Elementary Linear Algebra

Second Canadian Edition

Stewart Venit
California State University at Los Angeles

Wayne Bishop
California State University at Los Angeles

Jason Brown
Dalhousie University

NELSON EDUCATION

Elementary Linear Algebra, Second Canadian Edition

by Stewart Venit, Wayne Bishop, and Jason Brown

Vice President, Editorial Higher Education:
Anne Williams

Publisher:
Paul Fam

Executive Editor:
Jackie Wood

Executive Marketing Manager:
Sean Chamberland

Developmental Editor:
Rebecca Ryoji

Photo Researcher:
Sheila Hall

Permissions Coordinator:
Sheila Hall

Production Service:
Integra Software Services Pvt. Ltd.

Copy Editor:
Montgomery Kersell

Proofreader:
Integra Software Services Pvt. Ltd.

Indexer:
Integra Software Services Pvt. Ltd.

Manufacturing Manager:
Joanne McNeil

Design Director:
Ken Phipps

Managing Designer:
Franca Amore

Interior Design Modifications:
Integra Software Services Pvt. Ltd.

Cover Design:
Dianna Little

Cover Image:
© Duncan Walker/iStockphoto

Compositor:
Integra Software Services Pvt. Ltd.

Printer:
Edwards Brothers

Printed and bound in the United States
1 2 3 4 15 14 13 12

For more information contact Nelson Education Ltd., 1120 Birchmount Road, Toronto, Ontario, M1K 5G4. Or you can visit our Internet site at http://www.nelson.com

Library and Archives Canada Cataloguing in Publication Data

Venit, Stewart

Elementary linear algebra/Stewart Venit, Wayne Bishop, Jason Brown. — 2nd Canadian ed.

Includes index.
ISBN 978-0-17-650458-8

1. Algebras, Linear—Textbooks. I. Bishop, Wayne, 1942– II. Brown, Jason Ira, 1961– III. Title.

QA184.2.V46 2012 512′.5
C2011-908685-9

ISBN-13: 978-0-17-650458-8
ISBN-10: 0-17-650458-3

ABOUT THE AUTHORS

Stewart Venit

Stewart Venit received a Ph.D. in mathematics from the University of California, Berkeley, in 1971. For nearly 30 years, he has taught mathematics at California State University in Los Angeles. Professor Venit is a past department Chair and a recipient of the University's Outstanding Professor Award. He has authored numerous journal articles and 10 textbooks for mathematics and computer science.

Wayne Bishop

Wayne Bishop received a Ph.D. in mathematics from Western Michigan University in 1971, after which he began teaching at California State University in Los Angeles. He is a past Chair of the Department of Mathematics and spent one year teaching in Mexico City as a Fulbright Professor. Professor Bishop is very active in continuing efforts to improve mathematics education in California.

Jason Brown

Jason I. Brown received a Ph.D. in mathematics from the University of Toronto in 1987 and is currently a professor of mathematics at Dalhousie University in Halifax, Nova Scotia. He has written over 60 refereed research papers in combinatorics and its connections to other branches of mathematics. In 2004, he employed mathematical research to uncover how The Beatles played the opening chord of *A Hard Day's Night* and garnered widespread media attention, appearing on CBC radio's *As It Happens* and The Discovery Channel's *Daily Planet.* His work has been featured on the front page of the *National Post,* in *Guitar Player* magazine, and on the website of *The Wall Street Journal.* Professor Brown is an avid guitar player and songwriter, and he recently released his first CD titled "Songs in the Key of Pi." To find out more, visit **www.jasonibrown.com**.

PREFACE

PEDAGOGY

Features:

- Spotlight Applications
- Chapter Objectives
- New Marginal Biographies with Photos
- Computational Exercises
- Computational Notes
- New Marginal Notes
- Warnings
- Exercise Sets
- Section Applications
- Chapter Summaries
- Keywords
- Self-Tests
- Review Exercises
- Challenge Yourself Exercises
- Algorithmic Procedures

***Elementary Linear Algebra,* Second Canadian Edition,** is designed primarily as a first course in linear algebra for majors in mathematics, computer science, engineering, or the sciences. Knowledge of calculus is not required, except for optional material in Chapters 6 and 7 (Sections 6.6 and 7.6).

What's New in the Second Canadian Edition

***Elementary Linear Algebra,* Second Canadian Edition,** provides instructors with the mathematical rigor and content required in a university-level mathematics course, while providing students with a modern approach, interesting applications, and clarity of explanation they require to be successful.

In this second Canadian edition of ***Elementary Linear Algebra***, the text has become even more engaging, and readable for students, while continuing to be flexible for instructors. The following list highlights what has been added or revised to the new edition:

- A new application to **Social Networks** like Facebook is included in Chapter 3; a topical topic for students and instructor's alike.
- A new exciting application featuring **Markov Chains and Google PageRank** appears in Chapter 4. Students will enjoy learning about the math behind Google.
- **New historical notes added on mathematicians.** Historical notes and accompanying photos of key historical mathematicians highlight contributions made by these individuals as they relate to the topics discussed in the chapters.
- **New! Marginal Notes** help explain tricky material, expanding on lessons or providing hints on how to complete the math.
- **Important theorems** are now singled out for students with a new design feature so students can identify easily those that are most important in the chapter.
- **Eigenvalues and eigenvectors of matrices continue to be covered earlier in the text** (in Chapter 4). These are treated once again in the context of linear transformations in Chapter 8.

Below are chapter-specific additions and changes made to individual chapters.

Chapter 1:

- Areas of parallelograms, via cross products, are now in a theorem (with areas of triangles), with a reorganization and an additional exercise in Section 1.3.
- Section 1.3 opens with explanations of lines and planes as solutions to a single linear equation, and the term determinant is left out of the cross product formula.
- Section 1.2 contains vector projections and some additional interesting examples added in both Section 1.2 and Section 1.3.

Chapter 2:

- New notation was added for row operations.
- A challenge problem looks at solving linear systems over complex numbers.
- Six problems moved from Section 2.2 to Section 2.3.

Chapter 3:

- Additional properties of inverses (vis à vis transposes and products) are introduced. The definition of a symmetric matrix now appears in this chapter.
- A new application to social networks provides a topical topic!
- An example in Section 3.3 makes clearer the procedure for finding inverses.
- An exercise proves that the inverse of a symmetric invertible matrix is symmetric.

Chapter 4:

- A restructured Section 4.3 discusses adjoint form of inverse before Cramer's rule.
- Now included is an example of a determinant with variable entries.
- Geometric interpretation of eigenvalues and eigenvectors was added.

Chapter 5:

- Emphasis changes from linear dependence to linear independence.
- A notation was added for span, dimension, and rank.
- Motivation behind linear independence, spanning sets, and bases is highlighted.
- More questions were added to the review exercises.

Chapter 6:

- Section 6.5 was upgraded to include a serious introduction to complex numbers and complex spaces. Section 6.5 includes a mention of the dot product (the most important inner product) on complex spaces, and highlights, through the exercises, the results from inner product spaces that carry over.
- A proof in Section 6.6 was altered to be matrix theoretic.
- We added a significant number of exercises to Section 6.5, in conjunction with upgrading the content.
- Notations were added for dimension, span, and rank.
- Compact notation was inserted in addition to the more "wordy" format.
- Motivation for concepts like linear independence, spanning sets, and bases was added, and the exposition was increased by these additions.

Chapter 7:

- More motivation was added as to why one should study linear transformations.
- Historical notes were added through the new marginal features.
- Important theorems are highlighted to help improve student comprehension.

Chapter 8:

- The new Section 8.5 covers powers of matrices and linear recurrence relations.
- Historical and biographical notes were added.
- Usage of eigenvalues is more restricted in Section 8.1.

Chapter 9:
- Historical notes were added through the new marginal feature.
- Important theorems are highlighted to help improve student comprehension.

It's Been a Hard Day's Night!

Along the way, we included one of the authors' own applications of linear algebra, that of Fourier analysis to the opening chord of the The Beatles, *A Hard Day's Night*. (...alright!). There has long been a controversy over how The Beatles played the opening chord of *A Hard Day's Night*, certainly one of the most famous (and instantly recognizable) sounds in rock 'n' roll. Jason Brown used a *Fourier transform* to uncover the frequencies that made up the sound, and at what amplitudes. This mathematical process can be used to discover the original frequencies that are present.

Other Pedagogical Features

- **Vectors before linear equation**. Seen as a more modern approach, the text introduces vectors before linear equation to help facilitate student comprehension and clarity.
- **Computational Notes** appear throughout the text to give students some insight into the use of calculators and computers in linear algebra.
- **Warnings** are placed wherever appropriate to help students avoid common pitfalls.
- **Exercises** begin with a comprehensive array of relatively mechanical problems, followed by more-challenging "theoretical" exercises and, in many sections, computational exercises.
- **Self-Tests** appear at the end of each chapter to help students gauge their understanding of the material. Selected answers to the Self-Tests are provided in Appendix B.
- **Review Exercises** at the end of each chapter contain a greater number and wider variety of problems than the Self-Tests. Answers to the odd-numbered exercises are provided in Appendix B.
- **Algorithmic "procedures"** are found throughout the text to guide students through important, sometimes complicated, processes. These procedures are intended to provide both immediate and long-term review of the related material. They are usually given after a general description of the topic and one or two examples have been presented.
- **Chapter Keywords** appear at the end of each chapter. Students will find the keywords helpful reminders of the most important concepts and definitions within that chapter.
- **More-challenging problems** were added to each chapter after the Review Exercises. There have been requests for a few problems in each chapter that would stretch students a little further in their understanding of the concepts. These new problems also touch on interesting applications of linear algebra to other areas of mathematics, such as combinatorics.
- **An appendix introduces students to the computer algebra system** (CAS) *Maple* and explains how it can be integrated into solving problems in linear algebra. Some problems can benefit from the use of a CAS such as *Maple* or *Matlab*, yet it can be daunting for students to learn how to use such powerful programs. The appendix gives a concise introduction to *Maple* and its linear algebra capabilities.
- **Proofs** are stressed throughout the text. A course in linear algebra is a perfect opportunity to blend theory, computations, and applications, and the text strives to do just that. In fact, some of the applications, such as fractals

and music, aim to mix the three as seamlessly as possible, just as mathematicians do.

- **Unique and engaging style of writing and presentation** will ensure students find this book easy to read and follow, improving student comprehension and learning.

Text Organization

This second Canadian edition is organized for flexibility and use in a variety of courses. When designing a particular course, the instructor should be aware of the following points regarding the material:

- Chapter 1, *Geometry of* $\mathbf{R}^2$ *and* $\mathbf{R}^3$, is optional. This allows an instructor whose students have previously studied this material either to omit the chapter or to cover it quickly.
- All application sections are optional. None is prerequisite material for other topics or other applications.
- The chapter-opening Spotlights need not be covered in class. Exercises that refer to these applications have been highlighted for easy recognition and can be omitted by the instructor.

Course Flexibility

This text can be used in either a matrix-oriented course or a more traditionally structured course. Following are the sections that reflect the core material for these courses, both of which require approximately 30 hours of lecture time. Covering either set of core material should leave time to present other topics or applications, at the discretion of the instructor.

Matrix-Oriented Course
Chapter 2: Sections 2.1–2.3
Chapter 3: Sections 3.1–3.5
Chapter 4: Sections 4.1–4.4
Chapter 5: Sections 5.1–5.4
Chapter 6: Sections 6.1–6.4
Chapter 7: Sections 7.1–7.3
Chapter 8: Section 8

Traditional Course
Chapter 2: Sections 2.1–2.3
Chapter 3: Sections 3.1–3.3
Chapter 4: Sections 4.1–4.4
Chapter 5: Sections: 5.1–5.3
Chapter 6: Sections 6.1 and 6.2

Ancillaries

- **Instructor's Solutions Manual (978-0-17-665168-8):** A revised *Instructor's Solutions Manual* contains answers to the section and chapter exercises and solutions to the Self-Test exercises. For each chapter, a detailed list of the topics covered is provided, as well as general remarks about the material.
- **Student's Solutions Manual (978-0-17-665308-8):** A revised *Student's Solutions Manual* contains the solutions to all odd-numbered questions and exercises.
- **Enhanced Web Assign:** *Enhanced Web Assign* is a robust, easy-to-use homework system for students and is a powerful instructional tool for faculty and students that delivers an automatic grading solution for math and science courses and reinforces student learning through practice and instant feedback. End-of-chapter exercises in the text are enhanced with algorithmic functionality. Instructors can choose to include solutions or hide them.

Acknowledgments

I would like to thank the many people who have assisted in this latest revision of the text, including the panel of reviewers. Rebecca Ryoji, freelance developmental editor, managed this project through development, providing great direction and advice as we planned and executed the second edition. I'd also like to thank the people at Nelson Education Ltd., including executive editor Jackie Wood who helped set the course, as well as Sean Chamberland, Susan Calvert, Montgomery Kersell, and Reddy SSN. I would like to express appreciation to Peter Zvengrowski, Mark Blenkinsop, Carl Hickman, and Karla Cisneros, who made some excellent suggestions, as well as Danielle Cox, who prepared the solutions manual. Finally, I would like to thank my wife, Sondra, for all of her encouragement and support along the way, as well as my two sons, Shael and Zane, who prove, as Larry of the *Newhart* show once said, "Sometimes that DNA goes so right."

Jason I. Brown

Nelson Education Ltd. would like to thank the following reviewers for their guidance with the revision: Jeff Babb, University of Winnipeg; Ganesh Bhandari, Mount Royal University; Dzung Minh Ha, Ryerson University; Alistair Savage, University of Ottawa; Benjamin Steinberg, Carleton University; and Vicki Olds, University of Western Ontario.

CONTENTS

Elementary
Linear Algebra

SPOTLIGHT

Stephen Strathdee/Shutterstock

One of the fundamental concepts of linear algebra, and the central topic of this chapter, is that of the *vector*. Perhaps the most common physical application of vectors involves the analysis of the forces acting on an object and the resulting movement (if any) of that object. Natural forces affect our lives in innumerable ways, from the mundane (the force of gravity keeps the Earth in orbit around the Sun and us firmly planted on the ground) to the extraordinary (earthquakes and hurricanes can unleash catastrophic forces with awesome powers of destruction).

As a simple example to illustrate the use of vectors, let's take a look at the forces acting on an airplane in flight: these are the *thrust* of its engines (driving the plane forward), its *weight* (the force of gravity, pulling the plane down), and an *aerodynamic force* caused by the air rushing past it. The aerodynamic force can be broken into two components: *drag*, the resistance produced by the air (acting in the direction opposite to that of travel) and *lift*, which keeps the plane aloft (acting upward, roughly perpendicular to the direction of travel). Each of these forces can be represented by a vector (an "arrow" in the diagram), whose length represents the magnitude of the force and whose direction represents (as you may have guessed) its direction.

Similarly, the *velocity* of the airplane can be represented by a vector, giving the plane's speed and direction. Moreover, if the plane encounters a wind blowing at a certain speed in a certain direction, *its* velocity can be represented by another vector. In this situation, the question arises, "At what speed (relative to the ground) and in what direction does the plane actually fly?" In other words, what is the resultant *ground speed* and *true course* of its flight? The situation might look something like the figure.

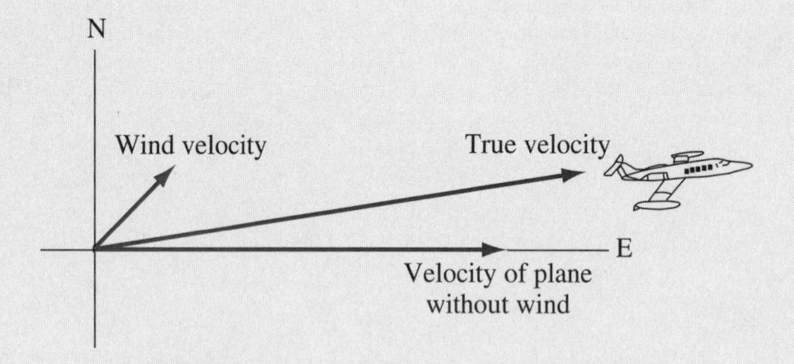

We return to this question and discuss the relation between forces and vectors further in the exercise sets of Sections 1.1 and 1.2.

CHAPTER 1

Geometry of $\mathbf{R}^2$ and $\mathbf{R}^3$

OBJECTIVES

You will learn to:

- Compute the length of a vector and the distance between vectors.
- Add and subtract vectors and multiply vectors by scalars.
- Determine the vector equivalent to a given directed line segment.
- Find the dot product and cross product (in $\mathbf{R}^3$) of two vectors.
- Determine if two vectors are orthogonal and find a vector orthogonal to two given vectors.
- Use vectors to determine the equations of lines in $\mathbf{R}^2$ and lines and planes in $\mathbf{R}^3$.

In the physical sciences, some quantities, such as mass and pressure, can be adequately described by simply giving their magnitude (size). Others, such as force and velocity, require not only magnitude but also direction. The latter quantities are usually represented by *vectors*. In this chapter, we develop the properties of vectors in the plane ($\mathbf{R}^2$) and 3-space ($\mathbf{R}^3$).

1.1 Vectors in $\mathbf{R}^2$ and $\mathbf{R}^3$

Each point in a plane can be identified with a unique ordered pair of real numbers by using a pair of perpendicular lines called *coordinate axes* (Figure 1). In a similar manner each point in 3-space can be represented by a unique ordered triple of real numbers (Figure 2). Throughout this text we denote the set of real numbers by $\mathbf{R}$, the set of ordered pairs of real numbers by $\mathbf{R}^2$, and the set of ordered triples by $\mathbf{R}^3$.

In addition to identifying a point in 2-space or 3-space, an ordered pair or triple is used to identify the *directed line segment* from the origin to the point indicated (Figures 3 and 4). This interpretation of an ordered pair or triple is called a **vector** in $\mathbf{R}^2$ or $\mathbf{R}^3$. Vectors are identified by boldfaced, lowercase, English letters. For example, $\mathbf{u} = (u_1, u_2, u_3)$ represents the vector in $\mathbf{R}^3$ with **components** u_1, u_2, and u_3. When we wish to speak of a *point* in $\mathbf{R}^2$ or $\mathbf{R}^3$, we use uppercase English letters to denote it, as in "the point P" or "the point $P(x_1, x_2, x_3)$." By convention, a point denoted by U has coordinates (u_1, u_2) in $\mathbf{R}^2$ or (u_1, u_2, u_3) in $\mathbf{R}^3$. Thus the point P and the vector $\mathbf{p}$ are represented by the same ordered pair or triple. In other words, the *endpoint* of the vector $\mathbf{p}$ is the point P.

The Bridgeman Art Library*

The use of ordered pairs and triples to study geometry dates back to René Descartes (1596–1650), a French mathematician and philosopher. His introduction of Cartesian coordinates allowed one to apply algebraic techniques to geometric problems.

NOTE Sometimes vectors are defined as directed line segments between *any* two points. We reserve the term *vector* for the special case of a directed

*Portrait of Rene Descartes (1596–1650) c.1649 (oil on canvas) (detail of 32939) by Frans Hals (1582/83–1666) (after) Louvre, Paris, France/ Giraudon/ The Bridgeman Art Library

FIGURE 1

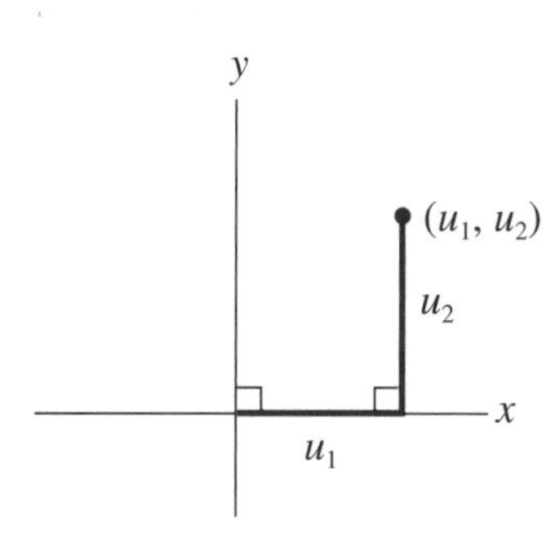

FIGURE 2

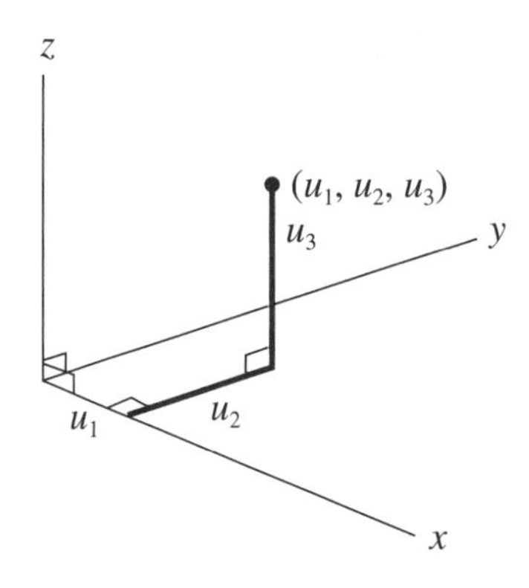

line segment that emanates from (begins at) the origin. More general directed line segments are important in geometric and physical applications. They are discussed later in this section and used throughout this chapter.

Two vectors **u** and **v** are **equal** if their *corresponding components* are equal; that is, $u_1 = v_1$ and $u_2 = v_2$ (and in $\mathbf{R}^3$ $u_3 = v_3$). The **zero vector** is the origin of the coordinate

FIGURE 3

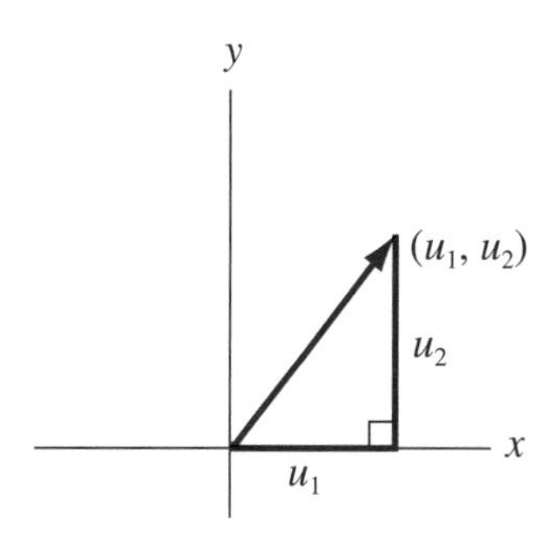

FIGURE 4

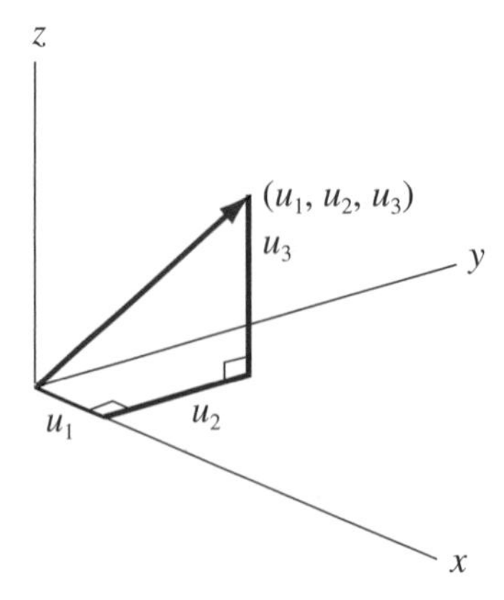

FIGURE 5

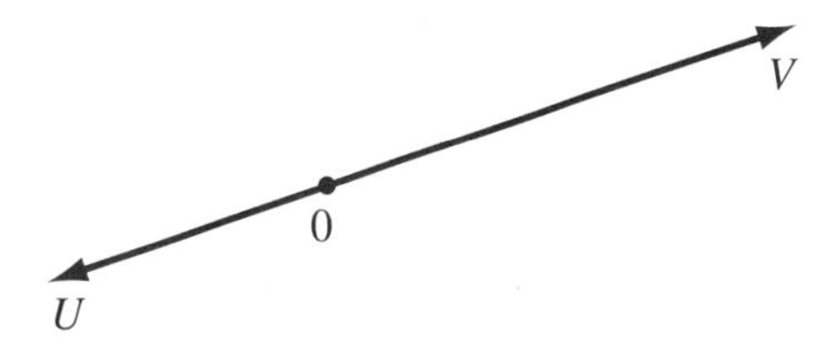

system, $\mathbf{0} = (0, 0)$ (or $(0, 0, 0)$). Thus the statement $\mathbf{u} \neq \mathbf{0}$ is equivalent to saying that at least one component of **u** is nonzero.

Two nonzero vectors **u** and **v** are **collinear** if the points U, V, and 0 are collinear points (Figure 5).

Length of a Vector

Recall that the distance between any two points in $\mathbf{R}^2$ or $\mathbf{R}^3$ can be given by the *distance formula.* The **distance** between two vectors **u** and **v** is defined to be the distance between their endpoints. Thus, in $\mathbf{R}^2$,

$$d(\mathbf{u}, \mathbf{v}) = \sqrt{(v_1 - u_1)^2 + (v_2 - u_2)^2}.$$

Similarly, for **u** and **v** in $\mathbf{R}^3$,

$$d(\mathbf{u}, \mathbf{v}) = \sqrt{(v_1 - u_1)^2 + (v_2 - u_2)^2 + (v_3 - u_3)^2}.$$

By considering the right triangles indicated in Figures 6 and 7, we see that the distance formula is an application of the Pythagorean theorem. For example, in Figure 6 the length of the hypotenuse of the right triangle is $d(\mathbf{u}, \mathbf{v})$, and the legs have lengths $v_1 - u_1$ and $v_2 - u_2$ (or, more generally, $|v_1 - u_1|$ and $|v_2 - u_2|$). Thus,

$$[d(\mathbf{u}, \mathbf{v})]^2 = (v_1 - u_1)^2 + (v_2 - u_2)^2,$$

and, by taking square roots, we obtain the first of the distance formulas. (For the second formula, this process is repeated twice.)

FIGURE 6

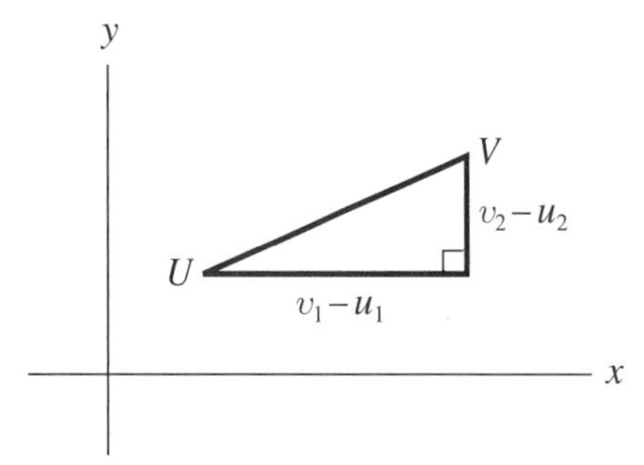

FIGURE 7

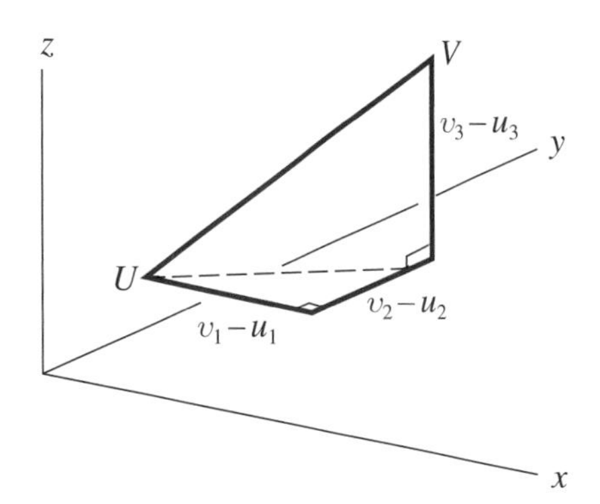

DEFINITION

The **length (norm, magnitude)** of a vector **u** is the distance from the origin to the endpoint of **u** and is denoted $\|\mathbf{u}\|$:

$$\|\mathbf{u}\| = d(\mathbf{0}, \mathbf{u}) = \begin{cases} \sqrt{u_1^2 + u_2^2}, & \text{if } \mathbf{u} \text{ is in } \mathbf{R}^2; \\ \sqrt{u_1^2 + u_2^2 + u_3^2}, & \text{if } \mathbf{u} \text{ is in } \mathbf{R}^3. \end{cases}$$

NOTE If $\|\mathbf{u}\| = 0$, then $\mathbf{u} = \mathbf{0}$ (and conversely). If $\|\mathbf{u}\| = 1$, the vector **u** is called a **unit vector**.

EXAMPLE 1

Find $\|\mathbf{u}\|$, where $\mathbf{u} = (2, -1, 3)$.

Solution

$$\|\mathbf{u}\| = \sqrt{2^2 + (-1)^2 + 3^2} = \sqrt{4 + 1 + 9} = \sqrt{14}$$ ■

Scalar Multiplication

An element of the set of real numbers **R** is called a **scalar**.

DEFINITION

Let c be a scalar and **u** be a vector in $\mathbf{R}^2$ or $\mathbf{R}^3$. The **scalar multiple** of **u** by c is the vector

$$c\mathbf{u} = \begin{cases} (cu_1, cu_2), & \text{if } \mathbf{u} \text{ is in } \mathbf{R}^2; \\ (cu_1, cu_2, cu_3), & \text{if } \mathbf{u} \text{ is in } \mathbf{R}^3. \end{cases}$$

In other words, $c\mathbf{u}$ is the vector obtained by multiplying each component of **u** by the scalar c.

EXAMPLE 2

Find $c\mathbf{u}$ for each scalar c and vector **u**.

a. $c = 3$, $\mathbf{u} = (1, -4, 4)$ **b.** $c = 0$, $\mathbf{u} = (2, 4)$

Solution

a. $c\mathbf{u} = (3, -12, 12)$ **b.** $c\mathbf{u} = (0, 0) = \mathbf{0}$ ■

Examples of the geometric interpretation of $c\mathbf{u}$ are given in Figure 8 and are justified by the next theorem.

FIGURE 8

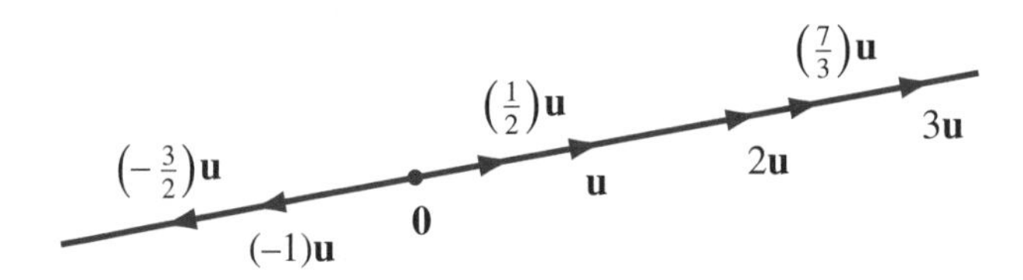

THEOREM 1

Let **u** be a nonzero vector in $\mathbf{R}^2$ or $\mathbf{R}^3$, and let c be a scalar. Then **u** and $c\mathbf{u}$ are collinear, and

a. if $c > 0$, then **u** and $c\mathbf{u}$ have the same direction;

b. if $c < 0$, then **u** and $c\mathbf{u}$ have opposite directions;

c. $\|c\mathbf{u}\| = |c|\,\|\mathbf{u}\|$.

Recall that two triangles are similar if they have the same angles.

Proof We prove this theorem for vectors in $\mathbf{R}^2$. (The proof is similar in $\mathbf{R}^3$.) Figures 9 and 10 illustrate the situation when $c > 0$ (part (a)) and $c < 0$ (part (b)), respectively. In each figure, the two right triangles are *similar*, and hence their corresponding angles are equal because their corresponding sides are in proportion; that is, the slopes of the lines determined by $\mathbf{u} = (u_1, u_2)$ and $c\mathbf{u} = (cu_1, cu_2)$,

$$\frac{u_2}{u_1} \quad \text{and} \quad \frac{cu_2}{cu_1},$$

are equal. Moreover, both lines have a point in common (the origin) and thus are identical. This shows that $\mathbf{u}$ and $c\mathbf{u}$ are collinear.

FIGURE 9

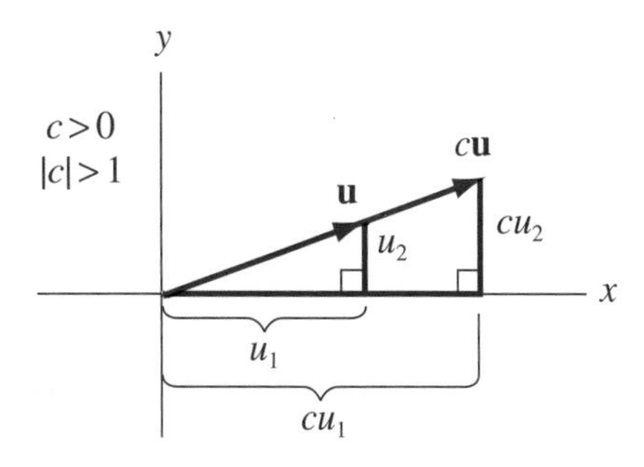

FIGURE 10

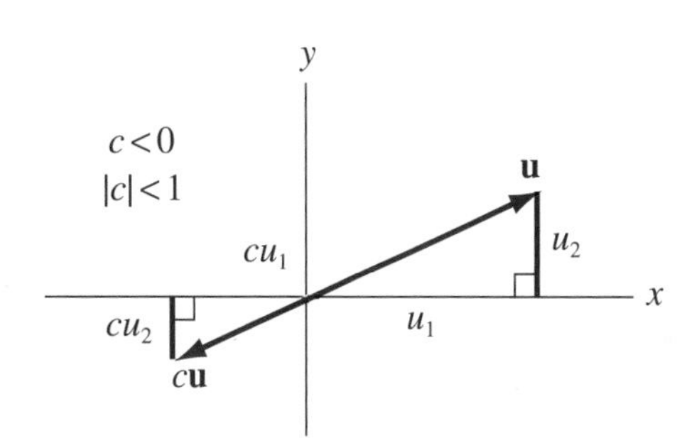

If $c > 0$ (part (a)), then the signs of the corresponding components of $\mathbf{u}$ and $c\mathbf{u}$ are the same, and the vectors have the same direction. If $c < 0$ (part (b)), the signs are reversed, so the vectors have opposite directions.

To verify part (c), we apply the distance formula:

$$\begin{aligned}\|c\mathbf{u}\| &= \|(cu_1, cu_2)\| = \sqrt{(cu_1)^2 + (cu_2)^2} \\ &= \sqrt{c^2(u_1^2 + u_2^2)} = \sqrt{c^2}\sqrt{u_1^2 + u_2^2} = |c|\,\|\mathbf{u}\| \end{aligned}$$ ■

EXAMPLE 3

Find the *midpoint* of the vector $\mathbf{u} = (2, 1, -4)$.

Solution Since $(\frac{1}{2})\mathbf{u}$ has the same direction as $\mathbf{u}$ and has half its length, the point corresponding to $(\frac{1}{2})\mathbf{u}$ is the midpoint of $\mathbf{u}$. Thus, the midpoint of the given vector is

$$\left(\frac{1}{2}\right)(2, 1, -4) = \left(1, \frac{1}{2}, -2\right).$$ ■

EXAMPLE 4

For a nonzero vector $\mathbf{u}$, find a unit vector with the same direction.

Solution We show that the desired vector is $(1/\|\mathbf{u}\|)\mathbf{u}$. It has the same direction as $\mathbf{u}$ by part (a) of Theorem 1, and it has length 1 (it is a unit vector) because

$$\left\| \frac{1}{\|\mathbf{u}\|}\mathbf{u} \right\| = \frac{1}{\|\mathbf{u}\|}\|\mathbf{u}\|$$

by part (c) of Theorem 1. ■

NOTE The scalar multiplier result (Theorem 1) holds in reverse as well: if **u** and **v** are nonzero collinear vectors, then $\mathbf{v} = c\mathbf{u}$ for some scalar c. The argument can be based on similar triangles, where c is taken as the constant of proportionality (or its negative if **u** and **v** have opposite directions).

Vector Addition

DEFINITION

Let **u** and **v** be vectors in $\mathbf{R}^2$ or $\mathbf{R}^3$. The **sum** of **u** and **v** is the vector

$$\mathbf{u} + \mathbf{v} = \begin{cases} (u_1 + v_1, u_2 + v_2), & \text{if } \mathbf{u}, \mathbf{v} \text{ are in } \mathbf{R}^2; \\ (u_1 + v_1, u_2 + v_2, u_3 + v_3), & \text{if } \mathbf{u}, \mathbf{v} \text{ are in } \mathbf{R}^3. \end{cases}$$

In other words, $\mathbf{u} + \mathbf{v}$ is the vector obtained by adding the corresponding components.

EXAMPLE 5

Find the sum of each pair of vectors.

a. $\mathbf{u} = (2, 1, -1), \mathbf{v} = (1, 2, 3)$ b. $\mathbf{u} = (3, -1), \mathbf{v} = (1, 3)$

Solution

a. $\mathbf{u} + \mathbf{v} = (3, 3, 2)$ b. $\mathbf{u} + \mathbf{v} = (4, 2)$ ■

Recall that a four-sided figure is a parallelogram if opposite sides are parallel.

We now look at the *geometric interpretation* of the sum of two vectors. The endpoint of $\mathbf{u} + \mathbf{v}$ is located by the endpoint of the directed line segment with the same direction and length as **v**, but originating at the endpoint of **u** (Figure 11). Another way to view $\mathbf{u} + \mathbf{v}$ (for noncollinear vectors) is as the diagonal from **0** of the parallelogram determined by **u** and **v** (Figure 12).

FIGURE 11

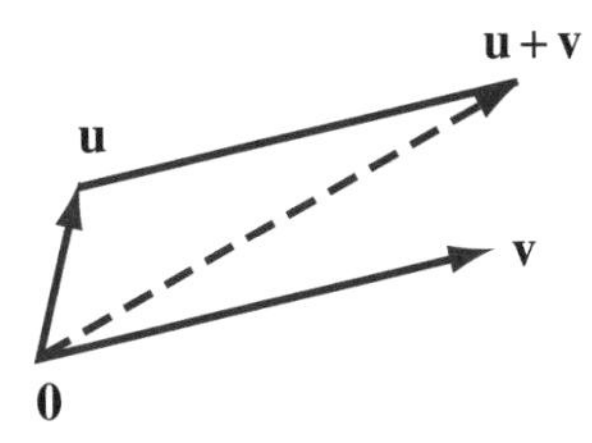

FIGURE 12

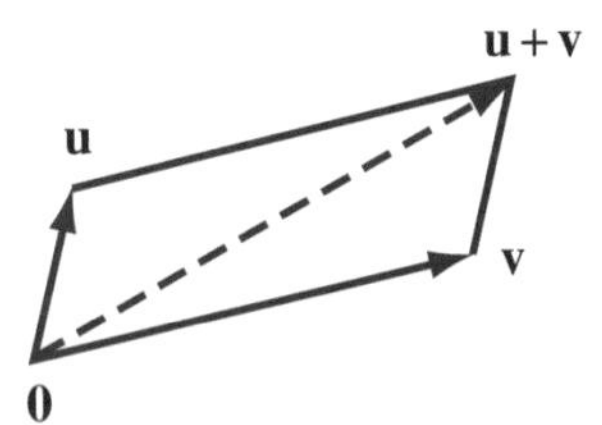

This theorem states that to add vectors, you add them "head-to-tail." That is, to add vectors **u** and **v**, place a copy of vector **v** with its tail at the head of vector **u**; **u** + **v** is the vector that runs from the tail of **u** to the head of the copy of **v**.

The preceding remarks are justified by Theorem 2.

THEOREM 2

For nonzero vectors **u** and **v**, the directed segment from the endpoint of **u** to the endpoint of $\mathbf{u} + \mathbf{v}$ is parallel and equal in length to the vector **v**.

Proof We show that the figure determined by 0, U, P, and V, where P is the endpoint of $\mathbf{u} + \mathbf{v}$, is a parallelogram. In a plane it is sufficient to show that the two pairs of opposite sides are equal in length. By the distance formula,

$$\begin{aligned} d(\mathbf{u}, \mathbf{u} + \mathbf{v}) &= d[(u_1, u_2), (u_1 + v_1, u_2 + v_2)] \\ &= \sqrt{[(u_1 + v_1) - u_1]^2 + [(u_2 + v_2) - u_2]^2} \\ &= \sqrt{v_1^2 + v_2^2} \\ &= \|\mathbf{v}\| \\ &= d(\mathbf{0}, \mathbf{v}). \end{aligned}$$

Similarly, since $d(\mathbf{v}, \mathbf{u} + \mathbf{v}) = d(\mathbf{0}, \mathbf{u})$, the figure is a parallelogram. In $\mathbf{R}^3$ these equations are as easy to check, but this is not quite enough to prove the assertion. It must also be confirmed that $\mathbf{u} + \mathbf{v}$ is *coplanar* with $\mathbf{0}$, $\mathbf{u}$, and $\mathbf{v}$. (See Exercise 37 of Section 1.2.) ■

For each vector $\mathbf{u}$, the vector that has the same length as $\mathbf{u}$ but points in the opposite direction is called the **negative** of $\mathbf{u}$ and is denoted $-\mathbf{u}$. By Theorem 1, $(-1)\mathbf{u}$ has these properties, so $(-1)\mathbf{u} = -\mathbf{u}$. It follows from the definition of scalar multiplication that

$$-\mathbf{u} = \begin{cases} (-u_1, -u_2), & \text{if } \mathbf{u} \text{ is in } \mathbf{R}^2; \\ (-u_1, -u_2, -u_3), & \text{if } \mathbf{u} \text{ is in } \mathbf{R}^3. \end{cases}$$

Now we can define subtraction of vectors in terms of addition by "adding the negative." That is, we define the **difference** of vectors $\mathbf{u}$ and $\mathbf{v}$ to be

$$\mathbf{u} - \mathbf{v} = \mathbf{u} + (-\mathbf{v}).$$

For example, if $\mathbf{u} = (0, -1, 2)$ and $\mathbf{v} = (-3, 0, 5)$, then $-\mathbf{u} = (0, 1, -2)$ and $\mathbf{u} - \mathbf{v} = (3, -1, -3)$.

The properties of vector addition and subtraction and scalar multiplication are given in Theorem 3 and follow directly from the definitions of these operations. To demonstrate how the arguments can be constructed, the proofs of parts (a), (b), and (f) of Theorem 3 are included, and the other proofs are left for exercises. In the proofs, we assume that the vectors are in $\mathbf{R}^2$. (For $\mathbf{R}^3$ just add a third component.)

THEOREM 3

Let $\mathbf{u}$, $\mathbf{v}$, and $\mathbf{w}$ be vectors in $\mathbf{R}^2$ or $\mathbf{R}^3$, and let c and d be scalars. Then

a. $\mathbf{u} + \mathbf{v} = \mathbf{v} + \mathbf{u}$ (Vector addition is commutative.)

b. $(\mathbf{u} + \mathbf{v}) + \mathbf{w} = \mathbf{u} + (\mathbf{v} + \mathbf{w})$ (Vector addition is associative.)

c. $\mathbf{u} + \mathbf{0} = \mathbf{u}$

d. $\mathbf{u} + (-\mathbf{u}) = \mathbf{0}$

e. $(cd)\mathbf{u} = c(d\mathbf{u})$

f. $(c + d)\mathbf{u} = c\mathbf{u} + d\mathbf{u}$ (Distributive laws)

g. $c(\mathbf{u} + \mathbf{v}) = c\mathbf{u} + c\mathbf{v}$ (Distributive laws)

h. $1\mathbf{u} = \mathbf{u}$

i. $(-1)\mathbf{u} = -\mathbf{u}$

j. $0\mathbf{u} = \mathbf{0}$

Proofs for (a), (b), and (f)

a. $$\begin{aligned} \mathbf{u} + \mathbf{v} &= (u_1 + v_1, u_2 + v_2) \\ &= (v_1 + u_1, v_2 + u_2) \\ &= \mathbf{v} + \mathbf{u} \end{aligned}$$

$$\begin{aligned}\textbf{b.}\quad (\mathbf{u}+\mathbf{v})+\mathbf{w} &= (u_1+v_1, u_2+v_2)+(w_1, w_2)\\ &= [(u_1+v_1)+w_1, (u_2+v_2)+w_2]\\ &= [u_1+(v_1+w_1), u_2+(v_2+w_2)]\\ &= (u_1, u_2)+(v_1+w_1, v_2+w_2)\\ &= \mathbf{u}+(\mathbf{v}+\mathbf{w})\end{aligned}$$

$$\begin{aligned}\textbf{f.}\quad (c+d)\mathbf{u} &= ((c+d)u_1, (c+d)u_2)\\ &= (cu_1+du_1, cu_2+du_2)\\ &= (cu_1, cu_2)+(du_1, du_2)\\ &= c\mathbf{u}+d\mathbf{u}\end{aligned}$$

■

Translation

Subtraction of vectors is a particularly useful way of discussing directed line segments that do not emanate from the origin. For points P and Q in $\mathbf{R}^2$ or $\mathbf{R}^3$, we denote the directed line segment from P to Q by the symbol $\overrightarrow{PQ}$. Two directed segments $\overrightarrow{PQ}$ and $\overrightarrow{RS}$ are said to be **equivalent** if they have the same direction and length (Figure 13). The next theorem relates equivalent directed segments to vector subtraction.

FIGURE 13

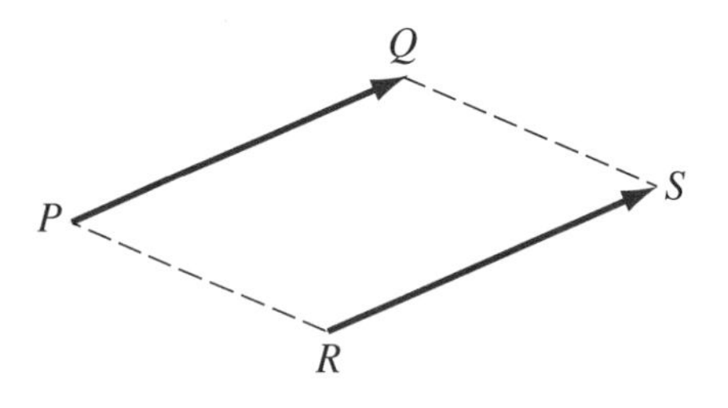

THEOREM 4

Let U and V be distinct points in $\mathbf{R}^2$ or $\mathbf{R}^3$. Then the vector $\mathbf{v}-\mathbf{u}$ is equivalent to the directed line segment from U to V. In other words, $\overrightarrow{UV}$ is parallel to the vector $\mathbf{v}-\mathbf{u}$, and $d(\mathbf{u}, \mathbf{v}) = \|\mathbf{v}-\mathbf{u}\|$.

You can view subtracting vector **u** from vector **v** as adding vector −**u** "head-to-tail" to vector **v**. That is, reverse the direction of vector **u** (to form vector −**u**) and place its tail at the head of vector **v**; **v** − **u** is the vector that runs from the tail of **v** to the head of −**u**.

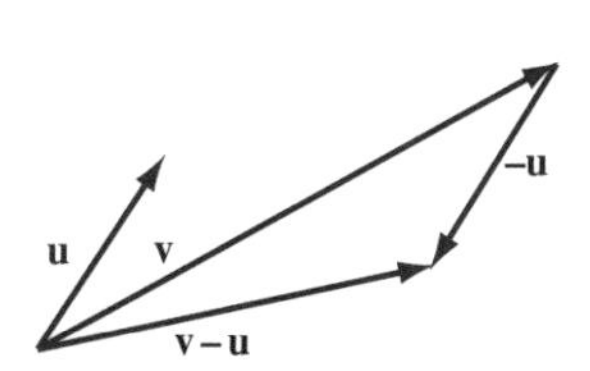

Proof We add **u** and **v** − **u**:

$$\begin{aligned}\mathbf{u}+(\mathbf{v}-\mathbf{u}) &= \mathbf{u}+(\mathbf{v}+(-\mathbf{u})) && \text{(Definition of subtraction)}\\ &= (\mathbf{u}+\mathbf{v})+(-\mathbf{u}) && \text{(Theorem 3(b))}\\ &= (\mathbf{v}+\mathbf{u})+(-\mathbf{u}) && \text{(Theorem 3(a))}\\ &= \mathbf{v}+(\mathbf{u}+(-\mathbf{u})) && \text{(Theorem 3(b))}\\ &= \mathbf{v}+\mathbf{0} && \text{(Theorem 3(d))}\\ &= \mathbf{v} && \text{(Theorem 3(c))}\end{aligned}$$

Since **v** is the sum of **u** and **v** − **u**, by Theorem 2, the vector **v** − **u** is parallel and equal in length (hence equivalent) to the directed line segment from U to V (Figure 14).

FIGURE 14

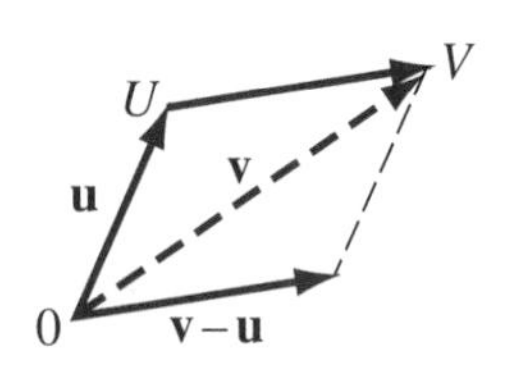

■

NOTE The process of replacing the directed line segment $\overrightarrow{UV}$ with the vector $\mathbf{v} - \mathbf{u}$ is often called *translating U to the origin.* In reverse, the process is called *translating to U.*

EXAMPLE 6

Let l_1 and l_2 be lines in $\mathbf{R}^3$. Suppose that l_1 is determined by the points $P(2, 2, 1)$ and $Q(-1, 0, 2)$ and that l_2 is determined by the points $R(2, 1, 1)$ and $S(8, 5, -1)$. Are the lines parallel?

Solution We translate the directed segments $\overrightarrow{PQ}$ and $\overrightarrow{RS}$ to the origin and compare them:

$$\mathbf{q} - \mathbf{p} = (-3, -2, 1)$$
$$\mathbf{s} - \mathbf{r} = (6, 4, -2)$$

Since one of these is a scalar multiple of the other, $\mathbf{s} - \mathbf{r} = -2(\mathbf{q} - \mathbf{p})$, they determine the same line through the origin. But, by Theorem 4, $\mathbf{q} - \mathbf{p}$ is parallel to l_1 and $\mathbf{s} - \mathbf{r}$ is parallel to l_2. Thus, l_1 and l_2 are parallel. ■

Alternative Notation

Notice that a vector (a, b) in $\mathbf{R}^2$ may be written as

$$(a, b) = a(1, 0) + b(0, 1).$$

Similarly, (a, b, c) in $\mathbf{R}^3$ may be written as

$$(a, b, c) = a(1, 0, 0) + b(0, 1, 0) + c(0, 0, 1).$$

In many engineering and physics books, the symbols reserved for these special vectors are $\mathbf{i} = (1, 0, 0)$, $\mathbf{j} = (0, 1, 0)$, and $\mathbf{k} = (0, 0, 1)$. Then the vector (a, b, c) is written as $a\mathbf{i} + b\mathbf{j} + c\mathbf{k}$. In $\mathbf{R}^2$, we have only $\mathbf{i}$ and $\mathbf{j}$, and we drop the last component.

EXAMPLE 7

Express $4\mathbf{i} - 3\mathbf{k}$ as an ordered triple.

Solution Since $\mathbf{j}$ is missing, its coefficient is 0. The result is $4\mathbf{i} - 3\mathbf{k} = (4, 0, -3)$. 

1.1 Exercises

In Exercises 1–16 let the vectors be given as follows:

$$\mathbf{r} = (-1, 1) \quad \mathbf{u} = (3, 1, 0)$$
$$\mathbf{s} = (2, 0) \quad \mathbf{v} = (2, 0, 1)$$
$$\mathbf{t} = (3, -2) \quad \mathbf{w} = (-1, -2, 3)$$

Perform each indicated operation.

1. $4\mathbf{w}$

2. $-3\mathbf{v}$

3. $-\mathbf{r}$

4. $\mathbf{s} + \mathbf{t}$

5. $\mathbf{u} + \mathbf{v}$

6. $\mathbf{t} - \mathbf{r}$

7. $\mathbf{w} - \mathbf{v}$

8. $3\mathbf{r} + 2\mathbf{s} - \mathbf{t}$

9. $\mathbf{u} - 5\mathbf{v} + 2\mathbf{w}$

10. $\|\mathbf{t}\|$

11. $\|\mathbf{w}\|$

12. $d(\mathbf{s}, \mathbf{t})$

13. $d(\mathbf{w}, \mathbf{v})$

14. $\|\mathbf{w} - \mathbf{v}\|$

15. $\|\mathbf{w}\| + \|\mathbf{v}\|$

16. $\|5(\mathbf{w} - \mathbf{v})\|$

In Exercises 17 and 18 find the unit vector with the same direction as the given vector.

17. $(3, -4)$

18. $(2, 1, -3)$

In Exercises 19 and 20 find the unit vector having the opposite direction of the given vector.

19. $(2, -1, 2)$

20. $(-1, 2)$

In Exercises 21–24 find the vector that is equivalent to the directed line segment from the first point to the second.

21. $(-1, 2), (3, 4)$

22. $(2, 1, 0), (-1, 3, 4)$

23. $(1, 5, 7), (2, 1, 0)$

24. $(1, -2), (-2, 3)$

25. Express $(2, 0, -3)$ in $\mathbf{i}, \mathbf{j}, \mathbf{k}$ form.

26. Express $3\mathbf{i} + 2\mathbf{j} - \mathbf{k}$ as an ordered triple.

27. Find the length of $2\mathbf{i} + 3\mathbf{j} - \mathbf{k}$.

28. Simplify $3(2\mathbf{i} + 7\mathbf{j} - \mathbf{k}) - 2(\mathbf{i} - 4\mathbf{j})$.

29. Prove the properties of Theorem 3 that remain unproved. Assume the vectors are in $\mathbf{R}^3$.

30. Prove that the midpoint of the line segment determined by the points P and Q is $(\frac{1}{2})\mathbf{p} + (\frac{1}{2})\mathbf{q}$. (*Hint:* This can be done directly by the distance formula, but using Theorem 4 is easier.)

31. Prove directly from the distance formula that $d(\mathbf{p}, \mathbf{q}) = \|\mathbf{q} - \mathbf{p}\|$.

32. Sketch a picture of the set of endpoints of all unit vectors in $\mathbf{R}^2$. What is the figure? What about in $\mathbf{R}^3$?

33. Describe the geometric figure consisting of the set of all solutions $\mathbf{x}$ in $\mathbf{R}^2$ to the equation $\|\mathbf{x} - \mathbf{p}\| = r$ for $\mathbf{p}$ in $\mathbf{R}^2$ and r in $\mathbf{R}$. What about in $\mathbf{R}^3$?

34. Let P and Q be distinct points and t a real number. Let $\mathbf{x}(t) = (1 - t)\mathbf{p} + t\mathbf{q}$. Use Theorem 4 to prove that

$$\frac{d(\mathbf{p}, \mathbf{x}(t))}{d(\mathbf{p}, \mathbf{q})} = |t|.$$

35. Use Exercise 34 to find a point one-third of the distance from P to Q.

36. Let l_1 be the line determined by $(2, 1, 3)$ and $(1, 2, -1)$, and l_2 be the line determined by $(0, 2, 3)$ and $(-1, 1, 2)$. Is l_1 parallel to l_2?

37. Let l_1 be the line determined by $(3, 1, 2)$ and $(4, 3, 1)$ and similarly l_2 be the line determined by $(1, 3, -3)$ and $(-1, -1, -1)$. Is l_1 parallel to l_2?

38. Let P, Q, and R be the three vertices of a triangle. Using Theorem 4, prove that the line segment that joins the midpoints of two of the sides is parallel to and half the length of the third side.

SPOTLIGHT *39.* If an airplane in flight encounters a wind (as we discussed in the Spotlight at the beginning of this chapter), the velocity of the plane without wind ($\mathbf{p}$) and the velocity of the wind ($\mathbf{w}$) can be represented by vectors (Figure 15). The magnitude of $\mathbf{p} + \mathbf{w}$ gives the *ground speed* of the plane, and the direction of $\mathbf{p} + \mathbf{w}$ gives its *true course.* Suppose a plane is flying due East at 600 km/h (kilometres per hour) when it is struck by a 80 km/h gust of wind blowing toward the Northeast. Then $\mathbf{p}$ can be represented by the vector $(600, 0)$, and $\mathbf{w}$ can be represented by a vector with magnitude 80 that makes an angle of 45° with both the North and East directions. Find the ground speed and true course of this plane.

FIGURE 15

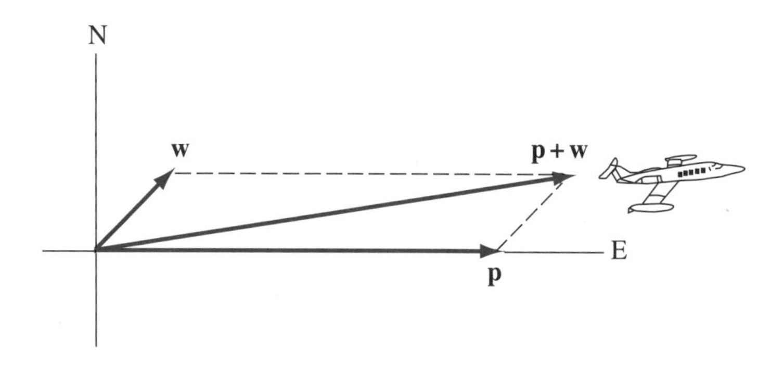

SPOTLIGHT ***40.*** Suppose that a box with mass 50 kg is sliding down an icy ramp inclined at an angle of 30° to the horizontal (Figure 16). What force **F** must a person exert parallel to the ramp to keep the box from sliding farther (assuming that no significant friction is present)? *Hint*: The weight of the box, **W**, is the vector sum of $\mathbf{W}_x$ and $\mathbf{W}_y$, its components parallel with and perpendicular to the ramp, respectively. To keep the box from sliding, the sum of **F** and $\mathbf{W}_x$ must be 0.

FIGURE 16

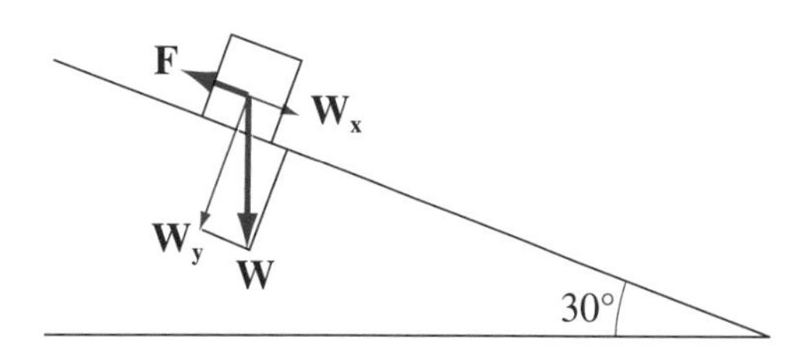

1.2 Dot and Cross Products

In this section, we define and develop the properties of two types of vector products: the *dot product*, which associates a scalar with every pair of vectors in $\mathbf{R}^2$ or $\mathbf{R}^3$, and the *cross product*, which yields another vector but is defined only in $\mathbf{R}^3$.

Dot Product

Since two nonzero vectors are assumed to have a common initial point (the origin), they form an angle. The following operation is useful in describing this angle.

DEFINITION

Let **u** and **v** be vectors in $\mathbf{R}^2$ or $\mathbf{R}^3$. The **dot product** (or **standard inner product**) of **u** and **v** is denoted by $\mathbf{u} \cdot \mathbf{v}$ and is defined as follows:

$$\mathbf{u} \cdot \mathbf{v} = \begin{cases} u_1v_1 + u_2v_2, & \text{if } \mathbf{u}, \mathbf{v} \text{ are in } \mathbf{R}^2; \\ u_1v_1 + u_2v_2 + u_3v_3, & \text{if } \mathbf{u}, \mathbf{v} \text{ are in } \mathbf{R}^3. \end{cases}$$

In words, $\mathbf{u} \cdot \mathbf{v}$ is the sum of the products of corresponding components of **u** and **v**.

EXAMPLE 1

Compute the dot product of each pair of vectors.

a. $\mathbf{u} = (1, 2, 3), \mathbf{v} = (-2, 4, 1)$

b. $\mathbf{u} = (1, 2), \mathbf{v} = (3, -4)$

c. $\mathbf{u} = (2, 1, -3), \mathbf{v} = (1, 1, 1)$

Solution

a. $\mathbf{u} \cdot \mathbf{v} = (1, 2, 3) \cdot (-2, 4, 1) = (1)(-2) + (2)(4) + (3)(1) = 9$

b. $\mathbf{u} \cdot \mathbf{v} = (1, 2) \cdot (3, -4) = (1)(3) + (2)(-4) = -5$

c. $\mathbf{u} \cdot \mathbf{v} = (2, 1, -3) \cdot (1, 1, 1) = (2)(1) + (1)(1) + (-3)(1) = 0$ ▣

Note that $\mathbf{u} \cdot \mathbf{v}$ is a *scalar* rather than a vector. For this reason, the dot product is often called the **scalar product**. The algebraic properties of the dot product are included in the next theorem.

THEOREM 1

Let **u**, **v**, and **w** be vectors in $\mathbf{R}^2$ or $\mathbf{R}^3$, and let c be a scalar. Then

a. $\mathbf{u} \cdot \mathbf{v} = \mathbf{v} \cdot \mathbf{u}$ (Dot product is commutative.)

b. $c(\mathbf{u} \cdot \mathbf{v}) = (c\mathbf{u}) \cdot \mathbf{v} = \mathbf{u} \cdot (c\mathbf{v})$ (Scalars factor out.)

c. $\mathbf{u} \cdot (\mathbf{v} + \mathbf{w}) = \mathbf{u} \cdot \mathbf{v} + \mathbf{u} \cdot \mathbf{w}$ (Distributive law)

d. $\mathbf{u} \cdot \mathbf{0} = 0$

e. $\mathbf{u} \cdot \mathbf{u} = \|\mathbf{u}\|^2$

Proofs for (a), (b), and (e) Assume that the vectors are in $\mathbf{R}^2$. The case of $\mathbf{R}^3$ is entirely analogous; just add a third component.

a.
$$\begin{aligned}\mathbf{u} \cdot \mathbf{v} &= u_1v_1 + u_2v_2\\ &= v_1u_1 + v_2u_2\\ &= \mathbf{v} \cdot \mathbf{u}\end{aligned}$$

b.
$$\begin{aligned}c(\mathbf{u} \cdot \mathbf{v}) &= c(u_1v_1 + u_2v_2)\\ &= cu_1v_1 + cu_2v_2 = \begin{cases}(cu_1)v_1 + (cu_2)v_2 = (c\mathbf{u}) \cdot \mathbf{v}\\ u_1(cv_1) + u_2(cv_2) = \mathbf{u} \cdot (c\mathbf{v})\end{cases}\end{aligned}$$

e.
$$\begin{aligned}\mathbf{u} \cdot \mathbf{u} &= u_1u_1 + u_2u_2\\ &= u_1^2 + u_2^2\\ &= \|\mathbf{u}\|^2\end{aligned}$$
▣

The proofs of (c) and (d) are left for exercises.

The geometric interpretation of $\mathbf{u} \cdot \mathbf{v}$ is particularly interesting. If **u** and **v** are nonzero vectors, they form an *angle*, since they both originate at the origin **0** (Figure 17). We use *radian* measure for the angle (although degree measure can be used as well), and we use the smallest nonnegative number possible. In other words, if θ is the measure of the angle, we have $0 \le \theta \le \pi$. The next theorem describes the relationship between this angle and $\mathbf{u} \cdot \mathbf{v}$.

FIGURE 17

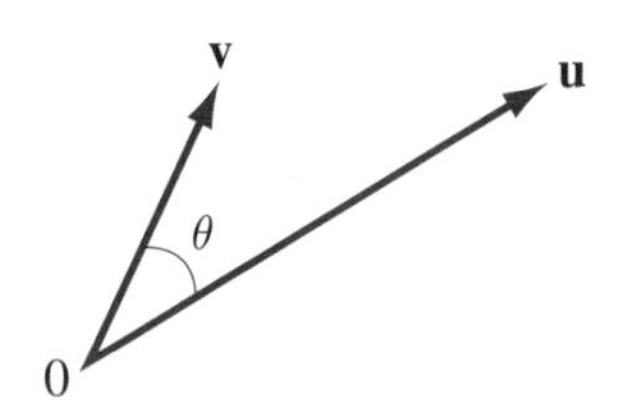

THEOREM 2

Let **u** and **v** be vectors in $\mathbf{R}^2$ or $\mathbf{R}^3$, and let θ be the angle between **u** and **v**. Then

$$\mathbf{u} \cdot \mathbf{v} = \|\mathbf{u}\|\,\|\mathbf{v}\| \cos\theta$$

or, equivalently, for nonzero $\mathbf{u}$ and $\mathbf{v}$,

$$\cos\theta = \frac{\mathbf{u} \cdot \mathbf{v}}{\|\mathbf{u}\|\,\|\mathbf{v}\|}.$$

Proof Consider the triangle determined by $\mathbf{u}$ and $\mathbf{v}$ (Figure 18). Let $a = \|\mathbf{u}\|$, $b = \|\mathbf{v}\|$, and $c = \|\mathbf{v} - \mathbf{u}\|$. Recall, from Section 1.1, that $\mathbf{v} - \mathbf{u}$ is equivalent to the directed line segment $\overrightarrow{UV}$, so c is the length of the third side of the triangle. To this triangle, we apply the law of cosines, $c^2 = a^2 + b^2 - 2ab\cos\theta$. Substituting for a, b, and c, we have

$$\|\mathbf{v} - \mathbf{u}\|^2 = \|\mathbf{u}\|^2 + \|\mathbf{v}\|^2 - 2\,\|\mathbf{u}\|\,\|\mathbf{v}\|\cos\theta. \quad \textbf{(1)}$$

FIGURE 18

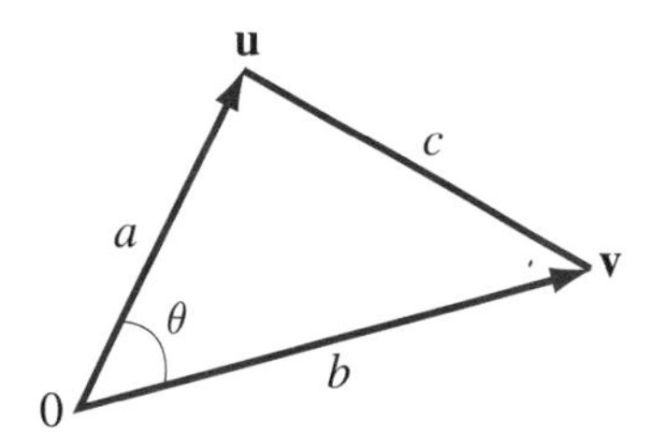

We need a more convenient form for $\|\mathbf{v} - \mathbf{u}\|^2$.

$$\begin{aligned}
\|\mathbf{v} - \mathbf{u}\|^2 &= (\mathbf{v} - \mathbf{u}) \cdot (\mathbf{v} - \mathbf{u}) && \text{(Theorem 1(e))}\\
&= (\mathbf{v} - \mathbf{u}) \cdot \mathbf{v} - (\mathbf{v} - \mathbf{u}) \cdot \mathbf{u} && \text{(Theorem 1(c))}\\
&= \mathbf{v} \cdot \mathbf{v} - \mathbf{u} \cdot \mathbf{v} - \mathbf{v} \cdot \mathbf{u} + \mathbf{u} \cdot \mathbf{u} && \text{(Theorem 1(c))}\\
&= \mathbf{v} \cdot \mathbf{v} - \mathbf{u} \cdot \mathbf{v} - \mathbf{u} \cdot \mathbf{v} + \mathbf{u} \cdot \mathbf{u} && \text{(Theorem 1(a))}\\
&= \|\mathbf{v}\|^2 - 2(\mathbf{u} \cdot \mathbf{v}) + \|\mathbf{u}\|^2 && \text{(Theorem 1(e))}
\end{aligned}$$

Returning now to Equation (1),

$$\begin{aligned}
\|\mathbf{v}\|^2 - 2(\mathbf{u} \cdot \mathbf{v}) + \|\mathbf{u}\|^2 &= \|\mathbf{u}\|^2 + \|\mathbf{v}\|^2 - 2\,\|\mathbf{u}\|\,\|\mathbf{v}\|\cos\theta\\
-2(\mathbf{u} \cdot \mathbf{v}) &= -2\,\|\mathbf{u}\|\,\|\mathbf{v}\|\cos\theta\\
\mathbf{u} \cdot \mathbf{v} &= \|\mathbf{u}\|\,\|\mathbf{v}\|\cos\theta.
\end{aligned}$$

■

EXAMPLE 2

Find the cosine of the angle between each pair of vectors.

a. $\mathbf{u} = (1, 2, 3)$, $\mathbf{v} = (-2, 4, 1)$

b. $\mathbf{u} = (2, 1, -3)$, $\mathbf{v} = (1, 1, 1)$

Solution

a.
$$\begin{aligned}
\cos\theta &= \frac{\mathbf{u} \cdot \mathbf{v}}{\|\mathbf{u}\|\,\|\mathbf{v}\|}\\
&= \frac{(1)(-2) + (2)(4) + (3)(1)}{\sqrt{1^2 + 2^2 + 3^2}\,\sqrt{(-2)^2 + 4^2 + 1^2}}\\
&= \frac{9}{\sqrt{14}\,\sqrt{21}}\\
&= \frac{3\sqrt{6}}{14}
\end{aligned}$$

(Using a calculator, θ is approximately 1.02 rad.)

b. $$\cos\theta = \frac{\mathbf{u}\cdot\mathbf{v}}{\|\mathbf{u}\|\,\|\mathbf{v}\|} = \frac{0}{\|\mathbf{u}\|\,\|\mathbf{v}\|} = 0$$

(Here the angle θ is $\pi/2$ rad, so θ is a right angle.) ■

EXAMPLE 3

Find the cosine of the angle at P determined by the directed line segments $\overrightarrow{PQ}$ and $\overrightarrow{PR}$, for points $P(1, 2)$, $Q(1+\sqrt{3}, -1)$, and $R(1-\sqrt{3}, 3)$.

Solution We need vectors to apply Theorem 2, so we translate $\overrightarrow{PQ}$ and $\overrightarrow{PR}$ to the origin by subtracting $\mathbf{p}$ from $\mathbf{q}$ and $\mathbf{r}$. This results in an angle at the origin determined by the vectors $\mathbf{q}-\mathbf{p}$ and $\mathbf{r}-\mathbf{p}$ (Figure 19). Since these vectors are respectively equivalent (and hence parallel) to $\overrightarrow{PQ}$ and $\overrightarrow{PR}$, the angle they form is the same, and we may use Theorem 2:

$$\cos\theta = \frac{(\mathbf{q}-\mathbf{p})\cdot(\mathbf{r}-\mathbf{p})}{\|\mathbf{q}-\mathbf{p}\|\,\|\mathbf{r}-\mathbf{p}\|} = \frac{(\sqrt{3},-3)\cdot(-\sqrt{3},1)}{\|(\sqrt{3},-3)\|\,\|(-\sqrt{3},1)\|} = \frac{-6}{\sqrt{12}\,\sqrt{4}} = -\frac{\sqrt{3}}{2}$$

(This time the angle θ is $5\pi/6$ rad.)

FIGURE 19

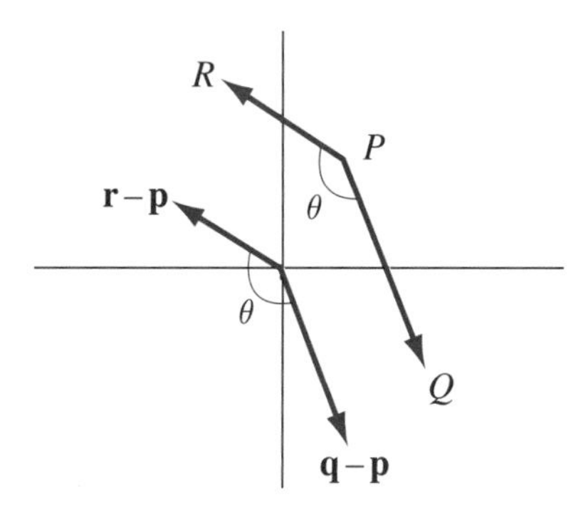

■

We investigate further the relationship between dot product and angle after introducing the following concept.

DEFINITION

Vectors $\mathbf{u}$ and $\mathbf{v}$ in $\mathbf{R}^2$ or $\mathbf{R}^3$ are **orthogonal** if $\mathbf{u}\cdot\mathbf{v} = 0$.

For example, the vectors $\mathbf{u} = (2, 1, -3)$ and $\mathbf{v} = (1, 1, 1)$ of Example 1 are orthogonal. In Example 2, we saw that these same vectors form a right angle. The words *orthogonal* (from Greek), *perpendicular* (from Latin), and *normal* (also Latin) mean essentially the same thing, "meet at right angles." However, usage sometimes makes one word preferable to another. As examples, two vectors are usually said to be *orthogonal*, two lines or planes are *perpendicular*, and a vector is *normal* to a plane.

THEOREM 3

Let $\mathbf{u}$ and $\mathbf{v}$ be nonzero vectors in $\mathbf{R}^2$ or $\mathbf{R}^3$, and let θ be the angle they form. Then θ is

An angle is acute if it is less than 90°, right if it equals 90°, and obtuse if it is greater than 90°.

a. An acute angle if $\mathbf{u} \cdot \mathbf{v} > 0$,

b. A right angle if $\mathbf{u} \cdot \mathbf{v} = 0$,

c. An obtuse angle if $\mathbf{u} \cdot \mathbf{v} < 0$.

Proof From Theorem 2, we have the equation

$$\mathbf{u} \cdot \mathbf{v} = \|\mathbf{u}\| \, \|\mathbf{v}\| \cos \theta.$$

Since $\|\mathbf{u}\|$ and $\|\mathbf{v}\|$ are positive, it follows that $\mathbf{u} \cdot \mathbf{v}$ is zero, positive, or negative as $\cos \theta$ is zero, positive, or negative. However, $\cos \theta > 0$ implies that θ is an acute angle since θ is between 0 and π. Similarly, $\cos \theta = 0$ implies that θ is a right angle, and $\cos \theta < 0$ implies that θ is obtuse. ■

In part (a) of Example 2, we calculated $\cos \theta$ to be $\frac{3\sqrt{6}}{14}$ and gave the approximate measure of the angle. Without a calculator, θ is seen to be an acute angle by Theorem 3. Similarly, the angle in Example 3 must be obtuse, since $(\mathbf{q} - \mathbf{p}) \cdot (\mathbf{r} - \mathbf{p}) = -6$.

There is one more concept related to dot products, which we shall revisit in Chapter 6. Let $\mathbf{u}$ and $\mathbf{v}$ be vectors in $\mathbf{R}^2$ or $\mathbf{R}^3$, with $\mathbf{v} \neq \mathbf{0}$. The **projection** of $\mathbf{u}$ onto $\mathbf{v}$, $\text{proj}_{\mathbf{v}}\mathbf{u}$, is the vector along $\mathbf{v}$ with length $\|\mathbf{u}\|\cos \theta$; this vector can be thought of as the part of $\mathbf{u}$ in the direction of $\mathbf{v}$ if a light is shone perpendicular to $\mathbf{v}$ in the plane determined by $\mathbf{u}$ and $\mathbf{v}$ (Figure 20). We multiply the scalar $\|\mathbf{u}\|\cos \theta$ by the unit vector $\frac{1}{\|\mathbf{v}\|}\mathbf{v}$ in the direction of $\mathbf{v}$ (see Example 4 of Section 1.1), so using Theorem 2, we have

$$\text{proj}_{\mathbf{v}}\mathbf{u} = (\|\mathbf{u}\|\cos \theta)\frac{1}{\|\mathbf{v}\|}\mathbf{v} = \left(\|\mathbf{u}\|\frac{\mathbf{u} \cdot \mathbf{v}}{\|\mathbf{u}\| \, \|\mathbf{v}\|}\right)\frac{1}{\|\mathbf{v}\|}\mathbf{v} = \frac{\mathbf{u} \cdot \mathbf{v}}{\|\mathbf{v}\|^2}\mathbf{v} = \left(\frac{\mathbf{u} \cdot \mathbf{v}}{\mathbf{v} \cdot \mathbf{v}}\right)\mathbf{v}.$$

It is straightforward to check that the length of the projection is $\dfrac{|\mathbf{u} \cdot \mathbf{v}|}{\|\mathbf{v}\|}$.

FIGURE 20

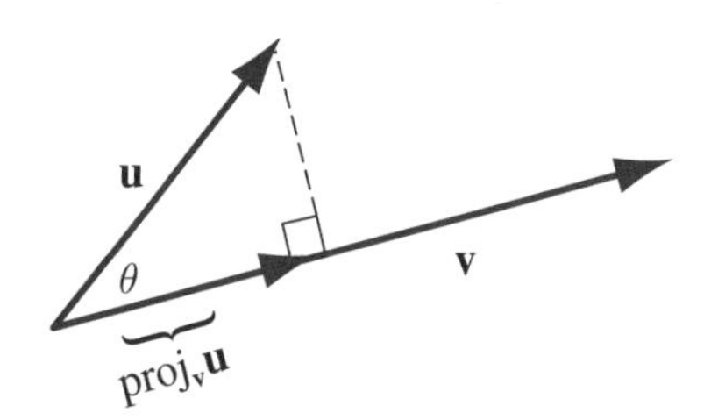

EXAMPLE 4

Find the projection of $\mathbf{u} = (-1, -1, 2)$ along $\mathbf{v} = (-3, 0, 4)$.

Solution

$$\begin{aligned}
\text{proj}_{\mathbf{v}}\mathbf{u} &= \left(\frac{\mathbf{u} \cdot \mathbf{v}}{\mathbf{v} \cdot \mathbf{v}}\right)\mathbf{v} \\
&= \left(\frac{(-1, -1, 2) \cdot (-3, 0, 4)}{(-3, 0, 4) \cdot (-3, 0, 4)}\right)(-3, 0, 4) \\
&= \frac{11}{25}(-3, 0, 4) \\
&= \left(-\frac{33}{25}, 0, \frac{44}{25}\right).
\end{aligned}$$

So the projection of $\mathbf{u}$ along $\mathbf{v}$ is $\left(-\frac{33}{25}, 0, \frac{44}{25}\right)$. ■

Cross Product

It may have seemed surprising that the dot product of two vectors is a scalar instead of a vector. There is, in $\mathbf{R}^3$, another type of product of two vectors that produces a vector.

DEFINITION

Let $\mathbf{u} = (u_1, u_2, u_3)$ and $\mathbf{v} = (v_1, v_2, v_3)$ be vectors in $\mathbf{R}^3$. The **cross product** of $\mathbf{u}$ and $\mathbf{v}$ is the vector

$$\mathbf{u} \times \mathbf{v} = (u_2v_3 - u_3v_2, u_3v_1 - u_1v_3, u_1v_2 - u_2v_1).$$

Fortunately, there is a convenient way of remembering this formula. Given an array of numbers

$$\begin{bmatrix} a & b \\ c & d \end{bmatrix},$$

we change the brackets to vertical lines to indicate the difference of the diagonal products,

$$\begin{vmatrix} a & b \\ c & d \end{vmatrix} = ad - bc.$$

With this notation, the expression for $\mathbf{u} \times \mathbf{v}$ becomes

$$\mathbf{u} \times \mathbf{v} = \left(\begin{vmatrix} u_2 & u_3 \\ v_2 & v_3 \end{vmatrix}, -\begin{vmatrix} u_1 & u_3 \\ v_1 & v_3 \end{vmatrix}, \begin{vmatrix} u_1 & u_2 \\ v_1 & v_2 \end{vmatrix} \right).$$

(Note the negative sign in the second component! There's more about this in Chapter 4.)

The arrays that appear in this expression are easy to remember by using the following device. Write the components of $\mathbf{u}$ above those of $\mathbf{v}$, forming the rectangular array

$$\begin{bmatrix} u_1 & u_2 & u_3 \\ v_1 & v_2 & v_3 \end{bmatrix}.$$

Then the square arrays that appear in the expression for $\mathbf{u} \times \mathbf{v}$ are obtained by successively deleting the first, second, and third columns of this rectangular array.

EXAMPLE 5

Find $\mathbf{u} \times \mathbf{v}$, where $\mathbf{u} = (1, 2, -1)$ and $\mathbf{v} = (0, 2, 3)$.

Solution Construct the rectangular array

$$\begin{bmatrix} 1 & 2 & -1 \\ 0 & 2 & 3 \end{bmatrix}.$$

Then

$$\begin{aligned} \mathbf{u} \times \mathbf{v} &= \left(\begin{vmatrix} 2 & -1 \\ 2 & 3 \end{vmatrix}, -\begin{vmatrix} 1 & -1 \\ 0 & 3 \end{vmatrix}, \begin{vmatrix} 1 & 2 \\ 0 & 2 \end{vmatrix} \right) \\ &= ((2)(3) - (-1)(2), -(1)(3) + (-1)(0), (1)(2) - (2)(0)) \\ &= (6 + 2, -3-0, 2-0) \\ &= (8, -3, 2). \end{aligned}$$

■

COMPUTATIONAL NOTE

In the course of your study of linear algebra, you will encounter many tedious computational problems, such as the calculation of cross products. The Computational Notes that appear from time to time throughout the text describe how a computer or calculator can be used to solve the problem under discussion and, occasionally, the pitfalls of doing so.

Corresponding Computational Exercises appear at the end of every section that contains these notes. In the Computational Exercises, you are asked to use computer software (possibly supplied by your instructor or computer lab) or a calculator to help you solve them. For example, an ordinary calculator can take some of the drudgery out of computing dot and cross products when the components of the given vectors aren't "nice" round numbers. Moreover, more sophisticated graphing calculators can compute dot products directly (once the given vectors are entered as "lists") and can aid in the computation of cross products by finding the three determinants for you.

On the other hand, you should always be able to carry out calculations by hand. Sometimes there are computational tricks and shortcuts that can be seen only by the human eye.

For our purposes, the most important property of the cross product is given by the following theorem.

THEOREM 4

The vector $\mathbf{u} \times \mathbf{v}$ is orthogonal to both $\mathbf{u}$ and $\mathbf{v}$.

Proof We compute $\mathbf{u} \cdot (\mathbf{u} \times \mathbf{v})$:

$$\begin{aligned}(u_1, u_2, u_3) \cdot (u_2v_3 - u_3v_2, u_3v_1 - u_1v_3, u_1v_2 - u_2v_1) \\ = u_1u_2v_3 - u_1u_3u_2 + u_2u_3v_1 - u_2u_1v_3 + u_3u_1v_2 - u_3u_2v_1 \\ = 0\end{aligned}$$

Thus, by definition, $\mathbf{u}$ is orthogonal to $\mathbf{u} \times \mathbf{v}$. Similarly, $\mathbf{v} \cdot (\mathbf{u} \times \mathbf{v}) = 0$, so $\mathbf{v}$ is orthogonal to $\mathbf{u} \times \mathbf{v}$. ■

EXAMPLE 6

Find a vector that is orthogonal to both $\mathbf{u} = (1, 2, -1)$ and $\mathbf{v} = (0, 2, 3)$.

Solution Taking the cross product of these vectors, we get $\mathbf{u} \times \mathbf{v} = (8, -3, 2)$, as in Example 5. By Theorem 4, this vector is orthogonal to both $\mathbf{u}$ and $\mathbf{v}$. ■

Other properties of cross products are given in the following theorem:

!

You might wonder whether a similar cross product exists in two dimensions, but it doesn't. In fact, cross products only exist in 1-, 3-, and 7-dimensional spaces, for reasons we won't go into here.

THEOREM 5

Let $\mathbf{u}$, $\mathbf{v}$ be vectors in $\mathbf{R}^3$, and c a scalar. Then

a. $\mathbf{u} \times \mathbf{v} = -(\mathbf{v} \times \mathbf{u})$

b. $\mathbf{u} \times (\mathbf{v} + \mathbf{w}) = (\mathbf{u} \times \mathbf{v}) + (\mathbf{u} \times \mathbf{w})$ (Distributive laws)

c. $(\mathbf{u} + \mathbf{v}) \times \mathbf{w} = (\mathbf{u} \times \mathbf{w}) + (\mathbf{v} \times \mathbf{w})$

d. $c(\mathbf{u} \times \mathbf{v}) = (c\mathbf{u}) \times \mathbf{v} = \mathbf{u} \times (c\mathbf{v})$ (Scalars factor out.)

e. $\mathbf{u} \times \mathbf{0} = \mathbf{0} \times \mathbf{u} = \mathbf{0}$

f. $\mathbf{u} \times \mathbf{u} = \mathbf{0}$

g. $\|\mathbf{u} \times \mathbf{v}\| = \|\mathbf{u}\| \, \|\mathbf{v}\| \sin \theta = \sqrt{\|\mathbf{u}\|^2 \, \|\mathbf{v}\|^2 - (\mathbf{u} \cdot \mathbf{v})^2}$

where θ is the angle determined by $\mathbf{u}$ and $\mathbf{v}$.

Proofs for (a), (b), and (g)

a. Let $\mathbf{u} = (u_1, u_2, u_3)$ and $\mathbf{v} = (v_1, v_2, v_3)$, and simply compute both $\mathbf{u} \times \mathbf{v}$ and $\mathbf{v} \times \mathbf{u}$. Notice that one vector is the negative of the other.

(The important fact here is the geometric interpretation: $\mathbf{u} \times \mathbf{v}$ and $\mathbf{v} \times \mathbf{u}$ have the same length but point in opposite directions.)

b. Let $\mathbf{u}$ and $\mathbf{v}$ be given as above, and $\mathbf{w} = (w_1, w_2, w_3)$. Then

$$\begin{aligned}\mathbf{u} \times (\mathbf{v} + \mathbf{w}) &= (u_1, u_2, u_3) \times (v_1 + w_1, v_2 + w_2, v_3 + w_3) \\ &= (u_2(v_3 + w_3) - u_3(v_2 + w_2), u_3(v_1 + w_1) \\ &\quad - u_1(v_3 + w_3), u_1(v_2 + w_2) - u_2(v_1 + w_1)) \\ &= (u_2v_3 + u_2w_3 - u_3v_2 - u_3w_2, u_3v_1 + u_3w_1 \\ &\quad - u_1v_3 - u_1w_3, u_1v_2 + u_1w_2 - u_2v_1 - u_2w_1) \\ &= (u_2v_3 - u_3v_2, u_3v_1 - u_1v_3, u_1v_2 - u_2v_1) \\ &\quad + (u_2w_3 - u_3w_2, u_3w_1 - u_1w_3, u_1w_2 - u_2w_1) \\ &= \mathbf{u} \times \mathbf{v} + \mathbf{u} \times \mathbf{w}.\end{aligned}$$

The proofs of (c) through (f) are straightforward computations, so they are left for exercises.

g. Note the similarity of statement (g) to the rule relating dot product and the cosine of the angle, $\|\mathbf{u}\| \, \|\mathbf{v}\| \cos \theta = \mathbf{u} \cdot \mathbf{v}$. Starting with this equation, we square both sides to obtain

$$\|\mathbf{u}\|^2 \, \|\mathbf{v}\|^2 \cos^2 \theta = (\mathbf{u} \cdot \mathbf{v})^2.$$

Since $\cos^2 \theta = 1 - \sin^2 \theta$,

$$\begin{aligned}\|\mathbf{u}\|^2 \, \|\mathbf{v}\|^2 (1 - \sin^2 \theta) &= (\mathbf{u} \cdot \mathbf{v})^2, \\ \|\mathbf{u}\|^2 \, \|\mathbf{v}\|^2 \sin^2 \theta &= \|\mathbf{u}\|^2 \, \|\mathbf{v}\|^2 - (\mathbf{u} \cdot \mathbf{v})^2.\end{aligned}$$

Taking square roots,

$$\|\mathbf{u}\| \, \|\mathbf{v}\| \sin \theta = \sqrt{\|\mathbf{u}\|^2 \, \|\mathbf{v}\|^2 - (\mathbf{u} \cdot \mathbf{v})^2}.$$

The computation to verify that $\|\mathbf{u} \times \mathbf{v}\| = \sqrt{\|\mathbf{u}\|^2 \, \|\mathbf{v}\|^2 - (\mathbf{u} \cdot \mathbf{v})^2}$ is tedious but straightforward. Simply expand each expression and verify that they are the same. The details are left for an exercise. ■

WARNING Although the cross product is distributive over addition (parts (b) and (c) of Theorem 5), it is *not* commutative (part (a) of Theorem 5). Moreover, it is easily shown that the cross product is not associative either (see Exercise 29).

We remark that from part (g) of Theorem 5, it follows that vectors $\mathbf{u}$ and $\mathbf{v}$ are collinear if and only if $\mathbf{u} \times \mathbf{v} = \mathbf{0}$, since vectors are collinear just in case the angle θ between them is 0 or π (and hence, just in case $\sin \theta = 0$). Geometrically, part (g) of Theorem 5 tells us that the length of $\mathbf{u} \times \mathbf{v}$ is the area of the parallelogram determined by $\mathbf{u}$ and $\mathbf{v}$ (Figure 21).

FIGURE 21

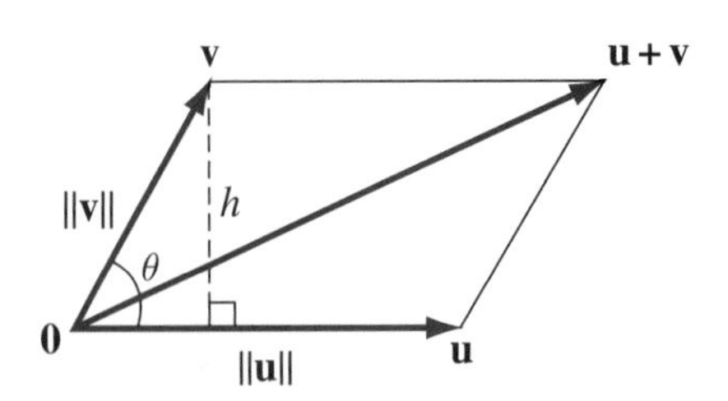

To see why this is so, let h be the height of the parallelogram from the endpoint of $\mathbf{v}$ to the side determined by $\mathbf{u}$. Then the area A of the parallelogram is given by $A = \|\mathbf{u}\| \, h$.

However, $\sin\theta = h/\|\mathbf{v}\|$, or $h = \|\mathbf{v}\|\sin\theta$. Making the substitution, we conclude that $A = \|\mathbf{u}\|\,\|\mathbf{v}\|\sin\theta$. Note that the triangle determined by the vectors **u** and **v** has half the area of the parallelogram determined by **u** and **v** (as the line segment from the end of **u** to the end of **v** divides the parallelogram into two congruent triangles). We conclude the following.

THEOREM 6

Suppose **u** and **v** are two nonzero vectors in $\mathbf{R}^3$. Then the area of the parallelogram determined by **u** and **v** is $\|\mathbf{u} \times \mathbf{v}\|$, and the area of the triangle determined by **u** and **v** is $\frac{1}{2}\|\mathbf{u} \times \mathbf{v}\|$. ■

EXAMPLE 7

Find the area of the triangle determined by the three points $P(1, 2, 3)$, $Q(-3, 2, 1)$, and $R(2, 4, 5)$.

Solution We translate $\overrightarrow{PQ}$ and $\overrightarrow{PR}$ to the origin to obtain the vectors $\mathbf{u} = \mathbf{q} - \mathbf{p}$ and $\mathbf{v} = \mathbf{r} - \mathbf{p}$. By Theorem 6, the area of the triangle determined by **u** and **v** is $\frac{1}{2}\|\mathbf{u} \times \mathbf{v}\|$ (Figure 22). Since $\mathbf{u} = (-4, 0, -2)$, and $\mathbf{v} = (1, 2, 2)$, we have

$$\begin{aligned}\frac{1}{2}\mathbf{u} \times \mathbf{v} &= \frac{1}{2}\left(\begin{vmatrix} 0 & -2 \\ 2 & 2 \end{vmatrix}, -\begin{vmatrix} -4 & -2 \\ 1 & 2 \end{vmatrix}, \begin{vmatrix} -4 & 0 \\ 1 & 2 \end{vmatrix}\right) \\ &= \frac{1}{2}\|(4, 6, -8)\| \\ &= \|(2, 3, -4)\| \\ &= \sqrt{29}.\end{aligned}$$

■

FIGURE 22

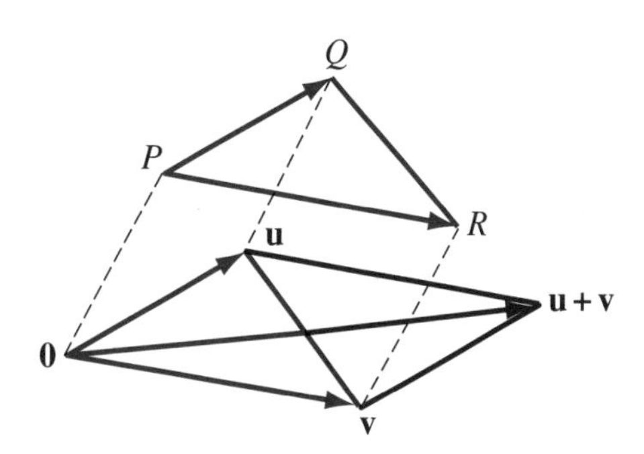

1.2 Exercises

In Exercises 1–10 let the vectors be given as follows:

$$\mathbf{s} = (-1, 2) \quad \mathbf{u} = (3, -1, 4)$$
$$\mathbf{t} = (2, 3) \quad \mathbf{v} = (-1, -3, 1)$$
$$\mathbf{w} = (-1, 1, 2)$$

Perform each indicated operation.

1. $\mathbf{s} \cdot \mathbf{t}$

2. $\mathbf{s} \cdot (3\mathbf{t})$

3. $\mathbf{u} \cdot \mathbf{v}$

4. $(3\mathbf{v}) \cdot (-2\mathbf{u})$

5. $\mathbf{u} \times \mathbf{v}$

6. $(\mathbf{v} \times \mathbf{u}) \cdot (-\mathbf{w})$

7. $\mathbf{v} \times (2\mathbf{u})$

8. $(2\mathbf{u} + \mathbf{v}) \times (3\mathbf{w})$

9. $(\mathbf{u} \times \mathbf{v}) \cdot \mathbf{w}$

10. $\mathbf{u} \cdot (\mathbf{w} \times \mathbf{v})$

In Exercises 11 and 12 determine if the angle formed by the two vectors is acute, right, or obtuse.

11. $(-2, 1), (5, 2)$

12. $(-1, 2, 3), (2, 0, 4)$

In Exercises 13–16 determine the cosine of the angle formed by the two vectors.

13. $(-2, 1), (5, 2)$

14. $(-1, 2, 3), (2, 0, 4)$

15. $(1, 0, 1), (-1, -1, 0)$

16. $(1, -1), (2, 0)$

In Exercises 17 and 18 find a vector that is orthogonal to both vectors.

17. $(1, 1, 3), (2, 1, -1)$

18. $(-1, 2, 3), (2, 0, 4)$

19. Find the area of the triangle with vertices at $(2, 1, 3)$, $(1, 0, 2)$, and $(-1, 1, 2)$.

20. Find the area of the parallelogram determined by the vectors $\mathbf{u} = (-1, 1, 0)$ and $\mathbf{v} = (2, 3, -1)$.

21. Prove that for any vector $\mathbf{u} = (a, b)$ in $\mathbf{R}^2$ the vector $\mathbf{r} = (-b, a)$ is orthogonal to $\mathbf{u}$.

In Exercises 22–25 find the projection of u on v and its length.

22. $\mathbf{u} = (1, 2), \mathbf{v} = (2, 1)$

23. $\mathbf{u} = (1, 2), \mathbf{v} = (-1, -2)$

24. $\mathbf{u} = (1, 0, -1), \mathbf{v} = (0, 1, 1)$

25. $\mathbf{u} = (1, 0, -1), \mathbf{v} = (0, -2, 0)$

The expressions in Exercises 26–28 are undefined. Explain why.

26. $\mathbf{u} \times (\mathbf{v} \cdot \mathbf{w})$

27. $\mathbf{u} \cdot (\mathbf{v} \cdot \mathbf{w})$

28. $\|\mathbf{u} \cdot \mathbf{v}\|$

29. Show that the cross product is not associative. That is, it is not always true that $(\mathbf{u} \times \mathbf{v}) \times \mathbf{w} = \mathbf{u} \times (\mathbf{v} \times \mathbf{w})$. (*Hint*: Try $(\mathbf{i} \times \mathbf{i}) \times \mathbf{j}$ and $\mathbf{i} \times (\mathbf{i} \times \mathbf{j})$, where $\mathbf{i} = (1, 0, 0)$ and $\mathbf{j} = (0, 1, 0)$.)

30. Use Theorem 5(g) to prove that $\mathbf{u}$ and $\mathbf{v}$ are collinear if and only if $\mathbf{u} \times \mathbf{v} = \mathbf{0}$.

31. Prove the properties of Theorem 1 that remain unproved.

32. Use Theorem 1(e) to prove the *parallelogram law*,

$$\|\mathbf{u} + \mathbf{v}\|^2 + \|\mathbf{u} - \mathbf{v}\|^2 = 2\|\mathbf{u}\|^2 + 2\|\mathbf{v}\|^2.$$

Interpret this result geometrically to conclude, "The sum of the squares of the sides of a parallelogram is equal to the sum of the squares of the diagonals."

33. Use Theorem 2 and the trigonometric identity, $\cos(\pi - \theta) = -\cos\theta$, to conclude that the angles formed by $\mathbf{u}$ and $\mathbf{v}$ and by $-\mathbf{u}$ and $\mathbf{v}$ are supplementary (add to π rad).

34. Prove the properties of Theorem 5 that remain unproved.

35. Let $\mathbf{u} = (u_1, u_2, u_3)$ be a unit vector in $\mathbf{R}^3$ in the direction of $\mathbf{v}$, that is, $\mathbf{u} = \left(\frac{1}{\|\mathbf{v}\|}\right)\mathbf{v}$. Use Theorem 2 to prove that u_1, u_2, and u_3 are the cosines of the angles formed by $\mathbf{v}$ with $(1, 0, 0)$, $(0, 1, 0)$, and $(0, 0, 1)$, respectively. The components of $\mathbf{u}$ are called the *direction cosines* of $\mathbf{v}$.

36. Use the result of Exercise 35 to find the direction cosines of the vector $(1, 2, -2)$.

A parallelepiped is a solid with six sides, each side being a parallelogram and opposite sides parallel.

37. Use Figure 23 to show that the *volume of the parallelepiped* determined by the vectors $\mathbf{u}$, $\mathbf{v}$, and $\mathbf{w}$ in $\mathbf{R}^3$ is given by

$$V = |(\mathbf{u} \times \mathbf{v}) \cdot \mathbf{w}|$$

(Recall that $V = Ah$, the area of the base times the perpendicular height.)

FIGURE 23

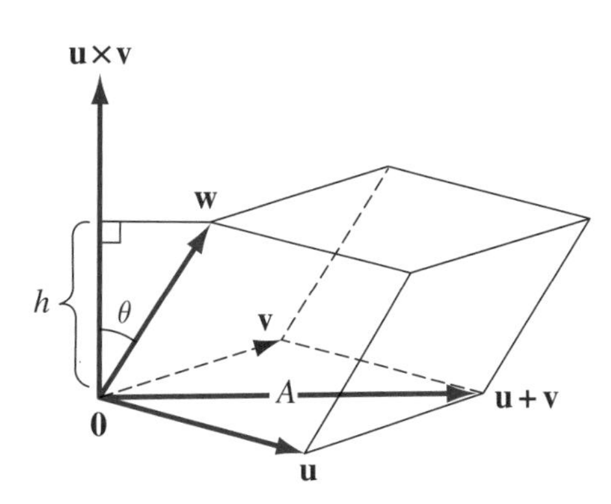

38. Use the result of Exercise 37 to find the volume of the parallelepiped determined by the vectors $\mathbf{u} = (2, 1, -1)$, $\mathbf{v} = (3, -1, 0)$, and $\mathbf{w} = (1, 1, -1)$.

39. Show that for all vectors **u**, **v**, and **w** in $\mathbf{R}^3$,

$$\mathbf{u} \bullet (\mathbf{v} - \mathbf{w}) = \mathbf{u} \bullet \mathbf{v} - \mathbf{u} \bullet \mathbf{w}.$$

40. Let **u**, $\mathbf{v}_1$, and $\mathbf{v}_2$ be vectors in $\mathbf{R}^3$, and let c_1 and c_2 be scalars. If **u** is orthogonal to both $\mathbf{v}_1$ and $\mathbf{v}_2$, prove that **u** is orthogonal to the vector $c_1\mathbf{v}_1 + c_2\mathbf{v}_2$.

41. Explain why the result of Exercise 40 implies that the sum of two vectors in $\mathbf{R}^3$ is coplanar with the vectors.

SPOTLIGHT *42.* If a force **F** (like those discussed in the Spotlight at the beginning of this chapter) is applied to an object and moves it an amount **s** (a certain distance and direction), then the *work* done is given by the dot product, $W = \mathbf{F} \bullet \mathbf{s}$. (If **F** is measured in newtons and **s** in metres, then the unit of measurement for W is the newton-metre.) Suppose that we drag a box across the floor by pulling on a rope that is attached to it and inclined at a 30° angle to the horizontal (Figure 24). What is the work done if we exert a constant force **F** of 200 N (newtons) on the rope in dragging the box 6 m?

FIGURE 24

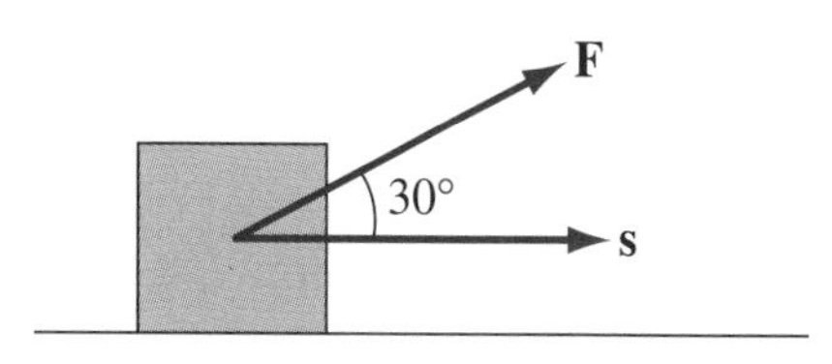

SPOTLIGHT *43.* Find the work done moving the box in Figure 24 if the force exerted, **F**, is 400 N and the distance **s** the box is moved is 5 m.

SPOTLIGHT *44.* When a twisting force **F** is applied to an object (for example, in using a wrench to loosen a nut, as in Figure 25), the magnitude of the resultant *torque* is given by $\|\mathbf{F} \times \mathbf{r}\|$, where **r** is the directed distance from the force to the object. (If **F** is measured in newtons and **r** in metres, the unit of measurement for torque is the newton-metre.) Suppose that a force of 70 N is applied at right angles to the wrench in Figure 25, which is 50 cm long. What is the magnitude of the torque on the nut?

FIGURE 25

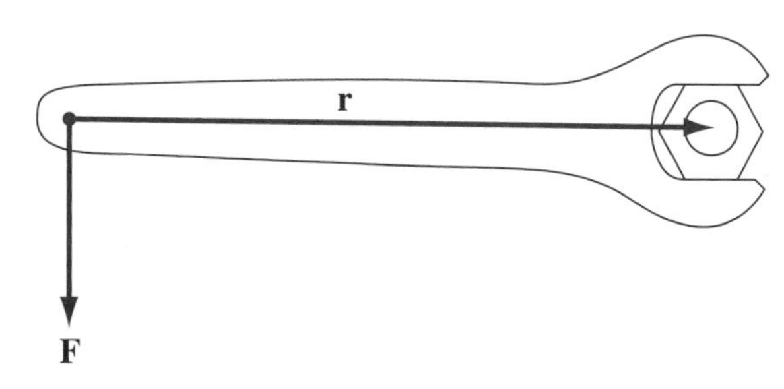

SPOTLIGHT *45.* Using the scenario described in Exercise 44, suppose the force is applied so that the angle between **F** and **r** is 80°. What is the magnitude of the torque in this case?

Computational Exercises

Use a calculator to help solve each problem in Exercises 46 and 47.

46. Find the cross product of the given vectors.
(a) $\mathbf{u} = (0.32, -1.41, 0.13)$, $\mathbf{v} = (5.19, 0.01, -3.68)$
(b) $\mathbf{u} = (1/2, 1/3, 1/4)$, $\mathbf{v} = (1/5, 1/6, 1/7)$

47. Find the angle formed by the following vectors.
(a) $\mathbf{u} = (-2, 1)$, $\mathbf{v} = (5, 2)$
(b) $\mathbf{u} = (-1, 2, 3)$, $\mathbf{v} = (2, 0, 4)$

1.3 Lines and Planes

A line in $\mathbf{R}^2$ can be written in the form $ax + by = c$ while a plane in $\mathbf{R}^3$ has the form $ax + by + cz = d$. Both share the commonality of being solutions to a single linear equation. But there are many ways to describe lines and planes, each with its own utility. In this section, we show how vector equations and functions can be used to describe lines and planes in $\mathbf{R}^3$.

Point-Parallel Form for a Line

Let $\mathbf{v}$ be a nonzero vector and let P be a point in $\mathbf{R}^3$. We wish to describe the line through P and parallel to $\mathbf{v}$ (Figure 26).

FIGURE 26

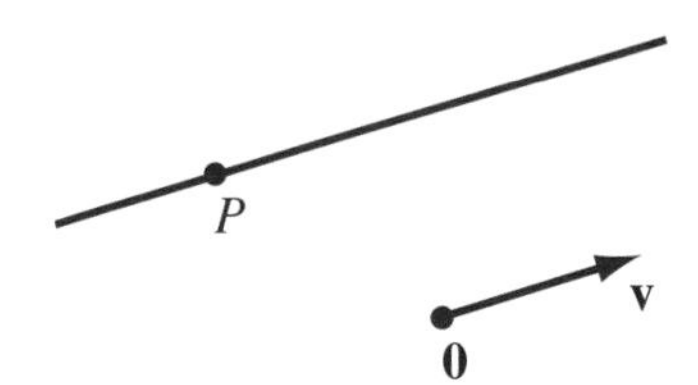

From Section 1.1, we know that a point is on the line determined by the vector $\mathbf{v}$ if and only if it has the form $t\mathbf{v}$ for some scalar t. From the properties of vector addition, the directed segment from P to the endpoint of $\mathbf{p} + t\mathbf{v}$ is parallel and equal in length to the vector $t\mathbf{v}$. But then the endpoint of $\mathbf{p} + t\mathbf{v}$ must lie on the line determined by P and the endpoint of $\mathbf{p} + \mathbf{v}$ (Figure 27). Thus, any point X on the line through P and parallel to $\mathbf{v}$ is the endpoint of a vector of the form $\mathbf{p} + t\mathbf{v}$. We write this as $\mathbf{x} = \mathbf{p} + t\mathbf{v}$ or, since $\mathbf{x}$ depends upon the value of t, as

$$\mathbf{x}(t) = \mathbf{p} + t\mathbf{v}. \tag{1}$$

FIGURE 27

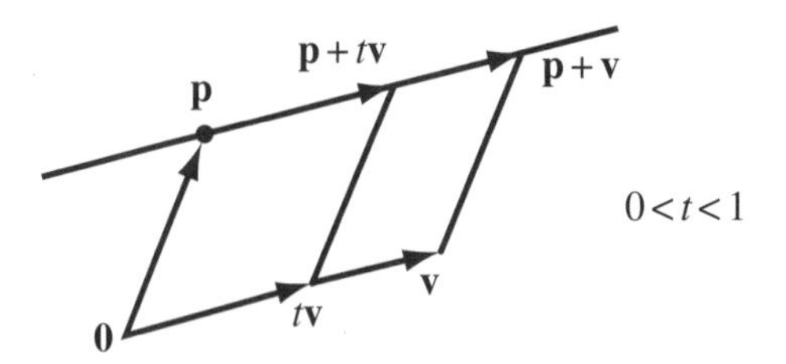

Equation (1) is called the **point-parallel form** for the line through the point P and parallel to the vector $\mathbf{v}$. Note that this equation is not an equation in the usual sense. It is instead a *function*, in which different vectors $\mathbf{x}(t)$ are obtained for different values of the variable t. For instance, $\mathbf{x}(0) = \mathbf{p}$ and $\mathbf{x}(1) = \mathbf{p} + \mathbf{v}$. Similarly, the endpoint of $\mathbf{x}(\frac{1}{2})$ is the midpoint of the segment from P to the endpoint of $\mathbf{p} + \mathbf{v}$, and $\mathbf{x}(2) = \mathbf{p} + 2\mathbf{v}$ and $\mathbf{x}(-1) = \mathbf{p} - \mathbf{v}$ are as pictured in Figure 28.

FIGURE 28

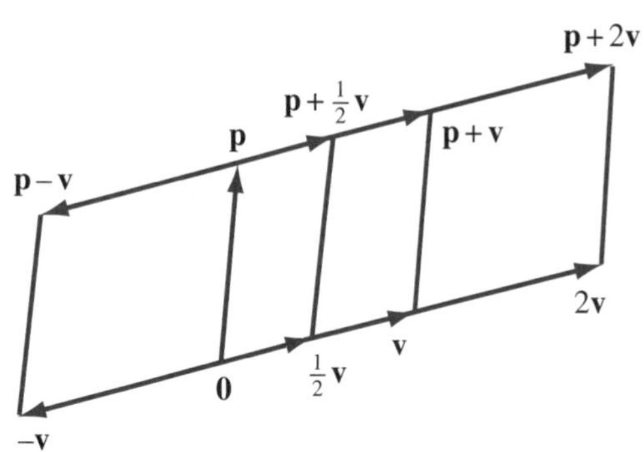

It is also important to note that, although we were discussing lines in $\mathbf{R}^3$, no special use of $\mathbf{R}^3$ was made in the discussion leading to Equation (1). Therefore, the interpretation of the equation holds in both $\mathbf{R}^2$ and $\mathbf{R}^3$.

EXAMPLE 1

Give a point-parallel form for the line through the point $P(2, 2, 0)$ and parallel to the vector $\mathbf{v} = (-3, -1, 2)$.

Solution From Equation (1), we have

$$\mathbf{x}(t) = (2, 2, 0) + t(-3, -1, 2).$$ ■

EXAMPLE 2

Give a point-parallel form for the line in $\mathbf{R}^2$ given by the equation $y = 2x - 1$.

Solution We need to find a point on the given line and a vector parallel to it. The given line has the y-intercept $(0, -1)$, so let $\mathbf{p} = (0, -1)$. Moreover, the line has slope $2 = 2/1$, so the vector $\mathbf{v} = (1, 2)$ is parallel to it. Thus, we have

$$\mathbf{x}(t) = (0, -1) + t(1, 2).$$ ■

It is sometimes convenient to write Equation (1) in terms of its coordinates. To do this, let $P(p_1, p_2, p_3)$, $V(v_1, v_2, v_3)$, and $X(x, y, z)$ be points in $\mathbf{R}^3$. Then Equation (1) becomes

$$\begin{aligned}(x, y, z) &= (p_1, p_2, p_3) + t(v_1, v_2, v_3)\\ &= (p_1 + tv_1, p_2 + tv_2, p_3 + tv_3).\end{aligned}$$

Equating coordinates, we have

$$\begin{aligned}x &= p_1 + tv_1,\\ y &= p_2 + tv_2,\\ z &= p_3 + tv_3.\end{aligned} \tag{2}$$

These equations are called **parametric equations** for the line. Taken together, they are equivalent to the original Equation (1). Of course, there are only equations for x and y in the case of $\mathbf{R}^2$.

EXAMPLE 3

Find parametric equations for the line through the point $(2, 2, 0)$ and parallel to the vector $(-3, -1, 2)$.

Solution From Example 1, we have the point-parallel form for the line

$$\mathbf{x}(t) = (2, 2, 0) + t(-3, -1, 2).$$

Equating coordinates, we have the parametric equations

$$\begin{aligned}x &= 2 - 3t,\\ y &= 2 - t,\\ z &= 2t.\end{aligned}$$ ■

Two-Point Form for a Line

Another way of determining a line geometrically is by two distinct points. Let P and Q be distinct points in $\mathbf{R}^2$ or $\mathbf{R}^3$. The **two-point form** for the line determined by P and Q is

$$\mathbf{x}(t) = (1 - t)\mathbf{p} + t\mathbf{q}, \tag{3}$$

where t ranges over all real numbers.

We now show that the set of all points determined by Equation (3) is the line that contains P and Q. Since $\mathbf{x}(0) = \mathbf{p}$ and $\mathbf{x}(1) = \mathbf{q}$, the set does contain the points P and Q. It can be shown that the set is a line by converting it to the form of Equation (1). Using properties of scalar multiplication and vector addition, we have

$$\begin{aligned}\mathbf{x}(t) &= (1 - t)\mathbf{p} + t\mathbf{q}\\ &= \mathbf{p} - t\mathbf{p} + t\mathbf{q}\\ &= \mathbf{p} + t(\mathbf{q} - \mathbf{p}).\end{aligned}$$

Thus the set of endpoints of all such $\mathbf{x}(t)$ is the line that contains the point P and is parallel to $\mathbf{q} - \mathbf{p}$ (Figure 29).

FIGURE 29

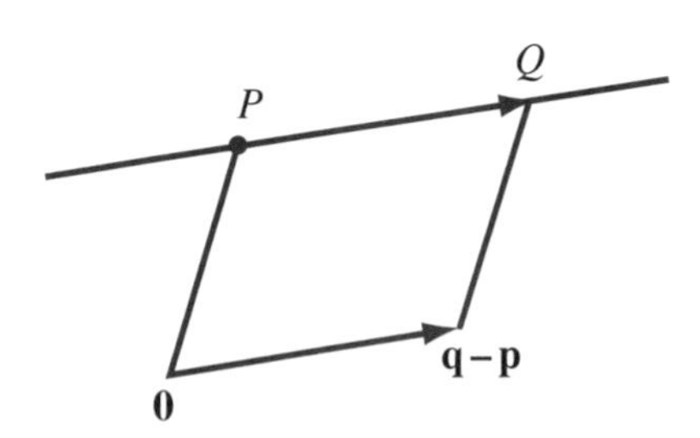

EXAMPLE 4 Describe the line through the points (1, 2, 3) and (−1, 1, 2) in both the two-point form and parametrically.

Solution From Equation (3), we have the two-point form,

$$\mathbf{x}(t) = (1 - t)(1, 2, 3) + t(-1, 1, 2).$$

To find the parametric equations, let $\mathbf{x} = (x, y, z)$ and perform the indicated operations on the right:

$$\begin{aligned}(x, y, z) &= (1 - t, 2 - 2t, 3 - 3t) + (-t, t, 2t)\\ &= (1 - 2t, 2 - t, 3 - t).\end{aligned}$$

Equating coordinates, we have the parametric equations,

$$\begin{aligned}x &= 1 - 2t,\\ y &= 2 - t,\\ \text{and}\quad z &= 3 - t.\end{aligned}$$

■

Point-Normal Form for a Plane

We turn our attention now to planes. Intuition tells us that the set of all directed line segments in $\mathbf{R}^3$ from a given point and perpendicular to a given directed line segment is a plane (Figure 30). We use this idea to develop an algebraic description of a plane.

FIGURE 30

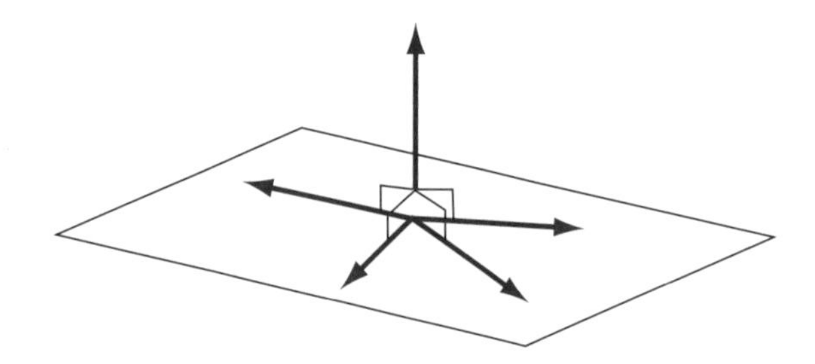

Let P be a point in $\mathbf{R}^3$, let $\mathbf{n}$ be a nonzero vector, and let Π denote the plane that contains the point P and is perpendicular to the line determined by $\mathbf{n}$ (Figure 31). Let X be an arbitrary point in Π, and consider the directed segment $\overrightarrow{PX}$. The vector $\mathbf{x} - \mathbf{p}$ is equivalent to $\overrightarrow{PX}$ (Theorem 4 of Section 1.1), and therefore is orthogonal to $\mathbf{n}$. In terms of dot product, we have

$$\mathbf{n} \cdot (\mathbf{x} - \mathbf{p}) = 0. \tag{4}$$

FIGURE 31

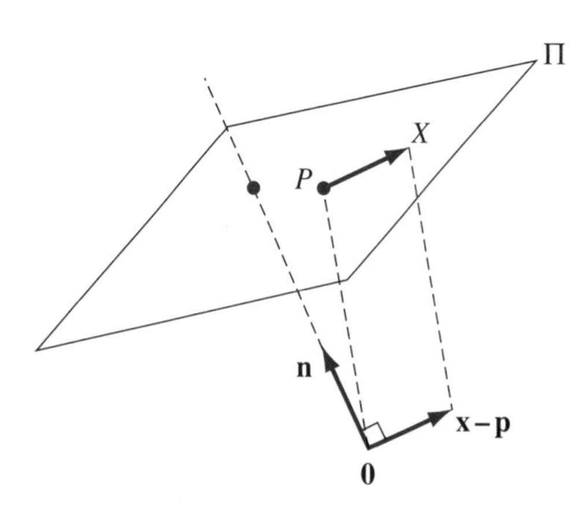

In other words, any point X in the plane Π satisfies Equation (4). Furthermore, if a point Y is not in Π, then $\mathbf{p} - \mathbf{y}$ cannot be orthogonal to $\mathbf{n}$. Thus, Equation (4) completely describes the plane Π. Equation (4) is called an equation in **point-normal form** for the plane, and $\mathbf{n}$ is called a **normal** for the plane. Note that any nonzero scalar multiple of $\mathbf{n}$ and any other point in the plane yield an equally valid equation, so the point-normal form is by no means unique.

EXAMPLE 5

Find a point-normal form for the plane that contains the point $P(2, 1, 3)$ with normal vector $\mathbf{n} = (1, 2, -2)$.

Solution From Equation (4) we have

$$\mathbf{n} \cdot (\mathbf{x} - \mathbf{p}) = 0, \quad \text{or} \quad (1, 2, -2) \cdot (\mathbf{x} - (2, 1, 3)) = 0. \quad \blacksquare$$

Sometimes an expanded form of Equation (1) is taken as the point-normal form. Specifically, let $\mathbf{n} = (n_1, n_2, n_3)$, $\mathbf{p} = (p_1, p_2, p_3)$, and $\mathbf{x} = (x, y, z)$. Then $\mathbf{n} \cdot (\mathbf{x} - \mathbf{p}) = 0$ becomes

$$(n_1, n_2, n_3) \cdot ((x, y, z) - (p_1, p_2, p_3)) = 0$$
$$(n_1, n_2, n_3) \cdot (x - p_1, y - p_2, z - p_3) = 0$$
$$n_1(x - p_1) + n_2(y - p_2) + n_3(z - p_3) = 0. \tag{5}$$

Equation (5) is equivalent to Equation (4), and it often goes by the same name. For instance, by letting $\mathbf{x} = (x, y, z)$ and performing the subtraction and dot product operations, the equation of Example 1 may be written

$$(x - 2) + 2(y - 1) - 2(z - 3) = 0.$$

The most efficient procedure for writing a "clean" equation for a plane through P with normal vector $\mathbf{n}$ results from expanding Equation (4) in a slightly different manner,

$$\begin{aligned}\mathbf{n}\cdot(\mathbf{x}-\mathbf{p}) &= 0\\ \mathbf{n}\cdot\mathbf{x}-\mathbf{n}\cdot\mathbf{p} &= 0\\ \mathbf{n}\cdot\mathbf{x} &= \mathbf{n}\cdot\mathbf{p}.\end{aligned}$$

Letting $\mathbf{n} = (a, b, c)$, $\mathbf{x} = (x, y, z)$, and $\mathbf{n}\cdot\mathbf{p} = d$, we obtain an equation for the plane in **standard form**:

$$ax + by + cz = d \tag{6}$$

EXAMPLE 6

Find an equation in standard form for the plane through the given point with normal $\mathbf{n}$.

a. $(-1, 1, 2)$, $\mathbf{n} = (2, 3, 4)$ b. the origin, $\mathbf{n} = (2, -1, 3)$

Solution

a. Since $(2, 3, 4)\cdot(-1, 1, 2) = 9$,

Equation (6) yields

$$2x + 3y + 4z = 9.$$

b. Here $\mathbf{p} = \mathbf{0}$ and $\mathbf{n}\cdot\mathbf{p} = 0$, so we have simply

$$2x - y + 3z = 0.$$

■

Suppose that we are given an equation for a plane in standard form

$$ax + by + cz = d, \quad (a, b, c) \neq \mathbf{0},$$

and we wish to transform it into point-normal form. The plane in question must be orthogonal to the vector $\mathbf{n} = (a, b, c)$, but we still need a point in the plane to get a point-normal form for it. One such point is easily found. If $a \neq 0$, then $\left(\frac{d}{a}, 0, 0\right)$ satisfies the equation. If $b \neq 0$, we can use $\left(0, \frac{d}{b}, 0\right)$. Similarly, if $c \neq 0$, we can use $\left(0, 0, \frac{d}{c}\right)$. Letting P be any of these, it is a simple matter to show that $\mathbf{n}\cdot(\mathbf{x}-\mathbf{p}) = 0$ reduces to the original equation $ax + by + cz = d$.

EXAMPLE 7

Find an equation in point-normal form for the plane described by the equation $3x - y + 2z = 5$.

Solution A normal vector is $(3, -1, 2)$, and the point $(0, -5, 0)$ satisfies the equation. Thus,

$$(3, -1, 2)\cdot(\mathbf{x} - (0, -5, 0)) = 0$$

is one such equation. ■

EXAMPLE 8

Find the point of intersection of the line $\mathbf{x}(t) = (2, 1, 3) + t(2, -2, 1)$ with the plane $x + 2y - z = 7$.

Solution We substitute the parametric equations for the line,

$$\begin{aligned}x &= 2 + 2t,\\ y &= 1 - 2t,\\ z &= 3 + t,\end{aligned}$$

into the equation for the plane,

$$(2 + 2t) + 2(1 - 2t) - (3 + t) = 7.$$

Solving for t, we obtain $t = -2$. Thus, the point is

$$\begin{aligned}\mathbf{x}(-2) &= (2, 1, 3) + (-2)(2, -2, 1)\\ &= (-2, 5, 1).\end{aligned}$$

■

If two planes are not parallel, then they intersect in a line. Finding that line requires solving a system of two equations; we'll investigate this further in the next chapter.

EXAMPLE 9

Find the parametric questions for the line of intersection of the planes $x - 2y + z = 3$ and $-x + y + z = -1$.

Solution We need to solve the system of equations

$$\begin{aligned} x - 2y + z &= 3 \\ -x + y + z &= -1. \end{aligned}$$

Adding the two equations, we find that $-y + 2z = 2$, that is, $y = 2z - 2$. Substituting this into the first equation, we find that $x - 2(2z - 2) + z = 3$, and we find that $x = 3z - 1$. By setting $z = t$ for some arbitrary number t, the solution to the two equations is the line

$$\begin{aligned} x &= 3t - 1, \\ y &= 2t - 2, \\ z &= t. \end{aligned}$$

As a check, we can substitute the line into the equation of each plane and ensure it does indeed lie on both planes (for *all* values of t):

$$(3t - 1) - 2(2t - 2) + t = 3 \text{ and } -(3t - 1) + (2t - 2) + t = -1. \quad \blacksquare$$

Plane Determined by Three Points

A plane Π in $\mathbf{R}^3$ is determined by three noncollinear points P, Q, and R. If three such points are specified, the cross product of two vectors can be used to find a normal vector. Specifically, the vector $\mathbf{q} - \mathbf{p}$ is equivalent to $\overrightarrow{PQ}$, and $\mathbf{r} - \mathbf{p}$ is equivalent to $\overrightarrow{PR}$ (Figure 32). Since P, Q, and R are not collinear, $\mathbf{q} - \mathbf{p}$ and $\mathbf{r} - \mathbf{p}$ are noncollinear vectors, and by Theorem 4 of Section 1.2,

$$\mathbf{n} = (\mathbf{q} - \mathbf{p}) \times (\mathbf{r} - \mathbf{p})$$

is a nonzero vector orthogonal to both $\mathbf{q} - \mathbf{p}$ and $\mathbf{r} - \mathbf{p}$. Thus, $\mathbf{n}$ is normal to the plane determined by $\mathbf{q} - \mathbf{p}$ and $\mathbf{r} - \mathbf{p}$, and hence to Π as well. Using this vector, $\mathbf{n}$, the equation

$$\mathbf{n} \cdot (\mathbf{x} - \mathbf{p}) = 0$$

describes the plane Π.

FIGURE 32

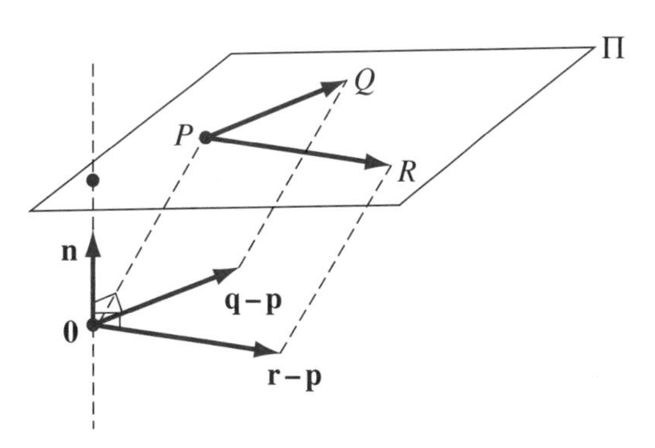

EXAMPLE 10

Find an equation for the plane determined by the points $P(2, 1, 3)$, $Q(4, 2, -1)$, and $R(6, 2, 2)$.

Solution For a normal vector, we have

$$\mathbf{n} = (\mathbf{q} - \mathbf{p}) \times (\mathbf{r} - \mathbf{p}) = (2, 1, -4) \times (4, 1, -1) = (3, -14, -2).$$

Thus a point-normal form for the plane is

$$(3, -14, -2) \cdot (\mathbf{x} - (2, 1, 3)) = 0$$

or, in standard form,

$$3x - 14y - 2z = -14. \quad \blacksquare$$

Parametric Form of a Plane

Lines were defined in point-parallel form by a point P and a vector on the line. Like lines, planes also have a parametric form, though with two variables rather than one. Suppose that a plane Π contains the point $P(p_1, p_2, p_3)$ and is parallel to the noncollinear vectors $\mathbf{u} = (u_1, u_2, u_3)$ and $\mathbf{v} = (v_1, v_2, v_3)$, that is, the line determined by $\mathbf{u}$ and the line determined by $\mathbf{v}$ each do not intersect the plane (or both lines lie wholly in the plane). Then a point $X(x, y, z)$ lies in the plane Π if and only if $\mathbf{x} = \mathbf{p} + s\mathbf{u} + t\mathbf{v}$ where s and t are some scalars (see Figure 33), and this means that

$$\begin{aligned} x &= p_1 + su_1 + tv_1, \\ y &= p_2 + su_2 + tv_2, \\ z &= p_3 + su_3 + tv_3. \end{aligned} \tag{7}$$

These are the parametric equations for the plane, in parameters s and t.

FIGURE 33

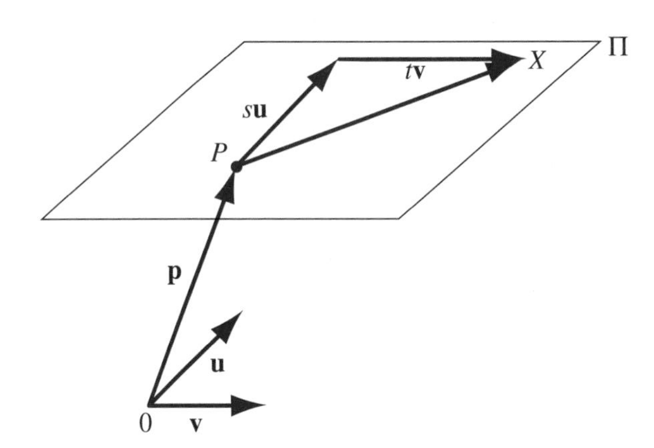

EXAMPLE 11

Find the parametric equations for the planes containing the point $(1, 2, -1)$ and parallel to the vectors $\mathbf{u} = (0, 1, 1)$ and $\mathbf{v} = (1, 0, 2)$.

Solution From Equation (7) the plane has parametric form

$$\begin{aligned} \mathbf{x} = \mathbf{p} + s\mathbf{u} + t\mathbf{v} &= (1, 2, -1) + s(0, 1, 1) + t(1, 0, 2) \\ &= (1 + t, 2 + s, -1 + s + 2t), \end{aligned}$$

that is,

$$\begin{aligned} x &= 1 + t, \\ y &= 2 + s, \\ z &= -1 + s + 2t. \end{aligned} \quad \blacksquare$$

We point out that using the methods of Example 10, a normal to the plane of Example 11 is

$$\mathbf{u} \times \mathbf{v} = (0, 1, 1) \times (1, 0, 2) = (2, 1, -1)$$

and hence point-normal form of the plane is

$$(2, 1, -1) \cdot (x - 1, y - 2, z - (-1)) = 0.$$

that is, $2x + y - z = 5$. You can easily verify that substituting the parametric equations $x = 1 + t$, $y = 2 + s$, $z = -1 + s + 2t$ satisfies this equation.

Remark **Determining the Distance between a Point and a Plane**
Determining the (perpendicular) distance between a given point P and a given plane Π is a standard problem in calculus and analytic geometry, but it can be done more easily using linear algebra. Let the equation of the plane Π be $\mathbf{n} \cdot (\mathbf{x} - \mathbf{q}) = 0$ and let P' be the point of intersection of the plane and the line through P that is perpendicular to it. Then the required distance is just $d(P, P') = \|\mathbf{p} - \mathbf{p}'\|$.

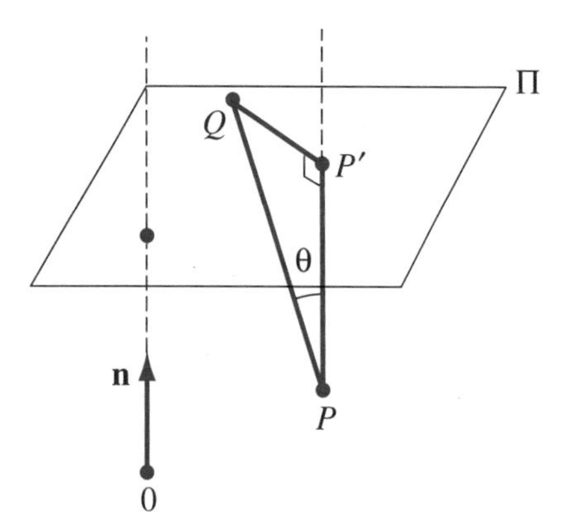

Now in the right triangle with vertices Q, P, and P',

$$\cos\theta = \frac{\|\mathbf{p} - \mathbf{p}'\|}{\|\mathbf{q} - \mathbf{p}\|},$$

so $$\|\mathbf{p} - \mathbf{p}'\| = \|\mathbf{q} - \mathbf{p}\| \cos\theta = \frac{\|\mathbf{n}\|\,\|\mathbf{q} - \mathbf{p}\| \cos\theta}{\|\mathbf{n}\|}.$$

But $\|\mathbf{n}\|\,\|\mathbf{q} - \mathbf{p}\| \cos\theta = |\mathbf{n} \cdot (\mathbf{q} - \mathbf{p})|$. (The absolute value sign is needed here because the angle between $\mathbf{n}$ and $\mathbf{q} - \mathbf{p}$ is either θ or $180° - \theta$, depending on whether or not $\mathbf{n}$ has the same direction as $\mathbf{q} - \mathbf{p}$.) Thus, the distance between the point P and the plane Π is given by

$$\|\mathbf{p} - \mathbf{p}'\| = \frac{|\mathbf{n} \cdot (\mathbf{q} - \mathbf{p})|}{\|\mathbf{n}\|}$$

For example, to find the distance from the point $P(2, 1, 3)$ to the plane $2x - 3y - z = 1$, we first need to express $2x - 3y - z = 1$ in point-normal form. One such form is $(2, -3, -1) \cdot (\mathbf{x} - (0, 0, -1)) = 0$. Thus, we can take $\mathbf{n} = (2, -3, -1)$ and $\mathbf{q} = (0, 0, -1)$. Substituting into the equation derived above, we have

$$\frac{|\mathbf{n} \cdot (\mathbf{q} - \mathbf{p})|}{\|\mathbf{n}\|} = \frac{|(2, -3, -1) \cdot (-2, -1, -4)|}{\|(2, -3, -1)\|} = \frac{3}{\sqrt{14}}.$$

1.3 Exercises

In Exercises 1–9 give a point-parallel or two-point form, and parametric equations for the $\mathbf{R}^2$ or $\mathbf{R}^3$ line described.

1. Through $P(2, 1, -3)$ and parallel to $\mathbf{v} = (1, 2, 2)$

2. Through $P(3, -1)$ and parallel to $\mathbf{v} = (2, 3)$

3. Through $P(2, -3, 1)$ and parallel to the x-axis

4. Through $P(1, 2, -1)$ and $Q(2, -1, 3)$

5. Through $P(2, 0, -2)$ and $Q(1, 4, 2)$

6. Through $P(1, 2, 3)$ and perpendicular to the plane $2x - y - 2z = 4$

7. Through $P(2, 4, 5)$ and perpendicular to the plane $5x - 5y - 10z = 2$

8. Through $P(3, -2)$ and perpendicular to the line $3x - 5y = 1$

9. Through $P(1, -1)$ and perpendicular to the line $x - 3y = 4$

In Exercises 10–13 give equations in both point-normal and standard form of the plane described.

10. Through $P(1, 2, 3)$ with normal $\mathbf{n} = (-3, 0, 1)$

11. Through the origin with normal $\mathbf{n} = (2, 1, 3)$

12. Through $P(-1, 2, 3)$, $Q(3, 2, 5)$, and $R(-1, 1, 2)$

13. Through $P(-1, 2, 3)$, $Q(0, 2, 0)$, and $R(1, 1, 3)$

In Exercises 14–16 convert the equation in standard form to point-normal form for a plane in $\mathbf{R}^3$.

14. $2x - y + z = 5$ *15.* $2x + 3y = 1$ *16.* $z = 0$

If, in $\mathbf{R}^2$, *P* is a point and n is a nonzero vector, then $\mathbf{n} \cdot (\mathbf{x} - \mathbf{p}) = 0$ is an equation of the line through *P* perpendicular to the direction n. Borrowing terminology from $\mathbf{R}^3$, we call this the *point-normal* form for the given line. Give the point-normal form for each $\mathbf{R}^2$ line described in Exercises 17–20.

17. Through $P(-1, 2)$ with normal $\mathbf{n} = (2, 1)$

18. Through $P(2, -1)$ with normal $\mathbf{n} = (0, 2)$

19. Through $P(-2, 5)$ and perpendicular to the line $x + 2y = 5$

20. Through $P(1, -3)$ and perpendicular to the line $3x - 4y = 1$

21. For the plane containing the point $(-1, 2, 3)$, $(0, 1, -1)$, and $(1, 0, 2)$, find both the standard and parametric forms of the plane.

22. Let $\mathbf{p} = (p_1, p_2)$, $\mathbf{u} = (u_1, u_2)$, and $\mathbf{x}(t) = \mathbf{x} = (x, y)$. Show that the point-parallel form $\mathbf{x}(t) = \mathbf{p} + t\mathbf{u}$ describes the same line in $\mathbf{R}^2$ as the point-normal form $\mathbf{n} \cdot (\mathbf{x} - \mathbf{p}) = 0$, where $\mathbf{n} = (u_2, -u_1)$.

23. Convert the line $\mathbf{x}(t) = (2, 3) + t(1, -3)$ into point-normal form. (See Exercise 22.)

24. Convert the line $2x - y = 5$ into point-parallel form.

25. Find the distance from the point (3, 2, 3) to the plane $x - 2y + z = 5$. (See the *Remark* earlier in this section.)

26. Show that this section's formula for the distance from a point to a plane in $\mathbf{R}^3$ yields the distance from a point to a line in $\mathbf{R}^2$.

27. Find the distance from the point $(2, -3)$ to the line $x + 2y = 5$. (See Exercise 26.)

28. Find the point of intersection of the line $\mathbf{x}(t) = (3, 0, 1) + t(-2, 1, 0)$ with the plane $3x + 2y - 3z = 3$.

29. Find the point of intersection of the line $\mathbf{x}(t) = (2, 1, 1) + t(-1, 0, 4)$ with the plane $x - 3y - z = 1$.

30. Let Π be the plane described by $\mathbf{n} \cdot (\mathbf{x} - \mathbf{q}) = 0$, and let l be the line $\mathbf{x}(t) = \mathbf{p} + t\mathbf{u}$ $(\mathbf{n}, \mathbf{u} \neq \mathbf{0})$. Show the following.

(a) If $\mathbf{n} \cdot \mathbf{u} = 0$, then l is contained in Π or *parallel* to it (they have no points in common).

(b) If $\mathbf{n} \cdot \mathbf{u} \neq 0$, then l and Π intersect in the point with coordinates

$$\mathbf{p} + \left(\frac{\mathbf{n} \cdot (\mathbf{q} - \mathbf{p})}{\mathbf{n} \cdot \mathbf{u}}\right)\mathbf{u},$$

which is at a distance of

$$\left|\frac{\mathbf{n} \cdot (\mathbf{q} - \mathbf{p})}{\mathbf{n} \cdot \mathbf{u}}\right| \|\mathbf{u}\|$$

from P.

31. Use Exercise 30 to show that the line $\mathbf{x}(t) = (2, 4, 5) + t(1, 0, 1)$ and the plane $2x - y - 2z = 5$ are parallel.

32. Two planes are *parallel* if they have no points in common. Show that the planes $2x - 3y + z = 1$ and $4x - 6y + 2z = 5$ are parallel. (*Hint*: First show that if they have one point in common, they are one and the same plane.)

33. Show that the planes $3x - 2y + z = 1$ and $x - y - 2z = 1$ are not parallel *without* finding any point of intersection.

34. Find the distance from the point $P(1, -1, 3)$ to the line l given by $\mathbf{x}(t) = (-1, 0, 2) + t(2, 1, 4)$. (*Hint*: Find the point of intersection of l with with plane through P and perpendicular to l.)

35. The lines $\mathbf{x}(t) = (2, 1, 3) + t(0, -2, -1)$ and $\mathbf{y}(s) = (3, 1, 5) + s(1, 0, 2)$ intersect at (2, 1, 3) (let $t = 0$ and $s = -1$). Find the cosine of the angle of intersection. (*Hint*: Just use the parallel vectors. The angle θ is restricted to $0 \leq \theta \leq \pi/2$ radians, so use Exercise 29 of Section 1.2.)

36. By considering complementary angles (ones whose sum is $\pi/2$ radians), it can be shown that the angle between two intersecting lines has the same measure as the angle formed by normal vectors to the lines (or its supplement). Find the cosine of the angle between the $\mathbf{R}^2$ lines $2x - y = 3$ and $x + 5y = 9$.

37. The *angle between two intersecting planes* is that of the lines formed by intersecting the planes with a third plane perpendicular to the line of intersection of the planes. As in Exercise 36, it suffices to use normal vectors to the planes. Find the cosine of the angle between the planes $3x - y + z = 3$ and $2x + 4y - 2z = 1$.

1.4 APPLICATION—Vectors and Interactive Multimedia

We have seen that vectors in $\mathbf{R}^2$ can be used to denote the location of a point in the plane. Vectors can be used to derive formulas for locating and moving objects when animating on the computer. But to do so, we need to talk about how programming languages typically assign the locations to objects on the computer screen.

A *window* is simply a rectangular area that will be drawn on the computer screen. It is made up of very tiny smaller rectangles, called *pixels*; when drawn on the screen, these pixels are so small that we can think of them as being points in the plane $\mathbf{R}^2$ of the window. The origin is often taken to be one of the corners of the window; for our discussion here, we shall assume that the origin is the lower left corner, with the x- and y-axes located as usual. Our association of vectors to points in the plane now allows

FIGURE 34

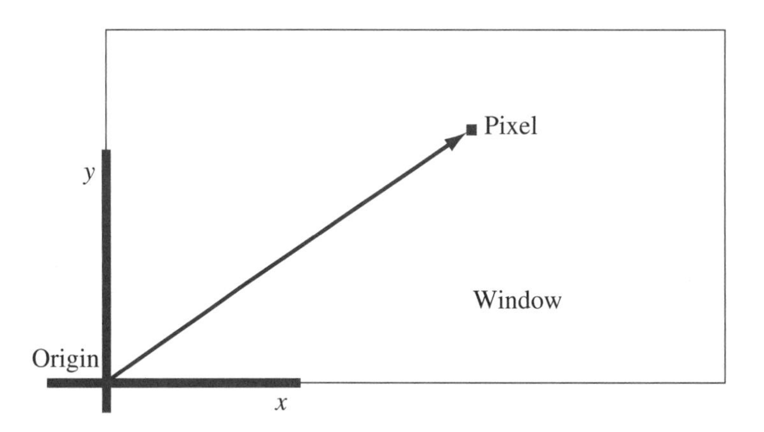

us to associate a vector to each pixel in a window. In the diagram below, if the pixel is 330 pixels to the right of the lower left corner and 235 pixels up, then it is associated with the vector (330, 235).

FIGURE 35

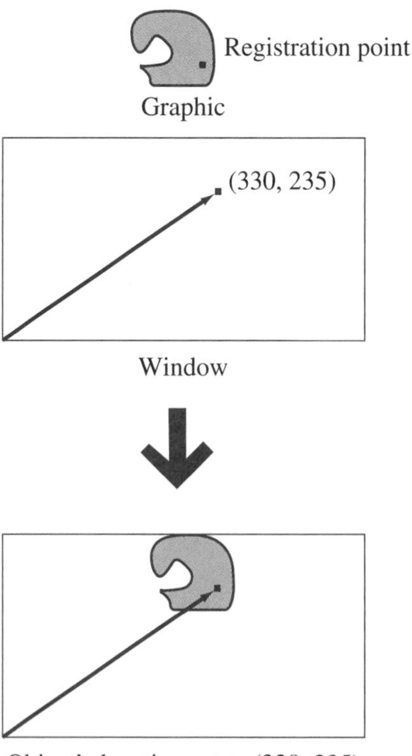

Graphics consist of pixels, in their own window. To assign a location on screen to a graphic, typically a special pixel of the object (the object's *registration point*) is chosen, whose placement on the screen will act as the location of the object. So, for example, if we set the location of an object called "graphic.bmp" to point (330, 235), the object is drawn in the main window with its registration point at (330, 235).

There is one other object whose location is often needed—the location of the cursor (as determined by the mouse or the trackpad). The cursor is like any other object on the screen, and its registration point is often called the *hot spot* of the cursor. We use $\mathbf{m}$ and $\mathbf{g}$ to denote the vectors that represent the current location of the registration points of the cursor and the graphic, respectively. We also let $\mathbf{m}_0$ and $\mathbf{g}_0$ denote the initial location of the cursor and object, respectively.

FIGURE 36

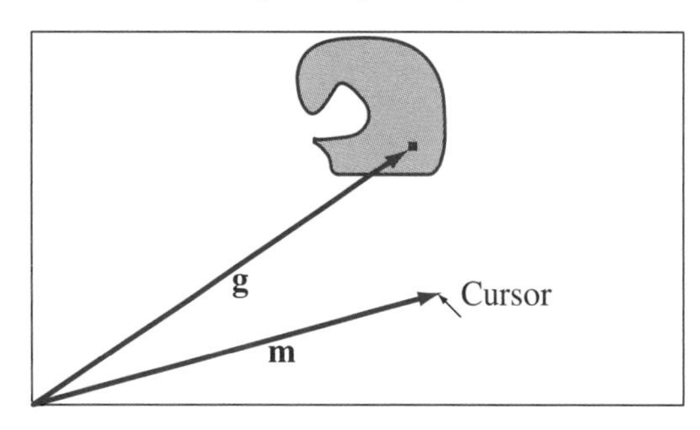

Vector operations are extremely useful for describing and manipulating the location of objects.

EXAMPLE 1

A user clicks the mouse on an object and, while holding down the mouse, moves the cursor around the screen. Using vectors, explain how to reset the location of the object as it is dragged.

Solution What we do not want to do is to reset the location of the object to the location of the cursor. The problem is that the location of the cursor and the object's location (as defined by the location of its registration point) are likely not the same initially, and so if the program resets the location of the object to the cursor location, that is, if it sets $\mathbf{g} = \mathbf{m}$, then the object will jump immediately to the location of the cursor's hot spot—not at all what is desired.

FIGURE 37

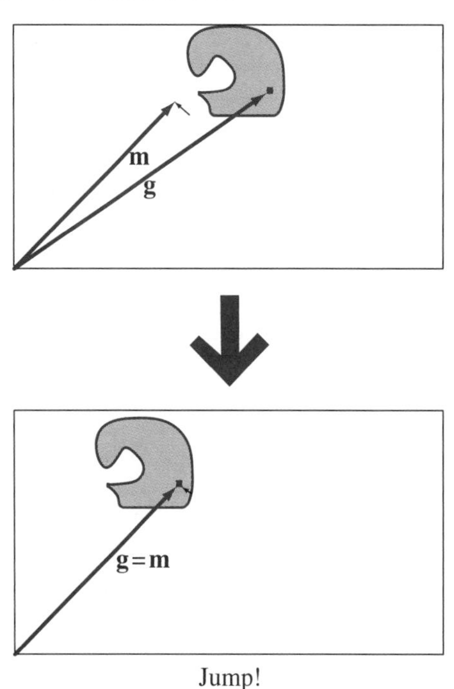

Notice that what we need to do is to find the location of the object's registration point with respect to the initial mouse click and then, when we move the cursor, keep the object's registration point offset by the same amount. Let $\mathbf{d}_0$ be the vector that starts at the mouse click's hot spot and ends at the registration point of the object. Clearly $\mathbf{d}_0 = \mathbf{g}_0 - \mathbf{m}_0$.

FIGURE 38

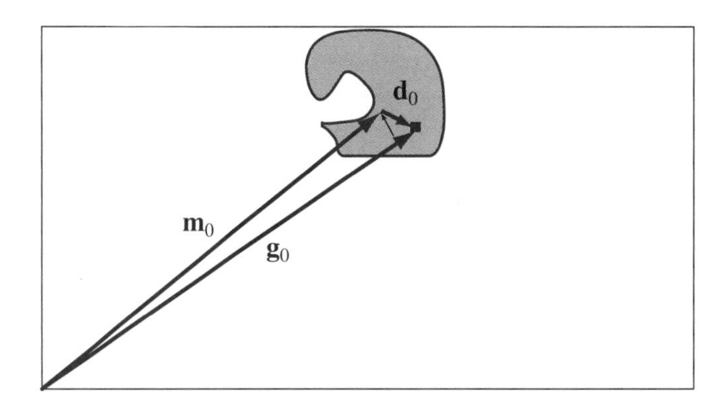

When we move the cursor to a new location (and thus have a new value for $\mathbf{m}$), we want to move the object's registration point to the new cursor's hot spot, offset by $\mathbf{d}_0$. This is denoted in vector notation by updating $\mathbf{g}$ to the current value of $\mathbf{m} + \mathbf{d}_0$.

FIGURE 39

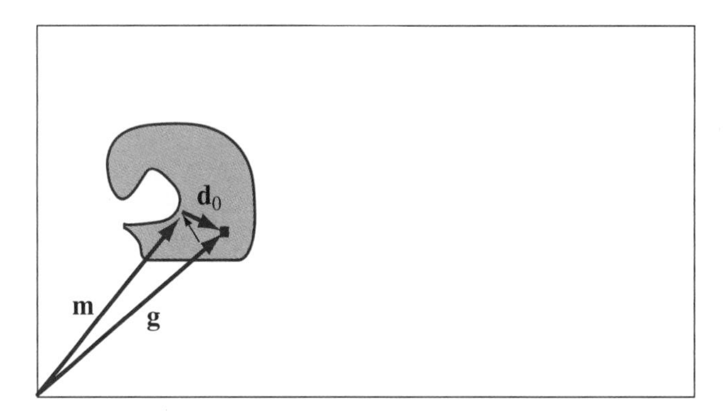

Notice that when the mouse is clicked and not yet moved, then the value of $\mathbf{m} + \mathbf{d}_0$ is $\mathbf{m} + (\mathbf{g}_0 - \mathbf{m}_0) = \mathbf{g}_0$, the initial location of the object, so that $\mathbf{g}$ is not updated (and the object is not yet relocated). ■

The use of vectors for interactive multimedia is not limited to movements of objects across the screen. Here is another example, a dial that rotates under the user's control.

EXAMPLE 2

We have a graphic on the screen that looks like a dial (Figure 40). We want to allow the user to rotate the dial by clicking on the notch at the top of the dial and dragging. How do we determine how many degrees to rotate the dial from the location of the cursor?

Solution We need to determine how many degrees to rotate the dial from the location of the cursor. We shall label the centre of the dial by X, the point at the top of the notch by Z, and the current cursor location by M (which we assume is not X, as then the angle is not well-defined). As well, we label the vectors $\overrightarrow{XZ}$ and $\overrightarrow{XM}$ as in Figure 41. We denote the angle between these vectors by θ.

Note that the angle θ can be found from the formula

$$\mathbf{z} \cdot \mathbf{p} = \|\mathbf{z}\| \, \|\mathbf{p}\| \cos \theta,$$

that is,

$$\cos \theta = \frac{\mathbf{z} \cdot \mathbf{p}}{\|\mathbf{z}\| \, \|\mathbf{p}\|}.$$

FIGURE 40

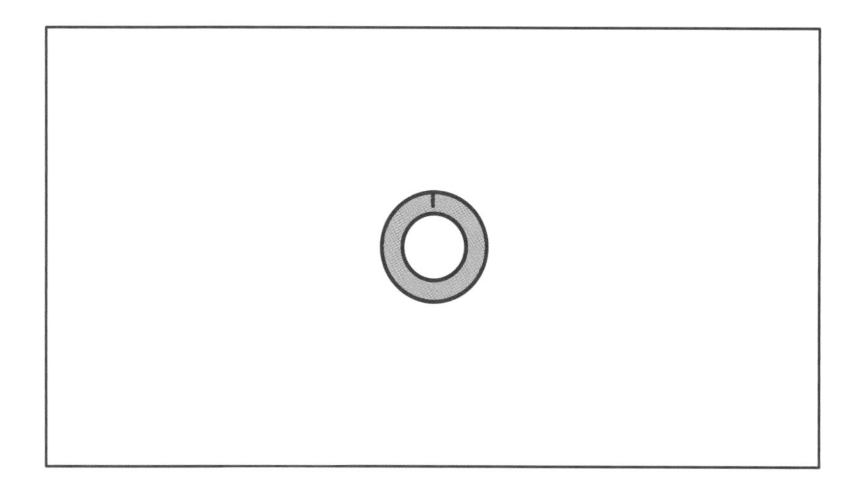

To determine the angle, we can use the *inverse trigonometric function* $\cos^{-1} x$ (sometimes written as $\arccos x$). But this only gives us an angle between 0 and π rad (or 0° and 180°). We still need to determine whether the angle is to the left or right of **z**, but if $X = (a, b)$ and $M = (c, d)$, then M is to the right of Z if $c > a$ and to the left of Z if $c < a$.

FIGURE 41

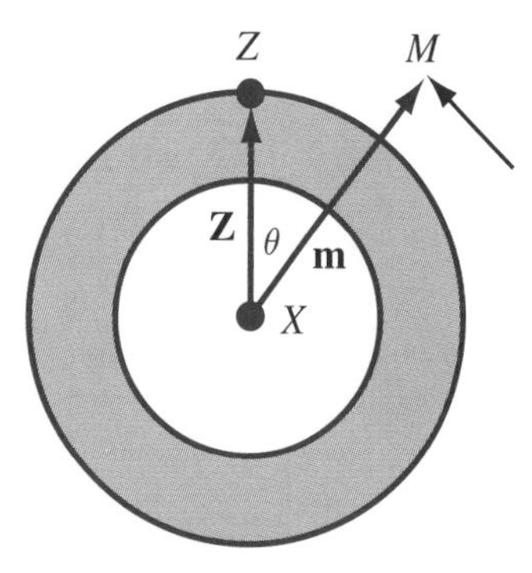

We now have all we need to calculate how many degrees to rotate the dial. We first calculate **z** and **p**. Then we use these and the formula for the dot product to calculate $\cos \theta$. Using the inverse cosine function, we find the angle θ between the vectors and, finally, we compare the x-coordinates a and c of X and M, respectively. If $c \geq a$, then we rotate the dial by θ clockwise (that is, to the right), and if $c < a$, then we rotate the dial by θ counterclockwise (that is, to the left). ■

These two examples only scratch the surface of how vectors can be used in interactive animation. The exercises provide you an opportunity to explore some others.

1.4 Exercises

1. In Example 2, if $c = a$, show that you rotate the dial 180° if $d < b$. What happens if $d \geq b$?

In Exercises 2–6 you are given an object whose registration point is at $P = (a, b)$. Explain, using vectors, how to reset the location of the object under the given user interaction.

2. Drag the object, provided that the mouse click is on the object but at least 5 pixels in distance away from the registration point of the object.
3. Keep the object always 10 pixels to the right and 20 pixels up from the cursor's location, as long as the mouse is down.
4. Move the object halfway toward the cursor when the mouse is down.
5. Drag the object together with a second object on the screen (at the appropriate location relative to the first object).
6. Drag a dial where the notch is horizontal at the right edge of the dial.

CHAPTER SUMMARY

Operations on Vectors in R^3

(For $\mathbf{R}^2$, delete the third component)

Let $\mathbf{u} = (u_1, u_2, u_3)$ and $\mathbf{v} = (v_1, v_2, v_3)$. Then

(a) $\|\mathbf{u}\| = \sqrt{u_1^2 + u_2^2 + u_3^2}$

(b) $d(\mathbf{u}, \mathbf{v}) = \|\mathbf{u} - \mathbf{v}\|$

(c) $\mathbf{u} + \mathbf{v} = (u_1 + v_1, u_2 + v_2, u_3 + v_3)$

(d) $\mathbf{u} - \mathbf{v} = (u_1 - v_1, u_2 - v_2, u_3 - v_3)$

(e) $c\mathbf{u} = (cu_1, cu_2, cu_3)$

(f) $\mathbf{u} \cdot \mathbf{v} = u_1v_1 + u_2v_2 + u_3v_3$

(g) $\mathbf{u} \times \mathbf{v} = (w_1, w_2, w_3)$

where $w_1 = \det\begin{bmatrix} u_2 & u_3 \\ v_2 & v_3 \end{bmatrix}$, $w_2 = -\det\begin{bmatrix} u_1 & u_3 \\ v_1 & v_3 \end{bmatrix}$, $w_3 = \det\begin{bmatrix} u_1 & u_2 \\ v_1 & v_2 \end{bmatrix}$

(only in $\mathbf{R}^3$).

Orthogonality

(a) DEFINITION: Vectors $\mathbf{u}$ and $\mathbf{v}$ are *orthogonal* if $\mathbf{u} \cdot \mathbf{v} = 0$.

(b) The vector $\mathbf{u} \times \mathbf{v}$ is orthogonal to both $\mathbf{u}$ and $\mathbf{v}$.

Equations of Lines and Planes

(a) Equation of the line through P and parallel to $\mathbf{v}$: $\mathbf{x} = \mathbf{p} + t\mathbf{v}$

(b) Equation of the line through P and Q: $\mathbf{x} = (1 - t)\mathbf{p} + t\mathbf{q}$

(c) Parametric equations of a line: $x_1 = p_1 + tv_1, x_2 = p_2 + tv_2, x_3 = p_3 + tv_3$

(d) Equation of the plane through P and normal to $\mathbf{n}$: $\mathbf{n} \cdot (\mathbf{x} - \mathbf{p}) = 0$

(e) Equation of a plane in standard form: $ax_1 + bx_2 + cx_3 = d$, where (a, b, c) is a vector normal to the plane

(f) Parametric equations of a plane: $x_1 = p_1 + su_1 + tv_1$, $x_2 = p_2 + su_2 + tv_2$, $x_3 = p_3 + su_3 + tv_3$.

KEYWORDS

vector	components
collinear	distance
norm	dot product
orthogonal	cross product
line	plane
normal	parametric equations
interactive multimedia	*window*
pixel	*registration point*

SELF-TEST

Answers to the Self-Tests are in Appendix B.

1. Perform each operation, given $\mathbf{y} = (3, 2, 1)$ and $\mathbf{z} = (-2, 3, -1)$.

(a) $2\mathbf{y} \cdot \mathbf{z}$
(b) $\mathbf{y} \times \mathbf{z}$
(c) $\|3\mathbf{z}\|$
(d) $d(3\mathbf{y}, 6\mathbf{z})$
(e) $\text{proj}_{\mathbf{z}}\mathbf{y}$
(f) $\text{proj}_{\mathbf{y}}\mathbf{z}$ **[Sections 1.1, 1.2]**

2. Give the two-point form and the point-parallel form for a line in $\mathbf{R}^3$ that contains the points $P(1, -1, 1)$ and $Q(4, 3, 2)$. **[Section 1.3]**

3. Determine the point (if any) of intersection of $\mathbf{x}(t) = t(1, 2, 4) + (-1, 0, 1)$ with the plane in $\mathbf{R}^3$ given by $2x - y + z = 5$. **[Section 1.3]**

4. Find the line of all vectors in $\mathbf{R}^3$ that are perpendicular to the plane $2x - 3y + z = 6$. **[Section 1.3]**

5. Find the line in $\mathbf{R}^3$ that is perpendicular to the plane $2x - 3y + z = 6$ and passes through the point $(2, 1, -3)$. **[Section 1.3]**

6. Express the plane $x - y + 2z = 6$ in $\mathbf{R}^3$ in point-normal form. **[Section 1.3]**

7. Determine whether or not the following lines in $\mathbf{R}^3$ are parallel:

$$\mathbf{x}(t) = t(2, 1, 0) + (1, -1, 1)$$
$$\mathbf{x}(t) = t(1, -1, 1) + (2, 1, 0)$$
[Section 1.3]

8. Find a standard-form equation for the plane that contains the three points $(1, 0, 1)$, $(1, 1, 2)$, $(3, 2, 0)$. **[Section 1.3]**

REVIEW EXERCISES

In Exercises 1–12 let $\mathbf{u} = (1, 2, 0)$ and $\mathbf{v} = (-1, 0, 3)$, and perform the indicated operations.

1. $\mathbf{u} - 3\mathbf{v}$

2. $2\mathbf{u} + 4\mathbf{v}$

3. $d(2\mathbf{u}, \mathbf{v})$

4. $d(\mathbf{u} + \mathbf{v}, \mathbf{u})$

5. $\|-2\mathbf{u}\|$

6. $\|\mathbf{u} - \mathbf{v}\|$

7. $\mathbf{u} \times \mathbf{v}$

8. $\left(\frac{1}{\|\mathbf{v}\|}\right)\mathbf{v}$

9. $\mathbf{u} \cdot \mathbf{v}$

10. $(\mathbf{u} + \mathbf{v}) \cdot (\mathbf{u} - \mathbf{v})$

11. Determine the unit vector that has a direction opposite to that of **v**.

12. Find a vector that has the same direction as that of **u** but whose length is 2.

13. Find an equation of the plane (in $\mathbf{R}^3$) that is perpendicular to the z-axis and that passes through the point (1, 2, 3).

14. Find an equation of the plane (in $\mathbf{R}^3$) that contains the points (1, 0, 0), (0, 1, 0), and (0, 0, 1).

15. The equation of a plane in $\mathbf{R}^3$ is $3x + y - 2z = 1$. Give a point-normal form for this equation.

16. A plane (in $\mathbf{R}^3$) has equation $(2, -1, 3) \cdot ((x, y, z) - (-1, 3, 0)) = 0$. Give this equation in standard form.

17. Find an equation of the line in $\mathbf{R}^3$ that completely contains the vector $\mathbf{v} = (1, 0, 2)$.

18. Determine the point (in $\mathbf{R}^3$) of intersection of the line $\mathbf{x}(t) = (1, 0, 1) + t(0, 1, 0)$ and the plane $x + y + z = 0$.

19. Determine all planes (in $\mathbf{R}^3$) orthogonal to the vector (1, 1, 1).

20. Let $\mathbf{u}$ and $\mathbf{v}$ be vectors in $\mathbf{R}^3$, and let c be a scalar. Which of the following operations are *not* defined?

(a) $3\mathbf{u} - c\mathbf{v}$ *(b)* $5c - \mathbf{u}$ *(c)* $\mathbf{u} \times \mathbf{v}$

(d) $\mathbf{u} \cdot (5\mathbf{v})$ *(e)* $\|c\mathbf{v}\|$ *(f)* $\|\mathbf{u} \cdot \mathbf{v}\|$

CHALLENGE YOURSELF

21. Find the point on the plane $x + 2y - z = 5$ that is closest to the point $P(1, -1, 0)$.

22. The parametric form of any plane Π has the form

$$\mathbf{x}(s, t) = (a, b, c) + s(x_1, y_1, z_1) + t(x_2, y_2, z_2) \quad (*)$$

for some real numbers $a, b, c, x_1, y_1, z_1, x_2, y_2,$ and z_2. What conditions must hold on $a, b, c, x_1, y_1, z_1, x_2, y_2, z_2$ for (*) to define a plane?

23. Suppose that $\mathbf{v} = (a, b) \neq (0, 0)$. Determine a unit vector perpendicular to $\mathbf{v}$ that points to the *left* side of $\mathbf{v}$ (if you are facing in the direction of $\mathbf{v}$).

24. Suppose that a rectangular box has sides of lengths a, b, and c. Find the angle that a diagonal makes with each of the three faces incident with its tail.

25. Consider the lines

$$\mathbf{x}_1(t) = (1, 1, 0) + (-1, 1, 1)$$

and

$$\mathbf{x}_2(s) = (0, 2, 1) + (2, 0, -1).$$

Show that the two lines lie on a common plane, and find the equation of this plane.

SPOTLIGHT

ARTSILENSEcom/Shutterstock

Systems of linear equations, the subject matter of most of this chapter, have important applications in many varied disciplines. To take one example from engineering, the analysis of fluid flow within or around an object eventually leads to a very large system of linear equations. This type of problem occurs when studying water flow within pipes (see Section 2.4), the flow of air past an airplane wing, and the movement of weather systems around the Earth.

As a simple example, consider the way heat "flows"—distributes itself—within a heat-conducting object. More specifically, suppose that the edges of a flat, rectangular metal plate have been heated to certain fixed temperatures. In this case, the interior of the plate eventually reaches an *equilibrium* temperature distribution—the temperature at any given point then remains unchanged as time passes. To approximate this temperature distribution, proceed as follows: construct an evenly spaced grid of points on the plate, and assume that at any interior point the equilibrium temperature is the average of the temperatures of its four neighbours. For example, letting u_i be the temperature at point i, we have (see the figure at the bottom of the page)

$$u_6 = \frac{(u_2 + u_5 + u_7 + u_{10})}{4} \quad \text{and} \quad u_7 = \frac{(u_3 + u_6 + u_8 + u_{11})}{4}.$$

As a specific example, suppose that at the point i on the boundary of this rectangular plate, the temperature is fixed at $10i$ degrees (so, at point 1, the temperature is 10°, at point 2 it is 20°, etc.). Then, applying the averaging principle at each of the two interior points (6 and 7), we obtain the following system of linear equations:

$$\begin{aligned} u_6 - \left(\frac{1}{4}\right)u_7 &= \frac{170}{4} \\ -\left(\frac{1}{4}\right)u_6 + \quad u_7 &= \frac{220}{4} \end{aligned}$$

(In Section 2.2, you will be asked to find the solution of this linear system.)

In a more realistic situation, thousands of grid points may be necessary to obtain a solution that has a high degree of accuracy. Such a grid results in thousands of equations and unknowns. To solve such a system, we need efficient, computer-capable methods of solution—like those discussed in Section 2.3.

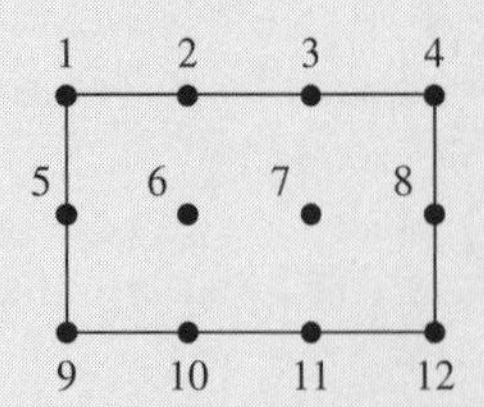

CHAPTER

2

Euclidean *m*-Space and Linear Equations

OBJECTIVES

You will learn to:

- Find the length of a vector in $\mathbf{R}^m$.
- Multiply a vector in $\mathbf{R}^m$ by a scalar.
- Find the sum, difference, and dot product of two vectors.
- Determine the equations of lines and hyperplanes in $\mathbf{R}^m$.
- Identify systems of linear equations and form their augmented matrices.
- Transform a matrix to row-reduced echelon form.
- Solve linear systems using the Gauss–Jordan and Gaussian elimination procedures.

In this chapter, we introduce basic linear algebra concepts that lay the groundwork for the remainder of the text. We begin by discussing *Euclidean m-space* ($\mathbf{R}^m$), made up of all vectors with m real components. Next we use the notation and geometry of $\mathbf{R}^m$ in describing solutions to systems of linear equations. We then present an efficient way to solve these systems and close the chapter with two detailed applications.

2.1 Euclidean *m*-Space

As you know from Chapter 1 vectors in the plane ($\mathbf{R}^2$) and 3-space ($\mathbf{R}^3$) are usually represented by ordered pairs (x, y), or ordered triples (x, y, z), as shown in Figures 1 and 2. These vectors can be combined in different ways (forming, for example, sums and dot products) and can be used to represent equations of lines and planes. In this section, we generalize many vector concepts from $\mathbf{R}^2$ and $\mathbf{R}^3$ to Euclidean m-space, $\mathbf{R}^m$.

Vectors in $\mathbf{R}^m$

We use $\mathbf{R}^m$ to denote the set of all ordered m-tuples,

$$\mathbf{u} = (u_1, u_2, \ldots, u_m),$$

where the u_i are in $\mathbf{R}$, the set of real numbers. An element of $\mathbf{R}^m$ is called a **vector**, or ***m*-vector**. The numbers $u_1, \ldots, u_m$ are called the **components** of **u** and, when unspecified, are denoted by the same letter as the vector. In other words, "m-vector **v**" means the vector $\mathbf{v} = (v_1, v_2, \ldots, v_m)$. In the case of $m = 2$ or $m = 3$, however, we sometimes continue to use $\mathbf{x} = (x, y)$ or $\mathbf{x} = (x, y, z)$.

FIGURE 1

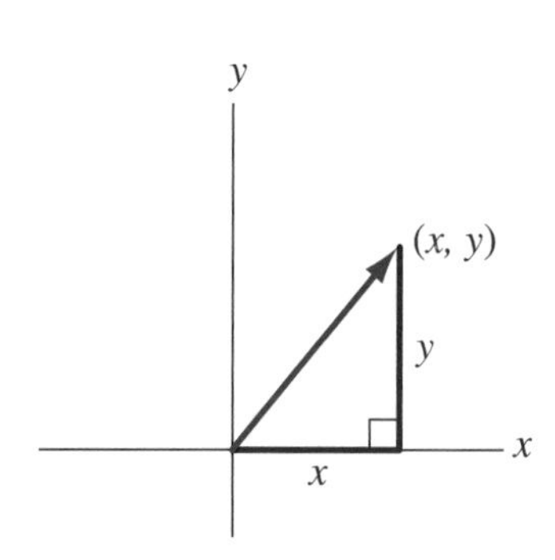

FIGURE 2

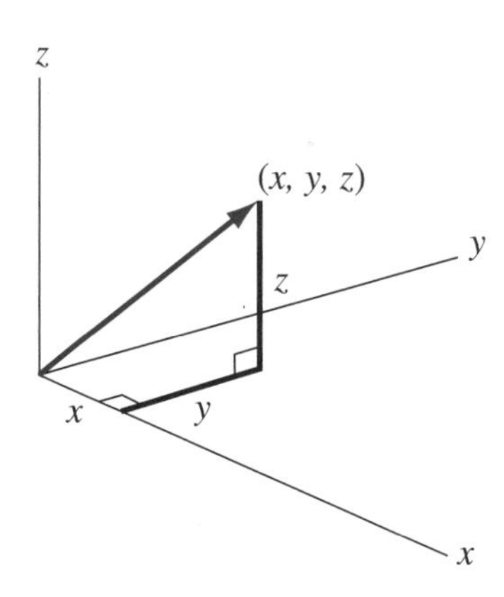

Two vectors **u** and **v** are **equal** if their *corresponding components* are equal—if $u_i = v_i$ for each i. The **zero vector** in $\mathbf{R}^m$ is the m-vector $\mathbf{0} = (0, 0, \ldots, 0)$. For a vector **u** in $\mathbf{R}^m$, the **negative** of **u** is the vector $-\mathbf{u} = (-u_1, -u_2, \ldots, -u_m)$.

Just as in $\mathbf{R}^2$ and $\mathbf{R}^3$, vectors in $\mathbf{R}^m$ have a magnitude and can be added, subtracted, and multiplied by scalars (constants). Moreover, by introducing the notions of dot product and orthogonality, we can speak of lines, planes, and hyperplanes in $\mathbf{R}^m$, even though we cannot actually picture them when $m \geq 4$.

Length of a Vector

DEFINITION

The **distance** between vectors **u** and **v** in $\mathbf{R}^m$ is given by the **distance formula,**

$$d(\mathbf{u}, \mathbf{v}) = \sqrt{(v_1 - u_1)^2 + (v_2 - u_2)^2 + \cdots + (v_m - u_m)^2}.$$

The **length (norm, magnitude)** of a vector **u** is its distance from **0**,

$$\|\mathbf{u}\| = d(\mathbf{0}, \mathbf{u}) = \sqrt{u_1^2 + u_2^2 + \cdots + u_m^2}.$$

NOTE If $\|\mathbf{u}\| = 0$, then $\mathbf{u} = \mathbf{0}$ (and conversely). A vector **u** is called a **unit vector** if $\|\mathbf{u}\| = 1$.

EXAMPLE 1

Find $\|\mathbf{u}\|$ where $\mathbf{u} = (1, 2, 3, -2)$.

Solution

$$\|\mathbf{u}\| = \sqrt{1^2 + 2^2 + 3^2 + (-2)^2} = \sqrt{18} = 3\sqrt{2}$$ ■

Scalar Multiplication

DEFINITION

Let c be a scalar and $\mathbf{u}$ a vector in $\mathbf{R}^m$. The **scalar multiple** of $\mathbf{u}$ by c is the vector

$$c\mathbf{u} = (cu_1, cu_2, \ldots, cu_m).$$

That is, $c\mathbf{u}$ is the vector obtained by multiplying each component of $\mathbf{u}$ by the scalar c.

If $\mathbf{u}$ and $\mathbf{v}$ are nonzero vectors, and there exists a scalar c such that $\mathbf{v} = c\mathbf{u}$, we say $\mathbf{u}$ and $\mathbf{v}$ are **collinear**. If this is the case, we say $\mathbf{u}$ and $\mathbf{v}$ *have the same direction* if c is positive and *have opposite directions* if c is negative.

THEOREM 1

Let $\mathbf{u}$ be a vector in $\mathbf{R}^m$ and c a scalar. Then

$$\|c\mathbf{u}\| = |c|\,\|\mathbf{u}\|.$$

Proof

$$\begin{aligned}\|c\mathbf{u}\| = \|(cu_1, \ldots, cu_m)\| &= \sqrt{(cu_1)^2 + \cdots + (cu_m)^2} \\ &= \sqrt{c^2(u_1^2 + \cdots + u_m^2)} = |c|\,\|\mathbf{u}\| \quad \blacksquare\end{aligned}$$

Note that this proof is exactly the same (with more components) as that given for Theorem 1 of Section 1.1. Furthermore, the interpretation of the result is exactly the same. Multiplication of a vector by a scalar changes the length by the same factor (in absolute value).

Vector Addition

DEFINITION

Let $\mathbf{u}$ and $\mathbf{v}$ be vectors in $\mathbf{R}^m$. The **sum** of $\mathbf{u}$ and $\mathbf{v}$ is the vector

$$\mathbf{u} + \mathbf{v} = (u_1 + v_1, u_2 + v_2, \ldots, u_m + v_m).$$

In words, $\mathbf{u} + \mathbf{v}$ is the vector obtained by adding the corresponding components.

The **negative** of $\mathbf{u}$ is the vector

$$-\mathbf{u} = (-u_1, -u_2, \ldots, -u_m),$$

and the **difference** of $\mathbf{u}$ and $\mathbf{v}$ is

$$\mathbf{u} - \mathbf{v} = \mathbf{u} + (-\mathbf{v}) = (u_1 - v_1, u_2 - v_2, \ldots, u_m - v_m).$$

EXAMPLE 2

Letting $c = 3$, $\mathbf{u} = (1, 3, 5, -1)$, and $\mathbf{v} = (1, 0, 2, 1)$, find $c\mathbf{u} - \mathbf{v}$.

Solution

$$\begin{aligned}c\mathbf{u} - \mathbf{v} &= 3(1, 3, 5, -1) - (1, 0, 2, 1) \\ &= (3, 9, 15, -3) - (1, 0, 2, 1) \\ &= (2, 9, 13, -4) \quad \blacksquare\end{aligned}$$

The following properties of vector addition and scalar multiplication follow directly from the definitions of these operations. The proofs are exactly the same as those given for Theorem 3 of Section 1.1 (with more components) and are left as exercises.

THEOREM 2

Let **u**, **v**, and **w** be vectors in $\mathbf{R}^m$ and c and d be scalars. Then

a. $\mathbf{u} + \mathbf{v} = \mathbf{v} + \mathbf{u}$ (Vector addition is commutative.)

b. $(\mathbf{u} + \mathbf{v}) + \mathbf{w} = \mathbf{u} + (\mathbf{v} + \mathbf{w})$ (Vector addition is associative.)

c. $\mathbf{u} + \mathbf{0} = \mathbf{u}$

d. $\mathbf{u} + (-\mathbf{u}) = \mathbf{0}$

e. $(cd)\mathbf{u} = c(d\mathbf{u})$

f. $(c + d)\mathbf{u} = c\mathbf{u} + d\mathbf{u}$ (Distributive laws)

g. $c(\mathbf{u} + \mathbf{v}) = c\mathbf{u} + c\mathbf{v}$

h. $1\mathbf{u} = \mathbf{u}$

i. $(-1)\mathbf{u} = -\mathbf{u}$

j. $0\mathbf{u} = \mathbf{0}$ ■

Dot Product

DEFINITION

Let **u** and **v** be vectors in $\mathbf{R}^m$. The **dot product** (or **standard inner product**) of **u** and **v** is

$$\mathbf{u} \cdot \mathbf{v} = u_1v_1 + u_2v_2 + \cdots + u_mv_m.$$

In words, **u** • **v** is the sum of the products of the corresponding components of **u** and **v**.

NOTE The definition of **u** • **v** can be given more concisely using *summation notation.* Given a set of numbers $a_1, a_2, \ldots, a_m$, the **summation notation** $\sum_{k=1}^{m} a_k$ represents their sum. In other words,

$$\sum_{k=1}^{m} a_k = a_1 + a_2 + \cdots + a_m.$$

Here k is called the *index of summation.* Summation notation behaves well with respect to addition and scalar multiplication:

$$\sum_{k=1}^{m} (a_k + b_k) = \sum_{k=1}^{m} a_k + \sum_{k=1}^{m} b_k \quad \text{and} \quad \sum_{k=1}^{m} ca_k = c\sum_{k=1}^{m} a_k.$$

Using summation notation, the dot product of **u** and **v** in $\mathbf{R}^m$ can be written as

$$\mathbf{u} \cdot \mathbf{v} = \sum_{k=1}^{m} u_kv_k.$$

Variables other than k can be used for the index of summation, so $\sum_{i=1}^{m} u_iv_i$ and $\sum_{j=1}^{m} u_jv_j$ also represent **u** • **v**. Throughout the rest of this text, whenever appropriate, we use summation notation in addition to the longer, expanded form of the sum.

Because the dot product of two vectors is a scalar, it is often called the **scalar product**. The algebraic properties of the dot product are included in the next theorem.

The proofs are exactly the same as those given in Theorem 1 of Section 1.2 (with more components) and are left as exercises.

THEOREM 3

Let **u**, **v**, and **w** be vectors in $\mathbf{R}^m$ and c be a scalar. Then

a. $\mathbf{u} \cdot \mathbf{v} = \mathbf{v} \cdot \mathbf{u}$ (Dot product is commutative.)

b. $c(\mathbf{u} \cdot \mathbf{v}) = (c\mathbf{u}) \cdot \mathbf{v} = \mathbf{u} \cdot (c\mathbf{v})$ (Scalars factor out.)

c. $\mathbf{u} \cdot (\mathbf{v} + \mathbf{w}) = \mathbf{u} \cdot \mathbf{v} + \mathbf{u} \cdot \mathbf{w}$ (Distributive law)

d. $\mathbf{u} \cdot \mathbf{0} = 0$

e. $\mathbf{u} \cdot \mathbf{u} = \|\mathbf{u}\|^2$

DEFINITION

Vectors **u** and **v** in $\mathbf{R}^m$ are **orthogonal** if $\mathbf{u} \cdot \mathbf{v} = 0$.

Lines in $\mathbf{R}^m$

The geometry of $\mathbf{R}^m$ for $m \geq 4$ is even more interesting than that of $\mathbf{R}^2$ and $\mathbf{R}^3$. The problem, of course, is that we cannot draw pictures of all of it at once. We must focus on subregions for pictures, and often this device is of little help.

Generalizing the $\mathbf{R}^2$ and $\mathbf{R}^3$ cases, we define *points* in $\mathbf{R}^m$ to be m-tuples of real numbers. We denote a point by, for example, $P(u_1, u_2, \ldots, u_m)$ to distinguish it from the vector $\mathbf{p} = (u_1, u_2, \ldots, u_m)$ represented by the same m-tuple. Although we have no intuitive feel for what a *line* might be in $\mathbf{R}^m$ when $m \geq 4$, we can again generalize the $\mathbf{R}^2$ and $\mathbf{R}^3$ cases to make a reasonable definition. To preserve the idea that two points determine a line, we therefore follow the *two-point form* for a line in $\mathbf{R}^3$, Equation (3) of Section 1.3, to define the geometric terms that follow.

DEFINITION

Let P and Q be distinct points in $\mathbf{R}^m$ and let

$$\mathbf{x}(t) = (1 - t)\mathbf{p} + t\mathbf{q}. \quad \textbf{(1)}$$

Then

a. The set of all $\mathbf{x}(t)$ for t in the real numbers is the **line** determined by P and Q.

b. The set of all $\mathbf{x}(t)$ for $0 \leq t \leq 1$ is the **line segment** determined by P and Q. By taking the $\mathbf{x}(t)$ in their natural order (start at $\mathbf{x}(0)$ and finish at $\mathbf{x}(1)$), we have the **directed line segment** from P to Q.

We can now speak of a nonzero vector **u** in $\mathbf{R}^m$ as a directed line segment $\overrightarrow{PQ}$ from the origin by letting $\mathbf{p} = \mathbf{0}$ and $\mathbf{q} = \mathbf{u}$. By letting t range over all real numbers, we have the *line determined* by **u**. Two nonzero vectors **u** and **v** are collinear if the lines they determine are one and the same.

Since we can view a vector **u** as a directed segment from **0**, we can generalize the *point-parallel form* for a line. The set of all vectors

$$\mathbf{x}(t) = \mathbf{p} + t\mathbf{v} \quad \textbf{(2)}$$

is the line that contains **P** and is parallel to **v**, where t is a real number and $\mathbf{v} \neq \mathbf{0}$.

EXAMPLE 3

Give a point-parallel form for the line through $P(2, 1, -1, 3)$ and parallel to $\mathbf{v} = (1, 0, -2, 1)$.

Solution

$$\mathbf{x}(t) = (2, 1, -1, 3) + t(1, 0, -2, 1)$$

We can also view this function parametrically as before. Let

$$\mathbf{x}(t) = (x, y, z, w).$$

Equating coordinates, we have the *parametric equations*,

$$x = 2 + t,$$
$$y = 1,$$
$$z = -1 - 2t,$$
$$\text{and} \quad w = 3 + t.$$

■

Thus, we can deal with lines just as easily in $\mathbf{R}^m$ with $m \geq 4$ as in $\mathbf{R}^3$.

Linear Equations

Let P be a point and $\mathbf{n}$ be a nonzero vector in $\mathbf{R}^m$. The **point-normal form** for P and $\mathbf{n}$ is the equation

$$\mathbf{n} \cdot (\mathbf{x} - \mathbf{p}) = 0. \tag{3}$$

Recall that in $\mathbf{R}^3$ (see Section 1.3) this is the equation of the *plane* through P with normal $\mathbf{n}$. In the general case $\mathbf{R}^m$, the set of all solutions to Equation (3) is called the **hyperplane** through P with normal $\mathbf{n}$.

EXAMPLE 4

Find a point-normal form for the hyperplane that contains the point $P(2, 1, 0, 3, 1)$ with normal $\mathbf{n} = (2, 1, -1, 3, 0)$.

Solution From Equation (3), we have the point-normal form,

$$(2, 1, -1, 3, 0) \cdot (\mathbf{x} - (2, 1, 0, 3, 1)) = 0.$$

■

Other forms of the equation of a plane are obtained by letting $\mathbf{n} = (a_1, a_2, \ldots, a_m)$, $\mathbf{p} = (p_1, p_2, \ldots, p_m)$, $\mathbf{x} = (x_1, x_2, \ldots, x_m)$, $b = \mathbf{n} \cdot \mathbf{p}$, and expanding Equation (3). The alternative point-normal form is

$$a_1(x_1 - p_1) + a_2(x_2 - p_2) + \cdots + a_m(x_m - p_m) = 0. \tag{4}$$

The **standard form** is obtained from Equation (3) by distributing $\mathbf{n} \cdot \mathbf{x} - \mathbf{n} \cdot \mathbf{p} = 0$ and transposing to obtain

$$a_1x_1 + a_2x_2 + \cdots + a_mx_m = b. \tag{5}$$

An equation that can be put into the form of Equation (5) is called a **linear equation** in the variables $x_1, x_2, \ldots, x_m$. This is a generalization of the equation $ax_1 + bx_2 = c$ (which describes a line in $\mathbf{R}^2$) and $ax_1 + bx_2 + cx_3 = d$ (which describes a plane in $\mathbf{R}^3$). (See Section 1.3.)

Note that a linear equation contains no products of variables (such as x_1^2 or x_1x_2) or functions of variables (such as $\sin x_1$, e^{x_2}, $\sqrt{x_4}$) other than multiplication by a constant scalar. For example,

$$3x_1 - 5x_1x_2 + 2x_2 = 1 \quad \text{and} \quad 3x - y^2 = 7$$

are *not* linear equations.

Given a linear equation in standard form (Equation 5), it is an easy matter to put it in point-normal form. Let $\mathbf{n} = (a_1, \ldots, a_m)$ be the vector of coefficients of the variables. For any i from 1 to m with $a_i \neq 0$, let $\mathbf{p} = (0, \ldots, 0, b/a_i, 0, \ldots, 0)$, where the coordinate b/a_i is in the ith position. Then $\mathbf{n} \cdot (\mathbf{x} - \mathbf{p}) = 0$ is equivalent to the original equation, as can easily be confirmed by expansion of the dot product.

EXAMPLE 5

Express the linear equation

$$2x_1 + 3x_2 - x_3 - x_4 = 5$$

in point-normal form for a hyperplane in $\mathbf{R}^4$.

Solution

Letting $\mathbf{n} = (2, 3, -1, -1)$, we have $a_3 = -1 \neq 0$, so the point $P(0, 0, 5/(-1), 0) = (0, 0, -5, 0)$ is in the hyperplane. The desired equation is

$$(2, 3, -1, -1) \cdot (\mathbf{x} - (0, 0, -5, 0)) = 0.$$ ■

Plane Determined by Two Vectors

We can discuss lines in $\mathbf{R}^m$ (point-parallel form) and hyperplanes in $\mathbf{R}^m$ (point-normal form), but so far we have no means of discussing other types of regions. This can be accomplished in a manner analogous to the point-parallel form by adding more terms—one more for a plane, two more for a three-dimensional region, and so on. As an example, we describe a plane in $\mathbf{R}^m$ determined by two noncollinear vectors.

Let $\mathbf{u}$ and $\mathbf{v}$ be noncollinear vectors; that is, one is not a scalar multiple of the other. Let X be any point in the plane determined by $\mathbf{u}$ and $\mathbf{v}$. The line through X parallel to the vector $\mathbf{v}$ must intersect the line determined by $\mathbf{u}$. Since any point on the latter line is a scalar multiple of $\mathbf{u}$, the point of intersection is $s\mathbf{u}$ for some real number s (Figure 3). Similarly, the line through X parallel to $\mathbf{u}$ must intersect the line determined by $\mathbf{v}$ in some point $t\mathbf{v}$. Then the quadrilateral $\mathbf{0}$, $s\mathbf{u}$, $\mathbf{x}$, $t\mathbf{v}$ is a parallelogram, and by Theorem 2 of Section 1.1,

$$\mathbf{x} = s\mathbf{u} + t\mathbf{v}.$$

FIGURE 3

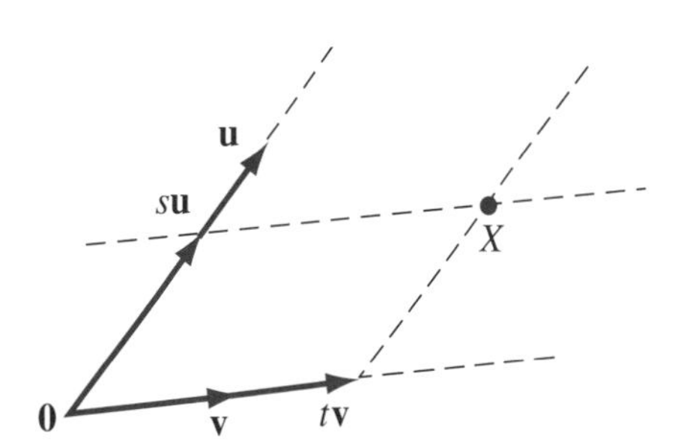

Thus, every point in the plane determined by $\mathbf{u}$ and $\mathbf{v}$ is representable as the sum of a multiple of $\mathbf{u}$ and a multiple of $\mathbf{v}$. A representation for a vector $\mathbf{x}$ of this type is called a **linear combination** of $\mathbf{u}$ and $\mathbf{v}$. Conversely, every linear combination $s\mathbf{u} + t\mathbf{v}$ lies in the plane of $\mathbf{u}$ and $\mathbf{v}$. We therefore may view the plane of $\mathbf{u}$ and $\mathbf{v}$ as the set of all points given by the function $\mathbf{x} = \mathbf{x}(s, t)$, where

$$\mathbf{x} = s\mathbf{u} + t\mathbf{v},$$

and s and t range independently over all real numbers.

If we start with three noncoplanar vectors $\mathbf{u}$, $\mathbf{v}$, and $\mathbf{w}$, we have an analogous situation. The set of all linear combinations $\mathbf{x} = \mathbf{x}(r, s, t)$ of $\mathbf{u}$, $\mathbf{v}$, and $\mathbf{w}$ given by

$$\mathbf{x} = r\mathbf{u} + s\mathbf{v} + t\mathbf{w}$$

is the 3-space determined by $\mathbf{u}$, $\mathbf{v}$, and $\mathbf{w}$.

Even if planes or 3-spaces do not contain the origin, a similar procedure can be used to obtain their equations: We first "translate" one of the points to the origin, creating vectors with the same magnitude and direction as the directed line segments beginning at

FIGURE 4

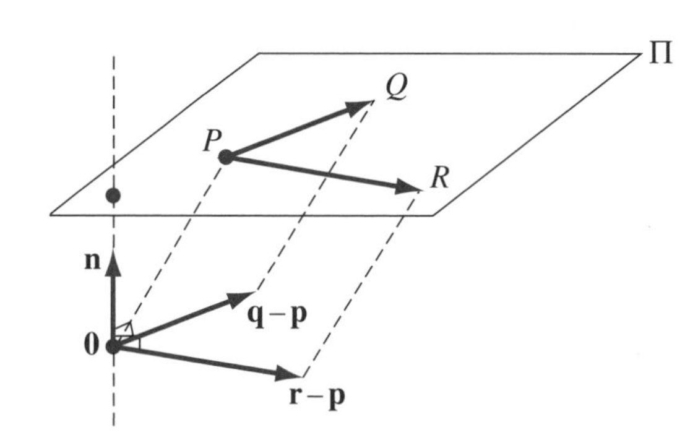

this point (Figure 4). For example, if P, Q, and R are the given points, we can translate P to the origin by determining the vectors $\mathbf{q} - \mathbf{p}$ and $\mathbf{r} - \mathbf{p}$. We then form the appropriate linear combination of these vectors and "translate back" to the original point as illustrated in the next example. (For more detail about this process, see Section 1.1.)

EXAMPLE 6

Describe the plane determined by $P(1, 2, 1, 1)$, $Q(3, -1, 4, 0)$, and $R(1, -1, 0, 2)$ in $\mathbf{R}^4$.

Solution We translate P to the origin. The vectors $\mathbf{q} - \mathbf{p}$ and $\mathbf{r} - \mathbf{p}$ then determine a parallel plane consisting of all linear combinations $s(\mathbf{r} - \mathbf{p}) + t(\mathbf{q} - \mathbf{p})$. Adding $\mathbf{p}$ to each such point, we translate back, and the desired plane is given by the set of all points of the form

$$\mathbf{x} = \mathbf{p} + s(\mathbf{r} - \mathbf{p}) + t(\mathbf{q} - \mathbf{p})$$

where s and t are real numbers. That is,

$$\mathbf{x} = (1, 2, 1, 1) + s(0, -3, -1, 1) + t(2, -3, 3, -1)$$

for all real numbers s and t. ■

2.1 Exercises

In Exercises 1–12 let the vectors be given as follows:

$$\mathbf{u} = (2, 1, 3, 0), \quad \mathbf{v} = (-1, 1, 2, 1), \quad \mathbf{w} = (2, -2, 0, 6).$$

Perform each indicated operation.

1. $\mathbf{u} + \mathbf{v}$

2. $3\mathbf{u} - \mathbf{v} + 2\mathbf{w}$

3. $2(\mathbf{u} + 2\mathbf{v} + 3\mathbf{w})$

4. $d(\mathbf{v}, \mathbf{w})$

5. $\mathbf{v} \cdot \mathbf{w}$

6. $\|\mathbf{w}\|$

7. $\|3\mathbf{w}\|$

8. $\dfrac{1}{\|\mathbf{w}\|}\mathbf{w}$

9. $d(\mathbf{v}, -\mathbf{u})$

10. $(\mathbf{u} \cdot \mathbf{v})(-\mathbf{w})$

11. $\|2(\mathbf{v} - \mathbf{w})\|$

12. $(3\mathbf{v}) \cdot (5\mathbf{w})$

13. Which pairs of the following vectors are orthogonal?

$$\mathbf{u}_1 = (2, 1, 3, -1), \quad \mathbf{u}_2 = (3, 1, 0, 2),$$
$$\mathbf{u}_3 = (0, -2, 1, 1), \quad \mathbf{u}_4 = (2, 0, 3, -3).$$

14. Prove that, for $\mathbf{u}$, $\mathbf{v}$ in $\mathbf{R}^m$, $d(\mathbf{u}, \mathbf{v}) = \|\mathbf{u} - \mathbf{v}\|$.

15. Prove that, for a nonzero vector $\mathbf{u}$ in $\mathbf{R}^m$, $(1/\|\mathbf{u}\|)\mathbf{u}$ is the unit vector in the direction of $\mathbf{u}$.

In Exercises 16 and 17 find the unit vector with the same direction as the one given.

16. $(1, 1, 3, 0, 5)$

17. $(2, 1, -1, 0, 3, 4)$

In Exercises 18 and 19 find all hyperplanes orthogonal to the given vector.

18. $(1, 0, -1, 0, 1)$

19. $(-2, -1, 0, 1)$

20. Give two-point and point-parallel forms and parametric equations for the line in $\mathbf{R}^5$ determined by $P(2, 1, 0, 3, 1)$ and $Q(1, -1, 3, 0, 5)$.

21. Give two-point and point-parallel forms and parametric equations for the line in $\mathbf{R}^4$ determined by $P(-1, 0, 3, 2)$ and $Q(-1, 0, 4, 5)$.

22. Give point-normal and standard forms for the hyperplane through $(-2, 1, 4, 0)$ with normal $(1, 2, -1, 3)$.

23. Give point-normal and standard forms for the hyperplane through $(3, 4, 5, 6, 7)$ with normal $(1, -1, 1, -1, 1)$.

24. Find a point-normal form for the linear equation $2x_1 - 3x_2 + x_4 - x_5 = 2$ in $\mathbf{R}^5$.

25. Repeat Exercise 24 viewing the equation in $\mathbf{R}^6$.

In Exercise 26–29 explain why each equation fails to be linear.

26. $x(2y + 3z) = 5$

27. $x_1 + 2x_2 + x_3 - x_4x_5 = 3$

28. $x + 2 \sin y + 3z = 0$

29. $x^2 + y^2 = 0$

30. Find the midpoint of the line segment that joins $(2, 1, 3, 5)$ with $(6, 3, 2, 1)$.

31. Prove the properties of Theorem 2.

32. The *angle between two nonzero vectors* $\mathbf{u}$ and $\mathbf{v}$ in $\mathbf{R}^m$ is defined to be the unique number θ, $0 \le \theta \le \pi$ radians such that $\cos \theta = \mathbf{u} \cdot \mathbf{v}/(\|\mathbf{u}\|\ \|\mathbf{v}\|)$. Find the cosine of the angle between the vectors $(3, 1, 1, 2, 1)$ and $(0, 2, 1, -2, 0)$.

33. Prove the properties of Theorem 3.

34. Describe the plane determined by the points $(3, 1, 0, 2, 1)$, $(2, 1, 4, 2, 0)$, and $(-1, 2, 1, 3, 1)$.

35. Describe the 3-space determined by the points $(3, 1, 0, 2, 1)$, $(2, 1, 4, 2, 0)$, $(-1, 2, 1, 3, 1)$, and $(0, 2, 0, 1, 0)$.

36. Prove the *parallelogram law* in $\mathbf{R}^m$, $\|\mathbf{u} + \mathbf{v}\|^2 + \|\mathbf{u} - \mathbf{v}\|^2 = 2\,\|\mathbf{u}\|^2 + 2\,\|\mathbf{v}\|^2$.

37. In metric geometry, a point B is *between* A and C if and only if $d(A, B) + d(B, C) = d(A, C)$. For distinct $\mathbf{p}$ and $\mathbf{q}$ in $\mathbf{R}^m$, prove that $\mathbf{x}(t) = (1 - t)\mathbf{p} + t\mathbf{q}$ is between $\mathbf{p}$ and $\mathbf{q}$ if $0 \le t \le 1$.

38. Use Exercise 14 to show that $d(\mathbf{p}, \mathbf{p} + t\mathbf{u}) = |t|\ \|\mathbf{u}\|$.

39. Find a point one-third of the distance from $(2, 1, 3, -4)$ to $(2, -2, 0, 2)$.

40. Let P, Q, and R be the vertices of a triangle in $\mathbf{R}^m$. Prove that the segment that joins the midpoints of two of the sides is parallel to and half the length of the third side.

41. Let $\mathbf{u}, \mathbf{v}_1, \mathbf{v}_2, \ldots, \mathbf{v}_n$ be vectors in $\mathbf{R}^m$ and $c_1, c_2, \ldots, c_n$ be scalars. If $\mathbf{u}$ is orthogonal to every $\mathbf{v}_i$, prove that $\mathbf{u}$ is orthogonal to $c_1\mathbf{v}_1 + c_2\mathbf{v}_2 + \cdots + c_n\mathbf{v}_n$.

2.2 Systems of Linear Equations

Linear equations were introduced in Section 2.1, with an emphasis on their geometrical interpretation in Euclidean m-space. In this section we recall some terminology associated with systems of linear equations and present a technique for solving them.

A **linear equation** in the variables $x_1, x_2, \ldots, x_n$ is one that can be put in the form

$$a_1x_1 + a_2x_2 + \cdots + a_nx_n = b,$$

where b and the coefficients a_i are constants and not all a_i equal zero. A **system of linear equations** (or **linear system**) is simply a finite set of linear equations. An n-vector $(s_1, s_2, \ldots, s_n)$ is a **solution** to a linear system (in n variables) if it satisfies every equation in the system. The variables in a linear system are also referred to as *unknowns*; of course, they may be denoted by other symbols such as x, y, z, or w.

Recall from Section 2.1 that the set of solutions to a linear equation is a line in $\mathbf{R}^2$, a plane in $\mathbf{R}^3$, and, generally speaking, a **hyperplane** in $\mathbf{R}^n$. From a geometric point of view, a solution to a system of linear equations represents a point that lies in the *intersection* of the regions described by the individual equations.

EXAMPLE 1

The linear system

$$2x = y - 4z$$
$$y = 2z$$

is a system of two equations in the three unknowns x, y, and z. One solution of this system is $(-1, 2, 1)$, since $x = -1$, $y = 2$, and $z = 1$ satisfies *both* equations. This is not the only solution, however. For example, $(0, 0, 0)$ and $(\frac{1}{2}, -1, -\frac{1}{2})$ are also solutions. In fact, any vector of the form $(-t, 2t, t)$ is a solution, where t represents an arbitrary real number. Consequently there are infinitely many solutions to this system—one for each value of t. For example, the solutions $(-1, 2, 1)$, $(0, 0, 0)$, and $(\frac{1}{2}, -1, -\frac{1}{2})$ are obtained be setting t equal to 1, 0, and $-\frac{1}{2}$, respectively. A solution set of this type is called a *one-parameter family* of solutions; here t is the *parameter*.

Like planes, hyperplanes can be written in parametric form. Moreover, the intersection of any number of planes can also be written in parametric form, as we shall see from our solutions to linear systems.

To check that all vectors of the form $(-t, 2t, t)$ are solutions, simply substitute the components for the corresponding variables in each equation. This substitution results in an identity in both cases, namely, $2(-t) = (2t) - 4(t)$ and $(2t) = 2(t)$. Consequently all such vectors are indeed solutions. ■

Note that the solution in Example 1 may be written as $(x, y, z) = t(-1, 2, 1)$, the point-parallel form of the line through the origin in $\mathbf{R}^3$ determined by the vector $(-1, 2, 1)$. Since each of the two given equations represents a plane in $\mathbf{R}^3$, we see that their intersection is this line (Figure 5).

FIGURE 5

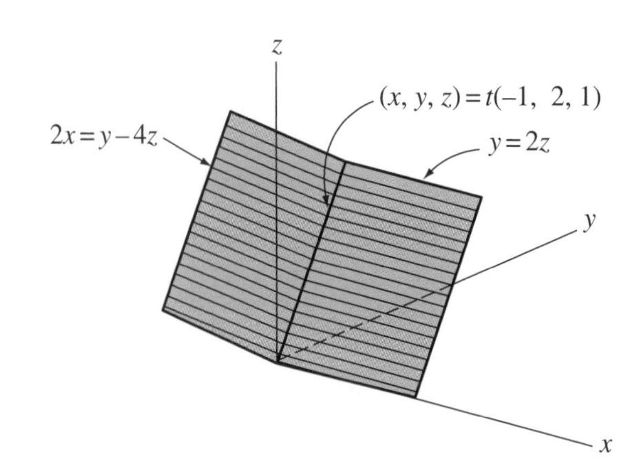

EXAMPLE 2

The linear system

$$x + y = 0$$
$$x + y = 1$$

is a system of two equations in two unknowns that does not have a solution. This can be seen by the following argument. Suppose (s_1, s_2) were a solution.

Then the first equation states that the two components s_1 and s_2 add up to 0, whereas the second states that they add up to 1. Both statements cannot be true at the same time, and consequently no pair of numbers (s_1, s_2) can satisfy both equations simultaneously.

Geometrically speaking, the given equations represent lines in $\mathbf{R}^2$. We have thus shown that these lines do not intersect—they must be parallel (Figure 6).

FIGURE 6

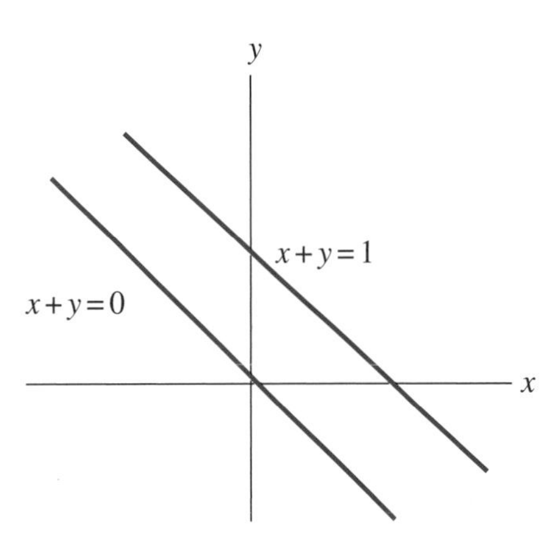

■

NOTE If a linear system has no solution, we say that it is **inconsistent.** If it has at least one solution, it is **consistent.** For example, the system of Example 1 is consistent, whereas that of Example 2 is inconsistent.

EXAMPLE 3

The system of equations

$$\begin{aligned} x_1 + x_2 \quad + x_3 + \quad x_4 &= 4 \\ x_1 + x_1 \qquad\quad + 2x_4 &= 2 \\ x_1 + x_1x_2 \qquad\quad - \quad x_4 &= 1 \end{aligned}$$

is *not* a linear system, because the third equation is not linear. (Remember, from Section 2.1, that if a product of variables occurs in an equation, the equation is not linear.) Although the first and second equations are linear, a set of equations forms a linear system if and only if *all* equations in the set are linear. ■

In solving a system of linear equations, it is convenient to begin by putting the system into **standard form**, which means writing it in such a way that

i. Each equation is in *standard form* (Section 2.1), and

ii. Like variables are aligned in vertical columns.

Note that the system of Example 2 is in standard form, but that of Example 1 is not. To put the latter into standard form, we simply transpose the variables to the left-hand side, obtaining

$$\begin{aligned} 2x - y + 4z &= 0 \\ y - 2z &= 0. \end{aligned}$$

We now proceed to describe a technique known as an *elimination procedure* for solving linear systems of equations. The basic idea of this technique is to transform the original linear system into simpler and simpler systems that have the same solutions as the one given. The final system should be so simple that its solution set can be easily determined.

Each step of this transformation process is accomplished by applying one of the three following **elementary operations**:

Note that we do not allow multiplication of an equation by 0, since it will always lead to the equation $0 = 0$ and may introduce *extraneous* solutions, that is, solutions that do not belong to the original linear system.

Likewise, operation (iii) insists that we replace one equation with the sum of that equation plus a multiple of *another* equation. If a multiple of that same equation were allowed, then we could possibly add an equation with (-1) times itself, yielding again $0 = 0$.

i. Multiply an equation in the system by a nonzero scalar.

ii. Interchange the positions of two equations in the system.

iii. Replace an equation with the sum of itself and a multiple of another equation of the system.

As we will see at the end of this section, performing any finite sequence of these elementary operations on a linear system of equations results in an **equivalent system**, one with the same set of solutions. Although a description of a *systematic* solution procedure based on these elementary operations is deferred until Section 2.3, we use this procedure to solve the examples that follow. In reading through these examples, you may begin to see the structure of this systematic procedure. In working the exercises for this section, however, feel free to use the elementary operations in any order you wish so as to arrive at a system with an "obvious" solution. It is often better to be clever than to be systematic!

EXAMPLE 4

Solve

$$\begin{aligned} -y+z&=3 &&\text{(A1)}\\ x-y-z&=0 &&\text{(B1)}\\ -x\qquad -z&=-3. &&\text{(C1)}\end{aligned}$$

Solution Interchange Equations (A1) and (B1):

$$\begin{aligned} x-y-z&=0 &&\text{(A2)}\\ -y+z&=3 &&\text{(B2)}\\ -x\qquad -z&=-3 &&\text{(C2)}\end{aligned}$$

Replace (C2) with the sum of (C2) and 1 times (A2):

$$\begin{aligned} x-y-z&=0 &&\text{(A3)}\\ -y+z&=3 &&\text{(B3)}\\ -y-2z&=-3 &&\text{(C3)}\end{aligned}$$

Multiply (B3) by -1:

$$\begin{aligned} x-y-z&=0 &&\text{(A4)}\\ y-z&=-3 &&\text{(B4)}\\ -y-2z&=-3 &&\text{(C4)}\end{aligned}$$

Replace (A4) with the sum of (A4) and 1 times (B4). Then replace (C4) with the sum of (C4) and 1 times (B4):

$$\begin{aligned} x\qquad -2z&=-3 &&\text{(A5)}\\ y-z&=-3 &&\text{(B5)}\\ -3z&=-6 &&\text{(C5)}\end{aligned}$$

Multiply (C5) by $-\frac{1}{3}$:

$$\begin{aligned} x \qquad - 2z &= -3 && \textbf{(A6)} \\ y - z &= -3 && \textbf{(B6)} \\ z &= 2 && \textbf{(C6)} \end{aligned}$$

Replace (A6) with the sum of (A6) and 2 times (C6). Then replace (B6) with the sum of (B6) and 1 times (C6):

$$\begin{aligned} x \qquad &= 1 && \textbf{(A7)} \\ y \qquad &= -1 && \textbf{(B7)} \\ z &= 2 && \textbf{(C7)} \end{aligned}$$

The only solution of this final system, and hence of the original one, is $(1, -1, 2)$.

Geometrically, we have shown that the intersection of the three planes given by Equations (A1), (B1), and (C1) is the single point $(1, -1, 2)$ (Figure 7).

FIGURE 7

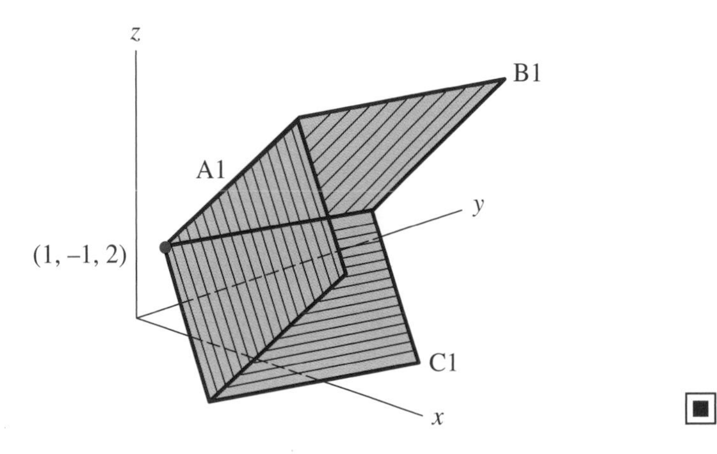

EXAMPLE 5

Solve

$$\begin{aligned} x_1 \qquad + 3x_3 + x_4 &= 0 && \textbf{(A1)} \\ -x_1 + 2x_2 + x_3 + x_4 &= 0 && \textbf{(B1)} \\ -x_1 + x_2 - x_3 \qquad &= 0. && \textbf{(C1)} \end{aligned}$$

Solution To solve the given system, replace (B1) with the sum of (B1) and 1 times (A1). Then replace (C1) with the sum of (C1) and 1 times (A1):

$$\begin{aligned} x_1 \qquad + 3x_3 + x_4 &= 0 && \textbf{(A2)} \\ 2x_2 + 4x_3 + 2x_4 &= 0 && \textbf{(B2)} \\ x_2 + 2x_3 + x_4 &= 0 && \textbf{(C2)} \end{aligned}$$

Multiply (B2) by $\frac{1}{2}$:

$$\begin{aligned} x_1 \qquad + 3x_3 + x_4 &= 0 && \textbf{(A3)} \\ x_2 + 2x_3 + x_4 &= 0 && \textbf{(B3)} \\ x_2 + 2x_3 + x_4 &= 0 && \textbf{(C3)} \end{aligned}$$

Replace (C3) with the sum of (C3) and -1 times (B3):

$$\begin{aligned} x_1 \qquad + 3x_3 + x_4 &= 0 && \textbf{(A4)} \\ x_2 + 2x_3 + x_4 &= 0 && \textbf{(B4)} \\ 0 &= 0 && \textbf{(C4)} \end{aligned}$$

Equation (C4) is an identity. It places no restriction on the values of the unknowns, and we discard it:

$$x_1 \quad + 3x_3 + x_4 = 0 \qquad \textbf{(A4)}$$
$$x_2 + 2x_3 + x_4 = 0 \qquad \textbf{(B4)}$$

Now solve (A4) for x_1 and (B4) for x_2:

$$x_1 = -3x_3 - x_4$$
$$x_2 = -2x_3 - x_4$$

Finally, let $x_3 = s$ and $x_4 = t$, obtaining the solution

$$x_1 = -3s - t$$
$$x_2 = -2s - t$$
$$x_3 = s$$
$$x_4 = t$$

or, in vector form, $(-3s - t, -2s - t, s, t)$, where the parameters s and t range independently over the real numbers. Once again we get a linear system with infinitely many solutions, but here they form a *two-parameter family.*

Among the solutions are

$$(0, 0, 0, 0), \qquad \text{when } s = 0, t = 0;$$
$$(-1, -1, 0, 1), \qquad \text{when } s = 0, t = 1;$$
$$(1, \tfrac{1}{2}, -\tfrac{1}{2}, \tfrac{1}{2}), \qquad \text{when } s = -\tfrac{1}{2}, t = \tfrac{1}{2}.$$

Geometrically, the solution set consists of the common points of intersection of the given hyperplanes in $\mathbf{R}^4$: $x_1 + 3x_3 + x_4 = 0$, $-x_1 + 2x_2 + x_3 + x_4 = 0$, and $-x_1 + x_2 - x_3 = 0$. Since any solution may be expressed as

$$\mathbf{x} = s(-3, -2, 1, 0) + t(-1, -1, 0, 1),$$

when $\mathbf{x} = (x_1, x_2, x_3, x_4)$, the set of all such points is the plane in $\mathbf{R}^4$ determined by the vectors $(-3, -2, 1, 0)$ and $(-1, -1, 0, 1)$. ■

COMPUTATIONAL NOTE

As you can see, solving a linear system—even if it has relatively few equations—can be a tedious process. Fortunately, computers (with the proper software) and sophisticated calculators can be used to take the drudgery out of this task and to speed up the process considerably. In Section 2.3, we discuss one of the methods used by computers and calculators to solve systems of linear equations.

THEOREM 1

If any one of the three elementary operations is performed on a given system of linear equations, an equivalent linear system is obtained.

Proof We show this to be true for elementary operation (iii) only. The statement is certainly true for elementary operation (ii), since this operation does not alter the system, and we leave the part of the proof concerning operation (i) as an exercise.

Let two equations of the given linear system be

$$a_1x_1 + a_2x_2 + \cdots + a_nx_n = b \qquad \textbf{(1)}$$
$$c_1x_1 + c_2x_2 + \cdots + c_nx_n = d. \qquad \textbf{(2)}$$

Suppose that we replace Equation (2) with the sum of itself and k times Equation (1), and leave all other equations in the system as they stand. We need to verify that $(s_1, s_2, \ldots, s_n)$ is a solution of the given system if and only if it is a solution of the transformed system. Since all the other equations remain unchanged, we need to only verify this for Equation (2), which has been transformed into

$$(c_1 + ka_1)x_1 + (c_2 + ka_2)x_2 + \cdots + (c_n + ka_n)x_n = d + kb. \quad \mathbf{(3)}$$

First, assume $(s_1, s_2, \ldots, s_n)$ satisfies the original system. Then, in particular, it satisfies Equations (1) and (2), so that

$$\begin{aligned} a_1s_1 + a_2s_2 + \cdots + a_ns_n &= b \\ c_1s_1 + c_2s_2 + \cdots + c_ns_n &= d. \end{aligned}$$

Multiplying the first of these by k, we have

$$k(a_1s_1 + a_2s_2 + \cdots + a_ns_n) = kb,$$

which when added to the second yields

$$k(a_1s_1 + a_2s_2 + \cdots + a_ns_n) + (c_1s_1 + c_2s_2 + \cdots + c_ns_n) = kb + d.$$

Finally, rearranging terms, we obtain

$$(c_1 + ka_1)s_1 + (c_2 + ka_2)s_2 + \cdots + (c_n + ka_n)s_n = d + kb, \quad \mathbf{(4)}$$

which says that $(s_1, s_2, \ldots, s_n)$ satisfies Equation (3).

Now assume that $(s_1, s_2, \ldots, s_n)$ satisfies the transformed system. This implies that Equation (4) holds. We also know that Equation (1) is satisfied, so that $a_1s_1 + a_2s_2 + \cdots + a_ns_n = b$. This implies that $k(a_1s_1 + a_2s_2 + \cdots + a_ns_n) = kb$. Subtracting this last equation from Equation (4) yields

$$c_1s_1 + c_2s_2 + \cdots + c_ns_n = d,$$

as desired. ▣

2.2 Exercises

In Exercises 1–6 determine whether or not the given system of equations is linear.

1. $\begin{aligned} x_1 - 3x_2 &= x_3 - 4 \\ x_4 &= 1 - x_1 \\ x_1 + x_4 + x_3 - 2 &= 0 \end{aligned}$

2. $\begin{aligned} 2x - \sqrt{y} + 3z &= -1 \\ x + 2y - z &= 2 \\ 4x - y \quad &= -1 \end{aligned}$

3. $\begin{aligned} 3x - xy \quad &= 1 \\ x + 2xy - y &= 0 \end{aligned}$

4. $\begin{aligned} y &= 2x - 1 \\ y &= -x \end{aligned}$

5. $\begin{aligned} 2x_1 - \sin x_2 &= 3 \\ x_2 &= x_1 + x_3 \\ -x_1 + x_2 - 3x_3 &= 0 \end{aligned}$

6. $\begin{aligned} x_1 + x_2 + x_3 &= 1 \\ -2x_1^2 \quad - 2x_3 &= -1 \\ 3x_2 - x_3 &= 2 \end{aligned}$

In Exercises 7–10 put the given system of linear equations into standard form.

7. $\begin{aligned} z &= 6 - y \\ z &= x + y \\ y + z - 3 &= x \end{aligned}$

8. $\begin{aligned} x_2 + x_1 &= 3 - x_3 + x_4 \\ x_2 &= 0 \\ x_3 &= 1 - x_1 - x_2 \end{aligned}$

9. $$\begin{aligned} x_2 + x_1 &= x_4 - x_3 \\ x_1 - 1 &= 0 \\ x_4 - 3 &= x_2 \end{aligned}$$

10. $$\begin{aligned} x + y &= 0 \\ x - z &= 1 \\ x - 1 &= y \end{aligned}$$

In Exercises 11–14 show that the given linear system has the solution indicated.

11. $$\begin{aligned} x - 3y + z &= 0 \\ x + y - 3z &= 0 \\ x - y - z &= 0 \end{aligned}$$
Solution: $(2s, s, s)$

12. $$\begin{aligned} x_1 - 3x_2 - x_3 + x_4 &= 1 \\ -x_1 + 2x_2 + x_3 + x_4 &= 0 \end{aligned}$$
Solution:
$(-2 + s + 5t, -1+2t, s, t)$

13. $$\begin{aligned} v - 2w + z &= 1 \\ 2u - v \qquad - z &= 0 \\ 4u + v - 6w + z &= 3 \end{aligned}$$
Solution: $(\frac{1}{2} + s, 1 + 2s - t, s, t)$

14. $$\begin{aligned} 2x - 3y + z &= 1 \\ -x \qquad + 2z &= 0 \\ 3x - 3y - z &= 1 \end{aligned}$$
Solution: $(2s, -\frac{1}{3} + \frac{5}{3}s, s)$

In Exercises 15–18 solve the given linear system.

15. $$\begin{aligned} x - 4y &= 1 \\ -2x + 8y &= -2 \end{aligned}$$

16. $$\begin{aligned} x_1 + x_2 - x_3 + x_4 &= 2 \\ -2x_1 \qquad + x_3 \qquad &= -4 \\ x_1 - x_2 - 2x_3 + 2x_4 &= -1 \end{aligned}$$

17. $$\begin{aligned} x_1 + x_2 + x_3 + x_4 &= 1 \\ 2x_1 + 3x_2 + 3x_3 \qquad &= 1 \\ -x_1 - 2x_2 - 2x_3 + x_4 &= 0 \\ - x_2 - x_3 + 2x_4 &= 1 \end{aligned}$$

18. $$\begin{aligned} 2u - v &= 0 \\ -3u + 2v &= 0 \\ 3u - v &= 0 \end{aligned}$$

19. Prove that, if we multiply any equation of an arbitrary linear system by a nonzero scalar, the resultant linear system is equivalent to the original one.

20. Show that the linear system

$$\begin{aligned} ax + by &= e \\ cx + dy &= f, \end{aligned}$$

where a, b, c, d, e, and f are constants, has a unique solution if $ad - bc \neq 0$. Express this solution in terms of a, b, c, d, e, and f.

SPOTLIGHT *21.* Solve the linear system that resulted from analyzing the problem posed in the Spotlight at the beginning of this chapter:

$$\begin{aligned} u_6 - \left(\frac{1}{4}\right)u_7 &= \frac{170}{4} \\ -\left(\frac{1}{4}\right)u_6 + \qquad u_7 &= \frac{220}{4} \end{aligned}$$

What does the solution represent? (Reread the Spotlight, if necessary.) In light of what it represents, does the solution seem reasonable?

SPOTLIGHT *22.* Consider a rectangular metal plate, similar to the one in the Spotlight at the beginning of this chapter, but having six interior grid points (Figure 8). Suppose that at point i on the boundary of the plate the temperature is fixed at $10i$ degrees. Set up a system of six linear equations in six unknowns $(u_7, u_8, u_9, u_{12}, u_{13}, u_{14})$ whose solution approximates the equilibrium temperature distribution at the interior grid points.

FIGURE 8

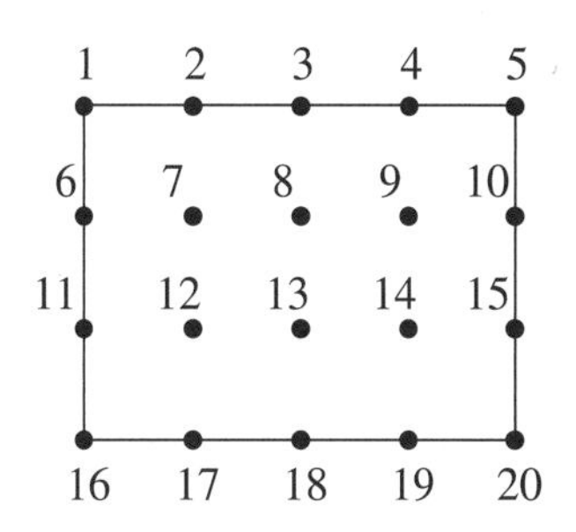

Computational Exercises

23. Consider the following system of linear equations:

$$\left(\frac{1}{3}\right)x + \left(\frac{1}{7}\right)y \ = 1$$
$$\left(\frac{1}{7}\right)x + \left(\frac{1}{11}\right)y = 0$$

(a) Use a calculator to solve this system, rounding all numbers (including intermediate results) to two decimal places.

(b) Use a graphing calculator to graph the lines whose equations are given in this linear system. Estimate the coordinates of the point of intersection to two decimal places.

(c) Solve the given system "by hand," without converting fractions to decimals. Compare your answers to parts (a) and (b). Do they agree to two decimal places?

Solve each of the linear systems in Exercises 24–26 by using a graphing calculator to graph the relevant lines and then finding their point(s) of intersection (if any).

24. $3x - y = 1$
$x - 2y = -3$

25. $y = 2x + 3$
$4x - 2y = 5$

26. $x - y = -1$
$y = 2x$
$x = -y - 3$

A *computer algebra system* (CAS), such as *Mathematica* and *Maple*, is a powerful software package that (in addition to its other functions) has the ability to manipulate symbolic expressions. A CAS can, for example, factor certain polynomials and solve equations that contain parameters. Exercises 27–29 make use of a CAS.

27. The equation of a circle can be written as $x^2 + y^2 + c_1x + c_2y + c_3 = 0$. Suppose the points (1, 1), (0, 2), and $(k, -1)$ lie on the same circle. Solve for the equation of the circle when $k = -1, 0, 1, 2$, and 3. (When solving a number of equations that involve changing a parameter, a CAS can help save time in performing repeated computations.) For what values of k is there no solution?

28. Use a CAS or a graphing calculator to plot the points and corresponding solution for the given values of k in Exercise 27.

29. Since a CAS can perform symbolic manipulation, solve directly for the equation of the circle with points $(-2, 0)$, $(1, 3)$, $(5, k)$. For what value of k is there no solution?

The term *matrix* was coined by the mathematician J.J. Sylvester (1814–1897), and comes from the latin word for "mother" (*mater*). Sylvester was Jewish, and confronted anti-Semitism throughout his life; he even refused to accept his degree from Cambridge University owing to the required pledge to the Church of England. In spite of it all, Sylvester taught on both sides of the Atlantic—at Johns Hopkins University in the United States and at the University of Oxford in England. His work in linear algebra was fundamental to the field. Sylvester was a colourful character with a love of verse, and even wrote a book setting out rules on the structure of poetry.

2.3 Row Reduction of Linear Systems

In Section 2.2, systems of linear equations were solved by means of an elimination procedure. In this section, we introduce the notion of a *matrix* and then use *matrices* (the plural of matrix) to write each step of this solution technique in a more concise form.

A **matrix** is a rectangular array of numbers, each of which is called an **entry** of the matrix. We enclose the rectangular array in brackets. For example,

$$\begin{bmatrix} 2 & 1 \\ -1 & 2 \\ 0 & 5 \end{bmatrix}, \tag{1}$$

$$\begin{bmatrix} 1 & -1 & 0 \\ 0 & 5 & 0 \end{bmatrix}, \tag{2}$$

$$\text{and} \quad \begin{bmatrix} 1 & \sqrt{2} & \pi \end{bmatrix} \tag{3}$$

are all matrices. Each horizontal line of numbers is called a **row** of the matrix; each vertical one is called a **column**. For example, in Matrix (1) the third row consists of the entries 0 and 5, whereas the first column is 2, -1, 0. In other books, you may see the rectangular array enclosed in parentheses. For example, Matrix (2) might be written

$$\begin{pmatrix} 1 & -1 & 0 \\ 0 & 5 & 0 \end{pmatrix}.$$

Matrices arise naturally in the study of systems of linear equations. If we take a given linear system, put it in standard form, and then delete all unknowns and any plus and equals signs, we obtain a rectangular array of numbers. This array is called the **augmented matrix** of the system. (When writing an augmented matrix, it is customary to draw a vertical line separating its last column—the one that corresponds to the right side constants of the system—from the rest of the matrix.)

EXAMPLE 1

Find the augmented matrices for the following linear systems.

a.
$$\begin{aligned} 2x - 3y + z &= 1 \\ x \qquad + 2z &= 0 \\ - y - z &= -5 \end{aligned}$$

b.
$$\begin{aligned} x_1 - 3x_2 - 1 &= x_4 - 4x_3 \\ - x_1 + 2x_3 &= 0 \end{aligned}$$

Solution

a. The augmented matrix is

$$\left[\begin{array}{ccc|c} 2 & -3 & 1 & 1 \\ 1 & 0 & 2 & 0 \\ 0 & -1 & -1 & -5 \end{array}\right].$$

b. We first rewrite the system in standard form to yield

$$\begin{aligned} x_1 - 3x_2 + 4x_3 - x_4 &= 1 \\ -x_1 \qquad + 2x_3 \qquad &= 0. \end{aligned}$$

Thus, the augmented matrix is

$$\left[\begin{array}{cccc|c} 1 & -3 & 4 & -1 & 1 \\ -1 & 0 & 2 & 0 & 0 \end{array}\right].$$

■

*James Joseph Sylvester, engraved by George J. Stodart (engraving) by English photographer (19th century) (after) Private Collection/ The Bridgeman Art Library

WARNING When an unknown is "missing" in the given system (when its coefficient is zero), be sure to put a zero in the appropriate place in the augmented matrix corresponding to this system.

In solving the linear systems of Examples 4 and 5 of Section 2.2, we made use of the *elementary operations*:

i. Multiply an equation by a nonzero scalar.

ii. Interchange the positions of two equations.

iii. Replace an equation by the sum of itself and a multiple of another equation.

In an analogous manner, we define for matrices the **elementary row operations**:

i. Multiply a row by a nonzero scalar.

ii. Interchange the position of two rows.

iii. Replace a row by the sum of itself and a multiple of another row.

It will be useful to introduce some notation for each elementary row operation. We will denote by cR_i the result of multiplying row i by the nonzero scalar c; by $R_i \leftrightarrow R_j$, the result of interchanging the positions of rows i and j, and by $R_i + cR_j$, the result of replacing row i by the sum of itself and c times row j. We will sometimes indicate to the right of the matrix the elementary row operation carried out.

NOTE When we say "multiply a row by," we mean more precisely, "multiply each entry of a row by"; and when we say "the sum of two rows," we mean "the sum of the corresponding entries in the two rows."

EXAMPLE 2

Given the matrix

$$\begin{bmatrix} 1 & -1 & 0 & 4 \\ 2 & 0 & -3 & 1 \\ 5 & -2 & 3 & 0 \end{bmatrix},$$

if we multiply the third row by 3 and replace the second row by the sum of itself and -2 times the first, we obtain the matrix

$$\begin{bmatrix} 1 & -1 & 0 & 4 \\ 0 & 2 & -3 & -7 \\ 15 & -6 & 9 & 0 \end{bmatrix}. \begin{matrix} \\ (R_2 + (-2)R_1) \\ (3R_3) \end{matrix}$$ ■

Performing elementary row operations on the augmented matrix of a given linear system has the same effect as performing the analogous elementary operations on the system itself. In other words, the matrix resulting from the elementary row operations is the same as the augmented matrix of the linear system resulting from the analogous elementary operations.

EXAMPLE 3

Consider the linear system that follows, with the augmented matrix at the right:

$$\begin{aligned} x_1 - 2x_2 \qquad &= 1 \\ 2x_2 + x_3 &= -4 \\ 3x_1 - 4x_2 + x_3 &= 0 \end{aligned} \qquad \left[\begin{array}{ccc|c} 1 & -2 & 0 & 1 \\ 0 & 2 & 1 & -4 \\ 3 & -4 & 1 & 0 \end{array}\right]$$

Let us perform the following elementary operations on the system. Replace the third equation with the sum of itself and -3 times the first, and multiply the second equation by $\frac{1}{2}$. Now perform the analogous elementary row operations on the matrix: replace the third row with the sum of itself and -3 times the first, and multiply the second row by $\frac{1}{2}$. Then the resulting system and matrix are

$$\begin{aligned} x_1 - 2x_2 \quad &= 1 \\ x_2 + \tfrac{1}{2}x_3 &= -2 \\ 2x_2 + x_3 &= -3 \end{aligned} \qquad \left[\begin{array}{ccc|c} 1 & -2 & 0 & 1 \\ 0 & 1 & \frac{1}{2} & -2 \\ 0 & 2 & 1 & -3 \end{array}\right]. \begin{array}{l} \\ (\tfrac{1}{2}R_2) \\ (R_3 + (-3)R_1) \end{array}$$

Notice that the resulting matrix is the augmented matrix for the resulting system. ■

Consequently, in implementing the elimination procedure, we can perform elementary row operations on the augmented matrices instead of elementary operations on the system itself. This reduces the amount of writing necessary, as well as the possibilities for error, and lends itself nicely to calculator and computer solution of linear systems. As an illustration of this process, we solve Example 4 of Section 2.2, using the same sequence of operations, but this time performing them on matrices.

EXAMPLE 4

Use elementary row operations on matrices to solve the linear system below (see Example 4 of Section 2.2).

$$\begin{aligned} -y + z &= 3 \\ x - y - z &= 0 \\ -x \qquad - z &= -3 \end{aligned}$$

Solution We first write the augmented matrix for the system:

$$\left[\begin{array}{ccc|c} 0 & -1 & 1 & 3 \\ 1 & -1 & -1 & 0 \\ -1 & 0 & -1 & -3 \end{array}\right]$$

Now we proceed as in Example 4 of Section 2.2. Interchange the first and second rows:

$$\left[\begin{array}{ccc|c} 1 & -1 & -1 & 0 \\ 0 & -1 & 1 & 3 \\ -1 & 0 & -1 & -3 \end{array}\right] \begin{array}{l} (R_1 \leftrightarrow R_2) \\ \\ \\ \end{array}$$

Replace the third row with the sum of itself and 1 times the first row:

$$\left[\begin{array}{ccc|c} 1 & -1 & -1 & 0 \\ 0 & -1 & 1 & 3 \\ 0 & -1 & -2 & -3 \end{array}\right] \begin{array}{l} \\ \\ (R_3 + 1R_1) \end{array}$$

Multiply the second row by -1:

$$\left[\begin{array}{ccc|c} 1 & -1 & -1 & 0 \\ 0 & 1 & -1 & -3 \\ 0 & -1 & -2 & -3 \end{array}\right] \begin{array}{l} \\ ((-1)R_2) \\ \\ \end{array}$$

Replace the first row with the sum of itself and 1 times the second row. Then replace the third row with the sum of itself and 1 times the second row:

$$\left[\begin{array}{ccc|c} 1 & 0 & -2 & -3 \\ 0 & 1 & -1 & -3 \\ 0 & 0 & -3 & -6 \end{array}\right] \begin{array}{l} (R_1 + 1R_2) \\ \\ (R_3 + 1R_2) \end{array}$$

Multiply the third row by $-\frac{1}{3}$:

$$\left[\begin{array}{ccc|c} 1 & 0 & -2 & -3 \\ 0 & 1 & -1 & -3 \\ 0 & 0 & 1 & 2 \end{array}\right] \quad ((-\tfrac{1}{3})R_2)$$

Replace the first row with the sum of itself and 2 times the third row. Then replace the second row with the sum of itself and 1 times the third row:

$$\left[\begin{array}{ccc|c} 1 & 0 & 0 & 1 \\ 0 & 1 & 0 & -1 \\ 0 & 0 & 1 & 2 \end{array}\right] \quad \begin{array}{l} (R_1 + 2R_3) \\ (R_2 + 1R_3) \\ \end{array}$$

Change back to equation form (write the linear system for which this is the augmented matrix):

$$\begin{aligned} x \qquad\qquad &= 1 \\ y \qquad &= -1 \\ z &= 2 \end{aligned}$$

Thus, the solution is $(1, -1, 2)$. ■

NOTE When one matrix can be obtained from another by means of a finite sequence of elementary row operations, the two matrices are said to be **row-equivalent**. For example, any two of the matrices in Example 4 are row-equivalent.

The following theorem is a restatement of Theorem 1 of Section 2.2. It tells us that although the augmented matrices are changing as we perform elementary row operations on them, the underlying linear systems all have the same solutions.

THEOREM 1

If two augmented matrices are row-equivalent, the corresponding linear systems are equivalent—they have the same solutions. ■

The final matrix of Example 4 is an especially simple one. It is said to be in *row-reduced echelon form*. The basic goal of the elimination procedure is to transform the original augmented matrix (by means of elementary row operations) into one that has this special form.

DEFINITION

A matrix is in **row-reduced echelon form** if it satisfies the following:

i. In each row that does not consist entirely of zeros, the first nonzero entry is a 1 (we call such an element **a leading 1**).

ii. In each column that contains a leading 1 of some row, all other entries are zero.

iii. In any two rows with some nonzero entries, the leading 1 of the higher row is farther to the left.

iv. Any row that contains only zeros is lower than any row with some nonzero entries.

EXAMPLE 5

The augmented matrices corresponding to the final linear systems in Examples 4 and 5 of Section 2.2 are, respectively,

$$\left[\begin{array}{ccc|c} 1 & 0 & 0 & 1 \\ 0 & 1 & 0 & -1 \\ 0 & 0 & 1 & 2 \end{array}\right] \quad \text{and} \quad \left[\begin{array}{cccc|c} 1 & 0 & 3 & 1 & 0 \\ 0 & 1 & 2 & 1 & 0 \\ 0 & 0 & 0 & 0 & 0 \end{array}\right].$$

Each is in row-reduced echelon form. ▣

EXAMPLE 6

The matrix

$$\begin{bmatrix} 1 & 2 & 0 & -1 & 0 \\ 0 & 0 & 1 & 2 & 1 \\ 0 & 0 & 0 & 0 & 0 \\ 0 & 0 & 0 & 0 & 0 \end{bmatrix}$$

is in row-reduced echelon form. ▣

EXAMPLE 7

The matrices

$$\begin{bmatrix} 1 & 1 & 0 \\ 0 & 1 & 0 \\ 0 & 0 & 1 \end{bmatrix}, \quad \begin{bmatrix} 1 & 0 & 0 \\ 0 & 0 & 1 \\ 0 & 1 & 0 \end{bmatrix}, \quad \begin{bmatrix} 1 & 0 \\ 0 & 2 \end{bmatrix}, \quad \text{and} \quad \begin{bmatrix} 0 & 0 & 0 \\ 1 & 0 & 0 \end{bmatrix}$$

are *not* in row-reduced echelon form. (They violate conditions ii, iii, i, and iv, respectively, of the definition.) ▣

Carl Friedrich Gauss (1777–1855) is often referred to as the "prince of mathematicians," so great was his research output in so many areas of mathematics. Gauss created *Gaussian* elimination to solve certain problems involving the earth's surface. A book on geodesy (a branch of earth sciences dealing with surveying) by Wilhelm Jordan (1842–1899) was the first to describe what we now call *Gauss–Jordan elimination.*

As we have said, the basic goal of the matrix elimination procedure is to transform, by means of a systematic sequence of elementary row operations, the original augmented matrix into one that is in row-reduced echelon form. We then change back to the "equation form" of the linear system and complete the solution process as in Example 4. This procedure is often called **Gauss–Jordan elimination** and is described in the next two examples.

EXAMPLE 8

Gauss

Jordan

Solve the following system by Gauss–Jordan elimination:

$$\begin{aligned} x_3 + 2x_4 &= 3 \\ 2x_1 + 4x_2 - 2x_3 &= 4 \\ 2x_1 + 4x_2 - x_3 + 2x_4 &= 7 \end{aligned}$$

Solution

Step 0. Write the augmented matrix for the system:

$$\left[\begin{array}{cccc|c} 0 & 0 & 1 & 2 & 3 \\ 2 & 4 & -2 & 0 & 4 \\ 2 & 4 & -1 & 2 & 7 \end{array}\right]$$

Step 1(a). Obtain a leading 1 in the first row, first column. (If there is a 0 in this position, interchange the first row with a row below it so that a nonzero entry appears there.) Interchange the first and second rows:

$$\left[\begin{array}{cccc|c} 2 & 4 & -2 & 0 & 4 \\ 0 & 0 & 1 & 2 & 3 \\ 2 & 4 & -1 & 2 & 7 \end{array}\right] \; (R_1 \leftrightarrow R_2)$$

(If the nonzero entry in the first row, first column is not a 1, multiply the first row by the reciprocal of this entry.) Multiply the first row by $\frac{1}{2}$:

$$\left[\begin{array}{cccc|c} 1 & 2 & -1 & 0 & 2 \\ 0 & 0 & 1 & 2 & 3 \\ 2 & 4 & -1 & 2 & 7 \end{array}\right] \begin{array}{l} \left(\frac{1}{2}R_1\right) \\ \\ \\ \end{array}$$

Step 1(b). Obtain zeros in other positions in the first column by adding appropriate multiples of the first row to other rows. Replace the third row with the sum of itself and -2 times the first:

$$\left[\begin{array}{cccc|c} 1 & 2 & -1 & 0 & 2 \\ 0 & 0 & 1 & 2 & 3 \\ 0 & 0 & 1 & 2 & 3 \end{array}\right] \begin{array}{l} \\ \\ (R_3 + (-2)R_1) \end{array}$$

Step 2. Obtain a leading 1 in the second row, second column. (If there is a 0 in this position, interchange the second row with a row below it so that a nonzero entry appears there. If this is not possible, go to the next column.) It is not possible, since all entries in the second column below the first row are zero.

Step 3(a). Obtain a leading 1 in the second row, third column. There is one there already.

Step 3(b). Obtain zeros in the other positions in the third column by adding appropriate multiples of the second row to the other rows. Replace the first row with the sum of itself and 1 times the second row. Then replace the third row with the sum of itself and -1 times the second row:

$$\left[\begin{array}{cccc|c} 1 & 2 & 0 & 2 & 5 \\ 0 & 0 & 1 & 2 & 3 \\ 0 & 0 & 0 & 0 & 0 \end{array}\right] \begin{array}{l} (R_1 + 1R_2) \\ \\ (R_3 + (-1)R_2) \end{array}$$

Step 3(c). Place any newly created zero rows at the bottom of the matrix. It is already there. (The matrix is now in row-reduced echelon form.)

Step 4. Change back to a system of equations, ignoring any rows that contain only zeros:

$$\begin{aligned} x_1 + 2x_2 \qquad + 2x_4 &= 5 \\ x_3 + 2x_4 &= 3 \end{aligned}$$

Step 5. Solve each equation for the unknown whose coefficient is a leading 1 in the final augmented matrix

$$\begin{aligned} x_1 &= 5 - 2x_2 - 2x_4 \\ x_3 &= 3 - 2x_4. \end{aligned}$$

The unknowns appearing on the right-hand side are taken to be parameters. Set $x_2 = s$ and $x_4 = t$. We have here a "two-parameter family of solutions" (see example 5 of Section 2.2). All solutions are of the form $(5 - 2s - 2t, s, 3 - 2t, t)$ or, equivalently, $(5, 0, 3, 0) + s(-2, 1, 0, 0) + t(-2, 0, -2, 1)$. ■

EXAMPLE 9

Solve the following system by Gauss–Jordan elimination:

$$\begin{aligned} x_1 + 2x_2 \qquad\qquad\qquad &= 1 \\ x_1 + 2x_2 + 3x_3 + x_4 &= 0 \\ -x_1 - x_2 + x_3 + x_4 &= -2 \\ x_2 + x_3 + x_4 &= -1 \\ -x_2 + 2x_3 \qquad &= 0 \end{aligned}$$

Solution

Step 0. Write the augmented matrix for the system:

$$\left[\begin{array}{rrrr|r} 1 & 2 & 0 & 0 & 1 \\ 1 & 2 & 3 & 1 & 0 \\ -1 & -1 & 1 & 1 & -2 \\ 0 & 1 & 1 & 1 & -1 \\ 0 & -1 & 2 & 0 & 0 \end{array}\right]$$

Step 1(a). Obtain a leading 1 in the first row, first column. There is one there already.

Step 1(b). Obtain zeros in the other positions in the first column. Replace the second row with the sum of itself and -1 times the first row. Then replace the third row with the sum of itself and 1 times the first row:

$$\left[\begin{array}{rrrr|r} 1 & 2 & 0 & 0 & 1 \\ 0 & 0 & 3 & 1 & -1 \\ 0 & 1 & 1 & 1 & -1 \\ 0 & 1 & 1 & 1 & -1 \\ 0 & -1 & 2 & 0 & 0 \end{array}\right] \begin{array}{l} \\ (R_2 + (-1)R_1) \\ (R_3 + 1R_1) \\ \\ \\ \end{array}$$

Step 2(a). Obtain a leading 1 in the second row, second column. (If there is a 0 in this position, interchange the second row with a row below it to create a nonzero entry there.) Interchange the second and third rows:

$$\left[\begin{array}{rrrr|r} 1 & 2 & 0 & 0 & 1 \\ 0 & 1 & 1 & 1 & -1 \\ 0 & 0 & 3 & 1 & -1 \\ 0 & 1 & 1 & 1 & -1 \\ 0 & -1 & 2 & 0 & 0 \end{array}\right] \begin{array}{l} \\ (R_2 \leftrightarrow R_3) \\ \\ \\ \\ \end{array}$$

We now have a 1 in the required position.

Step 2(b). Obtain zeros in the other positions in the second column. Replace the first row with the sum of itself and -2 times the second row. Next replace the fourth row with the sum of itself and -1 times the second row. Then replace the fifth row with the sum of itself and 1 times the second row:

$$\left[\begin{array}{rrrr|r} 1 & 0 & -2 & -2 & 3 \\ 0 & 1 & 1 & 1 & -1 \\ 0 & 0 & 3 & 1 & -1 \\ 0 & 0 & 0 & 0 & 0 \\ 0 & 0 & 3 & 1 & -1 \end{array}\right] \begin{array}{l} (R_1 + (-2)R_2) \\ \\ \\ (R_4 + (-1)R_2) \\ (R_5 + 1R_2) \end{array}$$

Step 2(c). Place any newly created zero rows at the bottom of the matrix. Interchange the fourth and fifth rows:

$$\left[\begin{array}{rrrr|r} 1 & 0 & -2 & -2 & 3 \\ 0 & 1 & 1 & 1 & -1 \\ 0 & 0 & 3 & 1 & -1 \\ 0 & 0 & 3 & 1 & -1 \\ 0 & 0 & 0 & 0 & 0 \end{array}\right] \begin{array}{l} \\ \\ \\ (R_4 \leftrightarrow R_5) \\ \\ \end{array}$$

Step 3(a). Obtain a leading 1 in the third row, third column. (If there is a nonzero entry there and it is not a 1, multiply the third row by the reciprocal of this entry.) Multiply the third row by $\frac{1}{3}$:

$$\left[\begin{array}{cccc|c} 1 & 0 & -2 & -2 & 3 \\ 0 & 1 & 1 & 1 & -1 \\ 0 & 0 & 1 & \frac{1}{3} & -\frac{1}{3} \\ 0 & 0 & 3 & 1 & -1 \\ 0 & 0 & 0 & 0 & 0 \end{array}\right] \begin{array}{l} \\ \\ \left(\frac{1}{3}R_3\right) \\ \\ \\ \end{array}$$

Step 3(b). Obtain zeros in the other positions of the third column. Replace the first row with the sum of itself and 2 times the third row. Next replace the second row with the sum of itself and -1 times the third row. Then replace the fourth row with the sum of itself and -3 times the third row:

$$\left[\begin{array}{cccc|c} 1 & 0 & 0 & -\frac{4}{3} & \frac{7}{3} \\ 0 & 1 & 0 & \frac{2}{3} & -\frac{2}{3} \\ 0 & 0 & 1 & \frac{1}{3} & -\frac{1}{3} \\ 0 & 0 & 0 & 0 & 0 \\ 0 & 0 & 0 & 0 & 0 \end{array}\right] \begin{array}{l} (R_1 + 2R_3) \\ (R_2 + (-1)R_3) \\ \\ (R_4 + (-3)R_3) \\ \\ \end{array}$$

Step 3(c). Place any newly created zero rows at the bottom of the matrix. It is already there. (The matrix is now in row-reduced echelon form.)

Step 4. Change back to a system of equations ignoring any zero rows:

$$\begin{aligned} x_1 \qquad\qquad - \frac{4}{3}x_4 &= \frac{7}{3} \\ x_2 \qquad + \frac{2}{3}x_4 &= -\frac{2}{3} \\ x_3 + \frac{1}{3}x_4 &= -\frac{1}{3} \end{aligned}$$

Step 5. Solve each equation for the unknown whose coefficient is a leading 1:

$$\begin{aligned} x_1 &= \frac{7}{3} + \frac{4}{3}x_4 \\ x_2 &= -\frac{2}{3} - \frac{2}{3}x_4 \\ x_3 &= -\frac{1}{3} - \frac{1}{3}x_4 \end{aligned}$$

We set $x_4 = s$ and obtain the "one-parameter family of solutions":

$$\left(\frac{7}{3} + \frac{4}{3}s, -\frac{2}{3} - \frac{2}{3}s, -\frac{1}{3} - \frac{1}{3}s, s\right)$$

or, equivalently,

$$\left(\frac{7}{3}, -\frac{2}{3}, -\frac{1}{3}, 0\right) + s\left(\frac{4}{3}, -\frac{2}{3}, -\frac{1}{3}, 1\right)$$ ■

Before continuing, we summarize the steps involved at the heart of the Gauss–Jordan elimination procedure: the transformation of the original augmented matrix to row-reduced echelon form.

Procedure to Row-Reduce a Matrix

i. In the leftmost column whose entries are not all zero, obtain a *leading 1* at the top of this column by using elementary row operations (i) and (ii).

ii. Obtain zeros in all other positions in this column by using elementary operation (iii).

iii. Place any newly created zero rows at the bottom of the matrix by using elementary operation (ii).

iv. Repeat step (i) for the *submatrix* obtained by deleting all rows containing previously obtained leading ones, and then do steps (ii) and (iii) on the full matrix. Continue until the row-reduced echelon form is obtained.

COMPUTATIONAL NOTE

Graphing calculators and some linear algebra software packages can be used to transform a matrix to row-reduced echelon form. To do this using a typical graphing calculator, enter the given matrix and then instruct the calculator to perform the required sequence of elementary row operations. Using computer software (such as *Matlab*) and some other types of graphing calculators is usually much easier: just enter the given matrix and issue the command to row-reduce it!

It is interesting and important to note that from the Gauss–Jordan procedure, if all the entries in the augmented matrix of a linear system are rational numbers, then a solution (if one exists) can also be chosen to consist of all rational numbers.

Using the Gauss–Jordan procedure, we can transform any matrix into row-reduced echelon form by means of a sequence of elementary row operations. We can also obtain this form by performing elementary row operations in a different order. We still arrive at the same row-reduced echelon form, however. These facts are stated without proof in the following two theorems. The first of these theorems can be proved by giving a precise description of the Gauss–Jordan procedure, which has been informally described in the preceding pages. A proof of the second can be constructed with the aid of the tools introduced in Chapter 5.

THEOREM 2

Every matrix can be transformed by a finite sequence of elementary row operations into one that is in row-reduced echelon form. ■

THEOREM 3

The row-reduced echelon form of a matrix is unique. 

In Example 4, the given linear system had a unique solution, whereas in Examples 8 and 9 there were infinitely many solutions. As we know from Section 2.2, a linear system may also have no solution. The following example illustrates what happens in the Gauss–Jordan procedure in such a case.

EXAMPLE 10

Solve

$$\begin{aligned} x_1 \qquad + x_3 &= 1 \\ x_2 - x_3 &= -1 \\ 2x_1 + x_2 + x_3 &= 2. \end{aligned}$$

Solution The augmented matrix for this system is

$$\left[\begin{array}{ccc|c} 1 & 0 & 1 & 1 \\ 0 & 1 & -1 & -1 \\ 2 & 1 & 1 & 2 \end{array}\right].$$

To transform it into row-reduced echelon form, first replace the third row with the sum of itself and -2 times the first row:

$$\left[\begin{array}{ccc|c} 1 & 0 & 1 & 1 \\ 0 & 1 & -1 & -1 \\ 0 & 1 & -1 & 0 \end{array}\right] \begin{array}{l} \\ \\ (R_3 + (-2)R_1) \end{array}$$

Then replace the third row with the sum of itself and -1 times the second, obtaining

$$\left[\begin{array}{ccc|c} 1 & 0 & 1 & 1 \\ 0 & 1 & -1 & -1 \\ 0 & 0 & 0 & 1 \end{array}\right]. \begin{array}{l} \\ \\ (R_3 + (-1)R_2) \end{array}$$

Although this matrix is not in row-reduced echelon form, we see that the third row represents the equation

$$0x_1 + 0x_2 + 0x_3 = 1,$$

which cannot be satisfied by any choice of x_1, x_2, and x_3. Thus, this equation, and hence the system, has no solution.

Geometrically, we have shown here that the planes in $\mathbf{R}^3$ whose equations are the given ones have no common points of intersection (Figure 9).

FIGURE 9

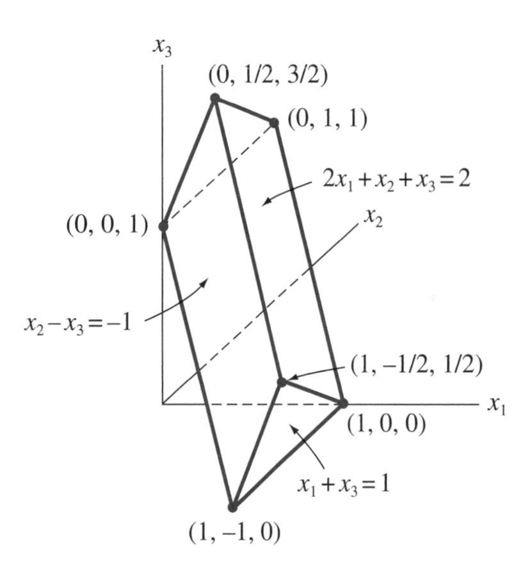

■

We may want to leave leading entries different from one as this allows us to solve a linear system whose augmented matrix contains only *integers* without having to introduce fractions until the very end.

COMPUTATIONAL NOTE

Gauss–Jordan elimination is not the most efficient matrix elimination procedure for use on computers. This honour belongs to a method known as **Gaussian elimination**. In the latter, elementary row operations are used to transform the original augmented matrix into **echelon form**, defined as follows:

i. The leftmost nonzero element in each row is called a **leading entry**.

ii. In each column that contains a leading entry of some row, all elements *below* the leading entry are zero.

iii. In any two rows with leading entries, the leading entry of the higher row is farther to the left.

iv. Any row that contains only zeros is lower than any row with some nonzero entries.

Note that compared with *row-reduced* echelon form, echelon form does not require the leading entry in a row to be a *one*, nor does it require *all* entries in a column with a leading entry, apart from the leading entry itself, to be zero. For example, the matrix

$$\begin{bmatrix} 2 & 1 & -1 & 3 & 5 \\ 0 & 0 & -3 & 6 & 1 \\ 0 & 0 & 0 & 2 & 4 \\ 0 & 0 & 0 & 0 & 0 \end{bmatrix}$$

is in echelon form. Once the original augmented matrix is transformed into echelon form, the linear system is solved by a process of *back-substitution*, as illustrated in Example 11.

EXAMPLE 11

Solve the system of Example 4 by Gaussian elimination:

$$\begin{aligned} -y + z &= 3 \\ x - y - z &= 0 \\ -x \qquad - z &= -3 \end{aligned}$$

Solution The augmented matrix is

$$\left[\begin{array}{rrr|r} 0 & -1 & 1 & 3 \\ 1 & -1 & -1 & 0 \\ -1 & 0 & -1 & -3 \end{array}\right].$$

Interchange the first and second rows:

$$\left[\begin{array}{rrr|r} 1 & -1 & -1 & 0 \\ 0 & -1 & 1 & 3 \\ -1 & 0 & -1 & -3 \end{array}\right] (R_1 \leftrightarrow R_2)$$

Replace the third row with the sum of itself and the first row:

$$\left[\begin{array}{rrr|r} 1 & -1 & -1 & 0 \\ 0 & -1 & 1 & 3 \\ 0 & -1 & -2 & -3 \end{array}\right] (R_3 + 1R_1)$$

Replace the third row with the sum of itself and -1 times the second row:

$$\left[\begin{array}{rrr|r} 1 & -1 & -1 & 0 \\ 0 & -1 & 1 & 3 \\ 0 & 0 & -3 & -6 \end{array}\right] (R_3 + (-1)R_2)$$

The matrix is now in echelon form. Convert back to equation form:

$$\begin{aligned} x - y - z &= 0 \\ -y + z &= 3 \\ -3z &= -6 \end{aligned}$$

We now solve this system by **back-substitution**. First, solve the last equation for z, obtaining $z = 2$. Next, substitute this value of z into the second equation, and solve it for y, obtaining $y = -1$. Finally, substitute $z = 2$, $y = -1$ into the first equation to get $x = 1$. As before, the solution is $x = 1$, $y = -1$, $z = 2$, or $(1, -1, 2)$. ■

2.3 Exercises

In Exercises 1–4 find the augmented matrix for each of the systems.

1. $$\begin{aligned} 2x - 3y + z &= 0 \\ x - 2z &= 1 \\ -4y + z &= -1 \end{aligned}$$

2. $$\begin{aligned} x_1 &= 3 - x_2 + x_3 \\ x_2 &= x_4 - x_3 \\ x_3 &= x_1 + x_4 + 1 \end{aligned}$$

3. $$\begin{aligned} u &= v \\ v &= w \end{aligned}$$

4. $$\begin{aligned} x + w &= 0 \\ y + z &= 0 \end{aligned}$$

In Exercises 5–10 determine which matrices are in row-reduced echelon form.

5. $\begin{bmatrix} 0 & 1 \\ 1 & 0 \end{bmatrix}$ *6.* $\begin{bmatrix} 1 & 2 & 0 \\ 0 & 0 & 1 \end{bmatrix}$ *7.* $\begin{bmatrix} 1 & 0 & 2 \\ 0 & 1 & 1 \end{bmatrix}$

8. $\begin{bmatrix} 1 & 2 & 0 & 3 \\ 0 & 1 & 0 & 2 \\ 0 & 0 & 1 & 0 \end{bmatrix}$ *9.* $\begin{bmatrix} 1 & 0 & 0 \\ 0 & 0 & 0 \\ 0 & 1 & 0 \end{bmatrix}$ *10.* $\begin{bmatrix} 2 & 0 & 0 \\ 0 & 2 & 0 \\ 0 & 0 & 2 \end{bmatrix}$

In Exercises 11–16 the given matrix is the augmented matrix for a linear system in the variables x_1, x_2, and x_3. Solve the system.

11. $\left[\begin{array}{ccc|c} 1 & 0 & 0 & 0 \\ 0 & 1 & 0 & 2 \\ 0 & 0 & 1 & -1 \end{array}\right]$ *12.* $\left[\begin{array}{ccc|c} 1 & 0 & 1 & 0 \\ 0 & 1 & -1 & 0 \\ 0 & 0 & 0 & 0 \end{array}\right]$

13. $\left[\begin{array}{ccc|c} 1 & 0 & 0 & 0 \\ 0 & 1 & 0 & 0 \\ 0 & 0 & 0 & 1 \end{array}\right]$ *14.* $\left[\begin{array}{ccc|c} 1 & 0 & 0 & 0 \\ 0 & 1 & 0 & 0 \\ 0 & 0 & 0 & 0 \end{array}\right]$

15. $\left[\begin{array}{ccc|c} 1 & 2 & 3 & 4 \\ 0 & 0 & 0 & 0 \end{array}\right]$ *16.* $\left[\begin{array}{ccc|c} 1 & 0 & 0 & 0 \\ 0 & 1 & 0 & 1 \\ 0 & 0 & 0 & 0 \end{array}\right]$

In Exercises 17–26 solve the linear systems by the Gauss–Jordan matrix elimination method.

17. $$\begin{aligned} x_1 - 2x_2 &= 1 \\ -2x_1 + 4x_2 &= -2 \end{aligned}$$

18. $$\begin{aligned} -x + y &= 0 \\ 2x - 2y &= 0 \end{aligned}$$

19. $$\begin{aligned} x - 2y + z &= 5 \\ -2x + 3y + z &= 1 \\ x + 3y + 2z &= 2 \end{aligned}$$

20. $$\begin{aligned} -x_1 + x_2 - 2x_3 &= 1 \\ x_1 + x_2 + 2x_3 &= -1 \\ x_1 + 3x_2 + 2x_3 &= -1 \end{aligned}$$

21. $$\begin{aligned} 2x_1 - 3x_2 + x_3 &= 1 \\ -x_1 + 2x_3 &= 0 \\ 3x_1 - 3x_2 - x_3 &= 1 \end{aligned}$$

22. $$\begin{aligned} x + y &= 0 \\ x + y &= -z \\ -x + 1 &= y \\ 1 + x + 2z &= 0 \end{aligned}$$

23. $$\begin{aligned} w &= x + y + z \\ w &= 2x - 3y + z - 1 \\ w &= -x + y - 2z + 2 \\ w &= 4x - 3y + 4z \end{aligned}$$

24. $$\begin{aligned} x_1 + 2x_2 - x_4 &= 0 \\ x_1 - 2x_2 + x_3 + x_4 &= 0 \\ 2x_1 - 3x_2 - x_4 &= 0 \end{aligned}$$

25. $$\begin{aligned} v - 2w + z &= 1 \\ 2u - v - z &= 0 \\ 4u + v - 6w + z &= 3 \end{aligned}$$

26. $$\begin{aligned} -x_1 + 3x_2 + x_4 &= 0 \\ 2x_1 - 5x_2 + x_3 - x_4 &= 1 \\ x_2 + x_3 + x_4 &= 1 \\ x_1 + x_2 + x_3 &= 0 \end{aligned}$$

27–36. Solve Exercises 17–26 by the Gaussian elimination procedure.

In Exercises 37–40 the given equations represent planes in $\mathbf{R}^3$. Describe the region (point, line, or plane), if any, that is the intersection of these planes.

37.
$$\begin{aligned} x + y + z &= -4 \\ x + 2y \quad\quad &= 1 \\ 2y + 3z &= -2 \end{aligned}$$

38.
$$\begin{aligned} y + z &= 1 \\ x + y + 2z &= -1 \\ x \quad + z &= 1 \end{aligned}$$

39.
$$\begin{aligned} x - y + z &= 0 \\ y - 2z &= 1 \\ 2x - y \quad\quad &= 1 \end{aligned}$$

40.
$$\begin{aligned} -x + 2y - z &= 1 \\ 2x - 4y + 2z &= -2 \\ x - 2y + z &= -1 \end{aligned}$$

41. Find the value of c such that the system

$$\begin{aligned} cx \quad + z &= 0 \\ 2y - 4z &= 0 \\ 2x - y \quad\quad &= 0 \end{aligned}$$

has a solution other than (0, 0, 0).

42. Find all values of k such that the system

$$\begin{aligned} x_1 - x_2 + 2x_3 &= 0 \\ x_2 - x_3 &= k \\ -x_1 + 2x_2 - 3x_3 &= 1 \end{aligned}$$

has no solution.

43. Given a system of linear equations in three variables with the augmented matrix

$$\left[\begin{array}{ccc|c} 1 & 0 & 1 & 1 \\ 0 & 1 & 1 & 2 \\ 0 & 2 & k & k \end{array}\right],$$

for what value of k is there no solution for the system? Is there any value of k for which there are infinitely many solutions?

44. The possible row-reduced echelon forms of the matrix

$$\begin{bmatrix} a & b \\ c & d \end{bmatrix}$$

are $\begin{bmatrix} 1 & 0 \\ 0 & 1 \end{bmatrix}$, $\begin{bmatrix} 1 & k \\ 0 & 0 \end{bmatrix}$, $\begin{bmatrix} 0 & 1 \\ 0 & 0 \end{bmatrix}$, and $\begin{bmatrix} 0 & 0 \\ 0 & 0 \end{bmatrix}$.

List the possible row-reduced echelon forms of the matrix

$$\begin{bmatrix} a & b & c \\ d & e & f \\ g & h & i \end{bmatrix}.$$

SPOTLIGHT *45* Consider the metal plate (see the Spotlight at the beginning of this chapter) pictured in Figure 10. Suppose that the temperatures at the upper left and lower right corners of this plate are fixed at 0° and 140°, respectively, and that the temperature on the boundary is evenly distributed between these points. (For example, the temperature in the upper right corner is 80°.)

(a) Construct the augmented matrix for a system of four linear equations in four unknowns whose solution approximates the equilibrium temperature distribution at the interior points.

(b) Solve the system in part (a) by using either Gauss–Jordan or Gaussian elimination.

FIGURE 10

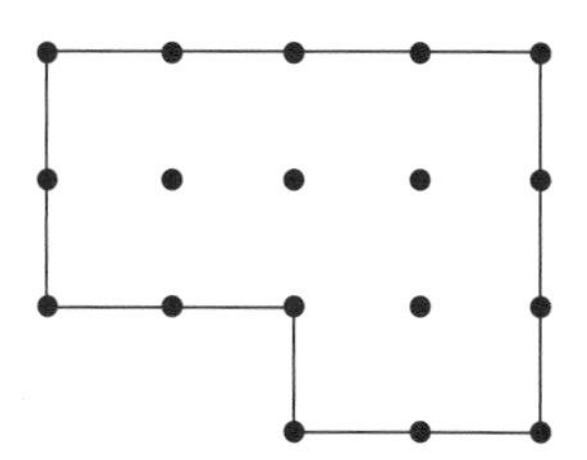

SPOTLIGHT *46.* Consider a metal plate shaped like the one in Figure 10 with its boundary held at the fixed temperature described in Exercise 45. Suppose, however, that we halve the distance between grid points both vertically and horizontally. (Now, for example, there are nine grid points along the top edge and seven down the right side.)

(a) How many equations and unknowns appear in the linear system that approximates the equilibrium temperature distribution at the interior grid points?

(b) How many unknowns (with nonzero coefficients) appear in *each* of the equations of this linear system?

COMPUTATIONAL NOTE

A computer or calculator rounds certain fractions (such as $\frac{1}{3}$) to a finite (relatively small) number of decimal places. Consequently "round-off errors" may be introduced at each step of an algorithm, such as Gauss–Jordan or Gaussian elimination. In "small" problems, like those in this exercise set, these errors are unlikely to have a significant effect on the accuracy of the result; however, it is possible. The cumulative effect of round-off errors is a function not only of the size of a linear system but also of its coefficients. Exercises 41–51 explore this phenomenon.

Computational Exercises

47–56. Solve Exercises 17–26 by using a graphing calculator or a computer software package. Compare your answers with the ones you obtained by hand.

57. Use a graphing calculator or computer software to solve the linear system that has the following augmented matrix:

$$\left[\begin{array}{cccccc|c} 1 & \frac{1}{2} & \frac{1}{3} & \frac{1}{4} & \frac{1}{5} & \frac{1}{6} & 1 \\ \frac{1}{2} & \frac{1}{3} & \frac{1}{4} & \frac{1}{5} & \frac{1}{6} & \frac{1}{7} & 0 \\ \frac{1}{3} & \frac{1}{4} & \frac{1}{5} & \frac{1}{6} & \frac{1}{7} & \frac{1}{8} & 0 \\ \frac{1}{4} & \frac{1}{5} & \frac{1}{6} & \frac{1}{7} & \frac{1}{8} & \frac{1}{9} & 0 \\ \frac{1}{5} & \frac{1}{6} & \frac{1}{7} & \frac{1}{8} & \frac{1}{9} & \frac{1}{10} & 0 \\ \frac{1}{6} & \frac{1}{7} & \frac{1}{8} & \frac{1}{9} & \frac{1}{10} & \frac{1}{11} & 0 \end{array}\right]$$

Compare your answer with the exact solution, which is $(36, -630, 3360, -7560, 7560, -2772)$.

58. Use a graphing calculator or computer software to solve the linear systems with the following augmented matrices. Give your answers to two decimal places.

(a) $\left[\begin{array}{ccc|c} 1 & \frac{1}{2} & \frac{1}{3} & 1 \\ \frac{1}{2} & \frac{1}{3} & \frac{1}{4} & 1 \\ \frac{1}{3} & \frac{1}{4} & \frac{1}{5} & 1 \end{array}\right]$ (b) $\left[\begin{array}{ccc|c} 1.00 & 0.33 & 0.20 & 1 \\ 0.33 & 0.20 & 0.15 & 1 \\ 0.20 & 0.15 & 0.11 & 1 \end{array}\right]$

59. Use a graphing calculator or computer software to solve the following systems of linear equations:

(a)
$$\begin{aligned} 3.56x_1 + 2.01x_2 + 3.10x_3 &= -4.19 \\ -1.01x_1 + 1.51x_2 - 0.05x_3 &= 8.15 \\ 2.07x_1 + 11.57x_2 + 5.93x_3 &= -3.14 \end{aligned}$$

(b)
$$\begin{aligned} 5x_1 + 2x_2 + 3x_3 + 2x_4 &= 5 \\ x_1 - 3x_2 + 2x_3 - 5x_4 &= 6 \\ -3x_1 + 4x_2 - 5x_3 - 7x_4 &= -3 \\ x_1 - 22x_2 + 25x_3 + 7x_4 &= -1 \end{aligned}$$

(c)
$$\begin{aligned} 212x_1 - 13x_2 + 5x_3 &= -101 \\ 52x_1 + 51x_2 - 13x_3 &= 10 \\ 108x_1 + 115x_2 + 31x_3 &= 53 \end{aligned}$$

In what way are the three systems similar?

60. Write the augmented matrices associated with each system in Exercise 59, and write the row-reduced form of each matrix. In what way are the row-reduced forms similar?

61. Replace the constant terms in each system of Exercise 59 by zeros. Solve the new systems. Are the solutions unique?

SPOTLIGHT *62.* Use a software package or graphing calculator to find the solution of the linear system described in Exercise 46.

2.4 APPLICATION—Electric Circuits and Pipe Networks

Linear algebra has many applications to physics. In this section we discuss two of the simplest: determining the current flowing in each wire of an electric circuit and finding the rate of flow of water in a network of pipes.

Current Flow in an Electric Circuit

We consider direct current (DC) electric circuits consisting of one *electromotive force* (emf), or more, and one or more *resistors*. Electromotive force, E, is measured in *volts* (denoted by V); *resistance*, R, is measured in *ohms* (Ω). The relationship between emf, resistance, and *current*, I (measured in *amperes* or A), in a simple circuit is given by *Ohm's law*:

$$E = IR$$

For example, in a circuit containing a simple emf of 6 V and a single resistance of 30 Ω, as in Figure 11, the current is calculated to be $6/30 = 1/5$ A. The direction of the current is by convention assumed to be from + to −, as indicated by the arrow in Figure 11.

FIGURE 11

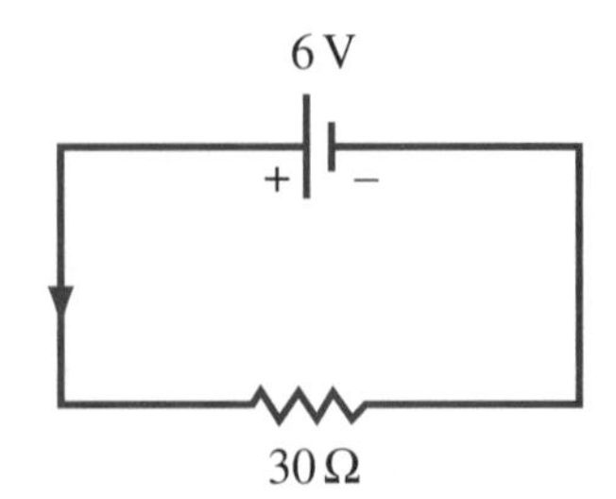

Assuming that the wires in a circuit have negligible resistance, the *voltage drop* across the resistor of Figure 11 must equal the applied emf. Since the current across the 30-Ω resistor is $\frac{1}{5}$ A, the voltage drop is $\frac{1}{5}(30) = 6$ V. In somewhat more complicated circuits, such as that of Figure 12, the sum of the voltage drops (across the resistors) must equal the *net* emf (the algebraic sum of the emfs). Here the net emf is 2 V (8 V − 6 V), since the two emfs are in opposition to each other. Equating total voltage drop to net emf, we obtain

$$6I + 4I + 2I = 2,$$

or $I = \frac{1}{6}$ A (flowing in the direction indicated).

FIGURE 12

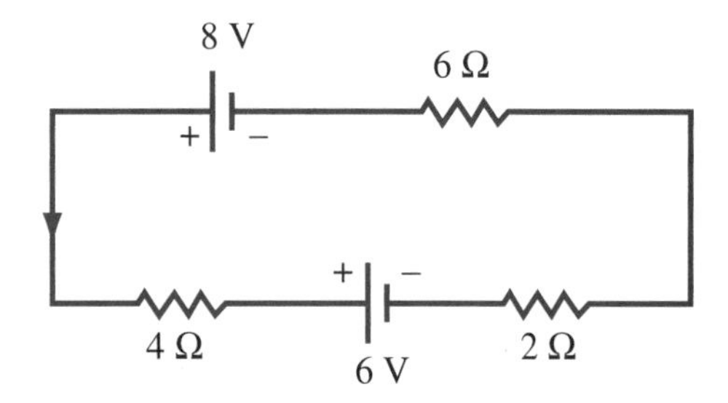

When a circuit consists of more than one *loop* (that is, a *closed path*), we use Kirchhoff's laws to determine the current in any branch. In such circuits there are points, called *nodes*, where three or more wires come together. For example, the circuit of Figure 13 has three loops (*ABCD*, *ADEF*, and *ABCDEF*) and two nodes (*A* and *D*).

FIGURE 13

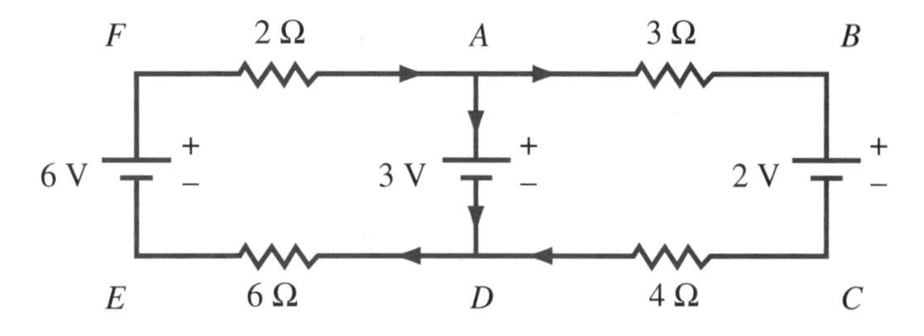

Kirchhoff's Laws

i. The algebraic sum of currents at any node in a circuit is zero.

ii. The algebraic sum of the voltage drops around any loop of the circuit is equal to the algebraic sum of the electromotive forces in that loop.

NOTE It is not necessary to guess correctly the direction of the current in any given wire of a complicated circuit. An arbitrary assignment is made, and if the current turns out to be negative, it simply means that the guess was wrong—the current flows in the opposite direction.

EXAMPLE 1

Find the current in each wire of the circuit of Figure 13.

Solution Let I_1 be the current from *D* to *E* to *F* to *A*, I_2 be the current from *A* to *B* to *C* to *D*, and I_3 be the current from *A* to *D*.

Since *A* is a node, and I_1 has been assigned to flow into *A*, whereas I_2 and I_3 flow out of *A*, we have, by the first law,

$$I_1 - I_2 - I_3 = 0. \quad \textbf{(1)}$$

Since *D* is a node, and I_1 flows out of *D*, whereas I_2 and I_3 flow into *D*, we have

$$I_2 + I_3 - I_1 = 0. \quad \textbf{(2)}$$

From loop *ABCD*, using the second law, we have

$$3I_2 + 4I_2 = 3 - 2$$

or, more simply,

$$7I_2 = 1. \tag{3}$$

From loop *ADEF*, we have

$$2I_1 + 6I_1 = 6 - 3$$

or, more simply,

$$8I_1 = 3. \tag{4}$$

From loop *ABCDEF*, we have

$$3I_2 + 4I_2 + 6I_1 + 2I_1 = 6 - 2$$

or, more simply,

$$8I_1 + 7I_2 = 4. \tag{5}$$

Equations (1) through (5) then provide the following system of linear equations:

$$\begin{aligned} I_1 - I_2 - I_3 &= 0 \\ -I_1 + I_2 + I_3 &= 0 \\ 7I_2 &= 1 \\ 8I_1 &= 3 \\ 8I_1 + 7I_2 &= 4 \end{aligned}$$

Following the methods of Section 2.3, we obtain the unique solution $I_1 = 3/8$, $I_2 = 1/7$, $I_3 = 13/56$. ■

Water Flow in a Network of Pipes

A situation very similar to electric circuits and Kirchhoff's laws is that of fluid flow through a plumbing network. In this situation, the equivalent of the voltage drop is the pressure drop along the length of a pipe due to internal friction, since the pipe does not move but the fluid does. The pressure drop d is assumed to obey the formula

$$d = \frac{vl}{\pi r^4} f = kf,$$

where f is the rate of flow in a pipe of length l and of radius r, and v is a constant based on the viscosity of the fluid. The constant k is a combination of the other constants,

$$k = \frac{vl}{\pi r^4},$$

which can be calculated for each pipe in the system. These constants k play the same role that resistances R did in a circuit analysis. Rather than list the pipe lengths, radii, and fluid viscosity, we simply assume these calculations have been made and proceed with known values k. The pressure itself can be generated by pumps in the network, or by gravity if the pipes are not all level with one another. We assume that the pressures and flows are represented by unspecified compatible units.

The following rules describe the flows:

i. The algebraic sum of flows at any node in the network is zero.

ii. The algebraic sum of the pressure drops around any loop of the network is equal to pressures in the loop. (Equivalently, taking pressure sources to be negative pressure drops, the sum around any loop is zero.)

It is easily seen that the first equation, $d = kf$, and conditions (i) and (ii) are very similar to Ohm's law and Kirchhoff's laws of circuit analysis. It is therefore not surprising that the solution, *the rate of flow in each pipe*, can be obtained in a similar fashion. It is assumed that the system is operating in a *stable* manner; that is, the flows are all underway and there is enough capacity in the system to keep them flowing. Flows in and out of a lake or a large cistern are examples of stability in this sense. Incidentally, the lake itself can be considered to be one of the pipes (with no pressure drop), to complete loops needed to solve the system.

EXAMPLE 2

In the network of pipes given in Figure 14, each node is given by a letter and each pipe by a number. Nodes *B* and *C* are pumps of pressure 200 and 300, respectively, that are driving the system, and *E* is a lake. Since the flows in 1 and 5 as well as in 3 and 4 must agree, they could be treated as one pipe. They do not look like one pipe, however, and may very well have different *k* values, so we do not make this assumption. Let the *k* values for the respective pipes be given by

Pipe i	1	2	3	4	5	6
k_i	5	4	2	4	2	1

FIGURE 14

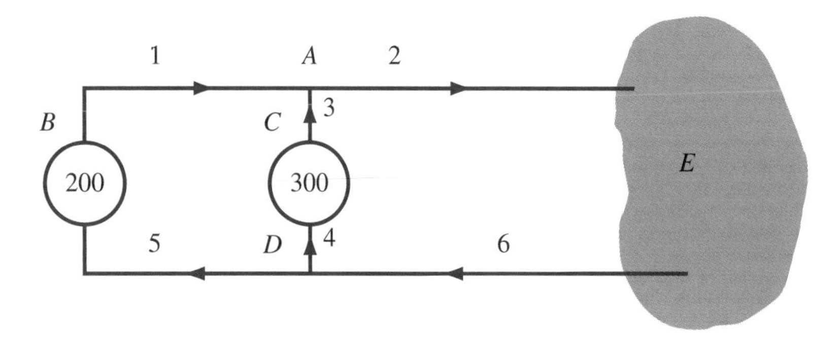

Let x_i represent the flow in pipe i. Then we have node equations at each of *A* through *E*, and loop equations corresponding to the loops *AEDBA*, *AEDCA*, and *ACDBA*. As was the case with Kirchhoff equations, there is some redundancy (dependence) in these equations, but that shows up in the solution process.

$$
\begin{aligned}
A:&\quad x_1 - x_2 + x_3 &= 0\\
B:&\quad x_1 - x_5 &= 0\\
C:&\quad x_3 - x_4 &= 0\\
D:&\quad x_4 + x_5 - x_6 &= 0\\
E:&\quad x_2 - x_6 &= 0\\
AEDBA:&\quad 5x_1 + 4x_2 + 2x_5 + x_6 &= 200\\
AEDCA:&\quad 4x_2 + 2x_3 + 4x_4 + x_6 &= 300\\
ACDBA:&\quad 5x_1 - 2x_3 - 4x_4 + 2x_5 &= -100
\end{aligned}
$$

Solving the system by Gauss–Jordan elimination, we have

$$
\begin{aligned}
x_1 &= 6.54\\
x_2 &= 30.84\\
x_3 &= 24.30\\
x_4 &= 24.30\\
x_5 &= 6.54\\
x_6 &= 30.84.
\end{aligned}
$$

■

2.4 Exercises

In Exercises 1 and 2 let the circuit be given by Figure 15. Calculate the current through each resistor, using the given values of resistances and emfs.

1. $a = 3\ \Omega$, $b = 4\ \Omega$, $c = 3\ \Omega$, $d = 2\ \Omega$, $e = 1\ \Omega$, $f = 6$ V, $g = 2$ V
2. $a = 4\ \Omega$, $b = 4\ \Omega$, $c = 5\ \Omega$, $d = 6\ \Omega$, $e = 4\ \Omega$, $f = 8$ V, $g = 10$ V

FIGURE 15

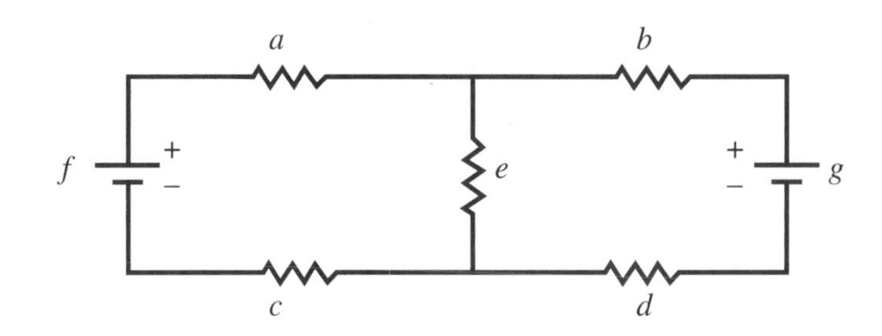

In Exercises 3 and 4 let the circuit be given by Figure 16. Calculate the current in each branch of the circuit, using the given values of resistances and emfs.

3. $a = 3\ \Omega$, $b = 4\ \Omega$, $c = 3\ \Omega$, $d = 2\ \Omega$, $e = 1\ \Omega$, $f = 6$ V, $g = 2$ V
4. $a = 4\ \Omega$, $b = 4\ \Omega$, $c = 5\ \Omega$, $d = 6\ \Omega$, $e = 4\ \Omega$, $f = 8$ V, $g = 10$ V

FIGURE 16

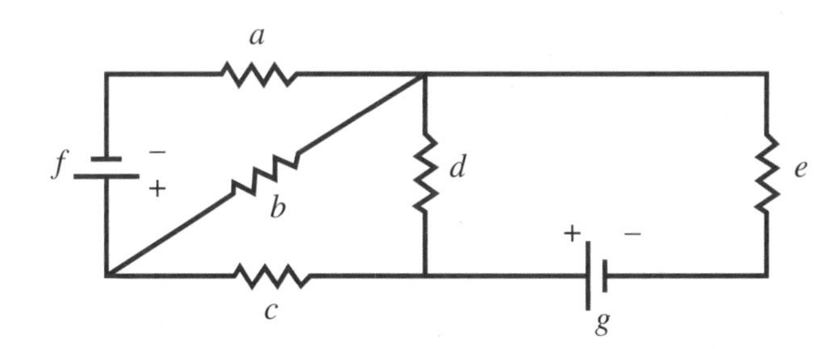

Induction ammeters can be used to measure the currents in live circuits. When the emf sources and these currents are known, Kirchhoff's laws can be used to calculate the resistances in a network. The system may have multiple solutions, since the resistances may not be completely determined, but additional information (for example, some known components in the system) can be used to complete the computations. When a parameterized family of solutions is all that is known, be sure to take into consideration the physical restriction that all resistances are nonnegative. In Exercises 5 and 6 use Figure 16 to calculate the resistance in each resistor if the current flow in amperes through each resistor is as indicated.

5. $a = 5$, $b = 3$, $c = 2$, $d = 4$, $e = 2$, $f = 6$ V, $g = 2$ V
6. $a = 6$, $b = 3$, $c = 3$, $d = 6$, $e = 3$, $f = 8$ V, $g = 10$ V

In Exercises 7 and 8 let a network of pipes be given by Figure 17. Calculate the flow in each branch, using the given pipe–viscosity constants k_i and pipe pressures.

7.

Pipe i	1	2	3	4	5	6	7	
k_i	3	1	5	1	2	1	3	$A = 200$ $B = 100$

8.

Pipe i	1	2	3	4	5	6	7
k_i	2	1	2	1	2	1	2

$A = 200$
$B = 400$

FIGURE 17

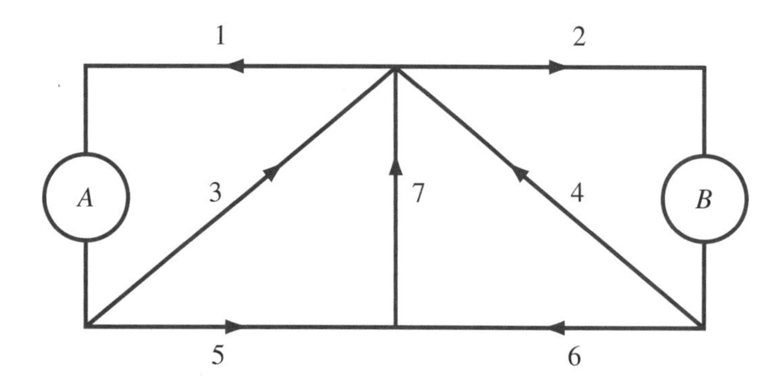

Analogously with Exercises 5 and 6, if we know the flow in each pipe of a network and the pressure at each source, the pipe-viscosity constants k_i can be computed. In Exercises 9 and 10 redraw Figure 16, interpreting the emf sources as pumps and the currents as flows. Then solve for the pipe constants, using the same numerical data as before.

9. $a = 5, b = 3, c = 2, d = 4, e = 2, f = 6, g = 2$

10. $a = 6, b = 3, c = 3, d = 6, e = 3, f = 8, g = 10$

Computational Exercise

11. Use a computer software package or graphing calculator to find the current in each wire of the circuit in Figure 18.

FIGURE 18

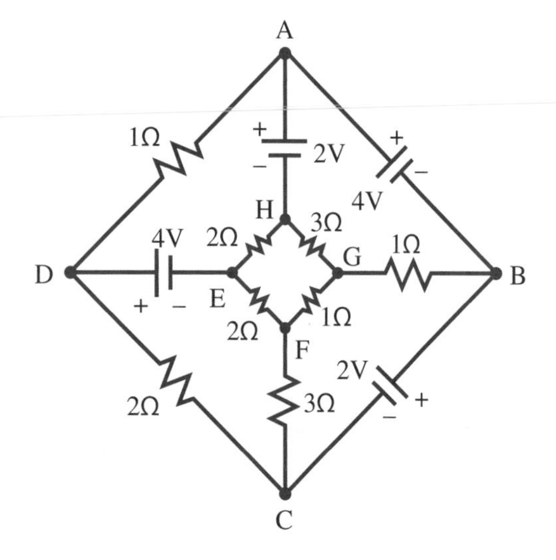

2.5 APPLICATION—Interpolating Polynomials

In this section, we develop a procedure to determine a polynomial that passes through a given set of points. Recall that a **polynomial** in the variable x is a function of the form

$$P(x) = a_0 + a_1x + a_2x^2 + \cdots + a_nx^n,$$

where the a_i are scalars and n is a nonnegative integer. If a_n is not equal to zero, the **degree** of the polynomial is equal to n. Polynomials of degree one, two, and three are called **linear**, **quadratic**, and **cubic**, respectively. The graph of a linear polynomial is a line in the plane, whereas that of a quadratic polynomial is a parabola.

We first consider the simple case of determining an equation of the line (the polynomial of degree one) that passes through two distinct points. This is a standard problem in elementary algebra, but here we solve it in a different way—one that illustrates the general process of passing a polynomial through a set of points.

Since a nonvertical line in the xy-plane may be represented by a linear polynomial $P(x) = a_0 + a_1x$, we can find the equation of the line through the points (x_0, y_0) and (x_1, y_1) by determining the coefficients a_0 and a_1. Each point must satisfy the desired equation, so we have

$$a_0 + a_1x_0 = y_0$$
$$\text{and} \quad a_0 + a_1x_1 = y_1.$$

But this is just a system of two linear equations in the two unknowns a_0 and a_1. If $x_0 \neq x_1$ (if the points do not lie on a vertical line), then this system has the unique solution:

$$a_0 = \frac{(y_0x_1 - y_1x_0)}{(x_1 - x_0)} \quad \text{and} \quad a_1 = \frac{(y_1 - y_0)}{(x_1 - x_0)}.$$

Following this lead, let us consider the situation in which we are given three points (x_0, y_0), (x_1, y_1), and (x_2, y_2), with x_0, x_1, and x_2 distinct. It is unlikely (although possible) that a line passes through all three points. On the other hand, it is reasonable to believe that we might be able to construct a quadratic polynomial $P(x) = a_0 + a_1x + a_2x^2$ through these points, since P has *three* undetermined coefficients. Requiring the given points to satisfy this polynomial equation, we obtain the linear system

$$\begin{aligned} a_0 + a_1x_0 + a_2x_0^2 &= y_0 \\ a_0 + a_1x_1 + a_2x_1^2 &= y_1 \\ a_0 + a_1x_2 + a_2x_2^2 &= y_2. \end{aligned} \qquad \textbf{(1)}$$

As in the two-point case, we can solve this linear system to obtain a (unique) solution for a_0, a_1, and a_2 and hence the formula for the desired polynomial.

EXAMPLE 1

Find a parabola that passes through the points $(0, -5)$, $(1, -1)$, and $(2, 5)$.

Solution We must find a quadratic polynomial $P(x) = a_0 + a_1x + a_2x^2$, where a_0, a_1, and a_2 satisfy

$$\begin{aligned} a_0 + 0a_1 + 0a_2 &= -5 \\ a_0 + 1a_1 + 1a_2 &= -1 \\ a_0 + 2a_1 + 4a_2 &= 5. \end{aligned}$$

Solving this linear system, we obtain $a_0 = -5$, $a_1 = 3$, and $a_2 = 1$. Thus, the parabola is given by $P(x) = -5 + 3x + x^2$ (or $y = x^2 + 3x - 5$; see Figure 19).

FIGURE 19

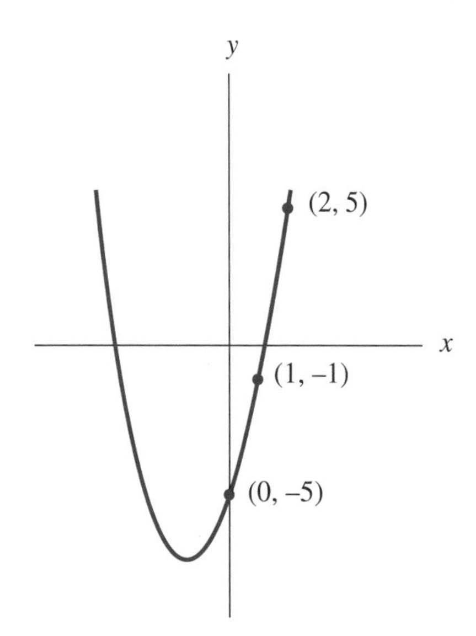

■

We now turn to the general case. We wish to find a polynomial of the form

$$P(x) = a_0 + a_1x + a_2x^2 + \cdots + a_nx^n$$

that passes through the given $n + 1$ points $(x_0, y_0), (x_1, y_1), \ldots, (x_n, y_n)$, where all x_i are distinct. Since each point lies on the curve $y = P(x)$, we obtain the following system of $n + 1$ linear equations in the $n + 1$ unknowns $a_0, a_1, \ldots, a_n$:

$$\begin{array}{ccccccccc} a_0 & + & a_1x_0 & + & a_2x_0^2 & + \cdots & + & a_nx_0^n & = y_0 \\ a_0 & + & a_1x_1 & + & a_2x_1^2 & + \cdots & + & a_nx_1^n & = y_1 \\ \vdots & & \vdots & & \vdots & & & \vdots & \vdots \\ a_0 & + & a_1x_n & + & a_2x_n^2 & + \cdots & + & a_nx_n^n & = y_n \end{array} \tag{2}$$

To solve this system, we form the augmented matrix

$$\left[\begin{array}{ccccc|c} 1 & x_0 & x_0^2 & \cdots & x_0^n & y_0 \\ 1 & x_1 & x_1^2 & \cdots & x_1^n & y_1 \\ \vdots & \vdots & \vdots & & \vdots & \vdots \\ 1 & x_n & x_n^2 & \cdots & x_n^n & y_n \end{array}\right]$$

and then perform the Gauss–Jordan elimination procedure of Section 2.3. The following theorem implies that the resulting solution is unique.

!

THEOREM 1

Let $(x_0, y_0), (x_1, y_1), \ldots, (x_n, y_n)$ be $n + 1$ points with all x_i distinct. Then there is a unique polynomial

$$P(x) = a_0 + a_1x + \cdots + a_nx^n$$

such that $P(x_i) = y_i$ for $i = 0, 1, \ldots, n$.

Proof We first show "existence"—that we can find *at least one* polynomial $P(x)$ that has the desired properties. For each $k = 0, 1, \ldots, n$, define the kth *Lagrange* polynomial:

$$L_k(X) = \frac{(x - x_0)(x - x_1)\cdots(x - x_{k-1})(x - x_{k+1})\cdots(x - x_n)}{(x_k - x_0)(x_k - x_1)\cdots(x_k - x_{k-1})(x_k - x_{k+1})\cdots(x_k - x_n)}$$

Notice that $L_k(x)$ is "missing" the factor $(x - x_k)$ in the numerator and hence has degree equal to n. Moreover, substituting x_i for x in the expression for $L_k(x)$, we see that

$$L_k(x_i) = \begin{cases} 1 & \text{if } i = k; \\ 0 & \text{if } i \neq k. \end{cases}$$

Thus, the polynomial

$$P(x) = y_0 L_0(x) + y_1 L_1(x) + \cdots + y_n L_n(x)$$

has degree n or less and satisfies $P(x_i) = y_i$ (for $i = 0, 1, \ldots, n$), as desired.

An alternate proof of uniqueness can be given with *Vandermonde determinants* — see Section 4.2.

To show "uniqueness"—that only one such polynomial exists—suppose that both $P(x)$ and $Q(x)$ have the desired properties. Define $R(x) = P(x) - Q(x)$. Then, $R(x_i) = P(x_i) - Q(x_i) = y_i - y_i = 0$, for $k = 0, 1, \ldots, n$. Since the x_i are distinct, this implies that $R(x)$ has at least $n + 1$ zeros. But the degree of R is less than or equal to n, so by the fundamental theorem of algebra, R must be the zero polynomial: $R(x) \equiv 0$. Thus, $P(x) - Q(x) \equiv 0$ or $P(x) \equiv Q(x)$, and our polynomial is unique. ■

DEFINITION

Let $(x_0, y_0), (x_1, y_1), \ldots, (x_n, y_n)$ be given points for which all x_i are distinct. The unique polynomial

$$P(x) = a_0 + a_1 x + a_2 x^2 + \cdots + a_n x^n$$

that passes through these points is called the **interpolating polynomial** for these points.

EXAMPLE 2

A table for e^x has the following entries:

x	1.0	1.1	1.2	1.3
e^x	2.7183	3.0042	3.3201	3.6693

Use these data to find an approximation to e^x at $x = 1.15$ by means of a cubic interpolating polynomial.

Solution Here (x_0, y_0), (x_1, y_1), (x_2, y_2), and (x_3, y_3) are (1.0, 2.7183), (1.1, 3.0042), (1.2, 3.3201), and (1.3, 3.6693), respectively. We shall find the cubic interpolating polynomial $P(x) = a_0 + a_1 x + a_2 x^2 + a_3 x^3$ for these points and then evaluate it at $x = 1.15$.

For these data, System (2) is

$$\begin{aligned} a_0 + (1.0)a_1 + (1.0)^2 a_2 + (1.0)^3 a_3 &= 2.7183 \\ a_0 + (1.1)a_1 + (1.1)^2 a_2 + (1.1)^3 a_3 &= 3.0042 \\ a_0 + (1.2)a_1 + (1.2)^2 a_2 + (1.2)^3 a_3 &= 3.3201 \\ a_0 + (1.3)a_1 + (1.3)^2 a_2 + (1.3)^3 a_3 &= 3.6693. \end{aligned}$$

Solving this linear system for a_0, a_1, a_2, and a_3, we obtain $a_0 = 0.7833$, $a_1 = 1.7000$, $a_2 = -0.3150$, and $a_3 = 0.5500$. Therefore, the desired approximation is $P(1.15)$, which rounds to 3.1582. (The correct value of $e^{1.15}$ to four decimal places is also 3.1582.) ■

NOTE The interpolating polynomial $P_n(x)$ for the $n + 1$ points $(x_0, y_0), (x_1, y_1), \ldots, (x_n, y_n)$ has degree less than or equal to n. Depending on the given data set, it may or may not have degree exactly equal to n. For example, in Examples 1 and 2, the interpolating polynomials for the given sets of three ($n = 2$) and four ($n = 3$) points did have degrees two and three, respectively. On the other hand, the unique polynomial of the form $P(x) = a_0 + a_1 x + a_2 x^2$ that passes through the three ($n = 2$) points (1, 10), (3, 10), and

(5, 10) is just $P(x) = 10$, a constant function. Here the degree of the interpolating polynomial is zero, strictly less than n.

2.5 Exercises

In Exercises 1–6 find a polynomial of the specified degree that passes through the given points.

1. $\{(0, -3), (1, 0), (2, 5)\}$, $n = 2$
2. $\{(-1, 4), (1, 2), (3, -4)\}$, $n = 2$
3. $\{(0, -1), (1, -1), (2, 3), (3, 17)\}$, $n = 3$
4. $\{(-1, 4), (0, 1), (1, 0), (3, 4)\}$, $n = 3$
5. $\{(-2, -15), (-1, -2), (0, 1), (1, 0), (2, 1)\}$, $n = 4$
6. $\{(0, -3), (1, -9), (2, -3), (3, 87), (4, 381)\}$, $n = 4$

A table of logarithms contains the following entries:

x	1.0	2.0	3.0	4.0
$\log_{10}x$	0.0000	0.3010	0.4771	0.6021

In Exercises 7 and 8 approximate $\log_{10}(2.5)$ by constructing an interpolating polynomial

7. of degree two, using the entries at $x = 1.0$, 2.0, and 3.0.
8. of degree three, using all the entries.
9. Find the (unique) solution of System (1) of three linear equations in the three unknowns a_0, a_1, and a_2.

Computational Exercises

Use a graphing calculator or a computer software package to solve the problems in Exercises 10–17.

10–15. Graph the interpolating polynomials you have constructed in Exercises 1–6.

16. Compare the value of $\log_{10}(2.5)$ given by your calculator to each of the approximations computed in Exercises 7 and 8.
17. Graph the interpolating polynomials of Exercises 7 and 8 and the function $y = \log_{10}x$ on the same set of axes. Does the higher degree interpolating polynomial seem to give a better approximation?
18. Use a computer algebra system (such as *Maple*) to do Exercise 9.

Use a computer algebra system to solve the problems in Exercises 19–22.

19. The equation of a conic section is given by $ax^2 + bxy + cy^2 + dx + ey + f = 0$. Do the points (2, 2), (3, 2), (3, 1), (1, 1), $(k, 2)$ lie on a conic section? If so, find the equation of the conic section. Find the values of k for which there is no solution. Plot the set of points for several values of k.
20. Do the points (2, 2), (3, 2), (3, 1), (1, 1), $(k, 0)$ lie on a conic section? If so, find the equation of the conic section. Find the values of k for which there is no solution. Plot the set of points for several values of k.

21. Solve for, plot, and identify the conic section given by the set of points $(-2, 2), (3, 1), (-5, 4), (1, 1), (4, 5)$.

22. It is sometimes possible to solve for the equation of a conic section passing through five given points. Make a conjecture as to what condition(s) must hold for these five points in order to find a solution. Experiment by trying to find the conic sections of arbitrary sets of five points.

CHAPTER SUMMARY

Operations on Vectors in $\mathbf{R}^m$

Let $\mathbf{u} = (u_1, u_2, \ldots, u_m)$ and $\mathbf{v} = (v_1, v_2, \ldots, v_m)$. Then

(a) *Length*: $\|\mathbf{u}\| = \sqrt{u_1^2 + u_2^2 + \cdots + u_m^2}$

(b) *Sum*: $\mathbf{u} + \mathbf{v} = (u_1 + v_1, u_2 + v_2, \ldots, u_m + v_m)$

(c) *Scalar multiple*: $c\mathbf{u} = (cu_1, cu_2, \ldots, cu_m)$

(d) *Dot product*: $\mathbf{u} \bullet \mathbf{v} = u_1v_1 + u_2v_2 + \cdots + u_mv_m$

Equations of Lines, Planes, and Hyperplanes in $\mathbf{R}^m$

(a) Equation of the line through P and parallel to $\mathbf{v}$: $\mathbf{x} = \mathbf{p} + t\mathbf{v}$

(b) Equation of the line through P and Q: $\mathbf{x} = (1 - t)\mathbf{p} + t\mathbf{q}$

(c) Equation of the hyperplane through P and normal to $\mathbf{n}$: $\mathbf{n} \bullet (\mathbf{x} - \mathbf{p}) = 0$

(d) Equation of a hyperplane in standard form: $a_1x_1 + a_2x_2 + \cdots + a_nx_n = b$, where $(a_1, a_2, \ldots, a_n)$ is a vector normal to the hyperplane.

(e) Equation of the plane determined by $\mathbf{u}$ and $\mathbf{v}$: $\mathbf{x} = r\mathbf{u} + s\mathbf{v}$

Solutions of Systems of Linear Equations

(a) DEFINITIONS: A *linear equation* in $x_1, x_2, \ldots, x_n$ has the form $a_1x_1 + a_2x_2 + \cdots + a_nx_n = b$. A *system of linear equations* is a finite set of linear equations.

(b) DEFINITION: A *solution* of a linear system is an n-vector, $(s_1, s_2, \ldots, s_n)$ that satisfies every equation in the system.

(c) There may be no solution, exactly one (a unique) solution, or infinitely many solutions to a system of linear equations.

Solving Linear Systems by Matrix Elimination

(a) DEFINITION: A matrix is in *row-reduced echelon form* if:

- The first nonzero entry in each row is a (leading) one (unless the row consists only of zeros, in which case it's at the bottom of the matrix).
- The other entries in a column that contains a leading one are zero.
- The leading ones "move" to the right as we move down the rows.

(b) Solve a linear system by Gauss–Jordan elimination as follows:

i. Form the augmented matrix for the system.

ii. Transform the augmented matrix to row-reduced echelon form.

iii. Convert back to a system of equations.

iv. Solve each equation for the unknown corresponding to the leading one.

(c) *Echelon form* differs from row-reduced echelon form in two ways: the first nonzero (leading) entry in a row need not be a one, and only entries in the column *below* a leading entry need be zero.

(d) *Gaussian elimination* refers to the procedure for solving a linear system that involves transforming the augmented matrix to echelon form.

KEYWORDS

length	scalar multiple
collinear	dot product
orthogonal	line
line segment	hyperplane
linear combination	linear system
consistent	elementary operation
matrix	augmented matrix
row-reduced echelon form	Gauss–Jordan elimination
Gaussian elimination	echelon form
back-substitution	*electromotive force*
resistor	*Ohm's law*
Kirchhoff's law	*interpolating polynomial*

SELF-TEST

1. Perform each operation on the vectors $\mathbf{u} = (2, 1, 0, -1)$ and $\mathbf{v} = (-3, 2, -1, 0)$:
(a) $(2\mathbf{u}) \cdot \mathbf{v}$ *(b)* $\|\mathbf{v} - \mathbf{u}\|$ **[Section 2.1]**

2. Give the two-point form and the point-parallel form for the line in $\mathbf{R}^4$ that contains the points $P(1, -1, 1, -1)$ and $Q(4, 3, 2, 1)$. **[Section 2.1]**

3. Express the hyperplane $x - y + 2z - w = 6$ in $\mathbf{R}^4$ in point-normal form. **[Section 2.1]**

4. Determine whether $(t, 2t, 3t)$, where t is an arbitrary real number, is a solution of the linear system

$$\begin{aligned} x + y - z &= 0 \\ 3x - 2y + z &= 1. \end{aligned}$$

[Section 2.2]

5. Construct the augmented matrix for the linear system in Exercise 4 and transform it to row-reduced echelon form. **[Section 2.3]**

In Exercise 6 and 7 solve the given linear system by Gauss–Jordan elimination.

6. $$\begin{aligned} x + 2y + z &= 1 \\ x + 2y + 3z &= 5 \\ 2y + 2z &= 2 \end{aligned}$$

7. $$\begin{aligned} x_1 - 2x_2 + x_3 - x_4 &= 1 \\ 2x_1 - 3x_2 + 2x_4 &= 2 \end{aligned}$$

[Section 2.3]

8. Solve the linear system of Exercise 6 by Gaussian elimination. **[Section 2.3]**

REVIEW EXERCISES

In Exercises 1–6 let $\mathbf{x} = (1, 0, 3, 0, 2)$ and $\mathbf{y} = (-1, 1, 0, 1, 4)$, and perform the indicated operation.

1. $\mathbf{x} - 3\mathbf{y}$

2. $d(\mathbf{x} + \mathbf{y}, \mathbf{x})$

3. $\|-2\mathbf{x}\|$

4. $\left(\frac{1}{\|\mathbf{y}\|}\right)\mathbf{y}$

5. $\mathbf{x} \cdot \mathbf{y}$

6. $(\mathbf{x} + \mathbf{y}) \cdot (\mathbf{x} - \mathbf{y})$

7. A hyperplane (in $\mathbf{R}^5$) has equation $(1, 2, 3, 4, 5) \bullet (\mathbf{x} - (1, 1, 1, 1, 1)) = 0$. Give this equation in standard form.

8. The equation of a hyperplane in $\mathbf{R}^4$ is $3x + y - 2z + w = 1$. Give a point-normal form for this equation.

9. Find an equation of the line in $\mathbf{R}^4$ that completely contains the vector $\mathbf{v} = (1, 0, 2, 0)$.

10. Determine the point (in $\mathbf{R}^4$) of intersection of the line $\mathbf{x}(t) = (1, 0, 1, 0) + t(0, 1, 0, 1)$ and the hyperplane $x + y + z + w = 0$.

11. Let $\mathbf{u}$ and $\mathbf{v}$ be vectors in $\mathbf{R}^5$ and c be a scalar. Which of the following operations are *not* defined?

(a) $3\mathbf{u} - c\mathbf{v}$ *(b)* $5c - \mathbf{u}$ *(c)* $\mathbf{u} \times \mathbf{v}$

(d) $\mathbf{u} \cdot (5\mathbf{v})$ *(e)* $\|c\mathbf{v}\|$ *(f)* $\|\mathbf{u} \cdot \mathbf{v}\|$

12. Find an equation of the plane in $\mathbf{R}^4$ that contains the points $(1, 0, 0, 1)$, $(0, 1, 0, 1)$, and $(0, 0, 1, 1)$.

In Exercises 13–20 solve the given linear system by the methods of Gauss–Jordan elimination and Gaussian elimination.

13.
$$\begin{aligned} x - 2y &= -1 \\ -2x + 4y &= 2 \end{aligned}$$

14.
$$\begin{aligned} -x + y - z &= 1 \\ x \quad\quad + z &= 3 \end{aligned}$$

15.
$$\begin{aligned} x_1 - 2x_2 + x_3 &= 0 \\ -x_1 + 2x_2 - x_3 &= 0 \end{aligned}$$

16.
$$\begin{aligned} x_1 + 2x_2 - 3x_3 &= -4 \\ 2x_1 - x_2 + x_3 &= 5 \\ 2x_2 + 3x_3 &= 1 \end{aligned}$$

17.
$$\begin{aligned} u + 2v - 3w &= 4 \\ -v + w &= -1 \\ 2u \quad\quad - w &= 3 \end{aligned}$$

18.
$$\begin{aligned} u + v - w &= 1 \\ v + 2w &= 0 \\ u + 4v + 5w &= 2 \end{aligned}$$

19.
$$\begin{aligned} x + y + 2z - w &= 1 \\ -x + 2y + z + w &= 2 \\ -x \quad\quad - z + w &= 3 \end{aligned}$$

20.
$$\begin{aligned} x - y + z - w &= 0 \\ y - z + 2w &= 0 \\ 2x - y + z \quad\quad &= 0 \end{aligned}$$

CHALLENGE YOURSELF

21. Find an equation of the hyperplane in $\mathbf{R}^4$ that completely contains the vectors $\mathbf{v}_1 = (1, 0, -1, 0)$, $\mathbf{v}_2 = (0, 1, 1, 1)$, and $\mathbf{v}_3 = (1, 1, -1, 0)$.

22. Solve the given linear system using Gauss–Jordan elimination. Under what conditions will it have a solution?

$$\begin{aligned} x - y + 2z &= 1 \\ ax - ay + 4z &= 2a \\ x + y + az &= 4 \end{aligned}$$

23. Find the equation of the line that is the intersection of the hyperplanes $2x - y - 2z - w = 0$, $2x - 3w = 2$, and $2y + 4z = 4$.

24. What relationship must hold among the constants a, b, and c for the following system to have a solution?

$$\begin{aligned} x + 2y - 3z &= a \\ -x - y + 2z &= b \\ x + 3y - 4z &= c \end{aligned}$$

25. Let $\mathbf{u}$ and $\mathbf{v}$ be nonzero vectors in $\mathbf{R}^m$. Find vectors $\mathbf{u}_1$ and $\mathbf{u}_2$ in $\mathbf{R}^m$, in terms of $\mathbf{u}$ and $\mathbf{v}$, such that $\mathbf{u} = \mathbf{u}_1 + \mathbf{u}_2$, $\mathbf{u}_1$ is parallel to $\mathbf{v}$, and $\mathbf{u}_2$ is perpendicular to $\mathbf{v}$.

26. Find the shortest distance between the lines $\mathbf{x}_1(t) = (-1, 0, 1, 0) + t(1, 1, 1, 0)$ and $\mathbf{x}_2(s) = (1, 2, 0, 0) + s(0, 1, -1, 0)$.

27. Recall that the complex number i satisfies $i = \sqrt{-1}$. Use Gauss–Jordan elimination to solve the following linear system:

$$\begin{aligned} ix + y &= 1 \\ x + iy &= -1 \end{aligned}$$

SPOTLIGHT

Peter Brett Charlton/Shutterstock

The notion of a *matrix*, a rectangular array of numbers, was used in Chapter 2 to develop a systematic and efficient method for solving systems of linear equations. Matrices have numerous other uses as well, often providing relatively simple ways of looking at seemingly complex problems. One of their most direct applications is to the analysis of networks. A *network* is a collection of locations (*nodes*) connected to one another by one- or two-way paths (*arcs*). Examples of networks include telephone and electric circuits, travel routes and job schedules, and systems of interlinked computers (computer networks).

As a specific (simple) example, suppose that a saleswoman has business in four cities, including her own. She can visit these cities in many different orders. For example, she might start from her home base (city 1), go to city 3, return home (city 1), and then go to cities 4 and 2 (in that order) before returning home again. We can describe this journey more simply by the sequence 1, 3, 1, 4, 2, 1. Using this notation, a second possible journey is 1, 3, 2, 4, 1.

We can depict each of these journeys as a *directed graph* consisting of *nodes* ("dots") representing the cities and *edges* (shown by "arrows") for the trips between them. Here are the directed graphs for the two journeys we just described:

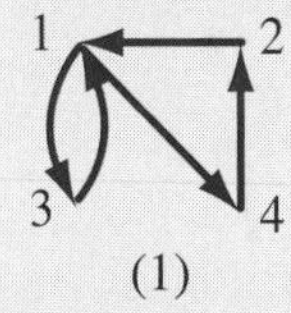

(1)

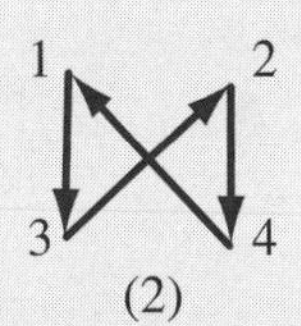

(2)

A standard way of analyzing a directed graph is to construct its *adjacency matrix*, which has one row and column corresponding to each node in the graph. If there *is* an edge from node i to node j, then the matrix entry in the ith row and jth column is 1; otherwise this entry is 0. For example, the adjacency matrices of the directed graphs shown previously are:

$$\textbf{(1)} \begin{bmatrix} 0 & 0 & 1 & 1 \\ 1 & 0 & 0 & 0 \\ 1 & 0 & 0 & 0 \\ 0 & 1 & 0 & 0 \end{bmatrix} \qquad \textbf{(2)} \begin{bmatrix} 0 & 0 & 1 & 0 \\ 0 & 0 & 0 & 1 \\ 0 & 1 & 0 & 0 \\ 1 & 0 & 0 & 0 \end{bmatrix}$$

In this chapter, you will study (among other things) how to perform operations, like addition and multiplication, on matrices. When applied to adjacency matrices, these operations help us to answer questions about the underlying graph and network. For example, in a certain network, is there a path from one given node to another? If so, what is the shortest path? The answers to these kinds of questions are studied in a branch of mathematics called *graph theory*. We return to this subject in Section 3.7.

CHAPTER

3

Matrices

OBJECTIVES

You will learn to:

- Add, subtract, and multiply matrices.
- Find scalar multiples, the transpose, and the inverse of a matrix.
- Use the matrix inverse to solve certain systems of linear equations.
- Find the rank of a matrix.
- Deduce, in some cases, information about solutions of linear systems without actually solving them.
- Determine the *LU* decomposition of a matrix.
- Use the *LU* decomposition to solve systems of linear equations.
- Determine the elementary matrices involved in a row reduction or *LU* decomposition.

In Section 2.3, we introduced matrices to help us solve systems of linear equations by means of an "elimination" procedure. In this chapter, we take a much more detailed look at matrices: first manipulating them in many of the same ways we did vectors, and then relating matrices, in new ways, to linear systems.

3.1 Operations on Matrices

In Chapter 2, you saw how vectors in $\mathbf{R}^m$ could be added, subtracted, multiplied by scalars, and (after a fashion) multiplied by one another. In this section, we define analogous operations on matrices.

Basic Terminology

Recall, from Section 2.3, that a **matrix** is simply a rectangular array of **entries**, which are usually real or complex numbers. If a given matrix has m rows and n columns, it is called an $\mathbf{m \times n}$ **matrix** (read, "m by n matrix"), and the numbers m and n are the **dimensions** of the matrix. The matrices

$$\begin{bmatrix} 1 & 3 \\ -1 & 0 \\ 2 & 4 \end{bmatrix}, \tag{1}$$

$$\begin{bmatrix} -1 & 0 & 1 \\ 4 & 2 & -3 \end{bmatrix}, \tag{2}$$

$$\text{and} \quad \begin{bmatrix} \frac{1}{2} \\ -6 \\ 4 \end{bmatrix} \tag{3}$$

In many ways, an $m \times n$ matrix is just like a vector in $\mathbf{R}^{mn}$, but there is much to be gained by writing it as a rectangular array rather than as a long vector. This is an example of an instance in mathematics where *how* you write something can be suggestive of so much more.

are, respectively, 3×2, 2×3, and 3×1 matrices. We sometimes refer to an $m \times 1$ matrix (with $m > 1$) as a **column vector** and a $1 \times n$ matrix ($n > 1$) as a **row vector**. For example, matrix (3) is called a column vector.

Capital (uppercase) letters, such as A, B, and C, are used to denote matrices, whereas the entries of a matrix are denoted by the corresponding lowercase letter with "double subscripts." For example, to describe a general 2×4 matrix, we write

$$A = \begin{bmatrix} a_{11} & a_{12} & a_{13} & a_{14} \\ a_{21} & a_{22} & a_{23} & a_{24} \end{bmatrix}.$$

Note that the first subscript gives the row number, and the second gives the column number. As another example, if

$$B = \begin{bmatrix} 3 & 6 & -1 \\ 4 & 2 & 0 \\ 1 & -3 & -5 \end{bmatrix},$$

then $b_{12} = 6$, $b_{21} = 4$, and $b_{33} = -5$.

Occasionally, when we want to make explicit this uppercase/lowercase convention, we write something like, "Let $A = [a_{ij}]$ and $B = [b_{ij}]$." This means that we wish to let A be a matrix with entries denoted a_{ij} and let B be a matrix with entries denoted b_{ij}. Moreover, if C is an $m \times n$ matrix, then the subscripts i and j of its entries take on the values $1, 2, \ldots, m$ and $1, 2, \ldots, n$, respectively. This is rarely stated explicitly, however.

Before proceeding to a description of matrix operations, we give one more definition. Given matrices A and B, the entries a_{ij} and b_{kp} are **corresponding entries** if and only if $i = k$ and $j = p$. In other words, an entry of A *corresponds* to an entry of B if and only if they occupy the same position in their respective matrices. For example, if

$$A = \begin{bmatrix} 1 & 2 & 3 \\ 4 & 5 & 6 \end{bmatrix} \quad \text{and} \quad B = \begin{bmatrix} 7 & 8 & 9 \\ -1 & -2 & -3 \end{bmatrix},$$

then 1 and 7, 2 and 8, and 4 and -1 are all pairs of corresponding entries. We shall speak of corresponding entries only when the two matrices in question have the same dimensions.

DEFINITION

Equality of matrices. Two matrices **A** and **B** are **equal**, written $A = B$, if they have the same dimensions and their corresponding entries are equal.

Formally, an $m \times n$ matrix A is equal to a $p \times q$ matrix B if and only if $m = p$, $n = q$, and $a_{ij} = b_{ij}$ for all i and j.

EXAMPLE 1

Which of the following matrices are equal?

2×3   2×2

$$A = \begin{bmatrix} 1 & 2 & -1 \\ 4 & 0 & 1+2 \end{bmatrix} \qquad B = \begin{bmatrix} 1 & 2 \\ 4 & 0 \end{bmatrix}$$

$$C = \begin{bmatrix} 1 & \frac{4}{2} & -1 \\ 4 & 0 & 3 \end{bmatrix} \qquad D = \begin{bmatrix} 1 & 2 & -1 \\ 4 & 0 & 1 \end{bmatrix}$$

Solution We have $A = C$, but this is the only pair of equal matrices. ∎

Sum, Difference, and Scalar Multiplication

DEFINITION

Sum of matrices: Let A and B be matrices with the same dimensions. The **sum** of A and B, written $A + B$, is the matrix obtained by adding corresponding entries of A and B.

Formally, we have $C = A + B$ if and only if $c_{ij} = a_{ij} + b_{ij}$ for all i and j.

EXAMPLE 2

Find the matrix C that is the sum of

$$A = \begin{bmatrix} 3 & 6 \\ -1 & 5 \\ 0 & 2 \end{bmatrix} \quad \text{and} \quad B = \begin{bmatrix} \frac{1}{2} & -4 \\ -2 & 0 \\ -3 & -2 \end{bmatrix}.$$

Solution

$$C = \begin{bmatrix} 3 + \frac{1}{2} & 6 + (-4) \\ -1 + (-2) & 5 + 0 \\ 0 + (-3) & 2 + (-2) \end{bmatrix} = \begin{bmatrix} \frac{7}{2} & 2 \\ -3 & 5 \\ -3 & 0 \end{bmatrix}$$

■

DEFINITION

Multiplication by a scalar: Let A be a matrix and c be a scalar. Then the **scalar multiple** cA is the matrix obtained by multiplying each entry of A by the scalar c.

Formally, we have $B = cA$ if and only if $b_{ij} = ca_{ij}$ for all i and j.

EXAMPLE 3

Let

$$A = \begin{bmatrix} 1 & -1 & 0 & 3 \\ 2 & 6 & -4 & 8 \end{bmatrix}.$$

Find $3A$ and $(-1)A$.

Solution We have

$$3A = \begin{bmatrix} 3 & -3 & 0 & 9 \\ 6 & 18 & -12 & 24 \end{bmatrix} \quad \text{and} \quad (-1)A = \begin{bmatrix} -1 & 1 & 0 & -3 \\ -2 & -6 & 4 & -8 \end{bmatrix}.$$

■

DEFINITION

The **negative** of a matrix B, written $-B$, is the matrix $(-1)B$. It is obtained from B by simply reversing the sign of every entry.

DEFINITION

Difference of matrices: Let A and B be matrices with the same dimensions. The **difference** of A and B, written $A - B$, is the matrix C given by $C = A + (-B)$.

NOTE The matrix $A - B$ can be obtained by simply subtracting each entry of B from the corresponding entry of A.

EXAMPLE 4

Find $C = 2A - B$, where

$$A = \begin{bmatrix} 1 & 2 \\ -1 & 0 \end{bmatrix} \quad \text{and} \quad B = \begin{bmatrix} 0 & 4 \\ -1 & 0 \end{bmatrix}.$$

Solution We have

$$2\begin{bmatrix} 1 & 2 \\ -1 & 0 \end{bmatrix} - \begin{bmatrix} 0 & 4 \\ -1 & 0 \end{bmatrix} = \begin{bmatrix} 2 & 4 \\ -2 & 0 \end{bmatrix} - \begin{bmatrix} 0 & 4 \\ -1 & 0 \end{bmatrix} = \begin{bmatrix} 2 & 0 \\ -1 & 0 \end{bmatrix}.$$ ■

DEFINITION

The matrix analog of the number 0 is called a **zero matrix**. A zero matrix, denoted by *0*, is a matrix of any dimensions consisting of only zero entries.

EXAMPLE 5

The matrices

$$\begin{bmatrix} 0 & 0 & 0 \\ 0 & 0 & 0 \\ 0 & 0 & 0 \end{bmatrix}, \quad \begin{bmatrix} 0 & 0 & 0 \\ 0 & 0 & 0 \end{bmatrix}, \quad \begin{bmatrix} 0 \\ 0 \\ 0 \end{bmatrix}, \quad \text{and} \quad [0 \quad 0 \quad 0 \quad 0]$$

are all zero matrices. ■

A zero matrix has the property that $A + 0 = A$ and $0 + A = A$ for any matrix A for which the sums are defined.

Matrix Multiplication

This product is suggested by the rectangular nature of matrices.

We now define the product of two matrices. From the way the other matrix operations have been defined, you might guess that we obtain the product of two matrices by simply multiplying corresponding entries. The definition of *product* given below is much more complicated than this but also considerably more useful in applications. Spotlight exercises at the end of this section provide a simple example of how our definition arises in a natural way. Another example is given in Section 3.2, and still others appear throughout the text.

Before getting to the definition, we introduce some notation. Given an $m \times n$ matrix A, the vector consisting of the entries of the ith row of A is denoted $\mathbf{a}_i$, whereas the vector consisting of the entries of the jth column of A is denoted $\mathbf{a}^j$. For example, if

$$A = \begin{bmatrix} 4 & 3 & -1 \\ 1 & 2 & 5 \\ 0 & 1 & 0 \end{bmatrix},$$

then $\mathbf{a}^1 = (4, 1, 0)$, $\mathbf{a}_1 = (4, 3, -1)$, and $\mathbf{a}_3 = (0, 1, 0)$. Also recall, from Section 2.1, that the dot product of the vectors $\mathbf{x} = (x_1, x_2, \ldots, x_r)$ and $\mathbf{y} = (y_1, y_2, \ldots, y_r)$, written $\mathbf{x} \bullet \mathbf{y}$, is given by the number $x_1y_1 + x_2y_2 + \cdots + x_ry_r$. For example, for the given matrix A, we have $\mathbf{a}^1 \bullet \mathbf{a}_1 = 4\,(4) + 1\,(3) + 0(-1) = 19$ and $\mathbf{a}_1 \bullet \mathbf{a}^3 = 11$.

DEFINITION

Matrix multiplication: Let A and B be matrices such that the number of columns of A is equal to the number of rows of B. Suppose that the number of rows of A is m and the number of columns of B is q. The **product** AB is an $m \times q$ matrix C, defined by

$$c_{ij} = \mathbf{a}_i \bullet \mathbf{b}^j.$$

In words, to obtain the entry in the ith row and jth column of C, we take the dot product of the ith row of A with the jth column of B.

NOTE If A is an $m \times n$ matrix and B is a $p \times q$ matrix, then the product AB is defined only if $n = p$ and, if so, AB is an $m \times q$ matrix. The entry of AB in the ith row and jth column is obtained by multiplying each entry in the ith row of A by the corresponding entry in the jth column of B, and adding up all such products. In other words,

$$c_{ij} = a_{i1}b_{1j} + a_{i2}b_{2j} + \cdots + a_{in}b_{nj}$$

or, using summation notation,

$$c_{ij} = \sum_{k=1}^{n} a_{ik}b_{kj}.$$

EXAMPLE 6

Given the pairs of matrices A and B, find the products AB and BA, if defined.

a. $A = \begin{bmatrix} 1 & 2 & 3 \\ -3 & -2 & -1 \end{bmatrix}, \quad B = \begin{bmatrix} -4 & 5 \\ 0 & 4 \\ -5 & 0 \end{bmatrix}$

b. $A = \begin{bmatrix} 1 & -1 \\ 1 & -1 \end{bmatrix}, \quad B = \begin{bmatrix} 2 & -3 \\ 2 & -3 \end{bmatrix}$

c. $A = \begin{bmatrix} 1 & 2 & 3 \\ 4 & 5 & 6 \\ 7 & 8 & 9 \end{bmatrix}, \quad B = \begin{bmatrix} x \\ y \\ z \end{bmatrix}$

Solution

a. Here A is a 2×3 matrix and B is a 3×2 matrix. Since the number of columns of A is equal to the number of rows of B, the product AB is defined. It has two rows (the same number as A) and two columns (the same number as B). Let $C = AB$. Then

$$\begin{aligned} c_{11} &= \mathbf{a}_1 \cdot \mathbf{b}^1 = (1, 2, 3) \cdot (-4, 0, -5) \\ &= 1(-4) + 2(0) + 3(-5) = -19, \\ c_{12} &= \mathbf{a}_1 \cdot \mathbf{b}^2 = (1, 2, 3) \cdot (5, 4, 0) \\ &= 1(5) + 2(4) + 3(0) = 13, \\ c_{21} &= \mathbf{a}_2 \cdot \mathbf{b}^1 = (-3, -2, -1) \cdot (-4, 0, -5) \\ &= (-3)(-4) + 0 + (-1)(-5) = 17, \\ \text{and} \quad c_{22} &= \mathbf{a}_2 \cdot \mathbf{b}^2 = (-3, -2, -1) \cdot (5, 4, 0) \\ &= (-3)(5) + (-2)(4) + 0 = -23. \end{aligned}$$

$$\text{Thus,} \qquad C = \begin{bmatrix} -19 & 13 \\ 17 & -23 \end{bmatrix}.$$

In terms of remembering the definition and actually doing the computations to get the ijth entry of AB, it is probably better to think, "go across the ith row of A and down the jth column of B," forming products of corresponding entries and adding the results.

We illustrate this device in finding BA. Now BA is defined, since B (now the "first matrix") has two columns, which is the same as the number of rows of A (the "second matrix"). Also, BA has three rows (as does B) and three columns (as does A). Letting $D = BA$, we have

$$D = \begin{bmatrix} -4 & 5 \\ 0 & 4 \\ -5 & 0 \end{bmatrix} \begin{bmatrix} 1 & 2 & 3 \\ -3 & -2 & -1 \end{bmatrix}.$$

We find

d_{11} by "going across first row of B and down first column of A," so
$d_{11} = (-4)(1) + 5(-3) = -19$;

d_{12} by "going across first row of B and down second column of A," so
$d_{12} = (-4)(2) + 5(-2) = -18$;

and d_{21} by "going across second row of B and down first column of A," so
$d_{21} = 0(1) + 4(-3) = -12$.

Similarly,

$$\begin{aligned} d_{13} &= (-4)(3) + (5)(-1) = -17, \\ d_{22} &= (0)(2) + (4)(-2) = -8, \\ d_{23} &= (0)(3) + (4)(-1) = -4, \\ d_{31} &= (-5)(1) + (0)(-3) = -5, \\ d_{32} &= (-5)(2) + (0)(-2) = -10, \\ \text{and } d_{33} &= (-5)(3) + (0)(-1) = -15. \end{aligned}$$

Thus, $$D = BA = \begin{bmatrix} -19 & -18 & -17 \\ -12 & -8 & -4 \\ -5 & -10 & -15 \end{bmatrix}.$$

Note that $AB \neq BA$ in this example. In fact, they do not even have the same dimensions.

b. Both products AB and BA are defined because both A and B are 2×2 matrices. Now

$$AB = \begin{bmatrix} 1 & -1 \\ 1 & -1 \end{bmatrix}\begin{bmatrix} 2 & -3 \\ 2 & -3 \end{bmatrix} = \begin{bmatrix} 0 & 0 \\ 0 & 0 \end{bmatrix} = 0,$$

the zero matrix, while

$$BA = \begin{bmatrix} 2 & -3 \\ 2 & -3 \end{bmatrix}\begin{bmatrix} 1 & -1 \\ 1 & -1 \end{bmatrix} = \begin{bmatrix} -1 & 1 \\ -1 & 1 \end{bmatrix}.$$

So we see that it is possible for both AB and BA to be defined and of the same dimensions, yet $AB \neq BA$. Moreover, we also see that it can happen that $AB = 0$ while $A \neq 0$ and $B \neq 0$ – such an occurrence does not happen with real numbers!

c. The product AB is defined, since A has three columns and B has three rows, and it is a 3×1 matrix. We have

$$AB = \begin{bmatrix} 1 & 2 & 3 \\ 4 & 5 & 6 \\ 7 & 8 & 9 \end{bmatrix}\begin{bmatrix} x \\ y \\ z \end{bmatrix} = \begin{bmatrix} x + 2y + 3z \\ 4x + 5y + 6z \\ 7x + 8y + 9z \end{bmatrix}.$$

However, the product BA is not defined, since the number of columns of B (one) is not equal to the number of rows of A (three).

Note that in this example, we can view AB as a function that takes a column vector B to another column vector. In fact, if A is an $m \times n$ matrix and X is an $n \times 1$ matrix (that is, a column vector), then AX is an $m \times 1$ column vector. So we can view multiplication on the left by A as a function that takes vectors in the Euclidean space $\mathbf{R}^n$ to vectors in $\mathbf{R}^m$. This viewpoint of matrices, as functions or *transformations*, is so important that we will devote all of Chapter 7 to the topic. ■

WARNING It should be apparent from Example 6 that multiplication of matrices "acts differently" from multiplication of numbers. The latter has the property of *commutativity* (for all numbers a and b, $ab = ba$), but matrix multiplication does not. For matrices, AB is not necessarily equal to BA.

One situation in which the products AB and BA are both defined (although they still need not be equal) is when A and B have the same dimensions and are *square*. A matrix is **square** if it has the same number of rows as columns. In this case, the number of rows (or columns) is called the **order** of the matrix. For example, matrix A of Example 6(b) is a square matrix of order 3.

DEFINITION

Powers of a matrix: Let A be a square matrix. Then,

$$A^1 = A$$
$$A^n = \underbrace{AA\cdots A}_{n}$$

EXAMPLE 7

Let

$$A = \begin{bmatrix} 1 & 0 \\ -1 & 2 \end{bmatrix}.$$

Find the matrices A^1, A^2, and A^3.

Solution

$$A^1 = A = \begin{bmatrix} 1 & 0 \\ -1 & 2 \end{bmatrix},$$

$$A^2 = AA = \begin{bmatrix} 1 & 0 \\ -1 & 2 \end{bmatrix}\begin{bmatrix} 1 & 0 \\ -1 & 2 \end{bmatrix} = \begin{bmatrix} 1 & 0 \\ -3 & 4 \end{bmatrix},$$

$$\text{and } A^3 = A^2A = \begin{bmatrix} 1 & 0 \\ -3 & 4 \end{bmatrix}\begin{bmatrix} 1 & 0 \\ -1 & 2 \end{bmatrix} = \begin{bmatrix} 1 & 0 \\ -7 & 8 \end{bmatrix}.$$

■

NOTE We found A^3 in Example 7 by forming the product A^2A^1, the exponents adding just as they do when the base is a number. In general, if A is a square matrix and m and n are positive integers, we have $A^mA^n = A^{m+n}$. It is also true that under the same conditions, $(A^m)^n = A^{mn}$.

The entries a_{ij} of a square matrix A for which $i = j$ form the **main diagonal** of A. For example, the main diagonal of

$$B = \begin{bmatrix} 3 & -1 & 4 \\ 0 & 1 & 2 \\ -2 & 5 & -3 \end{bmatrix}$$

consists of 3, 1, and -3. A square matrix in which every element *not* on the main diagonal (every "off diagonal" element) is zero is called a **diagonal matrix**. A special type of diagonal matrix is the **identity matrix**, denoted by I, in which each entry on the main diagonal is 1.

EXAMPLE 8

The matrices

$$\begin{bmatrix} 1 & 0 \\ 0 & 1 \end{bmatrix}, \quad \begin{bmatrix} 1 & 0 & 0 \\ 0 & 1 & 0 \\ 0 & 0 & 1 \end{bmatrix}, \quad \text{and} \quad \begin{bmatrix} 2 & 0 \\ 0 & -3 \end{bmatrix}$$

are all diagonal matrices. The first two are identity matrices of orders 2 and 3, respectively. The last is not an identity matrix. ■

The identity matrix is so named because it has the property that $AI = A$ and $IA = A$ for any matrix A for which the products are defined. Thus, it is the matrix analog of the number 1, the "multiplicative identity" for numbers. Also following the convention for numbers, we define

$$A^0 = I$$

for any square matrix A.

Transpose of a Matrix

It is sometimes necessary to interchange the rows and columns of a matrix. This operation is called *transposing* the matrix.

DEFINITION

Transpose of a matrix. The **transpose** of the $m \times n$ matrix A, written A^T, is the $n \times m$ matrix B whose jth column is the jth row of A. Equivalently, $B = A^T$ if and only if $b_{ij} = a_{ji}$ for each i and j. A square matrix A is **symmetric** if $A = A^T$.

EXAMPLE 9

Find the transpose of

a. $A = \begin{bmatrix} 1 & 2 \\ -1 & 3 \end{bmatrix}$ **b.** $B = \begin{bmatrix} 1 & -1 \\ 2 & 0 \\ 3 & 4 \end{bmatrix}$ **c.** $C = \begin{bmatrix} 1 & 2 & -1 \\ 2 & 3 & 0 \\ -1 & 0 & 4 \end{bmatrix}$

Solution

a. $A^T = \begin{bmatrix} 1 & -1 \\ 2 & 3 \end{bmatrix}$ **b.** $B^T = \begin{bmatrix} 1 & 2 & 3 \\ -1 & 0 & 4 \end{bmatrix}$ **c.** $C^T = \begin{bmatrix} 1 & 2 & -1 \\ 2 & 3 & 0 \\ -1 & 0 & 4 \end{bmatrix} = C,$

so C is symmetric. ■

COMPUTATIONAL NOTE

Most graphing calculators and linear algebra software packages are capable of performing all the matrix operations discussed in this section. To use one for this purpose, you first enter the elements of the given matrix or matrices and then either press the appropriate key (such as the "plus" key) or select the desired operation from a menu. Be aware, however, that for many of the simple examples and exercises presented in this text, the matrix operations can probably be performed more quickly "by hand"!

Rules Governing Operations on Matrices

We close this section with a theorem that states some of the rules governing the matrix operations already discussed.

THEOREM 1 (Properties of Matrices and Rules of Operation)

Let A, B, and C be matrices and a and b be scalars. Assume that the dimensions of the matrices are such that each operation is defined. Then

a. $A + B = B + A$ (Commutative law for addition)

b. $A + (B + C) = (A + B) + C$ (Associative law for addition)

c. $A + 0 = A$

d. $A + (-A) = 0$

e. $A(BC) = (AB)C$ (Associative law for multiplication)

f. $AI = A$, $IB = B$

g. $A(B + C) = AB + AC$, $(B + C)A = BA + CA$ (Distributive laws)

h. $a(B + C) = aB + aC$

i. $(a + b)C = aC + bC$

j. $(ab)C = a(bC)$

k. $1A = A$

l. $A0 = 0$, $0B = 0$

m. $a0 = 0$

n. $a(AB) = (aA)B = A(aB)$

o. $(A + B)^T = A^T + B^T$

p. $(AB)^T = B^T A^T$

Before we prove some of these assertions, we give a few illustrative examples.

EXAMPLE 10

Let

$$A = \begin{bmatrix} 1 & 0 & -1 \\ 2 & 1 & -1 \\ 3 & -2 & 0 \end{bmatrix}, \quad B = \begin{bmatrix} -1 & 2 & 3 \\ 0 & 1 & 0 \\ -2 & -1 & 0 \end{bmatrix}, \quad \text{and} \quad C = \begin{bmatrix} -2 & 1 & 1 \\ 1 & 1 & 0 \\ 3 & 0 & -3 \end{bmatrix}.$$

Find $A(BC)$, $(AB)C$, $A(B + C)$, and $AB + AC$.

Solution

$$A(BC) = \begin{bmatrix} 1 & 0 & -1 \\ 2 & 1 & -1 \\ 3 & -2 & 0 \end{bmatrix}\begin{bmatrix} 13 & 1 & -10 \\ 1 & 1 & 0 \\ 3 & -3 & -2 \end{bmatrix} = \begin{bmatrix} 10 & 4 & -8 \\ 24 & 6 & -18 \\ 37 & 1 & -30 \end{bmatrix}$$

$$(AB)C = \begin{bmatrix} 1 & 3 & 3 \\ 0 & 6 & 6 \\ -3 & 4 & 9 \end{bmatrix}\begin{bmatrix} -2 & 1 & 1 \\ 1 & 1 & 0 \\ 3 & 0 & -3 \end{bmatrix} = \begin{bmatrix} 10 & 4 & -8 \\ 24 & 6 & -18 \\ 37 & 1 & -30 \end{bmatrix}$$

(Note that $A(BC) = (AB)C$ for these choices of A, B, and C.)

$$A(B+C) = \begin{bmatrix} 1 & 0 & -1 \\ 2 & 1 & -1 \\ 3 & -2 & 0 \end{bmatrix} \begin{bmatrix} -3 & 3 & 4 \\ 1 & 2 & 0 \\ 1 & -1 & -3 \end{bmatrix} = \begin{bmatrix} -4 & 4 & 7 \\ -6 & 9 & 11 \\ -11 & 5 & 12 \end{bmatrix}$$

$$AB + AC = \begin{bmatrix} 1 & 3 & 3 \\ 0 & 6 & 6 \\ -3 & 4 & 9 \end{bmatrix} + \begin{bmatrix} -5 & 1 & 4 \\ -6 & 3 & 5 \\ -8 & 1 & 3 \end{bmatrix} = \begin{bmatrix} -4 & 4 & 7 \\ -6 & 9 & 11 \\ -11 & 5 & 12 \end{bmatrix}$$

(Note that $A(B + C) = AB + AC$ for these, A, B, and C.) ■

Of course, Example 10 does not constitute a *proof* of the fact that $A(BC) = (AB)C$ or that $A(B + C) = AB + AC$ for *all* matrices A, B, and C.

Proofs for Theorem 1 (b), (e), (n), and (p)

b. Let $M = A + (B + C)$ and $M' = (A + B) + C$.
Then $m_{ij} = a_{ij} + (b_{ij} + c_{ij}) = (a_{ij} + b_{ij}) + c_{ij} = m'_{ij}$ (since addition of numbers is associative).

e. This part of the proof illustrates the value of summation notation. Assume that A is $m \times n$, B is $n \times p$, and C is $p \times q$. Since

$$(BC)_{kj} = \sum_{s=1}^{p} b_{ks}c_{sj},$$

we have

$$\begin{aligned}(A(BC))_{ij} &= \sum_{k=1}^{n} a_{ik}(BC)_{kj} = \sum_{k=1}^{n} a_{ik}\left(\sum_{s=1}^{p} b_{ks}c_{sj}\right) = \sum_{k=1}^{n}\sum_{s=1}^{p} a_{ik}b_{ks}c_{sj} \\ &= \sum_{s=1}^{p}\left(\sum_{k=1}^{n} a_{ik}b_{ks}\right)c_{sj} = \sum_{s=1}^{p}(AB)_{is}c_{sj} \\ &= ((AB)C)_{ij},\end{aligned}$$

which proves the assertion.

n. Let $C = a(AB)$, $D = (aA)B$, and $E = A(aB)$. Then

$$c_{ij} = a(\mathbf{a}_i \cdot \mathbf{b}^j) = (a\mathbf{a}_i) \cdot \mathbf{b}^j = d_{ij}$$

(by Theorem 3b of Section 2.1). Moreover

$$d_{ij} = (a\mathbf{a}_i) \cdot \mathbf{b}^j = \mathbf{a}_i \cdot (a\mathbf{b}^j) = e_{ij}$$

(same theorem). Thus $c_{ij} = d_{ij} = e_{ij}$, as desired.

p. Again, as in part (c), assume that A is $m \times n$ and B is $n \times p$. Then AB is $m \times p$, and hence $(AB)^T$ is $p \times m$, while A^T is $n \times m$ and B^T is $p \times n$, so B^TA^T is also $p \times m$. We need only show now that corresponding entries are equal. On one hand, we have

$$\begin{aligned}\left((AB)^T\right)_{ij} &= (AB)_{ji} \\ &= \sum_{k=1}^{n} a_{jk}b_{ki} \\ &= \sum_{k=1}^{n} b_{ki}\,a_{jk} \\ &= \left(B^TA^T\right)_{ij}.\end{aligned}$$

We conclude that $(AB)^T = B^TA^T$. ■

3.1 Exercises

In Exercises 1–12 let

$$A = \begin{bmatrix} -1 & 0 & 1 \\ 2 & -1 & 3 \\ 0 & 1 & -2 \end{bmatrix} \quad \text{and} \quad B = \begin{bmatrix} 0 & 4 & -2 \\ 3 & 1 & 2 \\ -1 & 0 & 1 \end{bmatrix},$$

and let I be the identity matrix of order 3. Find each of the following.

1. $A + B$

2. $A - B$

3. $2A - 3B$

4. $A + I$

5. $A - \lambda I$ (λ, a scalar)

6. $A + \lambda B$ (λ, a scalar)

7. A^T

8. A^0

9. AB

10. $(AB)^T$

11. $B^T A^T$

12. $A^T B^T$

In Exercises 13–16 find each of the following, where

$$A = \begin{bmatrix} 2 & -1 & \frac{1}{2} \\ 1 & 0 & -2 \end{bmatrix}.$$

13. A^T

14. $\left(A^T\right)^T$

15. The columns $\mathbf{a}^1$ and $\mathbf{a}^3$ and the row $\mathbf{a}_2$

16. $\mathbf{a}^1 \cdot \mathbf{a}^3$ and $\mathbf{a}_1 \cdot \mathbf{a}_2$

In Exercises 17–22 find *AB* and *BA* when these products are defined.

17. $A = \begin{bmatrix} 1 & 2 & 0 \\ -2 & 0 & 1 \\ 0 & 0 & 3 \end{bmatrix}, \quad B = \begin{bmatrix} 1 & 1 & 1 \\ 3 & -1 & 0 \\ 2 & -2 & 0 \end{bmatrix}$

18. $A = \begin{bmatrix} 1 \\ 0 \\ 0 \end{bmatrix}, \quad B = \begin{bmatrix} -3 & 2 & 1 \end{bmatrix}$

19. $A = \begin{bmatrix} -1 & 1 & 0 \\ 2 & 3 & 5 \end{bmatrix}, \quad B = \begin{bmatrix} 0 & 4 & -2 \\ 3 & 6 & -1 \end{bmatrix}$

20. $A = \begin{bmatrix} -1 & 1 & 0 \\ 2 & 3 & 5 \end{bmatrix}, \quad B = \begin{bmatrix} 0 & 4 & -2 \\ 3 & 6 & -1 \\ 1 & 2 & -3 \end{bmatrix}$

21. $A = \begin{bmatrix} 2 & 1 & 0 \\ -1 & 4 & 3 \end{bmatrix}, \quad B = \begin{bmatrix} 1 & 0 & 0 \\ 0 & 1 & 0 \\ 0 & 0 & 1 \end{bmatrix}$

22. $A = \begin{bmatrix} 1 & 2 & 1 \\ -2 & 1 & -2 \end{bmatrix}, \quad B = \begin{bmatrix} 0 & 0 \\ 0 & 0 \\ 0 & 0 \\ 0 & 0 \end{bmatrix}$

23. Find

$$\begin{bmatrix} 1 & 3 \\ -1 & 1 \end{bmatrix}\begin{bmatrix} 0 & 1 & -1 \\ 1 & 0 & 1 \end{bmatrix} - 3\begin{bmatrix} 4 & 0 & -2 \\ 1 & \frac{2}{3} & 4 \end{bmatrix}.$$

24. Find $3A - 0B + 6I$ if

$$A = \begin{bmatrix} \frac{1}{2} & 2 \\ 5 & \frac{1}{4} \end{bmatrix}, \qquad B = \begin{bmatrix} \frac{1}{3} & \frac{1}{4} \\ \frac{1}{5} & \frac{1}{6} \end{bmatrix},$$

and I is the identity matrix of order 2.

In Exercises 25–28 find the powers A^2 and A^3 for the given matrix A.

25. $A = \begin{bmatrix} 1 & -1 \\ -1 & 2 \end{bmatrix}$

26. $A = \begin{bmatrix} 1 & 2 \\ 3 & 4 \end{bmatrix}$

27. $A = \begin{bmatrix} 1 & 0 & 0 \\ 2 & 1 & 0 \\ 3 & 2 & 1 \end{bmatrix}$

28. $A = \begin{bmatrix} 1 & 0 & 2 \\ -2 & -1 & 0 \\ 0 & 0 & 2 \end{bmatrix}$

In Exercises 29–32 let

$$A = \begin{bmatrix} 3 & 0 & -1 \\ 0 & 0 & 2 \\ 1 & 0 & -2 \end{bmatrix}, \quad B = \begin{bmatrix} 1 & 2 & 1 \\ 0 & 0 & 0 \\ -1 & 1 & 0 \end{bmatrix}, \quad \text{and} \quad C = \begin{bmatrix} 4 & 0 & -3 \\ -2 & -1 & 0 \\ 1 & 1 & 1 \end{bmatrix},$$

and verify each equation.

29. $(A + B)C = AC + BC$

30. $(2A)B = 2(AB)$

31. $A + C = C + A$

32. $(A + B) + C = (C + A) + B$

33. Prove the remaining parts of Theorem 1.

34. Use Theorem 1 to show that $(A + B) + C = (C + A) + B$ for all matrices A, B, and C for which the sums are defined.

For each matrix in Exercises 35 and 36, find a general formula for A^n.

35. $A = \begin{bmatrix} 1 & 1 \\ 0 & 1 \end{bmatrix}$

36. $A = \begin{bmatrix} 1 & -1 \\ -1 & 1 \end{bmatrix}$

37. A matrix A is called an *involution* if $A^2 = I$. Give an example of a 2×2 matrix (not equal to I) that is an involution.

38. A matrix A is *idempotent* if $A^2 = A$. Give an example of a 2×2 matrix (not equal to I or 0) that is idempotent.

39. Prove that $(kA)^T = kA^T$ for any matrix A and scalar k.

40. Prove that $(A^T)^T = A$ for all matrices A.

Computational Exercises

In Exercises 41 and 42 let *A* and *B* be square matrices of order 10 with

$$A = \begin{bmatrix} 1 & 2 & 3 & \dots & 10 \\ 2 & 3 & 4 & \dots & 11 \\ 3 & 4 & 5 & \dots & 12 \\ \vdots & \vdots & \vdots & & \vdots \\ 10 & 11 & 12 & \dots & 19 \end{bmatrix} \text{ and } B = \begin{bmatrix} 1 & -1 & 1 & -1 & \dots & -1 \\ 0 & 1 & -1 & 1 & \dots & 1 \\ 0 & 0 & 1 & -1 & \dots & -1 \\ \vdots & \vdots & \vdots & \vdots & & \vdots \\ 0 & 0 & 0 & 0 & \dots & 1 \end{bmatrix}.$$

Using a graphing calculator or computer software, find the following products

41. Find AB and BA. Are the two products equal?

42. Find $(AB)^T$, A^TB^T, and B^TA^T. Is any pair of these products equal?

In Exercises 43 and 44, let

$$A = \begin{bmatrix} 1 & \frac{1}{2} & \frac{1}{3} \\ \frac{1}{2} & \frac{1}{3} & \frac{1}{4} \\ \frac{1}{3} & \frac{1}{4} & \frac{1}{5} \end{bmatrix} \quad \text{and} \quad B = \begin{bmatrix} 1.00 & 0.33 & 0.20 \\ 0.33 & 0.20 & 0.15 \\ 0.20 & 0.15 & 0.11 \end{bmatrix}.$$

Find each answer to the nearest hundredth.

43. $A + B$ and $A - B$

44. AB and BA

45. Let $A = \begin{bmatrix} a & b & c \\ d & e & f \\ g & h & j \end{bmatrix}$ and $B = \begin{bmatrix} x_{11} & x_{12} & x_{13} \\ x_{21} & x_{22} & x_{23} \\ x_{31} & x_{32} & x_{33} \end{bmatrix}$.

Suppose that

$$AB = \begin{bmatrix} 1 & 0 & 0 \\ 0 & 1 & 0 \\ 0 & 0 & 1 \end{bmatrix}.$$

(a) Use a computer algebra system (CAS), such as *Mathematica*, to generate a system of nine equations with variables x_{ij}.

(b) Use a CAS to solve for the x_{ij}'s in terms of the constants a, b, c, d, e, f, g, h, and j.

(c) Under what conditions is there no solution?

(d) Suppose that

$$A = \begin{bmatrix} 1 & 0 & 3 \\ 0 & 2 & -1 \\ 2 & k & 3k \end{bmatrix}$$

for some constant k. Find values of k such that

$$AB = \begin{bmatrix} 1 & 0 & 0 \\ 0 & 1 & 0 \\ 0 & 0 & 1 \end{bmatrix},$$

for some matrix B.

46. Given the following matrices, use a computer software package or graphing calculator to find the 10th, 20th, 30th, and 40th powers of the matrices.

(a) $\begin{bmatrix} \frac{1}{2} & \frac{1}{3} & \frac{1}{6} \\ \frac{1}{4} & \frac{1}{2} & \frac{1}{4} \\ 0 & 1 & 0 \end{bmatrix}$ (b) $\begin{bmatrix} 0.25 & 0.30 & 0.45 \\ 0.85 & 0.10 & 0.05 \\ 0.15 & 0.50 & 0.35 \end{bmatrix}$

(c) $\begin{bmatrix} 0.13 & 0.24 & 0.31 & 0.32 \\ 0.51 & 0.19 & 0.23 & 0.07 \\ 0.19 & 0.31 & 0.48 & 0.02 \\ 0.22 & 0.29 & 0.25 & 0.24 \end{bmatrix}$

What do you notice? How are the matrices similar?

3.2 Matrix Equations and Inverses

In this section we develop a very concise way of writing a system of linear equations. This concise form suggests another solution procedure that in turn leads to the definition and investigation of the notion of the *inverse* of a square matrix.

Matrix Form of a Linear System

A system of m linear equations in the n unknowns $x_1, x_2, \ldots, x_n$ can be written in the following general form:

$$\begin{aligned} a_{11}x_1 + a_{12}x_2 + \cdots + a_{1n}x_n &= b_1 \\ a_{21}x_1 + a_{22}x_2 + \cdots + a_{2n}x_n &= b_2 \\ \vdots \qquad \vdots \qquad\qquad \vdots \quad &\quad \vdots \\ a_{m1}x_1 + a_{m2}x_2 + \cdots + a_{mn}x_n &= b_m \end{aligned}$$

The a_{ij} and the b_i are, respectively, the *coefficients* and *right-side constants* of the system. The "double subscript" notation is particularly convenient for the coefficients. Note that in the ith equation, the coefficient of the jth unknown is a_{ij}. For example, in the second equation the coefficient of the first unknown is denoted a_{21}.

Now recall, from Section 3.1, that two matrices are equal if and only if all pairs of corresponding entries are equal. For this reason, we can write the linear system just given as an equation between two $m \times 1$ matrices:

$$\begin{bmatrix} a_{11}x_1 + a_{12}x_2 + \cdots + a_{1n}x_n \\ a_{21}x_1 + a_{22}x_2 + \cdots + a_{2n}x_n \\ \vdots \qquad \vdots \qquad \vdots \\ a_{m1}x_1 + a_{m2}x_2 + \ldots + a_{mn}x_n \end{bmatrix} = \begin{bmatrix} b_1 \\ b_2 \\ \vdots \\ b_m \end{bmatrix}$$

However, by the definition of matrix multiplication, the matrix on the left can be written as the product of an $m \times n$ matrix and an $n \times 1$ matrix, so

$$\begin{bmatrix} a_{11} & a_{12} & \cdots & a_{1n} \\ a_{21} & a_{22} & \cdots & a_{2n} \\ \vdots & \vdots & & \vdots \\ a_{m1} & a_{m2} & \cdots & a_{mn} \end{bmatrix} \begin{bmatrix} x_1 \\ x_2 \\ x_3 \\ \vdots \\ x_n \end{bmatrix} = \begin{bmatrix} b_1 \\ b_2 \\ \vdots \\ b_m \end{bmatrix}.$$

Finally, denoting an $m \times n$ matrix with entries a_{ij} by A, an $n \times 1$ matrix with entries x_i by X, and an $m \times 1$ matrix with entries b_i by B, we can write the given system of linear equations in the concise matrix form

$$AX = B.$$

Here A is called the **coefficient matrix** of the system; X is the matrix, or (column) vector, of unknowns; and B is the matrix, or (column) vector, of constants. Since X and B are column vectors, we can write them using vector notation as $\mathbf{x}$ and $\mathbf{b}$. This leads to the conventional way of writing a linear system in **matrix form**:

$$A\mathbf{x} = \mathbf{b}$$

EXAMPLE 1

Write the following linear systems in matrix form.

a.
$$\begin{aligned} x + 2z &= 3 \\ 2x - 3y - 4z &= -1 \\ -\ y + \ z &= 0 \end{aligned}$$

b.
$$\begin{aligned} x_1 - 3x_2 + x_4 &= 0 \\ -\ x_2 + x_3 - x_4 &= 0 \end{aligned}$$

Solution

a.
$$\begin{bmatrix} 1 & 0 & 2 \\ 2 & -3 & -4 \\ 0 & -1 & 1 \end{bmatrix} \begin{bmatrix} x \\ y \\ z \end{bmatrix} = \begin{bmatrix} 3 \\ -1 \\ 0 \end{bmatrix}$$

b.
$$\begin{bmatrix} 1 & -3 & 0 & 1 \\ 0 & -1 & 1 & -1 \end{bmatrix} \begin{bmatrix} x_1 \\ x_2 \\ x_3 \\ x_4 \end{bmatrix} = \begin{bmatrix} 0 \\ 0 \end{bmatrix}$$

Both are written as $A\mathbf{x} = \mathbf{b}$, where in part (a),

$$A = \begin{bmatrix} 1 & 0 & 2 \\ 2 & -3 & -4 \\ 0 & -1 & 1 \end{bmatrix}, \quad \mathbf{x} = \begin{bmatrix} x \\ y \\ z \end{bmatrix}, \quad \mathbf{b} = \begin{bmatrix} 3 \\ -1 \\ 0 \end{bmatrix};$$

and in part (b),

$$A = \begin{bmatrix} 1 & -3 & 0 & 1 \\ 0 & -1 & 1 & -1 \end{bmatrix}, \quad \mathbf{x} = \begin{bmatrix} x_1 \\ x_2 \\ x_3 \\ x_4 \end{bmatrix}, \quad \mathbf{b} = \begin{bmatrix} 0 \\ 0 \end{bmatrix}.$$

■

NOTE It is often convenient to write a column vector on a horizontal line, so we establish the convention that $(c_1, c_2, \ldots, c_k)$ denotes the $k \times 1$ column vector

$$\begin{bmatrix} c_1 \\ c_2 \\ \vdots \\ c_k \end{bmatrix}.$$

Note that parentheses and commas distinguish our column vector written horizontally from the $1 \times k$ row vector $[c_1\ c_2\ \ldots\ c_k]$. This new convention also conforms with previous usage. In Chapter 2, we wrote solutions of a linear system with

n unknowns as n-tuples $(x_1, x_2, \ldots, x_n)$. We may still write solutions in this manner. For example, the solution of the system

$$\begin{aligned} 3x_1 - x_2 &= 0 \\ 2x_1 + x_2 &= 5 \end{aligned} \quad \text{or} \quad \begin{bmatrix} 3 & -1 \\ 2 & 1 \end{bmatrix}\begin{bmatrix} x_1 \\ x_2 \end{bmatrix} = \begin{bmatrix} 0 \\ 5 \end{bmatrix}$$

may be written as either

$$\begin{bmatrix} 1 \\ 3 \end{bmatrix} \quad \text{or} \quad (1, 3).$$

The Inverse of a Matrix

Writing a linear system in the form $A\mathbf{x} = \mathbf{b}$ suggests a deceptively simple means of solution. Were this a single (scalar) equation, $ax = b\ (a \neq 0)$, with unknown x, we can quickly solve it by multiplying both sides by the *multiplicative inverse* (the *reciprocal*) of a. The solution is $x = a^{-1}b\ (=(1/a)b = b/a)$. Taking our cue from this simpler situation, it is natural to ask if we can solve the matrix equation $A\mathbf{x} = \mathbf{b}$ in the same way. The answer is, "Yes, sometimes," but the process, which is developed next, takes considerably more work.

We must first define the (multiplicative) inverse of a matrix and then determine a procedure for finding it when it exists. The multiplicative inverse of a number a is a number b, such that $ab = 1$. (Since multiplication of numbers is commutative, we also have $ba = 1$.) Consequently a natural candidate for the inverse of a matrix A is a matrix B, such that $AB = I$. Matrix multiplication is not commutative, however, so we also require that $BA = I$. Note that if both products AB and BA are defined and equal to the same matrix I, then A and B must both be square and of the same order as I.

DEFINITION

Let A be a square matrix. If there exists a square matrix B of the same order as A such that $AB = BA = I$, we say that A is **invertible** (or **nonsingular**) and that B is the **inverse** of A. If A has no inverse, it is said to be **noninvertible** (or **singular**).

EXAMPLE 2

Show that the inverse of the matrix A is B, where

$$A = \begin{bmatrix} 2 & -1 \\ 0 & 1 \end{bmatrix} \quad \text{and} \quad B = \begin{bmatrix} \frac{1}{2} & \frac{1}{2} \\ 0 & 1 \end{bmatrix}.$$

Solution We have

$$AB = \begin{bmatrix} 2 & -1 \\ 0 & 1 \end{bmatrix}\begin{bmatrix} \frac{1}{2} & \frac{1}{2} \\ 0 & 1 \end{bmatrix} = \begin{bmatrix} 1 & 0 \\ 0 & 1 \end{bmatrix} = I$$

$$\text{and} \quad BA = \begin{bmatrix} \frac{1}{2} & \frac{1}{2} \\ 0 & 1 \end{bmatrix}\begin{bmatrix} 2 & -1 \\ 0 & 1 \end{bmatrix} = \begin{bmatrix} 1 & 0 \\ 0 & 1 \end{bmatrix} = I,$$

as desired. ▣

In the definition and in Example 2, we spoke of *the* inverse of a matrix. This terminology implies that a given matrix has at most one inverse. The following theorem justifies the use of this word.

THEOREM 1

If a matrix A is invertible, then the inverse is unique.

Proof Suppose that A has inverses B and C. We show that B and C must be equal as follows: $B = BI = B(AC) = (BA)C = IC = C$. ■

NOTE We denote the inverse of an invertible matrix A by the symbol A^{-1}. From the definition, it follows that if $A^{-1} = B$, then B is invertible, and $B^{-1} = A$ as well. Moreover, if A is a square matrix with inverse A^{-1}, then A^T is invertible with inverse $(A^{-1})^T$, since by Theorem 1 (p) of Section 3.1, we have $A^T(A^{-1})^T = ((A^{-1})A)^T = I^T = I$ and $(A^{-1})^T A^T = (A(A^{-1}))^T = I^T = I$. Similarly, if B is also a square invertible matrix of the same order as A, then AB is invertible with inverse $B^{-1}A^{-1}$ since, by Theorem 1 (e) of Section 3.1, $(AB)(B^{-1}A^{-1}) = A(BB^{-1})A^{-1} = AIA^{-1} = AA^{-1} = I$, and similarly, $(B^{-1}A^{-1})(AB) = I$, as well.

We now develop a technique for finding the inverse of a square matrix. First we state a theorem that simplifies this task somewhat. Its proof is given in Theorem 9 of Section 4.2.

THEOREM 2

Let A be a square matrix. If a square matrix B exists such that $AB = I$, then $BA = I$ as well, and thus $B = A^{-1}$. ■

Theorem 2 tells us that the inverse of an $n \times n$ matrix A is an $n \times n$ matrix X that solves the equation $AX = I$, where I is the identity matrix of order n. Before establishing a procedure to find this solution, we consider a particular example that illustrates the basic idea.

EXAMPLE 3

Find the inverse of

$$A = \begin{bmatrix} 1 & 1 & 1 \\ 0 & 2 & 1 \\ 1 & 0 & 1 \end{bmatrix}.$$

Solution We seek a 3×3 matrix X such that $AX = I$, that is,

$$\begin{bmatrix} 1 & 1 & 1 \\ 0 & 2 & 1 \\ 1 & 0 & 1 \end{bmatrix} \begin{bmatrix} x_{11} & x_{12} & x_{13} \\ x_{21} & x_{22} & x_{23} \\ x_{31} & x_{32} & x_{33} \end{bmatrix} = \begin{bmatrix} 1 & 0 & 0 \\ 0 & 1 & 0 \\ 0 & 0 & 1 \end{bmatrix}.$$

Now the first, second, and third columns of the matrix product on the left must be equal to, respectively, the first, second, and third columns of the identity matrix. Symbolically, we write this as $(AX)^1 = \mathbf{e}^1$, $(AX)^2 = \mathbf{e}^2$, and $(AX)^3 = \mathbf{e}^3$. By the definition of matrix multiplication,

$$(AX)^1 = \begin{bmatrix} \mathbf{a}_1 \cdot \mathbf{x}^1 \\ \mathbf{a}_2 \cdot \mathbf{x}^1 \\ \mathbf{a}_3 \cdot \mathbf{x}^1 \end{bmatrix} = \begin{bmatrix} (1, 1, 1) \cdot (x_{11}, x_{21}, x_{31}) \\ (0, 2, 1) \cdot (x_{11}, x_{21}, x_{31}) \\ (1, 0, 1) \cdot (x_{11}, x_{21}, x_{31}) \end{bmatrix} = \begin{bmatrix} x_{11} + x_{21} + x_{31} \\ 2x_{21} + x_{31} \\ x_{11} + x_{31} \end{bmatrix}.$$

This is exactly the result we get if we multiply the matrix A by the first column of X; that is,

$$A\mathbf{x}^1 = \begin{bmatrix} 1 & 1 & 1 \\ 0 & 2 & 1 \\ 1 & 0 & 1 \end{bmatrix} \begin{bmatrix} x_{11} \\ x_{21} \\ x_{31} \end{bmatrix} = \begin{bmatrix} x_{11} + x_{21} + x_{31} \\ 2x_{21} + x_{31} \\ x_{11} + x_{31} \end{bmatrix}.$$

Thus, $(AX)^1 = A\mathbf{x}^1$. Similarly, $(AX)^2 = A\mathbf{x}^2$ and $(AX)^3 = A\mathbf{x}^3$. Therefore, to find X, we can solve the three linear systems

$$A\mathbf{x}^1 = \mathbf{e}^1, \quad A\mathbf{x}^2 = \mathbf{e}^2, \quad \text{and} \quad A\mathbf{x}^3 = \mathbf{e}^3.$$

We do this by Gauss–Jordan elimination (Section 2.3), forming the augmented matrices

$$\left[\begin{array}{ccc|c} 1 & 1 & 1 & 1 \\ 0 & 2 & 1 & 0 \\ 1 & 0 & 1 & 0 \end{array}\right], \quad \left[\begin{array}{ccc|c} 1 & 1 & 1 & 0 \\ 0 & 2 & 1 & 1 \\ 1 & 0 & 1 & 0 \end{array}\right], \quad \left[\begin{array}{ccc|c} 1 & 1 & 1 & 0 \\ 0 & 2 & 1 & 0 \\ 1 & 0 & 1 & 1 \end{array}\right],$$

and transforming each to its row-reduced echelon form

$$\left[\begin{array}{ccc|c} 1 & 0 & 0 & 2 \\ 0 & 1 & 0 & 1 \\ 0 & 0 & 1 & -2 \end{array}\right], \quad \left[\begin{array}{ccc|c} 1 & 0 & 0 & -1 \\ 0 & 1 & 0 & 0 \\ 0 & 0 & 1 & 1 \end{array}\right], \quad \left[\begin{array}{ccc|c} 1 & 0 & 0 & -1 \\ 0 & 1 & 0 & -1 \\ 0 & 0 & 1 & 2 \end{array}\right].$$

Thus, $\mathbf{x}^1 = (2, 1, -2)$, $\mathbf{x}^2 = (-1, 0, 1)$, and $\mathbf{x}^3 = (-1, -1, 2)$, and we have

$$A^{-1} = \begin{bmatrix} 2 & -1 & -1 \\ 1 & 0 & -1 \\ -2 & 1 & 2 \end{bmatrix}.$$

■

The solution procedure of Example 3 contains a gross inefficiency. Since the coefficient matrices of the three linear systems $A\mathbf{x}^1 = \mathbf{e}^1$, $A\mathbf{x}^2 = \mathbf{e}^2$, and $A\mathbf{x}^3 = \mathbf{e}^3$ are identical, all three augmented matrices are transformed using the same row operations. We can therefore save a lot of work by doing all the row reductions at the same time. To do this, we form the 3×6 augmented matrix whose first three columns are those of A and whose last three columns are $\mathbf{e}^1$, $\mathbf{e}^2$, and $\mathbf{e}^3$ (that is, the last three columns are those of I):

$$\left[\begin{array}{ccc|ccc} 1 & 1 & 1 & 1 & 0 & 0 \\ 0 & 2 & 1 & 0 & 1 & 0 \\ 1 & 0 & 1 & 0 & 0 & 1 \end{array}\right]$$

We now transform this matrix to the row-reduced echelon form

$$\left[\begin{array}{ccc|ccc} 1 & 0 & 0 & 2 & -1 & -1 \\ 0 & 1 & 0 & 1 & 0 & -1 \\ 0 & 0 & 1 & -2 & 1 & 2 \end{array}\right].$$

Then the fourth, fifth, and sixth columns of the last matrix are $\mathbf{x}^1$, $\mathbf{x}^2$, and $\mathbf{x}^3$, the columns of A^{-1}. Thus (as before),

$$A^{-1} = \begin{bmatrix} 2 & -1 & -1 \\ 1 & 0 & -1 \\ -2 & 1 & 2 \end{bmatrix}.$$

NOTE In effect, to form the original 3×6 augmented matrix, we adjoined the identity matrix of order 3 to the coefficient matrix A. We denote this matrix by $[A \mid I]$. Similarly, the augmented matrix for the system $A\mathbf{x} = \mathbf{b}$ can be denoted $[A \mid \mathbf{b}]$.

We now turn to the general case of attempting to find the inverse of an $n \times n$ matrix A. This can be done by solving the matrix equation $AX = I$ (Theorem 2). As in Example 3, the latter is equivalent to solving the n linear systems

$$A\mathbf{x}^1 = \mathbf{e}^1, A\mathbf{x}^2 = \mathbf{e}^2, \ldots, A\mathbf{x}^n = \mathbf{e}^n$$

for the columns $\mathbf{x}^i$ of A^{-1}, where $\mathbf{e}^i$ is the column vector with a 1 in the i-th entry and 0's elsewhere. To solve these systems simultaneously, we transform the augmented matrix $[A \mid I]$ into row-reduced echelon form. The last n columns of the latter form the inverse of A, *provided that each of the systems* $A\mathbf{x}^i = \mathbf{e}^i$ *has a solution.* If one of them does not, the matrix equation $AX = I$ has no solution, and hence the inverse of A does not exist. We summarize this discussion in the form of a procedure.

Procedure for Finding a Matrix Inverse

Let A be a square matrix and I be the identity matrix of the same order. Form the augmented matrix $[A \mid I]$ and transform it into row-reduced echelon form $[C \mid D]$. Then

i. If C is the identity matrix, $D = A^{-1}$;

ii. If C is not the identity matrix (that is, C has a row of zeros), A is not invertible.

EXAMPLE 4

Find A^{-1} (if it exists) when

$$A = \begin{bmatrix} 1 & 0 & 1 & 1 \\ 0 & 0 & 1 & 0 \\ 1 & 1 & 1 & 0 \\ 1 & 0 & 0 & 2 \end{bmatrix}.$$

Solution We adjoin the (fourth-order) identity matrix to get $[A \mid I]$,

$$\left[\begin{array}{cccc|cccc} 1 & 0 & 1 & 1 & 1 & 0 & 0 & 0 \\ 0 & 0 & 1 & 0 & 0 & 1 & 0 & 0 \\ 1 & 1 & 1 & 0 & 0 & 0 & 1 & 0 \\ 1 & 0 & 0 & 2 & 0 & 0 & 0 & 1 \end{array}\right],$$

and transform this matrix to row-reduced echelon form,

$$\left[\begin{array}{cccc|cccc} 1 & 0 & 0 & 0 & 2 & -2 & 0 & -1 \\ 0 & 1 & 0 & 0 & -2 & 1 & 1 & 1 \\ 0 & 0 & 1 & 0 & 0 & 1 & 0 & 0 \\ 0 & 0 & 0 & 1 & -1 & 1 & 0 & 1 \end{array}\right].$$

Consequently, A^{-1} exists with

$$A^{-1} = \begin{bmatrix} 2 & -2 & 0 & -1 \\ -2 & 1 & 1 & 1 \\ 0 & 1 & 0 & 0 \\ -1 & 1 & 0 & 1 \end{bmatrix}.$$

■

EXAMPLE 5

Find the inverse of

$$A = \begin{bmatrix} 1 & -2 \\ -2 & 4 \end{bmatrix},$$

or show that no inverse exists.

Solution We form the augmented matrix

$$[A \mid I] = \left[\begin{array}{cc|cc} 1 & -2 & 1 & 0 \\ -2 & 4 & 0 & 1 \end{array}\right],$$

and transform it to the form

$$\left[\begin{array}{rr|rr} 1 & -2 & 1 & 0 \\ 0 & 0 & 2 & 1 \end{array}\right].$$

Since A cannot be transformed into the identity matrix, A has no inverse. ■

COMPUTATIONAL NOTE

Most graphing calculators and linear algebra software packages can be used to find the inverse of a matrix quickly and easily. After entering the given matrix, just press the appropriate key, enter the appropriate command, or select "inverse" from a menu. Using a calculator or computer to find an inverse is definitely a labour-saving process! On the other hand, the use of a calculator or computer may introduce rounding errors, so beware!

Method of Inverses

We now return to the idea of solving the linear system $A\mathbf{x} = \mathbf{b}$ by using inverse matrices. Let A be an $n \times n$ nonsingular matrix. We have

$$\begin{aligned} A\mathbf{x} &= \mathbf{b} & \\ A^{-1}A\mathbf{x} &= A^{-1}\mathbf{b} & &\text{(Multiplying both sides on the left by } A^{-1}\text{)} \\ I\mathbf{x} &= A^{-1}\mathbf{b} & &\text{(Theorem 1(e) of Section 3.1)} \\ \mathbf{x} &= A^{-1}\mathbf{b} & &\text{(Theorem 1(f) of Section 3.1)} \end{aligned}$$

Hence we can solve a linear system whose coefficient matrix is square and non-singular by finding the inverse and then multiplying it by the right-side vector. This type of solution procedure is called the **method of inverses**.

EXAMPLE 6

Solve the following system by the method of inverses.

$$\begin{aligned} x_1 + x_3 &= 2 \\ x_1 - x_2 &= -1 \\ 2x_2 + x_3 &= 1 \end{aligned}$$

Solution This linear system can be written in the form $A\mathbf{x} = \mathbf{b}$ by taking

$$A = \begin{bmatrix} 1 & 0 & 1 \\ 1 & -1 & 0 \\ 0 & 2 & 1 \end{bmatrix}, \quad \mathbf{x} = \begin{bmatrix} x_1 \\ x_2 \\ x_3 \end{bmatrix}, \quad \text{and} \quad \mathbf{b} = \begin{bmatrix} 2 \\ -1 \\ 1 \end{bmatrix}.$$

We must first find A^{-1}. We have

$$[A \mid I] = \left[\begin{array}{rrr|rrr} 1 & 0 & 1 & 1 & 0 & 0 \\ 1 & -1 & 0 & 0 & 1 & 0 \\ 0 & 2 & 1 & 0 & 0 & 1 \end{array}\right],$$

which has the row-reduced echelon form

$$\left[\begin{array}{rrr|rrr} 1 & 0 & 0 & -1 & 2 & 1 \\ 0 & 1 & 0 & -1 & 1 & 1 \\ 0 & 0 & 1 & 2 & -2 & -1 \end{array}\right].$$

Thus,
$$A^{-1} = \begin{bmatrix} -1 & 2 & 1 \\ -1 & 1 & 1 \\ 2 & -2 & -1 \end{bmatrix}.$$

Finally, we obtain $\mathbf{x} = A^{-1}\mathbf{b}$ by multiplying
$$\begin{bmatrix} -1 & 2 & 1 \\ -1 & 1 & 1 \\ 2 & -2 & -1 \end{bmatrix}\begin{bmatrix} 2 \\ -1 \\ 1 \end{bmatrix} = \begin{bmatrix} -3 \\ -2 \\ 5 \end{bmatrix}.$$

Hence, $x_1 = -3$, $x_2 = -2$, and $x_3 = 5$. ■

WARNING The method of inverses cannot be used to solve the linear system $A\mathbf{x} = \mathbf{b}$ if A is not square or if A is not invertible.

EXAMPLE 7

We wish to solve a number of linear systems of the form $A\mathbf{x} = \mathbf{b}$, all of which have the same fourth-order nonsingular coefficient matrix. If we know that
$$A^{-1} = \begin{bmatrix} 1 & 0 & -1 & 0 \\ 2 & 0 & 0 & 1 \\ 0 & -1 & -2 & 0 \\ 0 & 0 & 1 & 1 \end{bmatrix},$$

We can give a formula for the solution vector $\mathbf{x}$ in terms of the components of the right-side vector $\mathbf{b}$.

Solution We have $\mathbf{x} = A^{-1}\mathbf{b}$, so
$$\mathbf{x} = \begin{bmatrix} 1 & 0 & -1 & 0 \\ 2 & 0 & 0 & 1 \\ 0 & -1 & -2 & 0 \\ 0 & 0 & 1 & 1 \end{bmatrix}\begin{bmatrix} b_1 \\ b_2 \\ b_3 \\ b_4 \end{bmatrix} = \begin{bmatrix} b_1 - b_3 \\ 2b_1 + b_4 \\ -b_2 - 2b_3 \\ b_3 + b_4 \end{bmatrix}.$$

Thus, $\mathbf{x} = (b_1 - b_3, 2b_1 + b_4, -b_2 - 2b_3, b_3 + b_4)$. ■

COMPUTATIONAL NOTE

The method of inverses is not as efficient as Gauss–Jordan elimination for solving linear systems. It requires more operations—additions, subtractions, multiplications, and divisions—to complete the solution process. As you can see from these examples, however, if the inverse is somehow known beforehand, it is an extremely easy task to find the solution of the system by the method of inverses.

3.2 Exercises

In Exercises 1–4 write the given linear system in the matrix form $A\mathbf{x} = \mathbf{b}$.

1.
$$\begin{aligned} x - y + 3z &= 1 \\ x \quad - z &= 0 \\ -2x + y \quad &= -1 \end{aligned}$$

2.
$$\begin{aligned} 2x_1 - 3x_2 \quad + x_4 &= 6 \\ x_2 - x_3 + 3x_4 &= 0 \\ -x_1 \quad + x_4 &= 0 \end{aligned}$$

3.
$$\begin{aligned} x &= y - z \\ y &= x - z \end{aligned}$$

4.
$$\begin{aligned} x_1 &= 2 \\ x_2 &= 3 - x_1 \\ x_1 + x_2 + 1 &= 4 \end{aligned}$$

In Exercises 5–8 verify that matrices A and B are inverses of each other.

5. $A = \begin{bmatrix} 1 & 2 \\ 3 & 5 \end{bmatrix}, \quad B = \begin{bmatrix} -5 & 2 \\ 3 & -1 \end{bmatrix}$

6. $A = \begin{bmatrix} 1 & 0 \\ 4 & 2 \end{bmatrix}, \quad B = \begin{bmatrix} 1 & 0 \\ -2 & \frac{1}{2} \end{bmatrix}$

7. $A = \begin{bmatrix} 1 & 2 & -11 \\ 0 & -1 & 4 \\ 0 & 0 & 1 \end{bmatrix}, \quad B = \begin{bmatrix} 1 & 2 & 3 \\ 0 & -1 & 4 \\ 0 & 0 & 1 \end{bmatrix}$

8. $A = \begin{bmatrix} 1 & 0 & -1 \\ 3 & -1 & 0 \\ 0 & -1 & 1 \end{bmatrix}, \quad B = \frac{1}{2}\begin{bmatrix} -1 & 1 & -1 \\ -3 & 1 & -3 \\ -3 & 1 & -1 \end{bmatrix}$

In Exercises 9–16 find the inverse, if it exists, of the given matrix. If the matrix has no inverse, state this fact.

9. $\begin{bmatrix} 2 & 0 \\ -3 & 1 \end{bmatrix}$

10. $\begin{bmatrix} -1 & 2 \\ 3 & -4 \end{bmatrix}$

11. $\begin{bmatrix} 1 & 2 & -1 \\ 0 & 1 & 2 \\ 0 & 0 & 1 \end{bmatrix}$

12. $\begin{bmatrix} 1 & 0 & -2 \\ 2 & -1 & 0 \\ 1 & 0 & -2 \end{bmatrix}$

13. $\begin{bmatrix} 1 & -1 & 0 \\ -1 & 1 & 1 \\ 0 & -1 & 1 \end{bmatrix}$

14. $\begin{bmatrix} 1 & 0 & 0 \\ -1 & 2 & 0 \\ -2 & -1 & 1 \end{bmatrix}$

15. $\begin{bmatrix} 1 & 0 & 1 & 0 \\ 0 & 1 & 0 & 1 \\ 0 & 0 & 1 & 1 \\ 1 & 1 & 0 & 0 \end{bmatrix}$

16. $\begin{bmatrix} 1 & -1 & 0 & 0 \\ 0 & 1 & -1 & 0 \\ 0 & 0 & 1 & -1 \\ 1 & 0 & 0 & 1 \end{bmatrix}$

In Exercises 17 and 18 find the value of c for which A has no inverse.

17. $A = \begin{bmatrix} 1 & -2 \\ 3 & c \end{bmatrix}$

18. $A = \begin{bmatrix} 1 & -1 & c \\ 0 & 2 & 1 \\ -1 & 0 & 1 \end{bmatrix}$

In Exercises 19–22 solve the linear system $A\mathbf{x} = \mathbf{b}$ when the inverse, A^{-1}, and $\mathbf{b}$ are as given.

19. $A^{-1} = \begin{bmatrix} 1 & 2 \\ -2 & 3 \end{bmatrix}, \quad \mathbf{b} = \begin{bmatrix} 1 \\ -2 \end{bmatrix}$

20. $A^{-1} = \begin{bmatrix} 0 & 1 \\ 1 & 0 \end{bmatrix}, \quad \mathbf{b} = \begin{bmatrix} -1 \\ 1 \end{bmatrix}$

21. $A^{-1} = \begin{bmatrix} -1 & \frac{1}{2} & 0 \\ \frac{1}{2} & -1 & \frac{1}{2} \\ 0 & \frac{1}{2} & -1 \end{bmatrix}, \quad \mathbf{b} = \begin{bmatrix} -3 \\ 1 \\ 2 \end{bmatrix}$

22. $A^{-1} = \begin{bmatrix} 1 & -1 & 3 & 1 \\ 0 & 1 & 0 & 2 \\ -2 & -1 & -3 & -1 \\ 1 & -1 & 0 & 0 \end{bmatrix}, \quad \mathbf{b} = \begin{bmatrix} 1 \\ 2 \\ 3 \\ 4 \end{bmatrix}$

In Exercises 23–28 use the method of inverses, if possible, to solve the following systems.

23. $2x - 3y = 3$
$3x - 5y = 1$

24. $A\mathbf{x} = \mathbf{b}$, where

$$A = \begin{bmatrix} 1 & -3 \\ 2 & 4 \end{bmatrix}, \quad \mathbf{x} = \begin{bmatrix} x_1 \\ x_2 \end{bmatrix}, \quad \text{and} \quad \mathbf{b} = \begin{bmatrix} 3 \\ -1 \end{bmatrix}$$

25. $x - 3y + 4z = 4$
$2x + 2y \qquad = 0$
$\qquad y - 2z = 2$

26. $-x_1 + x_3 = 1$
$x_1 + x_2 = -1$
$x_2 - x_3 = 0$

27. $A\mathbf{x} = \mathbf{b}$, where

$$A = \begin{bmatrix} 1 & -1 & 2 \\ 2 & -1 & 1 \\ 0 & -1 & 3 \end{bmatrix}, \quad \mathbf{x} = \begin{bmatrix} x_1 \\ x_2 \\ x_3 \end{bmatrix}, \quad \text{and} \quad \mathbf{b} = \begin{bmatrix} 1 \\ 0 \\ 0 \end{bmatrix}$$

28. $x - 3y + z = 2$
$2x + y - 2z = 7$

29. Show that the inverse of the 2×2 matrix

$$\begin{bmatrix} a & b \\ c & d \end{bmatrix}$$

is the matrix

$$\frac{1}{ad - bc}\begin{bmatrix} d & -b \\ -c & a \end{bmatrix},$$

provided that $ad - bc \neq 0$.

In Exercises 30 and 31 use the result of Exercise 29 to find the inverses quickly.

30. $\begin{bmatrix} 3 & -2 \\ 1 & 4 \end{bmatrix}$

31. $\begin{bmatrix} 1 & -1 \\ 0 & 4 \end{bmatrix}$

32. Show that if A and P have inverses and are of the same order, then $P^{-1}AP$ has an inverse, and in fact $(P^{-1}AP)^{-1} = P^{-1}A^{-1}P$.

33. Show that the inverse of the nth-order diagonal matrix

$$D = \begin{bmatrix} a_1 & 0 & \cdots & 0 \\ 0 & a_2 & \cdots & 0 \\ \vdots & \vdots & \ddots & \vdots \\ 0 & 0 & \cdots & a_n \end{bmatrix} \quad \text{is} \quad D^{-1} = \begin{bmatrix} \frac{1}{a_1} & 0 & \cdots & 0 \\ 0 & \frac{1}{a_2} & \cdots & 0 \\ \vdots & \vdots & \ddots & \vdots \\ 0 & 0 & \cdots & \frac{1}{a_n} \end{bmatrix},$$

provided that the main diagonal entries a_i are all nonzero.

Computational Exercises

In Exercises 34–37 use a graphing calculator or software package to solve each problem.

34. Find the inverse of the following matrices.

$$(a)\ A = \begin{bmatrix} 1 & 0 & 0 & 0 & 0 \\ 0 & 0 & 0 & 1 & 0 \\ 0 & 1 & 0 & 0 & 0 \\ 0 & 0 & 0 & 0 & 1 \\ 0 & 0 & 1 & 0 & 0 \end{bmatrix} \qquad (b)\quad B = \begin{bmatrix} 1 & \frac{1}{2} & \frac{1}{3} & \frac{1}{4} & \frac{1}{5} \\ \frac{1}{2} & \frac{1}{3} & \frac{1}{4} & \frac{1}{5} & \frac{1}{6} \\ \frac{1}{3} & \frac{1}{4} & \frac{1}{5} & \frac{1}{6} & \frac{1}{7} \\ \frac{1}{4} & \frac{1}{5} & \frac{1}{6} & \frac{1}{7} & \frac{1}{8} \\ \frac{1}{5} & \frac{1}{6} & \frac{1}{7} & \frac{1}{8} & \frac{1}{9} \end{bmatrix}$$

35. Let A and B be the matrices in Exercise 34 and $\mathbf{b} = (1, 2, 3, 4, 5)$. Solve the following systems by the method of inverses.

(a) $A\mathbf{x} = \mathbf{b}$ *(b)* $B\mathbf{x} = \mathbf{b}$

(Notice that the system in part (a) can also be easily solved by hand.)

36. Solve the linear system $C\mathbf{x} = \mathbf{c}$ by the method of inverses, where C is the 10×10 matrix with ones on and above the main diagonal and zeros below it, and $\mathbf{b}$ is the 10-vector with all entries equal to 1. Compare your answer with the answer you obtain by solving this system by hand.

In Exercises 37 and 38 use a computer algebra system (such as *Mathematica*) to solve each problem.

37. Find the inverse of the matrix

$$A = \begin{bmatrix} 1 & 1 & k \\ 1 & k & 1 \\ k & 1 & 1 \end{bmatrix}$$

in terms of k. Are there any real values of k, other than 1, for which A^{-1} does not exist?

38. Find the inverse of the matrix

$$B = \begin{bmatrix} 1 & x & x^2 \\ 1 & y & y^2 \\ 1 & z & z^2 \end{bmatrix}$$

in terms of x, y, and z. For what values of x, y, and z does B^{-1} fail to exist?

39. Let A be an $n \times n$ matrix with zeros on the main diagonal and all other entries equal to -1. Use a graphing calculator or software package to find A^{-1} for $n = 3$, 4, and 5. Based on these results, find a formula (in terms of n) for A^{-1} that holds for all $n \geq 3$. (*Hint*: Interpret the decimals in A^{-1} as fractions.)

3.3 Theory of Linear Systems

In this section we investigate the nature of solutions of systems of linear equations rather than techniques for finding them. In doing so, we introduce an important special case: the *homogeneous* linear system.

We continue to write a linear system in the matrix form, $A\mathbf{x} = \mathbf{b}$, developed in Section 3.2. In such an equation, A is an $m \times n$ matrix, $\mathbf{x}$ is an n-vector, and $\mathbf{b}$ is an m-vector. If A is square, of course $m = n$.

THEOREM 1

Every system of linear equations $A\mathbf{x} = \mathbf{b}$ has no solution, exactly one solution, or infinitely many solutions.

Proof We have already seen examples of each of the three cases, so we need only show that there are no other possibilities. Assume that $\mathbf{y}$ and $\mathbf{z}$ are solutions with $\mathbf{y} \neq \mathbf{z}$. Since these are vectors in $\mathbf{R}^n$, the line they determine is given by the two-point form (Section 2.1),

$$\begin{aligned}\mathbf{x}(t) &= (1 - t)\mathbf{y} + t\mathbf{z}\\ &= \mathbf{y} + t(\mathbf{z} - \mathbf{y}).\end{aligned}$$

Each point $\mathbf{x}(t)$ on this line is also a solution of $A\mathbf{x} = \mathbf{b}$, as seen by

$$\begin{aligned}A(\mathbf{x}(t)) &= A(\mathbf{y} + t(\mathbf{z} - \mathbf{y}))\\ &= A\mathbf{y} + A(t(\mathbf{z} - \mathbf{y}))\\ &= \mathbf{b} + tA(\mathbf{z} - \mathbf{y})\\ &= \mathbf{b} + t(A\mathbf{z} - A\mathbf{y})\\ &= \mathbf{b} + t(\mathbf{b} - \mathbf{b}) = \mathbf{b}.\end{aligned}$$

Consequently, if there is more than one solution to $A\mathbf{x} = \mathbf{b}$, then there are infinitely many. ■

DEFINITION

The linear system $A\mathbf{x} = \mathbf{b}$ is said to be **homogeneous** if $\mathbf{b} = \mathbf{0}$, the vector consisting solely of zeros. If $\mathbf{b} \neq \mathbf{0}$, then the system is said to be **nonhomogeneous**.

For example, the linear systems

$$\begin{bmatrix} 1 & -1 & 2 & 0 \\ 3 & 0 & 1 & 5 \end{bmatrix}\begin{bmatrix} x_1 \\ x_2 \\ x_3 \\ x_4 \end{bmatrix} = \begin{bmatrix} 0 \\ 0 \end{bmatrix}$$

and

$$\begin{aligned}-x - 3y - z &= 0\\ 2x - 2z &= 0\\ 3y + z &= 0\end{aligned}$$

are *homogeneous*, but

$$\begin{aligned}x_1 - 2x_2 + x_3 &= 0\\ 3x_1 + x_2 - x_3 &= 1\end{aligned}$$

is *nonhomogeneous*.

NOTE The zero vector, $\mathbf{0}$, is a solution of every homogeneous linear system $A\mathbf{x} = \mathbf{0}$, since $A\mathbf{0} = \mathbf{0}$. The solution $\mathbf{x} = \mathbf{0}$ is called the **trivial solution** of the homogeneous system. Any other solution is called a **nontrivial** one.

THEOREM 2

Every homogeneous linear system $A\mathbf{x} = \mathbf{0}$ has either exactly one solution or infinitely many solutions.

Proof Since every homogeneous linear system has at least one solution (the trivial one), the result follows immediately from Theorem 1. ■

We can probe a little deeper into this subject by considering the structure of the row-reduced echelon form for the augmented matrix of the given system. The key to this line of investigation is the notion of *rank*.

DEFINITION

The **rank** of a matrix A is the number of nonzero rows in the row-reduced echelon form of A.

NOTE The rank of a matrix is *well defined* due to the uniqueness of the row-reduced echelon form (Theorem 3 of Section 2.3). No matter what sequence of elementary row operations is performed to put the given matrix in row-reduced echelon form, there is always the same number of nonzero rows.

EXAMPLE 1

Find the rank of

$$A = \begin{bmatrix} 1 & -1 & 2 & 1 \\ 0 & 1 & 1 & -2 \\ 1 & -3 & 0 & 5 \end{bmatrix}.$$

Solution The row-reduced echelon form of A is

$$\begin{bmatrix} 1 & 0 & 3 & -1 \\ 0 & 1 & 1 & -2 \\ 0 & 0 & 0 & 0 \end{bmatrix},$$

which has two nonzero rows. Hence the rank of A is 2. ■

NOTE Given the linear system $A\mathbf{x} = \mathbf{b}$, the rank of the coefficient matrix A is always less than or equal to the rank of the augmented matrix for the system $[A \mid \mathbf{b}]$. This follows from the fact that, when the latter is transformed into row-reduced echelon form, its first n columns are the row-reduced form of the former. It is consequently impossible for the row-reduced echelon form of A to have more nonzero rows than that of $[A \mid \mathbf{b}]$. In other words, it is impossible for the rank of the coefficient matrix to be greater than that of the augmented matrix.

THEOREM 3

For the linear system $A\mathbf{x} = \mathbf{b}$, where A is an $m \times n$ matrix, let p be the rank of A and q be the rank of $[A \mid \mathbf{b}]$. Then $A\mathbf{x} = \mathbf{b}$ has

a. No solution if $p < q$,

b. A unique solution if $p = q = n$,

c. Infinitely many solutions if $p = q$ and $p < n$.

Proof

a. If p is less than q, then there must be one row in the row-reduced echelon form of $[A \mid \mathbf{b}]$ that reads

$$[0 \quad 0 \quad \cdots \quad 0 \mid 1].$$

This corresponds to the equation $0 = 1$, and hence there is no solution of the system.

b. If p, q, and n are equal, then after discarding the zero rows of the row-reduced echelon form, we obtain the equivalent linear system

$$\begin{array}{cccc} x_1 & & & = d_1 \\ & x_2 & & = d_2 \\ & & \ddots & \vdots \\ & & x_n & = d_n, \end{array}$$

which has the unique solution $(d_1, d_2, \ldots, d_n)$.

c. If $p = q$ and $p < n$, then, after discarding the zero rows of the row-reduced echelon form and assuming that the first r columns of A are the ones transformed into those with leading ones, we obtain the equivalent linear system (with $r = p = q$):

$$\begin{array}{ccccccc} x_1 & & & + c_{1,r+1}x_{r+1} & + \cdots + & c_{1n}x_n & = d_1 \\ & x_2 & & + c_{2,r+1}x_{r+1} & + \cdots + & c_{2n}x_n & = d_2 \\ & & \ddots & \vdots & & \vdots & \vdots \\ & & x_r & + c_{r,r+1}x_{r+1} & + \cdots + & c_{rn}x_n & = d_r \end{array}$$

To see that this system has infinitely many solutions, take, for example, $x_{r+1} = t$, $x_{r+2} = \cdot \quad \cdot \quad \cdot = x_n = 0$ to get the solutions $(d_1 - c_{1,r+1}t, \ldots, d_r - c_{r,r+1}t, t, 0, 0, \ldots, 0)$. ■

EXAMPLE 2

Let's revisit examples 4, 8, and 10 from Section 2.3. Example 4 is the linear system

$$\begin{array}{rrrr} & -y & +z = & 3 \\ x & -y & -z = & 0 \\ -x & & -z = & -3, \end{array}$$

which can be written as $A\mathbf{x} = \mathbf{b}$, where $A = \begin{bmatrix} 0 & -1 & 1 \\ 1 & -1 & -1 \\ -1 & 0 & -1 \end{bmatrix}$. The row-reduced echelon form of the augmented matrix was found to be

$$\left[\begin{array}{ccc|r} 1 & 0 & 0 & 1 \\ 0 & 1 & 0 & -1 \\ 0 & 0 & 1 & 2 \end{array}\right].$$

It follows that the ranks of A, a 3×3 matrix, and the augmented matrix $[A \mid \mathbf{b}]$ are both 3, so by Theorem 3 (b) there is a unique solution (which was found to be $x = 1, y = -1, z = 2$).

In Example 8, the linear system

$$\begin{aligned} x_3 + 2x_4 &= 3 \\ 2x_1 + 4x_2 - 2x_3 \qquad &= 4 \\ 2x_1 + 4x_2 - x_3 + 2x_4 &= 7 \end{aligned}$$

can be written as $[A\mathbf{x} = \mathbf{b}]$, where $A = \begin{bmatrix} 0 & 0 & 1 & 2 \\ 2 & 4 & -2 & 0 \\ 2 & 4 & -1 & 2 \end{bmatrix}$. The row-reduced echelon form of the augmented matrix was found to be

$$\left[\begin{array}{cccc|c} 1 & 2 & 0 & 2 & 5 \\ 0 & 0 & 1 & 2 & 3 \\ 0 & 0 & 0 & 0 & 0 \end{array}\right],$$

so the ranks of A, a 3×4 matrix, and the augmented matrix $[A \mid \mathbf{b}]$ are both 2 (which is less than the number of rows of A), so by Theorem 3 (c) there are infinitely many solutions, agreeing with the solution back in Section 2.3

Finally, Example 10 considered the linear system

$$\begin{aligned} x_1 \qquad + x_3 &= 1 \\ x_2 - x_3 &= -1 \\ 2x_1 + x_2 + x_3 &= 2, \end{aligned}$$

which can be written as $A\mathbf{x} = \mathbf{b}$, where $A = \begin{bmatrix} 1 & 0 & 1 \\ 0 & 1 & -1 \\ 2 & 1 & 1 \end{bmatrix}$. The row-reduced echelon form of the augmented matrix was found to be

$$\left[\begin{array}{ccc|c} 1 & 0 & 1 & 1 \\ 0 & 1 & -1 & -1 \\ 0 & 0 & 0 & 1 \end{array}\right].$$

It follows that the rank of A, a 3×3 matrix, is 2, while the rank of the augmented matrix $[A \mid \mathbf{b}]$ is 3, so by Theorem 3 (a) there are no solutions. ■

For the homogeneous linear system $A\mathbf{x} = \mathbf{0}$, the augmented matrix is $[A \mid \mathbf{0}]$. Since the zeros in the last column of this matrix remain unchanged when it is transformed into row-reduced echelon form, the rank of A is always equal to the rank of $[A \mid \mathbf{0}]$. Thus, the following corollaries are immediate consequences of Theorem 3. We leave their proofs for exercises.

COROLLARY 1

The homogeneous linear system $A\mathbf{x} = \mathbf{0}$, where A is an $m \times n$ matrix, has

a. A unique solution (the trivial one) if the rank of A is equal to n;

b. Infinitely many solutions if the rank of A is less than n. ■

COROLLARY 2

A homogeneous linear system with more unknowns than equations has infinitely many solutions. ■

EXAMPLE 3

By inspection of the following systems, what can we say concerning the number of solutions they may have?

a.
$$\begin{aligned} x_1 + 2x_2 - x_3 &= 0 \\ 2x_1 - x_2 + 3x_3 &= 0 \end{aligned}$$

b.
$$\begin{aligned} 2x_1 + x_2 + x_3 &= 1 \\ -x_1 - x_2 + 2x_3 &= 0 \end{aligned}$$

c.
$$\begin{aligned} x_1 \qquad\qquad + 2x_4 &= 1 \\ x_2 \qquad - x_4 &= -1 \\ x_3 + x_4 &= 0 \end{aligned}$$

Solution System (a) is a homogeneous one with more unknowns than equations. By Corollary 2 of Theorem 3, it has infinitely many solutions. System (b) is a nonhomogeneous one with more unknowns than equations. By Theorem 3, it has either no solution or infinitely many solutions. System (c) is in row-reduced echelon form. The corresponding augmented matrix is

$$\left[\begin{array}{cccc|c} 1 & 0 & 0 & 2 & 1 \\ 0 & 1 & 0 & -1 & -1 \\ 0 & 0 & 1 & 1 & 0 \end{array}\right].$$

We see that the ranks of the coefficient and augmented matrices are both 3, whereas the number of variables n is 4. Hence, by Theorem 3, there are infinitely many solutions. ■

In Example 3, we can solve the system of part (c) to obtain the one parameter family of solutions $(1 - 2t, -1 + t, -t, t)$. The next theorem tells us that we could have deduced the number of parameters in the solution from the ranks of the coefficient and augmented matrices.

THEOREM 4

Consider the system $A\mathbf{x} = \mathbf{b}$, where A is an $m \times n$ matrix with rank equal to $r < n$. If the augmented matrix $[A \mid \mathbf{b}]$ also has rank $= r$, then the system has an $(n - r)$-parameter family of solutions.

Proof From Theorem 3, we know that $A\mathbf{x} = \mathbf{b}$ has infinitely many solutions. By discarding the zero rows of the row-reduced echelon form of $[A \mid \mathbf{b}]$ and assuming that the first r columns of A are the ones transformed into those that contain the leading ones, we obtain the equivalent system

$$\begin{array}{llll} x_1 & + c_{1,r+1}x_{r+1} & + \cdots + c_{1n}x_n & = d_1 \\ \quad x_2 & + c_{2,r+1}x_{r+1} & + \cdots + c_{2n}x_n & = d_2 \\ \qquad \ddots & \quad \vdots & \qquad \vdots & \quad \vdots \\ \qquad\quad x_r & + c_{r,r+1}x_{r+1} & + \cdots + c_{rn}x_n & = d_r. \end{array}$$

Since the variables $x_{r+1}, \ldots, x_n$ can be taken as parameters, we obtain an $(n - r)$-parameter family of solutions. ■

COMPUTATIONAL NOTE

You should be wary of using a computer or calculator to determine the rank of a matrix—you may get an erroneous answer. Because a computing device has to round off nonterminating decimals (such as the fraction $\frac{1}{3}$), small errors may be introduced in transforming a given matrix into row-reduced echelon form. These small errors could possibly result in nonzero entries occurring in the latter, where in fact there should be zeros, or vice-versa. It is therefore possible for the computed rank to be either smaller or larger than it actually is.

We now restrict ourselves to linear systems with the same number of equations as unknowns. Since in such cases the coefficient matrix is square, we call these **square linear systems**.

THEOREM 5

If A is an $n \times n$ matrix, the following conditions are equivalent:

a. A is invertible.

b. $A\mathbf{x} = \mathbf{b}$ has a unique solution for any $\mathbf{b}$.

c. $A\mathbf{x} = \mathbf{0}$ has only the trivial solution.

d. The rank of A is n.

e. The row-reduced echelon form of A is I, the identity matrix.

Proof We show the equivalence of these statements by showing that (a) implies (b), (b) implies (c), (c) implies (d), (d) implies (e), and (e) implies (a). Thus, any two statements imply each other.

***(a)* implies (b)**: We first note that the vector $A^{-1}\mathbf{b}$ is a solution, since $A(A^{-1}\mathbf{b}) = (AA^{-1})\mathbf{b} = I\mathbf{b} = \mathbf{b}$, as desired. Now suppose that $\mathbf{y}$ is also a solution. Then $A\mathbf{y} = \mathbf{b}$, so $A^{-1}(A\mathbf{y}) = A^{-1}\mathbf{b}$, or $(A^{-1}A)\mathbf{y} = A^{-1}\mathbf{b}$; but this yields $\mathbf{y} = A^{-1}\mathbf{b}$ as well. Thus, the solution is unique.

***(b)* implies (c)**: Since $A\mathbf{x} = \mathbf{0}$ is a special case of $A\mathbf{x} = \mathbf{b}$, the former has a unique solution. Since $\mathbf{0}$ is a solution, it is the only solution.

***(c)* implies (d)**: This is Corollary 1 of Theorem 3.

***(d)* implies (e)**: Since A has n rows and its rank is n, its row-reduced echelon form has n nonzero rows, each with a leading one and no zero rows. Since the matrix has n columns as well, however, the only nonzero element in column j is a 1 in the jth row. Thus, the row-reduced echelon form is the identity matrix.

***(e)* implies (a)**: This follows immediately from the procedure for matrix inverses presented in Section 3.2. ■

The next theorem gives a relationship between the solution of a homogeneous linear system, $A\mathbf{x} = \mathbf{0}$, and that of the corresponding nonhomogeneous system, $A\mathbf{x} = \mathbf{b}$. This result is of fundamental importance in the study of linear differential equations (see Section 7.4) and also appears in a variety of other applications.

THEOREM 6

Let A be an $m \times n$ matrix and $\mathbf{b}$ be an m-vector. Suppose that $\mathbf{x}_0$ is one *particular* solution of $A\mathbf{x} = \mathbf{b}$ and S is the set of *all* solutions to $A\mathbf{x} = \mathbf{0}$. Then all solutions of $A\mathbf{x} = \mathbf{b}$ have the form $\mathbf{x} = \mathbf{x}_0 + \mathbf{s}$, where $\mathbf{s}$ is in the set S.

Proof First we show that if a vector is of the form $\mathbf{x}_0 + \mathbf{s}$, then it must be a solution of $A\mathbf{x} = \mathbf{b}$. We have

$$A(\mathbf{x}_0 + \mathbf{s}) = A\mathbf{x}_0 + A\mathbf{s} = \mathbf{b} + \mathbf{0} = \mathbf{b},$$

so $\mathbf{x}_0 + \mathbf{s}$ satisfies the equation.

Conversely, if some vector (say, $\mathbf{y}$) is a solution of $A\mathbf{x} = \mathbf{b}$, then it must be equal to $\mathbf{x}_0 + \mathbf{s}$, for some $\mathbf{s}$ in S. To see this, for the given $\mathbf{y}$, define $\mathbf{s} = \mathbf{y} - \mathbf{x}_0$. Then $\mathbf{y} = \mathbf{x}_0 + \mathbf{s}$ and $\mathbf{s}$ is in S, as desired, because

$$A\mathbf{s} = A(\mathbf{y} - \mathbf{x}_0) = A\mathbf{y} - A\mathbf{x}_0 = \mathbf{b} - \mathbf{b} = \mathbf{0}.$$ ■

EXAMPLE 4

Consider the linear system of Example 3, part (c):

$$\begin{aligned} x_1 \qquad\qquad + 2x_4 &= 1 \\ x_2 \qquad - x_4 &= -1 \\ x_3 + x_4 &= 0 \end{aligned}$$

You can easily verify that $\mathbf{x}_0 = (1, -1, 0, 0)$ is a particular solution of this system. Moreover, the corresponding homogeneous system

$$\begin{aligned} x_1 \qquad\qquad + 2x_4 &= 0 \\ x_2 \qquad - x_4 &= 0 \\ x_3 + x_4 &= 0 \end{aligned}$$

has $(-2t, t, -t, t)$ as a one-parameter family of solutions. Thus, by Theorem 6, the general solution to the nonhomogeneous system is the set of all vectors of the form

$$(1, -1, 0, 0) + (-2t, t, -t, t) = (1 - 2t, -1 + t, -t, t).$$ ■

3.3 Exercises

In Exercises 1 and 2 determine if the system is homogeneous.

1. $$\begin{aligned} x_1 &= 2x_2 + x_3 \\ x_2 &= x_3 - x_4 \\ x_3 &= x_1 - 4x_4 \end{aligned}$$

2. $$\begin{aligned} x + y + 3z &= 0 \\ x - y + 2 &= 0 \\ z - 3 &= 0 \end{aligned}$$

In Exercises 3–6 find the rank of the given matrix.

3. $\begin{bmatrix} 1 & 2 & -1 \\ 3 & -6 & 2 \end{bmatrix}$

4. $\begin{bmatrix} 1 & 0 & -1 \\ 3 & 1 & 1 \\ -1 & -1 & -3 \end{bmatrix}$

5. $\begin{bmatrix} 1 & 0 & 1 \\ -2 & 1 & 1 \\ 1 & 1 & 2 \end{bmatrix}$

6. $\begin{bmatrix} 2 & 4 & 0 & -2 \\ 0 & 1 & -1 & 0 \\ 1 & 2 & -1 & -1 \end{bmatrix}$

In Exercises 7 and 8 consider the linear system $A\mathbf{x} = \mathbf{b}$, where

$$A = \begin{bmatrix} a & b \\ c & d \end{bmatrix}, \quad \mathbf{b} = \begin{bmatrix} e \\ f \end{bmatrix}, \quad \text{and} \quad A \neq 0.$$

7. Determine conditions on the constants a, b, c, d, e, and f so that

(a) The rank of A is 2,
(b) The rank of A is 1, but the rank of $[A \mid \mathbf{b}]$ is 2,
(c) The rank of A and the rank of $[A \mid \mathbf{b}]$ are both 1.

8. Using the results of Exercise 7, give conditions on a, b, c, d, e, and f so that the linear system has

(a) No solution,
(b) One solution,
(c) Infinitely many solutions.

In Exercises 9–14 determine whether the given statement is true or false.

9. A linear system with more unknowns than equations always has infinitely many solutions.

10. A homogeneous linear system with the same number of equations as unknowns always has a unique solution.

11. If a linear system has at least two solutions, it must have an infinite number of solutions.

12. If a linear system has no solution, the rank of the coefficient matrix must be less than the number of equations.

13. A homogeneous linear system with two equations and four unknowns must have a two-parameter family of solutions.

14. If a linear system has the same number of equations as unknowns and the coefficient matrix is invertible, the system has exactly one solution.

In Exercises 15–18 use the theorems and corollaries of Section 3.3 to say as much as possible concerning the number of solutions of the given linear system without actually solving it.

15.
$$\begin{aligned} 2x + 3y - z &= 0 \\ y + 2z - 3w &= 0 \\ 2x - 4z + w &= 0 \end{aligned}$$

16.
$$\begin{aligned} 2x + 3y - z &= 1 \\ y + 2z - 3w &= 0 \\ 2x - 4z + w &= 0 \end{aligned}$$

17.
$$\begin{aligned} 2x_1 + x_2 - x_3 &= 0 \\ 3x_1 - 2x_2 + x_3 &= 0 \\ x_1 - x_2 + x_3 &= 0 \end{aligned}$$

18.
$$\begin{aligned} 2x - 3y &= 2 \\ -x + 4y &= 3 \\ 4x + 3y &= 7 \end{aligned}$$

19. Let A, X, and B be $n \times n$ matrices with A nonsingular. If $AX = B$, show that $(AXA^{-1})A = B$.

20. Let A and B be invertible matrices with $AB = I$. Show that $\left(A^T\right)^{-1} = B^T$.

21. Suppose that the augmented matrix of a linear system is given by

$$\left[\begin{array}{ccc|c} 1 & 0 & 0 & 2 \\ 0 & 1 & 0 & 3 \\ 0 & 0 & x & y \end{array}\right].$$

For what values of x and y is there

(a) No solution?

(b) Exactly one solution?

(c) Infinitely many solutions?

Computational Exercises

22–25. Find the rank of the matrices in Exercises 3–6 by using a graphing calculator or computer software to transform each matrix to row-reduced echelon form. Do your results agree with the ones obtained by hand?

26. Find the rank of the matrix

$$A = \begin{bmatrix} 1 & \frac{1}{2} & \frac{1}{3} & \frac{1}{4} & \frac{1}{5} \\ \frac{1}{2} & \frac{1}{3} & \frac{1}{4} & \frac{1}{5} & \frac{1}{6} \\ \frac{1}{3} & \frac{1}{4} & \frac{1}{5} & \frac{1}{6} & \frac{1}{7} \\ \frac{1}{4} & \frac{1}{5} & \frac{1}{6} & \frac{1}{7} & \frac{1}{8} \\ \frac{1}{5} & \frac{1}{6} & \frac{1}{7} & \frac{1}{8} & \frac{1}{9} \end{bmatrix}$$

using a graphing calculator or computer software to transform A to row-reduced echelon form. (The rank of A is 5, but your answer may differ; see the Computational Note following Theorem 4.)

3.4 *LU* Decomposition

In Section 2.3, we discussed two procedures for solving systems of linear equations: Gauss–Jordan and Gaussian elimination. We noted that the second of these is more efficient than the first but left much unsaid for lack of knowledge of matrices. This section deals with an approach to solving systems of equations that will be seen to be just Gaussian elimination again but with a more powerful perspective. It is used in some computer software solutions to systems of linear equations and it is also an alternative way of looking at theoretical considerations about such systems.

Basics of *LU* Decomposition

The idea of ***LU* decomposition** is to replace the matrix A in a system of linear equations, $A\mathbf{x} = \mathbf{b}$, by the product of two matrices, $A = LU$; that is, to factor the matrix in a way that is somehow helpful. Here L stands for *lower* and U for *upper.* Matrix L is square, invertible, and its main diagonal has a 1 in every position (just like the identity matrix) and all of its other nonzero entries are below that diagonal. Matrix U is in echelon form as described in Section 2.3; that is, zeros below the leading nonzero entry of any row and "stair-stepping" down from left to right.

EXAMPLE 1

Show that L and U are an LU decomposition of A, where

$$A = \begin{bmatrix} 1 & 1 & 1 \\ 2 & 3 & 1 \\ 1 & 2 & 0 \end{bmatrix}, \quad L = \begin{bmatrix} 1 & 0 & 0 \\ 2 & 1 & 0 \\ 1 & 1 & 1 \end{bmatrix}, \quad \text{and} \quad U = \begin{bmatrix} 1 & 1 & 1 \\ 0 & 1 & -1 \\ 0 & 0 & 0 \end{bmatrix}.$$

Solution The matrix L is a lower matrix of the type described above, and U is in upper echelon form as required. Simple matrix multiplication confirms that $A = LU$. How to *find* L and U is addressed in Example 3. ■

The next example illustrates how we can use LU decomposition to solve a system of linear equations.

EXAMPLE 2

Solve the linear system of equations using LU decomposition:

$$\begin{aligned} x_1 + x_2 + x_3 &= 3 \\ 2x_1 + 3x_2 + x_3 &= 7 \\ x_1 + 2x_2 \qquad &= 4 \end{aligned}$$

Solution Notice that the coefficient matrix of this system is just the matrix A of Example 1. Therefore, the LU decomposition for A has already been provided. Let $\mathbf{y} = U\mathbf{x}$. Then solving $A\mathbf{x} = \mathbf{b}$ is equivalent to solving

$$A\mathbf{x} = (LU)\mathbf{x} = L(U\mathbf{x}) = L\mathbf{y} = \mathbf{b}.$$

Thus, we can first solve

$$L\mathbf{y} = \mathbf{b}$$

for the unknown column vector $\mathbf{y}$ and, since $\mathbf{y}$ is then known, next solve

$$U\mathbf{x} = \mathbf{y}$$

for the unknown column vector $\mathbf{x}$. This vector is the desired solution.

We solve $L\mathbf{y} = \mathbf{b}$ by writing it as the linear system of equations to which it corresponds,

$$\begin{aligned} y_1 \qquad\qquad &= 3 \\ 2y_1 + y_2 \qquad &= 7 \\ y_1 + y_2 + y_3 &= 4, \end{aligned}$$

and solving by *forward substitution*; that is, starting at the top and working our way down. This yields $y_1 = 3$, $y_2 = 1$, and $y_3 = 0$, or written as a column vector,

$$\mathbf{y} = \begin{bmatrix} 3 \\ 1 \\ 0 \end{bmatrix}$$

Now that $\mathbf{y}$ is known, we can solve $U\mathbf{x} = \mathbf{y}$ for $\mathbf{x}$, which is the system:

$$\begin{aligned} x_1 + x_2 + x_3 &= 3 \\ x_2 - x_3 &= 1 \end{aligned}$$

Letting $x_3 = t$, since x_3 doesn't correspond to a column containing a leading entry of some row, we solve for the other variables by backward substitution as in the Gaussian elimination procedure: $x_2 = 1 + t$ and $x_1 = 2 - 2t$. Thus, the desired solution is the vector $\mathbf{x} = (2 - 2t, 1 + t, t)$. ■

If another system of linear equations is posed that has the same coefficient matrix A but different right-side coefficient vector $\mathbf{b}$, the same factorization LU then leads to an easy solution procedure: a forward substitution followed by a backward substitution. Since most of the work in the method lies in finding the LU factorization, which has already been done, the new solution can be found very quickly. Although it might seem that having the same matrix of coefficients is never going to happen, in fact the situation is not all that rare. In many practical situations, the constraints or conditions that lead to the original system of linear equations are fixed, and the right-side coefficients change in later applications. Thus, having an LU factorization can be an important aid to computational speed and efficiency.

COMPUTATIONAL NOTE

It is somewhat surprising that an LU decomposition can be obtained with fewer computations than are needed to obtain *row-reduced* echelon form. In the exercises of the next section, we explore the actual number of computations needed, an important consideration for comparing different algorithms. Partly because of this efficiency, many linear algebra software packages, such as MAX and MATLAB, use LU decomposition in their algorithms.

Finding an *LU* Decomposition

Next we see how to find an LU decomposition for a given matrix. The method is really just that of Gaussian elimination (presented at the end of Section 2.3) in which we "remember" certain other information along the way. There is one consideration that we have been avoiding, however. It may be that a matrix *does not have* an LU decomposition unless some rows are interchanged (see Exercise 19). These interchanges occur naturally in the Gaussian elimination procedure but must be recorded somehow since the right-side coefficients of the linear system are not being "carried along" as they were in the Gaussian elimination procedure itself. These row operations are recorded by means of two matrices: a square matrix of zeros that is modified with each row reduction needed to bring the matrix to echelon form and a *permutation matrix*, an identity matrix that is modified to keep track of each row interchange that we make.

Procedure for LU Decomposition of a Matrix

Let A be an $m \times n$ matrix, O be an $m \times m$ matrix of zeros, and I be an $m \times m$ identity matrix. Row-reduce A to upper echelon form working from left to right.

i. Whenever two rows of A are interchanged, interchange the corresponding rows of the other two matrices, O and I, as well.

ii. Whenever a *pivot* (leading entry of a row of A) is used to place a zero in the same column below it, put the negative of the row multiplier in the corresponding position of the lower row of O.

iii. When A has been transformed into upper echelon form, call the result U and the transformed identity matrix P. Place a 1 in each position of the main diagonal of the transformed zero matrix and call the resulting matrix L. Then $LU = PA$.

Here P is just the matrix obtained from the identity matrix by interchanging some of its rows, those that were interchanged in A. Such a matrix is called a **permutation matrix** and contains all zeros except for exactly one 1 in each row and each column. It is customary to speak of an LU decomposition of the original matrix even though technically it may be the matrix PA that is actually factored. Sometimes permutations are unnecessary, but some simple examples show that they may be unavoidable (see Exercise 19). As the next example illustrates, this is not much of a problem.

EXAMPLE 3

Find an LU decomposition of A, where

$$A = \begin{bmatrix} 2 & 1 & 3 & 2 \\ 2 & 1 & -1 & 1 \\ 0 & 3 & -2 & 1 \\ 4 & 1 & 2 & 2 \end{bmatrix}.$$

Solution Following the procedure, we consider the three matrices A, O, and I:

$$A = \begin{bmatrix} 2 & 1 & 3 & 2 \\ 2 & 1 & -1 & 1 \\ 0 & 3 & -2 & 1 \\ 4 & 1 & 2 & 2 \end{bmatrix} \quad O = \begin{bmatrix} 0 & 0 & 0 & 0 \\ 0 & 0 & 0 & 0 \\ 0 & 0 & 0 & 0 \\ 0 & 0 & 0 & 0 \end{bmatrix} \quad I = \begin{bmatrix} 1 & 0 & 0 & 0 \\ 0 & 1 & 0 & 0 \\ 0 & 0 & 1 & 0 \\ 0 & 0 & 0 & 1 \end{bmatrix}$$

We perform Gaussian elimination on A. First, we multiply the first row by -1 and add it to the second row. We record this multiplier by placing its negative, $1 = -(-1)$, in position 2,1 of the second matrix:

$$A \rightarrow \begin{bmatrix} 2 & 1 & 3 & 2 \\ 0 & 0 & -4 & -1 \\ 0 & 3 & -2 & 1 \\ 4 & 1 & 2 & 2 \end{bmatrix} \quad O \rightarrow \begin{bmatrix} 0 & 0 & 0 & 0 \\ 1 & 0 & 0 & 0 \\ 0 & 0 & 0 & 0 \\ 0 & 0 & 0 & 0 \end{bmatrix} \quad I \rightarrow \begin{bmatrix} 1 & 0 & 0 & 0 \\ 0 & 1 & 0 & 0 \\ 0 & 0 & 1 & 0 \\ 0 & 0 & 0 & 1 \end{bmatrix}$$

Position 3,1 is already 0 so we move down to the fourth row. Multiply the first row by -2, and add it to the fourth row. At the same time, we place $2 = -(-2)$ in position 4,1 of the second matrix:

$$A \rightarrow \begin{bmatrix} 2 & 1 & 3 & 2 \\ 0 & 0 & -4 & -1 \\ 0 & 3 & -2 & 1 \\ 0 & -1 & -4 & -2 \end{bmatrix} \quad O \rightarrow \begin{bmatrix} 0 & 0 & 0 & 0 \\ 1 & 0 & 0 & 0 \\ 0 & 0 & 0 & 0 \\ 2 & 0 & 0 & 0 \end{bmatrix} \quad I \rightarrow \begin{bmatrix} 1 & 0 & 0 & 0 \\ 0 & 1 & 0 & 0 \\ 0 & 0 & 1 & 0 \\ 0 & 0 & 0 & 1 \end{bmatrix}$$

At this point, the 0 in position 2,2 makes a row interchange inevitable. We can interchange rows 2 and 3 or rows 2 and 4. The choice is arbitrary, although the final decomposition will reflect the decision made. A common choice in computer programs is the nonzero entry of *largest* absolute value, so-called "partial pivoting." In that case, the 3 of row 3 becomes the new leading entry. Working by hand, it is easier to avoid fractions as much as possible and, in that case, the -1 of row 4 is preferable. Using the latter strategy, we interchange rows 2 and 4 of all three matrices:

$$A \rightarrow \begin{bmatrix} 2 & 1 & 3 & 2 \\ 0 & -1 & -4 & -2 \\ 0 & 3 & -2 & 1 \\ 0 & 0 & -4 & -1 \end{bmatrix} \quad O \rightarrow \begin{bmatrix} 0 & 0 & 0 & 0 \\ 2 & 0 & 0 & 0 \\ 0 & 0 & 0 & 0 \\ 1 & 0 & 0 & 0 \end{bmatrix} \quad I \rightarrow \begin{bmatrix} 1 & 0 & 0 & 0 \\ 0 & 0 & 0 & 1 \\ 0 & 0 & 1 & 0 \\ 0 & 1 & 0 & 0 \end{bmatrix}$$

We now use the -1 in position 2,2 to eliminate the 3 in position 3,2. That is, we multiply row 2 by 3 and add it to row 3. Meanwhile, we record this by placing -3 in position 3,2 of the second matrix.

$$A \rightarrow \begin{bmatrix} 2 & 1 & 3 & 2 \\ 0 & -1 & -4 & -2 \\ 0 & 0 & -14 & -5 \\ 0 & 0 & -4 & -1 \end{bmatrix} \quad O \rightarrow \begin{bmatrix} 0 & 0 & 0 & 0 \\ 2 & 0 & 0 & 0 \\ 0 & -3 & 0 & 0 \\ 1 & 0 & 0 & 0 \end{bmatrix} \quad I \rightarrow \begin{bmatrix} 1 & 0 & 0 & 0 \\ 0 & 0 & 0 & 1 \\ 0 & 0 & 1 & 0 \\ 0 & 1 & 0 & 0 \end{bmatrix}$$

Finally we use the -14 in position 3,3 to eliminate the -4 in position 4,3. That is, multiply row 3 by $-(-4/-14) = -(2/7)$ and add it to row 4 while placing $-(-2/7) = 2/7$ in position 4,3 of the second matrix.

$$A \rightarrow \begin{bmatrix} 2 & 1 & 3 & 2 \\ 0 & -1 & -4 & -2 \\ 0 & 0 & -14 & -5 \\ 0 & 0 & 0 & \frac{3}{7} \end{bmatrix} \quad O \rightarrow \begin{bmatrix} 0 & 0 & 0 & 0 \\ 2 & 0 & 0 & 0 \\ 0 & -3 & 0 & 0 \\ 1 & 0 & \frac{2}{7} & 0 \end{bmatrix} \quad I \rightarrow \begin{bmatrix} 1 & 0 & 0 & 0 \\ 0 & 0 & 0 & 1 \\ 0 & 0 & 1 & 0 \\ 0 & 1 & 0 & 0 \end{bmatrix}$$

The matrix on the left is our U. Placing a 1 in each position on the main diagonal of the second, we have our matrix L. The third matrix is our permutation matrix P. Performing the multiplications of these matrices, we confirm that $LU = PA$. Notice that PA is just the matrix A with some of its rows interchanged:

$$LU = \begin{bmatrix} 1 & 0 & 0 & 0 \\ 2 & 1 & 0 & 0 \\ 0 & -3 & 1 & 0 \\ 1 & 0 & \frac{2}{7} & 1 \end{bmatrix} \begin{bmatrix} 2 & 1 & 3 & 2 \\ 0 & -1 & -4 & -2 \\ 0 & 0 & -14 & -5 \\ 0 & 0 & 0 & \frac{3}{7} \end{bmatrix}$$

$$= \begin{bmatrix} 1 & 0 & 0 & 0 \\ 0 & 0 & 0 & 1 \\ 0 & 0 & 1 & 0 \\ 0 & 1 & 0 & 0 \end{bmatrix} \begin{bmatrix} 2 & 1 & 3 & 2 \\ 2 & 1 & -1 & 1 \\ 0 & 3 & -2 & 1 \\ 4 & 1 & 2 & 2 \end{bmatrix} = \begin{bmatrix} 2 & 1 & 3 & 2 \\ 4 & 1 & 2 & 2 \\ 0 & 3 & -2 & 1 \\ 2 & 1 & -1 & 1 \end{bmatrix}$$ ■

We conclude this section with an example to show how to incorporate the permutation matrix into the LU decomposition solution process for a system of linear equations. The permutation matrix P just tells us how to reorder the right-side constraints; we then proceed exactly as in the case in which no row interchanges were necessary.

EXAMPLE 4

Solve the following linear system using LU decomposition.

$$\begin{aligned} 2x_1 + x_2 + 3x_3 + 2x_4 &= 1 \\ 2x_1 + x_2 - x_3 + x_4 &= 2 \\ 3x_2 - 2x_3 + x_4 &= 3 \\ 4x_1 + x_2 + 2x_3 + 2x_4 &= 4 \end{aligned}$$

Solution As usual, we use A to represent the coefficient matrix and $\mathbf{b}$ to represent the right-side column vector. Then the linear system has the matrix equation form

$$A\mathbf{x} = \mathbf{b}.$$

Since A is just the matrix of Example 3, we have already constructed an LU decomposition. That is, with L, U, and P from the solution of Example 3, we have $LU = PA$. Multiplying both sides of the matrix equation on the left by P yields

$$PA\mathbf{x} = P\mathbf{b}$$

and, since matrix multiplication is associative,

$$(PA)\mathbf{x} = P\mathbf{b},$$

where the right side simply amounts to a rearrangement of the entries of $\mathbf{b}$. We solve this system in exactly the same way as in Example 2. That is, let $\mathbf{y} = U\mathbf{x}$, and solve

$$(PA)\mathbf{x} = (LU)\mathbf{x} = L(U\mathbf{x}) = L\mathbf{y} = P\mathbf{b}.$$

First we solve the lower system $L\mathbf{y} = P\mathbf{b}$ by forward substitution. To be more specific, we write the matrix equation $L\mathbf{y} = P\mathbf{b}$ as a system of linear equations:

$$\begin{aligned} y_1 &= 1 \\ 2y_1 + y_2 &= 4 \\ -3y_2 + y_3 &= 3 \\ y_1 + \left(\frac{2}{7}\right)y_3 + y_4 &= 2 \end{aligned}$$

Solving by forward substitution, we have the solution $\mathbf{y} = (1, 2, 9, -\frac{11}{7})$. This is then the right-side column for the system $U\mathbf{x} = \mathbf{y}$:

$$\begin{aligned} 2x_1 + x_2 + 3x_3 + 2x_4 &= 1 \\ - x_2 - 4x_3 - 2x_4 &= 2 \\ - 14x_3 - 5x_4 &= 9 \\ \left(\frac{3}{7}\right)x_4 &= -\frac{11}{7} \end{aligned}$$

Solving by backward substitution, we have the desired solution to the original system: $x_4 = -\frac{11}{3}$, $x_3 = \frac{2}{3}$, $x_2 = \frac{8}{3}$, and $x_1 = \frac{11}{6}$. ■

3.4 Exercises

In Exercises 1 and 2 show that the matrices *L* and *U* yield an *LU* decomposition of *A*. That is, explain why *L* and *U* have the correct form, and verify that *A* = *LU*.

1. $A = \begin{bmatrix} 2 & -1 \\ -4 & 5 \end{bmatrix} \quad L = \begin{bmatrix} 1 & 0 \\ -2 & 1 \end{bmatrix} \quad U = \begin{bmatrix} 2 & -1 \\ 0 & 3 \end{bmatrix}$

2. $A = \begin{bmatrix} 1 & 2 & 1 \\ 2 & 1 & -1 \\ 2 & 1 & 1 \end{bmatrix} \quad L = \begin{bmatrix} 1 & 0 & 0 \\ 2 & 1 & 0 \\ 2 & 1 & 1 \end{bmatrix} \quad U = \begin{bmatrix} 1 & 2 & 1 \\ 0 & -3 & -3 \\ 0 & 0 & 2 \end{bmatrix}$

In Exercises 3–5 show that the matrices *L*, *U*, and *P* yield an *LU* decomposition of the matrix *A*. That is, explain why *L*, *U*, and *P* have the correct form, and verify that *PA* = *LU*.

3. $A = \begin{bmatrix} 2 & 1 & 1 \\ 2 & 1 & -1 \\ 4 & 3 & 2 \end{bmatrix} \quad L = \begin{bmatrix} 1 & 0 & 0 \\ 2 & 1 & 0 \\ 1 & 0 & 1 \end{bmatrix}$

$U = \begin{bmatrix} 2 & 1 & 1 \\ 0 & 1 & 0 \\ 0 & 0 & -2 \end{bmatrix} \quad P = \begin{bmatrix} 1 & 0 & 0 \\ 0 & 0 & 1 \\ 0 & 1 & 0 \end{bmatrix}$

4. $A = \begin{bmatrix} 1 & 2 & 1 & 2 \\ 1 & 2 & 3 & 3 \\ 2 & 1 & -1 & 1 \\ 3 & 0 & -1 & 2 \end{bmatrix} \quad L = \begin{bmatrix} 1 & 0 & 0 & 0 \\ 2 & 1 & 0 & 0 \\ 3 & 2 & 1 & 0 \\ 1 & 0 & 1 & 1 \end{bmatrix}$

$U = \begin{bmatrix} 1 & 2 & 1 & 2 \\ 0 & -3 & -3 & -3 \\ 0 & 0 & 2 & 2 \\ 0 & 0 & 0 & -1 \end{bmatrix} \quad P = \begin{bmatrix} 1 & 0 & 0 & 0 \\ 0 & 0 & 1 & 0 \\ 0 & 0 & 0 & 1 \\ 0 & 1 & 0 & 0 \end{bmatrix}$

5. $A = \begin{bmatrix} 1 & 2 & 1 & 2 \\ 3 & 5 & 1 & 2 \\ 2 & 1 & -1 & 1 \\ 2 & 1 & 2 & -2 \end{bmatrix} \quad L = \begin{bmatrix} 1 & 0 & 0 & 0 \\ 3 & 1 & 0 & 0 \\ 2 & 3 & 1 & 0 \\ 2 & 3 & 2 & 1 \end{bmatrix}$

$U = \begin{bmatrix} 1 & 2 & 1 & 2 \\ 0 & -1 & -2 & -4 \\ 0 & 0 & 3 & 9 \\ 0 & 0 & 0 & -12 \end{bmatrix} \quad P = \begin{bmatrix} 1 & 0 & 0 & 0 \\ 0 & 1 & 0 & 0 \\ 0 & 0 & 1 & 0 \\ 0 & 0 & 0 & 1 \end{bmatrix}$

6. Explain the significance of $P = I$ in Exercise 5 in terms of the Gaussian elimination procedure that created L, U, and P.

In Exercises 7–11 find an *LU* decomposition of the given matrix.

7. $\begin{bmatrix} 2 & -1 \\ 6 & 2 \end{bmatrix}$

8. $\begin{bmatrix} 3 & 1 & 2 \\ 3 & 1 & -2 \\ 6 & 1 & 3 \end{bmatrix}$

9. $\begin{bmatrix} 3 & 1 & 2 \\ 3 & -2 & 1 \\ 6 & 1 & 3 \end{bmatrix}$

10. $\begin{bmatrix} 1 & 2 & -1 & 1 \\ 2 & -1 & 2 & 1 \\ 3 & 2 & 1 & 0 \end{bmatrix}$

11. $\begin{bmatrix} 0 & 1 & -1 & 1 & 1 \\ 1 & -1 & 2 & -1 & 0 \\ 2 & -1 & 1 & 2 & 2 \end{bmatrix}$

In Exercises 12–18 solve each system of linear equations by *LU* decomposition. Note that all the coefficient matrices have been considered in Exercises 1–11.

12. $$\begin{aligned} 2x_1 - x_2 &= 3 \\ -4x_1 + 5x_2 &= -2 \end{aligned}$$

13. $$\begin{aligned} y_1 + 2y_2 + y_3 &= 1 \\ 2y_1 + y_2 - y_3 &= 2 \\ 2y_1 + y_2 + y_3 &= 3 \end{aligned}$$

14. $$\begin{aligned} 2x + y + z &= 3 \\ 2x + y - z &= 2 \\ 4x + 3y + 2z &= 1 \end{aligned}$$

15. $$\begin{aligned} w + 2x + y + 2z &= 2 \\ w + 2x + 3y + 3z &= 1 \\ 2w + x - y + z &= 1 \\ 3w - y + 2z &= 2 \end{aligned}$$

16. $$\begin{aligned} x_1 + 2x_2 + x_3 + 2x_4 &= 3 \\ 3x_1 + 5x_2 + x_3 + 2x_4 &= 2 \\ 2x_1 + x_2 - x_3 + x_4 &= 1 \\ 2x_1 + x_2 + 2x_3 - 2x_4 &= 0 \end{aligned}$$

17. $$\begin{aligned} 3x + y + 2z &= 2 \\ 3x + y - 2z &= 3 \\ 6x + y + 3z &= 4 \end{aligned}$$

18. $$\begin{aligned} x_2 - x_3 + x_4 + x_5 &= 1 \\ x_1 - x_2 + 2x_3 - x_4 &= 2 \\ 2x_1 - x_2 + x_3 + 2x_4 + 2x_5 &= 3 \end{aligned}$$

19. Show that row interchanges may be unavoidable by showing that there is no LU decomposition for the following matrix unless a permutation matrix P is involved.

$$A = \begin{bmatrix} 0 & 0 \\ 1 & 1 \end{bmatrix}$$

(*Hint*: Assign variables to the unknown entries of matrices L and U and set $A = LU$. Then solve for these variables.)

20. Redo Exercise 8 using different row interchanges to show that the LU decomposition of a matrix may not be unique.

21. If no row interchanges are made, the LU decomposition *is* unique (as opposed to the situation in the preceding exercise). Use the approach suggested in Exercise 19 to show that this is the case for matrix A below. If each variable has only one possible value, the decomposition is unique.

$$A = \begin{bmatrix} 1 & 1 & 3 & 4 \\ -1 & 0 & 1 & 3 \\ -2 & 1 & 2 & 4 \end{bmatrix}$$

SPOTLIGHT **22.** *(a)* Draw a directed graph (see the Spotlight at the beginning of this chapter) corresponding to each of the following permutation matrices:

$$M_1 = \begin{bmatrix} 0 & 1 & 0 & 0 \\ 0 & 0 & 0 & 1 \\ 1 & 0 & 0 & 0 \\ 0 & 0 & 1 & 0 \end{bmatrix} \quad \text{and} \quad M_2 = \begin{bmatrix} 0 & 0 & 0 & 1 \\ 0 & 0 & 1 & 0 \\ 0 & 1 & 0 & 0 \\ 1 & 0 & 0 & 0 \end{bmatrix}$$

(b) In a directed graph, the *out*-degree of a given node is defined to be the number of directed edges originating at that node; the *in*-degree of a given node is the number of directed edges terminating at this node. What is the in-degree and out-degree of each node in the graphs of part (a)?

(c) Let P be an arbitrary permutation matrix. Show that both the in-degree and out-degree of every node in the directed graph corresponding to P is equal to one.

Computational Exercises

23. Use a computer software package to find the LU-decomposition of the matrices of
(a) Exercise 11 *(b)* Exercise 21

24. Using computer software, solve the linear systems of the following exercises by performing an LU decomposition of their coefficient matrices and then working with the resulting L and U matrices.
(a) Exercise 16 *(b)* Exercise 18

25. By hand, find two different LU factorizations of the matrix of Exercise 8. Then, use your software package to factor this matrix. Which (if any) of the two factorizations does it display?

3.5 Elementary Matrices and Linear Systems

In the previous section we learned how to factor a matrix into an LU decomposition but not why the procedure works. In this section we introduce *elementary matrices* and use them to justify the LU decomposition process. They are also used to give a matrix interpretation of the entire row reduction process and provide an alternative means of proving some previous results.

Elementary Matrices

DEFINITION

A square matrix is called an **elementary matrix** if it can be obtained from the identity matrix of the same order by means of a single elementary row operation; that is, by means of

i. Multiplication of a row by a nonzero scalar, or
ii. Addition of a multiple of one row to a different row, or
iii. Interchange of two rows.

EXAMPLE 1

Show that each of the following is an elementary matrix:

$$E_1 = \begin{bmatrix} 1 & 0 & 0 \\ 0 & 1 & -2 \\ 0 & 0 & 1 \end{bmatrix} \quad E_2 = \begin{bmatrix} 0 & 0 & 0 & 1 \\ 0 & 1 & 0 & 0 \\ 0 & 0 & 1 & 0 \\ 1 & 0 & 0 & 0 \end{bmatrix} \quad E_3 = \begin{bmatrix} 1 & 0 & 0 & 0 \\ 0 & 1 & 0 & 0 \\ 0 & 0 & 4 & 0 \\ 0 & 0 & 0 & 1 \end{bmatrix}$$

Solution E_1 is obtained from the 3×3 identity matrix by multiplying row 3 by -2 and adding it to row 2. E_2 is obtained from the 4×4 identity matrix by interchanging rows 1 and 4. E_3 is obtained from the 4×4 identity matrix by multiplying row 3 by 4. ■

Much of the importance of elementary matrices comes from the following fact. Let I be the identity matrix with the same number of rows as a given matrix A. Then, the result of performing an elementary row operation on A is exactly the same as the result of multiplication on the left of A, or *premultiplication* of A, by the elementary matrix obtained by performing the same row operation on I. This fact is formalized in Theorem 1 after looking at another example.

EXAMPLE 2

Find the effect of premultiplication of a matrix A by each of the elementary matrices in Example 1.

Solution Since E_1 is a 3×3 matrix, for E_1A to be defined, A must have three rows. The number of columns doesn't matter at all. For the sake of variety, we'll use 4 with E_1, 1 with E_2, and 3 with E_3.

$$\begin{bmatrix} 1 & 0 & 0 \\ 0 & 1 & -2 \\ 0 & 0 & 1 \end{bmatrix}\begin{bmatrix} a_{11} & a_{12} & a_{13} & a_{14} \\ a_{21} & a_{22} & a_{23} & a_{24} \\ a_{31} & a_{32} & a_{33} & a_{34} \end{bmatrix} = \begin{bmatrix} a_{11} & a_{12} & a_{13} & a_{14} \\ a_{21} - 2a_{31} & a_{22} - 2a_{32} & a_{23} - 2a_{33} & a_{24} - 2a_{34} \\ a_{31} & a_{32} & a_{33} & a_{34} \end{bmatrix}$$

The resulting matrix product is the same as the matrix obtained by multiplying row 3 by -2 and adding it to row 2, the same description we had for E_1 in Example 1. Similarly, for the other two:

$$E_2\begin{bmatrix} a_{11} \\ a_{21} \\ a_{31} \\ a_{41} \end{bmatrix} = \begin{bmatrix} a_{41} \\ a_{21} \\ a_{31} \\ a_{11} \end{bmatrix} \quad \text{and} \quad E_3\begin{bmatrix} a_{11} & a_{12} & a_{13} \\ a_{21} & a_{22} & a_{23} \\ a_{31} & a_{32} & a_{33} \\ a_{41} & a_{42} & a_{43} \end{bmatrix} = \begin{bmatrix} a_{11} & a_{12} & a_{13} \\ a_{21} & a_{22} & a_{23} \\ 4a_{31} & 4a_{32} & 4a_{33} \\ a_{41} & a_{42} & a_{43} \end{bmatrix}$$

Again, the effect of each is exactly the same as the description for E_2 and E_3, respectively. ■

It is sometimes convenient to have names for each of the three types of elementary matrices that tell us the effect of premultiplication by that matrix at a glance. For any order n and any row i and column j, we introduce the following notation:

- $E_{ij}(c)$ is the matrix obtained from the identity matrix by replacing its (i,j)-entry by c.
- P_{ij} is the matrix obtained from the identity matrix by interchanging rows i and j. (Recall from the previous section that such a P_{ij} is an example of a *permutation* matrix.)

Using this notation, the matrices of Example 1 can be written as

$$E_1 = E_{23}(-2), \qquad E_2 = P_{14}, \qquad \text{and} \qquad E_3 = E_{33}(4).$$

The next theorem confirms the fact that performing an elementary row operation on a matrix has the same effect as premultiplication of that matrix by the corresponding elementary matrix. It also relates "undoing" the row operation to a corresponding matrix, the inverse of the elementary matrix.

THEOREM 1

Let A be an $m \times n$ matrix and E be an $m \times m$ elementary matrix. Then EA is the matrix obtained by performing on A the same row operation that converts I to E. Furthermore, E is invertible and E^{-1} is the matrix equivalent of a single elementary row operation of a similar type. More specifically, we have the following:

a. If $E = E_{ij}(c)$ for $i \neq j$, then EA is the matrix obtained from A by adding c times row j to row i. Furthermore, E is invertible and $E^{-1} = E_{ij}(-c)$; that is, repeat the same elementary row operation but with $-c$ as the row multiplier.

b. If $E = E_{ii}(c)$, then EA is obtained from A multiplying c times row i, where $c \neq 0$. Furthermore, E is invertible and $E^{-1} = E_{ii}(c^{-1})$; that is, repeat the same elementary row operation but with c^{-1} as the row multiplier.

c. If $E = P_{ij}$, then EA is obtained from A by interchanging rows i and j. In this case, E is its own inverse, $E^{-1} = P_{ij}$; that is, interchange the same two rows again.

The proofs of these facts involve little more than computations with very simple matrices. The details are considered in Exercises 8–11. ■

Multiple Row Operations

So far, we have only considered individual elementary row operations, but usually we are performing several in sequence. Theorem 1 and the fact that matrix multiplication is associative combine to make repeated row operations equivalent to multiplication on the left by a single matrix, the product of the corresponding elementary matrices.

EXAMPLE 3

Given a matrix A with at least three rows, find an invertible matrix M such that premultiplication of A by M yields the same result as the following three elementary row operations performed in sequence:

i. Interchange the first and second rows.

ii. Multiply the new first row by 4.

iii. Add -2 times the resulting first row to the third.

Solution This sequence of operations is equivalent to

$$E_{31}(-2)(E_{11}(4)(P_{12}A)) = (E_{31}(-2)E_{11}(4)P_{12})A,$$

by repeated use of Theorem 1 and associativity of multiplication. That is, the matrix

$$M = E_{31}(-2)E_{11}(4)P_{12}$$

is the desired matrix. ■

Although the product of a sequence of elementary matrices is a single matrix, M in the preceding example, it is often more convenient to think of it in its sequence form. In this form, it is easy to "undo" the corresponding row operations on a matrix.

We simply multiply their inverses together in reverse order. That is, if E_1, E_2, and E_3 are elementary matrices and A and B are such that

$$E_3E_2E_1A = B$$
$$\text{then} \quad A = (E_3E_2E_1)^{-1}B = E_1^{-1}E_2^{-1}E_3^{-1}B.$$

This fact is an immediate consequence of the following theorem.

THEOREM 2

If A and B are invertible matrices of the same order, then AB is also invertible and

$$(AB)^{-1} = B^{-1}A^{-1};$$

that is, the inverse of AB is the product of the inverses in the reverse order.

Proof To be an inverse of AB, a matrix X must have the property that $X(AB) = (AB)X = I$. All that is necessary is to let $X = B^{-1}A^{-1}$ and show that it multiplies correctly. First,

$$X(AB) = (B^{-1}A^{-1})(AB) = B^{-1}(A^{-1}A)B = B^{-1}(I)B = B^{-1}B = I.$$

$(AB)X$ is computed similarly. ■

Theorems 1 and 2 combine to give a nice matrix proof of an important fact that we learned in Section 3.3 and restate here as the next theorem.

THEOREM 3

A square matrix A is invertible if and only if its row-reduced echelon form is the identity matrix.

Proof First assume A row-reduces to the identity matrix. This means that a sequence of premultiplications by elementary matrices transforms A into the identity matrix

$$E_n \ldots E_2E_1A = I.$$

That is, the product of these elementary matrices, $E_n \ldots E_2E_1$, is the inverse of A.

Conversely, suppose that A is invertible, and let B be its row-reduced echelon form. That is, there is a sequence of elementary matrices $E_1, E_2, \ldots, E_n$, such that

$$E_n \ldots E_2E_1A = B.$$

Since A is invertible by hypothesis, and since the product matrix $E_n \ldots E_2E_1$ is as well, it follows from Theorem 2 that B is invertible. Since B is row-reduced and invertible, it must be the identity matrix. If it weren't, it would have to have a row of zeros, in which case so would BX for any matrix X of the same order. But then $BX = I$ could not have a solution, contradicting the fact that B is invertible. ■

NOTE The proof of Theorem 3 also gives another explanation for the validity of the procedure described in Section 3.2 for computing the inverse of a matrix A. Augment A by the identity matrix I, and transform it to row-reduced echelon form:

$$[A \mid I] \rightarrow [C \mid D]$$

Then C is the identity matrix if and only if A is invertible, in which case $D = A^{-1}$. Now that we can view the row reduction process as a sequence of premultiplications by elementary matrices, we see that if $C = I = E_n \ldots E_2E_1A$, then D is exactly the product $E_n \ldots E_2E_1$ premultiplying I, or just the product itself. But this product is the inverse of A, so D is just another name for that inverse.

LU Decomposition

We are finally ready to look at why the LU decomposition algorithm works. We do not present a formal proof, which involves a somewhat tedious proof by induction, but we do explore the essence of why the procedure is valid.

First, recall the procedure to obtain the factorization $PA = LU$ from a given matrix A. We create a zero matrix and an identity matrix with the same number of rows as A. As A is row-reduced to upper echelon form using elementary row operations in natural order from left to right to become U, the negative of each row multiplier is placed in the appropriate position of the zero matrix to obtain L, and row interchanges are carried through all three matrices to obtain the permutation matrix P.

We have now seen that each row operation can be viewed as premultiplication by an elementary matrix, so the row reduction process can be written as

$$E_n \ldots E_2E_1A = U,$$

where each E_i is an elementary matrix and U is in upper echelon form. Each of these elementary matrices has an inverse, so we can "recover" A by multiplying in order on the left by the sequence of inverses to write

$$A = E_1^{-1}E_2^{-1} \ldots E_n^{-1}U.$$

When each of the E_i is a lower matrix, then their product $L = E_1^{-1}E_2^{-1} \ldots E_n^{-1}$ is also a lower matrix, and we have A factored as LU, the kind of decomposition we sought. If none of the E_i are row interchanges, that is exactly what happens. Even more is true; the original row operations were done in an order that allows this product to be constructed simply by putting row multipliers in their corresponding positions. The following example illustrates this point.

EXAMPLE 4

Let $E_1 = E_{21}(1)$, $E_2 = E_{31}(2)$, and $E_3 = E_{32}(3)$. Show that $E_3E_2E_1$ is *not* just the identity matrix with the row multipliers from the given matrices placed appropriately. Also show that $(E_3E_2E_1)^{-1} = E_1^{-1}E_2^{-1}E_3^{-1} = E_{21}(-1)E_{31}(-2)E_{32}(-3)$ *is* just the identity matrix with the row multipliers placed appropriately.

Solution The computations are straightforward.

$$\begin{aligned} E_3E_2E_1 &= E_{32}(3)E_{31}(2)E_{21}(1) \\ &= \begin{bmatrix} 1 & 0 & 0 \\ 0 & 1 & 0 \\ 0 & 3 & 1 \end{bmatrix}\begin{bmatrix} 1 & 0 & 0 \\ 0 & 1 & 0 \\ 2 & 0 & 1 \end{bmatrix}\begin{bmatrix} 1 & 0 & 0 \\ 1 & 1 & 0 \\ 0 & 0 & 1 \end{bmatrix} = \begin{bmatrix} 1 & 0 & 0 \\ 1 & 1 & 0 \\ 5 & 3 & 1 \end{bmatrix} \end{aligned}$$

$$\begin{aligned} \text{and} \quad E_1^{-1}E_2^{-1}E_3^{-1} &= E_{21}(-1)E_{31}(-2)E_{32}(-3) \\ &= \begin{bmatrix} 1 & 0 & 0 \\ -1 & 1 & 0 \\ 0 & 0 & 1 \end{bmatrix}\begin{bmatrix} 1 & 0 & 0 \\ 0 & 1 & 0 \\ -2 & 0 & 1 \end{bmatrix}\begin{bmatrix} 1 & 0 & 0 \\ 0 & 1 & 0 \\ 0 & -3 & 1 \end{bmatrix} \\ &= \begin{bmatrix} 1 & 0 & 0 \\ -1 & 1 & 0 \\ -2 & -3 & 1 \end{bmatrix}. \end{aligned}$$

In the first matrix $E_3E_2E_1$, the 5 in the lower left corner, position 3,1, is not the row multiplier of the corresponding elementary matrix $E_2 = E_{31}(2)$. In the second matrix, however, each entry below the diagonal is the multiplier of the corresponding elementary matrix. ■

In the preceding example, the feature to note is that in $E_3E_2E_1 = E_{32}E_{31}E_{21}$, the row indices 3,3,2 decrease in moving from left to right but that in $(E_3E_2E_1)^{-1} = E_{21}E_{31}E_{32}$, they don't. The essence of the entire LU decomposition procedure is that as the Gaussian elimination is carried out in order from left to right, this feature is preserved. That is, the sequence of inverses of matrices E_{ij} is also of the form E_{ij} but in the right order to have this convenient product; that is, the row multipliers just need to be placed in the appropriate places.

In case some row interchanges occur during the Gaussian elimination procedure, we must deal with one more detail, but the idea is exactly the same. Again we assume that matrix A is row-reduced to upper echelon form by means of elementary row operations using Gaussian elimination in order from left to right. As before, this can be represented as the matrix equation

$$E_n \ldots E_2E_1A = U,$$

with each E_i being of the type $E_{ij}(c)$, addition of a multiple of one row to another, or P_{ij}, the interchange of two rows. Also as before, we know the inverse of each type of elementary matrix and can rewrite this equation as

$$A = E_1^{-1}E_2^{-1} \ldots E_n^{-1}U,$$

where each E_i^{-1} is either the corresponding $E_{ij}(-c)$ or the same row interchange P_{ij}, depending on which type of row operation was involved. The matrix U is still an upper matrix, but if any row interchanges occur, the product of inverses,

$$M = E_1^{-1}E_2^{-1} \ldots E_n^{-1},$$

fails to be a lower matrix. As good fortune would have it, the same row interchanges that were made along the way, when applied to M, leave us a lower matrix as desired. That is, we multiply both sides of the equation $A = E_1^{-1}E_2^{-1} \ldots E_n^{-1}U$ by P,

$$PA = P(E_1^{-1}E_2^{-1} \ldots E_n^{-1}U),$$

and regroup the product on the right to obtain

$$PA = (PE_1^{-1}E_2^{-1} \ldots E_n^{-1})U.$$

Now we let $L = PM = P(E_1^{-1}E_2^{-1} \ldots E_n^{-1}U)$, and we have the LU factorization

$$PA = LU.$$

3.5 Exercises

In Exercises 1–3 explain the effect of premultiplication by each elementary matrix in terms of an elementary row operation.

1. $\begin{bmatrix} 1 & 0 & 0 \\ 0 & 0 & 1 \\ 0 & 1 & 0 \end{bmatrix}$ *2.* $\begin{bmatrix} 2 & 0 & 0 \\ 0 & 1 & 0 \\ 0 & 0 & 1 \end{bmatrix}$ *3.* $\begin{bmatrix} 1 & 3 & 0 \\ 0 & 1 & 0 \\ 0 & 0 & 1 \end{bmatrix}$

4–6. Find the inverse of each of the matrices in Exercises 1–3 without doing any computation. Confirm your result by multiplying the resulting matrix by the original.

7. Show by example that the order reversal in Theorem 2 is essential. That is, find invertible matrices A and B such that $(AB)^{-1}$ is *not* $A^{-1}B^{-1}$. (Exercises 1–6 provide good examples.)

In Exercises 8–11 we explore why the inverse of each type of elementary matrix is as described in Theorem 1.

8. By multiplying them together, show that $E_{12}(c)$ and $E_{12}(-c)$ are inverses of each other, first when they are assumed to be 2×2 matrices and then again when they are assumed to be 3×3 matrices.

9. Repeat Exercise 8 for $E_{23}(c)$ and $E_{23}(-c)$, first as 3×3 matrices and then again as 4×4 matrices.

10. By multiplying them together, show that $E_{ii}(c)$ and $E_{ii}(c^{-1})$ are inverses of each other for any order n (and, of course, $1 \leq i \leq n$ and $c \neq 0$).

11. Explain why any P_{ij} is its own inverse.

12. Explain the effects of *postmultiplication* by an elementary matrix. That is, if A is an $m \times n$ matrix and E is an elementary matrix of order n, describe AE for each type of elementary matrix.

In Exercises 13–18 we explore the maximum number of computations (the total number of additions, subtractions, multiplications, and divisions) needed to use Gaussian elimination to transform a matrix *A* to upper echelon form, the *U* of its *LU* decomposition. We restrict consideration to the situation in which *A* is square of order *n* and in which no row interchanges occur.

13. Count the number of computations needed to place a zero below the 1,1 position in a 3×3 matrix. That is, how many computations are necessary to calculate the row multiplier and to then use the first row to "zero out" the first position of another row? How many rows require this number of computations?

14. Assuming the action of Exercise 13 has been completed (that is, each entry below the 1,1 position is zero), how many computations remain to complete the transformation to upper echelon form?

15. Repeat Exercises 13 and 14, assuming that A is a 4×4 matrix.

16. Following the idea of Exercise 15, calculate the number of computations needed to transform an $n \times n$ matrix A to upper echelon form by keeping track of the number of calculations at each step along the way and using ellipses (. . .) whenever necessary, for example, in expressions like $(n-1) + (n-2) + \cdots + 1$.

17. Simplify the expression of Exercise 16 by means of the summation formulas

$$1 + 2 + \cdots + n = \frac{n(n+1)}{2}$$

$$1^2 + 2^2 + \cdots + n^2 = \frac{n(n+1)(2n+1)}{6}$$

Conclude that Gaussian elimination is a cubic (n^3) algorithm. That is, as the number of rows n increases, the number of computations needed to transform a matrix to upper echelon form is a polynomial of degree 3.

18. Show that back- and forward-substitutions are quadratic (n^2) algorithms.

19. Suppose that A is a square invertible matrix. Show that A can be written as a product of elementary matrices. (*Hint*: The row-reduced echelon form of A is the identity matrix.)

3.6 APPLICATION—Least Square Polynomials

In Section 2.5, we developed a procedure for determining the polynomial that passes through a given set of points. This technique works well if the number of points given is small. However, in most experimental situations in which we are trying to obtain an equation relating a dependent and independent variable (say, y and x), the experiment is performed a relatively large number of times, resulting in many data points (x_i, y_i). If we tried to determine the interpolating polynomial for these points, we would have to do quite a bit of computing. Moreover, data obtained experimentally is subject to measurement errors, and these errors would be passed on to the approximating interpolating polynomial. Finally, finding such a high-degree polynomial makes it too dependent on the data chosen, and hides any underlying trend. In this situation it is better to construct a relatively low-degree polynomial that passes *near* the data points rather than through them. This is the idea behind *least square approximations.*

As an example, consider the simplest case—the *least square linear polynomial,* or **least square line**. Let $(x_0, y_0), (x_1, y_1), \ldots, (x_m, y_m)$ be a given set of points with all x_i distinct and d_k denote the vertical distance from the line $y = ax + b$ to the point (x_k, y_k) (Figure 1). That is, let $d_k = |y_k - (ax_k + b)|$. We wish to determine a and b such that the sum of the squares of the vertical distances d_k is minimized. In other words, we seek a linear polynomial $Q(x) = ax + b$ that minimizes the expression

$$S = [y_0 - (ax_0 + b)]^2 + [y_1 - (ax_1 + b)]^2 + \cdots + [y_m - (ax_m + b)]^2. \quad \textbf{(1)}$$

FIGURE 1

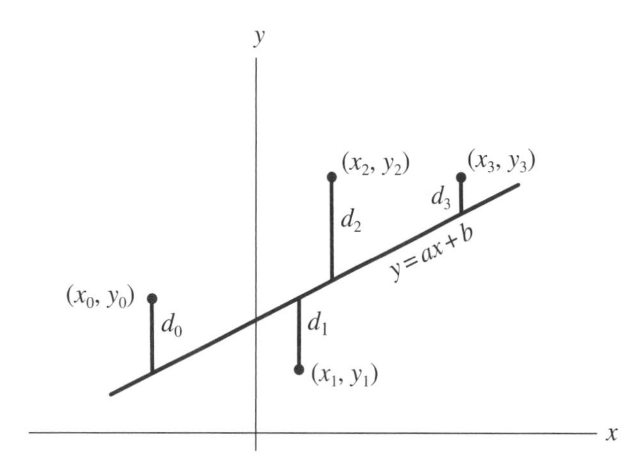

Here the ease of the mathematics drives the criterion chosen!

NOTE It may seem more logical to seek a polynomial that minimizes the sum of the vertical distances, or perhaps the sum of the perpendicular distances from each point to the line. These alternatives, however, result in a considerably more complicated analysis of the situation; the first requires absolute values, which are not as mathematically convenient as squaring. Minimizing the expression S of Equation (1) has a relatively simple as well as unique solution and works well in practical applications.

Instead of demonstrating how we can determine the least square line, we go directly to the more general case of the nth-degree least square polynomial. (The least square line is just the special case of the latter, when $n = 1$.)

Here we wish to determine a polynomial of the form

$$Q(x) = b_0 + b_1x + b_2x^2 + \cdots + b_nx^n$$

that minimizes the expression

$$S = [y_0 - Q(x_0)]^2 + [y_1 - Q(x_1)]^2 + \cdots + [y_m - Q(x_m)]^2. \qquad \textbf{(2)}$$

THEOREM 1

Let $(x_0, y_0), (x_1, y_1), \ldots, (x_m, y_m)$ be $m + 1$ points with all x_i distinct, and let n be a positive integer with $n \leq m$. Then there is a unique polynomial

$$Q(x) = b_0 + b_1x + b_2x^2 + \cdots + b_nx^n$$

that minimizes the expression S of Equation (2). The vector of coefficients $\mathbf{b} = (b_0, b_1, \ldots, b_n)$ satisfies the linear system

$$\left(U^TU\right)\mathbf{b} = U^T\mathbf{y}, \qquad \textbf{(3)}$$

where $\mathbf{y} = (y_0, y_1, \ldots, y_m)$ and

$$U = \begin{bmatrix} 1 & x_0 & x_0^2 & \cdots & x_0^n \\ 1 & x_1 & x_1^2 & \cdots & x_1^n \\ \vdots & \vdots & \vdots & & \vdots \\ 1 & x_m & x_m^2 & \cdots & x_m^n \end{bmatrix}.$$

The proof of Theorem 1 will appear in Section 6.6.

DEFINITION

Given the points, $(x_0, y_0), (x_1, y_1), \ldots, (x_m, y_m)$, the unique polynomial, $Q(x)$, of degree less than or equal to n that minimizes the expression S of Equation (2) is called the nth-degree **least square polynomial** for these points. System (3) is called the set of **normal equations** for the given data.

EXAMPLE 1

A student performs an experiment in an attempt to obtain an equation relating two physical quantities x and y. She performs the experiment six times, letting $x = 1, 2, 4, 6, 8$, and 10 units. These experiments result in corresponding y values of 2, 2, 3, 4, 6, and 8 units, respectively.

a. Find the least square linear polynomial (the least square line) for this data.

b. Find the least square quadratic polynomial (the least square parabola) for this data.

Solution Using the notation of Theorem 1, the given data are the set of six points (x_i, y_i), $i = 0, 1, \ldots, 5$: (1, 2), (2, 2), (4, 3), (6, 4), (8, 6), (10, 8). For part (a) we seek a polynomial of the form $Q_1(x) = b_0 + b_1x$, whereas for part (b) we seek one of the form $Q_2(x) = b_0 + b_1x + b_2x^2$.

a. Here the vectors $\mathbf{b}$ and $\mathbf{y}$ are $\mathbf{b} = (b_0, b_1)$ and $\mathbf{y} = (2, 2, 3, 4, 6, 8)$, and the matrices U and U^T are

$$U = \begin{bmatrix} 1 & 1 \\ 1 & 2 \\ 1 & 4 \\ 1 & 6 \\ 1 & 8 \\ 1 & 10 \end{bmatrix} \quad \text{and} \quad U^T = \begin{bmatrix} 1 & 1 & 1 & 1 & 1 & 1 \\ 1 & 2 & 4 & 6 & 8 & 10 \end{bmatrix}.$$

By computing $U^TU\mathbf{b}$ and $U^T\mathbf{y}$, we obtain the normal equations, System (3), $U^TU\mathbf{b} = U^T\mathbf{y}$, for part (a):

$$\begin{aligned} 6b_0 + 31b_1 &= 25 \\ 31b_0 + 221b_1 &= 170 \end{aligned}$$

Solving this linear system, we see that $b_0 = 0.6986$ and $b_1 = 0.6712$ (approximately), so that the least square line for the data is

$$Q_1(x) = 0.6986 + 0.6712x.$$

b. Here we have $\mathbf{b} = (b_0, b_1, b_2)$, $\mathbf{y} = (2, 2, 3, 4, 6, 8)$,

$$U = \begin{bmatrix} 1 & 1 & 1 \\ 1 & 2 & 4 \\ 1 & 4 & 16 \\ 1 & 6 & 36 \\ 1 & 8 & 64 \\ 1 & 10 & 100 \end{bmatrix}, \quad \text{and} \quad U^T = \begin{bmatrix} 1 & 1 & 1 & 1 & 1 & 1 \\ 1 & 2 & 4 & 6 & 8 & 10 \\ 1 & 4 & 16 & 36 & 64 & 100 \end{bmatrix}.$$

Therefore, the normal equations for part (b) are

$$\begin{aligned} 6b_0 + 31b_1 + 221b_2 &= 25 \\ 31b_0 + 221b_1 + 1801b_2 &= 170 \\ 221b_0 + 1801b_1 + 15665b_2 &= 1386, \end{aligned}$$

which has the approximate solution $b_0 = 1.836$, $b_1 = 0.0276$, and $b_2 = 0.0594$. Thus, the least square parabola for this data is

$$Q_2(x) = 1.836 + 0.0276x + 0.0594x^2.$$

(In Figure 2, we plotted the given data of this example and the least square line and parabola for the data.)

FIGURE 2

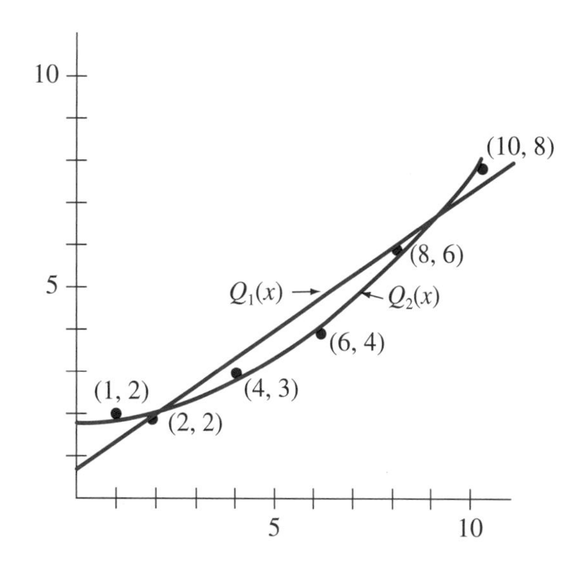

■

NOTE If $m = n$ in Theorem 1 (that is, if we seek a polynomial whose degree is 1 less than the number of data points), then the resulting least square polynomial is identical to the interpolating polynomial (see Section 2.5) through these points. To see why, notice that when $m = n$, the interpolating polynomial is defined and results in a minimum value (namely, 0) for the right side of Equation (2)—it passes *through* all the data points.

3.6 Exercises

In Exercises 1–4 find the least square polynomial of the specified degree for the given data points.

1. $\{(0, 0), (1, 0), (2, 1), (3, 3), (4, 5)\}$, $n = 1$

2. $\{(x, y) \mid y = x^3 \text{ and } x = 0, 1, 2, 3, 4, 5\}$, $n = 1$

3. $\{(x, y) \mid y = x^3 \text{ and } x = 0, 1, 2, 3, 4, 5\}$, $n = 2$

4. $\{(-2, 2), (-1, 1), (0, -1), (1, 0), (2, 3)\}$, $n = 2$

In Exercises 5 and 6 determine the system of normal equations that occurs in the process of finding the least square polynomial of specified degree for the given data.

5. $\{(x, y) \mid y = x^4 - 8x^2, x = -3, -2, -1, 0, 1, 2, 3\}$, $n = 3$

6. $\{(0, 0), (1, 0), (2, 1), (3, 3), (4, 5), (5, 4), (6, 5)\}$, $n = 4$

A firm that manufactures widgets has summarized the daily consumer demand $d(x)$ for widgets as a function of their price x in the following table:

x	1	1.5	2	2.5	3
$d(x)$	200	180	150	100	25

In Exercises 7 and 8, using least square polynomials, approximate the daily consumer demand as indicated.

7. By a linear function

8. By a quadratic function

Computational Exercises

Use a graphing calculator or computer software to help solve the following problems.

In Exercises 9 and 10 graph (on the same set of axes) the data points and least square polynomials of

9. Examples 5 and 6

10. Examples 7 and 8

For each of the given functions and intervals in Exercises 11 and 12:

(a) Approximate the given function by least square polynomials of degrees 3 and 5 at equally spaced points on the given interval.

(b) Compare the values obtained by the least square polynomials determined in part (a) at the midpoint of the given interval to that given by your calculator or software for the given function.

(c) Graph the given function and the approximating polynomials on the same set of axes. Does the higher degree polynomial appear to give a better approximation?

11. $y = \sin x, \quad 0 \le x \le \pi$

12. $y = e^x, \quad 0 \le x \le 2$

13. Given the following set of data points: $\{(-5, 6), (-4, 1), (-3, -1), (-2, 2), (-1, 11), (0, 12), (1, 4), (2, -1), (3, 1), (4, 2), (5, 8)\}$

(a) Plot these data points.

(b) Find an appropriate nth degree least square polynomial to fit the data.

(c) Graph the least square polynomial with the data points.

14. Repeat Exercise 13 for these data points: $\{(-2, -15), (-1, 0), (1, 1), (2, -2), (3, -5), (4, -1), (5, 10), (6, 25)\}$

3.7 APPLICATION—Social Networks

The spotlight at the beginning of the chapter highlighted matrices as a succinct way to model networks. In this section we'll talk about a specific application to social networks like Facebook, where the connections you make to others are of the essence.

Directed and Undirected Graphs

A **graph** (sometimes called a **directed graph** or **network**) G is a collection of objects, called **nodes** or **vertices**, and some ordered pairs of them (the ordered pairs in the graph are called **arcs**). An arc (u, v) is said to **start at** u and ***end at*** v. **Social networks** are simply graphs that arise in a social setting; the nodes (in this setting sometimes referred to as **actors**) are usually people, but can be groups, committees, or events.

For example, we could take as our set of actors a collection of people, say {Andrea, Bill, Carmen, David, Erin}. We then take pairs (A, B) where A likes B. So, for example, the arcs might be (Andrea, David), (Andrea, Erin), (Bill, Andrea), (Bill, Carmen), (Bill, Erin), (Carmen, Andrea), (Carmen, Bill), (David, Carmen), (Erin, Bill), and (Erin, Andrea), so

- Andrea likes David and Erin,
- Bill likes Andrea, Carmen, and Erin,
- Carmen likes Andrea and Bill,
- David likes only Carmen, and
- Erin likes Bill and Andrea.

We often visualize graphs by drawing labelled points for each node, and drawing an arrow from one node to another if they form an arc of the graph. A drawing of the graph just described is shown in Figure 3. Note that the actual points chosen for the nodes and the precise paths of the arrows drawn are of no importance; what matters is which points are joined to which. A drawing of a graph completely describes the graph.

FIGURE 3

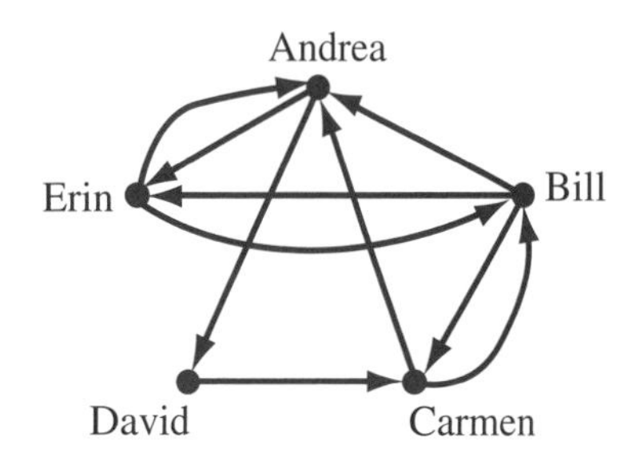

The **order** of a graph is its number of vertices. The graph of Figure 3 has order 5. A graph is **undirected** if the arc relation is *symmetric*—whenever there is an arc from node u to node v, there is also an arc from v to u as well. While the relationship of "liking" is not symmetric, the relation "is friends with" is. When the graph is undirected, one can omit the direction on the arcs when drawing the graph, and simply draw a line between two nodes. Figure 4 shows the (undirected) friendship graph of a group.

FIGURE 4

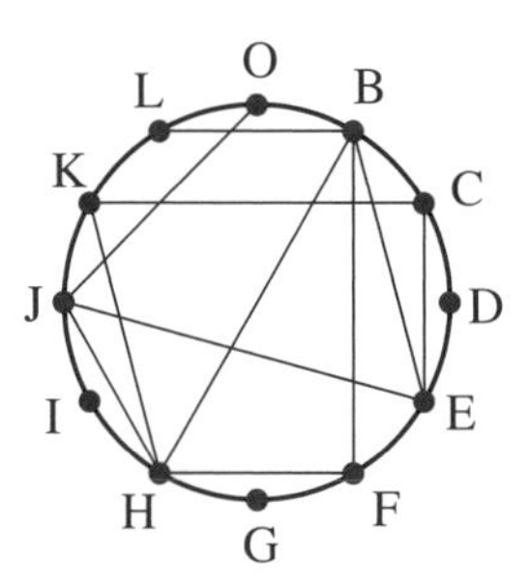

As mentioned in the chapter's spotlight, we can associate every graph with a matrix, called the graph's **adjacency matrix**. If a graph G has vertices $v_1, v_2, \ldots, v_n$ then the adjacency matrix of G is an $n \times n$ matrix A whose (i, j)th entry is 1 if there is an arc from v_i to v_j, and 0 otherwise. Note that we need to order the vertices of the graph first before we can write down the adjacency matrix. Often there is a "natural" order on the vertices, and once an ordering is chosen or assumed, we can talk about "the" adjacency matrix of the graph. For the example in Figure 4, the adjacency matrix, for the ordering

Andrea, Bill, Carmen, David, Erin,

is

$$A = \begin{bmatrix} 0 & 0 & 0 & 1 & 1 \\ 1 & 0 & 1 & 0 & 1 \\ 1 & 1 & 0 & 0 & 0 \\ 0 & 0 & 1 & 0 & 0 \\ 1 & 1 & 0 & 0 & 0 \end{bmatrix}$$

Some graph properties transfer over very nicely to matrix properties. For example, a graph is undirected if and only if its adjacency matrix is symmetric. Matrices are efficient ways to encode the information about a network, and algorithms are designed to use matrix operations to determine characteristics of the network.

Who Is the Most Social?

In a social network, we often seek out the most popular people. If the arc set is the symmetric relation of *friendship*, then the most popular people are those who have the most friends. For directed cases like the relationship of *liking*, active individuals might be defined to be those who like a lot of people or those who are liked by a lot of people. (Which do you think would be more impressive? We'll come back to this important point in Section 4.6.) The **in-degree** of a vertex v of a graph G is the number of arcs ending at v; the **out-degree** of v is the number of arcs starting at v. For undirected graphs, the common value of the in-degree and out-degree of a vertex is simply called its **degree**. In the graph of Figure 3, Andrea has in-degree 3 and out-degree 2, whereas in the graph of Figure 4, J has degree 5.

Calculating in-degrees and out-degrees turns out to be simple matrix multiplications on the adjacency matrix of a graph.

THEOREM 1

Let G be a graph of order n with vertices $v_1, v_2, \ldots, v_n$, A be the adjacency matrix of G, and $\mathbf{1}$ denote the $1 \times n$ column vector consisting of all 1's. Then

a. the in-degree of vertex v_i is the ith component of the row vector $\mathbf{1}^T A$,

b. the out-degree of vertex v_i is the ith component of the column vector $A\mathbf{1}$, and

c. the number of arcs in the graph is equal to $\mathbf{1}^T A \mathbf{1}$ (ignoring the brackets around the 1×1 matrix).

Proof Note that the result of multiplying A on the left by an all-1's row matrix sums the entries in each column of A:

$$[1 \quad 1 \quad \ldots \quad 1]\begin{bmatrix} a_{11} & \cdots & a_{1i} & \cdots & a_{1n} \\ a_{21} & \cdots & a_{2i} & \cdots & a_{2n} \\ \vdots & \vdots & \vdots & & \vdots \\ a_{n1} & \cdots & a_{ni} & \cdots & a_{nn} \end{bmatrix}$$

$$= \left[a_{11} + a_{21} + \cdots a_{n1}, \cdots, a_{1i} + a_{2i} + \cdots a_{ni}, \cdots, a_{1n} + a_{2n} + \cdots a_{nn}\right].$$

As the ith column corresponds to all arcs ending at v_i, its sum is the in-degree of vertex v_i. This proves (a). Parts (b) and (c) are proved analogously (see Exercise 16). ■

EXAMPLE 1

For the graph of Figure 4, its adjacency matrix (for the vertex ordering Andrea, Bill, Carmen, David, Erin) was shown to be

$$A = \begin{bmatrix} 0 & 0 & 0 & 1 & 1 \\ 1 & 0 & 1 & 0 & 1 \\ 1 & 1 & 0 & 0 & 0 \\ 0 & 0 & 1 & 0 & 0 \\ 1 & 1 & 0 & 0 & 0 \end{bmatrix}$$

Now

$$\mathbf{1}^T A = [1 \quad 1 \quad 1 \quad 1 \quad 1]\begin{bmatrix} 0 & 0 & 0 & 1 & 1 \\ 1 & 0 & 1 & 0 & 1 \\ 1 & 1 & 0 & 0 & 0 \\ 0 & 0 & 1 & 0 & 0 \\ 1 & 1 & 0 & 0 & 0 \end{bmatrix} = [3 \quad 2 \quad 2 \quad 1 \quad 2]$$

while

$$A\mathbf{1} = \begin{bmatrix} 0 & 0 & 0 & 1 & 1 \\ 1 & 0 & 1 & 0 & 1 \\ 1 & 1 & 0 & 0 & 0 \\ 0 & 0 & 1 & 0 & 0 \\ 1 & 1 & 0 & 0 & 0 \end{bmatrix}\begin{bmatrix} 1 \\ 1 \\ 1 \\ 1 \\ 1 \end{bmatrix} = \begin{bmatrix} 2 \\ 3 \\ 2 \\ 1 \\ 2 \end{bmatrix}.$$

We see from this that Andrea is liked by the most people (as the first entry of $\mathbf{1}^TA$ is 3, the largest value of $\mathbf{1}^TA$), while Bill likes the most people (since the second entry of $A\mathbf{1}$ is 3, the largest of $A\mathbf{1}^T$). Finally,

$$\mathbf{1}^TA\mathbf{1} = [1 \quad 1 \quad 1 \quad 1 \quad 1]\, A\mathbf{1} = [1 \quad 1 \quad 1 \quad 1 \quad 1]\begin{bmatrix} 2 \\ 3 \\ 2 \\ 1 \\ 2 \end{bmatrix} = [10]$$

and there are indeed 10 arcs in the graph. ■

Finding out who is the most social person can be vitally important in a social network if you want to control the spread of diseases, real or digital. It has been found that in web-like graphs, concentrating your focus on the most social individuals may make more sense than spending your energy on a larger group of less connected people.

Any Friend of a Friend ...

In many social networks, it is not only important who your friends are, but who your friends' friends are, and applications like Facebook make it easy to reach out to them. Finding out how many people you can reach is again a matrix multiplication, but now of just the adjacency matrix with itself. To see why, we prove an important general result on graphs.

A **walk of length *k*** from vertex u to vertex v of a graph G is simply a sequence of k arcs that start at u, end at v, with the end of each arc being the start of the subsequent arc. For example, for the graph of Figure 4,

(Carmen, Bill), (Bill, Andrea)

is a walk of length 2 from Carmen to Andrea, whereas

(Carmen Bill), (Bill, Andrea), (Andrea, David), (David, Carmen)

is a walk of length 4 from Carmen to herself.

From a social point of view, you might want to know how many walks there are of a certain length between two actors – are there even any at all? Matrices once again come to the rescue.

THEOREM 2

Let A be the adjacency matrix of a graph G of order n with vertices $v_1, v_2, \ldots, v_n$. Then for any $k \geq 1$ and any i and j, the number of walks of length k from node v_i to node v_j is the (i, j)th entry of A^k.

Proof We proceed by a famous proof technique called *induction*—we prove the result is true for the smallest case, $k = 1$, and then prove, assuming it holds for value k, that it holds for the next value, $k + 1$ (so that the validity of the result for $k = 1$ implies the validity of the result for $k = 2$, which implies the validity of the result for $k = 3$, and so on).

The result trivially holds when $k = 1$, as a walk of length 1 is just an arc, and thus the number of walks of length 1 from v_i to v_j is the number of arcs from v_i and v_j, which is exactly what the (i, j)th entry of $A = A^1$ counts.

We now assume that the result is true for value k, that is, for any i and j, the number of walks of length k from v_i to v_j is the (i, j)th entry of A^k. We will show that it follows that the result is true for $k + 1$ as well; that is, for any i and j, the number of walks of length $k + 1$ from v_i to v_j is the (i, j)th entry of A^{k+1}. The walks of length $k + 1$ from v_i to v_j arise as a walk of length k from v_i to some other vertex v_l followed by an arc from v_l to v_j. It follows that the number of walks of length $k + 1$ from v_i to v_j is equal to the sum, over all choices of l, of the number of walks from v_i to v_l, where there is an arc from v_l to v_j. Now there is an arc from v_l to v_j if the (i, j)th entry of A is 1, and there is no such arc if the (i, j)th entry of A is 0. It follows that the number of walks of length $k + 1$ from v_i to v_j is equal to

$$\sum_{l=1}^{n} (\text{number of walks of length } k \text{ from } v_i \text{ to } v_l) a_{lj}.$$

We now assume the result for k (for any i and j), so that the number of walks of length k from v_i to v_j is the (i, j)th entry of A^k. But then it follows that the number of walks of length $k + 1$ from v_i to v_j is equal to

$$\sum_{l=1}^{n} ((i, l)\text{th entry of } A^k) a_{lj},$$

and this is precisely the (i, j)th entry of A^{k+1} (think of multiplying across the ith row of A^k and down the jth column of A). So we have shown that the number of walks of length $k + 1$ from v_i to v_j is the (i, j)th entry of A^{k+1}, and this completes the argument. ■

EXAMPLE 2

For the graph of Figure 4, its adjacency matrix (for the vertex ordering Andrea, Bill, Carmen, David, Erin) was shown to be

$$A = \begin{bmatrix} 0 & 0 & 0 & 1 & 1 \\ 1 & 0 & 1 & 0 & 1 \\ 1 & 1 & 0 & 0 & 0 \\ 0 & 0 & 1 & 0 & 0 \\ 1 & 1 & 0 & 0 & 0 \end{bmatrix}.$$

A simple calculation shows that

$$A^3 = \begin{bmatrix} 2 & 1 & 1 & 1 & 2 \\ 3 & 1 & 3 & 2 & 4 \\ 3 & 3 & 1 & 1 & 1 \\ 1 & 0 & 1 & 1 & 2 \\ 3 & 3 & 1 & 1 & 1 \end{bmatrix}.$$

It follows that there are 4 walks of length 3 from Bill to Erin, as the (2,5)th entry of A^3 is 4. ■

EXAMPLE 3

Figure 4 describes the (undirected) friendship graph of a set of friends, and its adjacency matrix (with respect to the ordering O, B, C, D, E, F, G, H, I, J, K, L) is

$$A = \begin{bmatrix} 0 & 1 & 0 & 0 & 0 & 0 & 0 & 0 & 0 & 1 & 0 & 1 \\ 1 & 0 & 1 & 0 & 1 & 1 & 0 & 1 & 0 & 0 & 0 & 1 \\ 0 & 1 & 0 & 1 & 1 & 0 & 0 & 0 & 0 & 0 & 1 & 0 \\ 0 & 0 & 1 & 0 & 1 & 0 & 0 & 0 & 0 & 0 & 0 & 0 \\ 0 & 1 & 1 & 1 & 0 & 1 & 0 & 0 & 0 & 1 & 0 & 0 \\ 0 & 1 & 0 & 0 & 1 & 0 & 1 & 1 & 0 & 0 & 0 & 0 \\ 0 & 0 & 0 & 0 & 0 & 1 & 0 & 1 & 0 & 0 & 0 & 0 \\ 0 & 1 & 0 & 0 & 0 & 1 & 1 & 0 & 1 & 1 & 1 & 0 \\ 0 & 0 & 0 & 0 & 0 & 0 & 0 & 1 & 0 & 1 & 0 & 0 \\ 1 & 0 & 0 & 0 & 1 & 0 & 0 & 0 & 1 & 0 & 1 & 0 \\ 0 & 0 & 1 & 0 & 0 & 0 & 0 & 0 & 0 & 0 & 0 & 1 \\ 1 & 1 & 0 & 0 & 0 & 0 & 0 & 0 & 0 & 0 & 1 & 0 \end{bmatrix}.$$

Note that the number of common friends for two people is the number of walks of length two between them, and by Theorem 2, this is equal to the corresponding entry in A^2. Now

$$A^2 = \begin{bmatrix} 3 & 1 & 1 & 0 & 2 & 1 & 0 & 2 & 1 & 0 & 2 & 1 \\ 1 & 6 & 1 & 2 & 2 & 2 & 2 & 1 & 1 & 3 & 3 & 1 \\ 1 & 1 & 4 & 1 & 2 & 2 & 0 & 2 & 0 & 2 & 0 & 2 \\ 0 & 2 & 1 & 2 & 1 & 1 & 0 & 0 & 0 & 1 & 1 & 0 \\ 2 & 2 & 2 & 1 & 5 & 1 & 1 & 2 & 1 & 0 & 2 & 1 \\ 1 & 2 & 2 & 1 & 1 & 4 & 1 & 2 & 1 & 2 & 1 & 1 \\ 0 & 2 & 0 & 0 & 1 & 1 & 2 & 1 & 1 & 1 & 1 & 0 \\ 2 & 1 & 2 & 0 & 3 & 2 & 1 & 6 & 1 & 2 & 1 & 2 \\ 1 & 1 & 0 & 0 & 1 & 1 & 1 & 1 & 2 & 1 & 2 & 0 \\ 0 & 3 & 2 & 1 & 0 & 2 & 1 & 2 & 1 & 5 & 1 & 2 \\ 2 & 3 & 0 & 1 & 2 & 1 & 1 & 1 & 2 & 1 & 4 & 0 \\ 1 & 1 & 2 & 0 & 1 & 1 & 0 & 2 & 0 & 2 & 0 & 3 \end{bmatrix},$$

so we see, for example, that B and J have 3 common friends while C and G have none as the corresponding entries in A^2 are 3 and 0, respectively. ■

There is a well-known *six degrees of separation* principle that states that any two individuals in the world are connected via a chain of acquaintances of length at most 6, that is, if we form the social network whose nodes are all individuals in the world, with an arc connecting pairs who know each other, then in this huge graph, there is always a walk of length at most 6 between any two nodes. In matrix terms, this is equivalent to showing that for the corresponding adjacency matrix A, the matrix

$$I + A + A^2 + A^3 + A^4 + A^5 + A^6$$

has only positive entries (see Exercise 17).

More generally, you might want to know in a network whether you can eventually walk from any one individual to another; this property is known as *strongly connected* (in an undirected graph, the property is known as simply being *connected*). Connectivity can be checked from the adjacency matrix as well (see Exercise 18).

THEOREM 3

Suppose that G is a graph of order n with adjacency matrix A. Then G is strongly connected if and only if $I + A + A^2 + \cdots + A^{n-1}$ has all positive entries. ■

COMPUTATIONAL NOTE

The problem with verifying the six degrees of separation principle revolves around not only forming the huge matrix A (which has an order in the *billions*) but also carrying out the matrix powers. Fortunately, adjacency matrices are *sparse* (by far most entries are 0), and this allows for special algorithms to be created just for matrix multiplication. We will talk more about algorithms on very large matrices in Section 4.6.

We'll end by remarking that there are some applications for viewing your online social network. For example, once you have logged into your Facebook account, you can use a tool called *TouchGraph* (Figure 5). (Visit http://www.touchgraph.com/facebook to view the social network of you and your friends.)

FIGURE 5

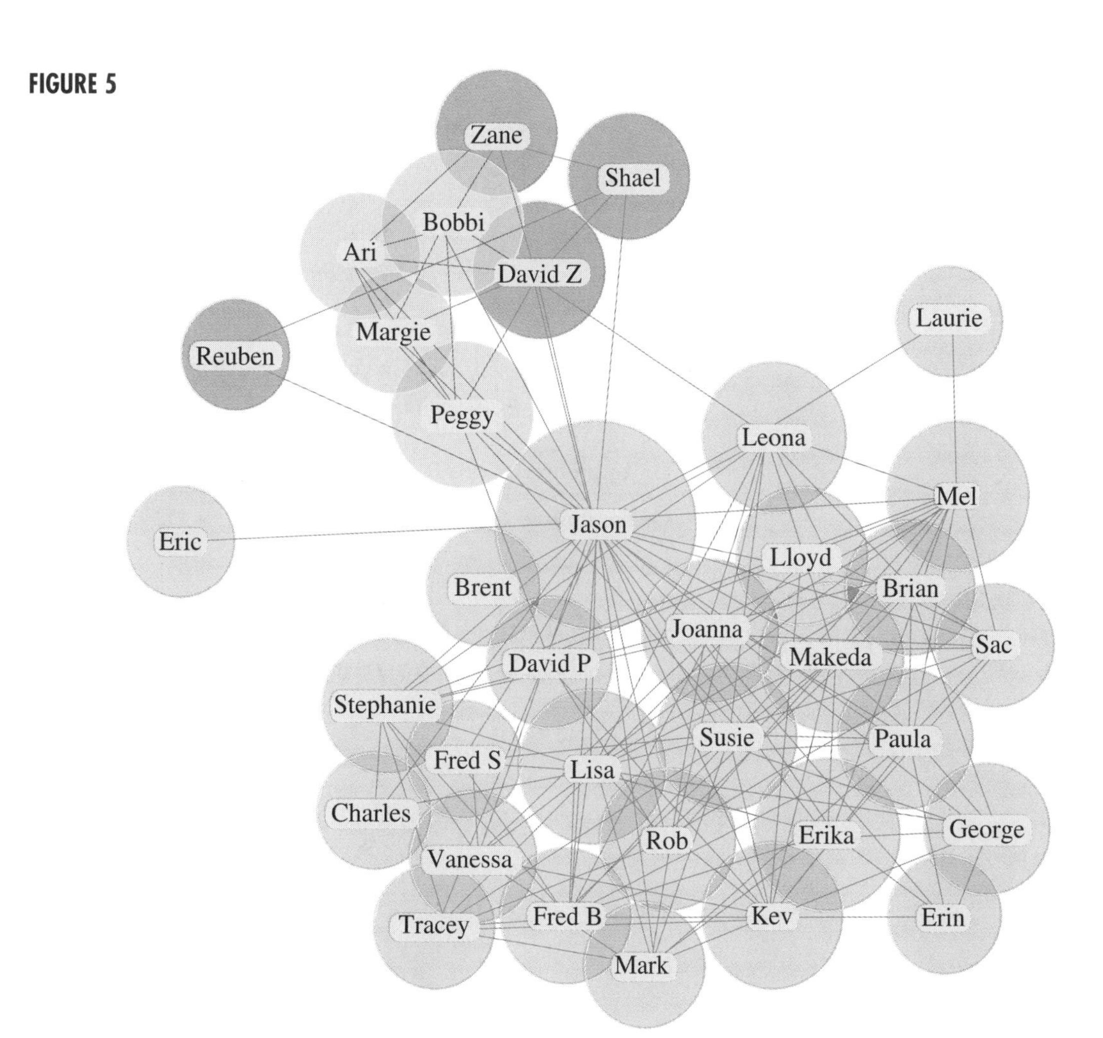

3.7 Exercises

In Exercises 1–5 let G be the graph with vertices {A, B, C, D, E, F}, and arcs {(A, B), (A, C), (A, E), (A, F), (B, A), (B, E), (C, A), (C, E), (D, A), (D, B), (D, E), (D, F), (E, B), (E, C), (F, A), (F, C), (F, E)}.

1. Draw G.

2. Write down the adjacency matrix of G (with respect to the natural ordering A, B, C, D, E, F).

3. List the in-degrees and out-degrees of each vertex.

4. If the arcs denote "likes," which node is the most popular? Why?

5. Is the graph strongly connected? Why or why not?

In Exercises 6–10 let G be the graph whose drawing is shown below.

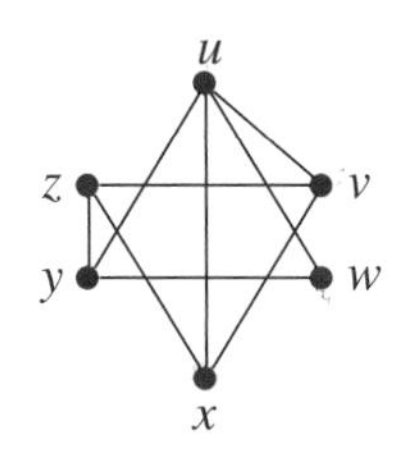

6. Is the graph undirected? Why or why not?

7. Write out the adjacency matrix.

8. How many walks are there of length 5 from u to v?

9. How many walks are there of length at most 5 from u to v?

10. What is the least value of k for which there is a walk of length k between *all* pairs of nodes?

For Exercises 11–15, suppose you have individuals A, B, D, E, F, H, J, K and committees C_1, C_2, C_3, with membership as follows:

On committee C_1: A, B, F, H, K

On committee C_2: A, E, H, J, K

On committee C_3: B, E, F, H, J, K.

Form a social network S on the nodes {A, B, D, E, F, H, J, K, C_1, C_2, C_3} with arcs (X, C_i) and (C_i, X) whenever person X belongs to committee C_i.

11. Draw S.

12. Is S undirected?

13. Write out the adjacency matrix of S with respect to the ordering A, B, D, E, F, H, J, K, C_1, C_2, C_3.

14. Explain how to use the adjacency matrix to find out how many common committees two individuals are on.

15. In this graph, what can you say about the number of walks between two individuals (or between two committees) of length k when k is odd? Why?

16. Prove parts (b) and (c) of Theorem 1.

17. Prove that if G is a graph with adjacency matrix A, then G has the "six degrees of separation" property if and only if $I + A + A^2 + A^3 + A^4 + A^5 + A^6$ has all positive entries.

18. Prove Theorem 3. (*Hint*: First show that if there is a walk from node u to another node v in a graph of length at least n, then there is a walk of length at most $n - 1$ from u to v).

19. The **distance** between two nodes u and v (written dist (u, v)) in an undirected graph G is the length of the shortest walk between them (if there is no walk between u and v, then the distance between them is defined to be ∞). Show that distance is a *metric* on the nodes of a graph; that is, for all nodes u, v, and w, we have

(a) $\text{dist}(u, u) = 0$,

(b) $\text{dist}(u, v) = \text{dist}(v, u)$, and

(c) $\text{dist}(u, v) + \text{dist}(v, w) \geq \text{dist}(u, w)$ (the *triangle inequality*).

20. Calculate the distance between all pairs of points in the graph whose drawing is given by Figure 4.

21. The **diameter** of an undirected graph G is the largest distance between pairs of points. How can you determine the diameter of a graph from its adjacency matrix?

22. Use *TouchGraph* to draw out your social network!

3.8 APPLICATION—Input–Output Analysis: Leontief Models

Wassily Leontief (1905–1999) was an economist who spent much of his research career at Harvard University. His work on input–output analysis won him a Nobel Prize in 1973. He applied his theory to study the flow of trade between countries.

In this section we consider *input–output analysis*, an application of linear algebra to economics. **Input–output analysis** is the study of the economics of situations involving production and consumption of various goods and services. The method of investigation described here is due to the famous economist Wassily Leontief. We consider two situations: first a closed model and then an open one.

Closed Leontief Model

A closed economic model is one in which each item (or value of service) produced by one segment of the economy is consumed by another segment of the same economy. In other words, there are no stored surpluses (or monetary equivalents) and no exports from the economy. We further assume that each segment of the economy requires a fixed fraction of the production of the other segments. The problem is to determine whether or not it is possible to have such a system with no "leftovers" and, if so, how much each segment should produce to achieve this state.

If we number the segments of the economy, say from 1 to n, the *input–output matrix* of the economy is defined to be an $n \times n$ matrix A in which each entry a_{ij} is the fraction of the total production of segment j required to supply the needs of segment i. (Since all of any segment's production must be consumed, the sum of the entries in any column must be 1.)

EXAMPLE 1

Assume the economy has three segments—farming, manufacturing, and service (such as repairing or transporting)—and assume that $\frac{1}{3}$ of the food produced by the farming segment is consumed by farmers themselves, the manufacturers consume $\frac{1}{2}$ of it, and the remaining $\frac{1}{6}$ is consumed by those offering services. Further assume that farmers requires $\frac{1}{2}$ of the manufactured goods, with the rest split evenly between the other two industries. Finally, assume that all three segments have equal need of services.

Solution By numbering the segments in order, we have

$$A = \begin{bmatrix} \frac{1}{3} & \frac{1}{2} & \frac{1}{3} \\ \frac{1}{2} & \frac{1}{4} & \frac{1}{3} \\ \frac{1}{6} & \frac{1}{4} & \frac{1}{3} \end{bmatrix}.$$

Let $\mathbf{x} = (x_1, x_2, x_3)$ be the dollar values associated with the output of each segment. Since the value produced by the farming segment must equal the sum of the amounts spent by farmers to fulfill their needs from the other segments (including other farmers), we have

$$x_1 = \frac{1}{3}x_1 + \frac{1}{2}x_2 + \frac{1}{3}x_3.$$

Similar equations for the expenditures by manufacturing and by service are

$$x_2 = \frac{1}{2}x_1 + \frac{1}{4}x_2 + \frac{1}{3}x_3$$

$$\text{and} \quad x_3 = \frac{1}{6}x_1 + \frac{1}{4}x_2 + \frac{1}{3}x_3.$$

In matrix form, these three equations can be written as

$$\mathbf{x} = A\mathbf{x},$$

and, since $\mathbf{x} = I\mathbf{x}$, we have the linear system

$$(I - A)\mathbf{x} = \mathbf{0}.$$

Solving it, we obtain

$$x_1 = \frac{5}{3}s, \qquad x_2 = \frac{14}{9}s, \quad \text{and} \quad x_3 = s.$$

Since these numbers are to represent the value of goods or services, only nonnegative solutions are realistic. Hence the system "balances" for any $s > 0$, and x_1, x_2, and x_3 as shown. ■

Returning to general considerations, let A be an input–output matrix and $\mathbf{x} = (x_1, \ldots, x_n)$ be the vector of values associated with the output of each segment. We wish to find the relative size of each segment of the economy.

Since the value produced by a given segment must equal the sum of the amounts spent by that segment to fulfill its needs from each of the other segments, we have the following system of equations:

$$\begin{aligned} x_1 &= a_{11}x_1 + a_{12}x_2 + \cdots + a_{1n}x_n \\ x_2 &= a_{21}x_1 + a_{22}x_2 + \cdots + a_{2n}x_n \\ &\vdots \\ x_n &= a_{n1}x_1 + a_{n2}x_2 + \cdots + a_{nn}x_n \end{aligned}$$

In matrix form, this becomes

$$\mathbf{x} = A\mathbf{x}$$

and, as in Example 1, we have

$$(I - A)\mathbf{x} = \mathbf{0}.$$

This system has the solution $\mathbf{x} = \mathbf{0}$ (nobody produces anything!), but it has other solutions as well. To be realistic, we can allow only nonnegative values for the entries of $\mathbf{x}$, since they are dollar values of output. The following theorem, which we assume without proof, tells us that there are realistic, nontrivial solutions.

THEOREM 1

Let A be the input–output matrix of a closed model. Then $(I - A)\mathbf{x} = \mathbf{0}$ has a solution in which all of its entries are positive. ■

Open Leontief Model

An **open economic model** is similar in structure to that of the closed model, except that one extra segment called *consumers* is added. The consumer segment is assumed to have a certain demand for each of the goods or services produced by the other segments. These demands are subject to change, however, so we wish to analyze the problem of what to do if the demands change. The other assumptions remain exactly as in the closed case—each segment requires a fixed fraction of the total output of each of the other segment's total production, and there are no surpluses (above consumer demand) produced.

Again we define the *input–output matrix* A of the economy in terms of fractions of production but with a slight change. We let a_{ij} be the fraction of the production of segment j required to supply the needs of segment i. Thus, the matrix A is the same as that of a closed economy in the case when all consumer demands equal zero. Letting $\mathbf{x} = (x_1, \ldots, x_n)$ be the production vector, $A\mathbf{x}$ is the vector of internal consumption. For instance, the first row of $A\mathbf{x}$ is

$$a_{11}x_1 + a_{12}x_2 + \cdots + a_{1n}x_n.$$

This represents the fraction of product 1 consumed by segment 1, plus the fraction of product 1 consumed by segment 2, and so on. The resulting sum is the amount of product 1 consumed by all of the producing segments. Similarly, the ith row of $A\mathbf{x}$ is the amount *produced* by the ith segment which is *consumed* by the producing segments.

Since there is an assumption that no surplus exists, the internal consumption plus the demand for each product must equal the total production of that product. Letting $\mathbf{d}$ be the vector of demands, with d_i being the demand for the ith product, we have, for each i,

$$a_{i1}x_1 + a_{i2}x_2 + \cdots + a_{in}x_n + d_i = x_i.$$

In terms of matrices, this system becomes

$$A\mathbf{x} + \mathbf{d} = \mathbf{x}.$$

Transposing $\mathbf{x}$, we have

$$(I - A)\mathbf{x} = \mathbf{d}$$

and, if $I - A$ is invertible,

$$\mathbf{x} = (I - A)^{-1}\mathbf{d}.$$

As in the case of the closed economy, we can accept only vectors $\mathbf{x}$ for which each entry is nonnegative. The following theorem (assumed without proof) gives fairly general conditions under which a solution can be obtained.

THEOREM 2

If a square matrix A has all positive entries and the sum of the entries in each of its columns is less than one, then $(I - A)$ is invertible. ■

EXAMPLE 2

In the economy of farming, manufacturing, and service industries of Example 1, suppose there is also a consumer segment. Table 1 gives the dollar value of each product consumed by a given segment. That is, an entry in the ith row of Table 1 gives the value (in millions of dollars) of the product associated with that row, which is consumed by the industry associated with that column. If the anticipated consumer demands for five years later are farming, 20, manufacturing, 30, and service, 20, find the amount each segment needs to produce in order to balance the system.

TABLE 1

Industry	Farming	Manufacturing	Service	Consumer	Total
Farming	10	15	5	10	40
Manufacturing	14	7	7	21	49
Service	6	6	12	6	30

Solution The input–output matrix A for the economy is computed as follows: a_{ij} is that fraction of the total of the jth segment consumed by the ith segment (ignoring the consumer segment). For example, the first column of A is $\mathbf{a}^1 = (\frac{10}{40}, \frac{14}{40}, \frac{6}{40})$, since farming consumes 10 of the total 40, manufacturing 14, and service 6. Computing the other entries, we have

$$A = \begin{bmatrix} \frac{10}{40} & \frac{15}{49} & \frac{5}{30} \\ \frac{14}{40} & \frac{7}{49} & \frac{7}{30} \\ \frac{6}{40} & \frac{6}{49} & \frac{12}{30} \end{bmatrix}.$$

Letting $\mathbf{d} = (20, 30, 20)$ be the vector of demands, we determine the solution to be

$$x = (I - A)^{-1}d = \begin{bmatrix} \frac{30}{40} & -\frac{15}{49} & -\frac{5}{30} \\ -\frac{14}{40} & \frac{42}{49} & -\frac{7}{30} \\ -\frac{6}{40} & -\frac{6}{49} & \frac{18}{30} \end{bmatrix}^{-1} \begin{bmatrix} 20 \\ 30 \\ 20 \end{bmatrix} = \begin{bmatrix} 77.18 \\ 85.60 \\ 70.10 \end{bmatrix}.$$

Thus, the approximate value of the farming segment should be $x_1 = 77.18$, manufacturing, $x_2 = 85.60$, and service, $x_3 = 70.10$. ■

3.8 Exercises

In Exercises 1 and 2 the given matrix *M* is the input–output matrix of a closed economy of three segments *A*, *B*, and *C*. Find the value of production for each segment.

1. $M = \begin{bmatrix} \frac{3}{4} & \frac{1}{5} & \frac{1}{3} \\ \frac{1}{8} & \frac{2}{5} & \frac{1}{6} \\ \frac{1}{8} & \frac{2}{5} & \frac{1}{2} \end{bmatrix}$

2. $M = \begin{bmatrix} 0.2 & 0.1 & 0.3 \\ 0.4 & 0.8 & 0.3 \\ 0.4 & 0.1 & 0.4 \end{bmatrix}$

3. Assume a closed economy has three segments *A*, *B*, and *C*, with *A* requiring $\frac{1}{2}$ of its own production and *B* and *C* splitting the rest evenly. All three segments require the same fraction of the production of *B*. Segment *A* requires $\frac{2}{3}$ of the production of *C*; *B* and *C* require the same amount of *C*. Find the value of production for each segment.

4. The closed Leontief model problem is sometimes viewed as a problem in *pricing* rather than *production*. The assumption is that in a closed model each segment requires not only a fixed *fraction* of the production of the other segments but a fixed *amount* of the production. Viewed in this manner, the problem is to decide how each segment should price its production in order to spend exactly what it earns. Explain the result of Exercise 3 in this context.

In Exercises 5 and 6 the given matrix M is the input–output matrix of an open economy of three segments A, B, and C, and the vector d is the predicted demand vector of the consumer segment. Find the production level for each segment required to balance the economy.

5. $M = \begin{bmatrix} \frac{1}{3} & \frac{1}{5} & \frac{1}{6} \\ \frac{1}{4} & \frac{2}{5} & \frac{1}{3} \\ \frac{1}{4} & \frac{1}{5} & \frac{1}{6} \end{bmatrix}, \quad \mathbf{d} = \begin{bmatrix} 3 \\ 5 \\ 1 \end{bmatrix}$

6. $M = \begin{bmatrix} 0.2 & 0.1 & 0.5 \\ 0.3 & 0.2 & 0.3 \\ 0.1 & 0.1 & 0.1 \end{bmatrix}, \quad \mathbf{d} = \begin{bmatrix} 20 \\ 15 \\ 5 \end{bmatrix}$

7. Assume that an open economy has three producing segments A, B, and C. Table 2 gives the current value of the production of the segment of each row that is consumed by the segment of each column. Calculate the production required of each industry to provide a balanced economy if the consumer demand rises to $\mathbf{d} = (15, 10, 6)$.

TABLE 2

	A	*B*	*C*	Consumers
A	12	4	8	10
B	8	9	4	6
C	2	4	3	3

Computational Exercise

Use a computer software package or graphing calculator to solve the following problem.

8. Assume an open economy with producing segments A, B, C, D, and E. Table 3 gives the value of the product associated with that row, which is consumed by the segment associated with that column. If the demand rises to 25 for A, 20 for B, 15 for C, 15 for D, and 20 for E, find the amount each segment needs to produce in order to balance the system.

TABLE 3

	A	*B*	*C*	*D*	*E*	Consumers
A	12	20	16	15	18	15
B	8	12	3	2	8	13
C	10	6	15	8	5	10
D	15	10	20	5	20	12
E	20	5	32	20	5	15

CHAPTER SUMMARY

Operations on Matrices

Let $A = [a_{ij}]$ and $B = [b_{ij}]$. Then, the i, j-entry of

(a) $A + B$ is $a_{ij} + b_{ij}$,

(b) $A - B$ is $a_{ij} - b_{ij}$,

(c) cA is ca_{ij},

(d) AB is $\mathbf{a}_i \cdot \mathbf{b}^j = a_{i1}b_{1j} + a_{i2}b_{2j} + \cdots + a_{in}b_{nj}$,

(e) A^T is a_{ji}.

Inverse of a Matrix

(a) DEFINITION: $B = A^{-1}$ if and only if $AB = BA = I$ (the identity matrix).

(b) To find A^{-1}, row-reduce $[A \mid I]$ to $[I \mid X]$, if possible. Then, $X = A^{-1}$. If A cannot be transformed to I, then A^{-1} does not exist.

(c) An $n \times n$ matrix A is invertible if and only if any of the following conditions hold:

- $A\mathbf{x} = \mathbf{b}$ has a unique solution for arbitrary b.
- The homogeneous linear system $A\mathbf{x} = \mathbf{0}$ has a unique solution.
- The row-reduced echelon form of A is the identity matrix.

Method of Inverses

(a) To solve $A\mathbf{x} = \mathbf{b}$, where A is square, find A^{-1}, and then multiply: $\mathbf{x} = A^{-1}\mathbf{b}$

(b) The method of inverses works if and only if $A\mathbf{x} = \mathbf{b}$ has a unique solution.

Rank of a Matrix

(a) DEFINITION: The *rank* of a matrix A is the number of nonzero rows in its row-reduced echelon form.

(b) Consider the linear system $A\mathbf{x} = \mathbf{b}$, where A is an $m \times n$ matrix. Let $p = \text{rank}(A)$ and $q = \text{rank}([A \mid \mathbf{b}])$. Then this system

- Has no solution if $p < q$.
- Has a unique solution if $p = q = n$.
- Has infinitely many solutions if $p = q$ and $p < n$.

LU Decomposition

(a) DEFINITION: The LU decomposition of A is $LU = PA$, where L is lower, U is upper, and P is a permutation matrix.

(b) To find the LU decomposition of A, transform A to echelon form (as in Gaussian elimination). Then

- U is the echelon form of A.
- L is the square matrix with zeros above the main diagonal, ones on the diagonal, and the negatives of the row multipliers used in the reduction process below the diagonal.

- P is the matrix obtained from the identity matrix by applying the row interchanges used in the reduction process to it.

(c) To solve $A\mathbf{x} = \mathbf{b}$ by an LU decomposition process:

i. Obtain the LU decomposition of A: $LU = PA$.

ii. Solve $L\mathbf{y} = P\mathbf{b}$ for $\mathbf{y}$ by forward-substitution.

iii. Solve $U\mathbf{x} = \mathbf{y}$ for the solution $\mathbf{x}$ by back-substitution.

Elementary Matrices

(a) DEFINITION: An *elementary matrix* is one that can be obtained from an identity matrix by performing an elementary row operation on it.

(b) Performing an elementary row operation on a matrix A is equivalent to multiplying A on the left by the corresponding elementary matrix.

(c) The inverse of an elementary matrix is an elementary matrix of the same form.

KEYWORDS

dimensions of a matrix
row vector
zero matrix
square matrix
transpose
matrix form of a linear system
homogeneous
rank
elementary matrix
normal equations
open economic model
column vector
scalar multiple
matrix multiplication
order
coefficient matrix
invertible
trivial solution
LU decomposition
least square polynomials
social network
closed economic model

SELF-TEST

In Problems 1–3, let

$$A = \begin{bmatrix} 1 & 0 \\ -1 & 2 \end{bmatrix} \quad \text{and} \quad B = \begin{bmatrix} 2 & 0 & 1 \\ 3 & -1 & 0 \end{bmatrix}.$$

Perform the indicated operations.

1. $3A - I$ *2.* AB *3.* B^T [Section 3.1]

4. Find the inverse of the matrix

$$C = \begin{bmatrix} 1 & 2 & 3 \\ 0 & 1 & 4 \\ 0 & 0 & 1 \end{bmatrix}.$$

[Section 3.2]

5. Solve the following linear system by the method of inverses.

$$\begin{aligned} 2x - y &= -1 \\ -x + y &= 2 \end{aligned}$$

[Section 3.2]

6. Suppose that the row-reduced echelon form of a linear system $A\mathbf{x} = \mathbf{b}$ has the following form, where a and b are arbitrary constants.

$$\left[\begin{array}{ccc|c} 1 & 0 & 0 & a \\ 0 & 1 & b & 0 \\ 0 & 0 & 0 & 0 \end{array}\right]$$

(a) What is the rank of A?
(b) How many solutions does $A\mathbf{x} = \mathbf{b}$ have?
(c) Is A invertible? [Section 3.3]

7. Is each of the following statements true or false? [Section 3.3]

(a) Every system of linear equations has at least one solution.
(b) If A^{-1} exists, then the linear system $A\mathbf{x} = \mathbf{b}$ has a unique solution for every right-side vector $\mathbf{b}$.

8. Find the LU decomposition of the matrix [Section 3.4]

$$A = \begin{bmatrix} 1 & 2 & 3 \\ -1 & -2 & 1 \\ 2 & 6 & 0 \end{bmatrix}.$$

9. Use the result of Problem 8 to solve the linear system $A\mathbf{x} = \mathbf{b}$, where A is the matrix of Problem 8 and $\mathbf{b} = (1, 3, 2)$. [Section 3.4]

10. For the matrix A of Problem 8, find elementary matrices $E_1, E_2, \ldots, E_n$, such that the matrix $R = E_n, \ldots, E_2E_1A$ is in row-reduced echelon form. [Section 3.5]

REVIEW EXERCISES

In Exercises 1–12 let

$$A = \begin{bmatrix} 1 & 0 & 1 \\ 0 & 1 & 3 \\ 0 & 0 & 3 \end{bmatrix} \quad \text{and} \quad B = \begin{bmatrix} -1 & 1 & 1 \\ 1 & 3 & 1 \\ 2 & 0 & -1 \end{bmatrix},$$

and determine each of the following.

1. $2A - 3B$
2. $2A + B - 3I$
3. AB
4. BA
5. A^{-1}
6. B^{-1}
7. $(A^TB)^{-1}$
8. $(A^TA)^{-1}$
9. rank of A
10. rank of B
11. $-A^2$
12. A^3

Write each of the linear systems in Exercises 13–18 in matrix form, and solve them by the method of inverses, if possible. If the method of inverses cannot be used, explain why.

13. $$\begin{aligned} x - 2y &= -1 \\ -2x + 4y &= 2 \end{aligned}$$

14. $$\begin{aligned} -x + y - z &= 1 \\ x \qquad + z &= 3 \end{aligned}$$

15. $$\begin{aligned} x_1 - 2x_2 + x_3 &= 0 \\ -x_1 + 2x_2 - x_3 &= 0 \end{aligned}$$

16. $$\begin{aligned} x_1 + 2x_2 - 3x_3 &= -4 \\ 2x_1 - x_2 + x_3 &= 5 \\ 2x_2 + 3x_3 &= 1 \end{aligned}$$

17. $$\begin{aligned} u + 2v - 3w &= 4 \\ -v + w &= -1 \\ 2u \qquad - w &= 3 \end{aligned}$$

18. $$\begin{aligned} u + v - w &= 1 \\ v + 2w &= 0 \\ u + 4v + 5w &= 2 \end{aligned}$$

In Exercises 19–22 the given matrix is the augmented matrix of a linear system. Without solving the system, determine how many solutions it has. If there are infinitely many solutions, determine the number of parameters present.

19. $\left[\begin{array}{ccc|c} 1 & 0 & 0 & 0 \\ 0 & 1 & 0 & 0 \\ 0 & 0 & 1 & 0 \end{array}\right]$

20. $\left[\begin{array}{cccc|c} 1 & 0 & 2 & 3 & 0 \\ 0 & 1 & -1 & 4 & 1 \end{array}\right]$

21. $\left[\begin{array}{ccc|c} 1 & 2 & 3 & 0 \\ 4 & 5 & 6 & 0 \end{array}\right]$

22. $\left[\begin{array}{ccc|c} 1 & 2 & 3 & 0 \\ 0 & 0 & 0 & 1 \\ 0 & 0 & 0 & 0 \end{array}\right]$

23. Confirm that L and U yield an LU decomposition of the matrix B at the beginning of these exercises where

$$L = \begin{bmatrix} 1 & 0 & 0 \\ -1 & 1 & 0 \\ -2 & \frac{1}{2} & 1 \end{bmatrix} \quad \text{and} \quad U = \begin{bmatrix} -1 & 1 & 1 \\ 0 & 4 & 2 \\ 0 & 0 & 0 \end{bmatrix}.$$

24. Find an LU decomposition for B different from that given in Exercise 23, this time with a permutation to avoid fractions.

25–30. Solve each of the systems in Exercises 13–18 using LU decomposition.

31. Give the 3×3 elementary matrix for which premultiplication corresponds to each row operation:

(a) Add 2 times the second row to the first.

(b) Interchange rows 1 and 3.

(c) Double the last row.

32. Find the inverse of each matrix of Exercise 31, and describe the row operation that corresponds to that matrix.

CHALLENGE YOURSELF

33. Suppose that A is an $n \times n$ matrix whose entries on and below the main diagonal are 0. Prove that A^n is an $n \times n$ zero matrix.

34. For what value(s) of a is the matrix A invertible?

$$A = \begin{bmatrix} 1 & 1 & 1 \\ -1 & 1 & a \\ -1 & a & -1 \end{bmatrix}$$

35. Two matrices A and B are said to *commute* if $AB = BA$. Find all matrices that commute with

$$A = \begin{bmatrix} 1 & 1 \\ -1 & 0 \end{bmatrix}.$$

36. Suppose that B is a 3×3 matrix that commutes with all other 3×3 matrices. What form does B have?

37. Suppose that A is an invertible $n \times n$ matrix. Let B be the $(n-1) \times (n-1)$ matrix formed from A by removing the last row and column. Prove that rank $(B) \geq n - 2$.

SPOTLIGHT

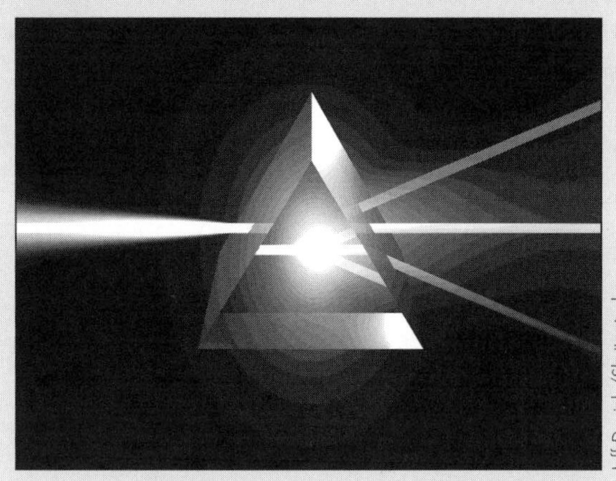
Jeff Daniels/Shutterstock

In Chapter 1, we used determinants to provide a simpler-looking formula for the cross product of two vectors. This elementary application of determinants illustrates one of their most important uses—determinants provide a convenient, easy-to-remember notation for expressing the solution to certain problems.

As another example, recall, from Section 2.1, that lines in $\mathbf{R}^2$, planes in $\mathbf{R}^3$, and hyperplanes in $\mathbf{R}^m$ can be represented by linear equations in two, three, and m variables, respectively. We can use determinants to express these equations in a simple way. For example, an equation of the line in $\mathbf{R}^2$ through the (distinct) points (x_1, y_1) and (x_2, y_2) can be written as

$$\det \begin{bmatrix} 1 & x_1 & y_1 \\ 1 & x_2 & y_2 \\ 1 & x & y \end{bmatrix} = 0$$

and an equation of the plane through the noncollinear points (x_1, y_1, z_1), (x_2, y_2, z_2), and (x_3, y_3, z_3) is given by

$$\det \begin{bmatrix} 1 & x_1 & y_1 & z_1 \\ 1 & x_2 & y_2 & z_2 \\ 1 & x_3 & y_3 & z_3 \\ 1 & x & y & z \end{bmatrix} = 0.$$

One more example: Exercise 37 of Section 1.2 tells us that the volume of a *parallelepiped* (see the figure) is given by the absolute value of $\mathbf{u} \cdot (\mathbf{v} \times \mathbf{w})$. This expression can in turn be expressed as $|\det(A)|$, where

$$A = \begin{bmatrix} u_1 & u_2 & u_3 \\ v_1 & v_2 & v_3 \\ w_1 & w_2 & w_3 \end{bmatrix}.$$

We return to these examples in the exercise sets of Sections 4.1 and 4.2. In Section 4.1, we show that determinants provide an easy way to check whether a matrix is invertible. Then, in Section 4.3, we present two more examples demonstrating the notational convenience of determinants—here, they provide formulas for the inverse of a matrix and the solution to certain linear systems.

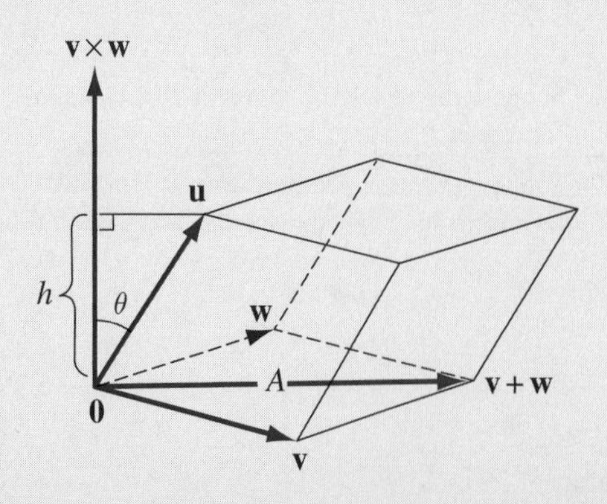

CHAPTER 4

Determinants

OBJECTIVES

You will learn to:

- Calculate the determinant of a square matrix of any order.
- Use elementary row (and column) operations to reduce the work involved in finding a determinant.
- Determine if a matrix is invertible by knowing the value of its determinant.
- Use determinants (in Cramer's rule) to solve certain systems of linear equations.
- Use determinants to find the inverse of a matrix.
- Calculate the eigenvalues and the corresponding eigenvectors of a square matrix.

Associated with each square matrix is a single number called the *determinant* of that matrix. In this chapter, we discuss the concept of a determinant, and describe its properties and some of its applications.

4.1 Definition of a Determinant

In Section 1.2, we defined the determinant of a 2×2 matrix for the sole purpose of providing a simple means of remembering the definition of cross product. In this section, we extend the definition to square matrices of any order. We first give a convention that makes this description easier.

NOTATION For an $m \times n$ matrix A, the *submatrix* A_{ij} is obtained by deleting the ith row and jth column of A.

The notion of a determinant was introduced by Gottfried Leibniz (1646–1716). Leibniz was a very prolific mathematician, who worked in many areas of mathematics. Both he and Sir Isaac Newton are credited (independently) with creating calculus.

EXAMPLE 1

Find the submatrix A_{23}, where

$$A = \begin{bmatrix} 1 & 1 & 0 & 2 \\ 2 & 1 & 1 & 1 \\ 3 & 0 & 0 & -1 \\ 1 & 1 & 2 & 1 \end{bmatrix}.$$

Solution We mentally line out the second row and the third column,

$$\begin{bmatrix} 1 & 1 & 0 & 2 \\ 2 & 1 & 1 & 1 \\ 3 & 0 & 0 & -1 \\ 1 & 1 & 2 & 1 \end{bmatrix},$$

obtaining

$$A_{23} = \begin{bmatrix} 1 & 1 & 2 \\ 3 & 0 & -1 \\ 1 & 1 & 1 \end{bmatrix}.$$

■

DEFINITION

Let $A = [a_{ij}]$ be an $n \times n$ matrix. The **determinant** of A is given by

i. $\det A = a_{11}$, if $n = 1$;

ii. $\det A = a_{11}a_{22} - a_{12}a_{21}$, if $n = 2$;

iii. $\det A = a_{11} \det A_{11} - a_{12} \det A_{12} + \cdots + (-1)^{1+n}a_{1n} \det A_{1n}$

$$= \sum_{j=1}^{n} (-1)^{1+j} a_{1j} \det \mathrm{A}_{1j},$$

if $n > 2$.

In words, case (iii) may be described as follows: Multiply each entry a_{1j} in the first row by $\det A_{1j}$, the determinant of the submatrix formed by deleting the row and the column that contain a_{1j}. Starting with a positive sign, alternate the signs of these terms—plus, minus, plus, minus, and so on. Finally, add all such terms.

NOTE The definition of determinant is a *recursive definition.* An explicit rule is given for $n = 1$ and $n = 2$. For $n = 3$ each A_{1j} is of order 2, so $\det A_{1j}$ has been defined. For $n = 4$ each A_{1j} is of order 3, so $\det A_{1j}$ can be computed by the $n = 3$ case. This process may be continued for all positive integers n.

EXAMPLE 2

Compute $\det A$, where A is given by

$$A = \begin{bmatrix} 2 & 1 & -3 \\ -3 & -2 & 0 \\ 2 & 1 & 2 \end{bmatrix}.$$

Solution Since A is order 3, the definition of determinant yields

$$\det A = a_{11} \det A_{11} - a_{12} \det A_{12} + a_{13} \det A_{13}$$
$$= (2) \det \begin{bmatrix} -2 & 0 \\ 1 & 2 \end{bmatrix} - (1) \det \begin{bmatrix} -3 & 0 \\ 2 & 2 \end{bmatrix} + (-3) \det \begin{bmatrix} -3 & -2 \\ 2 & 1 \end{bmatrix}$$

$$\begin{aligned}
&= (2)[(-2)(2) - (1)(0)] - (1)[(-3)(2) - (2)(0)] \\
&\quad + (-3)[(-3)(1) - (2)(-2)] \\
&= 2(-4) - (1)(-6) + (-3)(1) \\
&= -5.
\end{aligned}$$

■

EXAMPLE 3

Compute det A, where

$$A = \begin{bmatrix} 2 & 1 & -3 & 1 \\ -3 & -2 & 0 & 2 \\ 2 & 1 & 0 & -1 \\ 1 & 0 & 1 & 2 \end{bmatrix}.$$

Solution We have

$$\det A = (2)\det A_{11} - (1)\det A_{12} + (-3)\det A_{13} - (1)\det A_{14}.$$

This looks innocent enough, but note that each A_{1j} is of order 3, and computation of det A_{1j} requires repeating the process of Example 2. This must be done four times to finish the job. These values are as follows: det $A_{11} = 0$, det $A_{12} = 1$, det $A_{13} = 2$, and det $A_{14} = 1$. The final result is det $A = -8$. ■

We remark that the strange formula for the vector cross product in $\mathbf{R}^3$ can be succinctly written down via determinants. For by setting $\mathbf{i} = (1, 0, 0)$, $\mathbf{j} = (0, 1, 0)$, and $\mathbf{k} = (0, 0, 1)$ and allowing the definition for determinants to extend to vector entries, we have

$$\begin{aligned}
\begin{vmatrix} \mathbf{i} & \mathbf{j} & \mathbf{k} \\ u_1 & u_2 & u_3 \\ v_1 & v_2 & v_3 \end{vmatrix} &= \mathbf{i}\begin{vmatrix} u_2 & u_3 \\ v_2 & v_3 \end{vmatrix} - \mathbf{j}\begin{vmatrix} u_1 & u_3 \\ v_1 & v_3 \end{vmatrix} + \mathbf{k}\begin{vmatrix} u_1 & u_2 \\ v_1 & v_2 \end{vmatrix} \\
&= \begin{vmatrix} u_2 & u_3 \\ v_2 & v_3 \end{vmatrix}(1, 0, 0) - \begin{vmatrix} u_1 & u_3 \\ v_1 & v_3 \end{vmatrix}(0, 1, 0) + \begin{vmatrix} u_1 & u_2 \\ v_1 & v_2 \end{vmatrix}(0, 0, 1) \\
&= \left[\begin{vmatrix} u_2 & u_3 \\ v_2 & v_3 \end{vmatrix}, -\begin{vmatrix} u_1 & u_3 \\ v_1 & v_3 \end{vmatrix}, \begin{vmatrix} u_1 & u_2 \\ v_1 & v_2 \end{vmatrix}\right] \\
&= (u_2v_3 - u_3v_2, u_3v_1 - u_1v_3, u_1v_2 - u_2v_1) \\
&= \mathbf{u} \times \mathbf{v}.
\end{aligned}$$

Fortunately, there are easier ways to compute det A. In the remainder of this section and in Section 4.2, we discuss some ways to simplify this process considerably.

DEFINITION

Let A be a square matrix of order n, and let A_{ij} denote the submatrix (of order $n - 1$) obtained from A by deleting its ith row and its jth column. Then

i. the **i, j-minor** of A is given by $\det(A_{ij})$, and
ii. the **i, j-cofactor** of A is given by $(-1)^{i+j}\det(A_{ij})$.

EXAMPLE 4

Find the 2,3-minor and the 2,3-cofactor of

$$A = \begin{bmatrix} 2 & -2 & 1 & 1 \\ 1 & 3 & 3 & 2 \\ 1 & 0 & 9 & 1 \\ 3 & 4 & 2 & 0 \end{bmatrix}.$$

Solution The 2,3-minor is

$$\det A_{23} = \det\begin{bmatrix} 2 & -2 & 1 \\ 1 & 0 & 1 \\ 3 & 4 & 0 \end{bmatrix} = -10.$$

The 2,3-cofactor is

$$(-1)^{2+3}\det A_{23} = -\det\begin{bmatrix} 2 & -2 & 1 \\ 1 & 0 & 1 \\ 3 & 4 & 0 \end{bmatrix} = 10.$$

■

Note that Equation (iii) in the definition of the determinant may be written as

$$\begin{aligned}\det A &= a_{11}[(-1)^{1+1}\det A_{11}] + a_{12}[(-1)^{1+2}\det A_{12}] \\ &\quad + \cdots + a_{1n}[(-1)^{1+n}\det A_{1n}] \\ &= \sum_{j=1}^{n} a_{1j}[(-1)^{1+j}\det A_{1j}].\end{aligned}$$

In other words, det A is the sum of the products of each entry in the first row with the cofactor for that position. For this reason, Equation (iii) of the definition is called **cofactor expansion** along the first row (or simply **expansion** along the first row).

The next theorem states that cofactor expansion along *any* row (or down any column) of a square matrix also yields the determinant. Its proof is very cumbersome and is omitted here.

THEOREM 1

Let A be a matrix of order n with $n \geq 2$. Let i be a fixed row number. Then

$$\begin{aligned}\det A &= (-1)^{i+1}a_{i1}\det A_{i1} + (-1)^{i+2}a_{i2}\det A_{i2} \\ &\quad + \cdots + (-1)^{i+n}a_{in}\det A_{in} \\ &= \sum_{k=1}^{n} (-1)^{i+k}\, a_{ik}\det A_{ik}.\end{aligned}$$

Let j be a fixed column number. Then

$$\begin{aligned}\det A &= (-1)^{1+j}a_{1j}\det A_{1j} + (-1)^{2+j}a_{2j}\det A_{2j} \\ &\quad + \cdots + (-1)^{n+j}\, a_{nj}\det A_{nj} \\ &= \sum_{k=1}^{n} (-1)^{k+j}\, a_{kj}\det A_{kj}.\end{aligned}$$

■

In computations, it is convenient to choose a row or a column with as many zeros as possible, because it is not necessary to compute the cofactors of zero entries—the product always is zero.

EXAMPLE 5

Compute det A by expansion along the row or column with the maximum number of zeros, where

$$A = \begin{bmatrix} 2 & 1 & -3 & 1 \\ -3 & -2 & 0 & 2 \\ 2 & 1 & 0 & -1 \\ 1 & 0 & 1 & 2 \end{bmatrix}.$$

Solution This is just Example 3 again, but expanded down the third column:

$$\begin{aligned}\det A &= (-1)^{1+3}(-3)\det A_{13} + (-1)^{2+3}(0)\det A_{23}\\ &\quad + (-1)^{3+3}(0)\det A_{33} + (-1)^{4+3}(1)\det A_{43}\\ &= -3\det A_{13} - \det A_{43}\end{aligned}$$

The computation still requires evaluation of the determinants of two 3×3 matrices, but that is half as much work as was required before. Of course, the numerical value of det A is still -8. ■

NOTE It is easy to obtain the sign factors $(-1)^{i+j}$ in the terms of the expansion of a determinant, since the signs alternate when moving horizontally or vertically from one position to another (like the coloured squares on a checkerboard), as indicated in the following arrays for $n = 4$ and $n = 5$:

$$\begin{bmatrix} + & - & + & - \\ - & + & - & + \\ + & - & + & - \\ - & + & - & + \end{bmatrix} \quad \text{and} \quad \begin{bmatrix} + & - & + & - & + \\ - & + & - & + & - \\ + & - & + & - & + \\ - & + & - & + & - \\ + & - & + & - & + \end{bmatrix}$$

For example, if you wish to expand down the jth column, start in the upper left corner and proceed plus, minus, plus, minus, and so on, along the first row to the top of the jth column. That establishes the first position sign, and they alternate thereafter. If you are expanding along the ith row, alternate signs down the first column to the ith row, and proceed from there.

EXAMPLE 6

Compute det A, where

$$A = \begin{bmatrix} 2 & 9 & 3 & 2 \\ 4 & 0 & 0 & 6 \\ 3 & -1 & 1 & 2 \\ 5 & 0 & 0 & 1 \end{bmatrix}.$$

Solution For the simplest computation we should expand along the second row, the fourth row, the second column, or the third column, since all contain two zeros. We use the fourth row. Going down the first column—plus, minus, plus, minus—we see that the first position is negative, and therefore

$$\begin{aligned}\det A &= -(5)\det A_{41} + (0)\det A_{42} - (0)\det A_{43} + (1)\det A_{44}\\ &= -5\det\begin{bmatrix} 9 & 3 & 2 \\ 0 & 0 & 6 \\ -1 & 1 & 2 \end{bmatrix} + \det\begin{bmatrix} 2 & 9 & 3 \\ 4 & 0 & 0 \\ 3 & -1 & 1 \end{bmatrix}.\end{aligned}$$

The second row is the best for each of these third-order determinants,

$$\det\begin{bmatrix} 9 & 3 & 2 \\ 0 & 0 & 6 \\ -1 & 1 & 2 \end{bmatrix} = -(6)\det\begin{bmatrix} 9 & 3 \\ -1 & 1 \end{bmatrix} = -6(12) = -72$$

$$\text{and} \quad \det\begin{bmatrix} 2 & 9 & 3 \\ 4 & 0 & 0 \\ 3 & -1 & 1 \end{bmatrix} = -(4)\det\begin{bmatrix} 9 & 3 \\ -1 & 1 \end{bmatrix} = -4(12) = -48.$$

Therefore,

$$\det A = -5(-72) + (-48) = 312.$$ ■

The freedom to compute determinants along any row or column makes the computation of determinants of certain special matrices extremely simple.

THEOREM 2

If a square matrix has a zero row or column, its determinant is zero.

Proof Expand along the zero row or column. ■

THEOREM 3

If a square matrix has two equal rows (or two equal columns), its determinant is zero.

Proof Let n be the order of A. If $n = 2$, A has the form

$$A = \begin{bmatrix} a_{11} & a_{12} \\ a_{11} & a_{12} \end{bmatrix},$$

so that $\det A = a_{11}a_{12} - a_{11}a_{12} = 0$.

For $n > 2$, let A be a matrix of order n with two equal rows. Since n is at least 3, we can choose a row (say, the ith) that is not one of the two equal rows. Expand along this row to obtain

$$\det A = (-1)^{i+1}a_{i1} \det A_{i1} + (-1)^{i+2}a_{i2} \det A_{i2} + \cdots + (-1)^{i+n}a_{in} \det A_{in}.$$

Now each of the matrices A_{ij} has two equal rows (deleting a column does not disturb the equality of the rows). If $n = 3$, each of the A_{ij} is of order $n - 1 = 2$, and by the first case, $\det A_{ij} = 0$ for each j. But then $\det A = 0$ as well. If $n > 3$, repeat the process just described as often as needed until you get down to the 2×2 case in each term. Thus, $\det A = 0$.

A similar argument holds if two columns are equal. ■

EXAMPLE 7

Compute $\det B$ for

$$B = \begin{bmatrix} 2 & 1 & 3 & 2 \\ 1 & 4 & 2 & 1 \\ 3 & -2 & 1 & 3 \\ 2 & 1 & 0 & 2 \end{bmatrix}.$$

Solution The first and last columns are equal, so $\det B = 0$. ■

Recall, from Section 3.1, that the *main diagonal* of a square matrix A of order n consists of the entries a_{ii} for $i = 1$ to n. A square matrix is called **upper triangular** if all the entries below the main diagonal are zero and **lower triangular** if all the entries above the main diagonal are zero. Of course, a **diagonal matrix** (Section 3.1) is both upper and lower triangular.

As examples, let

$$A = \begin{bmatrix} 1 & 2 \\ 0 & 3 \end{bmatrix} \quad \text{and} \quad B = \begin{bmatrix} 2 & 0 & 0 \\ 1 & 2 & 0 \\ -1 & 1 & 3 \end{bmatrix}.$$

Here A is upper triangular and B is lower triangular.

THEOREM 4

If a matrix A of order n is upper triangular, lower triangular, or diagonal, then $\det A = a_{11}a_{22}\cdots a_{nn}$, the product of the entries on the main diagonal.

Proof Suppose that A is upper triangular. Then A has the form

$$\begin{bmatrix} a_{11} & a_{12} & \cdots & a_{1n} \\ 0 & a_{22} & \cdots & a_{2n} \\ \vdots & & \ddots & \vdots \\ 0 & 0 & \cdots & a_{nn} \end{bmatrix}.$$

Expansion down the first column gives at most one nonzero term

$$\det A = (-1)^{1+1}a_{11} \det A_{11} = a_{11} \det \begin{bmatrix} a_{22} & \cdots & a_{2n} \\ \vdots & \ddots & \vdots \\ 0 & \cdots & a_{nn} \end{bmatrix}.$$

The resulting matrix A_{11} is still upper triangular, so the same argument gives

$$\det A = a_{11}a_{22} \det \begin{bmatrix} a_{33} & \cdots & a_{3n} \\ \vdots & \ddots & \vdots \\ 0 & \cdots & a_{nn} \end{bmatrix}.$$

Continue this process until the submatrix is 2×2, so that

$$\begin{aligned} \det A &= a_{11}a_{22}\cdots a_{n-2,n-2} \det \begin{bmatrix} a_{n-1,n-1} & a_{n-1,n} \\ 0 & a_{nn} \end{bmatrix} \\ &= a_{11}a_{22}\cdots a_{nn}. \end{aligned}$$

Since a diagonal matrix is upper triangular, the same result holds for a diagonal matrix. In case A is lower triangular, the same argument applies except that expansion is done along first rows. ■

EXAMPLE 8

Compute $\det A$, where

$$A = \begin{bmatrix} 2 & 4 & 2 & 1 \\ 0 & 1 & 0 & 3 \\ 0 & 0 & 3 & 1 \\ 0 & 0 & 0 & -2 \end{bmatrix}.$$

Solution By Theorem 4, $\det A = (2)(1)(3)(-2) = -12$.

THEOREM 5

If I is an identity matrix of any order, then $\det I = 1$.

Proof Since I is diagonal, $\det I$ is the product of the main diagonal entries. Since each of these is 1, the product is also 1. ■

4.1 Exercises

1. Find the 3,1 and 3,2 minors and cofactors of

$$A = \begin{bmatrix} 2 & -1 & 3 \\ 1 & 2 & 1 \\ 0 & -3 & -4 \end{bmatrix}.$$

2. Find the 2,2 and 2,3 minors and cofactors of the matrix A in Exercise 1.

In Exercises 3–17 evaluate the determinant of each matrix.

3. $[3]$

4. $[-5]$

5. $\begin{bmatrix} 2 & 1 \\ 3 & -1 \end{bmatrix}$

6. $\begin{bmatrix} 4 & -6 \\ -2 & 3 \end{bmatrix}$

7. $\begin{bmatrix} 2 & 1 & 1 \\ 3 & 0 & -1 \\ 4 & 5 & 2 \end{bmatrix}$

8. $\begin{bmatrix} 1 & -2 & 3 \\ 2 & -2 & 4 \\ 0 & 3 & 2 \end{bmatrix}$

9. $\begin{bmatrix} 2 & 4 & 2 \\ 1 & 5 & 1 \\ 3 & -7 & 3 \end{bmatrix}$

10. $\begin{bmatrix} 1 & 3 & -1 & 2 \\ 2 & 1 & 0 & -4 \\ 3 & 1 & 0 & 4 \\ -1 & 2 & 0 & 3 \end{bmatrix}$

11. $\begin{bmatrix} 2 & 3 & 4 & 5 \\ 0 & 3 & 4 & 5 \\ 0 & 0 & 4 & 5 \\ 0 & 0 & 0 & 5 \end{bmatrix}$

12. $\begin{bmatrix} 2 & 0 & 2 & 0 \\ 1 & 4 & 0 & 0 \\ 2 & 3 & 1 & 0 \\ 2 & 5 & -1 & 0 \end{bmatrix}$

13. $\begin{bmatrix} a & ab \\ b & a^2 + b^2 \end{bmatrix}$

14. $\begin{bmatrix} ak & c \\ bk & d \end{bmatrix}$

15. $\begin{bmatrix} ak & ck \\ bk & dk \end{bmatrix}$

16. $\begin{bmatrix} \lambda - 2 & 3 \\ 4 & \lambda - 1 \end{bmatrix}$

17. $\begin{bmatrix} t - 3 & -1 \\ 3 & t \end{bmatrix}$

In Exercises 18–22 evaluate the determinant of the given matrix "by inspection."

18. $\begin{bmatrix} 1 & 2 & 3 \\ 0 & 4 & 5 \\ 0 & 0 & 6 \end{bmatrix}$

19. $\begin{bmatrix} 2 & 0 & 0 \\ 0 & 4 & 0 \\ 3 & 0 & -1 \end{bmatrix}$

20. $\begin{bmatrix} -1 & 0 & 0 \\ 0 & 3 & 0 \\ 0 & 0 & 5 \end{bmatrix}$

21. $\begin{bmatrix} 1 & 2 & 3 & 4 \\ 0 & 5 & 0 & 6 \\ 1 & 2 & 3 & 4 \\ 7 & 0 & 8 & 0 \end{bmatrix}$

22. $\begin{bmatrix} -1 & 1 & -3 & -1 \\ 3 & 2 & 2 & 3 \\ 0 & 1 & 1 & 0 \\ 2 & 1 & 0 & 2 \end{bmatrix}$

23. Verify one special case of Theorem 1 by proving that, for matrices of order 3, computation of the determinant by expansion down the second column yields the same result as expansion along the first row.

24. Determine a formula analogous to Theorem 4 for the determinant of a matrix A of order n such that $a_{ij} = 0$ unless $i + j = n + 1$.

25. Determine a formula analogous to Theorem 4 for the determinant of a matrix A of order n such that $a_{ij} = 0$ if $i + j > n + 1$. Such a matrix is upper triangular in appearance but not upper triangular with respect to the main diagonal.

26. Let a_{ij} be constants for all integers i and j from 1 to n, let x be a variable, let

$$A(x) = \begin{bmatrix} x + a_{11} & a_{12} & \cdots & a_{1n} \\ a_{21} & x + a_{22} & \cdots & a_{2n} \\ \vdots & \vdots & & \vdots \\ a_{n1} & a_{n2} & \cdots & x + a_{nn} \end{bmatrix},$$

and let $f(x) = det\, A(x)$ for each x. Prove that $f(x)$ is an nth degree polynomial.

27. Find all zeros (values of x such that $f(x) = 0$) of the polynomial

$$f(x) = \det \begin{bmatrix} x - 2 & 1 \\ 3 & x \end{bmatrix}.$$

28. Find all zeros of the polynomial

$$f(t) = \det \begin{bmatrix} t - 2 & 4 & -3 \\ 0 & t - 3 & 2 \\ 0 & 0 & t \end{bmatrix}.$$

29. For matrices of *order 3 only*, there is another convenient way to compute determinants that is sometimes called the "basket-weave method." First, construct a 3×5 array by writing down the entries of the given matrix and adjoining (on the right) its first and second columns (Figure 1). Then form the products of the three numbers on each of the six diagonals indicated, and set det A equal to the sum of the circled products 1, 2, and 3 minus the sum of circled products 4, 5, and 6. Show that this procedure does indeed yield the determinant of A.

FIGURE 1

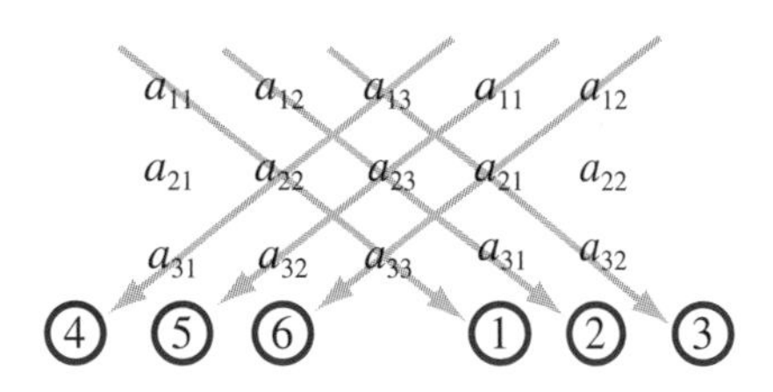

30. In Chapter 1, the *cross product* of two 3-vectors $\mathbf{v} = (a_1, a_2, a_3)$, $\mathbf{w} = (b_1, b_2, b_3)$ was defined to be

$$\mathbf{v} \times \mathbf{w} = (a_2b_3 - a_3b_2, a_3b_1 - a_1b_3, a_1b_2 - a_2b_1)$$

or, using **i, j, k** notation,

$$\mathbf{v} \times \mathbf{w} = (a_2b_3 - a_3b_2)\mathbf{i} + (a_3b_1 - a_1b_3)\mathbf{j} + (a_1b_2 - a_2b_1)\mathbf{k}.$$

Show that this vector can be conveniently remembered as the "determinant" of the matrix

$$\begin{bmatrix} \mathbf{i} & \mathbf{j} & \mathbf{k} \\ a_1 & a_2 & a_3 \\ b_1 & b_2 & b_3 \end{bmatrix}.$$

(The word *determinant* is in quotes because **i**, **j**, and **k** are vectors, so the result is a vector, not a scalar.)

31. Prove that $(A^T)_{ij} = (A_{ji})^T$, for any i, j, and square matrix A.

32. Let **u** and **v** be vectors in $\mathbf{R}^2$, and let A be the matrix with these vectors as rows:

$$A = \begin{bmatrix} u_1 & u_2 \\ v_1 & v_2 \end{bmatrix}$$

Show that $|\det A|$ is the *area of the parallelogram* determined by the vectors **u** and **v**.

SPOTLIGHT **33.** Let **u**, **v**, and **w** be vectors in $\mathbf{R}^3$, and let A be the matrix with these vectors as rows:

$$A = \begin{bmatrix} u_1 & u_2 & u_3 \\ v_1 & v_2 & v_3 \\ w_1 & w_2 & w_3 \end{bmatrix}$$

Show that $|\det A|$ is the *volume of the parallelepiped* determined by the vectors **u**, **v**, and **w** (see the Spotlight at the beginning of this chapter).

SPOTLIGHT **34.** Using the formula in Exercise 33, compute the volume of the parallelepiped determined by the vectors (1, 2, 3), (0, −4, 5), and (0, 0, 6).

SPOTLIGHT **35.** The set of points S given below is the set of vertices of a parallelepiped in $\mathbf{R}^3$. Verify this fact by translating a vertex to the origin, and then computing the volume of the parallelepiped.

$$S = \{(-1, 1, 3), (1, 1, 3), (0, 2, 3), (2, 2, 3), (0, 2, 6), (2, 2, 6), (1, 3, 6), (3, 3, 6)\}$$

4.2 Properties of Determinants

In this section, we examine some of the basic properties of determinants. Some of these properties enable us to compute determinants more quickly. Others provide important relationships between the value of a determinant and the invertibility of the corresponding matrix or the uniqueness of the solution to a related linear system.

THEOREM 1

Let A and A' be square matrices that are the same except that in one row (or in one column) of A' each entry of A is multiplied by a scalar c. Then $\det A' = c \det A$. (That is, c can be "factored out" of the row or column.)

Proof Suppose the difference occurs in the ith row. Then

$$A = \begin{bmatrix} a_{11} & \cdots & a_{1n} \\ \vdots & & \vdots \\ a_{i1} & \cdots & a_{in} \\ \vdots & & \vdots \\ a_{n1} & \cdots & a_{nn} \end{bmatrix} \quad \text{and} \quad A' = \begin{bmatrix} a_{11} & \cdots & a_{1n} \\ \vdots & & \vdots \\ ca_{i1} & \cdots & ca_{in} \\ \vdots & & \vdots \\ a_{n1} & \cdots & a_{nn} \end{bmatrix}.$$

Compute $\det A'$ by expansion along the ith row. Then

$$\begin{aligned} \det A' &= (-1)^{i+1} ca_{i1} \det A'_{i1} + (-1)^{i+2} ca_{i2} \det A'_{i2} \\ &\quad + \cdots + (-1)^{i+n} ca_{in} \det A'_{in} \\ &= \sum_{j=1}^{n} (-1)^{i+j} ca_{ij} \det A'_{ij} = c \sum_{j=1}^{n} (-1)^{i+j} a_{ij} \det A_{ij}, \end{aligned}$$

since c factors out of each term and since $A_{ij} = A'_{ij}$ for each j (the ith row has been deleted). The last expression is just $c \det A$. A similar argument holds for columns. ■

EXAMPLE 1

Compute $\det A$, given that

$$A = \begin{bmatrix} 2 & 0 & 4 \\ 1 & -6 & 2 \\ 3 & 9 & 12 \end{bmatrix}.$$

Solution In the computation of det A, 2 can be factored out of the first row, 3 out of the third row, 3 out of the second column, and 2 out of the third column:

$$\det\begin{bmatrix}2 & 0 & 4\\ 1 & -6 & 2\\ 3 & 9 & 12\end{bmatrix} = 2\det\begin{bmatrix}1 & 0 & 2\\ 1 & -6 & 2\\ 3 & 9 & 12\end{bmatrix} = 2(3)\det\begin{bmatrix}1 & 0 & 2\\ 1 & -6 & 2\\ 1 & 3 & 4\end{bmatrix}$$

$$= 2(3^2)\det\begin{bmatrix}1 & 0 & 2\\ 1 & -2 & 2\\ 1 & 1 & 4\end{bmatrix} = 2^2(3^2)\det\begin{bmatrix}1 & 0 & 1\\ 1 & -2 & 1\\ 1 & 1 & 2\end{bmatrix}$$

Expanding (say, along the first row), we still need to compute two 2×2 determinants, as in the original problem. Now, however, the numbers are easier to work with than before. The answer is $36(-2) = -72$. ■

The same idea can be used in reverse to clear the fractions in a matrix before computing its determinant. Because integers are easier to work with, there is less chance of error. The rule here is to examine each row (or column) for the smallest integer that clears the fractions (for the *least common denominator*), multiply the row by that number, and compensate by multiplying the determinant by the reciprocal of that number.

EXAMPLE 2

Compute det A, where

$$A = \begin{bmatrix}\frac{1}{3} & 0 & \frac{3}{4}\\ \frac{2}{5} & -1 & \frac{3}{2}\\ \frac{1}{8} & -\frac{3}{4} & \frac{5}{4}\end{bmatrix}.$$

Solution Since 12 clears the first row, 10 the second, and 8 the last, we multiply the determinant by $\frac{1}{12}$ and the first row by 12, and proceed similarly with the other rows.

$$\det A = \left(\frac{1}{12}\right)\left(\frac{1}{10}\right)\left(\frac{1}{8}\right)\det\begin{bmatrix}4 & 0 & 9\\ 4 & -10 & 15\\ 1 & -6 & 10\end{bmatrix}$$

Now -2 can be factored out of the second column:

$$\det A = \left(\frac{1}{12}\right)\left(\frac{1}{10}\right)\left(\frac{1}{8}\right)(-2)\det\begin{bmatrix}4 & 0 & 9\\ 4 & 5 & 15\\ 1 & 3 & 10\end{bmatrix}$$

The final value is $\det A = (-1/480)(83) = -83/480$. ■

WARNING It is *not* true in general that $\det(cA) = c\det(A)$. Since cA is formed by multiplying every row of A by the scalar c, we have

$$\det(cA) = c^n\det(A),$$

where n is the order of A. For example, if A is a 3×3 matrix, then $\det(2A) = 2^3\det(A) = 8\det(A)$.

EXAMPLE 3

Compute det A for the matrix

$$A = \begin{bmatrix}2 & 1 & 3 & -4\\ -1 & 0 & 1 & 2\\ -3 & 2 & -1 & 6\\ 4 & 1 & 4 & -8\end{bmatrix}.$$

Solution Your first thought might be to expand along the second row or column because of the zero. This is a reasonable approach, but if you notice that the last column is -2 times the first and use Theorem 1 to factor out -2, you obtain

$$\det A = -2 \det \begin{bmatrix} 2 & 1 & 3 & 2 \\ -1 & 0 & 1 & -1 \\ -3 & 2 & -1 & -3 \\ 4 & 1 & 4 & 4 \end{bmatrix}.$$

Now the first and last columns are equal, and by Theorem 3 of Section 4.1, $\det A = 0$. ■

The method of Example 3 is important enough to be identified as a theorem.

THEOREM 2

If a square matrix A has a row that is a scalar multiple of another row (or a column that is a scalar multiple of another column), then *det* $A = 0$.

Proof Use Theorem 1 to factor out the scalar multiple, and apply Theorem 3 of the previous section. ■

THEOREM 3

If two rows (or columns) of a square matrix are interchanged, then the determinant changes by a factor of -1; that is, its sign is reversed.

Proof Let A and A' be matrices of order n such that A' is obtained from A by interchanging two of its rows.

If $n = 2$, there are only two rows to interchange:

$$A = \begin{bmatrix} a_{11} & a_{12} \\ a_{21} & a_{22} \end{bmatrix} \quad \text{and} \quad A' = \begin{bmatrix} a_{21} & a_{22} \\ a_{11} & a_{12} \end{bmatrix}$$

Then $\det A' = a_{21}a_{12} - a_{11}a_{22} = -(a_{11}a_{22} - a_{21}a_{12}) = -\det A$.

For $n > 2$, expand $\det A'$ along any row not involved in the interchange. Suppose that the ith row is one such. Then

$$\det A' = (-1)^{i+1}a_{i1} \det A'_{i1} + (-1)^{i+2}a_{i2} \det A'_{i2} + \cdots + (-1)^{i+n}a_{in} \det A'_{in}.$$

Each submatrix A'_{ij} is the same as the corresponding submatrix A_{ij} of A, except that two rows of A_{ij} are interchanged to obtain A'_{ij}. (Taking out a row not involved in the interchange leaves two rows still interchanged.)

If $n = 3$, each submatrix A'_{ij} is order 2, and by the $n = 2$ case, each term in the expansion of $\det A'$ is the opposite of the corresponding term in the expansion of $\det A$ along the same row. But then $\det A' = -\det A$. If $n > 3$, repeat the process just given as often as necessary to reduce to the order-2 case. A similar argument holds if two columns are interchanged. ■

So far, we have seen how determinants of square matrices behave with respect to two of the three elementary row operations:

i. If a row is multiplied by a scalar, the determinant is multiplied by the same scalar.

ii. If two rows are interchanged, the sign of the determinant is reversed.

The third elementary row operation, addition of a scalar multiple of one row to another, has no effect at all on the determinant. It is also the most useful for simplifying computation of determinants.

THEOREM 4

If a scalar multiple of one row (or column) is added to another row (or column) of a square matrix, the determinant is unchanged.

Proof Let A and A' be matrices of order n, where A and A' agree except in the ith row, and $\mathbf{a}'_i = c\mathbf{a}_k + \mathbf{a}_i$; in other words, the ith row of A' is the sum of the ith row of A and c times the kth row of A, where $k \neq i$. We compute $\det A'$ by expansion along the ith row:

$$\begin{aligned}\det A' &= (-1)^{i+1}a'_{i1}\det A'_{i1} + \cdots + (-1)^{i+n}a'_{in}\det A'_{in}\\ &= \sum_{j=1}^{n}(-1)^{i+j}a'_{ij}\det A'_{ij}\end{aligned}$$

By hypothesis, $a'_{ij} = ca_{kj} + a_{ij}$ for each j, and since the ith row has been deleted, $A'_{ij} = A_{ij}$ for each j. Then

$$\begin{aligned}\det A' &= \sum_{j=1}^{n}(-1)^{i+j}(ca_{kj} + a_{ij})\det A_{ij}\\ &= c\sum_{j=1}^{n}(-1)^{i+j}a_{kj}\det A_{ij} + \sum_{j=1}^{n}a_{ij}(-1)^{i+j}\det A_{ij}\\ &= c\det A'' + \det A,\end{aligned}$$

where A'' is obtained from A by replacing the ith row by its own kth row. But then A'' has two equal rows (the ith and kth) and, by Theorem 3 of Section 4.1, $\det A'' = 0$. This leaves just $\det A' = \det A$.

A similar proof holds for columns. 

As we noted in the previous section, $\det A$ is easy to compute if a row or column has lots of zeros. The usefulness of Theorem 4 is that if A does not have many zeros, we can "create them"—we can find a new matrix with the same determinant that does have many zeros.

EXAMPLE 4

One last time, we compute the determinant of the matrix given in Example 3 of Section 4.1,

$$A = \begin{bmatrix} 2 & 1 & -3 & 1\\ -3 & -2 & 0 & 2\\ 2 & 1 & 0 & -1\\ 1 & 0 & 1 & 2\end{bmatrix}.$$

Solution We already noticed (Example 5 of Section 4.1) that it is most convenient to expand down the third column. Before doing so, however, the remaining work can be cut in half if we first add 3 times the last row to the first. Then we have

$$A' = \begin{bmatrix} 5 & 1 & 0 & 7 \\ -3 & -2 & 0 & 2 \\ 2 & 1 & 0 & -1 \\ 1 & 0 & 1 & 2 \end{bmatrix},$$

and, by Theorem 4,

$$\det A = \det A' = (-1)(1)\det A'_{43} = -\det \begin{bmatrix} 5 & 1 & 7 \\ -3 & -2 & 2 \\ 2 & 1 & -1 \end{bmatrix}.$$

To finish the computation as simply as possible, we use Theorem 4 again to obtain a new matrix with the same determinant as A'_{43} but with two zeros in some row or column. Any row or column will work, but a quick inspection shows that the second column or the third row requires minimal computation. We choose the third row. Adding -2 times the second column to the first, and just adding the second column to the third, we obtain

$$\det \begin{bmatrix} 5 & 1 & 7 \\ -3 & -2 & 2 \\ 2 & 1 & -1 \end{bmatrix} = \det \begin{bmatrix} 3 & 1 & 8 \\ 1 & -2 & 0 \\ 0 & 1 & 0 \end{bmatrix} = -(1)\det \begin{bmatrix} 3 & 8 \\ 1 & 0 \end{bmatrix} = 8.$$

Then $\det A = -8$. ■

One can calculate, using Theorems 1–4, the determinant of matrices with variable entries as well.

EXAMPLE 5

Compute $\det A$ where

$$A = \begin{bmatrix} 1 & a & -a \\ a & 1 & 0 \\ -a & 1 & a^2 \end{bmatrix}.$$

Solution We add $-a$ times the first row to the second row, and a times the first row to the third row to get matrix A'; by Theorem 4 these operations will not change the determinant.

$$A' = \begin{bmatrix} 1 & a & -a \\ 0 & 1-a^2 & a^2 \\ 0 & 1+a^2 & 0 \end{bmatrix}.$$

Expanding down the first column, we find that

$$\det A = \det A' = \det \begin{bmatrix} 1-a^2 & a^2 \\ 1+a^2 & 0 \end{bmatrix} = -(1+a^2)a^2.$$ ■

COMPUTATIONAL NOTE

It is somewhat surprising to realize that even with the aid of high-speed computers, simplifications similar to those in Example 4 must be made to compute $\det A$ when A is of even modest size. For example, the computation time of an order-25 matrix using only the definition of determinant is astronomically long, even using the fastest computers available (or on the drawing board). For this reason, computer (and calculator) programs to calculate the determinant of a matrix are based on the following idea. Use Theorems 3 and 4 to bring the matrix into upper triangular form, keeping track of the number of row interchanges used. By Theorem 4 of Section 4.1, the determinant is then just ε times the product of the main diagonal entries, where $\varepsilon = 1$ if there are an even number of row interchanges, and $\varepsilon = -1$ otherwise.

EXAMPLE 6

Find det A by using the process described in the preceding computational note, where

$$A = \begin{bmatrix} 1 & 2 & 0 & -2 \\ 0 & 0 & 2 & -1 \\ 0 & -1 & 1 & 0 \\ 1 & 3 & 4 & 1 \end{bmatrix}.$$

Solution By interchanging the second and third rows, and then adding multiples of each row to the bottom row, we obtain the sequence of equations

$$\det A = -\det \begin{bmatrix} 1 & 2 & 0 & -2 \\ 0 & -1 & 1 & 0 \\ 0 & 0 & 2 & -1 \\ 1 & 3 & 4 & 1 \end{bmatrix} = -\det \begin{bmatrix} 1 & 2 & 0 & -2 \\ 0 & -1 & 1 & 0 \\ 0 & 0 & 2 & -1 \\ 0 & 1 & 4 & 3 \end{bmatrix}$$

$$= -\det \begin{bmatrix} 1 & 2 & 0 & -2 \\ 0 & -1 & 1 & 0 \\ 0 & 0 & 2 & -1 \\ 0 & 0 & 5 & 3 \end{bmatrix} = -\det \begin{bmatrix} 1 & 2 & 0 & -2 \\ 0 & -1 & 1 & 0 \\ 0 & 0 & 2 & -1 \\ 0 & 0 & 0 & \frac{11}{2} \end{bmatrix}.$$

Finally, $\det A = -(1)(-1)(2)(11/2) = 11$. ■

Perhaps the most important single property of the determinant of a matrix is that it tells us whether or not the matrix in question is invertible.

THEOREM 5

A square matrix A is invertible if and only if $\det A \neq 0$.

Proof Let B be the row-reduced echelon form for A. Since B is obtained from A by elementary row operations, repeated application of Theorems 1, 3, and 4 implies that $\det B = c \det A$ for some nonzero scalar c. This implies that $\det A \neq 0$ if and only if $\det B \neq 0$. But B is a row-reduced square matrix and as such is either the identity matrix or has at least one row of zeros. In other words, $\det B = 1$ if B is the identity matrix, and $\det B = 0$ otherwise. Thus, we have $\det A \neq 0$ if and only if B is the identity matrix. By Theorem 5 of Section 3.3, this is equivalent to A being invertible. ■

EXAMPLE 7

Determine whether or not the matrix A is invertible, where A is given by

$$A = \begin{bmatrix} 2 & 1 & -3 & 1 \\ -3 & -2 & 0 & 2 \\ 2 & 1 & 0 & -1 \\ 1 & 0 & 1 & 2 \end{bmatrix}.$$

Solution Since A is the matrix of Example 4, $\det A = -8 \neq 0$, and thus A is invertible. ■

NOTE The fact that $\det A \neq 0$ implies that A is invertible does not tell us what the matrix A^{-1} looks like. Determinants can also be used for this purpose, as we will see in Section 4.3.

EXAMPLE 8

Let $x_1, x_2, \ldots, x_n$ be distinct real numbers. The matrix

$$\det \begin{bmatrix} 1 & x_1 & x_1^2 & \cdots & x_1^{n-2} & x_1^{n-1} \\ 1 & x_2 & x_2^2 & \cdots & x_2^{n-2} & x_2^{n-1} \\ \vdots & \vdots & \vdots & \ddots & \vdots & \vdots \\ 1 & x_n & x_n^2 & \cdots & x_n^{n-2} & x_n^{n-1} \end{bmatrix}$$

is called a *Vandermonde matrix*. Show that this matrix is invertible.

Solution We show that the determinant of the Vandermonde matrix is nonzero; by Theorem 5, this will immediately imply that the matrix is invertible.

Consider the following "trick" for calculating the determinant of the Vandermonde matrix: subtract x_1 times the second last column from the last column: This does not change the value of the determinant, but introduces a 0 in the last entry in the first row.

$$\det \begin{bmatrix} 1 & x_1 & x_1^2 & \cdots & x_1^{n-2} & x_1^{n-1} \\ 1 & x_2 & x_2^2 & \cdots & x_2^{n-2} & x_2^{n-1} \\ \vdots & \vdots & \vdots & \ddots & \vdots & \vdots \\ 1 & x_n & x_n^2 & \cdots & x_n^{n-2} & x_n^{n-1} \end{bmatrix}$$

$$= \det \begin{bmatrix} 1 & x_1 & x_1^2 & \cdots & x_1^{n-2} & 0 \\ 1 & x_2 & x_2^2 & \cdots & x_2^{n-2} & x_2^{n-1} - x_1 x_2^{n-2} \\ \vdots & \vdots & \vdots & \ddots & \vdots & \vdots \\ 1 & x_n & x_n^2 & \cdots & x_n^{n-2} & x_n^{n-1} - x_1 x_n^{n-2} \end{bmatrix}$$

We now proceed to do the same with the second and third last columns, and so on, all the way back to the first two columns. The determinant remains the same throughout. Thus, we derive that

$$\det \begin{bmatrix} 1 & x_1 & x_1^2 & \cdots & x_1^{n-2} & x_1^{n-1} \\ 1 & x_2 & x_2^2 & \cdots & x_2^{n-2} & x_2^{n-1} \\ \vdots & \vdots & \vdots & \ddots & \vdots & \vdots \\ 1 & x_n & x_n^2 & \cdots & x_n^{n-2} & x_n^{n-1} \end{bmatrix}$$

$$= \det \begin{bmatrix} 1 & 0 & 0 & \cdots & 0 & 0 \\ 1 & x_2 - x_1 & x_2^2 - x_1 x_2 & \cdots & x_2^{n-2} - x_1 x_2^{n-3} & x_2^{n-1} - x_1 x_2^{n-2} \\ \vdots & \vdots & \vdots & \ddots & \vdots & \vdots \\ 1 & x_n - x_1 & x_n^2 - x_1 x_n & \cdots & x_n^{n-2} - x_1 x_n^{n-3} & x_n^{n-1} - x_1 x_n^{n-2} \end{bmatrix}.$$

By carrying out the cofactor expansion along the top row, we get

$$\det\begin{bmatrix} 1 & x_1 & x_1^2 & \cdots & x_1^{n-2} & x_1^{n-1} \\ 1 & x_2 & x_2^2 & \cdots & x_2^{n-2} & x_2^{n-1} \\ \vdots & \vdots & \vdots & \ddots & \vdots & \vdots \\ 1 & x_n & x_n^2 & \cdots & x_n^{n-2} & x_n^{n-1} \end{bmatrix}$$

$$= \det\begin{bmatrix} x_2 - x_1 & x_2^2 - x_1x_2 & \cdots & x_2^{n-2} - x_1x_2^{n-3} & x_2^{n-1} - x_1x_2^{n-2} \\ \vdots & \vdots & \ddots & \vdots & \vdots \\ x_n - x_1 & x_n^2 - x_1x_n & \cdots & x_n^{n-2} - x_1x_n^{n-3} & x_n^{n-1} - x_1x_n^{n-2} \end{bmatrix}.$$

We observe that, in the determinant on the right side, we can factor out $(x_2 - x_1)$ from each entry in the first row, $(x_3 - x_1)$ from each entry in the second row, and so on. Thus

$$\det\begin{bmatrix} 1 & x_1 & x_1^2 & \cdots & x_1^{n-2} & x_1^{n-1} \\ 1 & x_2 & x_2^2 & \cdots & x_2^{n-2} & x_2^{n-1} \\ \vdots & \vdots & \vdots & \ddots & \vdots & \vdots \\ 1 & x_n & x_n^2 & \cdots & x_n^{n-2} & x_n^{n-1} \end{bmatrix}$$

$$= \det\begin{bmatrix} x_2 - x_1 & (x_2 - x_1)x_2 & \cdots & (x_2 - x_1)x_2^{n-3} & (x_2 - x_1)x_2^{n-2} \\ \vdots & \vdots & \ddots & \vdots & \vdots \\ x_n - x_1 & (x_n - x_1)x_n & \cdots & (x_n - x_1)x_n^{n-3} & (x_2 - x_1)x_n^{n-2} \end{bmatrix}.$$

By Theorem 1, we can factor out these common terms in each row:

$$\det\begin{bmatrix} 1 & x_1 & x_1^2 & \cdots & x_1^{n-2} & x_1^{n-1} \\ 1 & x_2 & x_2^2 & \cdots & x_2^{n-2} & x_2^{n-1} \\ \vdots & \vdots & \vdots & \ddots & \vdots & \vdots \\ 1 & x_n & x_n^2 & \cdots & x_n^{n-2} & x_n^{n-1} \end{bmatrix}$$

$$= (x_n - x_1)\cdots(x_2 - x_1)\det\begin{bmatrix} 1 & x_2 & \cdots & x_2^{n-3} & x_2^{n-2} \\ \vdots & \vdots & \ddots & \vdots & \vdots \\ 1 & x_n & \cdots & x_n^{n-3} & x_n^{n-2} \end{bmatrix}$$

The determinant on the right side is another Vandermonde determinant, of one size smaller. We proceed in the same way and continue to reduce the size of the Vandermonde determinant, and derive that

$$\det\begin{bmatrix} 1 & x_1 & x_1^2 & \cdots & x_1^{n-2} & x_1^{n-1} \\ 1 & x_2 & x_2^2 & \cdots & x_2^{n-2} & x_2^{n-1} \\ \vdots & \vdots & \vdots & \ddots & \vdots & \vdots \\ 1 & x_n & x_n^2 & \cdots & x_n^{n-2} & x_n^{n-1} \end{bmatrix}$$

$$= (x_n - x_1)\cdots(x_2 - x_1)(x_n - x_2)\cdots(x_3 - x_2)\cdots(x_n - x_{n-2})(x_{n-1} - x_{n-2})\det\begin{bmatrix} 1 & x_{n-1} \\ 1 & x_n \end{bmatrix}.$$

The last determinant is simply $\det\begin{bmatrix} 1 & x_{n-1} \\ 1 & x_n \end{bmatrix} = 1 \cdot x_n - x_{n-1} \cdot 1 = x_n - x_{n-1}$, so we derive that

$$\det\begin{bmatrix} 1 & x_1 & x_1^2 & \cdots & x_1^{n-2} & x_1^{n-1} \\ 1 & x_2 & x_2^2 & \cdots & x_2^{n-2} & x_2^{n-1} \\ \vdots & \vdots & \vdots & \ddots & \vdots & \vdots \\ 1 & x_n & x_n^2 & \cdots & x_n^{n-2} & x_n^{n-1} \end{bmatrix}$$
$$= (x_n - x_1)\cdots(x_2 - x_1)(x_n - x_2)\cdots(x_3 - x_2)\cdots(x_n - x_{n-2})(x_{n-1} - x_{n-2})(x_n - x_{n-1}).$$

This is the product of the differences $(x_i - x_j)$ over all $i > j$, and as $x_1, x_2, \ldots, x_n$ are distinct real numbers, none of these differences is zero. It follows that the determinant of the Vandermonde matrix is nonzero, and we are done. ■

Vandermonde determinants arise in a number of different applications (see Exercise 21 of Section 4.3).

The following theorem is an important direct consequence of Theorem 5. Its proof is left for an exercise.

THEOREM 6

Let A be a square matrix. Then the linear system $A\mathbf{x} = \mathbf{b}$ has a unique solution for every $\mathbf{b}$ if and only if $\det A \neq 0$. ■

The next theorem is basic in the study of linear algebra.

THEOREM 7

If A and B are square matrices of the same order, then

$$\det AB = (\det A)(\det B).$$ ■

Note that if A is an invertible square matrix, then $A^{-1}A = I$, which implies that $\det(A^{-1}A) = \det(I)$. Now by Theorem 7, $\det(A^{-1}A) = \det(A^{-1})\det(A)$, and by Theorem 5 of Section 4.1, $\det(I) = 1$. It follows that $\det(A^{-1}) = \dfrac{1}{\det(A)}$.

Before we prove Theorem 7, we show that it holds for a special case. This result is then used in the proof of the theorem itself.

LEMMA 1*

Let M be a square matrix, and let E be an elementary matrix of the same order. Then $\det(EM) = \det(E)\det(M)$.

Proof of Lemma 1 Recall, from Section 3.5, that an elementary matrix is a square matrix that can be obtained from the identity matrix by performing a single elementary row operation on it. Also recall that EM is the matrix obtained by performing on M the same elementary row operation that converts I into E. Thus, if E is obtained from I by, say, multiplication of one of its rows by the scalar c, then Theorem 2 tells us that

*A *lemma* is a theorem that is used solely for the purpose of proving another theorem.

$\det(E) = c\det(I) = c$ and that $\det(EM) = c\det(M)$. In other words, in this case, $\det(EM) = \det(E)\det(M)$. Moreover, a similar argument holds for the other two elementary row operations as well (using Theorems 3 and 4), which proves Lemma 1. ■

By repeatedly applying Lemma 1, we can prove the following result.

LEMMA 2

Let M be a square matrix, and let $E_1, E_2, \ldots, E_k$ be square elementary matrices of the same order as M. Then

$$\det(E_1E_2 \ldots E_kM) = \det(E_1)\det(E_2)\ldots\det(E_k)\det(M).$$ ■

Finally, setting $M = I$ in Lemma 2 yields one more useful fact.

LEMMA 3

If $E_1, E_2, \ldots, E_k$ are square elementary matrices of the same order, then $\det(E_1E_2 \ldots E_k) = \det(E_1)\det(E_2)\ldots\det(E_k)$.

It is now fairly easy to prove Theorem 7. ■

Proof of Theorem 7 We divide the proof into two cases.

Case i. Assume that A is invertible. Then, from Section 3.5 (Exercise 19), we see that A can be written as $A = E_1E_2 \ldots E_k$, where each of the E_i is an elementary matrix. Then, making use of Lemmas 2 and 3,

$$\begin{aligned}\det(AB) &= \det(E_1E_2\cdots E_kB) = \det(E_1)\det(E_2)\cdots\det(E_k)\det(B)\\ &= \det(E_1E_2\cdots E_k)\det(B)\\ &= \det(A)\det(B),\end{aligned}$$

as desired.

Case ii. Assume that A is not invertible. Then AB is noninvertible as well (see Exercise 33). Thus, by Theorem 5, $\det(A) = 0$ and $\det(AB) = 0$. But this means that $\det(AB) = \det(A)\det(B)$ in this case also. ■

NOTE We can extend Theorem 7 to cover products that involve more than two matrices. For example, if A, B, and C are square matrices of the same order, $\det(ABC) = \det((AB)C) = \det(AB)\det(C) = \det(A)\det(B)\det(C)$. More generally, if $A_1, A_2, \ldots, A_k$ are square matrices of the same order, then $\det(A_1A_2\cdots A_k) = \det(A_1)\det(A_2)\cdots\det(A_k)$.

Recall that the *transpose* of a matrix is the matrix obtained by interchanging its rows and columns. In other words, for each i, the ith *row* of a matrix A is the ith *column* of the matrix A^T, and the jth *column* of A is the jth *row* of A^T. The next theorem tells us that the determinant of a matrix does not change when we take its transpose.

THEOREM 8

If A is a square matrix, then $\det A^T = \det A$.

Proof Let n be the order of A. If $n = 2$,

$$A = \begin{bmatrix} a_{11} & a_{12} \\ a_{21} & a_{22} \end{bmatrix} \quad \text{and} \quad A^T = \begin{bmatrix} a_{11} & a_{21} \\ a_{12} & a_{22} \end{bmatrix}.$$

Hence, $\det A = a_{11}a_{22} - a_{21}a_{12} = \det A^T$.

For larger values of n, the proof is based on the fact that the expansion of det A^T repeatedly along first *rows* yields exactly the same terms, in the same order, as the expansion of det A repeatedly down first *columns*. ■

EXAMPLE 9

Suppose that A and B are square matrices of the same order, with det $A = 3$ and det $B = 5$. Find $\det(AB^T)$.

Solution

$$\begin{aligned}\det(AB^T) &= \det(A)\det(B^T) &&\text{(Theorem 7)}\\ &= \det(A)\det(B) &&\text{(Theorem 8)}\\ &= (3)(5) = 15\end{aligned}$$ ■

Careful analysis of Theorem 5 (and its supporting Theorem 5 of Section 3.3) shows only that if det $A \neq 0$, there exists a matrix B such that $AB = I$. This is because the left side of $[A \mid I]$ row-reduces to the identity matrix, and the resulting right side of $[I \mid B]$ must satisfy $AB = I$. Since matrix multiplication is in general *not* commutative, we cannot be sure that $BA = I$ as well, in order to conclude that B is the inverse of A. Theorems 7 and 8 can be used to prove this important fact, which completes the proof of Theorem 2 of Section 3.2. We restate Theorem 2 of Section 3.2 here and prove it.

THEOREM 9

Let A be a square matrix. If a square matrix B exists such that $AB = I$, then $BA = I$ as well, and thus $B = A^{-1}$.

Proof We first show that there is some square matrix C, such that $CA = I$. Consider the system

$$A^TX = I.$$

Since $(\det A)(\det B) = \det(AB) = \det I = 1$, we see that $\det A \neq 0$. However, $\det A^T = \det A$, so $\det A^T \neq 0$ as well. Thus, the row-reduced form of A^T is also equal to I, so we can row-reduce the system $[A^T \mid I]$ to obtain $[I \mid D]$ with $A^TD = I$. But then $(A^TD)^T = I^T$, and, from properties of transpose,

$$D^TA = I.$$

Thus, there is a matrix C, namely $C = D^T$, such that $CA = I$.

Finally, we show that $C = B$ (which implies that $B = A^{-1}$):

$$C = CI = C(AB) = (CA)B = IB = B$$ ■

Finally, we can use Theorem 7 to provide a different (and dare we say fun?) way to calculate determinants, replacing an $n \times n$ determinant by *just one* $(n-1)\times(n-1)$ determinant (in contrast to the definition of a determinant, that replaces an $n \times n$ determinant by n determinants of dimension $(n-1) \times (n-1)$).

The idea is the following. It is not hard to prove (see Exercise 27) that any $n \times n$ matrix

$$A = \begin{bmatrix} 1 & a_{12} & a_{13} & \cdots & a_{1n} \\ a_{21} & a_{22} & a_{23} & \cdots & a_{2n} \\ \vdots & \vdots & \vdots & \vdots & \vdots \\ a_{n1} & a_{n2} & a_{n3} & \cdots & a_{nn} \end{bmatrix}$$

with a 1 in the top left corner can be written as

$$A = \begin{bmatrix} 1 & 0 & 0 & \cdots & 0 \\ a_{21} & 1 & 0 & \cdots & 0 \\ \vdots & \vdots & \vdots & \vdots & \vdots \\ a_{n1} & 0 & 0 & \cdots & 1 \end{bmatrix} \begin{bmatrix} 1 & a_{12} & a_{13} & \cdots & a_{1n} \\ 0 & a_{22} - a_{21}a_{12} & a_{23} - a_{21}a_{13} & \cdots & a_{2n} - a_{21}a_{1n} \\ \vdots & \vdots & \vdots & \vdots & \vdots \\ 0 & a_{n2} - a_{n1}a_{12} & a_{n3} - a_{n1}a_{13} & \cdots & a_{nn} - a_{n1}a_{1n} \end{bmatrix}.$$

This can be written succinctly in **block matrix form** as

$$A = \begin{bmatrix} 1 & 0 \\ A_{21} & I \end{bmatrix} \begin{bmatrix} 1 & A_{12} \\ 0^T & A_{22} - A_{21}A_{12} \end{bmatrix}$$

where 0 is the zero matrix of size $1 \times (n-1)$, I is the identity matrix of size $(n-1) \times (n-1)$, and A_{21}, A_{12}, and A_{22} are **submatrices** of A of sizes $(n-1) \times 1$, $1 \times (n-1)$, and $(n-1) \times (n-1)$, respectively. Now it is easy to see that the determinant of the first matrix on the right side is 1 (as it is lower triangular with 1's on the diagonal), and by expanding down the first column of the second matrix on the right side, we find its determinant is

$$\det\begin{bmatrix} a_{22} - a_{21}a_{12} & a_{23} - a_{21}a_{13} & \cdots & a_{2n} - a_{21}a_{1n} \\ \vdots & \vdots & \vdots & \vdots \\ a_{n2} - a_{n1}a_{12} & a_{n3} - a_{n1}a_{13} & \cdots & a_{nn} - a_{n1}a_{1n} \end{bmatrix},$$

whose entries are formed by deleting the first row and column, and for each remaining (i,j)th entry, subtracting off the product of the ith in the first column and jth in the first row. We can think of this as subtracting from each a_{ij} (where $i, j \geq 2$) the product of its "feet," a_{i1} and a_{1j}, and we call the entire method the "feet" method for calculating determinants.

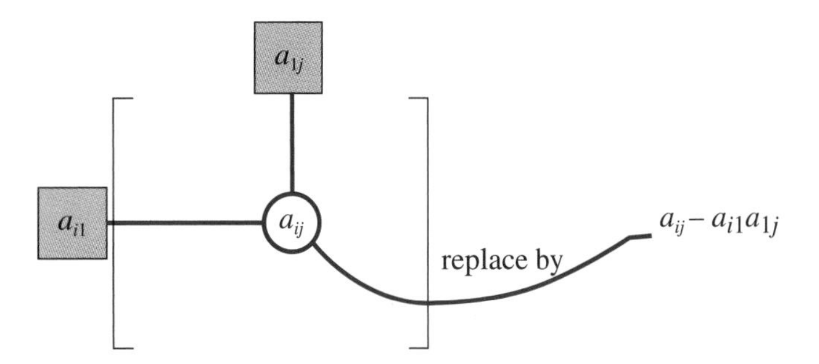

Thus

$$\det A = \det\begin{bmatrix} a_{22} - a_{21}a_{12} & a_{23} - a_{21}a_{13} & \cdots & a_{2n} - a_{21}a_{1n} \\ \vdots & \vdots & \vdots & \vdots \\ a_{n2} - a_{n1}a_{12} & a_{n3} - a_{n1}a_{13} & \cdots & a_{nn} - a_{n1}a_{1n} \end{bmatrix}$$

and we have replaced an $n \times n$ determinant by *just one* $(n-1) \times (n-1)$ determinant. We can continue until we have reduced it to a 2×2 determinant, which we already know how to calculate.

What do we do if the top left entry isn't a 1? If the entry isn't 0, we can factor it out (and multiply the ensuing determinant by it, by Theorem 1). Otherwise, if some other entry in the first column isn't 0, we can interchange rows (and multiply the determinant by -1, by Theorem 3) to make the top left entry nonzero, and proceed as described earlier. The only other case is that every element in the first column is 0, in which case, we can write down that the determinant is 0.

We illustrate the method with an example.

EXAMPLE 10

Find the determinant, using the "feet" method, of

$$A = \begin{bmatrix} 0 & 1 & -1 & 2 & 0 \\ 1 & -1 & 3 & 2 & 1 \\ 1 & 0 & 4 & -1 & 2 \\ -2 & 2 & 1 & 1 & 0 \\ -1 & 1 & 1 & 0 & 5 \end{bmatrix}.$$

Solution As the top left entry is 0, we interchange rows 1 and 2, thereby flipping the sign of the determinant. We then apply the "feet" procedure until we get down to a 2×2 determinant.

$$\begin{aligned} \det A &= -\det\begin{bmatrix} 1 & -1 & 3 & 2 & 1 \\ 0 & 1 & -1 & 2 & 0 \\ 1 & 0 & 4 & -1 & 2 \\ -2 & 2 & 1 & 1 & 0 \\ -1 & 1 & 1 & 0 & 5 \end{bmatrix} \\ &= -\det\begin{bmatrix} 1 & -1 & 2 & 0 \\ 1 & 1 & -3 & 1 \\ 0 & 7 & 5 & 2 \\ 0 & 4 & 2 & 6 \end{bmatrix} \\ &= -\det\begin{bmatrix} 2 & -5 & 1 \\ 7 & 5 & 2 \\ 4 & 2 & 6 \end{bmatrix} \\ &= \det\begin{bmatrix} 4 & 2 & 6 \\ 7 & 5 & 2 \\ 2 & -5 & 1 \end{bmatrix} \\ &= 4\det\begin{bmatrix} 1 & \frac{1}{2} & \frac{3}{2} \\ 7 & 5 & 2 \\ 2 & -5 & 1 \end{bmatrix} \\ &= 4\det\begin{bmatrix} \frac{3}{2} & -\frac{17}{2} \\ -6 & -2 \end{bmatrix} \\ &= 4(-3-51) \\ &= 4(-54) \\ &= -216 \end{aligned}$$

■

4.2 Exercises

Evaluate the determinant of each matrix in Exercises 1–9. Use the theorems of this section to simplify the work.

1. $\begin{bmatrix} 1 & 2 & 3 \\ 1 & 3 & 7 \\ 1 & 4 & 13 \end{bmatrix}$

2. $\begin{bmatrix} 1 & 0 & 0 & 3 \\ 0 & 3 & 9 & 21 \\ 0 & 0 & 2 & 1 \\ 2 & 0 & 0 & 7 \end{bmatrix}$

3. $\begin{bmatrix} 1 & 0 & 0 & 0 \\ 0 & 0 & 3 & 0 \\ 0 & 2 & 0 & 0 \\ 0 & 0 & 0 & 4 \end{bmatrix}$

4. $\begin{bmatrix} 2 & -1 & 3 & -4 \\ 1 & 0 & 2 & 1 \\ -4 & 2 & -6 & 8 \\ 3 & 1 & 2 & 0 \end{bmatrix}$ 5. $\begin{bmatrix} \frac{1}{3} & \frac{3}{5} & \frac{2}{5} \\ \frac{3}{8} & \frac{1}{2} & \frac{1}{4} \\ \frac{1}{3} & \frac{2}{3} & \frac{1}{2} \end{bmatrix}$ 6. $\begin{bmatrix} 0.02 & -0.10 & 0.08 \\ 0.03 & 0.06 & -0.09 \\ 0 & 0.07 & 0.20 \end{bmatrix}$

7. $\begin{bmatrix} x & 2x & -3x \\ x & x-1 & -3 \\ 0 & 0 & 2x-1 \end{bmatrix}$ 8. $\begin{bmatrix} t & -1 & 3 \\ 1 & t-3 & 1 \\ 2 & 3 & t-1 \end{bmatrix}$

9. $\begin{bmatrix} \lambda-1 & 0 & 0 & 0 \\ 2 & 0 & \lambda+1 & 0 \\ 1 & \lambda-2 & 0 & 0 \\ 2 & 3 & 9 & \lambda+2 \end{bmatrix}$

10. Determine which of the matrices in Exercises 1–6 are invertible. Do *not* compute the inverse.

11. For which values of the variables in Exercises 7–9 are the matrices invertible?

12. Prove Theorem 6: If A is a square matrix, the linear system $A\mathbf{x} = \mathbf{b}$ has a unique solution if and only if $\det A \neq 0$.

In Exercises 13–16 use determinants to determine whether or not the given linear system has a unique solution.

13. $\begin{aligned} x - y &= 4 \\ x + y &= 1 \end{aligned}$

14. $\begin{aligned} -x + y &= 0 \\ x - y &= 0 \end{aligned}$

15. $\begin{aligned} x + y + z &= 1 \\ x - y + 2z &= 0 \\ 2y - z &= 1 \end{aligned}$

16. $\begin{aligned} x + y &= 1 \\ y + z &= 0 \\ x + z &= 1 \end{aligned}$

17. Let A be a 3×3 matrix with $\det A = 4$. Find $\det(A^T)$ and $\det(2A)$.

18. Prove that, for a square matrix A of order n and scalar c, $\det(cA) = c^n \det A$.

19. Without expanding the determinant, show that for any values of s, t, and u,

$$\det \begin{bmatrix} s & s+1 & s+2 \\ t & t+1 & t+2 \\ u & u+1 & u+2 \end{bmatrix} = 0.$$

20. Let A' be obtained from the square matrix A by interchanging pairs of rows (columns) m times. Express $\det A'$ in terms of $\det A$ and m.

21. Use Exercise 20 to find a formula for $\det A'$ in terms of $\det A$, where A' is obtained from A by reversing the order in which the rows appear. That is, $\mathbf{a}'_i = \mathbf{a}_{n+1-i}$ *for each i.*

22. Compute and leave in simplest factored form:

$$\det \begin{bmatrix} 1 & s & s^2 \\ 1 & t & t^2 \\ 1 & u & u^2 \end{bmatrix}$$

23. We know that in general $AB \neq BA$. However, prove that $\det AB = \det BA$ for all square nth-order matrices A and B.

24. Prove that $\det A^m = (\det A)^m$, where A^m denotes the mth power of A.

25. If $A^T A = I$, prove that $det\,A = \pm 1$.

26. If S is an invertible matrix, and $B = S^{-1}AS$, prove that $\det A = \det B$.

27. Prove that any $n \times n$ matrix

$$A = \begin{bmatrix} 1 & a_{12} & a_{13} & \cdots & a_{1n} \\ a_{21} & a_{22} & a_{23} & \cdots & a_{2n} \\ \vdots & \vdots & \vdots & \vdots & \vdots \\ a_{n1} & a_{n2} & a_{n3} & \cdots & a_{nn} \end{bmatrix}$$

with a 1 in the top left corner can be written as

$$A = \begin{bmatrix} 1 & 0 & 0 & \cdots & 0 \\ a_{21} & 1 & 0 & \cdots & 0 \\ \vdots & \vdots & \vdots & \vdots & \vdots \\ a_{n1} & 0 & 0 & \cdots & 1 \end{bmatrix} \begin{bmatrix} 1 & a_{12} & a_{13} & \cdots & a_{1n} \\ 0 & a_{22} - a_{21}a_{12} & a_{23} - a_{21}a_{13} & \cdots & a_{2n} - a_{21}a_{1n} \\ \vdots & \vdots & \vdots & \vdots & \vdots \\ 0 & a_{n2} - a_{n1}a_{12} & a_{n3} - a_{n1}a_{13} & \cdots & a_{nn} - a_{n1}a_{1n} \end{bmatrix}.$$

28. Calculate the determinant of the following matrices using the "feet" method:

(a) $\begin{bmatrix} 2 & -1 & 3 & -4 \\ 1 & 0 & 2 & 1 \\ -4 & 2 & -6 & 8 \\ 3 & 1 & 2 & 0 \end{bmatrix}$

(b) $\begin{bmatrix} t & -1 & 3 \\ 1 & t-3 & 1 \\ 2 & 3 & t-1 \end{bmatrix}$

Exercises 29–31 give an extremely simple method of finding a linear equation of the hyperplane that contains *n* independent points in $\mathbf{R}^n$ (see the Spotlight at the beginning of this chapter).

SPOTLIGHT *29.* Let (a_1, a_2) and (b_1, b_2) be distinct points in $\mathbf{R}^2$. Prove that the equation

$$\det \begin{bmatrix} x & y & 1 \\ a_1 & a_2 & 1 \\ b_1 & b_2 & 1 \end{bmatrix} = 0$$

is a linear equation that describes the line that contains (a_1, a_2) and (b_1, b_2). (*Hint*: What happens when you make the substitutions $x = a_1$ and $y = a_2$? Similarly, what happens when you make the substitutions $x = b_1$ and $y = b_2$? Finally, expand along the top row.)

SPOTLIGHT *30.* Let (a_1, a_2, a_3), (b_1, b_2, b_3), and (c_1, c_2, c_3) be noncollinear points in $\mathbf{R}^3$. Prove that the equation

$$\det \begin{bmatrix} x & y & z & 1 \\ a_1 & a_2 & a_3 & 1 \\ b_1 & b_3 & b_3 & 1 \\ c_1 & c_3 & c_3 & 1 \end{bmatrix} = 0$$

is a linear equation that describes the plane containing (a_1, a_2, a_3), (b_1, b_2, b_3), and (c_1, c_2, c_3).

SPOTLIGHT **31.** Generalize the result of Exercises 29 and 30 to the case of n independent points in $\mathbf{R}^n$.

32. Compute the determinant of the following matrix by three different methods: (a) the definition, (b) triangularization, and (c) any other method. Which seems to be the best method?

$$A = \begin{bmatrix} 2 & 1 & -1 & 3 \\ 1 & 2 & 0 & 3 \\ -1 & 2 & -2 & 1 \\ 0 & 3 & 1 & 4 \end{bmatrix}$$

33. *(a)* Let A and B be square matrices of the same order. Show that if AB is invertible, then so is A with $A^{-1} = B(AB)^{-1}$. (*Hint*: Multiply A^{-1} by A.)
(b) Use the result of part (a) to show that if A is not invertible, then neither is AB.

Computational Exercises

In Exercises 34 and 35 use a computer or graphing calculator to find the determinant of the given matrix.

34. $\begin{bmatrix} 0.21 & -2.31 & 2.01 & 3.00 \\ 0.03 & -1.20 & 1.02 & 0.00 \\ 1.21 & 1.00 & -0.64 & -0.54 \\ 1.02 & 0.04 & 0.57 & 0.19 \end{bmatrix}$

35. $\begin{bmatrix} 1 & \frac{1}{2} & \frac{1}{3} & \frac{1}{4} & \frac{1}{5} & \frac{1}{6} \\ \frac{1}{2} & \frac{1}{3} & \frac{1}{4} & \frac{1}{5} & \frac{1}{6} & \frac{1}{7} \\ \frac{1}{3} & \frac{1}{4} & \frac{1}{5} & \frac{1}{6} & \frac{1}{7} & \frac{1}{8} \\ \frac{1}{4} & \frac{1}{5} & \frac{1}{6} & \frac{1}{7} & \frac{1}{8} & \frac{1}{9} \\ \frac{1}{5} & \frac{1}{6} & \frac{1}{7} & \frac{1}{8} & \frac{1}{9} & \frac{1}{10} \\ \frac{1}{6} & \frac{1}{7} & \frac{1}{8} & \frac{1}{9} & \frac{1}{10} & \frac{1}{11} \end{bmatrix}$

36. Let A be the $n \times n$ matrix with zeros on the main diagonal and all other entries equal to -1. Use a graphing calculator or computer software package to find det(A) for $n = 3$, 4, and 5. Based on these results, find a formula (in terms of n) for det(A) that holds for all $n \geq 3$.

In Exercises 37 and 38 use a computer algebra system to find the determinant of the given matrix.

37. $A = \begin{bmatrix} 1 & 1 & k \\ 1 & k & 1 \\ k & 1 & 1 \end{bmatrix}$ in terms of k

38. $B = \begin{bmatrix} 1 & x & x^2 \\ 1 & y & y^2 \\ 1 & z & z^2 \end{bmatrix}$ in terms of x, y, and z

39. Use the results of Exercises 37 and 38, together with the fact that a square matrix M is nonsingular if and only if det(M) $\neq 0$, to determine

(a) The values for k for which A^{-1} fails to exist, where A is the matrix of Exercise 37.
(b) The values of x, y, and z for which B^{-1} fails to exist, where B is the matrix of Exercise 38.

Use a computer algebra system to do Exercises 40 and 41.

40. Recall that certain parabolas can be described by the equation $ax^2 + bx + c + dy = 0$. Given three points on a parabola, (3, 1), (2, 5), (1, 1), substituting into this equation yields the homogeneous system

$$\begin{aligned} ax^2 + \ \ bx + c + \ \ dy &= 0 \\ 9a \ \ + 3b \ \ + c + \ \ d &= 0 \\ 4a \ \ + 2b \ \ + c + 5d &= 0 \\ a \ \ + \ \ b \ \ + c + \ \ d &= 0 \end{aligned}$$

with unknowns a, b, c, d. This system can be expressed as

$$\begin{bmatrix} x^2 & x & 1 & y \\ 9 & 3 & 1 & 1 \\ 4 & 2 & 1 & 5 \\ 1 & 1 & 1 & 1 \end{bmatrix} \begin{bmatrix} a \\ b \\ c \\ d \end{bmatrix} = 0.$$

Since the solution is nontrivial, the coefficient matrix is not invertible. Therefore, the determinant is zero. Find the equation of the parabola.

41. Proceed as in Exercise 40 to find an equation of the conic section fitting these points: (10, 1), (1, −6), (−2, 11), (−5, −18), (4, −1).

4.3 Cramer's Rule

In this section, we present new techniques for solving two old problems: inverting a matrix and solving a square linear system. In Sections 3.2 and 2.3 these problems were dealt with by using row reduction. Here we solve them with determinants.

Adjoint Form of Inverse

From Theorem 5 of Section 4.2, we know that if a square matrix A is invertible, then $\det A \neq 0$. If $\det A \neq 0$, we not only can invert A, we can also obtain an explicit formula for the inverse with the aid of determinants. Toward this goal we introduce the following matrix.

DEFINITION

For a square matrix A, the **(classical) adjoint** of A is the matrix

$$\text{Adj } A = [c_{ij}], \quad \text{where} \quad c_{ij} = (-1)^{i+j} \det A_{ji}.$$

In other words, the ijth entry of Adj A is the jith cofactor of A.

EXAMPLE 1

Find Adj A, where

$$A = \begin{bmatrix} 1 & -1 & 1 \\ 2 & 0 & 2 \\ 1 & 2 & 3 \end{bmatrix}.$$

Solution We compute three of the entries of $C = \text{Adj } A$ explicitly:

$$c_{11} = (-1)^{1+1} \det A_{11} = (-1)^{1+1} \det \begin{bmatrix} 0 & 2 \\ 2 & 3 \end{bmatrix} = (-1)^2(-4) = -4,$$

$$c_{12} = (-1)^{1+2} \det A_{21} = (-1)^{1+2} \det \begin{bmatrix} -1 & 1 \\ 2 & 3 \end{bmatrix} = (-1)^3(-5) = 5,$$

and $$c_{23} = (-1)^{2+3} \det A_{32} = (-1)^{2+3} \det \begin{bmatrix} 1 & 1 \\ 2 & 2 \end{bmatrix} = (-1)^5(0) = 0$$

The other entries are computed in a similar manner to give

$$\text{Adj } A = \begin{bmatrix} -4 & 5 & -2 \\ -4 & 2 & 0 \\ 4 & -3 & 2 \end{bmatrix}.$$

■

The next theorem provides another method for computing the adjoint. You'll probably find this way easier to remember.

THEOREM 1

For a square matrix A, Adj A can be formed by replacing each entry of A^T with the corresponding cofactor of A^T.

Proof This fact is an immediate consequence of Exercise 31 of Section 4.1 and Theorem 8 of Section 4.2. That is, $\det(A^T)_{ij} = \det(A_{ji})^T = \det A_{ji}$. ■

As an example of the preceding theorem, we look again at Example 1. First form the transpose

$$A^T = \begin{bmatrix} 1 & 2 & 1 \\ -1 & 0 & 2 \\ 1 & 2 & 3 \end{bmatrix}.$$

Then replace each entry with its cofactor:

$$\text{Adj } A = \begin{bmatrix} -4 & 5 & -2 \\ -4 & 2 & 0 \\ 4 & -3 & 2 \end{bmatrix}$$

The major importance of Adj A stems from the very simple product that results when it is multiplied by the original matrix A. As a numerical example, we again use the matrix of Example 1. We have

$$A(\text{Adj } A) = \begin{bmatrix} 1 & -1 & 1 \\ 2 & 0 & 2 \\ 1 & 2 & 3 \end{bmatrix} \begin{bmatrix} -4 & 5 & -2 \\ -4 & 2 & 0 \\ 4 & -3 & 2 \end{bmatrix} = \begin{bmatrix} 4 & 0 & 0 \\ 0 & 4 & 0 \\ 0 & 0 & 4 \end{bmatrix}$$

$$= 4\begin{bmatrix} 1 & 0 & 0 \\ 0 & 1 & 0 \\ 0 & 0 & 1 \end{bmatrix} = 4I.$$

Also,

$$(\text{Adj } A)A = \begin{bmatrix} -4 & 5 & -2 \\ -4 & 2 & 0 \\ 4 & -3 & 2 \end{bmatrix} \begin{bmatrix} 1 & -1 & 1 \\ 2 & 0 & 2 \\ 1 & 2 & 3 \end{bmatrix} = \begin{bmatrix} 4 & 0 & 0 \\ 0 & 4 & 0 \\ 0 & 0 & 4 \end{bmatrix}$$

$$= 4\begin{bmatrix} 1 & 0 & 0 \\ 0 & 1 & 0 \\ 0 & 0 & 1 \end{bmatrix} = 4I.$$

Since det $A = 4$, the preceding equations can be written

$$(\text{Adj } A)A = A(\text{Adj } A) = (\det A)I.$$

Dividing through by det A, we see that the inverse of A is given by (1/det A) Adj A. This fact suggests the following theorem, which yields the *adjoint form* of the inverse of a matrix.

THEOREM 2

For a square matrix A,

$$A(\text{Adj } A) = (\text{Adj } A)A = (\det A)I.$$

Moreover, if det $A \neq 0$, then A is invertible, and

$$A^{-1} = \frac{1}{\det A}\text{Adj } A.$$

Proof We compute (Adj A)A and leave A(Adj A) as an exercise. Let $A = [a_{ij}]$ be of order n, and let $C =$ (Adj A)A. Then, by the definition of matrix multiplication,

$$\begin{aligned} c_{ij} &= (-1)^{i+1}(\det A_{1i})a_{1j} + \cdots + (-1)^{i+n}(\det A_{ni})a_{nj} \\ &= (-1)^{i+1}a_{1j}\det A_{1i} + \cdots + (-1)^{i+n}a_{nj}\det A_{ni}. \end{aligned}$$

If $i = j$ (for an element on the main diagonal),

$$\begin{aligned} c_{ii} &= (-1)^{i+1}a_{1i}\det A_{1i} + \cdots + (-1)^{i+n}a_{ni}\det A_{ni} \\ &= \det A, \end{aligned}$$

since this is just the expansion of det A down the ith column. If $i \neq j$ (for an off-diagonal element),

$$\begin{aligned} c_{ij} &= (-1)^{i+1}a_{1j}\det A_{1i} + \cdots + (-1)^{i+n}a_{nj}\det A_{ni} \\ &= \det A', \end{aligned}$$

where A' is obtained from A by replacing the ith column by the jth column. (The expansion of det A' is down the ith column.) However, A' has two equal columns (the ith and jth), and therefore det $A' = 0$. That is, $c_{ij} = 0$ for any off-diagonal element. ■

We have already seen one example of Theorem 2. As another, we show how to compute the inverse of a 2 × 2 matrix by inspection.

EXAMPLE 2

Find the inverse of A, where

$$A = \begin{bmatrix} a_{11} & a_{12} \\ a_{21} & a_{22} \end{bmatrix}.$$

Solution If $\det A = a_{11}a_{22} - a_{21}a_{12} = 0$, A has no inverse. If $\det A \neq 0$, we proceed to compute $C = \text{Adj } A$:

$$\begin{aligned} c_{11} &= (-1)^{1+1} \det A_{11} = (1) \det [\, a_{22}] = a_{22}, \\ c_{12} &= (-1)^{1+2} \det A_{21} = (-1) \det [\, a_{12}] = -a_{12}, \\ c_{21} &= (-1)^{2+1} \det A_{12} = (-1) \det [\, a_{21}] = -a_{21}, \\ \text{and } c_{22} &= (-1)^{2+2} \det A_{22} = (1) \det [\, a_{11}] = a_{11} \end{aligned}$$

Clearly, all these entries can be found with no computation. Now, by Theorem 2,

$$A^{-1} = \frac{1}{\det A} \text{Adj } A = \frac{1}{\det A} \begin{bmatrix} a_{22} & -a_{12} \\ -a_{21} & a_{11} \end{bmatrix}.$$

Thus, to find the inverse of a 2×2 matrix A, interchange its diagonal entries, take the negatives of its off-diagonal entries, and divide each entry in the resulting matrix by $\det A$. ■

Cramer's Rule

Gabriel Cramer (1704–1752) was born in Geneva, Switzerland, and was gifted mathematically from childhood. His major work was on the theory of algebraic curves.

MIDDLE TEMPLE LIBRARY/SCIENCE PHOTO LIBRARY

Now we turn our attention to solving a system of linear equations $A\mathbf{x} = \mathbf{b}$, where A is a square matrix of order n. We denote by $A(i)$ the matrix obtained from A by replacing its ith column $\mathbf{a}^i$ by $\mathbf{b}$. For example, if the given linear system is

$$\begin{aligned} x_1 - x_2 + x_3 &= 6 \\ 2x_1 + 3x_3 &= 1 \\ 2x_2 - x_3 &= 0, \end{aligned}$$

then

$$A = \begin{bmatrix} 1 & -1 & 1 \\ 2 & 0 & 3 \\ 0 & 2 & -1 \end{bmatrix}, \qquad A(1) = \begin{bmatrix} 6 & -1 & 1 \\ 1 & 0 & 3 \\ 0 & 2 & -1 \end{bmatrix},$$

$$A(2) = \begin{bmatrix} 1 & 6 & 1 \\ 2 & 1 & 3 \\ 0 & 0 & -1 \end{bmatrix}, \quad \text{and} \quad A(3) = \begin{bmatrix} 1 & -1 & 6 \\ 2 & 0 & 1 \\ 0 & 2 & 0 \end{bmatrix}.$$

The following procedure for solving a square linear system is called **Cramer's rule** and is justified shortly.

THEOREM 3

Let $A\mathbf{x} = \mathbf{b}$ be a square linear system of n equations with $\det A \neq 0$. Then the solution to this system is given by

$$x_1 = \frac{\det A(1)}{\det A}, \quad x_2 = \frac{\det A(2)}{\det A}, \quad \ldots, \quad x_n = \frac{\det A(n)}{\det A}.$$

EXAMPLE 3

Solve the linear system of equations

$$\begin{aligned} 2x_1 + 3x_2 - x_3 &= 1 \\ x_1 + 4x_2 + 2x_3 &= 2 \\ 3x_1 - x_2 - x_3 &= 3. \end{aligned}$$

Solution Letting A be the coefficient matrix,

$$A = \begin{bmatrix} 2 & 3 & -1 \\ 1 & 4 & 2 \\ 3 & -1 & -1 \end{bmatrix},$$

and, noting that $\det A = 30 \neq 0$, Cramer's rule asserts that

$$x_1 = \frac{\det A(1)}{\det A} = \frac{\det \begin{bmatrix} 1 & 3 & -1 \\ 2 & 4 & 2 \\ 3 & -1 & -1 \end{bmatrix}}{\det A} = \frac{36}{30} = \frac{6}{5},$$

$$x_2 = \frac{\det A(2)}{\det A} = \frac{\det \begin{bmatrix} 2 & 1 & -1 \\ 1 & 2 & 2 \\ 3 & 3 & -1 \end{bmatrix}}{\det A} = -\frac{6}{30} = -\frac{1}{5},$$

$$\text{and } x_3 = \frac{\det A(3)}{\det A} = \frac{\det \begin{bmatrix} 2 & 3 & 1 \\ 1 & 4 & 2 \\ 3 & -1 & 3 \end{bmatrix}}{\det A} = \frac{24}{30} = \frac{4}{5}.$$

■

EXAMPLE 4

Solve the linear system of equations

$$\begin{aligned} x_1 + 2x_2 + x_3 &= 3 \\ 3x_1 + x_2 - x_3 &= -1 \\ 5x_1 \qquad - 3x_3 &= -5. \end{aligned}$$

Solution Letting A be the coefficient matrix, we see that $\det A = 0$. Cramer's rule does not apply. The row reduction methods of Section 2.3 should be used. ■

In a square linear system $A\mathbf{x} = \mathbf{b}$, the notation $A(i)$ used in Cramer's rule is sometimes not explicit enough. When there is any doubt, we write the matrix explicitly as a matrix of columns:

$$A(i) = [\mathbf{a}^1 \cdots \mathbf{a}^{i-1}\, \mathbf{b}\, \mathbf{a}^{i+1} \cdots \mathbf{a}^n]$$

Expressed as columns in this manner, Theorem 4 of Section 4.2 can be written as follows. Let A be an $n \times n$ matrix, and let d be a scalar. Then, for any $k \neq i$,

$$\det[\mathbf{a}^1 \cdots \mathbf{a}^{i-1}\, \mathbf{a}^i + d\mathbf{a}^k\, \mathbf{a}^{i+1} \cdots \mathbf{a}^n] = \det A.$$

Combined with Theorem 1 of Section 4.2, we can extend this as follows: Let A be an $n \times n$ matrix and c and d be scalars. Then, for any $k \neq i$,

$$\det[\mathbf{a}^1 \cdots \mathbf{a}^{i-1} c\mathbf{a}^i + d\mathbf{a}^k \mathbf{a}^{i+1} \cdots \mathbf{a}^n] = c \det A.$$

Repeated application of this fact gives the proof of the following result.

LEMMA 1

Let A be a square matrix of order n, and let $\mathbf{b}$ be a column vector with $\mathbf{b} = c_1\mathbf{a}^1 + \cdots + c_n\mathbf{a}^n$, a linear combination of the columns of A. Then, for each i,

$$\det A(i) = c_i \det A.$$

■

As an illustration of Lemma 1, let

$$A = \begin{bmatrix} 2 & 3 & -1 \\ 1 & 2 & 4 \\ 3 & -1 & 0 \end{bmatrix},$$

and notice that

$$\begin{bmatrix} 1 \\ -5 \\ 11 \end{bmatrix} = 3\begin{bmatrix} 2 \\ 1 \\ 3 \end{bmatrix} - 2\begin{bmatrix} 3 \\ 2 \\ -1 \end{bmatrix} - \begin{bmatrix} -1 \\ 4 \\ 0 \end{bmatrix}.$$

Then, by Lemma 1,

$$\det\begin{bmatrix} 2 & 1 & -1 \\ 1 & -5 & 4 \\ 3 & 11 & 0 \end{bmatrix} = -2\det\begin{bmatrix} 2 & 3 & -1 \\ 1 & 2 & 4 \\ 3 & -1 & 0 \end{bmatrix}.$$

Consider the square system of linear equations $A\mathbf{x} = \mathbf{b}$:

$$\begin{aligned} a_{11}x_1 + a_{12}x_2 + \cdots + a_{1n}x_n &= b_1 \\ a_{21}x_1 + a_{22}x_2 + \cdots + a_{2n}x_n &= b_2 \\ \vdots \qquad \vdots \qquad\quad \vdots \quad & \quad \vdots \\ a_{n1}x_1 + a_{n2}x_2 + \cdots + a_{nn}x_n &= b_n \end{aligned} \tag{1}$$

Another way of viewing this system is to note that each element in the first column is multiplied by x_1, each in the second by x_2, and so on. The system can thus be written as

$$\mathbf{a}^1x_1 + \mathbf{a}^2x_2 + \cdots + \mathbf{a}^nx_n = \mathbf{b}. \tag{2}$$

In words, $\mathbf{x} = (x_1, x_2, \ldots, x_n)$ is a solution to the system if and only if it is a vector such that x_1 times the first column of A plus x_2 times the second column, and so on, yields the column $\mathbf{b}$. We now prove Cramer's rule.

Proof From Equation (2), any solution $\mathbf{x}$ gives the column vector $\mathbf{b}$ as a linear combination of the columns of A,

$$\mathbf{b} = x_1\mathbf{a}^1 + \cdots + x_n\mathbf{a}^n.$$

From Lemma 1, $\det A(i) = x_i \det A$. If $\det A \neq 0$, divide both sides by $\det A$ to obtain the final equation. That the solution is *unique* in case $\det A \neq 0$ follows from Theorem 6 of Section 4.2. ■

COMPUTATIONAL NOTE

Solving a square system of equations by Cramer's rule and likewise computing the inverse of a matrix by using the classical adjoint of the matrix are numerically inefficient methods. Although the situation is not nearly as bad as computing determinants directly from the definition, the number of multiplications required to solve an $n \times n$ system by Cramer's rule is approximately n times the number needed to do this by row operations. For this reason, Cramer's rule and the adjoint form of the inverse are more useful as theoretical tools than for computational purposes.

4.3 Exercises

In Exercises 1–8 determine whether or not Cramer's rule can be used to solve the given system. If it can be used, solve the system by Cramer's rule. If it cannot be used, explain why not.

1. $\begin{aligned} x - 2y &= 3 \\ 2x + y &= 1 \end{aligned}$

2. $\begin{aligned} -x + y &= 1 \\ x - y &= 2 \end{aligned}$

3. $\begin{aligned} x + y - z &= 1 \\ 2x - 3z &= 0 \\ 2y + z &= 1 \end{aligned}$

4. $\begin{aligned} x - y + z &= 1 \\ 2x + 3y + z &= 2 \\ x + 2y &= 3 \end{aligned}$

5. $\begin{aligned} 3x - y + z &= 2 \\ 2x + y - z &= 1 \\ x + 5y - 3z &= 3 \end{aligned}$

6. $\begin{aligned} x + 2y - z &= 3 \\ 2x - z &= 1 \\ x + 6y - z &= 5 \end{aligned}$

7. $\begin{aligned} 3w + x + y + z &= 2 \\ x + 2y + 3z &= 1 \\ 3w - 2z &= 2 \\ 2x + 4y + 7z &= 1 \end{aligned}$

8. $\begin{aligned} 2w + x - z &= 1 \\ x + y &= 2 \\ y + z &= 2 \end{aligned}$

In Exercises 9–14 compute the adjoint of each matrix, and find the inverse of the matrix if it exists.

9. $\begin{bmatrix} 1 & 2 \\ -3 & 4 \end{bmatrix}$

10. $\begin{bmatrix} 1 & -1 \\ 1 & -2 \end{bmatrix}$

11. $\begin{bmatrix} 1 & -1 & 0 \\ 0 & 2 & -2 \\ 0 & 0 & 1 \end{bmatrix}$

12. $\begin{bmatrix} 2 & 1 & 0 \\ 1 & 2 & 2 \\ 3 & -1 & 4 \end{bmatrix}$

13. $\begin{bmatrix} 1 & 2 & -1 \\ 2 & 1 & 4 \\ 1 & 5 & -7 \end{bmatrix}$

14. $\begin{bmatrix} 1 & -2 & 0 \\ -2 & 1 & -2 \\ 0 & -2 & 1 \end{bmatrix}$

15. Use the adjoint form of the inverse to derive an explicit formula for computing the inverse of a 3×3 matrix.

16. Prove directly that $A(\text{Adj } A) = (\det A)I$ for any square matrix A.

17. A matrix A is *symmetric* if $A = A^T$. Prove that if A is symmetric and invertible, then A^{-1} is also symmetric and invertible.

18. Prove that the inverse of an invertible upper triangular matrix of order 3 is invertible and upper triangular. (The statement is in fact true for invertible upper triangular matrices in general.)

19. Prove that $\det A \neq 0$ if and only if $\det(\text{Adj } A) \neq 0$. (*Hint*: Use Theorem 2.)

20. Prove that $\det(\text{Adj } A) = (\det A)^{n-1}$, $n > 1$. (*Note*: You need Exercise 19 to prove the $\det A = 0$ case.)

21. Use Vandermonde determinants and Cramer's Rule to prove Theorem 1 of Section 2.5.

Computational Exercises

22. Solve the system of equations by Cramer's rule using a computer or calculator:

$$\begin{aligned} 0.02w - 0.23x + 0.40y - z &= 3.02 \\ 2.10w + x - 1.30y &= 4.11 \\ 1.16w - 0.15x - 2.35z &= 0 \\ 3.91x + 0.63y - z &= 2.34 \end{aligned}$$

23. Find the adjoint of the matrix using a computer or calculator:

$$\begin{bmatrix} 2.11 & 2.11 & -3.04 & 1.11 \\ -0.02 & 1.23 & 2.22 & 1.02 \\ 0.14 & -0.06 & 1.21 & -1.08 \\ 1.32 & 0.20 & 0 & 3.90 \end{bmatrix}$$

(*Hint*: If A is invertible, Adj $A = (\det A)A^{-1}$. Otherwise, use the definition of Adj A.)

24. Use a computer or calculator to find the following.
(a) The adjoint of the coefficient matrix of the linear system in Exercise 22
(b) The solution, by Cramer's rule, of the linear system $A\mathbf{x} = \mathbf{b}$, where A is the matrix of Exercise 23 and $\mathbf{b} = (1, 0, 0, 0)$

25. Use a computer algebra system and the adjoint form of the inverse to obtain a formula for finding the inverse of a general 3×3 matrix (see Exercise 15).

4.4 An Introduction to Eigenvalues

We can view an $n \times n$ matrix A as acting on a (column) vector $\mathbf{x} = \begin{bmatrix} x_1 \\ \vdots \\ x_n \end{bmatrix}$ in $\mathbf{R}^n$ by matrix multiplication:

$$\mathbf{x} \longrightarrow A\mathbf{x} = A\begin{bmatrix} x_1 \\ \vdots \\ x_n \end{bmatrix}$$

This again gives us a (column) vector in $\mathbf{R}^n$.

EXAMPLE 1

The matrix

$$A = \begin{bmatrix} 1 & -1 \\ 2 & 4 \end{bmatrix}$$

acts on vectors in $\mathbf{R}^2$. For example, A acts on vector $\begin{bmatrix} 1 \\ 0 \end{bmatrix}$ by taking it to

$$\begin{bmatrix} 1 & -1 \\ 2 & 4 \end{bmatrix}\begin{bmatrix} 1 \\ 0 \end{bmatrix} = \begin{bmatrix} 1 \\ 2 \end{bmatrix}$$

and on vector $\begin{bmatrix} -1 \\ 1 \end{bmatrix}$ by taking it to

$$\begin{bmatrix} 1 & -1 \\ 2 & 4 \end{bmatrix}\begin{bmatrix} -1 \\ 1 \end{bmatrix} = \begin{bmatrix} -2 \\ 2 \end{bmatrix}.$$

This is interesting, as the resulting vector is parallel to the original (since $\begin{bmatrix} -2 \\ 2 \end{bmatrix} = 2\begin{bmatrix} -1 \\ 1 \end{bmatrix}$ is a scalar multiple of $\begin{bmatrix} -1 \\ 1 \end{bmatrix}$).

The class of vectors **x** that satisfy the equation

λ is the Greek symbol lambda

$$A\mathbf{x} = \lambda\mathbf{x} \qquad \text{for some scalar } \lambda$$

are of particular interest. Of course, the zero vector **0** always satisfies this equation, so we are interested only in nonzero vectors that satisfy the equation.

DEFINITION

Let A be an $n \times n$ matrix. A scalar λ is called an **eigenvalue** of A if there is a *nonzero* vector $\mathbf{x}$ in $\mathbf{R}^n$ such that $A\mathbf{x} = \lambda\mathbf{x}$. The vector $\mathbf{x}$ is said to be an **eigenvector** corresponding to the eigenvalue λ.

We can view an $n \times n$ matrix A as a function that takes a vector $\mathbf{x}$ in $\mathbf{R}^n$ to another vector, $A\mathbf{x}$, also in $\mathbf{R}^n$. The eigenvectors of A are those vectors that, while not fixed, are taken to vectors parallel to themselves, and the eigenvalue is a measure of the "stretching" factor for these special vectors. Figure 2 shows an eigenvector $\mathbf{u}$ of a matrix A, as $A\mathbf{u}$ is parallel to $\mathbf{u}$, while vector $\mathbf{x}$ is not an eigenvector as $A\mathbf{x}$ is not parallel to $\mathbf{x}$.

FIGURE 2

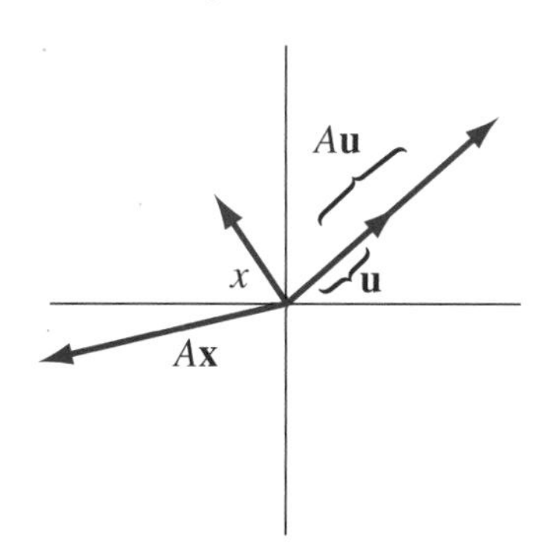

We now develop a systematic procedure for determining the eigenvalues and eigenvectors of an $n \times n$ matrix A. First, suppose we know that λ_0 is an eigenvalue of A. How do we find corresponding eigenvectors $\mathbf{x}$? By definition, $A\mathbf{x} = \lambda_0\mathbf{x}$, which may be rewritten as $A\mathbf{x} = \lambda_0 I\mathbf{x}$ or $(A - \lambda_0 I)\mathbf{x} = \mathbf{0}$, where I is the $n \times n$ identity matrix. Thus, $\mathbf{x}$ is an eigenvector corresponding to λ_0 if and only if it is a nonzero solution of the homogeneous linear system $(A - \lambda_0 I)\mathbf{x} = \mathbf{0}$.

EXAMPLE 2

The matrix

$$A = \begin{bmatrix} 1 & 1 \\ 6 & 2 \end{bmatrix}$$

has eigenvalues $\lambda_1 = -1$ and $\lambda_2 = 4$. Find the corresponding eigenvectors.

Solution All eigenvectors corresponding to $\lambda_1 = -1$ are nonzero solutions of the homogeneous linear system $(A - \lambda_1 I)\mathbf{x} = \mathbf{0}$ or $(A + I)\mathbf{x} = \mathbf{0}$, where

$$A + I = \begin{bmatrix} 2 & 1 \\ 6 & 3 \end{bmatrix}.$$

The solution of this system is found to be all vectors of the form $(s, -2s)$. So all eigenvectors corresponding to $\lambda_1 = -1$ are of the form $(s, -2s)$ where $s \neq 0$.

Similarly, to find the eigenvectors corresponding to $\lambda_2 = 4$, we solve the system $(A - 4I)\mathbf{x} = \mathbf{0}$. This yields eigenvectors of the form $(t, 3t)$, $t \neq 0$. ■

In Example 2, we saw how to find eigenvectors corresponding to a particular known eigenvalue. The next question is how to find the eigenvalues of an $n \times n$ matrix A in the first place.

Consider the equation $(A - \lambda I)\mathbf{x} = \mathbf{0}$. We seek scalars λ for which there exists *some nonzero* $\mathbf{x}$ satisfying the equation. By Theorem 5 of Section 3.3, there exists a nonzero solution $\mathbf{x}$ of the linear homogeneous system $(A - \lambda I)\mathbf{x} = \mathbf{0}$ if and only if the matrix $A - \lambda I$ is noninvertible. So the problem of finding the eigenvalues of A reduces to that of determining all scalars λ for which $A - \lambda I$ is noninvertible. But by Theorem 5 of Section 4.2, $A - \lambda I$ is noninvertible if and only if $\det(A - \lambda I) = 0$. Thus, λ is an eigenvalue of A if and only if $\det(A - \lambda I) = 0$.

EXAMPLE 3

Find the eigenvalues of

$$A = \begin{bmatrix} 3 & -1 \\ 2 & 0 \end{bmatrix}.$$

Solution We must determine all scalars λ for which $\det(A - \lambda I) = 0$. Now

$$A - \lambda I = \begin{bmatrix} 3 & -1 \\ 2 & 0 \end{bmatrix} - \lambda \begin{bmatrix} 1 & 0 \\ 0 & 1 \end{bmatrix} = \begin{bmatrix} 3 - \lambda & -1 \\ 2 & -\lambda \end{bmatrix},$$

so $\det(A - \lambda I) = (3 - \lambda)(-\lambda) + 2 = \lambda^2 - 3\lambda + 2$. Thus, the eigenvalues of A are those scalars λ that satisfy $\lambda^2 - 3\lambda + 2 = 0$. Solving this quadratic equation, we obtain $\lambda_1 = 1$ and $\lambda_2 = 2$. ■

The discussions that preceded Examples 2 and 3 can be formalized to provide the proof of the following theorem.

THEOREM 1

Let A be an $n \times n$ matrix, and let I be the identity matrix of order n. Then

a. The eigenvalues of A are those scalars λ that satisfy the equation $\det(A - \lambda I) = 0$;

b. The eigenvectors of A corresponding to an eigenvalue λ_0 are the nonzero solutions $\mathbf{x}$ of the homogeneous linear system $(A - \lambda_0 I)\mathbf{x} = \mathbf{0}$. ■

To facilitate the implementation of Theorem 1, we restate it as the following procedure.

Finding Eigenvalues and Corresponding Eigenvectors of a Square Matrix A

i. Form the matrix $A - \lambda I$ by subtracting λ from each diagonal entry of A.

ii. Solve the equation $\det(A - \lambda I) = 0$. The real solutions are the eigenvalues of A.

iii. For each eigenvalue λ_0, form the matrix $A - \lambda_0 I$.

iv. Solve the homogeneous linear system $(A - \lambda_0 I)\mathbf{x} = \mathbf{0}$. The nonzero solutions are the eigenvectors of A corresponding to λ_0.

EXAMPLE 4

Find the eigenvalues and corresponding eigenvectors of the matrix

$$A = \begin{bmatrix} 1 & -1 & 0 \\ 0 & 2 & -1 \\ 0 & 0 & -1 \end{bmatrix}.$$

Solution Following the procedure for finding the eigenvalues of A, we compute

$$\det(A - \lambda I) = \det \begin{bmatrix} 1-\lambda & -1 & 0 \\ 0 & 2-\lambda & -1 \\ 0 & 0 & -1-\lambda \end{bmatrix} = (1-\lambda)(2-\lambda)(-1-\lambda).$$

We now set this expression equal to zero, obtaining the equation

$$(1-\lambda)(2-\lambda)(-1-\lambda) = 0.$$

The eigenvalues of A are the roots of the equation:

$$\lambda_1 = 1, \quad \lambda_2 = 2, \quad \text{and} \quad \lambda_3 = -1$$

To find the eigenvectors corresponding to $\lambda_1 = 1$, we form the matrix

$$(A - 1I) = \begin{bmatrix} 0 & -1 & 0 \\ 0 & 1 & -1 \\ 0 & 0 & -2 \end{bmatrix}$$

and solve the homogeneous linear system $(A - 1I)\mathbf{x} = \mathbf{0}$. Doing so, we see that all eigenvectors corresponding to $\lambda_1 = 1$ take the form $(t, 0, 0)$, where $t \neq 0$.

To find the eigenvectors corresponding to $\lambda_2 = 2$ and $\lambda_3 = -1$, we work in a similar manner with the matrices $A - 2I$ and $A-(-1)I = A + I$, respectively. This yields eigenvectors of the form $(r, -r, 0)$, $r \neq 0$ for λ_2 and $(s, 2s, 6s)$, $s \neq 0$ for λ_3. ■

The equation $\det(A - \lambda I) = 0$, whose roots are the eigenvalues of A, is called the **characteristic equation** of A. If A is $n \times n$, the expression $\det(A - \lambda I)$ is a polynomial of degree equal to n (see Exercise 26 of Section 4.1), known as the **characteristic polynomial** for A. For example, the characteristic polynomials of Examples 3 and 4 are, respectively, $\lambda^2 - 3\lambda + 2$ (of degree 2) and $(1-\lambda)(2-\lambda)(-1-\lambda) = -\lambda^3 + 2\lambda^2 + \lambda - 2$ (of degree 3).

COMPUTATIONAL NOTE

It is not very easy to determine the eigenvalues and eigenvectors of an $n \times n$ matrix by the procedure of Theorem 1, even when n is small. Moreover, this technique is not a very efficient one to use in computer applications and may lead to inaccurate answers, because it is somewhat susceptible to the accumulation of round-off errors. For an outline of a better way to use a computer to approximate eigenvalues, see the Computational Note of Exercise 33 of Section 8.2.

You may have noticed in Example 4 that the eigenvalues of A turned out to be identical to the diagonal entries of this matrix. As you can see from Example 3, however, this is certainly not always true. The following theorem gives three special cases in which the diagonal entries of a matrix *are* its eigenvalues. Its proof follows from Theorem 4 of Section 4.1 and is left for an exercise.

THEOREM 2

The eigenvalues of an upper triangular, a lower triangular, or a diagonal matrix are identical to its diagonal entries. ■

EXAMPLE 5

Find the eigenvalues of the matrices

$$A = \begin{bmatrix} 8 & 0 & 0 \\ 0 & 0 & 0 \\ 0 & 0 & 7 \end{bmatrix} \quad \text{and} \quad B = \begin{bmatrix} 1 & 0 & 0 & 0 \\ 3 & -2 & 0 & 0 \\ 2 & -1 & 4 & 0 \\ 6 & 0 & 5 & -3 \end{bmatrix}.$$

Solution The matrices A and B are diagonal and lower triangular, respectively. So applying Theorem 2, the eigenvalues of A are 8, 0, and 7, whereas those of B are 1, -2, 4, and -3. ■

4.4 Exercises

In Exercises 1–12 find the characteristic equation and eigenvalues of the given matrix.

1. $\begin{bmatrix} 1 & 4 \\ 2 & -1 \end{bmatrix}$

2. $\begin{bmatrix} 1 & 2 \\ -2 & -1 \end{bmatrix}$

3. $\begin{bmatrix} 0 & 1 \\ -1 & 2 \end{bmatrix}$

4. $\begin{bmatrix} 0 & 0 \\ 0 & 0 \end{bmatrix}$

5. $\begin{bmatrix} 0 & -1 & 0 \\ -1 & 0 & -1 \\ 0 & -1 & 0 \end{bmatrix}$

6. $\begin{bmatrix} 2 & -1 & 0 \\ -1 & 2 & -1 \\ 0 & -1 & 2 \end{bmatrix}$

7. $\begin{bmatrix} 1 & 1 & 1 \\ 1 & 1 & 1 \\ 1 & 1 & 1 \end{bmatrix}$

8. $\begin{bmatrix} 1 & 0 & -1 \\ 1 & 1 & 2 \\ 2 & 1 & 1 \end{bmatrix}$

9. $\begin{bmatrix} 1 & 0 & 3 \\ 0 & -2 & 1 \\ 0 & 0 & 1 \end{bmatrix}$

10. $\begin{bmatrix} 1 & 0 & 0 & 0 \\ 0 & 2 & 0 & 0 \\ 3 & 1 & -1 & 0 \\ -2 & 0 & 0 & 2 \end{bmatrix}$

11. $\begin{bmatrix} 1 & 1 & 1 & 1 \\ 1 & 1 & 1 & 1 \\ 1 & 1 & 1 & 1 \\ 1 & 1 & 1 & 1 \end{bmatrix}$

12. $\begin{bmatrix} 1 & 0 & 0 & 0 \\ 0 & 0 & 1 & 0 \\ 0 & 0 & 0 & 1 \\ 0 & 1 & 0 & 0 \end{bmatrix}$

In Exercises 13–16 find the eigenvalues of the given matrix by inspection.

13. $\begin{bmatrix} 1 & 2 & 0 \\ 0 & 3 & -1 \\ 0 & 0 & 4 \end{bmatrix}$

14. $\begin{bmatrix} 1 & 0 & 0 \\ 0 & 5 & 0 \\ 0 & 0 & -6 \end{bmatrix}$

15. $\begin{bmatrix} 1 & 0 & 0 & 0 \\ 2 & 3 & 0 & 0 \\ 4 & 5 & 6 & 0 \\ 7 & 8 & 9 & 10 \end{bmatrix}$

16. $\begin{bmatrix} -1 & 0 & 1 & 0 \\ 0 & 0 & 0 & -2 \\ 0 & 0 & 2 & 0 \\ 0 & 0 & 0 & 0 \end{bmatrix}$

17. Prove Theorem 2.

18. Prove that 0 is an eigenvalue of a square matrix A if and only if A is noninvertible. (*Hint*: Use Theorem 5 of Section 3.3 or Theorem 5 of Section 4.2.)

19. Prove that A and A^T (the *transpose* of A) have the same eigenvalues.

20. Prove that λ is an eigenvalue of a nonsingular matrix A if an only if $1/\lambda$ is an eigenvalue of A^{-1}. What relationship holds between the eigenvectors of A and A^{-1}?

21. For the matrix

$$A = \begin{bmatrix} a & b \\ c & d \end{bmatrix},$$

show that A has (a) two eigenvalues if $(a - d)^2 + 4bc > 0$, (b) one eigenvalue if $(a - d)^2 + 4bc = 0$, and (c) no eigenvalues if $(a - d)^2 + 4bc < 0$.

4.5 APPLICATION—Markov Chains I: Theory

In this section, we demonstrate how multiplication of a vector by a matrix can be used to compute the probability of certain types of future events.

Markov Chains—The Basics

We first need to recall a few facts about probabilities in general. The probability of a random event is a number between 0 and 1 (inclusive) that measures the likelihood that the event will occur. For example, the probability of rolling the number 1 with a single die is $\frac{1}{6}$, since there is only one "successful" side out of the six sides of the die, and all outcomes are equally likely. The probability of rolling an even number, however, is $\frac{1}{2}$, since any of 2, 4, or 6 will be successful out of the six possible sides. Probability 0 implies that the event cannot occur, and probability 1 implies that the event must occur. As examples, the probability of rolling 7 with one die is 0, and the probability of rolling a number less than 10 with a single die is 1.

To compute probabilities of more complicated events, we use the following two facts. If two events are *independent*, the probability of both occuring is the *product* of the individual probabilities. For example, the probability of flipping a coin twice and obtaining heads both times is $\left(\frac{1}{2}\right)\left(\frac{1}{2}\right) = \frac{1}{4}$. (The coin cannot remember what it did before, so the events are independent.) If two events are *disjoint* (that is, they cannot both occur), the probability of one *or* the other occuring is the *sum* of the individual probabilities. For example, the probability of rolling an even number on a die was shown to be $\frac{3}{6} = \frac{1}{2}$. This can be broken up into three disjoint events: roll a 2, roll a 4, or roll a 6. Each of these has probability $\frac{1}{6}$, and $\frac{1}{6} + \frac{1}{6} + \frac{1}{6} = \frac{1}{2}$.

We now describe the setting for Markov chains. Repeated observations of a situation are made at regular intervals, and the current state of the situation is recorded. As an example, suppose the weather is observed every 6 hours, and the possible states of the weather are chosen to be fair (F), partly cloudy (P), cloudy (C), and rain (R). Based on past information, we assign to each state a probability of moving to some other state (or remaining the same). An important (and sometimes unrealistic) assumption is that these probabilities depend *only* on the current state, not on any previous ones. Continuing with the example, suppose the probabilities of moving from one weather state to another are as given in Table 1.

For instance, if it is cloudy now, the probability that it will be raining at the next observation is 0.2. Notice that each entry is nonnegative (probabilities must lie between 0 and 1), and that the sum of the entries in each row is 1 (no matter what the initial state is, *some* state must occur at the next observation, and the events are disjoint). Any such row is called a **probability vector**, a vector with nonnegative entries the sum of which is 1.

Our prediction for the weather will also be stated in terms of a probability vector, $\mathbf{x} = (x_1, x_2, x_3, x_4)$, where x_1 is the probability that the weather will be fair, x_2 is the probability of partly cloudy, and so on. Since *some* weather must occur, $x_1 + x_2 + x_3 + x_4 = 1$. For instance, if it is raining now, the prediction for 6 hours hence is (0.0, 0.2, 0.3, 0.5). What is the prediction for 12 hours hence? More generally, if $\mathbf{x}$ is the prediction for one time period, what is the prediction for the period after that?

TABLE 1

	Next State			
Current State	*F*	*P*	*C*	*R*
F	0.6	0.3	0.1	0.0
P	0.2	0.4	0.2	0.2
C	0.1	0.4	0.3	0.2
R	0.0	0.2	0.3	0.5

To answer this question, we reason as follows: let $\mathbf{y} = (y_1, y_2, y_3, y_4)$ be the prediction of probabilities in the next time period. That is, y_1 is the probability that the weather will be fair, y_2 is the probability that it will be partly cloudy, and so on. Looking just at y_1, there are four different ways of getting fair weather in the next time period. It could be fair now and stay fair, or it could change from one of three other states to fair. If it is fair, then the probability that it stays fair is 0.6. There is, however, a probability of x_1 that it is fair and these events are independent. Thus, the probability that it is fair and stays fair (*FF*) is $(x_1)(0.6)$. Similarly, the probability that it is partly cloudy and changes to fair (*PF*) is $(x_2)(0.2)$. Listing all the possibilities, we have

$$\begin{aligned}&\text{Probability of } FF \text{ is } 0.6x_1,\\&\text{Probability of } PF \text{ is } 0.2x_2,\\&\text{Probability of } CF \text{ is } 0.1x_3,\\&\text{Probability of } RF \text{ is } 0.0x_4.\end{aligned}$$

These four events are disjoint, so the probability that the weather is fair in the next time period is

$$y_1 = 0.6x_1 + 0.2x_2 + 0.1x_3 + 0.0x_4.$$

Note that y_1 is exactly the dot product of the first column of Table 1 with the vector $\mathbf{x}$. Carrying out a similar analysis for each of y_2, y_3, and y_4, we obtain y_i as the dot product of the ith column of the table with the vector $\mathbf{x}$. To make this conform to our usual convention for matrix multiplication, let T (for **transition matrix**) be the *transpose* of the entries in the table. Then this discussion yields the result that

$$\mathbf{y} = T\mathbf{x}. \tag{1}$$

Using Equation (1), we can answer the question posed earlier. If it is raining now, what is the probability vector for two periods in the future? Letting $\mathbf{x} = (0.0, 0.2, 0.3, 0.5)$, the prediction for the first time period, we have

$$\mathbf{y} = T\mathbf{x} = \begin{bmatrix} 0.6 & 0.2 & 0.1 & 0.0 \\ 0.3 & 0.4 & 0.4 & 0.2 \\ 0.1 & 0.2 & 0.3 & 0.3 \\ 0.0 & 0.2 & 0.2 & 0.5 \end{bmatrix} \begin{bmatrix} 0.0 \\ 0.2 \\ 0.3 \\ 0.5 \end{bmatrix} = \begin{bmatrix} 0.07 \\ 0.30 \\ 0.28 \\ 0.35 \end{bmatrix}.$$

Since $\mathbf{x}$ is equal to $T(0, 0, 0, 1)$, we have another way of computing $\mathbf{y}$, $\mathbf{y} = T^2(0, 0, 0, 1)$. Similarly, if we wanted the probability vector for the following time period, Equation (1) tells us that this prediction is given by

$$T\mathbf{y} = T(T^2(0, 0, 0, 1)) = T^3(0, 0, 0, 1).$$

Thus, we see that to go from any given state to any later state, we simply multiply the given vector by a power of T. The results of this specific example are generalized as the following procedure.

Markov Chain Procedure Suppose that there are n possible states:

i. Prepare an $n \times n$ matrix where the ijth entry is the probability of moving from state i to state j.

ii. Let T be the transpose of this matrix.

iii. For any probability vector $\mathbf{x}$ in $\mathbf{R}^n$ and for any natural number k, the probability vector for k time periods in the future is given by $\mathbf{y} = T^k\mathbf{x}$.

NOTE The vector $\mathbf{x}$ in the procedure is called an **initial vector** and is presented as a probability vector. If in fact we *know* the initial state (say, the ith state), then $\mathbf{x} = \mathbf{e}_i$, the ith elementary vector. Of course, $\mathbf{e}_i$ is itself a probability vector; the probability of the ith state is 1 and of any other is 0. As an example, in the preceding discussion, if the initial state is cloudy, the probability vector is $(0, 0, 1, 0)$.

EXAMPLE 1

Suppose that a particular telephone wire is capable of supporting up to four calls at any one time and that the number of calls that will be in progress 1 minute later than at a given time depends only on the number of calls in progress at the given time. Suppose further that the probability of moving from i calls at a given time to j calls 1 minute later is given in Table 2. If at some particular time all possible numbers of calls are equally likely, what is the probability that the line will be unavailable (four calls in progress) 3 minutes later?

TABLE 2

	j				
i	0	1	2	3	4
0	0.3	0.3	0.2	0.2	0.0
1	0.2	0.3	0.3	0.1	0.1
2	0.1	0.2	0.4	0.2	0.1
3	0.1	0.1	0.2	0.3	0.3
4	0.0	0.0	0.3	0.3	0.4

Solution We follow the Markov chain procedure stated earlier. Step (i) is already done. The matrix T is the transpose of the given array:

$$T = \begin{bmatrix} 0.3 & 0.2 & 0.1 & 0.1 & 0.0 \\ 0.3 & 0.3 & 0.2 & 0.1 & 0.0 \\ 0.2 & 0.3 & 0.4 & 0.2 & 0.3 \\ 0.2 & 0.1 & 0.2 & 0.3 & 0.3 \\ 0.0 & 0.1 & 0.1 & 0.3 & 0.4 \end{bmatrix}$$

Since all numbers of calls are equally likely, the initial vector is

$$\mathbf{x} = (0.2, 0.2, 0.2, 0.2, 0.2).$$

The probability vector for 3 minutes hence is then

$$\mathbf{y} = T^3\mathbf{x} = \begin{bmatrix} 0.3 & 0.2 & 0.1 & 0.1 & 0.0 \\ 0.3 & 0.3 & 0.2 & 0.1 & 0.0 \\ 0.2 & 0.3 & 0.4 & 0.2 & 0.3 \\ 0.2 & 0.1 & 0.2 & 0.3 & 0.3 \\ 0.0 & 0.1 & 0.1 & 0.3 & 0.4 \end{bmatrix}^3 \begin{bmatrix} 0.2 \\ 0.2 \\ 0.2 \\ 0.2 \\ 0.2 \end{bmatrix} = \begin{bmatrix} 0.1246 \\ 0.1712 \\ 0.2942 \\ 0.2232 \\ 0.1868 \end{bmatrix}.$$

The probability of having four calls in progress at that time is the last coordinate of $\mathbf{y}$, $y_5 = 0.1868$. ■

Stationary Distribution

DEFINITION

A probability vector $\mathbf{x}$ is a **stationary distribution** if $\mathbf{x} = T\mathbf{x}$, where T is the transition matrix of a Markov chain.

In other words, a stationary distribution is such that moving from one time period to the next results in no change. Every Markov chain has a stationary distribution and some have more than one. The following theorem shows how to determine them.

THEOREM 1

Let T be the transition matrix of a Markov chain. Then $\lambda = 1$ is an eigenvalue of T. If $\mathbf{x}$ is an eigenvector corresponding to $\lambda = 1$ with nonnegative components, and s is the sum of the components of $\mathbf{x}$, then $(1/s)\mathbf{x}$ is a stationary distribution.

Proof Let there be n states so that T is an $n \times n$ matrix. Since the sum of the entries in each column is 1, the sum of the entries in each column of $T - I$ is 0. This implies that if each row is added to the last row, the result is a zero row, and hence the resulting matrix has determinant zero. But adding one row to another does not change the value of the determinant (Theorem 4 of Section 4.2), so the determinant of $T - I$ is 0. Since $T - I = T - 1I$, $\lambda = 1$ is an eigenvalue of T.

If $\mathbf{x}$ is an eigenvector corresponding to $\lambda = 1$, then $T\mathbf{x} = \mathbf{x}$. Assuming further that $\mathbf{x}$ has nonnegative components, and that $s = x_1 + x_2 + \ldots + x_n$, the sum of the components of $\mathbf{x}$, we have $T((1/s)\mathbf{x}) = (1/s)T\mathbf{x} = (1/s)\mathbf{x}$. Now $(1/s)\mathbf{x}$ has nonnegative components, and the sum of the components is $(1/s)(x_1 + \ldots + x_n) = (1/s)s = 1$. Therefore $(1/s)\mathbf{x}$ is a probability vector, and consequently it is a stationary distribution. ■

EXAMPLE 2

Determine a stationary distribution for a Markov chain with transition matrix

$$T = \begin{bmatrix} \frac{1}{3} & \frac{1}{3} & \frac{1}{3} \\ \frac{1}{2} & \frac{1}{6} & 0 \\ \frac{1}{6} & \frac{1}{2} & \frac{2}{3} \end{bmatrix}.$$

Solution We first determine the eigenvectors of T corresponding to 1. To do so, we solve the homogeneous system $(T - I)\mathbf{x} = \mathbf{0}$ by row-reducing $T - I$,

$$T - I = \begin{bmatrix} -\frac{2}{3} & \frac{1}{3} & \frac{1}{3} \\ \frac{1}{2} & -\frac{5}{6} & 0 \\ \frac{1}{6} & \frac{1}{2} & -\frac{1}{3} \end{bmatrix} \rightarrow \begin{bmatrix} 1 & 0 & -\frac{5}{7} \\ 0 & 1 & -\frac{3}{7} \\ 0 & 0 & 0 \end{bmatrix}.$$

From the row-reduced echelon form we see that $\left(\frac{5}{7}, \frac{3}{7}, 1\right)$ is an eigenvector that corresponds to the eigenvalue 1. A (in fact, *the*) stationary distribution is therefore

$$\left(\frac{1}{\frac{5}{7} + \frac{3}{7} + 1}\right)\left(\frac{5}{7}, \frac{3}{7}, 1\right) = \left(\frac{1}{3}, \frac{1}{5}, \frac{7}{15}\right).$$ ■

Often there is a *unique* stationary distribution for a Markov chain, and succeeding stages of the chain give resulting stationary distributions that are successively closer to this stationary distribution and converge to it. The following theorem offers a very simple test for this situation. The proof is omitted.

THEOREM 2

Let T be the transition matrix of a Markov chain. If some power of T has strictly positive entries, then there is a unique stationary distribution $\mathbf{x}$, and for any probability vector $\mathbf{y}$, $T^n\mathbf{y}$ converges to $\mathbf{x}$ as n becomes large. ■

EXAMPLE 3

Show that some power of the transition matrix T of Example 1,

$$T = \begin{bmatrix} 0.3 & 0.2 & 0.1 & 0.1 & 0.0 \\ 0.3 & 0.3 & 0.2 & 0.1 & 0.0 \\ 0.2 & 0.3 & 0.4 & 0.2 & 0.3 \\ 0.2 & 0.1 & 0.2 & 0.3 & 0.3 \\ 0.0 & 0.1 & 0.1 & 0.3 & 0.4 \end{bmatrix},$$

has strictly positive entries so that there is a unique stationary distribution. Find an approximation to the stationary distribution by computing values of $T^n\mathbf{y}$ for some $\mathbf{y}$ and various values of n.

Solution The second power of T, T^2, has strictly positive entries, since

$$T^2 = \begin{bmatrix} 0.19 & 0.16 & 0.13 & 0.10 & 0.06 \\ 0.24 & 0.22 & 0.19 & 0.13 & 0.09 \\ 0.27 & 0.30 & 0.31 & 0.28 & 0.30 \\ 0.19 & 0.19 & 0.21 & 0.25 & 0.27 \\ 0.11 & 0.13 & 0.16 & 0.24 & 0.28 \end{bmatrix}.$$

Estimating the stationary distribution by raising T to successively higher powers (with the aid of a computer), we see that T^7 and T^8 agree to the nearest hundredth.

$$T^7 = \begin{bmatrix} 0.124 & 0.123 & 0.123 & 0.122 & 0.121 \\ 0.170 & 0.170 & 0.169 & 0.168 & 0.167 \\ 0.295 & 0.295 & 0.295 & 0.295 & 0.295 \\ 0.224 & 0.224 & 0.224 & 0.225 & 0.226 \\ 0.188 & 0.188 & 0.189 & 0.191 & 0.192 \end{bmatrix}$$

$$T^8 = \begin{bmatrix} 0.123 & 0.123 & 0.123 & 0.122 & 0.122 \\ 0.170 & 0.169 & 0.169 & 0.168 & 0.168 \\ 0.295 & 0.295 & 0.295 & 0.295 & 0.295 \\ 0.224 & 0.224 & 0.224 & 0.225 & 0.225 \\ 0.189 & 0.189 & 0.189 & 0.190 & 0.190 \end{bmatrix}$$

Letting $\mathbf{y} = (1, 0, 0, 0, 0)$, we take $T^8\mathbf{y} = (0.123, 0.170, 0.295, 0.224, 0.189)$ as our approximation to the stationary distribution. ■

COMPUTATIONAL NOTE

Estimating the stationary distribution $\mathbf{x}$ by setting $\mathbf{x}$ approximately equal to $T^n\mathbf{y}$ for larger values of n and an arbitrary vector $\mathbf{y}$ (as in Example 3) is commonly used in place of computing the actual stationary distribution when the conditions of Theorem 2 are met. By using the elementary vectors $\mathbf{e}_i$ for $\mathbf{y}$, we see that the columns of T^n must all equal $\mathbf{x}$ in the limit. Thus, the procedure is to raise T to successively higher powers until all of the columns agree within the desired accuracy. Any column of this matrix is then a good approximation to the stationary distribution.

4.5 Exercises

In Exercises 1–10 use the information given in the table of probabilities for a Markov chain, Table 3.

TABLE 3

	Next State		
Current State	1	2	3
1	0.2	0.5	0.3
2	0.1	0.8	0.1
3	0.2	0.4	0.4

1. Find the probability of moving from state 3 to state 2.
2. Find the probability of moving from state 2 to state 3.
3. Find the transition matrix for the Markov chain.
4. Find the transition matrix for moving from the current state to two states in the future.

5. What is the initial vector if all states are equally likely?

6. What is the initial vector if state 1 is excluded but the other two are equally likely?

7. If all states are equally likely, find the probability vector for the next state.

8. If state 1 is excluded, but the other two are equally likely, find the probability vector for the next state.

9. If the initial state is state 1, find the probability of state 2 two time periods in the future.

10. If the initial state is state 1, find the probability of state 3 three time periods in the future.

In Exercises 11–15 use the following information. Suppose that a man has three modes of transportation to work: he can walk, drive his car, or take the bus. He never walks two days in a row or drives two days in a row. If he walked the last time, he is twice as likely to drive as take the bus. In all other cases, he shows no preference.

11. Prepare the table of probabilities for this Markov chain, and find the transition matrix.

12. If he drives on Monday, what is the probability that he will walk on Tuesday? That he will drive on Tuesday?

13. If he is equally likely to choose any mode of transportation on Monday, what is the prediction for all modes of transportation for Tuesday?

14. If he takes the bus on Monday, what is the probability that he will take the bus on Wednesday?

15. If he takes the bus on Monday, what is the probability that he will take the bus all week (five days in succession)?

16. Find all stationary distributions of the transition matrix

$$T = \begin{bmatrix} \frac{1}{2} & \frac{1}{3} & 0 \\ \frac{1}{2} & \frac{1}{3} & 0 \\ 0 & \frac{1}{3} & 1 \end{bmatrix}.$$

17. Find all stationary distributions of the transition matrix

$$T = \begin{bmatrix} \frac{1}{2} & \frac{1}{2} & 0 \\ \frac{1}{2} & \frac{1}{2} & 0 \\ 0 & 0 & 1 \end{bmatrix}.$$

18. Find all stationary distributions of the transition matrix of Exercise 11.

19. If T is a transition matrix of a Markov chain, prove that the sum of the entries in each column of T^2 is also equal to 1.

20. Extend Exercise 19 to T^n.

Computational Exercises

In Exercises 21 and 22 use computer software or a graphing calculator to estimate the components of each stationary distribution to the nearest hundredth.

21. Find the stationary distribution for the transition matrix of Exercise 3.

22. Find the stationary distribution for the transition matrix T obtained from Table 1.

23. Suppose that you have only four moods: happy, sad, angry, and confused. Let t_{ij} be the probability of being in mood i tomorrow if you are in mood j today. Suppose further that these probabilities do not change from one day to another. Let T be the matrix of entries t_{ij},

$$T = \begin{bmatrix} 0.90 & 0.40 & 0.20 & 0.03 \\ 0.05 & 0.40 & 0.20 & 0.08 \\ 0.03 & 0.10 & 0.40 & 0.13 \\ 0.02 & 0.10 & 0.20 & 0.76 \end{bmatrix}.$$

Use the power method of Theorem 2 to estimate the stationary distribution for the transition matrix T. How would people be most apt to describe your personality based on this result?

24. Use a computer software package or graphing calculator to solve the following problem. Given Table 4 and the initial data in Table 5:

(a) Construct a transition matrix and an initial probability vector based on the empirical data.
(b) Find the population distribution (as percentages) in each of the regions in 10 years, 20 years.
(c) Find the stationary distribution. Express the populations (as percentages) for the stationary distribution.

TABLE 4 Movement Between Regions from Year One to Year Two (Number of People)

Region Moved to	Region Moved from: South	West	North	East
South	—	34	56	10
West	76	—	23	91
North	32	43	—	76
East	12	8	14	—

TABLE 5 Population in Year One

Region	Population
South	569
West	981
North	345
East	289

Larry Page (left) and Sergey Brin were undergraduate students at Stanford University. In the 1980s Page and Brin came up with an idea in a research project that proposed a new type of search engine. A little bit of linear algebra formed the basis for a startling new method of web searching.

Bloomberg via Getty Images

4.6 APPLICATION—Markov Chains II: Google PageRank

You might ask, what can you do with Markov chains? Well, one thing you can do is roll it into a billion dollar idea! It is such a great "math-to-riches" story that we'll spend the section recounting it.

Ranking the Web

One can look at the World Wide Web as a large graph (called a **web graph**). Each node is a Web page, and the arcs indicate links: if Web page A links to Web page B, then we have an arc from A to B.

EXAMPLE 1

Suppose that we have a group of 8 Web pages, labelled A through H, with links as follows:

- page A links to pages B, E, F, and G
- page B links to pages C, D, and E
- page C links to pages B and D
- page D links to no pages
- page E links to page G
- page F links to page B
- page G links to pages A and F
- page H links to pages F and G

A drawing of the associated web graph is shown in Figure 3.

FIGURE 3

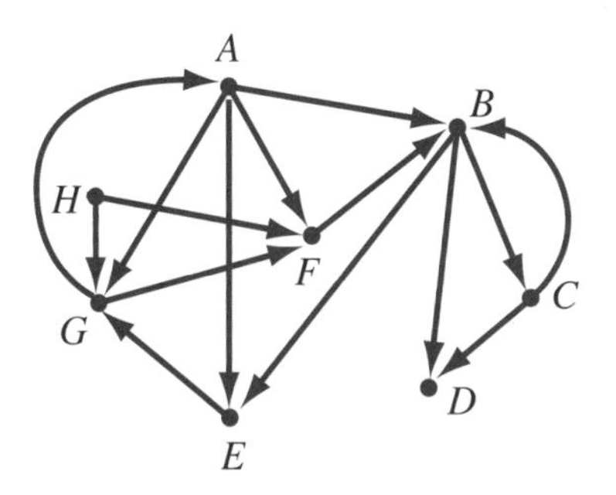

Like the question we posed in Section 3.7 about popular individuals in social networks, what constitutes a popular and important Web page? Not a Web page that has a lot of links; after all, a Web page can choose to link to a lot of other pages, and this should not affect its importance. A much better guide to a page's significance is the number of links coming *into* the page. The creator of a page has no control over its in-links, and it seems likely that a page with a lot of in-links is an *authority* on its subject. Moreover, a Web page is even more important if other important pages have links into it (this seems rather circular, but we'll make it precise shortly). Finally, the importance of a link from a page X to a page Y depends on how many links X has coming out of it; if it has few, each link is rather important, but if it has many, the importance of each link is diluted.

This observation is the basis of Brin and Page's function $r(P)$ for ranking pages on the Web—deciding a pecking order of relative importance. It seems reasonable that the importance of page X should be the sum of the importance of the pages that link into it, but by also somehow modifying the importance of each page Y by the number of links that come out of that page (more links should drop the importance; fewer should increase it). Borrowing the terminology from Section 3.7, for a page X, $N^{in}(X)$ denotes the *in-neighbours* of X (that is, the set of pages that have arcs *into* X), and $N^{out}(X)$, the *out-neighbours* of X (that is, the set of pages that have arcs *out of* X). Brin and Page suggested that the rank of page P (its **PageRank**) be given by

$$r(X) = \sum_{Y \in N^{in}(X)} \frac{r(Y)}{d^{out}(Y)} \tag{1}$$

where $d^{out}(Y)$ is the out-degree of Y, that is, the number of out-neighbours of Y. (Trivially, if all of the PageRanks were 0, (1) would be satisfied, so we insist that all of the ranks are at least 0 and at least one rank is positive.) You can think of formula (1) as stating that a page divides its PageRank equally among its out-neighbours.

Let's Go Surfin' Now

Now you might think that the formula doesn't really define the PageRank, as the PageRank of every page is defined in terms of the PageRanks of other pages. To see how to get around this, let's formulate the problem with matrices and probabilities. For a directed graph, we'll consider the nodes as states, and when you are at any given node, you move with equal probability to one of its out-neighbours (think of this as *random Web surfing*, where you would jump randomly via links from page to page). The probability that you move from node i to node j is equal to $1/d^{out}(i)$.

For the graph in Figure 3 (which corresponds to the web graph described in Example 1), the transition matrix (using the terminology of the previous section) is the transpose of the matrix

$$\begin{bmatrix} 0 & \frac{1}{4} & 0 & 0 & \frac{1}{4} & \frac{1}{4} & \frac{1}{4} & 0 \\ 0 & 0 & \frac{1}{3} & \frac{1}{3} & \frac{1}{3} & 0 & 0 & 0 \\ 0 & \frac{1}{2} & 0 & \frac{1}{2} & 0 & 0 & 0 & 0 \\ 0 & 0 & 0 & 0 & 0 & 0 & 0 & 0 \\ 0 & 0 & 0 & 0 & 0 & 0 & 1 & 0 \\ 0 & 1 & 0 & 0 & 0 & 0 & 0 & 0 \\ \frac{1}{2} & 0 & 0 & 0 & 0 & \frac{1}{2} & 0 & 0 \\ 0 & 0 & 0 & 0 & 0 & \frac{1}{2} & \frac{1}{2} & 0 \end{bmatrix},$$

that is,

$$T = \begin{bmatrix} 0 & 0 & 0 & 0 & 0 & 0 & \frac{1}{2} & 0 \\ \frac{1}{4} & 0 & \frac{1}{2} & 0 & 0 & 1 & 0 & 0 \\ 0 & \frac{1}{3} & 0 & 0 & 0 & 0 & 0 & 0 \\ 0 & \frac{1}{3} & \frac{1}{2} & 0 & 0 & 0 & 0 & 0 \\ \frac{1}{4} & \frac{1}{3} & 0 & 0 & 0 & 0 & 0 & 0 \\ \frac{1}{4} & 0 & 0 & 0 & 0 & 0 & \frac{1}{2} & \frac{1}{2} \\ \frac{1}{4} & 0 & 0 & 0 & 1 & 0 & 0 & \frac{1}{2} \\ 0 & 0 & 0 & 0 & 0 & 0 & 0 & 0 \end{bmatrix}.$$

The PageRank of Web page B is, by formula (1),

$$r(B) = \frac{r(A)}{4} + \frac{r(C)}{2} + \frac{r(F)}{1}$$

since A divides up its PageRank equally among its 4 out-neighbours (so each gets $r(A)/4$ of A's PageRank). C divides up its PageRank equally among its 2 out-neighbours (so each gets $r(C)/2$ of C's PageRank), and F, as it only has one out-neighbour, namely B, sends all of its PageRank to B.

But here is the key observation—if we form a column vector of the PageRanks,

$$\mathbf{r} = \begin{bmatrix} r(A) \\ r(B) \\ r(C) \\ r(D) \\ r(E) \\ r(F) \\ r(G) \\ r(H) \end{bmatrix},$$

then

$$T\mathbf{r} = \begin{bmatrix} 0 & 0 & 0 & 0 & 0 & 0 & \frac{1}{2} & 0 \\ \frac{1}{4} & 0 & \frac{1}{2} & 0 & 0 & 1 & 0 & 0 \\ 0 & \frac{1}{3} & 0 & 0 & 0 & 0 & 0 & 0 \\ 0 & \frac{1}{3} & \frac{1}{2} & 0 & 0 & 0 & 0 & 0 \\ \frac{1}{4} & \frac{1}{3} & 0 & 0 & 0 & 0 & 0 & 0 \\ \frac{1}{4} & 0 & 0 & 0 & 0 & 0 & \frac{1}{2} & \frac{1}{2} \\ \frac{1}{4} & 0 & 0 & 0 & 1 & 0 & 0 & \frac{1}{2} \\ 0 & 0 & 0 & 0 & 0 & 0 & 0 & 0 \end{bmatrix} \begin{bmatrix} r(A) \\ r(B) \\ r(C) \\ r(D) \\ r(E) \\ r(F) \\ r(G) \\ r(H) \end{bmatrix} = \begin{bmatrix} \frac{r(G)}{2} \\ \frac{r(A)}{4} + \frac{r(C)}{2} + r(F) \\ \frac{r(B)}{3} \\ \frac{r(B)}{3} + \frac{r(C)}{2} \\ \frac{r(A)}{4} + \frac{r(B)}{3} \\ \frac{r(A)}{4} + \frac{r(G)}{2} + \frac{r(H)}{2} \\ \frac{r(A)}{4} + r(E) + \frac{r(H)}{2} \\ 0 \end{bmatrix}.$$

The second entry in the column vector on the right corresponds to the PageRank $r(B)$ of the second node, B. You can check that all of the components correspond to the PageRank of the associated node in the graph, as calculated by formula (1). It follows that from formula (1), $T\mathbf{r}$ is the same as vector $\mathbf{r}$, that is, $T\mathbf{r} = \mathbf{r}$. This holds for any graph, so we have the following crucial result.

THEOREM 1

Let T be the transition matrix of a web graph G. Then the PageRank column vector $\mathbf{r}$ of G satisfies $T\mathbf{r} = \mathbf{r}$, that is, $\mathbf{r}$ is an eigenvector of T with eigenvalue 1. ■

EXAMPLE 2

For the web graph of Example 1 (as shown in Figure 3), we have seen that the transition matrix is

$$T = \begin{bmatrix} 0 & 0 & 0 & 0 & 0 & 0 & \frac{1}{2} & 0 \\ \frac{1}{4} & 0 & \frac{1}{2} & 0 & 0 & 1 & 0 & 0 \\ 0 & \frac{1}{3} & 0 & 0 & 0 & 0 & 0 & 0 \\ 0 & \frac{1}{3} & \frac{1}{2} & 0 & 0 & 0 & 0 & 0 \\ \frac{1}{4} & \frac{1}{3} & 0 & 0 & 0 & 0 & 0 & 0 \\ \frac{1}{4} & 0 & 0 & 0 & 0 & 0 & \frac{1}{2} & \frac{1}{2} \\ \frac{1}{4} & 0 & 0 & 0 & 1 & 0 & 0 & \frac{1}{2} \\ 0 & 0 & 0 & 0 & 0 & 0 & 0 & 0 \end{bmatrix}.$$

A calculation shows that the only vector $\mathbf{r}$ that satisfies $T\mathbf{r} = \mathbf{r}$ is the zero vector, so there is *no* PageRank function. ■

If the story ended there, there would be no pot of gold for Brin and Page. But there is even a bigger mystery here. If T is the transition of a Markov chain, then by Theorem 1 of Section 4.5, there *has* to be a non-zero vector $\mathbf{r}$ that satisfies $T\mathbf{r} = \mathbf{r}$ (namely an eigenvector of T with eigenvalue 1) but there isn't one! So what gives?

Making the Markov Connection

The problem we glossed over is that the probabilities we defined aren't quite a Markov chain. To be a Markov chain, the sum of all the probabilities out of each state has to be 1, that is, in T, the column sums must all be 1. This holds true in the transition matrices we defined in this section *except* for those nodes that have no outgoing links (which are often called *sinks*). You can't travel *anywhere* from sinks, and the sum of all the probabilities leading from such a node is 0.

Here is Brin and Page's fix. What they did was to allow, with a small probability, a jump from a sink to another node (even itself!) at random. That is, for each sink, with probability $1/n$ you jump to another node (possibly staying where you are), where n is number of nodes in the graph. This corresponds to changing in T each entry in the *column* corresponding to a sink to $1/n$. We'll call this new matrix the **altered transition matrix**, $\widetilde{T}$.

EXAMPLE 3

Continuing with Examples 1 and 2, the only sink is node D, so replacing each entry in D's column by 1/8 we have the new altered transition matrix

$$\widetilde{T} = \begin{bmatrix} 0 & 0 & 0 & \frac{1}{8} & 0 & 0 & \frac{1}{2} & 0 \\ \frac{1}{4} & 0 & \frac{1}{2} & \frac{1}{8} & 0 & 1 & 0 & 0 \\ 0 & \frac{1}{3} & 0 & \frac{1}{8} & 0 & 0 & 0 & 0 \\ 0 & \frac{1}{3} & \frac{1}{2} & \frac{1}{8} & 0 & 0 & 0 & 0 \\ \frac{1}{4} & \frac{1}{3} & 0 & \frac{1}{8} & 0 & 0 & 0 & 0 \\ \frac{1}{4} & 0 & 0 & \frac{1}{8} & 0 & 0 & \frac{1}{2} & \frac{1}{2} \\ \frac{1}{4} & 0 & 0 & \frac{1}{8} & 1 & 0 & 0 & \frac{1}{2} \\ 0 & 0 & 0 & \frac{1}{8} & 0 & 0 & 0 & 0 \end{bmatrix}.$$

Note that matrix $\widetilde{T}$ is indeed the transition matrix of a Markov chain, as *all* of the row sums in the transpose are 1. We can now look for a PageRank vector, knowing that by Theorem 1 of Section 4.5, there has to be such a vector (namely an eigenvector corresponding to eigenvalue 1). A calculation shows that the eigenvectors corresponding to eigenvalue 1 are all scalar multiples of

$$[212, 468, 192, 288, 245, 283, 352, 36]^T.$$

The sum of its coordinates is 2076, so the corresponding stationary distribution is

$$\left[\frac{212}{2076}, \frac{468}{2076}, \frac{192}{2076}, \frac{288}{2076}, \frac{245}{2076}, \frac{283}{2076}, \frac{352}{2076}, \frac{36}{2076}\right]^T,$$

which is approximately

$$[0.1021, 0.2254, 0.0925, 0.1387, 0.1180, 0.1363, 0.1696, 0.0173].$$

This gives a PageRank for each page—larger numbers correspond to higher ranked pages. The biggest value is 0.2254, the PageRank of page B. Ordering by PageRank, from most important to least, we have the ranking B, G, D, F, E, A, C, H. ■

Now we have a way to find the PageRank of a web graph, namely to form the altered transition matrix, which happens to be a Markov chain, and the stationary distribution will yield our ranking.

Brin and Page should have been done, but they recognized a few problems. First of all, there is no guarantee that there is only one stationary distribution of the Markov chain. Moreover, even if there were, calculating eigenvectors can be difficult for as large matrices as they needed to consider (remember, the actual web graph has *billions* of nodes, and thus the altered transition matrix is HUGE!). Even if they formed the matrix and tried solving $\widetilde{T}\mathbf{r} = \mathbf{r}$ as a linear system (which it is!), they wouldn't have enough time and resources to solve it.

If You Can't Make the Math Fit the Problem ...

The solution again lies in the theory of Markov chains. Theorem 2 of Section 4.5 says that *provided that some power of the transition matrix of a Markov chain has all positive entries*, you could start with *any* probability vector $\mathbf{y}$, and $T^m\mathbf{y}$ would converge to the stationary distribution (this is called the **power method** for finding the stationary distribution). This held out the promise that instead of solving for the stationary vector of the altered transition matrix of a web graph, we could use successive matrix multiplications with column vectors to approximate the PageRank vector (that is, provided that all of the entries in some power of the altered transition matrix are positive). But sadly, even when we have dealt with the sinks in our alteration, we still may have lots of zeros, and sometimes, raising the matrix to higher powers will still leave zeros (see Exercise 4).

Again, if you can't make the math fit the problem, fit the problem to the math! Brin and Page again altered the matrix, this time to ensure that all the entries are positive. So what we do is quite simple—we add the same tiny bit to each entry of the matrix (Brin and Page thought of this as *teleporting*, à la Star Trek, to a random page). But of course we want to keep the matrix a transition matrix of a Markov chain; that is, all of the column sums need to be 1. So letting J be a matrix, of the same $n \times n$ size as the altered transition matrix, with all entries $1/n$, we pick a number α strictly between 0 and 1 and form the new matrix (depending on α)

$$G_\alpha = \alpha\widetilde{T} + (1 - \alpha)J;$$

that is, we multiply each entry of the altered transition matrix by α and add $(1 - \alpha)$. This matrix G is a **google matrix** of the web graph, and is the crowning glory. No matter what α is, all of the entries of G_α are positive and the column sums are 1 (see Exercise 5), so G_α is the transition matrix of a new Markov chain. If α is really close to 1, then G_α is close to $\widetilde{T}$, the altered transition matrix of the web graph, and we feel that the PageRank vector of G_α should be close to that of $\widetilde{T}$ (which, after all, is what we desire). We can now use the power method, starting with any probability distribution (a vector with all entries equal to $1/n$ will do, for example, and is used in practice).

Any choice of α between 0 and 1 will do, but different choices will yield different PageRank vectors. Which one does Google actually choose? The actual number is a carefully guarded company secret, but it's believed to be around 0.85. Why did they settle on this number, rather than something closer to 1? Well, because the computation of vectors in the power method is so expensive, resource-wise, Google needs to conserve the number of iterations, and yet be close enough to the PageRank vector in order to determine the ranking. It can be shown mathematically that the kth iteration under the power method is within about α^k of the true PageRank vector. For $\alpha = 0.85$, $\alpha^{50} \approx 0.0003$, and so the PageRank vector would be accurate to about 3 decimal places, which is probably good enough for ranking purposes.

EXAMPLE 4

Continuing with Examples 1, 2, and 3, for $\alpha = 0.85$, the google matrix is (up to 5 decimal places)

$$G = 0.85\begin{bmatrix} 0 & 0 & 0 & \frac{1}{8} & 0 & 0 & \frac{1}{2} & 0 \\ \frac{1}{4} & 0 & \frac{1}{2} & \frac{1}{8} & 0 & 1 & 0 & 0 \\ 0 & \frac{1}{3} & 0 & \frac{1}{8} & 0 & 0 & 0 & 0 \\ 0 & \frac{1}{3} & \frac{1}{2} & \frac{1}{8} & 0 & 0 & 0 & 0 \\ \frac{1}{4} & \frac{1}{3} & 0 & \frac{1}{8} & 0 & 0 & 0 & 0 \\ \frac{1}{4} & 0 & 0 & \frac{1}{8} & 0 & 0 & \frac{1}{2} & \frac{1}{2} \\ \frac{1}{4} & 0 & 0 & \frac{1}{8} & 1 & 0 & 0 & \frac{1}{2} \\ 0 & 0 & 0 & \frac{1}{8} & 0 & 0 & 0 & 0 \end{bmatrix} +.0.15\begin{bmatrix} \frac{1}{8} & \frac{1}{8} & \frac{1}{8} & \frac{1}{8} & \frac{1}{8} & \frac{1}{8} & \frac{1}{8} & \frac{1}{8} \\ \frac{1}{8} & \frac{1}{8} & \frac{1}{8} & \frac{1}{8} & \frac{1}{8} & \frac{1}{8} & \frac{1}{8} & \frac{1}{8} \\ \frac{1}{8} & \frac{1}{8} & \frac{1}{8} & \frac{1}{8} & \frac{1}{8} & \frac{1}{8} & \frac{1}{8} & \frac{1}{8} \\ \frac{1}{8} & \frac{1}{8} & \frac{1}{8} & \frac{1}{8} & \frac{1}{8} & \frac{1}{8} & \frac{1}{8} & \frac{1}{8} \\ \frac{1}{8} & \frac{1}{8} & \frac{1}{8} & \frac{1}{8} & \frac{1}{8} & \frac{1}{8} & \frac{1}{8} & \frac{1}{8} \\ \frac{1}{8} & \frac{1}{8} & \frac{1}{8} & \frac{1}{8} & \frac{1}{8} & \frac{1}{8} & \frac{1}{8} & \frac{1}{8} \\ \frac{1}{8} & \frac{1}{8} & \frac{1}{8} & \frac{1}{8} & \frac{1}{8} & \frac{1}{8} & \frac{1}{8} & \frac{1}{8} \\ \frac{1}{8} & \frac{1}{8} & \frac{1}{8} & \frac{1}{8} & \frac{1}{8} & \frac{1}{8} & \frac{1}{8} & \frac{1}{8} \end{bmatrix}$$

$$= \begin{bmatrix} 0.01875 & 0.01875 & 0.01875 & 0.125 & 0.01875 & 0.01875 & 0.44375 & 0.01875 \\ 0.23125 & 0.01875 & 0.44375 & 0.125 & 0.01875 & 0.86875 & 0.01875 & 0.01875 \\ 0.01875 & 0.30208 & 0.01875 & 0.125 & 0.01875 & 0.01875 & 0.01875 & 0.01875 \\ 0.01875 & 0.30208 & 0.44375 & 0.125 & 0.01875 & 0.01875 & 0.01875 & 0.01875 \\ 0.23125 & 0.30208 & 0.01875 & 0.125 & 0.01875 & 0.01875 & 0.01875 & 0.01875 \\ 0.23125 & 0.01875 & 0.01875 & 0.125 & 0.01875 & 0.01875 & 0.44375 & 0.44375 \\ 0.23125 & 0.01875 & 0.01875 & 0.125 & 0.86875 & 0.01875 & 0.01875 & 0.44375 \\ 0.01875 & 0.01875 & 0.01875 & 0.125 & 0.01875 & 0.01875 & 0.01875 & 0.01875 \end{bmatrix}.$$

Starting with the probability vector

$$[0.125, 0.125, 0.125, 0.125, 0.125, 0.125, 0.125, 0.125]^T$$

we use the power method, and after 50 iterations, we get (to 3 decimal places) the vector

$$[0.104, 0.214, 0.093, 0.133, 0.116, 0.140, 0.167, 0.033]^T.$$

Comparing this to the PageRank vector, for the altered transition matrix

$$[0.1021, 0.2254, 0.0925, 0.1387, 0.1180, 0.1363, 0.1696, 0.0173].$$

found in Example 3, we see that we are indeed pretty close, without having to solve a linear system. More importantly, the ranking we get from the google matrix, *B, G, F, D, E, A, C, H*, is almost the same as the ranking we got from the altered transition matrix - only *D* and *F* have switched places. ■

And that's the idea in a mathematical nutshell. For searches on Google's site, there is also a component of the ranking that is topic-specific, but the most important part of the ranking is from the google matrix. In the end, even the power method needs some tweaking, as the matrices are so large, but luckily, the fact that the original transition matrix is so sparse (that is, has so few nonzero elements) makes for easier matrix multiplication. But we leave the details to another time.

4.6 Exercises

1. Explain why the PageRanks of Example 3 make sense.

FIGURE 4

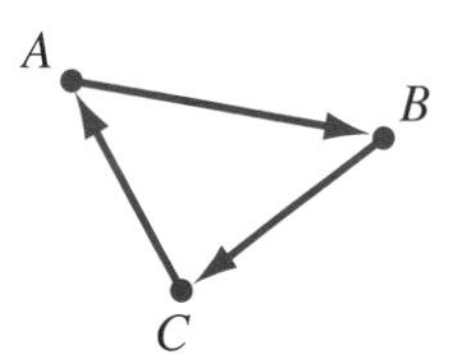

2. For the graph in Figure 4, find the PageRanks of G_α (for any α).

FIGURE 5

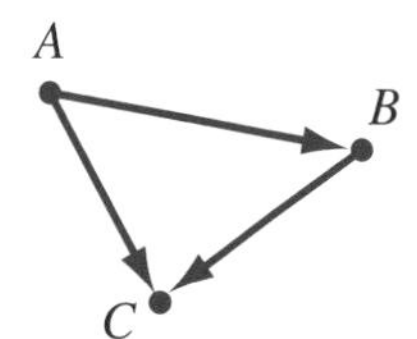

3. For the graph in Figure 5, find the PageRanks of $G_{0.5}$.
4. Find a graph such that, while you can walk from any node to any other, there is no k such that the transition matrix has all positive entries.

5. Prove that the google matrix G_α always has positive entries with a column sum of 1 (and so is the transition matrix of a Markov chain).

6. Prove that the *largest* eigenvalue of a transition matrix of a Markov chain is 1. (*Hint*: Consider the transpose of the transition matrix and show that it has a largest eigenvalue of 1. Then argue that this implies that the transition matrix also has a largest eigenvalue of 1 as well.)

Computational Exercises

7. Generate the first 5 vectors under the power method for Example 4.

8. Repeat Example 4 for the google matrices for $G_{0.75}$, $G_{0.5}$, and $G_{0.25}$.

CHAPTER SUMMARY

Computation of a Determinant

Let A be an $n \times n$ matrix and let A_{ij} denote the submatrix of A obtained by deleting its ith row and jth column. Then

(a) DEFINITION: The i, j-minor of A is $M_{ij} = \det(A_{ij})$.
DEFINITION: The i, j-cofactor of A is $C_{ij} = (-1)^{i+j}M_{ij}$.

(b) $\det(A)$ can be obtained by *cofactor expansion* along any row or down any column; that is, for any fixed row i or fixed column j,

$$\begin{aligned}\det(A) &= a_{i1}C_{i1} + a_{i2}C_{i2} + \cdots + a_{in}C_{in} \\ &= a_{1j}C_{1j} + a_{2j}C_{2j} + \cdots + a_{nj}C_{nj}\end{aligned}$$

Properties of Determinants

(a) Let A be a square matrix. If A' is the matrix obtained from A by

- Interchanging two rows of A, then $\det(A') = -\det(A)$,
- Multiplying a row of A by the scalar c, then $\det(A') = c\det(A)$,
- Adding a multiple of one row of A to another row of A, then $\det(A') = \det(A)$.

(b) A square matrix A is invertible if and only if $\det(A) \neq 0$. The linear system $A\mathbf{x} = \mathbf{b}$, where A is square, has a unique solution if and only if $\det(A) \neq 0$.

(c) If A and B are square matrices of the same order, then $\det(AB) = \det(A)\det(B)$. If A is a square matrix, then $\det(A^T) = \det(A)$.

Cramer's Rule

Given the linear system $A\mathbf{x} = \mathbf{b}$, where A is a square matrix with $\det(A) \neq 0$, let $A(k)$ be the matrix obtained from A by replacing its kth column by the elements of $\mathbf{b}$. Then the components of the solution are given by $x_i = \det A(i)/\det A$.

The Adjoint Matrix

(a) DEFINITION: Let A be a square matrix. Then the *adjoint* of A, $\text{Adj}(A)$, is the transpose of the matrix obtained from A by replacing each entry by its cofactor; that is, letting C_{ij} be the i, j-cofactor of A; the i, j-entry of $\text{Adj}(A)$ is C_{ji}.

(b) Let A be a square matrix with $\det(A) \neq 0$. Then $A^{-1} = (1/\det A)\,\text{Adj}(A)$.

Eigenvalues and Eigenvectors

(a) DEFINITION: A scalar λ is an *eigenvalue* and the nonzero vector $\mathbf{x}$ is an *eigenvector* of the square matrix A if $A\mathbf{x} = \lambda\mathbf{x}$.

(b) The eigenvalues of the matrix A are the roots of the *characteristic equation* $\det(A - \lambda I) = 0$.

(c) The eigenvectors corresponding to eigenvalue λ_0 are the nonzero solutions of the homogeneous linear system $(A - \lambda_0 I)\mathbf{x} = \mathbf{0}$.

KEYWORDS

determinant
minor
cofactor
cofactor expansion
upper triangular
lower triangular
diagonal matrix
Vandermonde matrix
Cramer's rule
adjoint
eigenvalue
eigenvector
diagonalizable
characteristic equation
characteristic polynomial
Markov chain
probability vector
initial vector
stationary distribution
transition matrix
web graph
google matrix

SELF-TEST

Find the determinant of each matrix in Exercises 1–3. Use any facts about determinants that simplify the work.

1. $\begin{bmatrix} 3 & 2 & 1 \\ 1 & 4 & 2 \\ 2 & 5 & 3 \end{bmatrix}$ *2.* $\begin{bmatrix} 0.02 & 0.03 & 0.01 \\ -0.01 & 0.06 & 0.02 \\ 0.02 & -0.09 & -0.03 \end{bmatrix}$ *3.* $\begin{bmatrix} 0 & 2 & 0 & 0 \\ 3 & 0 & 0 & 0 \\ 0 & 0 & 0 & 2 \\ 0 & 0 & 3 & 0 \end{bmatrix}$

4. Solve the equation

$$\det\begin{bmatrix} 1-t & 2 & 0 \\ 0 & 2-t & 2 \\ 0 & 0 & 3-t \end{bmatrix} = 0.$$

[Sections 4.1, 4.2]

Solve the linear systems in Exercises 5 and 6 by Cramer's rule.

5. $\begin{aligned} x - y &= 2 \\ 3x + 4y &= 3 \end{aligned}$

6. $\begin{aligned} 2x + y + z &= 3 \\ 3x \quad - z &= 2 \\ 4x + 5y + 2z &= 1 \end{aligned}$

[Section 4.3]

7. Find all values of t for which the matrix A is invertible:

$$A = \begin{bmatrix} 2 & 1 & 1 \\ 3 & 1 & t \\ 1 & t & -2 \end{bmatrix}$$

[Section 4.2]

8. If $\det(A^2) = 7$, show that A is invertible.

[Section 4.2]

9. Determine Adj A, where A is the matrix of Exercise 1. **[Section 4.3]**

10. If A and B are *similar* (that is, if there exists an invertible matrix P such that $B = P^{-1}AP$), show that det A = det B. **[Section 4.2]**

For each matrix in Exercises 11 and 12, find

a. Its characteristic equation,

b. All eigenvalues,

c. All eigenvectors.

11. $\begin{bmatrix} 1 & -1 \\ -1 & 1 \end{bmatrix}$ *12.* $\begin{bmatrix} 1 & -1 & 0 \\ 0 & 2 & 1 \\ 0 & 0 & 1 \end{bmatrix}$ **[Section 4.4]**

REVIEW EXERCISES

In Exercises 1–8 evaluate the determinant of the given matrix in the easiest way you can.

1. $\begin{bmatrix} 1 & 1 & 0 \\ 1 & 0 & 1 \\ 0 & 1 & 1 \end{bmatrix}$ *2.* $\begin{bmatrix} 1 & -1 & 1 \\ -1 & 0 & -1 \\ 0 & 1 & -1 \end{bmatrix}$ *3.* $\begin{bmatrix} 1 & 0 & 0 \\ 1 & 2 & 0 \\ 1 & 2 & 3 \end{bmatrix}$

4. $\begin{bmatrix} 1 & 2 & 3 \\ 3 & 2 & 1 \\ 2 & 4 & 6 \end{bmatrix}$ *5.* $\begin{bmatrix} 1 & x & y \\ 1 & x^2 & y^2 \\ 1 & x^3 & y^3 \end{bmatrix}$ *6.* $\begin{bmatrix} 1 & x & y \\ 1 & x^2 & y^2 \\ 0 & 2x & 2y \end{bmatrix}$

7. $\begin{bmatrix} 1 & 0 & 2 & 1 \\ 0 & 1 & -1 & 0 \\ 2 & 0 & 5 & 2 \\ -1 & 0 & -2 & 0 \end{bmatrix}$ *8.* $\begin{bmatrix} 1 & 1 & 1 & 1 \\ 0 & 1 & 2 & 0 \\ 1 & 0 & 2 & 1 \\ -1 & -1 & -1 & 0 \end{bmatrix}$

9. Use the results of Exercises 1–4 to determine in which cases the inverse of the given matrix exists.

10. Let A be a square matrix such that $A^3 = I$. Find the value of det A.

11. Let A be an invertible matrix with det $A = 2$. Find $\det(A^{-1})$ and $\det(A^T)^{-1}$.

12. Can you give an example of a linear system that can be solved by Cramer's rule but not by the method of inverses? If not, why not?

13–16. For each matrix A in Exercises 1–4, determine Adj A and, if it exists, A^{-1}.

In Exercises 17–22 solve the given linear system by Cramer's rule, if possible.

17. $\begin{aligned} x - 2y &= 1 \\ 2x - y &= 1 \end{aligned}$

18. $\begin{aligned} x + y &= 2 \\ 2x - 3y &= -1 \end{aligned}$

19. $\begin{aligned} x_1 + x_2 &= 1 \\ x_1 + x_3 &= 0 \\ x_2 + x_3 &= 0 \end{aligned}$

20. $\begin{aligned} x_1 - x_2 \qquad &= 1 \\ x_1 \qquad - x_3 &= 2 \\ x_2 - x_3 &= 3 \end{aligned}$

21. $\begin{aligned} x + z \qquad &= 0 \\ y \qquad + w &= 0 \\ x - z \qquad &= 0 \\ y + 2z + w &= 0 \end{aligned}$

22. $\begin{aligned} x + y \qquad &= 0 \\ y + z \qquad &= 0 \\ z + w &= 0 \\ x \qquad + w &= 0 \end{aligned}$

In Exercises 23–26 find the characteristic equation and eigenvalues of the given matrix.

23. $\begin{bmatrix} 1 & 2 \\ 2 & 1 \end{bmatrix}$ *24.* $\begin{bmatrix} 0 & 3 \\ -2 & 0 \end{bmatrix}$

25. $\begin{bmatrix} 2 & 0 & 0 \\ 0 & 1 & 3 \\ 0 & 3 & 1 \end{bmatrix}$

26. $\begin{bmatrix} -1 & 0 & 1 & 2 \\ 0 & 1 & -1 & -3 \\ 0 & 0 & 2 & 4 \\ 0 & 0 & 0 & 5 \end{bmatrix}$

27. Find all eigenvectors for each eigenvalue in the matrix of Exercise 23.

28. Find all eigenvectors for each eigenvalue in the matrix of Exercise 25.

CHALLENGE YOURSELF

29. Let $r > \lambda > 0$ and consider the $n \times n$ matrix

$$A = \begin{bmatrix} r & \lambda & \lambda & \cdots & \lambda & \lambda \\ \lambda & r & \lambda & \cdots & \lambda & \lambda \\ \vdots & \vdots & & \ddots & & \vdots \\ \lambda & \lambda & \lambda & \cdots & \lambda & r \end{bmatrix}.$$

Show that A is invertible by calculating its determinant.

30. For what values of λ is the matrix *not* invertible?

$$A = \begin{bmatrix} \lambda - 1 & 0 & 0 \\ 1 & \lambda & 1 \\ 2 & 2 & 1 + \lambda \end{bmatrix}$$

31. Suppose that (a, b) and (c, d) are two distinct points in $\mathbf{R}^2$. Prove that the equation of the straight line through these two points can be given as

$$\det \begin{bmatrix} 1 & x & y \\ 1 & a & b \\ 1 & c & d \end{bmatrix} = 0.$$

32. Prove that if $\det(A) = 1$ or -1 and A has all integer entries, then A^{-1} also has all integer entries.

33. Suppose that A is a diagonal matrix and that all eigenvalues of A are strictly between -1 and 1. Show that $\lim_{n \to \infty} A^n = 0$, that is, each entry of A^n tends to 0.

SPOTLIGHT

Terrance Emerson/Shutterstock

The central theme of this chapter is the notion of *dimension*, as applied to certain special subsets of $\mathbf{R}^m$ called *subspaces.* Although it will take us some time (and a lot of words) to define this term formally, *dimension* is a fairly intuitive concept.

When we look at a vector in $\mathbf{R}^m$, $(v_1, v_2, \ldots, v_m)$, we see that it has m independent components; no constraints tie any of these components to the others. For this reason, we say that $\mathbf{R}^m$ has "m degrees of freedom," or that it has "dimension equal to m." For example, $\mathbf{R}^3$ (as you might expect) has dimension 3. On the other hand, consider the subset (subspace) of $\mathbf{R}^m$ consisting of all vectors (x, y, z) for which $x + y + z = 0$. This subspace possesses just two degrees of freedom; we are free to choose any two of the components in its vectors, but the third must then be determined by the constraint $x + y + z = 0$. (For example, if we arbitrarily choose x and y, then $z = -x - y$.) Hence, we say that the dimension of this subspace is 2.

The idea of dimension is used in physics to help analyze the motions within certain mechanical systems, ranging from a collection of atomic or subatomic particles to the orbiting bodies within our solar system. If a system consists of n objects, then (since each one has three coordinates) we can describe the *configuration* of that system (the positions of all objects within it) by specifying $3n$ coordinates. If n is relatively large, describing the configuration of a system can be very cumbersome. Due to interactions among the objects within the system or external constraints, however, there may be far fewer *independent* coordinates—far fewer degrees of freedom. These independent *generalized coordinates,* as they are known in physics, can be used to reduce the complexity of the system, enabling us to analyze it more easily.

As a simple physical example, consider a system of two weights, W_1 and W_2, connected by a wire that passes over a pulley (see the figure below). As the pulley turns, the weights move up and down. We could give the position of each weight by specifying its three coordinates relative to a coordinate system in 3-space. This requires six numbers to describe the configuration of the coordinate system. The weights can only move in the vertical direction, however, so a single coordinate can describe the position of each (the other two coordinates remain fixed). Moreover, as one weight moves up or down a certain distance, the other moves the same distance in the opposite direction. Thus, a single coordinate (specifying, say, the distance of W_1 from the top of the pulley) completely describes the location of both weights. This system has only one degree of freedom!

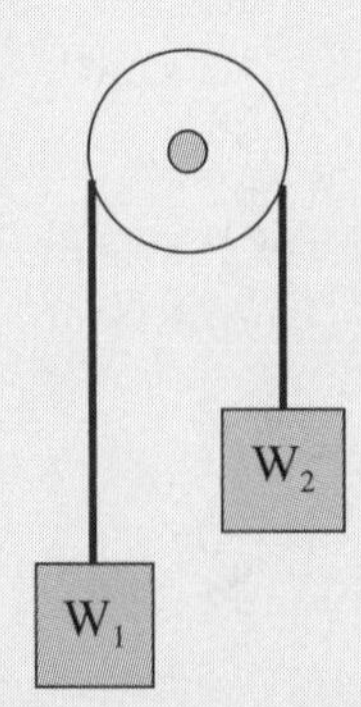

We will return to the notions of *dimension* and *degrees of freedom* in Section 5.3.

CHAPTER 5

Independence and Basis in $\mathbf{R}^m$

OBJECTIVES

You will learn to:

- Determine if a set of vectors in $\mathbf{R}^m$ is linearly independent and, if not, to express one of the vectors as a linear combination of the others.
- Determine if a subset of $\mathbf{R}^m$ is a subspace.
- Determine if a set of vectors spans a given subspace of $\mathbf{R}^m$.
- Determine if a set of vectors is a basis for a given subspace of $\mathbf{R}^m$.
- Find a basis for and the dimension of a given subspace of $\mathbf{R}^m$.
- Relate the rank of a matrix to its row space and column space.

We have sometimes referred to $\mathbf{R}^m$, the set of all m-vectors, as "m-dimensional space." In this chapter, we introduce concepts that not only make the notion of dimension more precise, but also allow us to view $\mathbf{R}^m$ in a different light.

5.1 Linear Independence and Dependence

As we saw in Section 2.1, two 3-vectors, $\mathbf{v}_1$ and $\mathbf{v}_2$, determine a plane in $\mathbf{R}^3$, as long as one of them is not a scalar multiple of the other. Moreover, a third 3-vector, $\mathbf{w}$, lies in the plane of $\mathbf{v}_1$ and $\mathbf{v}_2$ if in some sense it is "dependent" on these vectors. More precisely, this occurs if (and only if) we can find scalars c_1 and c_2 such that $\mathbf{w} = c_1\mathbf{v}_1 + c_2\mathbf{v}_2$ (Figure 1). In Section 5.1, we investigate such dependence relations in the more general setting of $\mathbf{R}^m$.

Many of the essential ideas in vector space theory, such as linear independence, spanning sets, and dimension, are due to the mathematician Hermann Grassman (1809–1877). Among his other research works is "Grassman's law" which describes how we perceive colour.

Courtesy of Mac Tutor, St. Andrews University

FIGURE 1

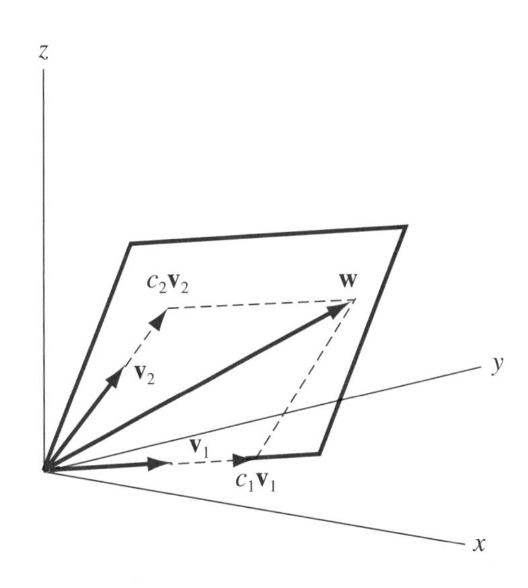

DEFINITION

A **linear combination** of the vectors $\mathbf{v}_1, \mathbf{v}_2, \ldots, \mathbf{v}_n$ in $\mathbf{R}^m$ is an expression of the form

$$c_1\mathbf{v}_1 + c_2\mathbf{v}_2 + \cdots + c_n\mathbf{v}_n,$$

where the c_i are scalars.

EXAMPLE 1

Let $\mathbf{v} = (1, 2, 3)$, $\mathbf{v}_1 = (1, 0, 0)$, $\mathbf{v}_2 = (1, 1, 0)$, and $\mathbf{v}_3 = (1, 1, 1)$. Since

$$(1, 2, 3) = (-1)(1, 0, 0) + (-1)(1, 1, 0) + 3(1, 1, 1),$$

we can express $\mathbf{v}$ as a linear combination of the given vectors, $\mathbf{v} = c_1\mathbf{v}_1 + c_2\mathbf{v}_2 + c_3\mathbf{v}_3$, by taking $c_1 = -1$, $c_2 = -1$, and $c_3 = 3$. ■

EXAMPLE 2

Let $\mathbf{v}_1 = (-2, 0, 1)$, $\mathbf{v}_2 = (1, -1, 2)$, and $\mathbf{v}_3 = (4, -2, 3)$. Express $\mathbf{v}_3$ as a linear combination of the other two vectors.

Solution We need to find constants c_1 and c_2 such that

$$(4, -2, 3) = c_1(-2, 0, 1) + c_2(1, -1, 2).$$

To determine c_1 and c_2, we first perform the scalar multiplication and addition on the right, obtaining

$$\begin{aligned}(4, -2, 3) &= (-2c_1, 0, c_1) + (c_2, -c_2, 2c_2)\\ &= (-2c_1 + c_2, -c_2, c_1 + 2c_2).\end{aligned}$$

Now, equating corresponding components yields the linear system

$$\begin{aligned}4 &= -2c_1 + c_2\\ -2 &= - c_2\\ 3 &= c_1 + 2c_2.\end{aligned}$$

This system has solution: $c_1 = -1$, $c_2 = 2$. Thus,

$$(4, -2, 3) = (-1)(-2, 0, 1) + (2)(1, -1, 2).$$

(Figure 2 gives a picture of the situation.)

FIGURE 2

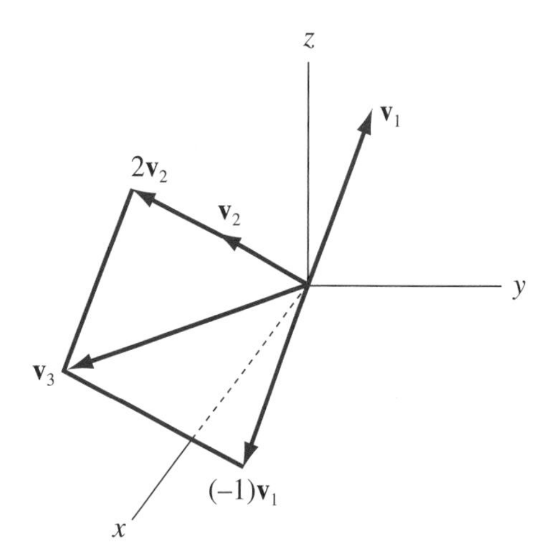

When one vector $\mathbf{v}$ in a set is a linear combination of the others, the set is, from one point of view, redundant because you derive $\mathbf{v}$ from the rest of the set via the operations of scalar multiplication and vector addition. In the following definition we make more precise what redundancy amongst vectors means.

DEFINITION

A set of vectors $\{\mathbf{v}_1, \mathbf{v}_2, \ldots, \mathbf{v}_n\}$ in $\mathbf{R}^m$ is **linearly independent** if the only way to write the zero vector as a linear combination of the vectors is with all of the coefficients being 0; that is, the only solution to the vector equation

$$c_1\mathbf{v}_1 + c_2\mathbf{v}_2 + \cdots + c_n\mathbf{v}_n = \mathbf{0}$$

is the **trivial solution** $c_1 = c_2 = \cdots = c_n = 0$. A set consisting of a single vector is linearly independent if that vector is not equal to $\mathbf{0}$. If the set is *not* linearly independent, it is said to be **linearly dependent**.

EXAMPLE 3

Determine whether or not the set of vectors $\{\mathbf{v}_1, \mathbf{v}_2, \mathbf{v}_3\}$ is linearly independent, where

$$\mathbf{v}_1 = (1, 0, -1), \qquad \mathbf{v}_2 = (0, 1, 2), \quad \text{and} \quad \mathbf{v}_3 = (1, 1, 3).$$

Solution We wish to determine whether or not the vector equation $c_1\mathbf{v}_1 + c_2\mathbf{v}_2 + c_3\mathbf{v}_3 = \mathbf{0}$ has a nontrivial solution. This equation is equivalent to

$$c_1(1, 0, -1) + c_2(0, 1, 2) + c_3(1, 1, 3) = (0, 0, 0),$$
$$(c_1, 0, -c_1) + (0, c_2, 2c_2) + (c_3, c_3, 3c_3) = (0, 0, 0),$$
$$\text{or} \qquad (c_1 + c_3, c_2 + c_3, -c_1 + 2c_2 + 3c_3) = (0, 0, 0).$$

Now, equating corresponding components, we get the system of equations

$$\begin{aligned} c_1 \qquad + c_3 &= 0 \\ c_2 + c_3 &= 0 \\ -c_1 + 2c_2 + 3c_3 &= 0. \end{aligned}$$

In matrix form, this system is

$$\begin{bmatrix} 1 & 0 & 1 \\ 0 & 1 & 1 \\ -1 & 2 & 3 \end{bmatrix}\begin{bmatrix} c_1 \\ c_2 \\ c_3 \end{bmatrix} = \begin{bmatrix} 0 \\ 0 \\ 0 \end{bmatrix}.$$

So we may restate the original problem as follows: "Does the linear system, $A\mathbf{c} = \mathbf{0}$, where

$$A = \begin{bmatrix} 1 & 0 & 1 \\ 0 & 1 & 1 \\ -1 & 2 & 3 \end{bmatrix}, \quad \mathbf{c} = \begin{bmatrix} c_1 \\ c_2 \\ c_3 \end{bmatrix}, \quad \text{and} \quad \mathbf{0} = \begin{bmatrix} 0 \\ 0 \\ 0 \end{bmatrix},$$

have a nontrivial solution?" If so, the set $\{\mathbf{v}_1, \mathbf{v}_2, \mathbf{v}_3\}$ is linearly dependent.

Transforming the augmented matrix for this system to row-reduced echelon form, we obtain

$$\left[\begin{array}{ccc|c} 1 & 0 & 0 & 0 \\ 0 & 1 & 0 & 0 \\ 0 & 0 & 1 & 0 \end{array}\right].$$

Hence, the system has the unique solution $c_1 = 0$, $c_2 = 0$, $c_3 = 0$, and the set $\{\mathbf{v}_1, \mathbf{v}_2, \mathbf{v}_3\}$ is linearly independent. ■

EXAMPLE 4

Show that the vectors

$$\mathbf{v}_1 = (-2, 0, 1), \quad \mathbf{v}_2 = (1, -1, 2), \quad \text{and} \quad \mathbf{v}_3 = (4, -2, 3)$$

are linearly dependent.

Solution We have already shown, in Example 2, that $\mathbf{v}_3 = (-1)\mathbf{v}_1 + 2\mathbf{v}_2$. (Geometrically, this means that $\mathbf{v}_3$ lies in the plane determined by $\mathbf{v}_1$ and $\mathbf{v}_2$, as shown in Figure 2.) Rearranging, we see that $1\mathbf{v}_1 + (-2)\mathbf{v}_2 + 1\mathbf{v}_3 = 0$, so the set is linearly dependent. ■

NOTE In Example 4, we also can express $\mathbf{v}_1$ as a linear combination of $\mathbf{v}_2$ and $\mathbf{v}_3$, and $\mathbf{v}_2$ as a linear combination of $\mathbf{v}_1$ and $\mathbf{v}_3$. You might try to find the appropriate scalars involved.

The following theorem gives another characterization of linear dependence. This characterization explains why linear independence corresponds to no redundancy amongst vectors.

THEOREM 1

Let $\mathcal{S} = \{\mathbf{v}_1, \mathbf{v}_2, \ldots, \mathbf{v}_n\}$. Then $\mathcal{S}$ is a linearly dependent set if and only if one of the vectors is a linear combination of the others.

Proof First assume that the vector equation $c_1\mathbf{v}_1 + c_2\mathbf{v}_2 + \cdots + c_n\mathbf{v}_n = \mathbf{0}$ has a solution with at least one $c_i \neq 0$. Say it is c_1 that is not zero. Then dividing both sides by c_1 yields

$$\mathbf{v}_1 + \left(\frac{c_2}{c_1}\right)\mathbf{v}_2 + \cdots + \left(\frac{c_n}{c_1}\right)\mathbf{v}_n = \mathbf{0},$$

and transposing all terms except $\mathbf{v}_1$ to the right-hand side, we obtain

$$\mathbf{v}_1 = -\left(\frac{c_2}{c_1}\right)\mathbf{v}_2 - \cdots - \left(\frac{c_n}{c_1}\right)\mathbf{v}_n.$$

Thus, we have expressed one of the vectors as a linear combination of the others. (The same argument holds if some other c_i is not zero.)

Now assume that we can express one of the vectors (say, $\mathbf{v}_1$) as a linear combination of the others. In other words, there exist $k_2, \ldots, k_n$ such that $\mathbf{v}_1 = k_2\mathbf{v}_2 + \cdots + k_n\mathbf{v}_n$. This implies that $1\mathbf{v}_1 - k_2\mathbf{v}_2 - \cdots - k_n\mathbf{v}_n = \mathbf{0}$. By letting $c_1 = 1$ and $c_i = -k_i$ for $i > 1$, we have a solution of the vector equation $c_1\mathbf{v}_1 + c_2\mathbf{v}_2 + \cdots + c_n\mathbf{v}_n = \mathbf{0}$ with at least one $c_i \neq 0$, namely, $c_1 = 1$. That is, the set is linearly dependent. ■

In Example 3, notice that the first, second, and third columns of matrix A are, respectively, the given vectors $\mathbf{v}_1$, $\mathbf{v}_2$, and $\mathbf{v}_3$. This is no coincidence and suggests the next procedure, which is justified by Theorem 4 (given toward the end of this section). We first introduce the following terminology.

DEFINITION

An **elementary vector** in $\mathbf{R}^m$ is a vector that has one component equal to 1 and all other components equal to 0. If the 1 occurs as the ith component, the elementary vector is denoted $\mathbf{e}_i$. A column of a matrix whose entries form an elementary vector is called an **elementary column**.

For example, if

$$A = \begin{bmatrix} 1 & 2 & 0 & 0 & 1 \\ 0 & 0 & 1 & 0 & 2 \\ 0 & 0 & 0 & 1 & 3 \end{bmatrix},$$

then the first, third, and fourth columns of A are *elementary columns*. Specifically $\mathbf{a}^1 = \mathbf{e}_1$, $\mathbf{a}^3 = \mathbf{e}_2$, and $\mathbf{a}^4 = \mathbf{e}_3$.

Test for Linear Independence/Dependence Let $\mathcal{S} = \{\mathbf{v}_1, \mathbf{v}_2, \ldots, \mathbf{v}_n\}$ be a set of m-vectors. Construct the $m \times n$ matrix A, whose ith column $\mathbf{a}^i$ consists of the components of $\mathbf{v}_i$. The set $\mathcal{S}$ is linearly independent if and only if the row-reduced echelon form of A contains only distinct elementary columns.

EXAMPLE 5

Determine whether or not the set

$$\{(-1, 0, 2, 1), (3, 1, -2, 0), (0, 1, 4, 3)\}$$

is linearly independent.

Solution The 4×3 matrix A, with these vectors as columns, is

$$A = \begin{bmatrix} -1 & 3 & 0 \\ 0 & 1 & 1 \\ 2 & -2 & 4 \\ 1 & 0 & 3 \end{bmatrix}.$$

The row-reduced echelon form of A is

$$\begin{bmatrix} 1 & 0 & 3 \\ 0 & 1 & 1 \\ 0 & 0 & 0 \\ 0 & 0 & 0 \end{bmatrix}.$$

Since the third column is not an elementary column, this set is linearly dependent. ■

COMPUTATIONAL NOTE

It is possible to use computer software or a graphing calculator to help you determine whether a set of vectors is or is not linearly independent. To do so, form the matrix A described in the test for linear independence/dependence and transform A into row-reduced echelon form. Then inspect the resulting matrix. If it consists entirely of distinct elementary columns, the original set is linearly independent; otherwise the set is linearly dependent. Unfortunately, rounding errors inherent in computer calculations (see the computational note of Section 3.3) can result in nonzero entries appearing in the row-reduced echelon form where, in fact, zeros should occur, or vice versa. As a result, it is possible to obtain an incorrect answer.

Since the set of vectors of Example 5 is linearly dependent, one of them can be expressed as a linear combination of the others. In fact, we have

$$(0, 1, 4, 3) = (3)(-1, 0, 2, 1) + (1)(3, 1, -2, 0).$$

Notice that the coefficients (3 and 1) of the linear combination appear in the third column of the row-reduced echelon matrix. When a set of m-vectors is linearly dependent, these *coefficients of dependence* can always be found in this way. To be more explicit, we state the following procedure (justified by Theorem 5 at the end of this section).

Determining Coefficients of Dependence Let $\mathcal{S} = \{\mathbf{v}_1, \mathbf{v}_2, \ldots, \mathbf{v}_n\}$ be a set of vectors in $\mathbf{R}^m$. Construct an $m \times n$ matrix A with $\mathbf{a}^i = \mathbf{v}_i$. Transform A to its row-reduced echelon form, B. Suppose a column of B, say $\mathbf{b}^j$, is either a non-elementary column or is a duplication of a preceding elementary column. Then we have

$$\mathbf{v}_j = b_{1j}\mathbf{w}_1 + b_{2j}\mathbf{w}_2 + \cdots + b_{kj}\mathbf{w}_k,$$

where $\mathbf{w}_1, \mathbf{w}_2, \ldots, \mathbf{w}_k$ are the vectors in $\mathcal{S}$ that have been transformed into the distinct elementary columns of B that precede the jth column.

EXAMPLE 6

Let $\mathbf{v}_1 = (1, -1, 2)$, $\mathbf{v}_2 = (-2, 3, 0)$, $\mathbf{v}_3 = (0, 1, 4)$, $\mathbf{v}_4 = (1, 0, 1)$, and $\mathbf{v}_5 = (2, 1, 1)$. Show that the set $\{\mathbf{v}_1, \mathbf{v}_2, \mathbf{v}_3, \mathbf{v}_4, \mathbf{v}_5\}$ is linearly dependent and express at least one $\mathbf{v}_i$ as a linear combination of the others.

Solution The matrix A with $\mathbf{a}^i = \mathbf{v}_i$ is

$$A = \begin{bmatrix} 1 & -2 & 0 & 1 & 2 \\ -1 & 3 & 1 & 0 & 1 \\ 2 & 0 & 4 & 1 & 1 \end{bmatrix}.$$

The row-reduced echelon form of A is the matrix

$$B = \begin{bmatrix} 1 & 0 & 2 & 0 & -1 \\ 0 & 1 & 1 & 0 & 0 \\ 0 & 0 & 0 & 1 & 3 \end{bmatrix}.$$

We see that $\mathbf{v}_1$, $\mathbf{v}_2$, and $\mathbf{v}_4$ have been transformed into the elementary vectors $\mathbf{e}_1$, $\mathbf{e}_2$, and $\mathbf{e}_3$, respectively. (Thus, $\mathbf{v}_1$, $\mathbf{v}_2$, and $\mathbf{v}_4$ become $\mathbf{w}_1$, $\mathbf{w}_2$, and $\mathbf{w}_3$ in the terminology of the procedure.) We conclude that

$$\mathbf{v}_3 = 2\mathbf{v}_1 + 1\mathbf{v}_2 \quad \text{(From the third column of } B\text{)}$$

$$\text{and} \quad \mathbf{v}_5 = (-1)\mathbf{v}_1 + (0)\mathbf{v}_2 + 3\mathbf{v}_4 \quad \text{(From the fifth column of } B\text{)}.$$

This can easily be checked by computing

$$2(1, -1, 2) + (-2, 3, 0) = (0, 1, 4)$$

$$\text{and} \qquad (-1)(1, -1, 2) + 3(1, 0, 1) = (2, 1, 1). \qquad \blacksquare$$

If we phrase Example 6 simply as, "Is the set $\{\mathbf{v}_1, \mathbf{v}_2, \mathbf{v}_3, \mathbf{v}_4, \mathbf{v}_5\}$ linearly independent?" the following theorem enables us to answer immediately, "No." We leave the proof for Exercise 27.

THEOREM 2

Let $\mathcal{S}$ be a set of n vectors in $\mathbf{R}^m$ with $n > m$. Then $\mathcal{S}$ is linearly dependent. ■

EXAMPLE 7

Show that the vectors $\mathbf{v}_1 = (1, -1)$, $\mathbf{v}_2 = (0, 1)$, and $\mathbf{v}_3 = (2, 1)$ form a linearly dependent set.

Solution Here the given set consists of three 2-vectors. By Theorem 2 (with $m = 2$, $n = 3$), it is linearly dependent. ■

If we are given a subset of $\mathbf{R}^m$ containing exactly m vectors, the following theorem supplies us with a relatively simple alternative to our previous test for linear independence. It is a consequence of Theorem 4 (following Example 8), since for a square matrix A, $\det A \neq 0$ if and only if A is invertible, which is true if and only if its row-reduced form is the identity matrix.

THEOREM 3

Let $\mathcal{S} = \{\mathbf{v}_1, \mathbf{v}_2, \ldots, \mathbf{v}_m\}$ be a set of vectors in $\mathbf{R}^m$ and A be an $m \times m$ matrix with $\mathbf{a}^i = \mathbf{v}_i$. Then $\mathcal{S}$ is linearly independent if and only if $\det A \neq 0$. ■

EXAMPLE 8

Determine whether or not the set $\mathcal{S} = \{\mathbf{v}_1, \mathbf{v}_2, \mathbf{v}_3\}$ is linearly dependent, where $\mathbf{v}_1 = (1, 2, 3)$, $\mathbf{v}_2 = (-1, 0, 1)$, and $\mathbf{v}_3 = (0, 0, 1)$.

Solution We form the matrix A with $\mathbf{a}^i = \mathbf{v}_i$:

$$A = \begin{bmatrix} 1 & -1 & 0 \\ 2 & 0 & 0 \\ 3 & 1 & 1 \end{bmatrix}$$

Now $\det A = 2 \neq 0$, so, by Theorem 3, $\mathcal{S}$ is linearly independent. ■

To summarize our discussion so far, suppose that we are given a set of n vectors in $\mathbf{R}^m$.

i. If $n > m$, then the set is linearly dependent.

ii. If $n = m$, we can test for linear independence by using the row-reduction procedure or by using determinants.

iii. If $n < m$, we can test for linear independence by using the row-reduction procedure.

We close this section with two theorems, the results of which have already been used in our computational work.

THEOREM 4

Let $\mathcal{S} = \{\mathbf{v}_1, \mathbf{v}_2, \ldots, \mathbf{v}_n\}$ be a set of m-vectors, A be the $m \times n$ matrix with $\mathbf{a}^i = \mathbf{v}_i$, and B be the row-reduced echelon form of A. Then the set $\mathcal{S}$ is linearly independent if and only if the columns of B are the elementary vectors $\mathbf{e}_1, \mathbf{e}_2, \ldots, \mathbf{e}_n$.

Proof The set $\mathcal{S}$ is linearly independent if and only if the vector equation $c_1\mathbf{v}_1 + c_2\mathbf{v}_2 + \cdots + c_n\mathbf{v}_n = \mathbf{0}$ has a unique solution, the trivial one. By the definition of matrix multiplication and equality of vectors, this is true if and only if $A\mathbf{c} = \mathbf{0}$, where $\mathbf{c} = (c_1, c_2, \ldots, c_n)$, has a unique solution. But, by Theorem 3 of Section 3.3, the latter is true if and only if the rank of A is equal to n—that is, if and only if the number of nonzero rows of B is equal to n. Since, in row-reduced echelon form, each nonzero row has a leading 1, and each column with a leading 1 has all other entries equal to 0, the rank of A is n if and only if the columns of B are the m-vectors $\mathbf{e}_1, \mathbf{e}_2, \ldots, \mathbf{e}_n$, as desired. ■

THEOREM 5

Let $\mathcal{S} = \{\mathbf{v}_1, \mathbf{v}_2, \ldots, \mathbf{v}_n\}$ be a set of m-vectors, A be the $m \times n$ matrix with columns $\mathbf{a}^i = \mathbf{v}_i$, and B be the row-reduced echelon form of A, and suppose that one of its columns, $\mathbf{b}^j$, is not an elementary vector or is a duplication of a preceding elementary vector. Then $\mathcal{S}$ is linearly dependent, and we have $\mathbf{v}_j = b_{1j}\mathbf{w}_1 + b_{2j}\mathbf{w}_2 + \cdots + b_{kj}\mathbf{w}_k$, where $\mathbf{w}_1, \mathbf{w}_2, \ldots, \mathbf{w}_k$ are vectors in $\mathcal{S}$ that have been transformed into $\mathbf{e}_1, \mathbf{e}_2, \ldots, \mathbf{e}_k$, the distinct elementary columns of B that precede the jth column.

Proof By Theorem 4, the set $\mathcal{S}$ is linearly dependent. Now, the vector equation $c_1\mathbf{v}_1 + c_2\mathbf{v}_2 + \cdots + c_n\mathbf{v}_n = \mathbf{0}$ is equivalent to $A\mathbf{c} = \mathbf{0}$, which, in turn, is equivalent to $B\mathbf{c} = \mathbf{0}$. For the sake of notational convenience, suppose that $\mathbf{w}_i = \mathbf{v}_i$ ($i = 1, 2, \ldots, k$). Then the first k equations of $B\mathbf{c} = \mathbf{0}$ become

$$\begin{array}{llll} c_1 & & + \cdots + b_{1j}c_j + \cdots = 0 \\ & c_2 & + \cdots + b_{2j}c_j + \cdots = 0 \\ & \ddots & \qquad\vdots \qquad\qquad \vdots \\ & & c_k + \cdots + b_{kj}c_j + \cdots = 0. \end{array}$$

Letting $c_j = 1$ and $c_i = 0$ for all variables other than $c_1, \ldots, c_k$, we have a solution $c_1 = -b_{1j}$, $c_2 = -b_{2j}, \ldots, c_k = -b_{kj}$, $c_j = 1$, and $c_i = 0$ for $i > k$ and $i \neq j$. But the c_i were the coefficients needed to express

$$c_1\mathbf{v}_1 + c_2\mathbf{v}_2 + \cdots + c_n\mathbf{v}_n = \mathbf{0}.$$

On replacing the c_i with the solution values, we have

$$-b_{1j}\mathbf{v}_1 - b_{2j}\mathbf{v}_2 - \cdots - b_{kj}\mathbf{v}_k + 0\mathbf{v}_{k+1} + \cdots + 0\mathbf{v}_{j-1} + 1\mathbf{v}_j + 0\mathbf{v}_{j+1} + \cdots + 0\mathbf{v}_n = \mathbf{0},$$

or, solving for $\mathbf{v}_j$,

$$\mathbf{v}_j = b_{1j}\mathbf{v}_1 + b_{2j}\mathbf{v}_2 + \cdots + b_{kj}\mathbf{v}_k,$$

as desired. ■

5.1 Exercises

In Exercises 1–4 let $\mathbf{u} = (1, 0, -1)$ and $\mathbf{v} = (-2, 1, 1)$.

1. Write $\mathbf{w}_1 = (-1, 2, -1)$ as a linear combination of $\mathbf{u}$ and $\mathbf{v}$.

2. Show that $\mathbf{w}_2 = (-1, 1, 1)$ *cannot* be written as a linear combination of $\mathbf{u}$ and $\mathbf{v}$.

3. For what value of c is the vector $(1, 1, c)$ a linear combination of $\mathbf{u}$ and $\mathbf{v}$?

4. If the vector (x_1, x_2, x_3) is a linear combination of $\mathbf{u}$ and $\mathbf{v}$, find an equation relating x_1, x_2, and x_3.

In Exercises 5–14 determine whether the given set of vectors is linearly dependent or independent. If it is linearly dependent, express one of the vectors in the set as a linear combination of the others.

5. $\{(1, 2, 3), (-1, 0, 1), (0, 1, 2)\}$

6. $\{(-1, 1, 2), (3, 3, 1), (1, 2, 2)\}$

7. $\{(0, 1, 2, 3), (0, -1, 2, -1), (0, 1, 0, 1)\}$

8. $\{(-1, -2, 2, 1), (0, 0, 0, 0), (1, 2, 3, 4)\}$

9. $\{(1, 2), (-1, 2)\}$

10. $\{(1, -1, 0), (2, 4, 0)\}$

11. $\{(1, 0), (0, 1), (1, 1)\}$

12. $\{(2, 0, 1), (1, -2, 0), (4, -4, 1), (1, 1, 1)\}$

13. $\{(1, -1, 1, -1), (0, 1, 0, 1) (1, 1, 0, 0), (2, 1, 1, 1)\}$

14. $\{(1, 1, 1, 1), (0, 1, 2, 3), (0, -1, -2, -3), (1, 0, -1, -2)\}$

In Exercises 15–18 let $\mathcal{S} = \{\mathbf{v}_1, \mathbf{v}_2, \mathbf{v}_3, \mathbf{v}_4, \mathbf{v}_5\}$ be vectors in $\mathbf{R}^3$ and A be the 3×5 matrix with the ith column $\mathbf{a}^i = \mathbf{v}_i$. Suppose that the row-reduced echelon form of A is

$$\begin{bmatrix} 1 & 2 & 0 & -1 & 0 \\ 0 & 0 & 1 & 3 & 0 \\ 0 & 0 & 0 & 0 & 1 \end{bmatrix}.$$

Are the indicated sets linearly dependent or independent? If a set is linearly dependent, express one vector as a linear combination of the others.

15. $\{\mathbf{v}_1, \mathbf{v}_2, \mathbf{v}_3\}$

16. $\{\mathbf{v}_1, \mathbf{v}_3, \mathbf{v}_4\}$

17. $\{\mathbf{v}_1, \mathbf{v}_4, \mathbf{v}_5\}$

18. $\{\mathbf{v}_3, \mathbf{v}_4, \mathbf{v}_5\}$

In Exercises 19–26 use the definition or Theorem 2 or 3 to determine whether or not the given set is linearly dependent.

19. $\{(1, 2), (0, 2), (1, 0), (-1, 1)\}$

20. $\{(1, -1, 0), (1, 1, 1), (0, -2, 1), (1, 4, -2)\}$

21. $\{(1, 2), (3, -1)\}$

22. $\{(1, 0, 1), (-1, 1, 0), (0, 1, 1)\}$

23. $\{(1, -1, 3, 0), (1, 0, 1, 0), (0, 0, 0, 0), (4, 2, 3, -6)\}$

24. $\{(2, 0, 0, 0), (2, 1, 0, 0), (-1, 3, -2, 0), (1, -2, 4, -3)\}$

25. $\{(1, 1, 1), (2, 2, 2)\}$

26. $\{(1, 0, -1, 0), (3, 6, 1, 2), (-2, 0, 2, 0)\}$

27. Prove Theorem 2: If $\mathcal{S}$ is a set of n vectors in $\mathbf{R}^m$ with $n > m$, then $\mathcal{S}$ is linearly dependent. (*Hint*: Apply Theorem 3, Corollary 2, of Section 3.3.)

28. Let $\mathbf{v}_1, \mathbf{v}_2, \ldots, \mathbf{v}_n$ be arbitrary vectors in $\mathbf{R}^m$. Show that $\{\mathbf{0}, \mathbf{v}_1, \mathbf{v}_2, \ldots, \mathbf{v}_n\}$ is linearly dependent.

29. Show that any set of distinct elementary vectors in $\mathbf{R}^m$ is linearly independent.

30. Let $\{\mathbf{v}_1, \mathbf{v}_2, \mathbf{v}_3, \mathbf{v}_4\}$ be a linearly independent set in $\mathbf{R}^m$. Show that $\{\mathbf{v}_1, \mathbf{v}_2, \mathbf{v}_3\}$ is also linearly independent.

31. Let $\mathcal{S}$ be a linearly independent set in $\mathbf{R}^m$. Show that any subset of $\mathcal{S}$ is also linearly independent.

32. Let $\{\mathbf{v}_1, \mathbf{v}_2, \ldots, \mathbf{v}_n\}$ be a linearly dependent set in $\mathbf{R}^m$ and $\mathbf{x}$ be any vector in $\mathbf{R}^m$. Show that $\{\mathbf{v}_1, \ldots, \mathbf{v}_n, \mathbf{x}\}$ is linearly dependent.

33. Let $\mathbf{u}$ and $\mathbf{v}$ be nonzero orthogonal vectors in $\mathbf{R}^m$. Show that they are linearly independent. (*Hint*: Use the fact that $\mathbf{u} \cdot (c\mathbf{u} + d\mathbf{v}) = c\mathbf{u} \cdot \mathbf{u}$ and $\mathbf{v} \cdot (c\mathbf{u} + d\mathbf{v}) = d\mathbf{v} \cdot \mathbf{v}$.)

34. Show that a set $\{\mathbf{u}_1, \mathbf{u}_2, \ldots, \mathbf{u}_n\}$ of mutually orthogonal nonzero vectors in $\mathbf{R}^m$ is linearly independent. (See Exercise 33.)

Computational Exercises

In Exercises 35–37 using computer software or a graphing calculator to transform a matrix to row-reduced echelon form, determine whether or not the following sets of vectors are linearly dependent. If a set is linearly dependent, express one of its vectors as a linear combination of the others.

35. $\{(1, 2, 3, 4, 5), (3, 4, 5, 2, 1), (2, 3, 1, 5, 4), (4, 5, 2, 1, 3), (5, 1, 4, 3, 2)\}$

36. $\{(1, 1, 1, 1, 1, 1), (2, 2, 2, 2, 2, 3), (3, 3, 3, 3, 3, 5), (4, 4, 4, 4, 4, 8)\}$

37. $\{\mathbf{v}_1, \mathbf{v}_2, \mathbf{v}_3, \mathbf{v}_4\}$, where $\mathbf{v}_1 = (0.216, 0.531, 0.870, 1.213)$, $\mathbf{v}_2 = (0.760, 1.432, 1.614, 0.666)$, $\mathbf{v}_3 = (0.228, 0.370, -1.026, -1.760)$, $\mathbf{v}_4 = (-0.050, 0, -0.450, 0)$

In Exercises 38–40 use a computer algebra system to find all values of the variables, if any, for which the vectors are linearly independent.

38. $\{(1, 3, 4), (2, -1, 1), (1, x, 2x)\}$

39. $\{(2, 1, 4, 0), (a, 0, b, 0), (1, 1, -2, 2), (0, a, 0, b)\}$

40. $\{(1, a, 2, a), (2, b, 3, b), (1, 1, 1, 1)\}$

In Exercises 41–42 use a computer algebra system to find all values of the variables, if any, for which the vector (1, 2, 3, 4) can be expressed as a linear combination of the vectors in the set.

41. $\{(1, -1, 0, 1), (2, 1, -1, 0), (2, -1, x, y)\}$

42. $\{(1, 1, 1, 1), (-2, 1, 0, x), (x, 1, y, -1)\}$

5.2 Subspaces of $\mathbf{R}^m$

As we saw in Chapters 1 and 2, lines, planes, and certain other regions in $\mathbf{R}^m$ can be described as linear combinations of vectors. In this section, we see that these regions are *subspaces* of $\mathbf{R}^m$ formed by the *span* of a set of vectors.

Subspaces of $\mathbf{R}^m$

DEFINITION

A **subspace** of $\mathbf{R}^m$ is a nonempty set, $\mathbf{S}$, of vectors satisfying the following two conditions:

i. If $\mathbf{v}_1$ and $\mathbf{v}_2$ are both in $\mathbf{S}$, then $\mathbf{v}_1 + \mathbf{v}_2$ is in $\mathbf{S}$.
ii. If $\mathbf{v}$ is in $\mathbf{S}$, then $c\mathbf{v}$ is in $\mathbf{S}$ for any scalar c.

NOTE If $\mathbf{v}_1$ and $\mathbf{v}_2$ are vectors in a subspace $\mathbf{S}$, and c_1 and c_2 are scalars, then $c_1\mathbf{v}_1 + c_2\mathbf{v}_2$ is also in $\mathbf{S}$. This results from the fact that both $c_1\mathbf{v}_1$ and $c_2\mathbf{v}_2$ are in $\mathbf{S}$ due to condition (ii) of the definition, and therefore their sum is in $\mathbf{S}$ from condition (i). More generally, if $\mathbf{v}_1, \mathbf{v}_2, \ldots, \mathbf{v}_n$ are vectors in a subspace $\mathbf{S}$, then every linear combination of the $\mathbf{v}_i$ is in $\mathbf{S}$ (Exercise 27 at the end of this section).

Taking $c = 0$ in condition (ii), we see that every subspace must contain the vector $\mathbf{0}$. In fact, it is easily checked (Exercise 4) that $\mathbf{S} = \{\mathbf{0}\}$, the set containing only the zero vector, is itself a subspace of $\mathbf{R}^m$. Moreover, since $\mathbf{R}^m$ is a subset of itself, $\mathbf{R}^m$ is a subspace as well. Thus, for each m, $\mathbf{R}^m$ has at least two subspaces, $\{\mathbf{0}\}$ and $\mathbf{R}^m$. Other examples of subspaces are given in Examples 1 through 3.

EXAMPLE 1

Show that the set $\mathbf{S}$ consisting of all 4-vectors whose first and last components are equal to zero is a subspace of $\mathbf{R}^4$.

Solution All vectors in $\mathbf{S}$ are of the form $(0, a, b, 0)$, where a and b are scalars. To verify that $\mathbf{S}$ is a subspace, we check conditions (i) and (ii) of the definition.

i. We have $(0, a_1, b_1, 0) + (0, a_2, b_2, 0) = (0, a_1 + a_2, b_1 + b_2, 0)$. Since this is a 4-vector whose first and last components are zero, the sum of two vectors in $\mathbf{S}$ is in $\mathbf{S}$.

ii. We have $c(0, a, b, 0) = (0, ca, cb, 0)$. Thus, any scalar multiple of a vector in $\mathbf{S}$ is in $\mathbf{S}$.

Consequently, $\mathbf{S}$ is a subspace of $\mathbf{R}^4$. ■

EXAMPLE 2

Let A be an $m \times n$ matrix. Show that the set $\mathbf{S}$ of all solutions of the homogeneous linear system $A\mathbf{x} = \mathbf{0}$ is a subspace of $\mathbf{R}^n$.

Solution To verify that $\mathbf{S}$ is a subspace, we check conditions (i) and (ii) of the definition.

i. Let $\mathbf{x}_1$ and $\mathbf{x}_2$ be vectors in $\mathbf{R}^n$ that are solutions of $A\mathbf{x} = \mathbf{0}$. That is, $A\mathbf{x}_1 = \mathbf{0}$ and $A\mathbf{x}_2 = \mathbf{0}$. Now $A(\mathbf{x}_1 + \mathbf{x}_2) = A\mathbf{x}_1 + A\mathbf{x}_2 = \mathbf{0} + \mathbf{0} = \mathbf{0}$ (by part (g) of Theorem 1 of Section 3.1). So $\mathbf{x}_1 + \mathbf{x}_2$ is a solution of $A\mathbf{x} = \mathbf{0}$, and hence is in $\mathbf{S}$.

ii. Let $\mathbf{x}$ be a solution of $A\mathbf{x} = \mathbf{0}$ and c be a scalar. Then $A(c\mathbf{x}) = c(A\mathbf{x}) = c\mathbf{0} = \mathbf{0}$ (by part (n) of Theorem 1 of Section 3.1). So $c\mathbf{x}$ is a solution of $A\mathbf{x} = \mathbf{0}$, and hence is in $\mathbf{S}$.

Thus, $\mathbf{S}$ is a subspace of $\mathbf{R}^n$. ■

NOTE The subspace of Example 2 is called the **solution space** of the given homogeneous linear system. It is *not* true (Exercise 12) that the set of all solutions of a *nonhomogeneous* linear system forms a subspace.

EXAMPLE 3

Show that the set **S** of all linear combinations of the vectors $\mathbf{u}_1 = (1, 2, 3)$ and $\mathbf{u}_2 = (1, 0, -1)$ is a subspace of $\mathbf{R}^3$.

Solution A vector is in **S** if and only if it is of the form $a\mathbf{u}_1 + b\mathbf{u}_2$ for some scalars a and b. We check conditions (i) and (ii).

- **i.** Let $\mathbf{v}_1$ and $\mathbf{v}_2$ be vectors in **S**. Then there are scalars a_1, b_1 and a_2, b_2 such that $\mathbf{v}_1 = a_1\mathbf{u}_1 + b_1\mathbf{u}_2$ *and* $\mathbf{v}_2 = a_2\mathbf{u}_1 + b_2\mathbf{u}_2$. Now $\mathbf{v}_1 + \mathbf{v}_2 = (a_1 + a_2)\mathbf{u}_1 + (b_1 + b_2)\mathbf{u}_2$, a linear combination of $\mathbf{u}_1$ and $\mathbf{u}_2$, so $\mathbf{v}_1 + \mathbf{v}_2$ is in **S**.
- **ii.** Let $\mathbf{v}$ be a vector in **S**, $\mathbf{v} = a\mathbf{u}_1 + b\mathbf{u}_2$ and c be a scalar. Then $c\mathbf{v} = (ca)\mathbf{u}_1 + (cb)\mathbf{u}_2$, a linear combination of $\mathbf{u}_1$ and $\mathbf{u}_2$, so $c\mathbf{v}$ is in **S**.

Thus, **S** is a subspace of $\mathbf{R}^3$. ■

NOTE As we saw in Section 2.1, the set of all linear combinations of two non-collinear vectors in $\mathbf{R}^m$ describes a plane. The plane generated in Example 3 is pictured in Figure 3. Note that it contains the origin, as must all subspaces of $\mathbf{R}^m$.

In the solution to Example 3, we made no use of the fact that $\mathbf{u}_1$ and $\mathbf{u}_2$ were specific vectors in $\mathbf{R}^3$. Consequently, a similar argument can be used to prove the following theorem. We leave its proof for Exercise 28.

FIGURE 3

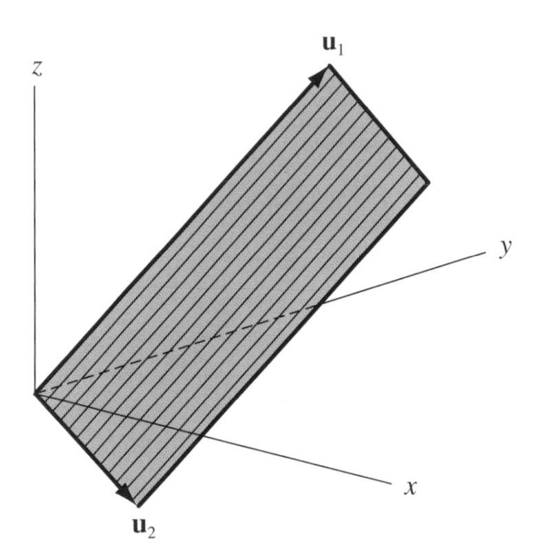

THEOREM 1

Let $\mathcal{T} = \{\mathbf{v}_1, \mathbf{v}_2, \ldots, \mathbf{v}_n\}$ be a set of vectors in $\mathbf{R}^m$. The set of all linear combinations, $c_1\mathbf{v}_1 + c_2\mathbf{v}_2 + \cdots + c_n\mathbf{v}_n$, of the vectors in $\mathcal{T}$ is a subspace of $\mathbf{R}^m$. ■

NOTE The subspace described in Theorem 1 is called the **subspace generated by** $\mathcal{T}$. For example, the subspace of $\mathbf{R}^3$ described in Example 3 is the *subspace generated by* $(1, 2, 3)$ *and* $(1, 0, -1)$.

We now give an example of a subset of $\mathbf{R}^m$ that is *not* a subspace.

EXAMPLE 4

Show that the set **S** of all vectors of the form $(a, 1)$ is not a subspace of $\mathbf{R}^2$.

Solution To show that condition (i) of the definition does not hold for **S**, we need only find two vectors in **S** whose sum is not in **S**. Let $\mathbf{v}_1 = (2, 1)$ and $\mathbf{v}_2 = (3, 1)$. Then $\mathbf{v}_1$ and $\mathbf{v}_2$ are both in **S** (since their second components are equal to 1), but $\mathbf{v}_1 + \mathbf{v}_2 = (5, 2)$ is not in **S**. (You can also show that condition (ii) fails to hold.) ■

Span of a Set of Vectors

DEFINITION

Let **S** be a subspace of $\mathbf{R}^m$. A subset $\mathcal{T}$ of **S** **spans S** if every vector in **S** can be written as a linear combination of elements of $\mathcal{T}$. In this case, we say that $\mathcal{T}$ is a **spanning set** for **S**, or $\mathcal{T}$ spans **S**, and write $\mathbf{S} = \text{Span } \mathcal{T}$.

EXAMPLE 5

Show that the set of m distinct elementary vectors in $\mathbf{R}^m$, $\mathcal{T} = \{\mathbf{e}_1, \mathbf{e}_2, \ldots, \mathbf{e}_m\}$, spans $\mathbf{R}^m$.

Solution Let $\mathbf{w} = (w_1, w_2, \ldots, w_m)$ be an arbitrary vector in $\mathbf{R}^m$. Then

$$\mathbf{w} = w_1\mathbf{e}_1 + w_2\mathbf{e}_2 + \cdots + w_m\mathbf{e}_m$$

is a linear combination of the $\mathbf{e}_i$, as desired. ■

EXAMPLE 6

Let **S** be the subspace of $\mathbf{R}^4$ consisting of all vectors whose first and last components are zero (see Example 1). Show that the vectors $\mathbf{v}_1 = (0, 1, -1, 0)$ and $\mathbf{v}_2 = (0, 2, -1, 0)$ span **S**.

Solution Let $\mathbf{w} = (0, a, b, 0)$ be an arbitrary vector in **S**. We must show that there are constants c_1 and c_2 such that $\mathbf{w} = c_1\mathbf{v}_1 + c_2\mathbf{v}_2$, that is,

$$(0, a, b, 0) = c_1(0, 1, -1, 0) + c_2(0, 2, -1, 0).$$

Performing the scalar multiplications and additions on the right side and equating corresponding components yields the two equations

$$\begin{aligned} c_1 + 2c_2 &= a \\ \text{and } -c_1 - c_2 &= b. \end{aligned}$$

(Equating the first, as well as the last, components just gives $0 = 0$.) Since this system has the solution $c_1 = -(a + 2b)$, $c_2 = a + b$, we know that $\{\mathbf{v}_1, \mathbf{v}_2\}$ spans **S**. For example, to express the vector $(0, 3, 7, 0)$ as a linear combination of $\mathbf{v}_1$ and $\mathbf{v}_2$, take $c_1 = -(3 + 2(7)) = -17$ and $c_2 = 3 + 7 = 10$:

$$(0, 3, 7, 0) = -17(0, 1, -1, 0) + 10(0, 2, -1, 0)$$ ■

If **S** is the subspace of $\mathbf{R}^m$ *generated by a set* $\mathcal{T}$ of vectors in **S**, then, by definition, $\mathcal{T}$ spans **S**. If linear dependence corresponds to *lack of redundancy*, then we can view spanning as *sufficiency*—the vectors are enough to generate the subspace by scalar multiplication and vector addition. The following theorem simplifies the task of determining proper subsets of $\mathcal{T}$ that also span **S**.

THEOREM 2

Let **S** be the subspace of $\mathbf{R}^m$ spanned by the set $\mathcal{T} = \{\mathbf{v}_1, \mathbf{v}_2, \ldots, \mathbf{v}_n\}$. If one of the vectors in $\mathcal{T}$ is a linear combination of the others, then the subset of $\mathcal{T}$ obtained by deleting that vector still spans **S**.

Proof For the sake of notational convenience, suppose that $\mathbf{v}_n$ is a linear combination of the other vectors in $\mathcal{T}$; that is, $\mathbf{v}_n = c_1\mathbf{v}_1 + c_2\mathbf{v}_2 + \cdots + c_{n-1}\mathbf{v}_{n-1}$. Let **w** be an arbitrary vector in **S**. Since $\mathcal{T}$ spans **S**, we have $\mathbf{w} = a_1\mathbf{v}_1 + a_2\mathbf{v}_2 + \cdots + a_n\mathbf{v}_n$ for some scalars $a_1, a_2, \ldots, a_n$. But then

$$\begin{aligned}\mathbf{w} &= a_1\mathbf{v}_1 + a_2\mathbf{v}_2 + \cdots + a_{n-1}\mathbf{v}_{n-1} + a_n(c_1\mathbf{v}_1 + c_2\mathbf{v}_2 + \cdots + c_{n-1}\mathbf{v}_{n-1}) \\ &= (a_1 + a_nc_1)\mathbf{v}_1 + (a_2 + a_nc_2)\mathbf{v}_2 + \cdots + (a_{n-1} + a_nc_{n-1})\mathbf{v}_{n-1}.\end{aligned}$$

Thus, $\mathbf{w}$ is a linear combination of $\{\mathbf{v}_1, \mathbf{v}_2, \ldots, \mathbf{v}_{n-1}\}$, the subset of $\mathcal{T}$ obtained by deleting $\mathbf{v}_n$, and this subset spans **S**, as desired. ■

EXAMPLE 7

Let **S** be the subspace of $\mathbf{R}^3$ generated by the set $\mathcal{T} = \{(1, -1, 0), (-2, 1, 2), (-1, -1, 4)\}$. Show that the subset $\mathcal{T}_0 = \{(1, -1, 0), (-2, 1, 2)\}$ spans **S**.

Solution To apply Theorem 2, we show that $(-1, -1, 4)$ is a linear combination of $(1, -1, 0)$ and $(-2, 1, 2)$ by using Theorem 5 of Section 5.1. We construct the matrix

$$\begin{bmatrix} 1 & -2 & -1 \\ -1 & 1 & -1 \\ 0 & 2 & 4 \end{bmatrix}$$

and transform it to the row-reduced echelon form

$$\begin{bmatrix} 1 & 0 & 3 \\ 0 & 1 & 2 \\ 0 & 0 & 0 \end{bmatrix}.$$

Consequently, $(-1, -1, 4) = 3(1, -1, 0) + 2(-2, 1, 2)$ (from the third column of the latter matrix). Thus, $\mathcal{T}_0$ spans **S**. ■

NOTE Geometrically, we have shown that the three vectors of Example 7 lie in a plane—the plane generated by $(1, -1, 0)$ and $(-2, 1, 2)$, illustrated in Figure 4.

FIGURE 4

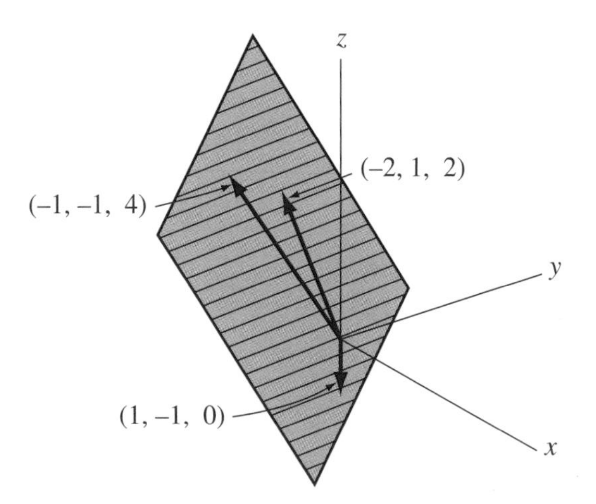

5.2 Exercises

In Exercises 1–6 show that the given sets of vectors are subspaces of $\mathbf{R}^m$.

1. The set of all scalar multiples of the vector $(1, 2, 3)$ (of $\mathbf{R}^3$)
2. The set of all m-vectors whose first component is 0 (of $\mathbf{R}^m$)
3. The set of all linear combinations of the vectors $(1, 0, 1, 0)$ and $(0, 1, 0, 1)$ (of $\mathbf{R}^4$)
4. The set consisting solely of the zero vector, $\{\mathbf{0}\}$ (of $\mathbf{R}^m$)

5. The set of all vectors of the form $(a, b, a, -b, a + b)$ (of $\mathbf{R}^4$)

6. The set of all vectors (x, y, z) such that $x + y + z = 0$ (of $\mathbf{R}^3$)

In Exercises 7–12 show that the given sets of vectors do *not* form subspaces of $\mathbf{R}^m$.

7. The set consisting of the single vector $(1, 1, 1, 1)$ (of $\mathbf{R}^4$)

8. The set of all m-vectors whose first component is 2 (of $\mathbf{R}^m$)

9. The set of all m-vectors *except* the m-vector $\mathbf{0}$ (of $\mathbf{R}^m$)

10. The set of all 4-vectors *except* the vector $(1, 1, 1, 1)$ (of $\mathbf{R}^4$)

11. The set of all 3-vectors whose components have the sum of 1 (of $\mathbf{R}^3$)

12. The set of all solutions of $A\mathbf{x} = \mathbf{b}, \mathbf{b} \neq \mathbf{0}$ (of $\mathbf{R}^n$, where A is $m \times n$)

In Exercises 13–16 determine which sets span $\mathbf{R}^3$.

13. $\{(1, 2, 3), (-1, 0, 1), (0, 1, 2)\}$

14. $\{(-1, 1, 2), (3, 3, 1), (1, 2, 2)\}$

15. $\{(1, 2, -1), (1, 0, 1)\}$

16. $\{(1, 0, 0), (0, 1, 0), (0, 0, 1), (1, 1, 1)\}$

17. Show that $\{\mathbf{e}_2, \mathbf{e}_3\}$ spans the subspace of $\mathbf{R}^4$ consisting of all vectors whose first and last components are zero.

18. Show that $\{(1, 1, 0), (1, 0, 1)\}$ spans the subspace of $\mathbf{R}^3$ consisting of all vectors (x, y, z) such that $x = y + z$.

19–24. Find a set that spans each of the subspaces in Exercises 1 through 6.

25. Show that $\{(0, 1, 1)\}$ spans the solution space of $A\mathbf{x} = \mathbf{0}$, where

$$A = \begin{bmatrix} 1 & -1 & 1 \\ 2 & 1 & -1 \\ 0 & -1 & 1 \end{bmatrix}.$$

26. Show that $\{(1, 1, 0), (-2, 0, 1)\}$ spans the solution space of $A\mathbf{x} = \mathbf{0}$, where

$$A = \begin{bmatrix} 1 & -1 & 2 \\ -2 & 2 & -4 \end{bmatrix}.$$

27. Show that if $\mathbf{v}_1, \mathbf{v}_2, \ldots, \mathbf{v}_n$ are vectors in a subspace $\mathbf{S}$ of $\mathbf{R}^m$, then $\mathbf{w} = c_1\mathbf{v}_1 + c_2\mathbf{v}_2 + \cdots + c_n\mathbf{v}_n$ is in $\mathbf{S}$ for all scalars $c_1, c_2, \ldots, c_n$.

28. Prove Theorem 1: Let $\mathcal{T} = \{\mathbf{v}_1, \mathbf{v}_2, \ldots, \mathbf{v}_n\}$ be a set of vectors in $\mathbf{R}^m$, and show that the set of all linear combinations of the vectors in $\mathcal{T}$ is a subspace of $\mathbf{R}^m$.

5.3 Basis and Dimension

In Section 5.1, we considered sets that are linearly independent. In Section 5.2, we considered sets that span a given subspace. We have observed that linear independence corresponds to no redundancy—not too much—while spanning corresponds to sufficiency—enough. Subsets of a subspace that are linearly independent *and* span the subspace are of particular importance in linear algebra; these are the basic building blocks of subspaces. They are both enough and not too much—just right! In this section, we investigate the properties of such sets.

Basis for a Subspace

DEFINITION

Let **S** be a subspace of $\mathbf{R}^m$. A set $\mathcal{T}$ of vectors in **S** is a **basis** for **S** if

i. $\mathcal{T}$ is linearly independent and
ii. $\mathcal{T}$ spans **S**.

EXAMPLE 1

Show that the set of elementary vectors in $\mathbf{R}^m$, $\mathcal{T} = \{\mathbf{e}_1, \mathbf{e}_2, \ldots, \mathbf{e}_m\}$, is a basis for $\mathbf{R}^m$.

Solution We must show that $\mathcal{T}$ is linearly independent and that $\mathcal{T}$ spans $\mathbf{R}^m$. To demonstrate the former, we show that the vector equation

$$c_1\mathbf{e}_1 + c_2\mathbf{e}_2 + \cdots + c_m\mathbf{e}_m = \mathbf{0}$$

has only the trivial solution. Now, each $\mathbf{e}_i$ has a 1 in its ith position and zeros everywhere else. Hence, if we perform the scalar multiplications and additions on the left side of this equation, we obtain

$$(c_1, c_2, \ldots, c_m) = (0, 0, \ldots, 0).$$

Equating corresponding components gives $c_i = 0$ for all i, so $\mathcal{T}$ is linearly independent. Since we have already shown, in Example 5 of Section 5.2, that $\mathcal{T}$ spans $\mathbf{R}^m$, this set is indeed a basis for $\mathbf{R}^m$. ■

NOTE The basis $\{\mathbf{e}_1, \mathbf{e}_2, \ldots, \mathbf{e}_m\}$ is called the **standard basis** for $\mathbf{R}^m$. For example, the standard basis for $\mathbf{R}^2$ is $\{\mathbf{e}_1, \mathbf{e}_2\} = \{(1, 0), (0, 1)\}$, and the standard basis for $\mathbf{R}^3$ is $\{(1, 0, 0), (0, 1, 0), (0, 0, 1)\}$.

EXAMPLE 2

Show that the set $\mathcal{T} = \{(1, 0, 0), (1, 2, 0), (1, 2, 3)\}$ is a basis for $\mathbf{R}^3$.

Solution We can show that $\mathcal{T}$ is linearly independent and that $\mathcal{T}$ spans $\mathbf{R}^3$ at the same time. Let $\mathbf{w} = (w_1, w_2, w_3)$ be an arbitrary vector in $\mathbf{R}^3$. Then $\mathcal{T}$ spans $\mathbf{R}^3$ if there exist scalars c_1, c_2, c_3 such that

$$c_1(1, 0, 0) + c_2(1, 2, 0) + c_3(1, 2, 3) = (w_1, w_2, w_3).$$

Performing the scalar multiplications and equating corresponding components yields the linear system

$$\begin{aligned} c_1 + c_2 + c_3 &= w_1 \\ 2c_2 + 2c_3 &= w_2 \\ 3c_3 &= w_3, \end{aligned}$$

which has the coefficient matrix

$$A = \begin{bmatrix} 1 & 1 & 1 \\ 0 & 2 & 2 \\ 0 & 0 & 3 \end{bmatrix}.$$

Now $\det A = 6$. Since $\det A$ is nonzero, Theorem 3 of Section 5.1 tells us that $\mathcal{T}$ is linearly independent. Theorem 6 of Section 4.2 tells us that our linear system has a (unique) solution. Hence, $\mathcal{T}$ spans $\mathbf{R}^3$ as well. ■

Now suppose that we wish to find a basis for the subspace **S** generated by a set $\mathcal{T}$ of vectors in **S**. By definition, $\mathcal{T}$ spans **S**, but $\mathcal{T}$ is not necessarily linearly independent. The following theorem and accompanying procedure show us how to *reduce a spanning set to a basis.*

THEOREM 1

Let **S** be the subspace of $\mathbf{R}^m$ generated by the set $\mathcal{T} = \{\mathbf{v}_1, \mathbf{v}_2, \ldots, \mathbf{v}_n\}$. Then there is a subset of $\mathcal{T}$ that is a basis for **S**.

Proof By Theorem 4 of Section 5.1, $\mathcal{T}_0$ is linearly independent. By Theorem 5 of Section 5.1, the other vectors in $\mathcal{T}$ are linear combinations of those in $\mathcal{T}_0$. But then, by Theorem 2 of Section 5.2, $\mathcal{T}_0$ spans **S** as well. Hence, $\mathcal{T}_0$ is a basis for **S**. ■

Procedure for Reducing a Spanning Set to a Basis
Let A be the $m \times n$ matrix with columns $\mathbf{a}^i = \mathbf{v}_i$ and B be its row-reduced echelon form. If $\mathbf{w}_1, \mathbf{w}_2, \ldots, \mathbf{w}_s$ are the vectors in $\mathcal{T}$ that are transformed into the distinct elementary columns of B, then $\mathcal{T}_0 = \{\mathbf{w}_1, \mathbf{w}_2, \ldots, \mathbf{w}_s\}$ is a basis for **S**.

EXAMPLE 3

Find a basis for the subspace, **S**, of $\mathbf{R}^4$ generated by $\mathcal{T} = \{\mathbf{v}_1, \mathbf{v}_2, \mathbf{v}_3, \mathbf{v}_4, \mathbf{v}_5\}$, where $\mathbf{v}_1 = (1, 0, -1, 1)$, $\mathbf{v}_2 = (-2, 0, 2, -2)$, $\mathbf{v}_3 = (1, 1, 1, 1)$, $\mathbf{v}_4 = (1, -1, -3, 1)$, and $\mathbf{v}_5 = (1, -1, 1, -1)$.

Solution We apply the procedure, constructing the matrix A with $\mathbf{a}^i = \mathbf{v}_i$,

$$A = \begin{bmatrix} 1 & -2 & 1 & 1 & 1 \\ 0 & 0 & 1 & -1 & -1 \\ -1 & 2 & 1 & -3 & 1 \\ 1 & -2 & 1 & 1 & -1 \end{bmatrix},$$

which has the row-reduced echelon form

$$B = \begin{bmatrix} 1 & -2 & 0 & 2 & 0 \\ 0 & 0 & 1 & -1 & 0 \\ 0 & 0 & 0 & 0 & 1 \\ 0 & 0 & 0 & 0 & 0 \end{bmatrix}.$$

We see that $\mathbf{v}_1$, $\mathbf{v}_3$, and $\mathbf{v}_5$ have been transformed into the distinct elementary vectors $\mathbf{e}_1$, $\mathbf{e}_2$, and $\mathbf{e}_3$ ($\mathbf{v}_1$, $\mathbf{v}_3$, and $\mathbf{v}_5$ are the $\mathbf{w}_1$, $\mathbf{w}_2$, and $\mathbf{w}_3$ of the procedure), so $\{\mathbf{v}_1, \mathbf{v}_3, \mathbf{v}_5\}$ is a basis for **S**. ■

In Theorem 1, we learned how to *reduce a spanning set to a basis.* This raises the question, "Is it possible to *extend a linearly independent set to a basis*?" More precisely, if $\mathcal{T}$ is a linearly independent subset of a subspace **S**, "Are there vectors in **S** that can be added to the set $\mathcal{T}$ to obtain a basis for **S**?" The answer is "Yes." If we have a spanning set for **S**, there is a straightforward computational technique to construct this basis. The next theorem and accompanying procedure describe the technique.

THEOREM 2

Let **S** be the subspace of $\mathbf{R}^m$ generated by the set $\mathcal{T} = \{\mathbf{v}_1, \mathbf{v}_2, \ldots, \mathbf{v}_n\}$ and $\mathcal{T}_1 = \{\mathbf{u}_1, \mathbf{u}_2, \ldots, \mathbf{u}_s\}$ be a linearly independent subset of **S**. Then there is a subset $\mathcal{T}_2$ of $\mathcal{T}$ such that the union of $\mathcal{T}_1$ and $\mathcal{T}_2$ is a basis for **S**. Thus every linearly independent subset of a subspace **S** can be extended to a basis of **S**.

Proof By the procedure of Theorem 1, the columns of A that correspond to the distinct elementary columns of B form a basis for **S**. Since the first s columns of A are linearly independent, they are transformed into $\mathbf{e}_1, \ldots, \mathbf{e}_s$, the first s columns of B. Thus, each element of $\mathcal{T}_1$ is in the basis. Let $\mathcal{T}_2$ be the set of other basis vectors. Then $\mathcal{T}_2$ is a subset of $\mathcal{T}$, since each element comes from the last n columns of A. So the union of $\mathcal{T}_1$ and $\mathcal{T}_2$ is just the set of all basis elements. ■

Procedure for Extending an Independent Set to a Basis

Let A be the $m \times (s + n)$ matrix, with columns $\mathbf{a}^i = \mathbf{u}_i$ for $i = 1, \ldots, s$ and $\mathbf{a}^{s+j} = \mathbf{v}_j$ for $j = 1, \ldots, n$. In other words, the first s columns of A are the vectors in $\mathcal{T}_1$ and the last n columns are the vectors in $\mathcal{T}$. Let B be the row-reduced echelon matrix for A and $\mathbf{w}_1, \mathbf{w}_2, \ldots, \mathbf{w}_t$ be those vectors in $\mathcal{T}$ that are transformed into distinct elementary columns of B other than $\mathbf{e}_1, \mathbf{e}_2, \ldots, \mathbf{e}_s$. Then $\mathcal{T}_2 = \{\mathbf{w}_1, \mathbf{w}_2, \ldots, \mathbf{w}_t\}$.

EXAMPLE 4

Let $\mathcal{T}_1 = \{\mathbf{v}_1, \mathbf{v}_2\}$ be the set of linearly independent vectors in $\mathbf{R}^4$ with $\mathbf{v}_1 = (1, 0, -1, 1)$ and $\mathbf{v}_2 = (1, 1, 1, -1)$. Extend this set to form a basis for $\mathbf{R}^4$.

Solution Since $\mathbf{R}^4$ is spanned by the set $\mathcal{T} = \{\mathbf{e}_1, \mathbf{e}_2, \mathbf{e}_3, \mathbf{e}_4\}$, we may find a subset $\mathcal{T}_2$ of this set $\mathcal{T}$ such that the union of $\mathcal{T}_1$ and $\mathcal{T}_2$ is a basis for $\mathbf{R}^4$. Following the procedure of Theorem 2, we construct the matrix

$$\left[\begin{array}{rr|rrrr} 1 & 1 & 1 & 0 & 0 & 0 \\ 0 & 1 & 0 & 1 & 0 & 0 \\ -1 & 1 & 0 & 0 & 1 & 0 \\ 1 & -1 & 0 & 0 & 0 & 1 \end{array}\right]$$

and transform it to row-reduced echelon form:

$$\left[\begin{array}{rr|rrrr} 1 & 0 & 0 & 1 & 0 & 1 \\ 0 & 1 & 0 & 1 & 0 & 0 \\ 0 & 0 & 1 & -2 & 0 & -1 \\ 0 & 0 & 0 & 0 & 1 & 1 \end{array}\right]$$

We see that $\mathbf{e}_1$ and $\mathbf{e}_3$ of the set $\mathcal{T}$ have been transformed into distinct elementary columns of the latter matrix, so the set $\{\mathbf{v}_1, \mathbf{v}_2, \mathbf{e}_1, \mathbf{e}_3\}$ is a basis for $\mathbf{R}^4$. ■

We now state an important theorem concerning bases, to be proved in Chapter 6.

THEOREM 3

Let $\mathbf{S}$ be a subspace of $\mathbf{R}^m$. Then any basis for $\mathbf{S}$ contains the same number of vectors as any other basis. ■

For example, we showed (Example 1) that there is a basis for $\mathbf{R}^m$ containing m vectors. As a result, Theorem 3 tells us that *every* basis for $\mathbf{R}^m$ must contain m vectors.

Dimension of a Subspace

DEFINITION

Let $\mathbf{S}$ be a subspace of $\mathbf{R}^m$. The **dimension** of $\mathbf{S}$, dim $\mathbf{S}$, is equal to the number of vectors in any basis for $\mathbf{S}$.

NOTE If we find one basis for a subspace that contains n vectors, and then determine another basis for the same subspace, Theorem 3 guarantees that it too contains n vectors. Consequently, we say that the concept of dimension is "well defined."

EXAMPLE 5

Find the dimension of $\mathbf{R}^2$, $\mathbf{R}^3$, and $\mathbf{R}^m$.

Solution Bases for these subspaces are given by $\{\mathbf{e}_1, \mathbf{e}_2\}$, $\{\mathbf{e}_1, \mathbf{e}_2, \mathbf{e}_3\}$, and $\{\mathbf{e}_1, \mathbf{e}_2, \ldots, \mathbf{e}_m\}$, respectively. Thus, $\mathbf{R}^2$ has dimension 2, $\mathbf{R}^3$ has dimension 3, and, in general, $\mathbf{R}^m$ has dimension m. ■

EXAMPLE 6

Find the dimension of the subspace $\mathbf{S}$ of $\mathbf{R}^4$ consisting of all vectors whose first and last components are equal to 0 (see Example 1 of Section 5.2).

Solution In Example 6 of Section 5.2 we showed that the vectors $\mathbf{v}_1 = (0, 1, -1, 0)$ and $\mathbf{v}_2 = (0, 2, -1, 0)$ span $\mathbf{S}$. By transforming the matrix

$$\begin{bmatrix} 0 & 0 \\ 1 & 2 \\ -1 & -1 \\ 0 & 0 \end{bmatrix}$$

to row-reduced echelon form, we can show that they are also linearly independent. Hence, $\{\mathbf{v}_1, \mathbf{v}_2\}$ is a basis for $\mathbf{S}$. Thus, $\dim \mathbf{S} = 2$. ■

One of the more important subspaces of $\mathbf{R}^m$ is the *solution space* of a homogeneous linear system—the set of all solutions of $A\mathbf{x} = \mathbf{0}$ for a given matrix A (see Example 2 of Section 5.2). This subspace is called the *kernel* of matrix A, and is denoted by Ker A. The next theorem and its accompanying procedure describe how we can find a basis for, and hence the dimension of, these subspaces. This is a procedure we use several times in the chapters that follow.

THEOREM 4

Let A be an $m \times n$ matrix, B be its row-reduced echelon form, and the rank of A (the number of nonzero rows of B) be equal to r. Then the dimension of the solution space of $A\mathbf{x} = \mathbf{0}$ is $n - r$. (The dimension of this solution space is called the *nullity* of A.) ■

Before we prove Theorem 4, we present a procedure for determining a *basis* of $n - r$ vectors for the solution space and an example illustrating this procedure. The proof of the theorem is simply a verification that the set given in the procedure is in fact a basis.

Procedure for Finding a Basis for the Solution Space of $A\mathbf{x} = \mathbf{0}$

Let A be an $m \times n$ matrix with rank equal to r.

i. Solve $A\mathbf{x} = \mathbf{0}$ by Gauss–Jordan elimination (Section 2.3). There are $n - r$ parameters in the solution.

ii. Obtain $n - r$ vectors by setting, in turn, each parameter in the solution equal to 1 as the rest are set equal to 0.

iii. The $n - r$ vectors of step (ii) are a basis for the solution space.

EXAMPLE 7

Find a basis for and the dimension of the solution space of the homogeneous linear system $A\mathbf{x} = \mathbf{0}$, where

$$A = \begin{bmatrix} 1 & 1 & 5 & 0 & 1 \\ 1 & -1 & 1 & -2 & -1 \\ 1 & 1 & 5 & 1 & 1 \\ 0 & 2 & 4 & 3 & 2 \\ 1 & 1 & 5 & 2 & 1 \end{bmatrix}.$$

Solution Following the procedure just given, we first transform A to its row-reduced echelon form:

$$B = \begin{bmatrix} 1 & 0 & 3 & 0 & 0 \\ 0 & 1 & 2 & 0 & 1 \\ 0 & 0 & 0 & 1 & 0 \\ 0 & 0 & 0 & 0 & 0 \\ 0 & 0 & 0 & 0 & 0 \end{bmatrix}$$

The rank of A is 3, so the dimension of the solution space is $5 - 3 = 2$. Translating B into a system of equations and solving for x_1, x_2, and x_4 (the variables that correspond to the leading 1's of B), we obtain

$$\begin{aligned} x_1 &= -3x_3, \\ x_2 &= -2x_3 - x_5, \\ x_4 &= \ \ 0. \end{aligned}$$

Thus, the solution of $A\mathbf{x} = \mathbf{0}$ is

$$(-3s, -2s - t, s, 0, t).$$

Setting $s = 1$ and $t = 0$, we get $\mathbf{v}_1 = (-3, -2, 1, 0, 0)$; setting $t = 1$ and $s = 0$, we get $\mathbf{v}_2 = (0, -1, 0, 0, 1)$. Then $\{\mathbf{v}_1, \mathbf{v}_2\}$ is a basis for the solution space of $A\mathbf{x} = \mathbf{0}$. ■

We now prove Theorem 4.

Proof of Theorem 4 By Theorem 4 of Section 3.3, the solution set of $A\mathbf{x} = \mathbf{0}$ contains $(n - r)$ parameters. Recalling the setting with $d_1 = d_2 = \cdots = d_r = 0$, we assume that the first r columns of A are the ones transformed into a set of distinct elementary columns of B. This gives the following system of equations:

$$\begin{array}{cccccc} x_1 & & & + b_{1,r+1}x_{r+1} & + \cdots + b_{1,n}x_n & = 0 \\ & x_2 & & + b_{2,r+1}x_{r+1} & + \cdots + b_{2,n}x_n & = 0 \\ & & \ddots & \vdots & \vdots & \vdots \\ & & & x_r + b_{r,r+1}x_{r+1} & + \cdots + b_{r,n}x_n & = 0 \end{array}$$

Solving for $x_1, \ldots, x_r$, and setting $x_{r+1} = t_1, x_{r+2} = t_2, \ldots, x_n = t_{n-r}$, we obtain this solution:

$$\begin{array}{lcccccc} x_1 & = -b_{1,r+1}t_1 & - & b_{1,r+2}t_2 & - \cdots - & b_{1,n}t_{n-r} \\ x_2 & = -b_{2,r+1}t_1 & - & b_{2,r+2}t_2 & - \cdots - & b_{2,n}t_{n-r} \\ \vdots & \vdots & & \vdots & & \vdots \\ x_r & = -b_{r,r+1}t_1 & - & b_{r,r+2}t_2 & - \cdots - & b_{r,n}t_{n-r} \\ x_{r+1} & = \quad 1t_1 & + & 0t_2 & + \cdots + & 0t_{n-r} \\ x_{r+2} & = \quad 0t_1 & + & 1t_2 & + \cdots + & 0t_{n-r} \\ \vdots & \vdots & & \vdots & \ddots & \vdots \\ x_n & = \quad 0t_1 & + & 0t_2 & + \cdots + & 1t_{n-r} \end{array} \tag{1}$$

Letting $\mathbf{v}_i$ be the vector obtained by setting $t_i = 1$ and the rest of the parameters equal to 0, we may write the solution given by System (1) in vector form as

$$\mathbf{x} = t_1\mathbf{v}_1 + t_2\mathbf{v}_2 + \cdots + t_{n-r}\mathbf{v}_{n-r}. \tag{2}$$

Each vector $\mathbf{v}_i$ is in the solution space, and since any solution $\mathbf{x}$ can be written as a linear combination of $\mathcal{T} = \{\mathbf{v}_1, \ldots, \mathbf{v}_{n-r}\}$, $\mathcal{T}$ spans the solution space. Moreover, $\mathcal{T}$ is linearly independent. To see this, note that each vector $\mathbf{v}_i$ has 1 as its $(r + i)$th component and 0 for each of the other last $(n - r)$ components. Set $\mathbf{0}$ equal to a linear combination of the vectors in $\mathcal{T}$, say,

$$\mathbf{0} = c_1\mathbf{v}_1 + c_2\mathbf{v}_2 + \cdots + c_{n-r}\mathbf{v}_{n-r}.$$

Looking only at the $(r + i)$th component, we have

$$0 = c_1(0) + \cdots + c_{i-1}(0) + c_i(1) + c_{i+1}(0) + \cdots + c_{n-r}(0),$$

from which $0 = c_i$. Since this holds for each i, $\mathcal{T}$ is linearly independent. Thus, $\mathcal{T}$ spans the solution space and is linearly independent, so it is a basis. ■

We can generalize the procedure of Theorem 4 to provide us with a method for finding the dimension of, and a basis for, any subspace in which we can give the general form for its vectors. This technique is illustrated in the next example.

EXAMPLE 8

Find a basis for, and the dimension of, the subspace $\mathbf{S}$ of $\mathbf{R}^4$ of all vectors (a, b, c, d) for which $d = a + b + c$.

Solution The general form of a vector in $\mathbf{S}$ is $(a, b, c, a + b + c)$. Notice that three parameters, a, b, and c, appear in this general vector. Comparing this situation to that of Theorem 4 leads us to believe that the dimension of $\mathbf{S}$ is 3. Moreover, imitating the procedure associated with this theorem, we can obtain a basis $\{\mathbf{v}_1, \mathbf{v}_2, \mathbf{v}_3\}$ for $\mathbf{S}$ by setting

$$a = 1, b = 0, c = 0: \quad \mathbf{v}_1 = (1, 0, 0, 1);$$
$$a = 0, b = 1, c = 0: \quad \mathbf{v}_2 = (0, 1, 0, 1);$$
$$a = 0, b = 0, c = 1: \quad \mathbf{v}_3 = (0, 0, 1, 1).$$

It is easy to check that $\{\mathbf{v}_1, \mathbf{v}_2, \mathbf{v}_3\}$ *is* a basis for $\mathbf{S}$, so the dimension of $\mathbf{S}$ is indeed 3. ■

WARNING In Example 8, a "general vector" for the given subspace contained three parameters, and the dimension of that subspace was 3. Although it is often true, it is *not always* true that the dimension of a subspace $\mathbf{S}$ is equal to the number of parameters in a general vector for $\mathbf{S}$. For example, let $\mathbf{S}_0$ be the subspace of $\mathbf{R}^2$ consisting of all vectors of the form $(a + b, a + b)$. This general vector contains two parameters, a and b, but the dimension of $\mathbf{S}_0$ is 1; a basis for $\mathbf{S}_0$ is the set consisting of the single vector $(1, 1)$. Using the terminology introduced in the Spotlight at the beginning of this chapter, the vectors in $\mathbf{S}_0$ have only one "degree of freedom"—once one of their components is chosen, the other is determined. (Also, see Exercises 32 and 33.)

The following theorem gives some useful relationships among the notions of *linear dependence, span, basis,* and *dimension.* We only prove part (i), but the proofs of the other parts are just as easy.

!

THEOREM 5

Let $\mathbf{S}$ be a subspace of $\mathbf{R}^m$ with dimension equal to n and $\mathcal{T}$ be a subset of $\mathbf{S}$.

i. If the number of vectors in $\mathcal{T}$ is greater than n, then $\mathcal{T}$ is not linearly independent.

ii. If the number of vectors in $\mathcal{T}$ is less than n, then $\mathcal{T}$ does not span $\mathbf{S}$.

iii. If the number of vectors in $\mathcal{T}$ is equal to n, then $\mathcal{T}$ is linearly independent if and only if it spans $\mathbf{S}$.

Proof of (i) Suppose that the number of vectors in $\mathcal{T}$ is greater than n and $\mathcal{T}$ is linearly independent. By Theorem 2, $\mathcal{T}$ can be extended to a basis for $\mathbf{S}$. But this gives a basis for $\mathbf{S}$ with more than n vectors, which contradicts Theorem 3. Hence, $\mathcal{T}$ must be linearly dependent. ■

This theorem can sometimes be used to simplify the task of answering questions concerning linear dependence, span, or basis.

EXAMPLE 9

a. Do the vectors (0, 1, 2, 3), (1, 3, 2, 6), (−1, 4, 0, 1) span $\mathbf{R}^4$?

b. Show that the set {(0, 1, 2, 0), (0, 1, 0, 0), (0, 1, 1, 0)} is linearly dependent.

c. Show that the set {(1, 0, 0), (1, 2, 0), (1, 2, 3)} is a basis for $\mathbf{R}^3$.

Solution

a. Apply part (ii) of Theorem 5. We are given three vectors in $\mathbf{R}^4$, which has dimension 4 (Example 5). Consequently, the given vectors do not span $\mathbf{R}^4$.

b. The given vectors all lie in the subspace $\mathbf{S}$ of $\mathbf{R}^4$, consisting of all vectors with first and last components equal to zero. By Example 6, the dimension of $\mathbf{S}$ is 2. Consequently, by part (i) of Theorem 5, the given set is linearly dependent.

c. By part (iii) of Theorem 5, we need only show that the given set of three vectors is linearly independent (since the dimension of $\mathbf{R}^3$ is 3). Let A be the 3×3 matrix with columns formed by the given vectors. Since $\det A = 6 \neq 0$, the set is linearly independent by Theorem 3 of Section 5.1. ■

The technique in the solution for part (c) of Example 9 is important enough to state as a theorem.

THEOREM 6

Let $\mathcal{T} = \{\mathbf{v}_1, \mathbf{v}_2, \ldots, \mathbf{v}_m\}$ be a set of vectors in $\mathbf{R}^m$. Then $\mathcal{T}$ is a basis for $\mathbf{R}^m$ if and only if $\det A \neq 0$, where A is the $m \times m$ matrix with $\mathbf{a}^i = \mathbf{v}_i$.

Proof By Theorem 3 of Section 5.1, $\mathcal{T}$ is linearly independent if and only if det $A \neq 0$. But since $\mathbf{R}^m$ has dimension m, by part (iii) of Theorem 5, $\mathcal{T}$ is a basis for $\mathbf{R}^m$ if and only if it is linearly independent. Consequently, $\mathcal{T}$ a basis for $\mathbf{R}^m$ if and only if det $A \neq 0$. ■

EXAMPLE 10

Determine whether or not the set $\mathcal{T} = \{\mathbf{v}_1, \mathbf{v}_2, \mathbf{v}_3, \mathbf{v}_4\}$ is a basis for $\mathbf{R}^4$, where $\mathbf{v}_1 = (1, 0, 0, 0)$, $\mathbf{v}_2 = (-1, 1, 3, 2)$, $\mathbf{v}_3 = (1, 2, 3, 1)$, and $\mathbf{v}_4 = (1, 1, 2, 1)$.

Solution Here we have four vectors in $\mathbf{R}^4$, so we can apply Theorem 6. To do so, we construct the matrix A with $\mathbf{a}^i = \mathbf{v}_i$:

$$A = \begin{bmatrix} 1 & -1 & 1 & 1 \\ 0 & 1 & 2 & 1 \\ 0 & 3 & 3 & 2 \\ 0 & 2 & 1 & 1 \end{bmatrix}$$

Since $\det A = 0, \mathcal{T}$ is not a basis for $\mathbf{R}^4$. ■

5.3 Exercises

In Exercises 1–8 determine whether or not the given set forms a basis for the indicated subspace.

1. {(1, 2, 3), (−1, 0, 1), (0, 1, 2)} for $\mathbf{R}^3$

2. {(−1, 1, 2), (3, 3, 1), (1, 2, 2)} for $\mathbf{R}^3$

3. $\{(1, -1, 0), (0, 1, -1)\}$ for the subspace of $\mathbf{R}^3$ of all (x, y, z) such that $x + y + z = 0$

4. $\{(1, 1, 0), (1, 1, 1)\}$ for the subspace of $\mathbf{R}^3$ of all (x, y, z) such that $y = x + z$

5. $\{(1, 1, 0, 0), (0, 0, 1, 1)\}$ for the subspace of $\mathbf{R}^4$ of all vectors of the form $(a, a + b, b, b)$

6. $\{(1, 2, -1, 3), (0, 0, 0, 0)\}$ for the subspace of $\mathbf{R}^4$ of all vectors of the form $(a, b, a - b, a + b)$

7. $\{(1, -1), (-2, 3)\}$ for the subspace of $\mathbf{R}^2$ generated by the set $(1, 2)$ and $(2, 1)$

8. $\{(2, 0, 2), (1, 1, 1)\}$ for the subspace of $\mathbf{R}^3$ generated by $(1, 0, 1)$ and $(0, 1, 0)$

In Exercises 9–12 find a basis for the subspace generated by the given set of vectors.

9. $\{(1, -1, 1), (2, 0, 1), (1, 1, 0)\}$

10. $\{(1, 0, 1, 0), (0, -1, 1, 2), (2, -1, 3, 2), (1, 1, 1, 1)\}$

11. $\{(-1, 1, 2, -1), (1, 0, -1, 1), (-1, 2, 3, -1), (1, 1, 0, 1)\}$

12. $\{(1, -1, 0), (0, 1, 1), (1, 0, 2)\}$

In Exercises 13–16 extend the given set to form a basis for $\mathbf{R}^m$.

13. $\{(1, 1, 0), (1, -1, 1)\}$ for $\mathbf{R}^3$

14. $\{(-1, 1, 0, 0), (1, -1, 1, 0), (0, 0, 1, 2)\}$ for $\mathbf{R}^4$

15. $\{(1, 0, 0, 1), (0, 1, 1, 0), (1, 1, 1, 2)\}$ for $\mathbf{R}^4$

16. $\{(-1, -1, 2, 1), (2, 1, -1, -2), (0, -1, 4, 0)\}$ for $\mathbf{R}^4$

In Exercises 17–20 explain why the given statement is true "by inspection."

17. The set $\{(1, 0, 3), (-1, 1, 0), (1, 2, 4), (0, -1, -2)\}$ is linearly dependent.

18. The set $\{(1, -1, 2), (0, 1, 1)\}$ does not span $\mathbf{R}^3$.

19. If the set $\{\mathbf{v}_1, \mathbf{v}_2, \mathbf{v}_3, \mathbf{v}_4\}$ of vectors in $\mathbf{R}^4$ is linearly independent, then it spans $\mathbf{R}^4$.

20. The set $\{(0, 1, -1, 0), (0, -1, 2, 0)\}$ is linearly independent, so it spans the subspace of $\mathbf{R}^4$ of all vectors of the form $(0, a, b, 0)$.

In Exercises 21–28 find a basis for, and the dimension of, the indicated subspace.

21. The solution space of the homogeneous linear system

$$\begin{aligned} x_1 - 2x_2 + x_3 &= 0 \\ x_2 - x_3 + x_4 &= 0 \\ x_1 - x_2 + x_4 &= 0 \end{aligned}$$

22. The solution space of

$$\begin{aligned} x_1 - 3x_2 + x_3 - x_5 &= 0 \\ x_1 - 2x_2 + x_3 - x_4 &= 0 \\ x_1 - x_2 + x_3 - 2x_4 + x_5 &= 0 \end{aligned}$$

23. The subspace of $\mathbf{R}^3$ of all vectors of the form $(x, 2x, z)$

24. The subspace of $\mathbf{R}^3$ of all vectors of the form $(x + y, y, x)$

25. The subspace of $\mathbf{R}^4$ of all vectors of the form $(x, -y, x - 2y, 3y)$

26. The subspace of $\mathbf{R}^4$ of all vectors of the form $(0, x, y, x - y)$

27. The subspace of $\mathbf{R}^4$ generated by the set

$$\{(1, -1, 3, 1), (-2, 0, -2, 0), (0, -1, 2, 1)\}$$

28. The subspace of $\mathbf{R}^3$ generated by the set

$$\{(1, -1, 0), (0, 1, 1), (2, -1, 1), (1, 0, 1)\}$$

29. Let $\mathbf{v}$ be a nonzero vector in $\mathbf{R}^m$. Find the dimension of the subspace $\mathbf{S}$ of $\mathbf{R}^m$ consisting of all scalar multiples of $\mathbf{v}$. (Geometrically, $\mathbf{S}$ is the line through the origin containing $\mathbf{v}$.)

30. Let $\mathbf{v}_1$ and $\mathbf{v}_2$ be two noncollinear vectors in $\mathbf{R}^m$. Find the dimension of the subspace $\mathbf{S}$ of $\mathbf{R}^m$ generated by $\mathbf{v}_1$ and $\mathbf{v}_2$. (Geometrically, $\mathbf{S}$ is the plane containing the given vectors.)

31. Let $\{\mathbf{v}_1, \mathbf{v}_2, \ldots, \mathbf{v}_n\}$ be a basis for a subspace $\mathbf{S}$ of $\mathbf{R}^m$. Show that every vector $\mathbf{x}$ in $\mathbf{S}$ may be written as a *unique* linear combination of the vectors $\mathbf{v}_i$.

In Exercises 32 and 33 find the dimension of the given subspace by viewing dimension as the number of "degrees of freedom" allowed for the components of the vectors in that subspace. (See the Spotlight at the beginning of this chapter.)

SPOTLIGHT *32.* The subspace of $\mathbf{R}^3$ consisting of all vectors of the form

(a) (a, b, c) *(b)* $(a + b + c, a + b + c, 0)$

SPOTLIGHT *33.* The subspace of $\mathbf{R}^4$ consisting of all vectors of the form

(a) $(a, b, a + b, a - b)$ *(b)* $(a + b, c + d, 0, 0)$

SPOTLIGHT *34.* How many degrees of freedom does each of the following physical systems have; that is, what is the fewest number of coordinates needed to describe the positions of all the objects in the system? (See the Spotlight at the beginning of this chapter.)

(a) The system in Figure 5, which contains three "objects": the weights W_1, W_2, and W_3. As the top pulley turns, weight W_3 and the lower pulley move up or down. As the lower pulley turns, weights W_1 and W_2 move up and down.

FIGURE 5

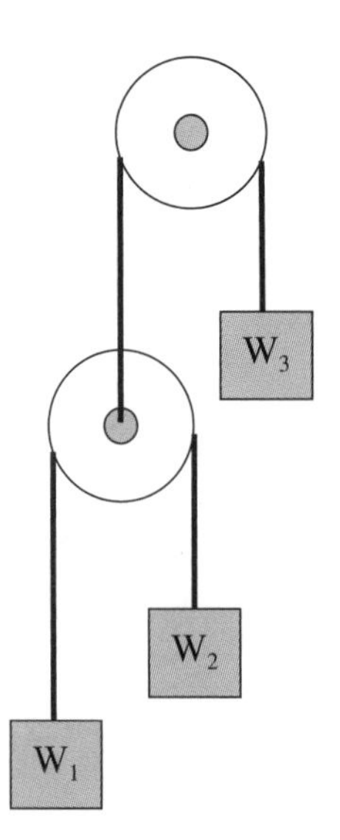

(b) The system in Figure 6, which contains a single object: a box sliding straight down an inclined plane as the plane itself moves along a horizontal track

FIGURE 6

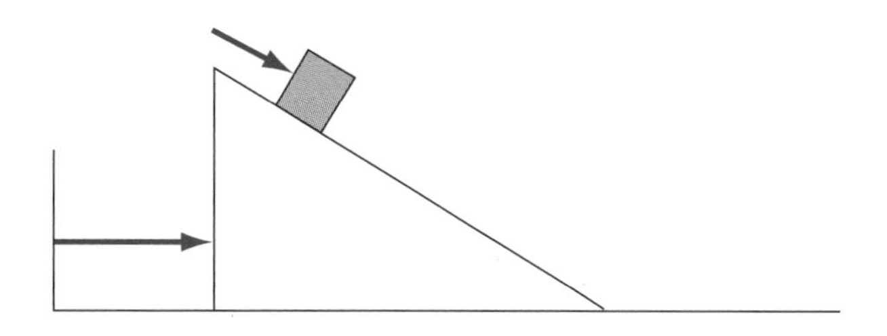

Computational Exercises

In Exercises 35–37 use a computer algebra system to find all values of the variables, if any, for which the given set is a basis for $\mathbf{R}^m$.

35. $\{(1, 2, 3), (1, 2, x), (1, x, x + y)\}$, $n = 3$

36. $\{(2, 1, -1, 1), (2, 2, 1, 1), (x, 2, y, 2), (1, x + y, -1, x - y)\}$, $n = 4$

37. $\{(2, 1, 1, 0), (1, 2, 0, 1), (x + 2, 1, y, 2), (1, x - y, 0, x + y)\}$, $n = 4$

5.4 Rank of a Matrix

In Section 3.3, the notion of the *rank* of a matrix was defined as the number of nonzero rows in its row-reduced echelon form. In this section, we introduce the notions of *column rank* and *row rank* of a matrix and show that these three types of rank are, in fact, one and the same.

DEFINITION

Let A be an $m \times n$ matrix. The subspace of $\mathbf{R}^m$ generated by the columns of the matrix, $\mathbf{a}^1, \mathbf{a}^2, \ldots, \mathbf{a}^n$, is called the **column space** of A and is denoted by $\mathrm{CS}(A)$. Its dimension is called the **column rank** of A.

EXAMPLE 1

Let A be given by

$$A = \begin{bmatrix} 1 & -1 & 3 & -1 & 2 \\ 2 & 2 & 1 & -1 & 1 \\ 1 & 0 & 1 & -15 & -1 \\ 1 & 0 & 2 & 4 & 2 \end{bmatrix}.$$

Describe the column space of A, and find a basis for it. Determine the column rank of A.

Solution The column space of A is the subspace of $\mathbf{R}^4$ generated by the vectors $\mathbf{v}_1 = (1, 2, 1, 1)$, $\mathbf{v}_2 = (-1, 2, 0, 0)$, $\mathbf{v}_3 = (3, 1, 1, 2)$, $\mathbf{v}_4 = (-1, -1, -15, 4)$, and $\mathbf{v}_5 = (2, 1, -1, 2)$. To find a basis for this subspace, we follow the procedure of Theorem 1 of Section 5.3 and transform A into the row-reduced echelon form

$$B = \begin{bmatrix} 1 & 0 & 0 & -34 & -4 \\ 0 & 1 & 0 & 24 & 3 \\ 0 & 0 & 1 & 19 & 3 \\ 0 & 0 & 0 & 0 & 0 \end{bmatrix}.$$

Since the vectors $\mathbf{v}_1$, $\mathbf{v}_2$, and $\mathbf{v}_3$ have been transformed into distinct elementary vectors, the set $\{\mathbf{v}_1, \mathbf{v}_2, \mathbf{v}_3\}$ is a basis for the column space of A. Moreover, the dimension (the number of vectors in a basis) of the latter is 3, and so the column rank of A is 3. ■

Note that, in Example 1, the column rank and the rank (the number of nonzero rows in the row-reduced echelon form) of the matrix are equal. The following theorem states that this is no coincidence.

THEOREM 1

The dimension of the column space of a matrix A is equal to the rank of A.

Proof Let B be the row-reduced echelon form of A. According to the procedure of Theorem 1 of Section 5.3, we need only show that the number of distinct elementary columns of B is equal to the number of nonzero rows of B. The last statement follows immediately from the definition of row-reduced echelon form. The number of nonzero rows is the same as the number of *leading ones*, and the number of leading ones is, in turn, the same as the number of distinct elementary columns. ■

By considering the rows of a matrix instead of its columns, we can develop a theorem similar to the one just given.

DEFINITION

Let A be an $m \times n$ matrix. The subspace of $\mathbf{R}^n$ generated by the rows of the matrix $\mathbf{a}_1, \mathbf{a}_2, \ldots, \mathbf{a}_m$ is called the **row space** of A, denoted by RS(A). Its dimension is called the **row rank** of A.

EXAMPLE 2

As in Example 1, let

$$A = \begin{bmatrix} 1 & -1 & 3 & -1 & 2 \\ 2 & 2 & 1 & -1 & 1 \\ 1 & 0 & 1 & -15 & -1 \\ 1 & 0 & 2 & 4 & 2 \end{bmatrix}.$$

Describe the row space of A, and find a basis for it. Determine the row rank of A.

Solution The row space of A is the subspace of $\mathbf{R}^5$ generated by the vectors $\mathbf{w}_1 = (1, -1, 3, -1, 2)$, $\mathbf{w}_2 = (2, 2, 1, -1, 1)$, $\mathbf{w}_3 = (1, 0, 1, -15, -1)$, and $\mathbf{w}_4 = (1, 0, 2, 4, 2)$. To find a basis for it, we again apply Theorem 1 of Section 5.3. We must first form the matrix whose ith column is $\mathbf{w}_i$. This matrix is simply the transpose of A,

$$A^T = \begin{bmatrix} 1 & 2 & 1 & 1 \\ -1 & 2 & 0 & 0 \\ 3 & 1 & 1 & 2 \\ -1 & -1 & -15 & 4 \\ 2 & 1 & -1 & 2 \end{bmatrix},$$

which has the row-reduced echelon form

$$B = \begin{bmatrix} 1 & 0 & 0 & \frac{2}{3} \\ 0 & 1 & 0 & \frac{1}{3} \\ 0 & 0 & 1 & -\frac{1}{3} \\ 0 & 0 & 0 & 0 \\ 0 & 0 & 0 & 0 \end{bmatrix}.$$

Since $\mathbf{w}_1$, $\mathbf{w}_2$, and $\mathbf{w}_3$ are transformed to $\mathbf{e}_1$, $\mathbf{e}_2$, and $\mathbf{e}_3$, the set $\{\mathbf{w}_1, \mathbf{w}_2, \mathbf{w}_3\}$ is a basis for the row space of A, and so the dimension of the latter is 3. Thus, the row rank of A is 3. ■

In Example 1, we found that the *column rank* of this matrix A is also 3. It is, in fact, always true that the row rank of a given matrix is the same as its column rank. As an intermediate step toward obtaining this result, we prove the following theorem.

THEOREM 2

Let A be an $m \times n$ matrix. If C is the matrix obtained from A by performing an elementary row operation on the latter, then the row space of A is the same as the row space of C.

Proof Let the rows of A be $\mathbf{a}_1, \mathbf{a}_2, \ldots, \mathbf{a}_m$ and those of C be $\mathbf{c}_1, \mathbf{c}_2, \ldots, \mathbf{c}_m$. Then the row space of A is the subspace $\mathbf{S}_1$ of $\mathbf{R}^n$ generated by $\mathcal{T}_1 = \{\mathbf{a}_1, \mathbf{a}_2, \ldots, \mathbf{a}_m\}$and the row space of C is the subspace $\mathbf{S}_2$ generated by $\mathcal{T}_2 = \{\mathbf{c}_1, \mathbf{c}_2, \ldots, \mathbf{c}_m\}$. We must show that under each of the three elementary row operations, $\mathbf{S}_1 = \mathbf{S}_2$.

i. Suppose that C is obtained from A by interchanging two rows. Then $\mathcal{T}_1 = \mathcal{T}_2$, and so $\mathbf{S}_1 = \mathbf{S}_2$.

ii. Suppose that C is obtained from A by multiplying row i by the nonzero scalar k. Then

$$\mathcal{T}_2 = \{\mathbf{a}_1, \ldots, \mathbf{a}_{i-1}, k\mathbf{a}_i, \mathbf{a}_{i+1}, \ldots, \mathbf{a}_m\}.$$

If $\mathbf{v}$ is in the row space of C, $\mathbf{S}_2$, then there exist scalars $c_1, c_2, \ldots, c_m$ such that

$$\begin{aligned}\mathbf{v} &= c_1\mathbf{a}_1 + \cdots + c_{i-1}\mathbf{a}_{i-1} + c_i(k\mathbf{a}_i) + c_{i+1}\mathbf{a}_{i+1} + \cdots + c_m\mathbf{a}_m \\ &= c_1\mathbf{a}_1 + \cdots + c_{i-1}\mathbf{a}_{i-1} + (c_ik)\mathbf{a}_i + c_{i+1}\mathbf{a}_{i+1} + \cdots + c_m\mathbf{a}_m,\end{aligned}$$

a linear combination of $\mathcal{T}_1$. Thus, $\mathbf{v}$ lies in the row space of A, $\mathbf{S}_1$. Similarly, we can show that if $\mathbf{w}$ is in $\mathbf{S}_1$, it must also be in $\mathbf{S}_2$. Hence, $\mathbf{S}_1 = \mathbf{S}_2$.

iii. Suppose that C is obtained from A by adding k times row j to row i. Then

$$\mathcal{T}_2 = \{\mathbf{a}_1, \ldots, \mathbf{a}_{i-1}, k\mathbf{a}_j + \mathbf{a}_i, \mathbf{a}_{i+1}, \ldots, \mathbf{a}_m\}.$$

By an argument similar to the one in part (ii), we can show that in this case as well, $\mathbf{S}_1 = \mathbf{S}_2$. ■

NOTE By applying Theorem 2 successively a finite number of times, we see that any finite sequence of row operations on a matrix A results in a matrix C that has the same row space as A. In other words, any two matrices that are *row-equivalent* have the same row space.

EXAMPLE 3

Show that the matrices

$$A = \begin{bmatrix} 5 & 3 & 4 \\ 2 & 2 & -4 \\ 3 & 2 & 1 \\ -1 & 2 & 1 \end{bmatrix} \quad \text{and} \quad C = \begin{bmatrix} 3 & 1 & 8 \\ 1 & 4 & -3 \\ -2 & -1 & -3 \\ 1 & 1 & -2 \end{bmatrix}$$

have the same row space, and find a basis for this subspace of $\mathbf{R}^3$.

Solution By performing the computations, we see that both A and C have the row-reduced echelon form

$$B = \begin{bmatrix} 1 & 0 & 0 \\ 0 & 1 & 0 \\ 0 & 0 & 1 \\ 0 & 0 & 0 \end{bmatrix}.$$

Thus, A and C are row-equivalent to the same matrix, so they are row-equivalent to each other and have the same row space.

This row space is also that of B, generated by $\{(1, 0, 0), (0, 1, 0), (0, 0, 1), (0, 0, 0)\}$. A basis for this subspace is clearly $\{(1, 0, 0), (0, 1, 0), (0, 0, 1)\}$. ■

NOTE Examples 2 and 3 demonstrate two essentially different methods of producing bases for the row space of a matrix. The first provides a basis that consists of a *subset* of the given rows. The second provides a basis that ordinarily is *not* a subset of the original rows. The elements of the latter, however, are in some sense simpler than those of the former.

We now give the main result of this section.

THEOREM 3

Let A be a matrix. The row rank, the column rank, and the rank of A are all equal (and denoted by rank A).

Proof By Theorem 1, the column rank of A is equal to the rank of A. Thus, it suffices to show that the row rank of A is equal to the rank of A.

Let B be the row-reduced echelon form of A. According to Theorem 2 and the definition of rank, we need only show that the nonzero rows of B form a basis for the row space of B. Since these nonzero rows span the row space, it is sufficient to prove that the nonzero rows of B form a linearly independent set. We leave the verification of this last statement for Exercise 19. ■

5.4 Exercises

In Exercises 1–6 verify that the row rank of each matrix is equal to the column rank by explicitly finding the dimensions of the row space and the column space.

1. $\begin{bmatrix} 1 & 2 & 0 \\ 0 & 0 & 1 \\ 0 & 0 & 0 \end{bmatrix}$

2. $\begin{bmatrix} 1 & 2 & 1 \\ 2 & 1 & -1 \end{bmatrix}$

3. $\begin{bmatrix} 1 & 0 & -1 \\ -1 & 0 & 1 \end{bmatrix}$

4. $\begin{bmatrix} 0 & 0 \\ -1 & 2 \\ 2 & -4 \end{bmatrix}$

5. $\begin{bmatrix} 1 & -1 & 3 \\ 0 & 1 & 1 \\ 1 & 1 & 0 \\ 2 & -1 & 1 \end{bmatrix}$

6. $\begin{bmatrix} 1 & -1 & 1 & 0 \\ 1 & 1 & 0 & 0 \\ 1 & 0 & 0 & 1 \end{bmatrix}$

7–12. Find a basis for the column space of each matrix in Exercises 1–6.

13–18. Find a basis for the row space of each matrix in Exercises 1–6.

19. Prove that the set of nonzero rows of the row-reduced echelon form of a matrix is linearly independent.

The method used in Example 2 to find a basis for the row space of matrix A has the advantage of introducing no new mathematical tools, simply finding a basis for the column space of A^T. An alternative approach parallels the development based on the idea of *column operations*. Exercises 20 through 30 explore this idea.

20. Define the three **elementary column operations** on a matrix (see Section 2.3).

21. Define the notion of a matrix being in **column-reduced echelon form.**

22. For a matrix A, prove that a column operation on A has exactly the same effect as the corresponding row operation on A^T followed by transposing the resulting matrix.

23. If A is a matrix and B is its column-reduced echelon form, prove that the rows of A that correspond to a distinct set of elementary rows of B form a basis for the row space of A.

24–29. Find a basis for the row space of each matrix in Exercises 1–6 using the method outlined in Exercise 23.

30. Recall that to solve the equation $A\mathbf{x} = \mathbf{b}$, we augment A by $\mathbf{b}$, obtaining $[A \mid \mathbf{b}]$, and row-reduce. We can use this technique to solve the equation $\mathbf{x}^T A = \mathbf{b}^T$ by solving $A^T\mathbf{x} = \mathbf{b}$ since $(\mathbf{x}^T A)^T = A^T\mathbf{x}$. Devise a technique (using column operations) to solve the problem $\mathbf{x}^T A = \mathbf{b}^T$ without using the transpose of A.

5.5 APPLICATION—Signal Compression

We have seen, in Example 1 of Section 5.3, that the set of elementary vectors $\mathcal{T} = \{\mathbf{e}_1, \mathbf{e}_2, \ldots, \mathbf{e}_m\}$ is a basis for $\mathbf{R}^m$, the standard basis for $\mathbf{R}^m$. It is such a natural and easy basis that you may wonder why you would want to work with any other basis of $\mathbf{R}^m$. But there is an application to image processing that relies on quite a different basis. We begin with a simple example that highlights how one basis might be better than another when we want to **compress** a given **signal** (or input) vector, that is, send the vector in a compact format, smaller than the original vector.

EXAMPLE 1

Consider $\mathbf{R}^4$ which consists of all 4-tuples. We define the vectors $\mathbf{v}_1 = (1, 1, 1, 1)$, $\mathbf{v}_2 = (1, 1, -1, -1)$, $\mathbf{v}_3 = (1, -1, 0, 0)$, and $\mathbf{v}_4 = (0, 0, 1, -1)$. We can check, via Theorem 6 of Section 5.3, that $B_2 = \{\mathbf{v}_1, \mathbf{v}_2, \mathbf{v}_3, \mathbf{v}_4\}$ is indeed a basis for $\mathbf{R}^4$ as

$$\det\begin{bmatrix} 1 & 1 & 1 & 0 \\ 1 & 1 & -1 & 0 \\ 1 & -1 & 0 & 1 \\ 1 & -1 & 0 & -1 \end{bmatrix} = 8.$$

It follows that every vector in $\mathbf{R}^4$ can be written (uniquely) as a linear combination of the vectors in B_2. So, in particular, the vector (14.5, 15.1, 3, 2.9) can be expressed as a linear combination of $\mathbf{v}_1 = (1, 1, 1, 1)$, $\mathbf{v}_2 = (1, 1, -1, -1)$, $\mathbf{v}_3 = (1, -1, 0, 0)$, and $\mathbf{v}_4 = (0, 0, 1, -1)$:

$$\begin{aligned}(14.5, 15.1, 3, 2.9) &= a(1, 1, 1, 1) + b(1, 1, -1, -1) + c(1, -1, 0, 0) + d(0, 0, 1, -1) \\ &= (a + b + c, a + b - c, a - b + d, a - b - d)\end{aligned}$$

Solving the linear system,

$$\begin{aligned} a + b + c &= 14.5 \\ a + b - c &= 15.1 \\ a - b + d &= 3 \\ a - b - d &= 2.9, \end{aligned}$$

we find that $a = 8.875$, $b = 5.925$, $c = -0.3$, and $d = 0.05$. Thus,

$$\begin{aligned}(14.5, 15.1, 3, 2.9) &= 8.875(1, 1, 1, 1) + 5.925(1, 1, -1, -1) \\ &\quad - 0.3(1, -1, 0, 0) + 0.05(0, 0, 1, -1).\end{aligned}$$

■

Note that the vector (14.5, 15.1, 3, 2.9) is rather unusual in that it consists of blocks of numbers that are roughly equal (14.5, 15.1 and 3, 2.9) with rather large differences between the blocks. We observe that, while none of the numbers 14.5, 15.1, 3, and 2.9 are close to zero, two of the coefficients in terms of the basis B_2 are close to zero, namely -0.3 and 0.05. What do we lose when we drop these coefficients down to 0? We change the vector (14.5, 15.1, 3, 2.9) into the vector

$$\begin{aligned} &8.875(1, 1, 1, 1) + 5.925(1, 1, -1, -1) + 0(1, -1, 0, 0) + 0(0, 0, 1, -1) \\ &= 8.875(1, 1, 1, 1) + 5.925(1, 1, -1, -1) \\ &= (14.8, 14.8, 2.95, 2.95), \end{aligned}$$

which is close to the original vector. On the other hand, if we keep track of the vector by its coefficients as a linear combination of the vectors in B_2, we have two coefficients that are zero. *Zero is always easier to maintain and transmit.*

If we choose a basis so that we can round down many consecutive coefficients to zero, then there is a savings to be had. We can compress the original "signal" vector by compressing the zeros—we can send the number of zeros rather than a description for each. For example, if every number we transmit uses 8 bits (8 binary digits), then a string of say 32 zeros takes, without compression, $8 \times 32 = 256$ bits. On the other hand, we could simply transmit the number 32 (using 5 bits) with the understanding that it refers to the number of zeros. This compresses the string by a factor of 8.

EXAMPLE 2

In this example, our input is a longer string of numbers:

24.18, 25.07, 23.70, 24.65, 25.82, 25.78, 5.29, 5.50,

0.06, 0.34, 1.13, −0.95, 99.50, 98.98, 99.21, 98.85

We cut this string up into 4-tuples, that is, into vectors in $\mathbf{R}^4$:

$$\begin{aligned} \mathbf{u}_1 &= (24.18, 25.07, 23.70, 24.65) \\ \mathbf{u}_2 &= (25.82, 25.78, 5.29, 5.50) \\ \mathbf{u}_3 &= (0.06, 0.34, 1.13, -0.95) \\ \mathbf{u}_4 &= (99.50, 98.98, 99.21, 98.85) \end{aligned}$$

For each of these vectors, in turn, we express (as in Example 1) the vector as a linear combination of the basis vectors $\mathbf{v}_1 = (1, 1, 1, 1)$, $\mathbf{v}_2 = (1, 1, -1, -1)$, $\mathbf{v}_3 = (1, -1, 0, 0)$, and $\mathbf{v}_4 = (0, 0, 1, -1)$ (rounding all the coefficients to two decimal places):

$$\begin{aligned} \mathbf{u}_1 &= 24.40\mathbf{v}_1 + 0.23\mathbf{v}_2 - 0.45\mathbf{v}_3 - 0.48\mathbf{v}_4 \\ \mathbf{u}_2 &= 15.60\mathbf{v}_1 + 10.20\mathbf{v}_2 + 0.02\mathbf{v}_3 - 0.11\mathbf{v}_4 \\ \mathbf{u}_3 &= 0.15\mathbf{v}_1 + 0.06\mathbf{v}_2 - 0.14\mathbf{v}_3 + 1.04\mathbf{v}_4 \\ \mathbf{u}_4 &= 99.14\mathbf{v}_1 + 0.11\mathbf{v}_2 + 0.26\mathbf{v}_3 + 0.18\mathbf{v}_4 \end{aligned}$$

Putting these coefficients together, we derive the following string:

24.40, 0.23, −0.45, −0.48, 15.60, 10.20, 0.02, −0.11,

0.15, 0.06, −0.14, 1.04, 99.14, 0.11, 0.26, 0.18

We now need to choose a rule for deciding which numbers can be rounded down to 0. Let us choose to round a number down to 0 if it is strictly between -0.5 and 0.5. Using this rule, the output string now becomes

24.40, 0, 0, 0, 15.60, 10.20, 0, 0, 0, 0, 0, 1.04, 99.14, 0, 0, 0.

This string has a number of substrings of consecutive zeros, and can be compressed.

Finally, from the output string, we can recover our *approximation* to the original input signal, by piecing together our approximations to each of $\mathbf{u}_1$, $\mathbf{u}_2$, $\mathbf{u}_3$, and $\mathbf{u}_4$:

$$\begin{aligned} \mathbf{u}_1 &\approx 24.40\mathbf{v}_1 + 0\mathbf{v}_2 + 0\mathbf{v}_3 + 0\mathbf{v}_4 = (24.40, 24.40, 24.40, 24.40) \\ \mathbf{u}_2 &\approx 15.60\mathbf{v}_1 + 10.20\mathbf{v}_2 + 0\mathbf{v}_3 + 0\mathbf{v}_4 = (25.80, 25.80, 5.40, 5.40) \end{aligned}$$

$$\mathbf{u}_3 \approx 0\mathbf{v}_1 + 0\mathbf{v}_2 + 0\mathbf{v}_3 + 1.04\mathbf{v}_4 = (0, 0, 1.04, 1.04)$$
$$\mathbf{u}_4 \approx 99.14\mathbf{v}_1 + 0\mathbf{v}_2 + 0\mathbf{v}_3 + 0\mathbf{v}_4 = (99.14, 99.14, 99.14, 99.14)$$

These yield the string

24.40, 24.40, 24.40, 24.40, 25.80, 25.80, 5.40, 5.40,
0, 0, 1.04, 1.04, 99.14, 99.14, 99.14, 99.14,

which is fairly close to the original input string. ■

Greyscale images subdivide the image into many small rectangles, called pixels, each having an associated number between 0 and 255, indicating the intensity of black at that location. This sequence of numbers can be viewed as a large vector, which can be cut up into a number of smaller vectors. Now the numbers associated with neighbouring pixels in an image are often related to one another, in that it is common to have large areas with roughly the same level of grey, while between these areas are rather large changes in the level of grey (corresponding to edges between regions). Also, bases such as B_2 (known as *wavelets*) well reflect such sequences, in that when we express such vectors in terms of the basis, typically there will be subsequences of coefficients that are zero or close to zero. By setting a *threshold* below which numbers are rounded to zero, we get a vector close to the original vector, but with many subsequences of zeros that can be compressed. This compression is known as *lossy* as some information is lost in the process (although we gain in terms of being able to compress the data).

With greater and greater amounts of data being transferred (whether graphics, video, or other data files), compression continues to play a vital role in communication.

5.5 Exercises

In Exercises 1 and 2 show that the given set is a basis for the indicated space.

1. $B_1 = \{(1,1),(1,-1)\}$ for $\mathbf{R}^2$

2. $B_3 = \{(1, 1, 1, 1, 1, 1, 1, 1), (1, 1, 1, 1, -1, -1, -1, -1), (1, 1, -1, -1, 0, 0, 0, 0), (0, 0, 0, 0, 1, 1, -1, -1), (1, -1, 0, 0, 0, 0, 0, 0), (0, 0, 1, -1, 0, 0, 0, 0), (0, 0, 0, 0, 1, -1, 0, 0), (0, 0, 0, 0, 0, 0, 1, -1)\}$ for $\mathbf{R}^8$

In Exercises 3 and 4 express the given vector v as a linear combination of the basis.

3. $\mathbf{v} = (-7, -6.5, 2.7, 3)$ in terms of B_2

4. $\mathbf{v} = (3.1, 3.2, 2.9, 3, -1, -0.9, 1, 1.1)$ in terms of B_3

In Exercises 5 through 7 use the given threshold rule to round the coefficients and produce the output vector.

5. $\mathbf{v} = (-7, -6, -6.5, 3)$ in terms of B_2
Use the threshold rule that rounds each number to its nearest integer.

6. $\mathbf{v} = (3.1, 3.2, 2.9, 3, -1, -0.9, 1, 1.1)$ in terms of B_4
Use the threshold rule that rounds each number to one decimal place.

7. $\mathbf{v} = (10.5, 11, 10.7, 10.6, 10.5, 10.3, 4.1, 3.9, 4, 4.1, -1.1, -1, -1, -1.2, 5, 5.1, 4.9, 4.8, 3, 3.1)$
Cut the vector into five vectors, each of length 4, and then process each vector with basis B_2. Finally, use the threshold rule that rounds each coefficient to the nearest integer.

CHAPTER SUMMARY

Linear Independence and Dependence in $\mathbf{R}^m$

(a) DEFINITION: A set of vectors, $\{\mathbf{v}_1, \mathbf{v}_2, \ldots, \mathbf{v}_n\}$, in $\mathbf{R}^m$ is *linearly independent* if and only if

- The vector equation $c_1\mathbf{v}_1 + c_2\mathbf{v}_2 + \ldots + c_n\mathbf{v}_n = \mathbf{0}$ has only the trivial solution with each $c_i = 0$;

or

- None of its vectors can be written as a linear combination of the others.

The set is *linearly dependent* if and only if it is not linearly independent.

(b) We can determine if a set of vectors, $\{\mathbf{v}_1, \mathbf{v}_2, \ldots, \mathbf{v}_n\}$, in $\mathbf{R}^m$ is linearly independent by constructing the $m \times n$ matrix A with ith column $\mathbf{a}^i = \mathbf{v}_i$ and transforming it to row-reduced echelon form. The set is linearly independent if and only if the columns of the row-reduced echelon form are the distinct elementary vectors, $\mathbf{e}_1, \mathbf{e}_2, \ldots, \mathbf{e}_n$.

Subspaces of $\mathbf{R}^m$

(a) DEFINITION: A nonempty subset $\mathbf{S}$ of $\mathbf{R}^m$ is a *subspace* if both of the following conditions hold:

i. Whenever two vectors are in $\mathbf{S}$, their sum is in $\mathbf{S}$;

and

ii. Whenever a vector is in $\mathbf{S}$, all scalar multiples of this vector are in $\mathbf{S}$.

(b) The set of all solutions of the homogeneous linear system $A\mathbf{x} = \mathbf{0}$, where A is an $m \times n$ matrix, is a subspace of $\mathbf{R}^m$ called the *solution space* of the system.

Span of a Set of Vectors in $\mathbf{R}^m$

(a) DEFINITION: A subset of vectors, $\mathcal{T}$, in a subspace $\mathbf{S}$ *spans* $\mathbf{S}$ if every vector in $\mathbf{S}$ can be written as a linear combination of the vectors in $\mathcal{T}$.

(b) DEFINITION: Let $\mathcal{T}$ be a set of vectors in $\mathbf{R}^m$. The set of all linear combinations of the vectors in $\mathcal{T}$ is a subspace called the *subspace generated* by $\mathcal{T}$.

Basis for a Subspace of $\mathbf{R}^m$

(a) DEFINITION: A subset $\mathcal{T}$ of a subspace $\mathbf{S}$ is a *basis* for $\mathbf{S}$ if $\mathcal{T}$ is linearly independent and $\mathcal{T}$ spans $\mathbf{S}$.

(b) The *standard basis* for $\mathbf{R}^m$ is $\{\mathbf{e}_1, \mathbf{e}_2, \ldots, \mathbf{e}_m\}$.

(c) If a set $\mathcal{T}$ spans a subspace $\mathbf{S}$, then there is a subset of $\mathcal{T}$ that is a basis for $\mathbf{S}$.

(d) If a set of vectors $\mathcal{T}$ in a subspace $\mathbf{S}$ is linearly independent, then there is a set containing $\mathcal{T}$ that is a basis for $\mathbf{S}$.

(e) Every basis for a given subspace $\mathbf{S}$ contains the same number of vectors.

Dimension of a Subspace of $\mathbf{R}^m$

(a) DEFINITION: The *dimension* of a subspace $\mathbf{S}$ is the number of vectors in a basis for $\mathbf{S}$.

(b) Let **S** be a subspace with dimension n and $\mathcal{T}$ be a subset of **S** containing s vectors.

i. If $s > n$, then $\mathcal{T}$ is linearly dependent;

ii. If $s < n$, then $\mathcal{T}$ does not span **S**;

iii. If $s = n$, then $\mathcal{T}$ is linearly independent if and only if it spans **S**.

Column Space, Row Space, and Rank of a Matrix

(a) DEFINITION: The subspace generated by the column vectors of a matrix A is called the *column space* of A. Its dimension is called the *column rank* of A.

(b) DEFINITION: The subspace generated by the row vectors of a matrix A is called the *row space* of A. Its dimension is called the *row rank* of A.

(c) For a given matrix A, its column rank, row rank, and rank (the number of nonzero rows in its row-reduced echelon form) are all equal.

KEYWORDS

linear combination
elementary vector
solution space
basis
dimension
row space
signal
linearly independent
subspace
spanning set
standard basis
column space
rank
compression

SELF-TEST

1. Determine whether the set of vectors

$$\{(1, 0, 1, 0), (0, 1, -1, 1), (2, 1, 1, 1)\}$$

is linearly independent or linearly dependent. If it is linearly dependent, express one of the vectors as a linear combination of the others. [Section 5.1]

2. Determine whether or not the set of vectors

$$\{(1, 1, 1), (0, 1, 2), (1, 2, 3)\}$$

spans $\mathbf{R}^3$. [Section 5.2]

3. Show that the set of all vectors of the form $(a + b, a - b, c, 0)$ is a subspace of $\mathbf{R}^4$. [Section 5.2]

4. Determine whether or not the set of all vectors with first component equal to 1 is a subspace of $\mathbf{R}^m$. [Section 5.2]

5. Show that the set

$$\{(1, 1 ,0, 0), (1, -1, 0, 0), (0, 0, 1, 0)\}$$

is a basis for the subspace of Exercise 3. [Section 5.3]

6. Use the result of Exercise 5 to determine the dimension of the subspace of Exercise 3. [Section 5.3]

7. Determine the dimension of, and a basis for, the solution space of the homogeneous linear system $A\mathbf{x} = \mathbf{0}$, where

$$A = \begin{bmatrix} 1 & 0 & 1 & 2 \\ 0 & 1 & 1 & 1 \end{bmatrix}.$$ [Section 5.3]

8. Let $\mathbf{S}$ be a subspace of $\mathbf{R}^5$ of dimension 3 and $\mathcal{T}$ be a subset of $\mathbf{S}$. Are the following statements true or false?

   (a) If $\mathcal{T}$ contains two vectors, then it cannot span $\mathbf{S}$.
   (b) If $\mathcal{T}$ contains three vectors, then it must be a basis for $\mathbf{S}$.
   (c) If $\mathcal{T}$ contains four vectors, then it must be linearly dependent. [Section 5.3]

9. Let

$$B = \begin{bmatrix} 1 & 2 & 3 & 4 \\ 0 & 1 & 2 & 3 \\ 1 & 1 & 1 & 1 \end{bmatrix}.$$

   Determine a basis for, and the dimension of, the following subspaces.

   (a) The column space of B *(b)* The row space of B [Section 5.4]

10. If $\mathcal{S} = \{\mathbf{u}, \mathbf{v}, \mathbf{w}\}$ is a basis for $\mathbf{R}^3$, show that $\mathcal{T} = \{\mathbf{u}, \mathbf{v}\}$ is a basis for the subspace $\mathbf{S}$ generated by $\mathbf{u}$ and $\mathbf{v}$. [Section 5.3]

REVIEW EXERCISES

In Exercises 1–4 determine whether the given set of vectors is linearly independent or dependent. If it is dependent, express one of the vectors as a linear combination of the others.

1. $\{(1, 0, 0), (1, 2, 0), (1, 2, 3)\}$
2. $\{(1, 2, -1), (0, 1, -1), (2, 1, 1)\}$
3. $\{(1, 0, 1, 0), (-1, 1, 2, -1), (3, -1, 0, 1)\}$
4. $\{(1, 2, 3, 4, 5), (5, 4, 3, 2, 1)\}$

In Exercises 5–8 determine whether or not the given set of vectors is a subspace of $\mathbf{R}^m$.

5. The set of all vectors in $\mathbf{R}^4$ of the form $(0, x, -2x, x + 2y)$
6. The set of all vectors in $\mathbf{R}^4$ with the first two components zero and the third component equal to the square of the fourth
7. The set of all vectors $\mathbf{x}$ in $\mathbf{R}^3$ for which $\mathbf{x} + (0, 0, 1)$ is not $\mathbf{0}$
8. The set of all vectors $\mathbf{x}$ in $\mathbf{R}^3$ such that $\mathbf{x} \cdot (0, 0, 1) = 0$

In Exercises 9 and 10 determine if the given vector x spans the indicated subspace S.

9. $\mathbf{x} = (-1, 1, 1)$, and $\mathbf{S}$ is the solution space of $A\mathbf{x} = \mathbf{0}$, where

$$A = \begin{bmatrix} 1 & 0 & 1 \\ 0 & 1 & -1 \end{bmatrix}.$$

10. $\mathbf{x} = (0, 1, 0)$, and $\mathbf{S}$ is the row space of the matrix

$$\begin{bmatrix} 1 & 0 & -1 \\ -1 & 0 & 1 \end{bmatrix}.$$

In Exercises 11–14 determine if the given set is a basis for the indicated subspace.

11. $\{(1, 0, 0), (1, 2, 0), (1, 2, 3)\}$ for $\mathbf{R}^3$
12. $\{(1, 0, 0, 0), (1, 1, 0, 0), (1, 1, 1, 0), (1, 1, 1, 1)\}$ for $\mathbf{R}^4$
13. $\{(1, 2, 0, 0), (1, 0, 2, 0)\}$ for the subspace of $\mathbf{R}^4$ of all vectors of the form $(x, x + y, x - y, 0)$
14. $\{(1, 0, 1, 0), (0, 1, 2, 1)\}$ for the subspace of $\mathbf{R}^4$ of all vectors of the form $(x, y, x + y + z, y - z)$

In Exercises 15–21 find a basis for, as well as the dimension of, each of the given subspaces.

15. The subspace of $\mathbf{R}^4$ of all vectors with the second component equal to the sum of the other components

16. The subspace of $\mathbf{R}^5$ of all vectors with the first and second components zero and the last component equal to the sum of the first four

17. The subspace of $\mathbf{R}^3$ consisting of all vectors $\mathbf{x}$ such that $\mathbf{x} \bullet (1, 2, 3) = 0$

18. The solution space of $A\mathbf{x} = \mathbf{0}$, where

$$A = \begin{bmatrix} 1 & 2 & 3 & 4 \\ 0 & 1 & 2 & 3 \end{bmatrix}$$

19. The row space of the matrix

$$B = \begin{bmatrix} 1 & 1 & 1 & 1 & 1 \\ 0 & 1 & 1 & 1 & 1 \\ 0 & 0 & 1 & 1 & 1 \end{bmatrix}$$

20. The column space of matrix B of Exercise 19

21. The subspace of $\mathbf{R}^4$ generated by the vectors $(1, -1, 2, 0)$, $(-1, -1, 4, -2)$, $(2, -1, 1, 1)$, and $(2, -3, 7, -1)$

22. Find a basis for $\mathbf{R}^3$ that contains the vectors $(1, 2, 3)$ and $(3, 2, 1)$.

In Exercises 23–28 find a basis for the row space, a basis for the column space, and the rank of the matrix.

23. $\begin{bmatrix} -1 & 1 \\ 2 & -2 \end{bmatrix}$

24. $\begin{bmatrix} -1 & 1 \\ 2 & 2 \end{bmatrix}$

25. $\begin{bmatrix} 1 & 0 & 1 \\ -2 & 0 & -2 \\ 0 & 1 & 3 \\ -1 & 1 & 2 \end{bmatrix}$

26. $\begin{bmatrix} 1 & -2 & 0 & -1 \\ 0 & 0 & 1 & 1 \\ 1 & -2 & 3 & 2 \end{bmatrix}$

27. $\begin{bmatrix} 1 & -1 & 1 & 1 \\ 0 & 1 & 2 & 0 \\ 3 & -2 & 5 & 3 \\ 0 & 0 & 1 & 0 \end{bmatrix}$

28. $\begin{bmatrix} 1 & 0 & 0 & 1 & 2 \\ 2 & -1 & 1 & 0 & 1 \\ 4 & 1 & 1 & -1 & -3 \\ 3 & -1 & 2 & 1 & 0 \\ 0 & -1 & 1 & -2 & 4 \end{bmatrix}$

CHALLENGE YOURSELF

29. Suppose that A is an $m \times n$ matrix and B is a $p \times m$ matrix. Prove that $\text{rank}(BA) \leq \text{rank}(A)$ and $\text{rank}(BA) \leq \text{rank}(B)$.

30. Let n, m, r, and λ be positive integers with $r > \lambda$. A *balanced incomplete block design* with parameters n, m, r, and λ (or an (n, m, r, λ)*-design*) is a collection of m subsets of the set $V_n = \{1, 2, \ldots, n\}$ such that each element of V_n is in r of the subsets and each pair of distinct elements of V_n is in λ of the subsets. For example, the collection

$$\{\{0, 1, 2\}, \{0, 3, 6\}, \{0, 4, 5\}, \{1, 4, 6\}, \{1, 3, 5\}, \{2, 5, 6\}, \{2, 3, 4\}\}$$

is a $(7, 7, 3, 1)$-design.

(a) For an (n, m, r, λ)-design, list the n elements for V_n and the collection of m subsets in some order. Form the $n \times m$ matrix A whose (i, j)th entry is 1 if element i is in the jth set and 0 otherwise. Show that the rank of matrix AA^T is n. (*Hint*: Use Exercise 29 from the Review Exercises for Chapter 4.)

(b) Conclude that for any (n, m, r, λ)-design, $m \geq n$.

31. What is the dimension of the subspace of $\mathbf{R}^5$ consisting of all vectors $\mathbf{v}$ in $\mathbf{R}^5$ such that all three of the following conditions hold?

i. $\mathbf{v} \cdot (1, 1, 1, 0, 0) = 0$
ii. $\mathbf{v} \cdot (1, 0, 0, -1, 1) = 0$
iii. $\mathbf{v} \cdot (-1, 2, 2, 3, -3) = 0$

32. Suppose that $\{\mathbf{v}_1, \mathbf{v}_2, \ldots, \mathbf{v}_k\}$ is a basis for a subspace $\mathbf{S}$ of $\mathbf{R}^m$. Prove that $\{\mathbf{v}_1, \mathbf{v}_1 + \mathbf{v}_2, \ldots, \mathbf{v}_1 + \mathbf{v}_2 + \ldots + \mathbf{v}_k\}$ is also a basis for $\mathbf{S}$.

33. Suppose that A is a matrix.

(a) Prove that if A is not square, then either the rows or the columns are linearly dependent.
(b) Prove that if A is square, then the rows are linearly independent if and only if the columns are linearly independent.

SPOTLIGHT

ASSOCIATED PRESS

As you study vector spaces, it will probably strike you that they are very abstract. Vector spaces have their practical side, too, with many concrete applications. One of the most interesting of these involves music.

All musical sound waves are modelled by linear combinations of sines and cosines, and are located in a vector space of such functions. The individual trigonometric terms encode both the frequency of a component pure tone together with its loudness. Fourier analysis breaks down the sound wave into its component parts. It is utterly amazing that each of us can inherently carry out the calculation, whether we know linear algebra or not!

The sampled data from a piece of music on a CD is really a large string of equally spaced data points from the sound wave. A mathematical process called a *Fourier transform* can be used to discover the original frequencies that are present. Computer programs are indispensable in carrying out this process. Fourier transforms can therefore be very useful in trying to uncover musical mysteries. In Section 6.6, we'll apply such a process, along with some deductive reasoning, to one of the most famous chords in rock 'n' roll history.

We'll also touch upon an application of vector spaces to graphics. Objects such as clouds, lightning, mountains, trees, and coastlines are all poorly described by the usual building blocks of geometry—lines, points, polygons, and circles. Landforms differ from these mathematical figures in that they are *self-similar*—part of a landform looks very much like the whole landform, except for the fact that it is smaller.

Fractals are mathematical objects that have self-similarity. One way to construct fractals involves taking a certain collection of line segments, called a *seed*, and repeatedly replacing each of the line segments by an appropriate copy of the seed. Vectors are an ideal language to describe such a procedure, and notions of the projections onto and perpendicular to given vectors, introduced in Section 6.4, play a central role.

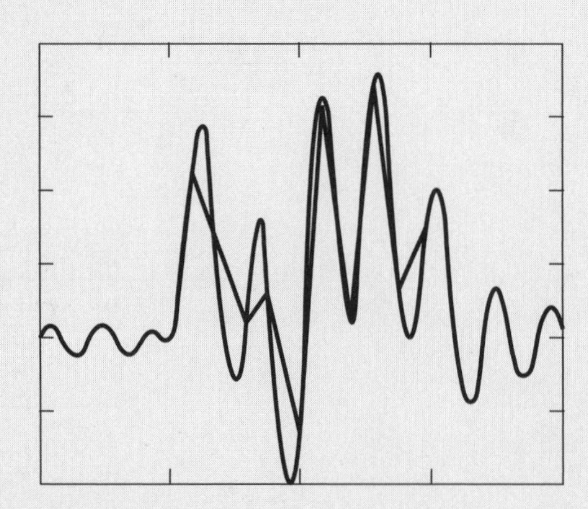

CHAPTER 6

Vector Spaces

OBJECTIVES

You will learn to:

- Define vector space and subspace of a vector space.
- Determine whether or not a given set of vectors in a vector space is linearly dependent and, if it is, express one of the vectors as a linear combination of the others.
- Determine if a given set of vectors spans a vector space.
- Determine if a given set of vectors is a basis for a vector space.
- Determine the dimension of a given vector space.
- Find the coordinate vector of a given vector relative to a given basis.
- Define an inner product space.
- Determine whether or not a given set of vectors is orthogonal (or orthonormal).
- Use the Gram–Schmidt process to construct an orthonormal basis for a given inner product space.

In the preceding chapters, we developed the properties of special vectors: those in Euclidean m-space, $\mathbf{R}^m$. Here we study more general collections of vectors—sets, with arithmetic properties similar to $\mathbf{R}^m$, called *vector spaces.* By collecting these diverse sets under one name, we can investigate their common properties all at once.

6.1 Vector Spaces and Subspaces

A *vector space* is intuitively a set that, together with operations of addition and multiplication by scalars, behaves in essentially the same manner as $\mathbf{R}^m$. The following definition makes this precise.

The general definition of a vector space was formulated by Giuseppe Peano (1858–1932) in 1888. He was born in Italy and was a prolific researcher in a number of areas of mathematics, including logic and set theory. He is famous for his construction of a curve that could fill a two-dimensional space, and this became one of the first examples of a fractal (see Section 6.8).

Courtesy of MacTutor, St. Andrews University

DEFINITION

A set **V**, together with operations called *addition* and *multiplication by scalars* (or *scalar multiplication*), is a **vector space** if the following properties are satisfied for every choice of elements **u**, **v**, and **w** in **V** and scalars c and d in **R**. The elements of **V** are called **vectors**.

Vector Space Axioms

a. $\mathbf{u} + \mathbf{v}$ is in **V**. (**V** is closed under addition.)

b. $c\mathbf{u}$ is in **V**. (**V** is closed under scalar multiplication.)

c. $\mathbf{u} + \mathbf{v} = \mathbf{v} + \mathbf{u}$ (Addition is commutative.)

d. $(\mathbf{u} + \mathbf{v}) + \mathbf{w} = \mathbf{u} + (\mathbf{v} + \mathbf{w})$ (Addition is associative.)

e. There is an element, denoted by **0**, in **V** such that $\mathbf{u} + \mathbf{0} = \mathbf{u}$ for all **u** in **V**. (**0** is the **zero** of **V**.)

f. For each **u** in **V**, there is an element, denoted by $-\mathbf{u}$, such that $\mathbf{u} + (-\mathbf{u}) = \mathbf{0}$. ($-\mathbf{u}$ is the **negative** of **u**.)

g. $(cd)\mathbf{u} = c(d\mathbf{u})$

h. $(c + d)\mathbf{u} = c\mathbf{u} + d\mathbf{u}$ (Distributive laws)

i. $c(\mathbf{u} + \mathbf{v}) = c\mathbf{u} + c\mathbf{v}$

j. $1\mathbf{u} = \mathbf{u}$

NOTE Since we have restricted the scalars to be the set of real numbers, **V** is called a vector space *over the real numbers*. If the scalars were allowed to be complex numbers, the same definition would apply, and the resulting structure would be called a vector space *over the complex numbers*. In Section 8.4, we study such vector spaces.

EXAMPLE 1

The set $\mathbf{R}^m$ of all m-vectors, with the usual operations of addition and multiplication by scalars, is a vector space.

This statement is a consequence of Theorem 2 of Section 2.1. ■

EXAMPLE 2

The set $\mathbf{M}^{m,n}$ of all $m \times n$ matrices, with the usual operations of addition and scalar multiplication, is a vector space.

This statement is a consequence of Theorem 1 of Section 3.1. ■

EXAMPLE 3

The set $\mathbf{R}^+$ of positive real numbers is *not* a vector space under the usual operations of addition and multiplication, since it is not closed under scalar multiplication. For example, 2 is in $\mathbf{R}^+$ and -1 is a real number, but $(-1)(2)$ is not in $\mathbf{R}^+$. Thus, property (b) is not satisfied. (Neither is property (e), but all we need to show is that *one* property fails to hold.) ■

EXAMPLE 4

The set **C** of all complex numbers, with the usual operations of addition of complex numbers and multiplication by real numbers, is a vector space (over the real numbers).

Solution Recall that **C** is the set of all elements of the form $z = a + bi$, where a and b are real numbers. Two complex numbers, $a + bi$ and $c + di$, are equal if and only if $a = c$ and $b = d$. Addition is defined by

$$(a + bi) + (c + di) = (a + c) + (b + d)i$$

and scalar multiplication by $k(a + bi) = (ka) + (kb)i$. Since each of these is an element of **C**, properties (a) and (b) of the definition of vector space are satisfied.

c. For $z_1 = a_1 + b_1 i$ and $z_2 = a_2 + b_2 i$,

$$\begin{aligned} z_1 + z_2 &= (a_1 + b_1 i) + (a_2 + b_2 i) = (a_1 + a_2) + (b_1 + b_2)i \\ &= (a_2 + a_1) + (b_2 + b_1)i = (a_2 + b_2 i) + (a_1 + b_1 i) \\ &= z_2 + z_1. \end{aligned}$$

d. For $z_1 = a_1 + b_1 i$, $z_2 = a_2 + b_2 i$, and $z_3 = a_3 + b_3 i$,

$$\begin{aligned} (z_1 + z_2) + z_3 &= [(a_1 + b_1 i) + (a_2 + b_2 i)] + (a_3 + b_3 i) \\ &= [(a_1 + a_2) + (b_1 + b_2)i] + (a_3 + b_3 i) \\ &= [(a_1 + a_2) + a_3] + [(b_1 + b_2) + b_3]i \\ &= [a_1 + (a_2 + a_3)] + [b_1 + (b_2 + b_3)]i \\ &= (a_1 + b_1 i) + [(a_2 + a_3) + (b_2 + b_3)i] \\ &= (a_1 + b_1 i) + [(a_2 + b_2 i) + (a_3 + b_3 i)] \\ &= z_1 + (z_2 + z_3). \end{aligned}$$

e. $0 + 0i$ serves as the zero element, since

$$(a + bi) + (0 + 0i) = (a + 0) + (b + 0)i = a + bi.$$

f. For $z = a + bi$,

$$-z = -a - bi,$$

since $(a + bi) + (-a - bi) = 0 + 0i$.

g. For $z = a + bi$ and real numbers c and d,

$$\begin{aligned} (cd)(a + bi) &= (cd)a + (cd)bi \\ &= c(da) + c(db)i \\ &= c(da + dbi) \\ &= c(d(a + bi)). \end{aligned}$$

h. For $z = a + bi$ and real numbers c and d,

$$\begin{aligned} (c + d)(a + bi) &= (c + d)a + (c + d)bi \\ &= (ca + da) + (cb + db)i \\ &= (ca + cbi) + (da + dbi) \\ &= c(a + bi) + d(a + bi). \end{aligned}$$

i. For $z_1 = a_1 + b_1 i$, $z_2 = a_2 + b_2 i$, and real number c,

$$\begin{aligned} c(z_1 + z_2) &= c[(a_1 + b_1 i) + (a_2 + b_2 i)] \\ &= c[(a_1 + a_2) + (b_1 + b_2)i] \\ &= c(a_1 + a_2) + c(b_1 + b_2)i \\ &= (ca_1 + ca_2) + (cb_1 + cb_2)i \\ &= (ca_1 + cb_1 i) + (ca_2 + cb_2 i) \\ &= c(a_1 + b_1 i) + c(a_2 + b_2 i) \\ &= cz_1 + cz_2. \end{aligned}$$

j. For $z = a + bi$,

$$1z = 1(a + bi) = (1a) + (1b)i = a + bi = z.$$

Thus, **C** is a vector space. ▣

NOTE Example 4 did not mention that two complex numbers can be *multiplied.* This added structure is ignored when **C** is being viewed as a vector space over the real numbers. You may notice that special characteristics of m-vectors and matrices have been ignored as well. In the study of vector spaces, only properties (a) through (j) and their consequences are examined.

The next example deals with *polynomials*. Recall that a **polynomial p** is an expression of the form

$$\mathbf{p}(x) = a_n x^n + a_{n-1}x^{n-1} + \cdots + a_1 x + a_0,$$

where each a_i is a real number called the *coefficient* of x^i. The *degree* of a nonzero polynomial **p** is n, provided a_n, its leading coefficient, is not equal to zero. Two polynomials are *equal* if all their corresponding coefficients are equal. The *sum* of two polynomials is the polynomial obtained by adding their corresponding coefficients, that is, adding "like" terms. The *product* of a polynomial by a scalar is the polynomial obtained by multiplying each of its coefficients by the scalar.

EXAMPLE 5

Let **P** be the set of all polynomials in x with the usual addition and scalar multiplication. Then **P** is a vector space.

Solution Clearly, properties (a) and (b) of the definition of vector space are satisfied by these operations. The **0** for **P** is the zero polynomial, the one whose graph is the x-axis. The negative of **p**, $-\mathbf{p}$, is the polynomial with each coefficient replaced by its negative. (The graph of $-\mathbf{p}$ is the reflection of the graph of **p** in the x-axis.) The other vector space properties follow easily; the arguments are very similar to those of Example 4. ■

EXAMPLE 6

(*From calculus*) Let $\mathbf{C}(-\infty, \infty)$ denote the set of all continuous real-valued functions of a real variable x. Addition of **f** and **g** in $\mathbf{C}(-\infty, \infty)$ and multiplication by a scalar c are defined in the usual way for functions, by

i. $(\mathbf{f} + \mathbf{g})(x) = \mathbf{f}(x) + \mathbf{g}(x)$, for all real numbers x,

ii. $(c\mathbf{f})(x) = c(\mathbf{f}(x))$, for all real numbers x.

Then $\mathbf{C}(-\infty, \infty)$ is a vector space over **R**.

Solution Vector space properties (a) and (b) follow from the fact that the sum of two continuous functions is continuous, and a scalar multiple of a continuous function is continuous. Further properties rest on the fact that two functions are equal if they agree at every point. That is, $\mathbf{f} = \mathbf{g}$ if and only if $\mathbf{f}(x) = \mathbf{g}(x)$ for each real number x. For example, to check property (c), let **f** and **g** be in $\mathbf{C}(-\infty, \infty)$. Then for each real number x,

$$(\mathbf{f} + \mathbf{g})(x) = \mathbf{f}(x) + \mathbf{g}(x) = \mathbf{g}(x) + \mathbf{f}(x) = (\mathbf{g} + \mathbf{f})(x).$$

Thus, $\mathbf{f} + \mathbf{g} = \mathbf{g} + \mathbf{f}$. The other properties are also easily checked. It should be noted that **0** is the zero function of Example 5, and that the negative of **f** in $\mathbf{C}(-\infty, \infty)$ is the function $-\mathbf{f}$, defined by $(-\mathbf{f})(x) = -(\mathbf{f}(x))$ for each x. As in Example 5, the graph of $-\mathbf{f}$ is the reflection of the graph of **f** in the x-axis. ■

NOTE In Example 6, we showed that $\mathbf{C}(-\infty, \infty)$, the set of all continuous real-valued functions defined on the real line, is a vector space. In a similar manner, it can be shown that the following sets of real-valued functions are also vector spaces:

$\mathbf{C}(a, b)$, the set of all continuous functions on the open interval (a, b);

$\mathbf{C}[a, b]$, the set of all continuous functions on the closed interval $[a, b]$;

$\mathbf{C}^{\infty}(a, b)$, the set of all functions that have derivatives of all orders on (a, b);

$\mathbf{C}^{\infty}(-\infty, \infty)$, the set of all functions that have derivatives of all orders on the entire real line.

Remark At this point, a logical question to ask is, "Why are abstract vector spaces so important?" As you read more about them, work on the exercises, and see their applications, you may be able to supply some specific reasons yourself. Here, in general terms, is an answer to the question posed above:

- Abstract vector spaces can be applied to help solve real-world problems. Two of these applications are given later: approximation of continuous functions (Section 6.6) and linear differential equations (Section 7.6).
- Vector spaces allow us to see the common underlying structure of certain seemingly different mathematical entities. For example, you have already seen that matrices and polynomials, which on the surface appear to have little in common, share many of the same properties.
- Some vector spaces, such as $\mathbf{R}^n$, are easier to study than others. By standard techniques (for example, using *coordinates*, as in Section 6.3), it may be possible to rephrase a problem that involves an abstract vector space in terms of a simpler vector space.
- In many areas of mathematics and its applications, one or more underlying vector spaces give the structure of the entity being studied. With this in mind, the mathematician (or other expert) can make use of the extensive information available about vector spaces to pursue difficult questions without needing to "reinvent the wheel."

To put it simply, vector spaces supply mathematicians and those who use mathematics with powerful tools to attack many problems.

In the previous examples, the operations of addition and scalar multiplication have been defined in a "natural" way—in the way you would expect. In describing a vector space, however, we can define these operations in any way we like. As long as the given set under these operations satisfies properties (a) through (j) of the definition, it is a vector space; if these properties are not satisfied, then it is not a vector space. The next example illustrates this point.

EXAMPLE 7

Let $\mathbf{V}$ be the set $\mathbf{R}^2$ of all ordered pairs of real numbers. Let addition be defined in the usual way, but define scalar multiplication, $*$, by

$$c * \mathbf{v} = |c|\, \mathbf{v}$$

for any scalar c and ordered pair $\mathbf{v}$. That is, $c * \mathbf{v}$ is the usual scalar product of the *absolute value* of c times $\mathbf{v}$. Show that $\mathbf{V}$ is *not* a vector space.

Solution All we need to show is that *one* of properties (a) through (j) fails to hold for *some* choice of elements of $\mathbf{V}$ and *some* choice of scalars. We show that property (h) fails when $c = 1$, $d = -1$, and $\mathbf{u} = (2, 3)$. Since

$$(c + d) * \mathbf{u} = (1 + (-1)) * (2, 3) = 0 * (2, 3) = 0(2, 3) = (0, 0),$$

but

$$c * \mathbf{u} + d * \mathbf{u} = 1 * \mathbf{u} + (-1) * \mathbf{u} = 1\mathbf{u} + 1\mathbf{u} = \mathbf{u} + \mathbf{u} = (4, 6),$$

we have that $(c + d) * \mathbf{u} \neq c * \mathbf{u} + d * \mathbf{u}$. Thus, $\mathbf{V}$ is not a vector space under these operations. (You can check that the rest of the properties *do* hold.) ■

The next theorem gives three additional properties common to all vector spaces. Although all of them may look "obvious," these properties must be proved from our existing knowledge of vector spaces.

THEOREM 1

Let $\mathbf{V}$ be a vector space, $\mathbf{v}$ an element of $\mathbf{V}$, and c a scalar, then

a. $0\mathbf{v} = \mathbf{0}$,

b. $(-1)\mathbf{v} = -\mathbf{v}$,

c. $c\mathbf{0} = \mathbf{0}$.

Proof

a. Since $0\mathbf{v} = (0 + 0)\mathbf{v} = 0\mathbf{v} + 0\mathbf{v}$ by property (h), and $0\mathbf{v} = 0\mathbf{v} + \mathbf{0}$ by property (e), we have

$$\mathbf{0} + 0\mathbf{v} = 0\mathbf{v} + 0\mathbf{v},$$

because addition is commutative. By property (f), every element has a negative, so there is an element $-(0\mathbf{v})$ that we add to both sides:

$$(\mathbf{0} + 0\mathbf{v}) + -(0\mathbf{v}) = (0\mathbf{v} + 0\mathbf{v}) + -(0\mathbf{v})$$

Applying property (d), and then (f) and (e), we have

$$\mathbf{0} + (0\mathbf{v} + -(0\mathbf{v})) = 0\mathbf{v} + (0\mathbf{v} + -(0\mathbf{v}))$$
$$\mathbf{0} + \mathbf{0} = 0\mathbf{v} + \mathbf{0}$$
$$\mathbf{0} = 0\mathbf{v}.$$

b. We first show that $(-1)\mathbf{v}$ "acts like" $-\mathbf{v}$; that is, $(-1)\mathbf{v} + \mathbf{v} = \mathbf{0}$. Using properties (h) and (j), we compute

$$(-1)\mathbf{v} + \mathbf{v} = (-1)\mathbf{v} + 1\mathbf{v} = (-1 + 1)\mathbf{v} = 0\mathbf{v} = \mathbf{0},$$

the last equality being part (a) of this theorem. But also, $-\mathbf{v} + \mathbf{v} = \mathbf{0}$, and thus we have

$$(-1)\mathbf{v} + \mathbf{v} = -\mathbf{v} + \mathbf{v}.$$

Adding $-\mathbf{v}$ to both sides of the equation, we obtain

$$(-1)\mathbf{v} = -\mathbf{v}.$$

c. Since $c(\mathbf{0}) = c(\mathbf{0} + \mathbf{0}) = c(\mathbf{0}) + c(\mathbf{0})$, we have

$$\mathbf{0} + c(\mathbf{0}) = c(\mathbf{0}) + c(\mathbf{0}).$$

Now adding $-c(\mathbf{0})$ to each side, we have the result. ■

Since polynomials are examples of continuous functions, the set $\mathbf{P}$ of Example 5 can be viewed as a subset of $\mathbf{C}(-\infty, \infty)$ of Example 6. Furthermore, the operations are "compatible," which means that addition of polynomials is the same as addition of functions, and multiplication by scalars is also the same in both vector spaces. We say that $\mathbf{P}$ is a *subspace* of $\mathbf{C}(-\infty, \infty)$.

DEFINITION

Let $\mathbf{V}$ be a vector space and $\mathbf{W}$ a subset of $\mathbf{V}$, then $\mathbf{W}$ is a **subspace** of $\mathbf{V}$ if $\mathbf{W}$ is a vector space using the operations of $\mathbf{V}$.

NOTE By virtue of the next theorem, this definition of subspace agrees with that of Section 5.2 in the case of subspaces of $\mathbf{R}^m$. Thus, all of the subspaces of $\mathbf{R}^m$ considered in Chapter 5 are subspaces under this new definition.

THEOREM 2

Let $\mathbf{V}$ be a vector space and $\mathbf{W}$ a nonempty subset of $\mathbf{V}$, then $\mathbf{W}$ is a subspace of $\mathbf{V}$ if and only if vector space properties (a) and (b) are satisfied; that is, if and only if $\mathbf{W}$ is closed under addition and scalar multiplication.

Proof If $\mathbf{W}$ is a subspace, *all* of properties (a) through (j) must be sa
particular, (a) and (b) must hold. Conversely, suppose that (a) and (b) a
Properties (c), (d), and (g) through (j) are "inherited" from $\mathbf{V}$. In other words, si
hold for all vectors in $\mathbf{V}$, and $\mathbf{W}$ is a subset of $\mathbf{V}$, they hold for all vectors in $\mathbf{W}$ as we

Since $\mathbf{V}$ is a vector space, there is an element $\mathbf{0}$ in $\mathbf{V}$, by property (e), and fo
each $\mathbf{w}$ in $\mathbf{W}$, $-\mathbf{w}$ is in $\mathbf{V}$, by property (f). We need to know that each of these is in $\mathbf{W}$ to complete the argument that $\mathbf{W}$ is a subspace. By property (b), for any $\mathbf{w}$ in $\mathbf{W}$, it follows that $0\mathbf{w}$ and $(-1)\mathbf{w}$ are in $\mathbf{W}$. But from Theorem 1, these are seen to be simply $\mathbf{0}$ and $-\mathbf{w}$. Hence $\mathbf{W}$ is a subspace. ■

We close this section with three additional examples: the first giving a subspace of the vector space $\mathbf{M}^{m,n}$, the second a subspace of $\mathbf{P}$, and the third a *subset* of $\mathbf{P}$ that is *not a subspace* of $\mathbf{P}$.

EXAMPLE 8

Let $\mathbf{D}^{2,2} = \{A \text{ in } \mathbf{M}^{2,2} \mid a_{12} = a_{21} = 0\}$ be the set of all 2×2 *diagonal* matrices. Then $\mathbf{D}^{2,2}$ is a subspace of $\mathbf{M}^{2,2}$, the set of *all* 2×2 matrices.

Solution By Theorem 2, all we need to show is that $\mathbf{D}^{2,2}$ satisfies properties (a) and (b). (In other words, we only need to demonstrate that $\mathbf{D}^{2,2}$ is *closed* under the addition and scalar multiplication of $\mathbf{M}^{2,2}$.) To do this, let A and B be diagonal matrices and c a scalar, then A and B have the form

$$A = \begin{bmatrix} a_1 & 0 \\ 0 & a_2 \end{bmatrix} \quad \text{and} \quad B = \begin{bmatrix} b_1 & 0 \\ 0 & b_2 \end{bmatrix},$$

and we have

$$A + B = \begin{bmatrix} a_1 + b_1 & 0 \\ 0 & a_2 + b_2 \end{bmatrix}$$

$$\text{and} \quad cA = \begin{bmatrix} ca_1 & 0 \\ 0 & ca_2 \end{bmatrix}.$$

Since $A + B$ and cA are diagonal matrices—that is, elements of $\mathbf{D}^{2,2}$—the set $\mathbf{D}^{2,2}$ is a subspace of $\mathbf{M}^{2,2}$. ■

NOTE By Theorem 2, often the easiest way to show that a set with two operations is a vector space is to show that it is a subspace of a known vector space. The following is an example of this technique.

EXAMPLE 9

Let $\mathbf{P}_n$ be the set of all polynomials of degree less than or equal to n, together with $\mathbf{0}$, and use the usual polynomial arithmetic, then $\mathbf{P}_n$ is a vector space for each positive integer n.

Solution We show that $\mathbf{P}_n$ is a subspace of $\mathbf{P}$ for each positive integer n. Since adding two polynomials does not introduce new terms of degree higher than those of the polynomials, property (a) is satisfied. Likewise, multiplication by a scalar leaves the degree unchanged unless the scalar is zero, which *lowers* the degree. Thus, property (b) is satisfied as well, and $\mathbf{P}_n$ is a subspace of $\mathbf{P}$. Since $\mathbf{P}_n$ is a subspace of $\mathbf{P}$, it is a vector space in its own right. ■

EXAMPLE 10

Let $\mathbf{P}_3'$ be the set of all polynomials of degree *exactly equal* to three together with $\mathbf{0}$ and use the usual polynomial arithmetic, then $\mathbf{P}_3'$ is *not* a subspace of $\mathbf{P}_3$, the set of all polynomials of degree *less than or equal* to 3.

Solution To show this, we need only show that either property (a) or property (b) *fails* to hold for a specific choice of polynomials or scalars (or both). For example, let $\mathbf{p}(x) = x^3 + 2x - 1$ and let $\mathbf{q}(x) = -x^3 + x^2 - 3$, then

$$(\mathbf{p} + \mathbf{q})(x) = x^2 + 2x - 4,$$

which is a polynomial of degree 2. Thus, $\mathbf{p}$ and $\mathbf{q}$ are two elements in $\mathbf{P}_3'$ whose sum is not in $\mathbf{P}_3'$, so $\mathbf{P}_3'$ is not a subspace of $\mathbf{P}_3$. ■

6.1 Exercises

In Exercises 1–12 determine whether or not the given set is a vector space under the given operations. If it is not a vector space, list all properties that fail to hold.

1. The set of all 2×3 matrices whose second column consists of zeros: the usual matrix operations
2. The set of all 2×3 matrices whose entries have the sum of 1: the usual matrix operations
3. The set of all polynomials of degree greater than or equal to 3 together with $\mathbf{0}$: the usual polynomial operations
4. The set of all polynomials of degree less than or equal to 2, whose coefficients have the sum of zero: the usual operations
5. The set of all polynomials with positive coefficients: the usual operations
6. The set of all vectors in $\mathbf{R}^3$: the usual scalar multiplication but addition of $\mathbf{u}$ and $\mathbf{v}$ defined by taking the cross product of $\mathbf{u}$ and $\mathbf{v}$

In Exercises 7 and 8 the *only* elements of the set are the formal symbols x and y, with no special properties other than those listed in the exercises.

7. The set $\{\mathbf{x}, \mathbf{y}\}$, where $\mathbf{x} \neq \mathbf{y}$, with addition $\mathbf{x} + \mathbf{x} = \mathbf{x}$, $\mathbf{x} + \mathbf{y} = \mathbf{y} + \mathbf{x} = \mathbf{y}$, $\mathbf{y} + \mathbf{y} = \mathbf{x}$, and scalar multiplication $c\mathbf{x} = \mathbf{x}$, $c\mathbf{y} = \mathbf{x}$ for all scalars c
8. Same as Exercise 7 except scalar multiplication: $c\mathbf{x} = \mathbf{x}$ and $c\mathbf{y} = \mathbf{y}$ for all scalars c
9. The set of all continuous functions with the property that the function evaluated at 0 is 0
10. The set of all continuous functions with the property that the function evaluated at 0 is 1
11. The set of all continuous functions with the property that the function is 0 at every integer; for example, $\mathbf{f}(x) = \sin(\pi x)$
12. The set consisting solely of a single element $\mathbf{0}$, with addition $\mathbf{0} + \mathbf{0} = \mathbf{0}$ and scalar multiplication $c\mathbf{0} = \mathbf{0}$ for all scalars c

In Exercises 13–16 determine whether or not the given set is a subspace of $\mathbf{M}^{2,2}$ (see Example 2).

13. The set of all 2×2 matrices whose entries have the sum of zero
14. The set of all 2×2 matrices with a determinant of zero
15. The set of all 2×2 matrices of the form
$$\begin{bmatrix} a & b \\ c & 1 \end{bmatrix},$$
where a, b, and c are real numbers
16. The set of all 2×2 matrices of the form
$$\begin{bmatrix} a & b \\ b & a + b \end{bmatrix},$$
where a and b are real numbers

In Exercises 17 and 18 determine whether or not the given set is a subspace of $\mathbf{M}^{n,n}$, the space of all square matrices of order *n*.

17. The diagonal matrices of order n

18. The matrices of order n with trace equal to 0 (The **trace** of a square matrix is the sum of the diagonal elements.)

In Exercises 19 and 20 determine whether or not the given set is a subspace of $C(-\infty, \infty)$ (see Example 6).

19. The set of all functions of the form $a \sin x + b \cos x$, where a and b are real numbers

20. The set $\mathbf{D}$ of all differentiable functions

21. If $\mathcal{S} = \{\mathbf{v}_1, \mathbf{v}_2, \ldots, \mathbf{v}_n\}$ is a set of vectors in a vector space $\mathbf{V}$, show that

$$c_1\mathbf{v}_1 + c_2\mathbf{v}_2 + \cdots + c_n\mathbf{v}_n$$

is also in $\mathbf{V}$, where all c_i are arbitrary scalars.

6.2 Linear Independence, Basis, and Dimension

In Chapter 5, we studied the concepts of *linear independence, span, basis,* and *dimension* in the context of Euclidean m-space, $\mathbf{R}^m$. In this section, we generalize these ideas to arbitrary vector spaces.

Linear Independence

We follow Section 5.1 for the extended definition of linear independence of a set of vectors. Consequently, the definition given here is compatible with the definition for the special case of vectors in $\mathbf{R}^m$.

DEFINITION

Let $\mathcal{S} = \{\mathbf{v}_1, \mathbf{v}_2, \ldots, \mathbf{v}_n\}$ be a set of vectors in a vector space $\mathbf{V}$. The set $\mathcal{S}$ is **linearly independent** if the equation

$$c_1\mathbf{v}_1 + c_2\mathbf{v}_2 + \cdots + c_n\mathbf{v}_n = \mathbf{0}$$

has *only* the trivial solution, $c_i = 0$, for all i. An *infinite* set $\mathcal{S}$ is **linearly independent** if every finite subset of it is linearly independent. If $\mathcal{S}$ is not linearly independent, we say that it is **linearly dependent**.

NOTE If $\mathcal{S} = \{\mathbf{v}_1, \mathbf{v}_2, \ldots, \mathbf{v}_n\}$ is a set of vectors in a vector space $\mathbf{V}$, a **linear combination** of the $\mathbf{v}_i$ is any expression $\mathbf{v}$ of the form

$$\mathbf{v} = c_1\mathbf{v}_1 + c_2\mathbf{v}_2 + \cdots + c_n\mathbf{v}_n,$$

where the c_i are scalars. By Exercise 21 of Section 6.1, all such expressions are in $\mathbf{V}$. Thus, $\mathcal{S}$ is linearly independent if and only if the only linear combination of the $\mathbf{v}_i$ equal to zero is the one in which every scalar c_i is zero.

EXAMPLE 1

Determine whether or not the set

$$\mathcal{S} = \left\{ \begin{bmatrix} 1 & 0 \\ 0 & -1 \end{bmatrix}, \begin{bmatrix} 0 & 1 \\ 0 & -1 \end{bmatrix}, \begin{bmatrix} 0 & 0 \\ 1 & -1 \end{bmatrix} \right\}$$

is linearly independent in $\mathbf{M}^{2,2}$ (see Example 2 of Section 6.1).

Solution We consider the matrix equation

$$c_1 \begin{bmatrix} 1 & 0 \\ 0 & -1 \end{bmatrix} + c_2 \begin{bmatrix} 0 & 1 \\ 0 & -1 \end{bmatrix} + c_3 \begin{bmatrix} 0 & 0 \\ 1 & -1 \end{bmatrix} = \begin{bmatrix} 0 & 0 \\ 0 & 0 \end{bmatrix}.$$

If the only solution is the trivial one, $c_1 = c_2 = c_3 = 0$, then $\mathcal{S}$ is linearly independent. To determine whether or not this is the case, we perform the scalar multiplications and additions on the left side to obtain

$$\begin{bmatrix} c_1 & c_2 \\ c_3 & -c_1 - c_2 - c_3 \end{bmatrix} = \begin{bmatrix} 0 & 0 \\ 0 & 0 \end{bmatrix}$$

and then equate corresponding entries. This yields the following homogeneous system of four linear equations:

$$\begin{aligned} c_1 \qquad\qquad &= 0 \\ c_2 \qquad &= 0 \\ c_3 &= 0 \\ -c_1 - c_2 - c_3 &= 0 \end{aligned}$$

From the first three equations, it is clear that the only solution of this system is the trivial one. Thus, $\mathcal{S}$ is linearly independent. ■

EXAMPLE 2

Determine whether or not the set

$$\mathcal{S} = \{x + 1, x - 1, 2x + 1\}$$

is linearly independent in $\mathbf{P}_1$, the space of polynomials of degree less than or equal to 1.

Solution We consider the equation

$$c_1(x + 1) + c_2(x - 1) + c_3(2x + 1) = 0.$$

Combining terms of the same degree, we have

$$(c_1 + c_2 + 2c_3)x + (c_1 - c_2 + c_3) = 0.$$

This yields the following set of linear equations:

$$\begin{aligned} c_1 + c_2 + 2c_3 &= 0 \\ c_1 - c_2 + c_3 &= 0 \end{aligned}$$

Since this system is a homogeneous system of linear equations with more unknowns than equations, it has (infinitely many) nonzero solutions. Hence, the set $\mathcal{S}$ is linearly dependent. ■

EXAMPLE 3

(*From calculus*) Determine whether or not the set $\mathcal{S} = \{1, x, \sin x\}$ is linearly independent in the vector space, $\mathbf{C}(-\infty, \infty)$, of continuous functions (see Example 6 of the previous section).

Solution Here we consider the equation

$$c_1(1) + c_2x + c_3 \sin x = 0,$$

where 0 is the zero function. Since this equation must hold for *all* x, it holds for $x = 0$, $\pi/2$, and π. Successively substituting these values for x into the previous equation yields

$$\begin{aligned} c_1 \qquad\qquad\qquad &= 0 \\ c_1 + \left(\frac{\pi}{2}\right)c_2 + c_3 &= 0 \\ c_1 + (\pi)c_2 \qquad\quad &= 0. \end{aligned}$$

You can easily check that this homogeneous linear system has only the trivial solution. Hence, $\mathcal{S}$ is linearly independent. ■

Basis for a Vector Space

We now investigate the notions of *span, basis,* and *dimension* of an arbitrary vector space. The definitions are virtually identical to the definitions given in the context of $\mathbf{R}^m$ in Sections 5.2 and 5.3.

DEFINITION

Let $\mathbf{V}$ be a vector space. A set $\mathcal{S}$ of vectors in $\mathbf{V}$ **spans** (or **generates**) $\mathbf{V}$ if every vector in $\mathbf{V}$ can be expressed as a linear combination of a finite subset of $\mathcal{S}$.

EXAMPLE 4

Show that the set $\mathcal{S} = \{1, x - 1, x^2 - 1\}$ spans $\mathbf{P}_2$, the subspace of $\mathbf{P}$ consisting of all polynomials of degree less than or equal to 2 (see Example 9 of Section 6.1).

Solution Let $\mathbf{p}(x) = a_0 + a_1x + a_2x^2$ be an arbitrary vector in $\mathbf{P}_2$. We must show that there exist scalars c_1, c_2, and c_3 such that

$$c_1(1) + c_2(x - 1) + c_3(x^2 - 1) = a_0 + a_1x + a_2x^2.$$

Collecting like terms on the left side, we obtain

$$(c_1 - c_2 - c_3) + c_2x + c_3x^2 = a_0 + a_1x + a_2x^2.$$

Since two polynomials are equal if and only if their corresponding coefficients are equal, this last equation implies that

$$\begin{aligned} c_1 - c_2 - c_3 &= a_0 \\ c_2 \qquad &= a_1 \\ c_3 &= a_2. \end{aligned}$$

Now this linear system has a solution for c_1, c_2, and c_3 no matter what a_0, a_1, and a_2 are. Hence, $\mathcal{S}$ spans $\mathbf{P}_2$. ■

DEFINITION

Let $\mathbf{V}$ be a vector space. A set $\mathcal{S}$ of vectors in $\mathbf{V}$ is a **basis** for $\mathbf{V}$ if

i. $\mathcal{S}$ is linearly independent, and

ii. $\mathcal{S}$ spans $\mathbf{V}$.

Just as the set $\{\mathbf{e}_1, \mathbf{e}_2, \ldots, \mathbf{e}_m\}$ provides a *stardard basis* for $\mathbf{R}^m$, there are standard bases for $\mathbf{P}_n$, $\mathbf{P}$, and $\mathbf{M}^{m,n}$. The next theorem describes these especially simple bases.

THEOREM 1

Each of the following is a basis for the respective vector space:

a. The set $\mathcal{S} = \{1, x, x^2, \ldots, x^n\}$ for $\mathbf{P}_n$, the vector space of all polynomials of degree less than or equal to n together with the zero polynomial

b. The set $\mathcal{S} = \{1, x, x^2, \ldots, x^n, \ldots\}$ for $\mathbf{P}$, the vector space of all polynomials

c. The set $\{E^{ij} \mid i = 1, \ldots, m;\ j = 1, \ldots, n\}$, where E^{ij} is an $m \times n$ matrix with one in the i, j position and zeros elsewhere, for $\mathbf{M}^{m,n}$, the vector space of all $m \times n$ matrices

Proof We prove only parts (a) and (b), leaving the proof of (c) as an exercise.

To prove part (a), we must show that $\mathcal{S}$ is linearly independent and spans $\mathbf{P}_n$. First consider the equation

$$c_1(1) + c_2(x) + \cdots + c_{n+1}(x^n) = \mathbf{0} = 0(1) + 0(x) + \cdots + 0(x^n).$$

Since two polynomials are equal if and only if their corresponding coefficients are equal, we must have $c_i = 0$ for $i = 1, \ldots, n + 1$. Hence, $\mathcal{S}$ is linearly independent.

Now suppose that $\mathbf{p}(x) = a_0 + a_1x + \cdots + a_nx^n$ is an arbitrary member of $\mathbf{P}_n$, then $\mathbf{p}$ is *already* expressed as a linear combination of $\mathcal{S}$ with the a_i as the scalars (since $a_0 = a_0 \cdot 1$). Thus, $\mathcal{S}$ spans $\mathbf{P}_n$ as well.

The proof that $\mathcal{S} = \{1, x, x^2, \ldots\}$ is a basis for $\mathbf{P}$ is almost exactly the same as the proof for part (a). The important fact is that each polynomial has *some* degree, so that we can choose n to be the degree of the polynomial and we can express the polynomial as a linear combination of the *finite* set $\{1, x, \ldots, x^n\}$. ▣

NOTE Usually we study situations in which the spanning set $\mathcal{S}$ is itself finite and *all* the elements of $\mathcal{S}$ can be used to represent a given vector as a linear combination of elements of $\mathcal{S}$. If $\mathcal{S}$ is infinite, however, we must find for each vector a finite subset of $\mathcal{S}$ to use to represent that vector, since sums of infinitely many terms have no meaning in general vector spaces.

EXAMPLE 5

Find a basis for (a) $\mathbf{P}_3$ and (b) $\mathbf{M}^{2,3}$.

Solution We apply Theorem 1 to these special cases:

a. $\mathcal{S} = \{1, x, x^2, x^3\}$ is a basis for $\mathbf{P}_3$.

b. $$\mathcal{S} = \left\{\begin{bmatrix}1 & 0 & 0\\0 & 0 & 0\end{bmatrix}, \begin{bmatrix}0 & 1 & 0\\0 & 0 & 0\end{bmatrix}, \begin{bmatrix}0 & 0 & 1\\0 & 0 & 0\end{bmatrix}, \begin{bmatrix}0 & 0 & 0\\1 & 0 & 0\end{bmatrix}, \begin{bmatrix}0 & 0 & 0\\0 & 1 & 0\end{bmatrix}, \begin{bmatrix}0 & 0 & 0\\0 & 0 & 1\end{bmatrix}\right\}$$

is a basis for $\mathbf{M}^{2,3}$. ▣

EXAMPLE 6

Show that the set

$$\mathcal{S} = \begin{bmatrix}1 & 0\\0 & -1\end{bmatrix}, \begin{bmatrix}0 & 1\\0 & -1\end{bmatrix}, \begin{bmatrix}0 & 0\\1 & -1\end{bmatrix}$$

is a basis for the subspace $\mathbf{S}$ of $\mathbf{M}^{2,2}$ consisting of all 2×2 matrices, whose entries have the sum of zero.

Solution We know that $\mathbf{S}$ is a subspace of $\mathbf{M}^{2,2}$ by Exercise 13 of Section 6.1 and that $\mathcal{S}$ is certainly a subset of $\mathbf{S}$. Therefore, all we need to do is verify that $\mathcal{S}$ is linearly independent and spans $\mathbf{S}$. We have already shown the first part in Example 1. To show that $\mathcal{S}$ spans $\mathbf{S}$, let

$$A = \begin{bmatrix}a_1 & a_2\\a_3 & a_4\end{bmatrix}, \qquad a_1 + a_2 + a_3 + a_4 = 0,$$

be an arbitrary vector in **S**. We must find scalars c_1, c_2, and c_3 such that

$$c_1\begin{bmatrix}1 & 0\\0 & -1\end{bmatrix} + c_2\begin{bmatrix}0 & 1\\0 & -1\end{bmatrix} + c_3\begin{bmatrix}0 & 0\\1 & -1\end{bmatrix} = \begin{bmatrix}a_1 & a_2\\a_3 & a_4\end{bmatrix}.$$

Performing the scalar multiplications and additions on the left side and equating corresponding entries yields the linear system

$$\begin{aligned} c_1 \qquad\qquad\quad &= a_1\\ c_2 \qquad\quad &= a_2\\ c_3 &= a_3\\ -c_1 - c_2 - c_3 &= a_4. \end{aligned}$$

From the first three equations, we see that $c_1 = a_1$, $c_2 = a_2$, and $c_3 = a_3$. Substituting these values for the c_i into the fourth equation yields $-a_1 - a_2 - a_3 = a_4$ or, equivalently, $a_1 + a_2 + a_3 + a_4 = 0$. But the latter is true by hypothesis, so A can be expressed as a linear combination of the vectors in $\mathcal{S}$ for all choices of the a_i; that is, the set $\mathcal{S}$ spans **S**. Thus, $\mathcal{S}$ is a basis for **S**. ■

The following theorems are the counterparts of theorems in Chapter 5. For example, the next theorem is a generalization of Theorem 2 of Section 5.2. That theorem was stated in terms of a *subspace* of $\mathbf{R}^m$, whereas the next theorem refers to a *vector space.* Since any subspace of a vector space is also a vector space, the following result does encompass the former one. The proof is exactly the same as was given before, and hence is omitted.

THEOREM 2

Let **V** be a vector space spanned by the set $\mathcal{T}$. If one of the vectors in $\mathcal{T}$ is a linear combination of the others, then the set obtained by deleting that vector still spans **V**. ■

If $\mathcal{T}$ in Theorem 2 is a finite set, repeated application of Theorem 2 results in a set that both spans and is linearly independent—a basis. The result is true even if $\mathcal{T}$ is infinite, and we state this as the next theorem.

THEOREM 3

Let **V** be a vector space spanned by a set $\mathcal{T}$. Then there is a subset of $\mathcal{T}$ that is a basis for **V**. In other words, a generating set may be reduced to a basis. ■

EXAMPLE 7

The set $\mathcal{T} = \{1, x + 1, x - 1\}$ generates $\mathbf{P}_1$, the vector space of polynomials of degree less than or equal to 1 (Exercise 15). Theorem 3 tells us that there is a subset of $\mathcal{T}$ that is a basis for $\mathbf{P}_1$. In fact, there are three of them. Each of $\mathcal{T}_1 = \{1, x + 1\}$, $\mathcal{T}_2 = \{1, x - 1\}$, and $\mathcal{T}_3 = \{x + 1, x - 1\}$ is a basis for $\mathbf{P}_1$ (Exercise 43).

THEOREM 4

Let **V** be a vector space spanned by a set $\mathcal{S}$ and $\mathcal{T}_1$ be a linearly independent subset of **V**. Then there is a subset $\mathcal{T}_2$ of $\mathcal{S}$ such that $\mathcal{T} = \mathcal{T}_1 \cup \mathcal{T}_2$, the union of $\mathcal{T}_1$ and $\mathcal{T}_2$, is a basis for **V**. In other words, a linearly independent set may be extended to a basis.

Proof We prove only the case in which $\mathcal{T}_1$ and $\mathcal{S}$ are finite sets. Consider the set $\mathcal{T}_1 \cup \mathcal{S}$. Fixing the order of the elements, we take first the elements of $\mathcal{T}_1$ and then those of $\mathcal{S}$, and we apply Theorem 2. Since $\mathcal{S}$ alone spans $\mathbf{V}$, the set $\mathcal{T}_1 \cup \mathcal{S}$ does as well. Since no element of $\mathcal{T}_1$ is a linear combination of the others ($\mathcal{T}_1$ is linearly independent), if $\mathcal{T}_1 \cup \mathcal{S}$ is not linearly independent, some element of $\mathcal{S}$ is a linearly dependent combination of the others and can be deleted. Repeated application of this process yields a set that spans $\mathbf{V}$, is linearly independent, and contains $\mathcal{T}_1$. The remaining elements constitute $\mathcal{T}_2$. ■

EXAMPLE 8

From Example 1, we know that

$$\mathcal{T}_1 = \left\{ \begin{bmatrix} 1 & 0 \\ 0 & -1 \end{bmatrix}, \begin{bmatrix} 0 & 1 \\ 0 & -1 \end{bmatrix}, \begin{bmatrix} 0 & 0 \\ 1 & -1 \end{bmatrix} \right\}$$

is a linearly independent subset of $\mathbf{M}^{2,2}$. From part (c) of Theorem 1, we know that the set

$$\mathcal{S} = \left\{ \begin{bmatrix} 1 & 0 \\ 0 & 0 \end{bmatrix}, \begin{bmatrix} 0 & 1 \\ 0 & 0 \end{bmatrix}, \begin{bmatrix} 0 & 0 \\ 1 & 0 \end{bmatrix}, \begin{bmatrix} 0 & 0 \\ 0 & 1 \end{bmatrix} \right\}$$

is a generating set (in fact a basis) for $\mathbf{M}^{2,2}$. Taking

$$\mathcal{T}_2 = \left\{ \begin{bmatrix} 0 & 0 \\ 0 & 1 \end{bmatrix} \right\}$$

to be the subset of $\mathcal{S}$, it can be shown that the union of $\mathcal{T}_1$ and $\mathcal{T}_2$ is a basis for $\mathbf{M}^{2,2}$ (Exercises 5 and 17). ■

Dimension of a Vector Space

THEOREM 5

Let $\mathbf{V}$ be a vector space with bases $\mathcal{B}$ and $\mathcal{B}'$. Then $\mathcal{B}$ and $\mathcal{B}'$ have the same number of elements.

Proof We consider only the case in which $\mathcal{B}$ and $\mathcal{B}'$ are finite sets. Let

$$\mathcal{B} = \{\mathbf{v}_1, \ldots, \mathbf{v}_n\} \quad \text{and} \quad \mathcal{B}' = \{\mathbf{w}_1, \ldots, \mathbf{w}_m\}.$$

We must show that $m = n$. It suffices to assume that $n \le m$ and show that $n = m$, since the two sets could be interchanged before starting the proof.

Since $\mathcal{B}$ spans $\mathbf{V}$ and $\mathbf{w}_1$ is in $\mathbf{V}$, there are scalars $c_1, \ldots, c_n$ such that

$$\mathbf{w}_1 = c_1\mathbf{v}_1 + \cdots + c_n\mathbf{v}_n.$$

If every $c_i = 0$, $\mathbf{w}_1$ is the zero vector, and thus $\mathbf{0}$ is in $\mathcal{B}'$. It is easy to show (Exercise 45) that if $\mathbf{0}$ is in a set, the set is not linearly independent. But $\mathcal{B}'$ is linearly independent, so $\mathbf{w}_1 \neq \mathbf{0}$, and thus at least one $c_i \neq 0$. By renumbering the elements of $\mathcal{B}$, if necessary, we may assume that $c_1 \neq 0$. Now we solve for $\mathbf{v}_1$:

$$\mathbf{v}_1 = \left(\frac{1}{c_1}\right)\mathbf{w}_1 - \left(\frac{c_2}{c_1}\right)\mathbf{v}_2 - \cdots - \left(\frac{c_n}{c_1}\right)\mathbf{v}_n$$

Thus, $\mathbf{v}_1$ is a linear combination of the elements of the set

$$\mathcal{B}_1 = \{\mathbf{w}_1, \mathbf{v}_2, \ldots, \mathbf{v}_n\}.$$

Letting $\mathcal{B}^*$ be the set $\mathcal{B}$ together with $\mathbf{w}_1$, we have a spanning set (since $\mathcal{B}$ alone is a spanning set), with the element $\mathbf{v}_1$ as a linear combination of the others. By Theorem 2, the set $\mathcal{B}_1$, obtained by deleting $\mathbf{v}_1$, still spans $\mathbf{V}$.

Since $\mathcal{B}_1$ spans $\mathbf{V}$ and $\mathbf{w}_2$ is in $\mathbf{V}$, there are scalars $d_1, \ldots, d_n$ such that

$$\mathbf{w}_2 = d_1\mathbf{w}_1 + d_2\mathbf{v}_2 + \cdots + d_n\mathbf{v}_n.$$

If $d_2 = d_3 = \cdots = d_n = 0$, we have $\mathbf{w}_2 = d_1\mathbf{w}_1$, from which it is easy to show th
set $\mathcal{B}'$ is not linearly independent. Thus, one of the $d_i \neq 0$ for $i = 2, \ldots, n.$
renumbering the remaining elements of $\mathcal{B}$, if necessary, we may assume that $d_2 \neq 0$. Now
we solve for $\mathbf{v}_2$:

$$\mathbf{v}_2 = -\left(\frac{d_1}{d_2}\right)\mathbf{w}_1 + \left(\frac{1}{d_2}\right)\mathbf{w}_2 - \left(\frac{d_3}{d_2}\right)\mathbf{v}_3 - \cdots - \left(\frac{d_n}{d_2}\right)\mathbf{v}_n$$

Thus, $\mathbf{v}_2$ is a linear combination of the elements of the set

$$\mathcal{B}_2 = \{\mathbf{w}_1, \mathbf{w}_2, \mathbf{v}_3, \ldots, \mathbf{v}_n\}.$$

Letting $\mathcal{B}^*$ be the set $\mathcal{B}_1$ together with $\mathbf{w}_2$, we have a spanning set for $\mathbf{V}$, with the element $\mathbf{v}_2$ as a linear combination of the others. By Theorem 2, the set $\mathcal{B}_2$ still spans $\mathbf{V}$.

Continuing this same idea, we can use each $\mathbf{w}_i$ to replace some $\mathbf{v}_i$ and obtain

$$\mathcal{B}_k = \{\mathbf{w}_1, \ldots, \mathbf{w}_k, \mathbf{v}_{k+1}, \ldots, \mathbf{v}_n\}$$

as a spanning set for $\mathbf{V}$ until $k = n$. In other words, we continue until we run out of elements of $\mathcal{B}$ and have

$$\mathcal{B}_n = \{\mathbf{w}_1, \ldots, \mathbf{w}_n\}$$

as a spanning set for $\mathbf{V}$. If $n < m$, we have $\mathbf{w}_{n+1}$ in the span of $\mathcal{B}_n$. It follows easily that $\mathcal{B}'$ cannot be linearly independent—a contradiction. Therefore, $n = m$. ■

DEFINITION

The **dimension** of a vector space $\mathbf{V}$ that has a finite spanning set is equal to the number of vectors in a basis for $\mathbf{V}$.

NOTE For the zero vector space $\{\mathbf{0}\}$ it is customary to take the empty set (no elements) as a basis. Thus, the dimension of the zero vector space is 0. If a vector space $\mathbf{V}$ has a finite basis or, more generally, a finite generating set, it can be shown by an argument similar to the proof of Theorem 5 that all bases for $\mathbf{V}$ are finite and hence have the same number of elements. Such a vector space is said to be **finite-dimensional**. If a basis for a vector space $\mathbf{V}$ is not finite, we say that $\mathbf{V}$ is **infinite-dimensional**.

As a direct consequence of Theorem 1 and the definition of dimension, we have the following theorem.

THEOREM 6

a. $\mathbf{P}_n$ has dimension $n + 1$.

b. $\mathbf{P}$ is infinite dimensional.

c. $\mathbf{M}^{m,n}$ has dimension mn. ■

EXAMPLE 9

Find the dimension of the subspace $\mathbf{S}$ of $\mathbf{M}^{3,3}$ consisting of all 3×3 diagonal matrices.

Solution It can be easily shown that a basis for $\mathbf{S}$ is given by

$$\left\{\begin{bmatrix} 1 & 0 & 0 \\ 0 & 0 & 0 \\ 0 & 0 & 0 \end{bmatrix}, \begin{bmatrix} 0 & 0 & 0 \\ 0 & 1 & 0 \\ 0 & 0 & 0 \end{bmatrix}, \begin{bmatrix} 0 & 0 & 0 \\ 0 & 0 & 0 \\ 0 & 0 & 1 \end{bmatrix}\right\}.$$

Therefore, $\mathbf{S}$ has dimension equal to 3. ■

We close this section with a theorem that is sometimes useful in quickly answering questions concerning a given subset of a vector space: "Is it linearly independent?" "Does it span?" "Is it a basis?" We leave the proof for an exercise.

at the
By

THEOREM 7

Let $\mathcal{S}$ be a subset of a vector space $\mathbf{V}$ of dimension n.

a. If $\mathcal{S}$ contains fewer than n vectors, it does not span $\mathbf{V}$.

b. If $\mathcal{S}$ contains more than n vectors, it is not linearly independent (it is linearly dependent).

c. If $\mathcal{S}$ contains exactly n vectors and is either linearly independent or spans $\mathbf{V}$, then it is a basis for $\mathbf{V}$. ■

EXAMPLE 10

Demonstrate each of the following:

a. The set $\{x, 1 + x, 1 - x\}$ is linearly dependent in $\mathbf{P}_1$.

b. The set

$$\left\{\begin{bmatrix} 1 & 1 & 1 \\ 0 & 0 & 0 \end{bmatrix}, \begin{bmatrix} 0 & 0 & 0 \\ 1 & 1 & 1 \end{bmatrix}\right\}$$

does not span $\mathbf{M}^{2,3}$.

Solution These follow immediately from conditions (b) and (a) of Theorem 7, respectively. In part (a), the set consists of three elements and is a subset of the two-dimensional vector space $\mathbf{P}_1$. Therefore, it is linearly dependent. In part (b), the given set contains two vectors in the six-dimensional $\mathbf{M}^{2,3}$, so it cannot span. ■

6.2 Exercises

In Exercises 1–12 determine whether or not the given set is linearly independent. If the set is linearly dependent, write one of its vectors as a linear combination of the others.

1. $\{1, 1 + x, 1 + x + x^2\}$ in $\mathbf{P}_2$

2. $\{1 + x^2, 1 + x + 2x^2, x + x^2\}$ in $\mathbf{P}_2$

3. $\{1, x - 1, x + 1\}$ in $\mathbf{P}_1$

4. $\{x - 1, x + 1\}$ in $\mathbf{P}_1$

5. $\left\{\begin{bmatrix} 1 & 0 \\ 0 & -1 \end{bmatrix}, \begin{bmatrix} 0 & 1 \\ 0 & -1 \end{bmatrix}, \begin{bmatrix} 0 & 0 \\ 1 & -1 \end{bmatrix}, \begin{bmatrix} 0 & 0 \\ 0 & 1 \end{bmatrix}\right\}$ in $\mathbf{M}^{2,2}$

6. $\left\{\begin{bmatrix} 1 & 2 \\ -1 & 0 \end{bmatrix}, \begin{bmatrix} 0 & -1 \\ 1 & 1 \end{bmatrix}, \begin{bmatrix} 1 & 0 \\ 1 & 2 \end{bmatrix}\right\}$ in $\mathbf{M}^{2,2}$

7. $\{1, \sin^2 x, \cos^2 x\}$ in $\mathbf{C}(-\infty, \infty)$

8. $\{1, e^x, e^{-x}\}$ in $\mathbf{C}(-\infty, \infty)$

9. $\left\{\begin{bmatrix} 2 & 0 & 0 \\ 0 & -1 & 0 \\ 0 & 0 & 1 \end{bmatrix}, \begin{bmatrix} -2 & 0 & 0 \\ 0 & -1 & 0 \\ 0 & 0 & -1 \end{bmatrix}\right\}$

in the vector space of 3×3 diagonal matrices

10. $\left\{\begin{bmatrix} 2 & 0 & 0 \\ 1 & -1 & 0 \\ 2 & 1 & -1 \end{bmatrix}, \begin{bmatrix} 2 & 0 & 0 \\ 0 & -1 & 0 \\ 2 & 1 & -1 \end{bmatrix}\right\}$

in the vector space of 3×3 matrices with trace 0

11. $\{2 + x - 3 \sin x + \cos x, x + \sin x - 3 \cos x, 1 - 2x + 3 \sin x + \cos x, 2 - x - \sin x - \cos x, 2 + x - 3 \cos x\}$ in the space of all linear combinations of the set $\{1, x, \sin x, \cos x\}$

12. All 2×2 matrices with exactly two 1's and two 0's for entries in the space of all 2×2 matrices

13–24. Determine whether or not each set in Exercises 1–12 generates (that is, spans) the indicated vector space. (Theorem 7 may come in handy here.)

25–36. Determine whether or not each set in Exercises 1–12 is a basis for the indicated vector space.

In Exercises 37–42 find the dimension of the given vector space.

37. $\mathbf{M}^{3,4}$

38. $\mathbf{P}_5$

39. The subspace of $\mathbf{M}^{2,2}$ consisting of all diagonal 2×2 matrices

40. The subspace of $\mathbf{M}^{2,2}$ consisting of all 2×2 matrices whose diagonal entries are zero

41. The subspace of $\mathbf{P}_1$ consisting of all polynomials $\mathbf{p}(x) = a_0 + a_1x$ such that $a_0 + a_1 = 0$

42. The subspace of $\mathbf{P}_3$ consisting of all polynomials

$$\mathbf{p}(x) = a_0 + a_1x + a_2x^2 + a_3x^3$$

with $a_2 = 0$

43. Show that $\mathcal{T}_1 = \{1, x + 1\}$, $\mathcal{T}_2 = \{1, x - 1\}$, and $\mathcal{T}_3 = \{x + 1, x - 1\}$ are all bases for $\mathbf{P}_1$.

44. Prove part (c) of Theorem 1.

45. If $\mathcal{T}$ is a subset of a vector space $\mathbf{V}$, and $\mathbf{0}$ is in $\mathcal{T}$, prove that $\mathcal{T}$ is linearly dependent.

46. Let $\mathcal{S} = \{\mathbf{v}_1, \mathbf{v}_2, \ldots, \mathbf{v}_n\}$ be a linearly independent set in a vector space $\mathbf{V}$.

(a) Show that the set $\{\mathbf{v}_1, \mathbf{v}_2, \ldots, \mathbf{v}_m\}$ is linearly independent if $m < n$.

(b) More generally, show that any subset of $\mathcal{S}$ is linearly independent.

47. Let $\mathcal{S} = \{\mathbf{v}_1, \mathbf{v}_2, \ldots, \mathbf{v}_n\}$ be a linearly dependent set in a vector space $\mathbf{V}$.

(a) If $\mathbf{w}$ is a vector in $\mathbf{V}$, show that the set $\{\mathbf{w}, \mathbf{v}_1, \mathbf{v}_2, \ldots, \mathbf{v}_n\}$ is linearly dependent.

(b) More generally, if $\mathcal{T}$ is any set in $\mathbf{V}$ that contains $\mathcal{S}$, show that $\mathcal{T}$ is linearly dependent.

48. If $\mathcal{B}$ is a basis for a vector space $\mathbf{V}$ and $\mathbf{v}$ is in $\mathbf{V}$, show that the representation of $\mathbf{v}$ as a linear combination of elements of $\mathcal{B}$ is unique.

6.3 Coordinate Vectors

In this section, we show how the theory of $\mathbf{R}^m$ developed in Chapters 2 through 5 can sometimes be usefully applied to more general finite dimensional vector spaces.

If $\mathcal{B}$ is a basis for a vector space $\mathbf{V}$, it follows from Exercise 48 of Section 6.2 that each vector $\mathbf{v}$ in $\mathbf{V}$ can be expressed as a linear combination of the elements of $\mathcal{B}$ in one and only one way.

DEFINITION

Let $\mathcal{B} = \{\mathbf{b}_1, \ldots, \mathbf{b}_n\}$ be a basis for a vector space $\mathbf{V}$ and $\mathbf{v}$ be a vector in $\mathbf{V}$. The unique vector $\mathbf{x} = (x_1, \ldots, x_n)$ in $\mathbf{R}^n$ such that $\mathbf{v} = x_1\mathbf{b}_1 + \cdots + x_n\mathbf{b}_n$ is called the **coordinate vector** of $\mathbf{v}$ with respect to the basis $\mathcal{B}$.

EXAMPLE 1

Find the coordinate vector of the polynomial $\mathbf{p}(x) = 1 + 2x - 3x^2$ with respect to the basis $\{1, x, x^2, x^3\}$ for $\mathbf{P}_3$, the vector space of polynomials of degree less than or equal to 3.

Solution Since $\mathbf{p}(x) = 1(1) + 2(x) + (-3)(x^2) + 0(x^3)$, the coordinate vector is $(1, 2, -3, 0)$. ■

NOTE If $\mathbf{V}$ is a vector space with basis $\mathcal{B} = \{\mathbf{b}_1, \ldots, \mathbf{b}_n\}$, then the coordinate vector of each $\mathbf{b}_i$ is $\mathbf{e}_i$, *the* ith standard basis vector for $\mathbf{R}^n$.

THEOREM 1

Let $\mathbf{V}$ be a vector space with basis $\mathcal{B} = \{\mathbf{b}_1, \ldots, \mathbf{b}_n\}$, $\mathbf{v}, \mathbf{v}_1, \ldots \mathbf{v}_m$ be elements of $\mathbf{V}$, and $\mathbf{x}, \mathbf{x}_1, \ldots, \mathbf{x}_m$ be their respective coordinate vectors in $\mathbf{R}^n$. Then $\mathbf{v}$ is a linear combination of $\mathbf{v}_1, \ldots, \mathbf{v}_m$ if and only if $\mathbf{x}$ is a linear combination of $\mathbf{x}_1, \ldots, \mathbf{x}_m$. That is, for scalars $d_1, \ldots, d_m$,

$$\mathbf{v} = d_1\mathbf{v}_1 + \cdots + d_m\mathbf{v}_m$$

if and only if

$$\mathbf{x} = d_1\mathbf{x}_1 + \cdots + d_m\mathbf{x}_m.$$

Proof For each $i = 1, \ldots, m$, let $\mathbf{x}_i = (x_{i1}, \ldots, x_{in})$ be the coordinate vector of $\mathbf{v}_i$ and $\mathbf{x} = (x_1, \ldots, x_n)$ be the coordinate vector of $\mathbf{v}$, all with respect to the basis $\mathcal{B}$. That is,

$$\mathbf{v}_i = x_{i1}\mathbf{b}_1 + \cdots + x_{in}\mathbf{b}_n \quad (i = 1, \ldots, m)$$
$$\text{and} \quad \mathbf{v} = x_1\mathbf{b}_1 + \cdots + x_n\mathbf{b}_n.$$

Assuming that $\mathbf{v}$ is a linear combination of the $\mathbf{v}_i$, there are scalars d_i such that $\mathbf{v} = d_1\mathbf{v}_1 + \cdots + d_m\mathbf{v}_m$. Replacing each $\mathbf{v}$ in this equation by its representation as a linear combination of the basis vectors, we obtain

$$\mathbf{v} = d_1(x_{11}\mathbf{b}_1 + \cdots + x_{1n}\mathbf{b}_n) + \cdots + d_m(x_{m1}\mathbf{b}_1 + \cdots + x_{mn}\mathbf{b}_n),$$

or by regrouping terms

$$\mathbf{v} = (d_1x_{11} + \cdots + d_mx_{m1})\mathbf{b}_1 + \cdots + (d_1x_{1n} + \cdots + d_mx_{mn})\mathbf{b}_n.$$

This last equation expresses $\mathbf{v}$ as a linear combination of basis vectors, and such a representation is unique. But we also have $\mathbf{v} = x_1\mathbf{b}_1 + \cdots + x_n\mathbf{b}_n$. Thus, $x_i = d_1x_{1i} + \cdots + d_mx_{mi}$ for each $i = 1, \ldots, n$.

Taken collectively,

$$\begin{aligned}\mathbf{x} &= (x_1, \ldots, x_n)\\ &= (d_1x_{11} + \cdots + d_mx_{m1}, \ldots, d_1x_{1n} + \cdots + d_mx_{mn})\\ &= d_1(x_{11}, \ldots, x_{1n}) + \cdots + d_m(x_{m1}, \ldots, x_{mn})\\ &= d_1\mathbf{x}_1 + \cdots + d_m\mathbf{x}_m.\end{aligned}$$

The statements in this proof are reversible, and thus the result is proved. ■

By letting $\mathbf{x} = \mathbf{0}$ in Theorem 1, we obtain the following corollary.

COROLLARY

Using the notation of Theorem 1, $\{\mathbf{v}_1, \ldots, \mathbf{v}_m\}$ is a linearly independent set in $\mathbf{V}$ if and only if $\{\mathbf{x}_1, \ldots, \mathbf{x}_m\}$ is a linearly independent set in $\mathbf{R}^n$. ■

EXAMPLE 2

Determine whether or not the set

$$\mathcal{S} = \left\{ \begin{bmatrix} 2 & 1 \\ 1 & 1 \end{bmatrix}, \begin{bmatrix} 1 & 0 \\ 1 & 0 \end{bmatrix}, \begin{bmatrix} 0 & 1 \\ -1 & 1 \end{bmatrix}, \begin{bmatrix} 5 & 1 \\ 4 & 3 \end{bmatrix} \right\}$$

is linearly independent in $\mathbf{M}^{2,2}$, the set of all 2×2 matrices.

Solution The coordinate vectors of the elements of $\mathcal{S}$ with respect to the basis $\{E^{11}, E^{12}, E^{21}, E^{22}\}$ (part (c) of Theorem 1 of Section 6.2) are

$$\begin{bmatrix} 2 & 1 \\ 1 & 1 \end{bmatrix} \leftrightarrow (2, 1, 1, 1),$$
$$\begin{bmatrix} 1 & 0 \\ 1 & 0 \end{bmatrix} \leftrightarrow (1, 0, 1, 0),$$
$$\begin{bmatrix} 0 & 1 \\ -1 & 1 \end{bmatrix} \leftrightarrow (0, 1, -1, 1),$$
$$\text{and } \begin{bmatrix} 5 & 1 \\ 4 & 3 \end{bmatrix} \leftrightarrow (5, 1, 4, 3).$$

As a consequence of the corollary to Theorem 1, $\mathcal{S}$ is linearly independent if and only if $\mathcal{S}' = \{(2, 1, 1, 1), (1, 0, 1, 0), (0, 1, -1, 1), (5, 1, 4, 3)\}$ is linearly independent. But these are vectors in $\mathbf{R}^4$, so we may use the methods of Chapter 5. Since

$$\det \begin{bmatrix} 2 & 1 & 0 & 5 \\ 1 & 0 & 1 & 1 \\ 1 & 1 & -1 & 4 \\ 1 & 0 & 1 & 3 \end{bmatrix} = 0,$$

$\mathcal{S}'$ is linearly dependent. Hence, $\mathcal{S}$ is as well. ■

NOTE Since the *order* in which basis elements are listed affects the coordinate vectors with respect to that basis, an order must be chosen and fixed throughout the computation. The order for $\{E^{ij}\}$ of Theorem 1 of Section 6.2 is taken *along rows*. That is, let $i = 1$, and take $j = 1, \ldots, n$; then let $i = 2$, and take $j = 1, \ldots, n$; and so on.

EXAMPLE 3

Show that $\mathcal{B} = \{1, x - 1, (x - 1)^2\}$ is a basis for $\mathbf{P}_2$, the set of all polynomials of degree less than or equal to 2, and express

$$\mathbf{p}(x) = 2x^2 + x - 1$$

as a linear combination of the elements of $\mathcal{B}$.

Solution We know, from part (a) of Theorem 1 of Section 6.2, that $\mathcal{B}' = \{1, x, x^2\}$ is a basis for $\mathbf{P}_2$. The coordinate vectors of each element of $\mathcal{B}$ with respect to the basis $\mathcal{B}'$ are

$$1 \leftrightarrow (1, 0, 0),$$
$$x - 1 = -1 + x \leftrightarrow (-1, 1, 0),$$
$$\text{and } (x - 1)^2 = 1 - 2x + x^2 \leftrightarrow (1, -2, 1).$$

Applying Theorem 1 to the coordinate vectors, we know that $\mathcal{B}$ is linearly independent, since

$$\det\begin{bmatrix}1 & -1 & 1\\ 0 & 1 & -2\\ 0 & 0 & 1\end{bmatrix} \neq 0.$$

Since $\mathcal{B}$ is linearly independent and $\mathbf{P}_2$ is a three-dimensional vector space, $\mathcal{B}$ is a basis.

For the second part, we need the coordinate vector of $\mathbf{p}$,

$$2x^2 + x - 1 = -1 + x + 2x^2 \leftrightarrow (-1, 1, 2).$$

According to Theorem 1, the coefficients needed to express $(-1, 1, 2)$ as a linear combination of $\{(1, 0, 0), (-1, 1, 0), (1, -2, 1)\}$ are exactly the coefficients needed to express $\mathbf{p}$ as a linear combination of elements of $\mathcal{B}$. But this is also a problem from Chapter 5. We perform the row reduction

$$\left[\begin{array}{ccc|c}1 & -1 & 1 & -1\\ 0 & 1 & -2 & 1\\ 0 & 0 & 1 & 2\end{array}\right] \rightarrow \left[\begin{array}{ccc|c}1 & 0 & 0 & 2\\ 0 & 1 & 0 & 5\\ 0 & 0 & 1 & 2\end{array}\right]$$

and, as a consequence of Theorem 5 of Section 5.1, we obtain from the last column

$$\mathbf{p}(x) = 2(1) + 5(x - 1) + 2(x - 1)^2.$$

■

EXAMPLE 4

Show that $\mathcal{S} = \{x^2 - 1, x^2 + 1\}$ is a linearly independent subset of $\mathbf{P}_3$, and extend $\mathcal{S}$ to a basis.

Solution Letting $\mathcal{B} = \{1, x, x^2, x^3\}$, we have the coordinate vectors

$$x^2 - 1 = -1 + 0x + x^2 + 0x^3 \leftrightarrow (-1, 0, 1, 0)$$
$$\text{and} \quad x^2 + 1 = 1 + 0x + x^2 + 0x^3 \leftrightarrow (1, 0, 1, 0).$$

Forming the matrix of columns and row reducing, we have

$$\begin{bmatrix}-1 & 1\\ 0 & 0\\ 1 & 1\\ 0 & 0\end{bmatrix} \rightarrow \begin{bmatrix}1 & 0\\ 0 & 1\\ 0 & 0\\ 0 & 0\end{bmatrix}.$$

Since the columns of the row-reduced matrix are distinct elementary columns, by Theorem 4 of Section 5.1, the coordinate vectors are linearly independent. Hence, the original polynomials are as well.

To extend $\mathcal{S}$ to a basis, we augment the matrix of coordinate vector columns by a basis for $\mathbf{R}^4$ (specifically, the standard basis) and row-reduce:

$$\left[\begin{array}{cc|cccc}-1 & 1 & 1 & 0 & 0 & 0\\ 0 & 0 & 0 & 1 & 0 & 0\\ 1 & 1 & 0 & 0 & 1 & 0\\ 0 & 0 & 0 & 0 & 0 & 1\end{array}\right] \rightarrow \left[\begin{array}{cc|cccc}1 & 0 & -\frac{1}{2} & 0 & \frac{1}{2} & 0\\ 0 & 1 & \frac{1}{2} & 0 & \frac{1}{2} & 0\\ 0 & 0 & 0 & 1 & 0 & 0\\ 0 & 0 & 0 & 0 & 0 & 1\end{array}\right]$$

From the row-reduced matrix, we see that $\mathbf{e}_2$ and $\mathbf{e}_4$ are needed to form a basis for $\mathbf{R}^4$. The corresponding polynomials are x and x^3. Thus, $\{x^2 - 1, x^2 + 1, x, x^3\}$ is a basis for $\mathbf{P}_3$.

■

COMPUTATIONAL NOTE

More abstract vector spaces may be so complicated that computer assistance is needed to study them. Is it necessary to develop new software to handle these vector spaces? The answer is, "Of course not!"

Coordinate vectors provide the mechanism for converting the original situation into the familiar $\mathbf{R}^n$ case, where row-reduction, inverse, and other programs exist to resolve the problems at hand. The results must then be translated back to the original setting to complete the solution.

Examples 2, 3, and 4 have small integer coefficients to make the arithmetic easy to do by hand, but that is not necessary for computer or calculator applications. See Exercises 22 through 25, which demonstrate the idea with concrete examples.

6.3 Exercises

In Exercises 1–5 find the coordinate vector of v with respect to the given basis $\mathscr{B}$ for the vector space V.

1. $\mathbf{v} = 2 - x + 3x^3$, $\mathscr{B} = \{1, x, x^2, x^3\}$, $\mathbf{V} = \mathbf{P}_3$

2. $\mathbf{v} = \begin{bmatrix} 1 & 2 & 1 \\ -1 & 1 & 2 \end{bmatrix}$, $\mathscr{B} = \{E^{ij} \mid i = 1, 2; j = 1, 2, 3\}$, $\mathbf{V} = \mathbf{M}^{2,3}$

3. $\mathbf{v} = 2 - 5x$, $\mathscr{B} = \{x + 1, x - 1\}$, $\mathbf{V} = \mathbf{P}_1$

4. $\mathbf{v} = \begin{bmatrix} -2 & 0 \\ 0 & 3 \end{bmatrix}$, $\mathscr{B} = \left\{\begin{bmatrix} 1 & 0 \\ 0 & 0 \end{bmatrix}, \begin{bmatrix} 0 & 0 \\ 0 & 1 \end{bmatrix}\right\}$

$\mathbf{V}$ is the vector space of all 2×2 diagonal matrices.

5. $\mathbf{v} = \begin{bmatrix} -2 & 0 \\ 0 & 3 \end{bmatrix}$, $\mathscr{B} = \left\{\begin{bmatrix} 3 & 1 \\ 0 & 2 \end{bmatrix}, \begin{bmatrix} 1 & -1 \\ 0 & 1 \end{bmatrix}, \begin{bmatrix} 2 & 2 \\ 0 & 0 \end{bmatrix}\right\}$

$\mathbf{V}$ is the vector space of all 2×2 matrices A such that $a_{21} = 0$.

In Exercises 6–12 use coordinate vectors to decide whether or not the given set is linearly independent. If it is linearly dependent, express one of the vectors as a linear combination of the others.

6. $\mathscr{S}_1 = \{x^2 + x - 1, x^2 - 2x + 3, x^2 + 4x - 5\}$ in $\mathbf{P}_2$

7. $\mathscr{S}_2 = \{x^2 + x - 1, x^2 - 2x + 3, x^2 + 4x - 3\}$ in $\mathbf{P}_3$

8. $\mathscr{S}_3 = \left\{\begin{bmatrix} 2 & -1 \\ 1 & 0 \end{bmatrix}, \begin{bmatrix} 1 & 0 \\ 2 & 1 \end{bmatrix}, \begin{bmatrix} 3 & 1 \\ -1 & 1 \end{bmatrix}, \begin{bmatrix} -1 & 4 \\ 2 & 0 \end{bmatrix}\right\}$ in $\mathbf{M}^{2,2}$

9. $\mathscr{S}_4 = \left\{\begin{bmatrix} 1 & 2 \\ -1 & 0 \end{bmatrix}, \begin{bmatrix} 0 & -1 \\ 1 & 1 \end{bmatrix}, \begin{bmatrix} 1 & 0 \\ 1 & 2 \end{bmatrix}\right\}$ in $\mathbf{M}^{2,2}$

10. $\mathscr{S}_5 = \left\{\begin{bmatrix} 2 & 1 \\ 1 & 3 \end{bmatrix}, \begin{bmatrix} 3 & -1 \\ -1 & 2 \end{bmatrix}, \begin{bmatrix} 1 & 0 \\ 0 & 1 \end{bmatrix}, \begin{bmatrix} 2 & 2 \\ 2 & 4 \end{bmatrix}\right\}$,

in the subspace of $\mathbf{M}^{2,2}$ of all matrices of the form

$$\begin{bmatrix} a & b \\ b & a + b \end{bmatrix}$$

11. $\{2 + x - 3 \sin x + \cos x, x + \sin x - 3 \cos x, 1 - 2x + 3 \sin x + \cos x, 2 - x - \sin x - \cos x, 2 + x - 3 \cos x\}$ in the space of all linear combinations of the set $\{1, x, \sin x, \cos x\}$

12. All 2×2 matrices with exactly two 1's and two 0's for entries in the space of all 2×2 matrices

In Exercises 13–17 the given set is linearly independent. Extend it to a basis; that is, include other elements (from the standard basis for the vector space) so that the resulting set is a basis.

13. $\{x^2 - x + 1, x - 1, x + 1\}$ in $\mathbf{P}_2$

14. $\{x + 1, x^2 + x, x^3 + x^2 + x\}$ in $\mathbf{P}_3$

15. $\left\{\begin{bmatrix} 1 & -1 \\ 1 & 0 \end{bmatrix}, \begin{bmatrix} 1 & 1 \\ 0 & 0 \end{bmatrix}, \begin{bmatrix} 0 & 0 \\ 1 & 1 \end{bmatrix}\right\}$ in $\mathbf{M}^{2,2}$

16. $\left\{\begin{bmatrix} 1 & 2 \\ 3 & 1 \end{bmatrix}, \begin{bmatrix} 1 & 3 \\ 2 & 1 \end{bmatrix}\right\}$ in $\mathbf{M}^{2,2}$

17. $\left\{\begin{bmatrix} 1 & 0 & -1 \\ 1 & -1 & 1 \end{bmatrix}, \begin{bmatrix} 0 & 1 & 1 \\ 2 & 1 & -1 \end{bmatrix}, \begin{bmatrix} 1 & -1 & 1 \\ 1 & -1 & 1 \end{bmatrix}, \begin{bmatrix} 0 & 1 & 0 \\ 0 & 1 & 0 \end{bmatrix}\right\}$ in $\mathbf{M}^{2,3}$

In Exercises 18–21 the given set spans the space. Reduce it to an independent set that has the same span.

18. $\{x^2 - x + 1, x^2 + 2x + 1, x - 2, x + 2, x^2 - 1, x^2 + 1\}$ in $\mathbf{P}_2$

19. $\{x + 1, x - 1, x^2 + 1, x^2 - 1, x^2 + x, x^2 - x\}$ in $\mathbf{P}_2$

20. $\left\{\begin{bmatrix} 0 & 0 & 0 \\ 0 & -1 & 0 \\ 0 & 0 & 1 \end{bmatrix}, \begin{bmatrix} 2 & 0 & 0 \\ 0 & 1 & 0 \\ 0 & 0 & 1 \end{bmatrix}, \begin{bmatrix} 1 & 0 & 0 \\ 0 & 1 & 0 \\ 0 & 0 & 1 \end{bmatrix}, \begin{bmatrix} 2 & 0 & 0 \\ 0 & 2 & 0 \\ 0 & 0 & -1 \end{bmatrix}\right\}$

in the vector space of 3×3 diagonal matrices

21. $\left\{\begin{bmatrix} 2 & 1 \\ 0 & 1 \end{bmatrix}, \begin{bmatrix} -1 & 1 \\ 0 & 0 \end{bmatrix}, \begin{bmatrix} 1 & 2 \\ 0 & 2 \end{bmatrix}, \begin{bmatrix} 1 & -1 \\ 0 & -1 \end{bmatrix}, \begin{bmatrix} 2 & -1 \\ 0 & -1 \end{bmatrix}\right\}$

in the space of all 2×2 matrices A with $a_{21} = 0$

Computational Exercises

In Exercises 22 and 23 decide if the functions in the given set are linearly independent or dependent. If they are dependent, express one of them as a linear combination of the others.

$$\begin{aligned} f_1(x) &= 1.1x + 6.3x^2 - 3.1 \sin x \\ f_2(x) &= 2.1x - 3.3x^2 + 5.2 \sin x \\ f_3(x) &= 1.3x + 1.2x^2 - 1.1 \sin x \\ f_4(x) &= 8.9x - 7.5x^2 + 13.4 \sin x \end{aligned}$$

22. $S = \{f_1, f_2, f_3\}$

23. $S = \{f_1, f_2, f_3, f_4\}$

In Exercises 24 and 25 use a computer algebra system to find all values for the coefficients *a* and *b* for which the set of vectors is dependent.

$$f_1(x) = 1 + x + x^2 + \sin x$$
$$f_2(x) = 2 + x + 3x^2 + \sin x$$
$$f_3(x) = 1 - x + x^2 + a\sin x$$
$$f_4(x) = 3 - (a + b)x + x^2 + b\sin x$$

24. $S = \{f_1, f_2, f_3, f_4\}$

25. $S = \{f_1, 2f_3, 3f_3, 4f_4\}$

6.4 Inner Product Spaces

In Section 2.1, we introduced the notion of the *dot product* (also called *inner product* or *scalar product*) of two vectors in $\mathbf{R}^m$. In this section, we generalize this notion to vectors in an arbitrary vector space. This, in turn, leads to a natural definition of *orthogonality* of such vectors and the study of bases that consist of orthogonal vectors.

In Theorem 3 of Section 2.1, we gave several important properties of the dot product in $\mathbf{R}^m$. We now use these properties to define our extended notion of the dot product. Following standard convention, we use the alternative terminology *inner product* and also change the previous notation in the general definition.

DEFINITION

Let $\mathbf{V}$ be a vector space over the real numbers. An **inner product** for $\mathbf{V}$ is a function that associates with every pair of vectors, $\mathbf{u}$ and $\mathbf{v}$, in $\mathbf{V}$, a real number, $(\mathbf{u}, \mathbf{v})$, satisfying the following properties. For all $\mathbf{u}$, $\mathbf{v}$, and $\mathbf{w}$ in $\mathbf{V}$ and scalars c,

a. $(\mathbf{u}, \mathbf{v}) = (\mathbf{v}, \mathbf{u})$

b. $(\mathbf{u}, \mathbf{v} + \mathbf{w}) = (\mathbf{u}, \mathbf{v}) + (\mathbf{u}, \mathbf{w})$

c. $(c\mathbf{u}, \mathbf{v}) = c(\mathbf{u}, \mathbf{v})$

d. $(\mathbf{u}, \mathbf{u}) \geq 0$ and $(\mathbf{u}, \mathbf{u}) = 0$, if and only if $\mathbf{u} = \mathbf{0}$.

A vector space $\mathbf{V}$ together with an inner product is called an **inner product space**.

EXAMPLE 1

Let $\mathbf{u} = (u_1, u_2, \ldots, u_m)$ and $\mathbf{v} = (v_1, v_2, \ldots, v_m)$ be arbitrary vectors in the vector space $\mathbf{R}^m$. Define

$$(\mathbf{u}, \mathbf{v}) = u_1v_1 + u_2v_2 + \cdots + u_mv_m,$$

the dot product of $\mathbf{u}$ and $\mathbf{v}$. Then $(\mathbf{u}, \mathbf{v})$ defines an inner product for $\mathbf{R}^m$.

Solution By Theorem 3 of Section 2.1, this function satisfies the four properties of the definition and is therefore an inner product. ■

Example 1 verifies that the dot product for $\mathbf{R}^m$, defined in Section 2.1, is an inner product for this vector space. The next example demonstrates that we can define other inner products for $\mathbf{R}^m$—there may be more than one inner product for a given vector space.

EXAMPLE 2

Let $\mathbf{u} = (u_1, u_2, \ldots, u_m)$ and $\mathbf{v} = (v_1, v_2, \ldots, v_m)$ be arbitrary vectors in $\mathbf{R}^m$. Define

$$(\mathbf{u}, \mathbf{v}) = u_1v_1 + 2u_2v_2 + 3u_3v_3 + \cdots + mu_mv_m.$$

Then $(\mathbf{u}, \mathbf{v})$ is an inner product for $\mathbf{R}^m$.

Solution We must verify properties (a) through (d) of the definition:

a. $$\begin{aligned}(\mathbf{u}, \mathbf{v}) &= u_1v_1 + 2u_2v_2 + \cdots + mu_mv_m \\ &= v_1u_1 + 2v_2u_2 + \cdots + mv_mu_m \\ &= (\mathbf{v}, \mathbf{u})\end{aligned}$$

b. $$\begin{aligned}(\mathbf{u}, \mathbf{v} + \mathbf{w}) &= u_1(v_1 + w_1) + 2u_2(v_2 + w_2) + \cdots + mu_m(v_m + w_m) \\ &= (u_1v_1 + u_1w_1) + (2u_2v_2 + 2u_2w_2) + \\ &\quad \cdots + (mu_mv_m + mu_mw_m) \\ &= (u_1v_1 + 2u_2v_2 + \cdots + mu_mv_m) + (u_1w_1 + 2u_2w_2 + \\ &\quad \cdots + mu_mw_m) \\ &= (\mathbf{u}, \mathbf{v}) + (\mathbf{u}, \mathbf{w})\end{aligned}$$

c. $$\begin{aligned}(c\mathbf{u}, \mathbf{v}) &= cu_1v_1 + 2cu_2v_2 + \cdots + mcu_mv_m \\ &= c(u_1v_1 + 2u_2v_2 + \cdots + mu_mv_m) \\ &= c(\mathbf{u}, \mathbf{v})\end{aligned}$$

d. $(\mathbf{u}, \mathbf{u}) = u_1^2 + 2u_2^2 + \cdots + mu_m^2$,

which is nonnegative, and if

$u_1^2 + 2u_2^2 + \cdots + mu_m^2 = 0$,

then $u_1 = u_2 = \cdots = u_m = 0$

Thus, $\mathbf{u} = \mathbf{0}$, as desired. ■

NOTE Example 2 shows that inner products other than the dot product may be defined on $\mathbf{R}^m$. If no inner product on $\mathbf{R}^m$ is explicitly mentioned, we assume that the usual dot product is the inner product intended.

EXAMPLE 3

Let $\mathbf{p}$ and $\mathbf{q}$ be arbitrary vectors in $\mathbf{P}_1$, the vector space of all polynomials in x of degree less than or equal to 1. Define

$$(\mathbf{p}, \mathbf{q}) = \mathbf{p}(0)\mathbf{q}(0) + \mathbf{p}(1)\mathbf{q}(1).$$

Then $(\mathbf{p}, \mathbf{q})$ is an inner product for $\mathbf{P}_1$.

Solution Properties (a), (b), and (c) are easily verified. For property (d), we have $(\mathbf{p}, \mathbf{p}) = [\mathbf{p}(0)]^2 + [\mathbf{p}(1)]^2$, which is nonnegative. Moreover, $(\mathbf{p}, \mathbf{p}) = 0$ if and only if $\mathbf{p}(0) = 0$ and $\mathbf{p}(1) = 0$. But the latter implies that $\mathbf{p}(x)$ is the zero polynomial, since the graph of $\mathbf{p}(x)$, a line, goes through the points (0, 0) and (1, 0) and is therefore the graph of the zero polynomial. ■

EXAMPLE 4

(*From calculus*) Let $\mathbf{f}$ and $\mathbf{g}$ be vectors in $\mathbf{C}[-1, 1]$, the vector space of all real-valued functions continuous on the interval $[-1, 1]$. (See the note following Example 6 of Section 6.1.) Define

$$(\mathbf{f}, \mathbf{g}) = \int_{-1}^{1} \mathbf{f}(x)\mathbf{g}(x)\, dx.$$

Then $(\mathbf{f}, \mathbf{g})$ is an inner product for $\mathbf{C}[-1, 1]$.

Solution The verification that $(\mathbf{f}, \mathbf{g})$ is an inner product rests on some properties of the definite integral learned in calculus. Specifically, properties (b), (c), and (d) hold because, respectively,

$$\int_{-1}^{1} \mathbf{f}(x)(\mathbf{g}(x) + \mathbf{h}(x))\, dx = \int_{-1}^{1} \mathbf{f}(x)\mathbf{g}(x)\, dx + \int_{-1}^{1} \mathbf{f}(x)\mathbf{h}(x)\, dx,$$

$$\int_{-1}^{1} c\mathbf{f}(x)\mathbf{g}(x)\,dx = c\int_{-1}^{1} \mathbf{f}(x)\mathbf{g}(x)\,dx,$$

$$\text{and} \quad \int_{-1}^{1} (\mathbf{f}(x))^2\,dx \geq 0 \qquad \text{with} \qquad \int_{-1}^{1} (\mathbf{f}(x))^2\,dx = 0$$

if and only if $\mathbf{f}(x)$ is the zero function on $[-1, 1]$. Property (a) holds simply because $\mathbf{f}(x)\mathbf{g}(x) = \mathbf{g}(x)\mathbf{f}(x)$ for all x in $[-1, 1]$. ■

From the defining properties of the inner product, we can derive several other properties satisfied by all inner products. Some of these are presented in the next two theorems.

THEOREM 1

Let $\mathbf{V}$ be an inner product space. Then for all vectors $\mathbf{u}_1$, $\mathbf{u}_2$, and $\mathbf{v}$ in $\mathbf{V}$ and scalars c_1 and c_2,

a. $(c_1\mathbf{u}_1 + c_2\mathbf{u}_2, \mathbf{v}) = (\mathbf{v}, c_1\mathbf{u}_1 + c_2\mathbf{u}_2) = c_1(\mathbf{u}_1, \mathbf{v}) + c_2(\mathbf{u}_2, \mathbf{v})$,

b. $(\mathbf{0}, \mathbf{v}) = (\mathbf{v}, \mathbf{0}) = 0$.

Proof

a.
$$\begin{aligned}
(c_1\mathbf{u}_1 + c_2\mathbf{u}_2, \mathbf{v}) &= (\mathbf{v}, c_1\mathbf{u}_1 + c_2\mathbf{u}_2) && \text{(Property (a))}\\
&= (\mathbf{v}, c_1\mathbf{u}_1) + (\mathbf{v}, c_2\mathbf{u}_2) && \text{(Property (b))}\\
&= (c_1\mathbf{u}_1, \mathbf{v}) + (c_2\mathbf{u}_2, \mathbf{v}) && \text{(Property (a))}\\
&= c_1(\mathbf{u}_1, \mathbf{v}) + c_2(\mathbf{u}_2, \mathbf{v}) && \text{(Property (c))}
\end{aligned}$$

b. We leave this proof for Exercise 32. 

In Section 2.1, we saw that for any $\mathbf{u}$ in $\mathbf{R}^m$, the *length*, or *norm*, of $\mathbf{u}$, $\|\mathbf{u}\|$, is equal to $\sqrt{\mathbf{u} \cdot \mathbf{u}}$. In a general inner product space, we use this for a definition. If $\mathbf{u}$ is a vector in an inner product space, then the **norm** of $\mathbf{u}$, denoted $\|\mathbf{u}\|$, is given by $\|\mathbf{u}\| = \sqrt{(\mathbf{u}, \mathbf{u})}$. The next theorem, known as the **Cauchy–Schwarz inequality**, relates the inner product of vectors $\mathbf{u}$ and $\mathbf{v}$ to the norms of $\mathbf{u}$ and $\mathbf{v}$.

THEOREM 2

Let $\mathbf{u}$ and $\mathbf{v}$ be vectors in an inner product space. Then

$$|(\mathbf{u}, \mathbf{v})| \leq \|\mathbf{u}\|\,\|\mathbf{v}\|,$$

where $|(\mathbf{u}, \mathbf{v})|$ denotes the *absolute value* of $(\mathbf{u}, \mathbf{v})$.

Proof First note that if $\mathbf{u} = \mathbf{0}$, then $(\mathbf{u}, \mathbf{v}) = 0$, by part (b) of Theorem 1, and

$$\|\mathbf{u}\| = \sqrt{(\mathbf{u}, \mathbf{u})} = 0$$

as well. Thus, the theorem holds if $\mathbf{u} = \mathbf{0}$.

Now assume that $\mathbf{u} \neq \mathbf{0}$. If x is an arbitrary real number, we have

$$\begin{aligned}
0 &\leq (x\mathbf{u} + \mathbf{v}, x\mathbf{u} + \mathbf{v}) && \text{(Property (d))}\\
&= x^2(\mathbf{u}, \mathbf{u}) + x(\mathbf{u}, \mathbf{v}) + x(\mathbf{v}, \mathbf{u}) + (\mathbf{v}, \mathbf{v}) && \text{(Part (a) of Theorem 1)}\\
&= x^2(\mathbf{u}, \mathbf{u}) + 2x(\mathbf{u}, \mathbf{v}) + (\mathbf{v}, \mathbf{v}). && \text{(Property (a))}
\end{aligned}$$

Now, setting $(\mathbf{u}, \mathbf{u}) = a$, $2(\mathbf{u}, \mathbf{v}) = b$, and $(\mathbf{v}, \mathbf{v}) = c$, this inequality becomes $ax^2 + bx + c \geq 0$, which is quadratic in x. In geometric terms, this implies that the parabola $y = ax^2 + bx + c$ must be contained in the region on and above the x-axis (Figures 1 and 2). But this, in turn, implies that the polynomial $ax^2 + bx + c$ cannot have two distinct real zeros, since, if it did, the curve would dip below the x-axis (Figure 3).

FIGURE 1
No real roots

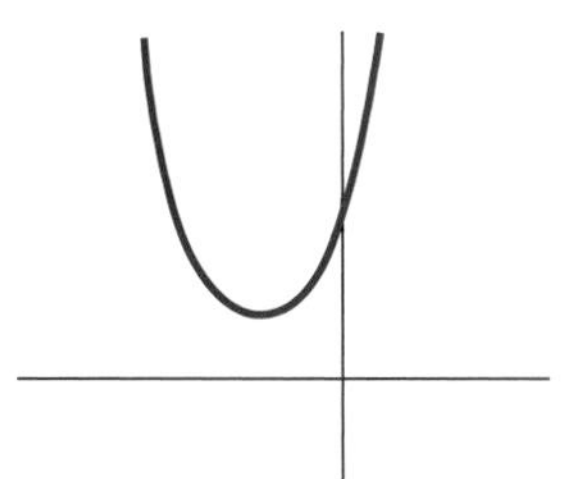

FIGURE 2
One real root

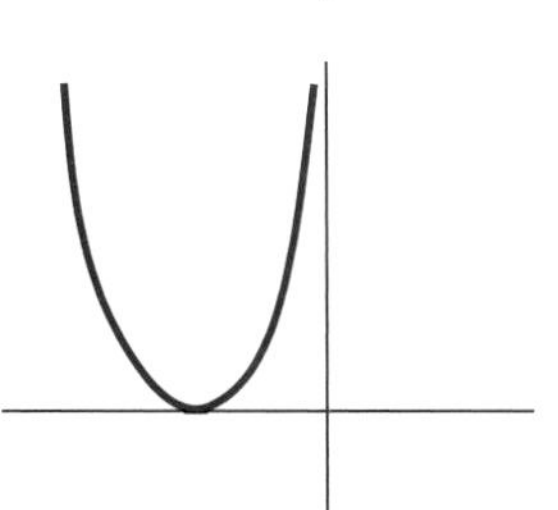

FIGURE 3
Two real roots

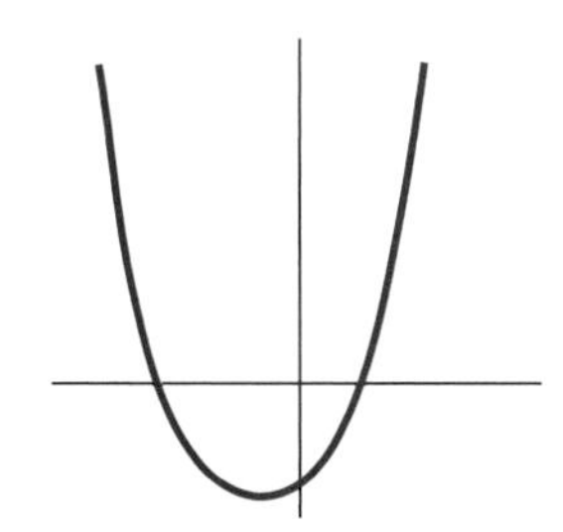

Now the equation $ax^2 + bx + c = 0$ fails to have two real roots (it has either no roots or one root) if and only if $b^2 - 4ac \leq 0$. But this implies that $b^2 \leq 4ac$, or

$$4(\mathbf{u}, \mathbf{v})^2 \leq 4(\mathbf{u}, \mathbf{u})(\mathbf{v}, \mathbf{v}),$$

$$\text{or} \qquad (\mathbf{u}, \mathbf{v})^2 \leq (\mathbf{u}, \mathbf{u})(\mathbf{v}, \mathbf{v}).$$

Finally, taking square roots of both sides, we obtain

$$\sqrt{(\mathbf{u}, \mathbf{v})^2} \leq \sqrt{(\mathbf{u}, \mathbf{u})}\,\sqrt{(\mathbf{v}, \mathbf{v})}$$

$$\text{or} \qquad |(\mathbf{u}, \mathbf{v})| \leq \|\mathbf{u}\|\,\|\mathbf{v}\|.$$

■

EXAMPLE 5

With the inner products of Examples 1 and 4, the Cauchy–Schwarz inequality becomes

$$|u_1v_1 + u_2v_2 + \cdots + u_mv_m| \leq \sqrt{u_1^2 + u_2^2 + \cdots + u_m^2}\,\sqrt{v_1^2 + v_2^2 + \cdots + v_m^2}$$

$$\text{and} \qquad \left|\int_{-1}^{1} \mathbf{f}(x)\mathbf{g}(x)\,dx\right| \leq \sqrt{\int_{-1}^{1} (\mathbf{f}(x))^2\,dx}\,\sqrt{\int_{-1}^{1} (\mathbf{g}(x))^2\,dx},$$

respectively.

■

Orthogonal Sets and Orthogonal Bases

The standard basis for $\mathbf{R}^m$, $\{\mathbf{e}_1, \mathbf{e}_2, \ldots, \mathbf{e}_m\}$, has several computational advantages over an arbitrary one. The standard basis vectors are pairwise orthogonal, each has unit length, and it is a simple matter to write an arbitrary vector in $\mathbf{R}^m$ as a linear combination of these basis vectors. Unfortunately, subspaces of $\mathbf{R}^m$ do not ordinarily have bases that are subsets of $\{\mathbf{e}_1, \mathbf{e}_2, \ldots, \mathbf{e}_m\}$. Nevertheless, we can construct bases for them—*orthonormal* ones—that do have the nice properties of the standard basis. The same situation applies to other inner product spaces as well. Here we investigate the properties of orthonormal bases and present a method for replacing an arbitrary basis by one orthonormal basis.

Recall that two vectors in $\mathbf{R}^m$ are *orthogonal* if their dot product is equal to zero. The next definition generalizes the concept of orthogonality to inner product spaces.

DEFINITION

Two vectors, $\mathbf{u}$ and $\mathbf{v}$, in an inner product space $\mathbf{V}$ are **orthogonal** if $(\mathbf{u}, \mathbf{v}) = 0$. A set of vectors, $\mathcal{S}$, is an **orthogonal set** if every pair of distinct vectors in $\mathcal{S}$ is orthogonal. A set $\mathcal{T}$ is an **orthonormal set** if it is orthogonal and every vector in $\mathcal{T}$ has norm equal to 1.

EXAMPLE 6

Define an inner product on the vector space $\mathbf{M}^{2,2}$, the set of all 2×2 matrices, by

$$(A, B) = a_{11}b_{11} + a_{12}b_{12} + a_{21}b_{21} + a_{22}b_{22}.$$

In this inner product space, show that the matrices

$$C_1 = \begin{bmatrix} 1 & -1 \\ 1 & 0 \end{bmatrix}, \quad C_2 = \begin{bmatrix} -1 & -1 \\ 0 & 2 \end{bmatrix}, \quad \text{and} \quad C_3 = \begin{bmatrix} 0 & 2 \\ 2 & 1 \end{bmatrix}$$

form an orthogonal set. (Exercise 4 asks you to show that (A, B) is an inner product.)

Solution We show that $(C_1, C_2) = (C_1, C_3) = (C_2, C_3) = 0$ by computing each of these inner products. We have

$$(C_1, C_3) = (1)(0) + (-1)(2) + (1)(2) + (0)(1) = 0,$$

as desired. Similarly, $(C_1, C_2) = 0$ and $(C_2, C_3) = 0$. Thus, $\{C_1, C_2, C_3\}$ is orthogonal. ■

EXAMPLE 7

Show that $\mathcal{T} = \{\mathbf{v}_1, \mathbf{v}_2, \mathbf{v}_3\}$ is an orthonormal set in $\mathbf{R}^3$, where

$$\mathbf{v}_1 = \left(\frac{1}{3}, \frac{2}{3}, -\frac{2}{3}\right),$$

$$\mathbf{v}_2 = \left(0, \frac{1}{\sqrt{2}}, \frac{1}{\sqrt{2}}\right),$$

$$\text{and} \quad \mathbf{v}_3 = \left(-\frac{4}{3\sqrt{2}}, \frac{1}{3\sqrt{2}}, -\frac{1}{3\sqrt{2}}\right).$$

Solution We first check that $\mathcal{T}$ is orthogonal by verifying that the dot products $\mathbf{v}_1 \cdot \mathbf{v}_2$, $\mathbf{v}_1 \cdot \mathbf{v}_3$, and $\mathbf{v}_2 \cdot \mathbf{v}_3$ are all equal to 0. We then check the "length = 1" condition by verifying that $\|\mathbf{v}_1\| = \|\mathbf{v}_2\| = \|\mathbf{v}_3\| = 1$. Thus, $\mathcal{T}$ is an orthonormal set. ■

The following theorem (which generalizes Exercise 34 of Section 5.1) gives an important property of orthogonal sets that do not contain the zero vector.

THEOREM 3

Let $\mathcal{T} = \{\mathbf{v}_1, \mathbf{v}_2, \ldots, \mathbf{v}_n\}$ be an orthogonal set of nonzero vectors in an inner product space $\mathbf{V}$. Then $\mathcal{T}$ is linearly independent.

Proof Consider the equation $c_1\mathbf{v}_1 + c_2\mathbf{v}_2 + \cdots + c_n\mathbf{v}_n = \mathbf{0}$. We need to show that all c_i must be zero. To accomplish this, form the inner product of both sides of this equation with each vector $\mathbf{v}_j$. For each j, this yields the equation

$$c_1(\mathbf{v}_1, \mathbf{v}_j) + c_2(\mathbf{v}_2, \mathbf{v}_j) + \cdots + c_j(\mathbf{v}_j, \mathbf{v}_j) + \cdots + c_n(\mathbf{v}_n, \mathbf{v}_j) = (\mathbf{0}, \mathbf{v}_j).$$

Because the set $\mathcal{T}$ is orthogonal, however, all inner products in this equation *except* $(\mathbf{v}_j, \mathbf{v}_j)$ are equal to zero. Thus, the equation reduces to

$$c_j(\mathbf{v}_j, \mathbf{v}_j) = 0.$$

By hypothesis, $\mathbf{v}_j \neq \mathbf{0}$ for $j = 1, 2, \ldots, n$, so $(\mathbf{v}_j, \mathbf{v}_j) \neq 0$, and the last equation implies that $c_j = 0$ $(j = 1, 2, \ldots, n)$, as desired. ■

Since $\mathbf{0}$ cannot be a member of an ortho*normal* set $(\|\mathbf{0}\| = 0 \neq 1)$, Theorem 3 has an immediate corollary.

COROLLARY

An orthonormal set of vectors is linearly independent. 

As a consequence of this theorem and corollary, if an orthonormal set, or an orthogonal set of nonzero vectors, spans an inner product space, it is a basis for this space.

DEFINITION

Let $\mathbf{V}$ be an inner product space. If a set $\mathcal{T}$ of vectors is a basis for $\mathbf{V}$, and

i. If $\mathcal{T}$ is *orthogonal*, it is called an **orthogonal basis** for $\mathbf{V}$;
ii. If $\mathcal{T}$ is *orthonormal*, it is called an **orthonormal basis** for $\mathbf{V}$.

For example, it is easy to check that the standard basis for $\mathbf{R}^m$, $\{\mathbf{e}_1, \mathbf{e}_2, \ldots, \mathbf{e}_m\}$, is an orthonormal set (Exercise 31), and consequently it is an orthonormal basis for $\mathbf{R}^m$. The standard basis is certainly not the *only* orthonormal basis for $\mathbf{R}^m$, however. For example, the set $\mathcal{T}$ of Example 7 is an orthonormal set containing three vectors. By Theorem 3 and part (iii) of Theorem 5 of Section 5.3, it is an orthonormal basis for $\mathbf{R}^3$.

The next theorem gives one of the major reasons why orthonormal bases are so desirable—they provide a simple way to write an arbitrary vector in a given vector space as a linear combination of the vectors in an orthonormal basis for that space.

THEOREM 4

Let $\mathcal{T} = \{\mathbf{v}_1, \mathbf{v}_2, \ldots, \mathbf{v}_n\}$ be an orthonormal basis for an inner product space $\mathbf{V}$.

Let $\mathbf{x}$ be an arbitrary vector in $\mathbf{V}$. Then $\mathbf{x} = c_1\mathbf{v}_1 + c_2\mathbf{v}_2 + \cdots + c_n\mathbf{v}_n$, where $c_j = (\mathbf{x}, \mathbf{v}_j)(j = 1, 2, \ldots, n)$.

Proof Since $\mathcal{T}$ is a basis for $\mathbf{V}$, we know that there exist scalars $c_1, c_2, \ldots, c_n$ such that $\mathbf{x} = c_1\mathbf{v}_1 + c_2\mathbf{v}_2 + \cdots + c_n\mathbf{v}_n$. Moreover,

$$\begin{aligned}(\mathbf{x}, \mathbf{v}_j) &= ((c_1\mathbf{v}_1 + c_2\mathbf{v}_2 + \cdots + c_j\mathbf{v}_j + \cdots + c_n\mathbf{v}_n), \mathbf{v}_j) \\ &= c_1(\mathbf{v}_1, \mathbf{v}_j) + c_2(\mathbf{v}_2, \mathbf{v}_j) + \cdots + c_j(\mathbf{v}_j, \mathbf{v}_j) + \cdots + c_n(\mathbf{v}_n, \mathbf{v}_j) \\ &= c_j(\mathbf{v}_j, \mathbf{v}_j),\end{aligned}$$

due to the orthogonality of $\mathcal{T}$. But $(\mathbf{v}_j, \mathbf{v}_j) = 1$, since $\mathcal{T}$ is orthonormal, so $(\mathbf{x}, \mathbf{v}_j) = c_j$, as desired. ■

EXAMPLE 8

Express $\mathbf{x} = (1, 2, 3)$ as a linear combination of the vectors

$$\mathbf{v}_1 = \left(\frac{1}{3}, \frac{2}{3}, -\frac{2}{3}\right),$$

$$\mathbf{v}_2 = \left(0, \frac{1}{\sqrt{2}}, \frac{1}{\sqrt{2}}\right),$$

$$\text{and} \quad \mathbf{v}_3 = \left(-\frac{4}{3\sqrt{2}}, \frac{1}{3\sqrt{2}}, -\frac{1}{3\sqrt{2}}\right).$$

Solution In Example 7, we showed that $\{\mathbf{v}_1, \mathbf{v}_2, \mathbf{v}_3\}$ is an orthonormal set. By Theorem 3, this set is linearly independent; by part (iii) of Theorem 5 of Section 5.3, it is a basis for $\mathbf{R}^3$. We may therefore apply Theorem 4 (with $\mathbf{V} = \mathbf{R}^3$) to obtain $\mathbf{x} = c_1\mathbf{v}_1 + c_2\mathbf{v}_2 + c_3\mathbf{v}_3$, with

$$c_1 = \mathbf{x} \cdot \mathbf{v}_1 = -\frac{1}{3},$$

$$c_2 = \mathbf{x} \cdot \mathbf{v}_2 = \frac{5}{\sqrt{2}},$$

$$\text{and} \quad c_3 = \mathbf{x} \cdot \mathbf{v}_3 = -\frac{5}{3\sqrt{2}}.$$ ■

In the preceding discussion, we defined *orthonormal basis* and gave some of its properties. Now we describe how to construct one from a known, but nonorthonormal, basis.

If the known basis is *orthogonal*, all we need to do to produce an orthonormal basis is to *normalize* each vector in the given basis—in other words, divide each vector by its length. This preserves the orthogonality of the given set and, at the same time, creates vectors of norm 1.

EXAMPLE 9

Find an orthonormal basis for the subspace **S** of $\mathbf{R}^4$ generated by the orthogonal set $\mathcal{T} = \{\mathbf{v}_1, \mathbf{v}_2, \mathbf{v}_3\}$, where

$$\mathbf{v}_1 = (0, 2, -1, 1), \quad \mathbf{v}_2 = (0, 0, 1, 1), \quad \text{and} \quad \mathbf{v}_3 = (-2, 1, 1, -1).$$

Solution Since $\mathcal{T}$ is orthogonal, it is linearly independent, and since $\mathcal{T}$ generates **S**, it spans **S**. Hence, $\mathcal{T}$ is an *orthogonal* basis for **S**. To obtain an ortho*normal* basis for **S**, all we need to do is normalize each vector in $\mathcal{T}$. Form the set

$$\begin{aligned}\mathcal{T}' &= \left\{\left(\frac{1}{\|\mathbf{v}_1\|}\right)\mathbf{v}_1, \left(\frac{1}{\|\mathbf{v}_2\|}\right)\mathbf{v}_2, \left(\frac{1}{\|\mathbf{v}_3\|}\right)\mathbf{v}_3\right\} \\ &= \left\{\left(0, \frac{2}{\sqrt{6}}, -\frac{1}{\sqrt{6}}, \frac{1}{\sqrt{6}}\right), \left(0, 0, \frac{1}{\sqrt{2}}, \frac{1}{\sqrt{2}}\right), \left(-\frac{2}{\sqrt{7}}, \frac{1}{\sqrt{7}}, \frac{1}{\sqrt{7}}, -\frac{1}{\sqrt{7}}\right)\right\}.\end{aligned}$$

Then $\mathcal{T}'$ is an orthonormal basis for **S**. ■

When the known basis is not orthogonal, the process of finding an orthonormal basis is more complicated but still mechanical. To motivate the procedure, we consider the special case of two vectors in $\mathbf{R}^m$.

Let **u** and **v** be nonzero vectors in $\mathbf{R}^m$, and let **x** be the vector along **v** such that the points 0, X, and U form a right triangle. The vector **x** is called the **orthogonal projection** of **u** onto **v** (Figure 4).

FIGURE 4

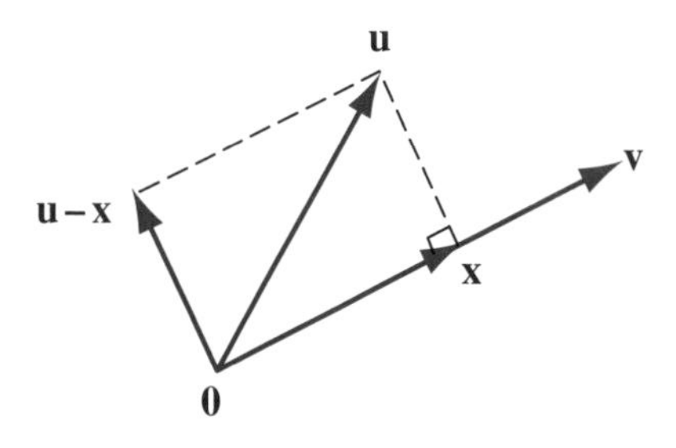

THEOREM 5

Let **u** and **v** be nonzero vectors in $\mathbf{R}^m$. The orthogonal projection of **u** onto **v** is given by

$$\mathbf{x} = \left(\frac{\mathbf{u} \cdot \mathbf{v}}{\mathbf{v} \cdot \mathbf{v}}\right)\mathbf{v}.$$

Proof Since $\mathbf{u} - \mathbf{x}$ is equivalent to $\overrightarrow{XU}$, we need only check that $\mathbf{u} - \mathbf{x}$ is orthogonal to **v**:

$$(\mathbf{u} - \mathbf{x}) \cdot \mathbf{v} = \left(\mathbf{u} - \left(\frac{\mathbf{u} \cdot \mathbf{v}}{\mathbf{v} \cdot \mathbf{v}}\right)\mathbf{v}\right) \cdot \mathbf{v} = \mathbf{u} \cdot \mathbf{v} - \left(\frac{\mathbf{u} \cdot \mathbf{v}}{\mathbf{v} \cdot \mathbf{v}}\right)\mathbf{v} \cdot \mathbf{v} = 0 \quad \blacksquare$$

Note that the sets {**v**, **u**} and {**v**, $\mathbf{u} - \mathbf{x}$} span the same set, because **u** is a linear combination of **v** and $\mathbf{u} - \mathbf{x}$, whereas $\mathbf{u} - \mathbf{x}$ is a linear combination of **v** and **u**. Specifically, we have

$$\mathbf{u} = \left(\frac{\mathbf{u} \cdot \mathbf{v}}{\mathbf{v} \cdot \mathbf{v}}\right)\mathbf{v} + (1)(\mathbf{u} - \mathbf{x})$$

and

$$\mathbf{u} - \mathbf{x} = (1)\mathbf{u} + (-1)\left(\frac{\mathbf{u} \cdot \mathbf{v}}{\mathbf{v} \cdot \mathbf{v}}\right)\mathbf{v}.$$

It follows that if {**v**, **u**} is a basis for a subspace of $\mathbf{R}^m$, then {**v**, $\mathbf{u} - \mathbf{x}$} is an *orthogonal* basis for the same subspace.

This process of replacing a basis by an orthogonal one for the same subspace can be generalized to any finite number of vectors in any inner product space. We present the proof of this procedure after we demonstrate its use.

Procedure for Constructing an Orthogonal Basis (Gram–Schmidt Process)

Compute the vectors in the basis as described in the following theorem.

THEOREM 6

Let $\mathcal{T} = \{\mathbf{u}_1, \mathbf{u}_2, \ldots, \mathbf{u}_n\}$ be a basis for an inner product space **V**. Let $\mathcal{T}' = \{\mathbf{v}_1, \mathbf{v}_2, \ldots, \mathbf{v}_n\}$ be defined as follows:

$$\mathbf{v}_1 = \mathbf{u}_1,$$

$$\mathbf{v}_2 = \mathbf{u}_2 - \frac{(\mathbf{u}_2, \mathbf{v}_1)}{(\mathbf{v}_1, \mathbf{v}_1)}\mathbf{v}_1,$$

$$\mathbf{v}_3 = \mathbf{u}_3 - \frac{(\mathbf{u}_3, \mathbf{v}_1)}{(\mathbf{v}_1, \mathbf{v}_1)}\mathbf{v}_1 - \frac{(\mathbf{u}_3, \mathbf{v}_2)}{(\mathbf{v}_2, \mathbf{v}_2)}\mathbf{v}_2,$$

$$\vdots$$

$$\mathbf{v}_n = \mathbf{u}_n - \frac{(\mathbf{u}_n, \mathbf{v}_1)}{(\mathbf{v}_1, \mathbf{v}_1)}\mathbf{v}_1 - \frac{(\mathbf{u}_n, \mathbf{v}_2)}{(\mathbf{v}_2, \mathbf{v}_2)}\mathbf{v}_2 - \cdots - \frac{(\mathbf{u}_n, \mathbf{v}_{n-1})}{(\mathbf{v}_{n-1}, \mathbf{v}_{n-1})}\mathbf{v}_{n-1}$$

Then the set $\mathcal{T}'$ is an orthogonal basis for **V**. An orthonormal basis for **V** is given by $\mathcal{T}'' = \{\mathbf{w}_1, \mathbf{w}_2, \ldots, \mathbf{w}_n\}$, where $\mathbf{w}_i = (1/\|\mathbf{v}_i\|)\mathbf{v}_i$ for $i = 1, 2, \ldots, n$. ■

Before proving this theorem, we present a couple of examples.

EXAMPLE 10

Let **S** be the subspace of $\mathbf{R}^5$ generated by the set $\mathcal{T} = \{\mathbf{u}_1, \mathbf{u}_2, \mathbf{u}_3\}$, where $\mathbf{u}_1 = (-1, -1, 1, 0, 0)$, $\mathbf{u}_2 = (0, -1, 0, 0, 1)$, and $\mathbf{u}_3 = (1, -1, 0, 1\ 0)$. Find an orthonormal basis for **S**.

Solution We apply the Gram–Schmidt process to $\mathcal{T}$.

$$\mathbf{v}_1 = \mathbf{u}_1 = (-1, -1, 1, 0, 0)$$

$$\begin{aligned}\mathbf{v}_2 &= \mathbf{u}_2 - \left(\frac{\mathbf{u}_2 \cdot \mathbf{v}_1}{\mathbf{v}_1 \cdot \mathbf{v}_1}\right)\mathbf{v}_1 \\ &= (0, -1, 0, 0, 1) - \left(\frac{1}{3}\right)(-1, -1, 1, 0, 0) \\ &= \left(\frac{1}{3}, -\frac{2}{3}, -\frac{1}{3}, 0, 1\right)\end{aligned}$$

$$\begin{aligned}\mathbf{v}_3 &= \mathbf{u}_3 - \left(\frac{\mathbf{u}_3 \cdot \mathbf{v}_1}{\mathbf{v}_1 \cdot \mathbf{v}_1}\right)\mathbf{v}_1 - \left(\frac{\mathbf{u}_3 \cdot \mathbf{v}_2}{\mathbf{v}_2 \cdot \mathbf{v}_2}\right)\mathbf{v}_2 \\ &= (1, -1, 0, 1, 0) - \left(\frac{0}{3}\right)(-1, -1, 1, 0, 0) - \left(\frac{1}{\frac{5}{3}}\right)\left(\frac{1}{3}, -\frac{2}{3}, -\frac{1}{3}, 0, 1\right) \\ &= \left(\frac{4}{5}, -\frac{3}{5}, \frac{1}{5}, 1, -\frac{3}{5}\right)\end{aligned}$$

The set $\mathcal{T}' = \{\mathbf{v}_1, \mathbf{v}_2, \mathbf{v}_3\}$ is an orthogonal basis for **S**. Normalizing each vector, we obtain an orthonormal basis $\mathcal{T}'' = \{\mathbf{w}_1, \mathbf{w}_2, \mathbf{w}_3\}$, where

$$\mathbf{w}_1 = \left(-\frac{\sqrt{3}}{3}, -\frac{\sqrt{3}}{3}, \frac{\sqrt{3}}{3}, 0, 0\right),$$

$$\mathbf{w}_2 = \left(\frac{\sqrt{15}}{15}, -\frac{2\sqrt{15}}{15}, -\frac{\sqrt{15}}{15}, 0, \frac{\sqrt{15}}{5}\right),$$

$$\text{and} \quad \mathbf{w}_3 = \left(\frac{2\sqrt{15}}{15}, -\frac{\sqrt{15}}{10}, \frac{\sqrt{15}}{30}, \frac{\sqrt{15}}{6}, -\frac{\sqrt{15}}{10}\right).$$ ■

EXAMPLE 11

(*From calculus*) Find an orthonormal basis for $\mathbf{P}_2$, the vector space of all polynomials of degree less than or equal to 2, under the inner product $(\mathbf{p}, \mathbf{q}) = \int_{-1}^{1} \mathbf{p}(x)\mathbf{q}(x)\,dx$.

Solution We know that a basis for $\mathbf{P}_2$ (the *standard basis* for $\mathbf{P}_2$) is given by $\mathcal{T} = \{\mathbf{u}_1, \mathbf{u}_2, \mathbf{u}_3\}$, where $\mathbf{u}_1 = 1$, $\mathbf{u}_2 = x$, and $\mathbf{u}_3 = x^2$ (Theorem 1 of Section 6.2). Applying Theorem 6, $\mathcal{T}' = \{\mathbf{v}_1, \mathbf{v}_2, \mathbf{v}_3\}$ is an *orthogonal* basis for $\mathbf{P}_2$ if

$$\mathbf{v}_1 = 1,$$

$$\mathbf{v}_2 = x - \left(\frac{\int_{-1}^{1}(x)(1)\,dx}{\int_{-1}^{1}(1)(1)\,dx}\right)(1) = x - \left(\frac{0}{2}\right)1 = x,$$

$$\text{and}\quad \mathbf{v}_3 = x^2 - \left(\frac{\int_{-1}^{1}(x^2)(1)\,dx}{\int_{-1}^{1}(1)(1)\,dx}\right)(1) - \left(\frac{\int_{-1}^{1}(x^2)(x)\,dx}{\int_{-1}^{1}(x)(x)\,dx}\right)(x)$$

$$= x^2 - \left(\frac{2/3}{2}\right)(1) - \left(\frac{0}{2/3}\right)(x) = x^2 - \frac{1}{3}.$$

Finally, $\mathcal{T}'' = \{\mathbf{w}_1, \mathbf{w}_2, \mathbf{w}_3\}$ is an ortho*normal* basis for $\mathbf{P}_2$ if we let

$$\mathbf{w}_1 = \left(\frac{1}{\sqrt{\int_{-1}^{1}(1)(1)\,dx}}\right)(1) = \frac{1}{\sqrt{2}},$$

$$\mathbf{w}_2 = \left(\frac{1}{\sqrt{\int_{-1}^{1}(x)(x)\,dx}}\right)(x) = \frac{1}{\sqrt{2/3}}x = \sqrt{\frac{3}{2}}x,$$

$$\text{and}\quad \mathbf{w}_3 = \left(\frac{1}{\sqrt{\int_{-1}^{1}(x^2 - 1/3)(x^2 - 1/3)\,dx}}\right)\left(x^2 - \frac{1}{3}\right)$$

$$= \frac{1}{2/3\sqrt{2/5}}\left(x^2 - \frac{1}{3}\right) = \frac{1}{2}\sqrt{\frac{5}{2}}(3x^2 - 1).$$ ■

The following two lemmas are used in the proof of Theorem 6.

LEMMA 1

Let $\mathcal{T}$ and $\mathcal{T}'$ be as in Theorem 6, and let

$$a_{ij} = \frac{(\mathbf{u}_i, \mathbf{v}_j)}{(\mathbf{v}_j, \mathbf{v}_j)}.$$

Then $(a_{ij}\mathbf{v}_j, \mathbf{v}_j) = (\mathbf{u}_i, \mathbf{v}_j)$, and, if $\mathbf{v}_i$ and $\mathbf{v}_j$ are orthogonal, $(a_{ij}\mathbf{v}_j, \mathbf{v}_i) = 0$.

Proof We have

$$(a_{ij}\mathbf{v}_j, \mathbf{v}_j) = \frac{(\mathbf{u}_i, \mathbf{v}_j)}{(\mathbf{v}_j, \mathbf{v}_j)}(\mathbf{v}_j, \mathbf{v}_j) = (\mathbf{u}_i, \mathbf{v}_j).$$

If $\mathbf{v}_i$ and $\mathbf{v}_j$ are orthogonal, we have $(a_{ij}\mathbf{v}_j, \mathbf{v}_i) = a_{ij}(\mathbf{v}_j, \mathbf{v}_i) = 0$, as desired. ■

LEMMA 2

Let $\mathcal{T}$ and $\mathcal{T}'$ be as in Theorem 6. Then every vector in $\mathcal{T}'$ is nonzero.

Proof Suppose that a vector in $\mathcal{T}'$ (say, $\mathbf{v}_k$) is zero. Then, using the notation of Lemma 1,

$$\mathbf{u}_k = a_{k1}\mathbf{v}_1 + a_{k2}\mathbf{v}_2 + \cdots + a_{k,k-1}\mathbf{v}_{k-1}.$$

Now each $\mathbf{v}_i$ is a linear combination of $\mathbf{u}_i$ and the $\mathbf{v}_j$ that precede it. By successive substitutions into this equation, we can write $\mathbf{u}_k$ as a linear combination of $\mathbf{u}_1, \mathbf{u}_2, \ldots, \mathbf{u}_{k-1}$. Hence, $\{\mathbf{u}_1, \mathbf{u}_2, \ldots, \mathbf{u}_k\}$ is linearly dependent; by Exercise 47 of Section 6.2, so is $\mathcal{T}$, because $\mathcal{T}$ contains this set. But this is a contradiction, since $\mathcal{T}$ is a basis for **V**, so all members of $\mathcal{T}'$ must be nonzero. ■

Proof of Theorem 6 We first show that $\mathcal{T}'$ is an orthogonal set and then that it is a basis for **V**. The fact that $\mathcal{T}''$ is an orthonormal basis for **V** follows easily, and we leave it for Exercise 33.

Using the notation and results of Lemma 1, we have

$$\begin{aligned}(\mathbf{v}_2, \mathbf{v}_1) &= (\mathbf{u}_2 - a_{21}\mathbf{v}_1, \mathbf{v}_1) = (\mathbf{u}_2, \mathbf{v}_1) - (a_{21}\mathbf{v}_1, \mathbf{v}_1)\\ &= (\mathbf{u}_2, \mathbf{v}_1) - (\mathbf{u}_2, \mathbf{v}_1) = 0.\end{aligned}$$

Therefore, $\mathbf{v}_2$ and $\mathbf{v}_1$ are orthogonal. Furthermore,

$$\begin{aligned}(\mathbf{v}_3, \mathbf{v}_1) &= (\mathbf{u}_3 - a_{31}\mathbf{v}_1 - a_{32}\mathbf{v}_2, \mathbf{v}_1)\\ &= (\mathbf{u}_3, \mathbf{v}_1) - (a_{31}\mathbf{v}_1, \mathbf{v}_1) - (a_{32}\mathbf{v}_2, \mathbf{v}_1)\\ &= (\mathbf{u}_3, \mathbf{v}_1) - (\mathbf{u}_3, \mathbf{v}_1) - 0 = 0.\end{aligned}$$

Therefore, $\mathbf{v}_3$ and $\mathbf{v}_1$ are orthogonal. In a similar manner, we can show that $\mathbf{v}_3$ and $\mathbf{v}_2$ are orthogonal and that $\mathbf{v}_4$ is orthogonal, successively, to $\mathbf{v}_1$, $\mathbf{v}_2$, and $\mathbf{v}_3$. For example,

$$\begin{aligned}(\mathbf{v}_4, \mathbf{v}_2) &= (\mathbf{u}_4 - a_{41}\mathbf{v}_1 - a_{42}\mathbf{v}_2 - a_{43}\mathbf{v}_3, \mathbf{v}_2)\\ &= (\mathbf{u}_4, \mathbf{v}_2) - (a_{41}\mathbf{v}_1, \mathbf{v}_2) - (a_{42}\mathbf{v}_2, \mathbf{v}_2) - (a_{43}\mathbf{v}_3, \mathbf{v}_2)\\ &= (\mathbf{u}_4, \mathbf{v}_2) - 0 - (\mathbf{u}_4, \mathbf{v}_2) - 0 = 0\end{aligned}$$

(by once again applying Lemma 1). We continue this process until we finally show that $\mathbf{v}_n$ is orthogonal, successively, to $\mathbf{v}_1, \mathbf{v}_2, \ldots, \mathbf{v}_{n-1}$. Thus, $\mathcal{T}'$ is an orthogonal set.

To show that $\mathcal{T}'$ is a basis for **V**, note that it contains n vectors, as does the known basis, $\mathcal{T}$. Thus, by part (iii) of Theorem 7 of Section 6.2, all we need to show is that $\mathcal{T}'$ is linearly independent. But since $\mathcal{T}'$ is an orthogonal set of nonzero vectors (Lemma 2), $\mathcal{T}'$ is linearly independent and therefore a basis—an orthogonal basis—for **V**, by Theorem 3. ■

6.4 Exercises

In Exercises 1–6 determine whether or not each of the following is an inner product for the indicated vector space. If it is not, list all properties violated.

1. $(\mathbf{p}, \mathbf{q}) = a_0b_0 + a_1b_1$; $\mathbf{p}(x) = a_0 + a_1x$, $\mathbf{q}(x) = b_0 + b_1x$ in $\mathbf{P}_1$
2. $(\mathbf{p}, \mathbf{q}) = \mathbf{p}(0)\mathbf{q}(0)$; $\mathbf{p}$, $\mathbf{q}$ in $\mathbf{P}_1$
3. $(A, B) = \det(A)\det(B)$; A, B in $\mathbf{M}^{2,2}$
4. $(A, B) = a_{11}b_{11} + a_{12}b_{12} + a_{21}b_{21} + a_{22}b_{22}$; A, B in $\mathbf{M}^{2,2}$
5. $(\mathbf{f}, \mathbf{g}) = \mathbf{f}'(0)\mathbf{g}'(0)$; $\mathbf{f}$, $\mathbf{g}$ in $\mathbf{C}^1(-1, 1)$, the vector space of all differentiable functions on the interval $(-1, 1)$
6. $(\mathbf{f}, \mathbf{g}) = \left(\max_{[-1,1]} |\mathbf{f}(x)|\right)\left(\max_{[-1,1]} |\mathbf{g}(x)|\right)$; $\mathbf{f}$, $\mathbf{g}$ in $\mathbf{C}[-1, 1]$

In Exercises 7–16 determine whether or not the given set of vectors is (a) orthogonal or (b) orthonormal for the indicated vector space and inner product.

7. $\{(1/\sqrt{5}, 2/\sqrt{5}, 0), (-2/\sqrt{5}, 1/\sqrt{5}, 0), (0, 0, 1)\}$ for $\mathbf{R}^3$

8. $\{(0, 1, 0), (1, 0, 1), (0, 0, 1)\}$ for $\mathbf{R}^3$

9. $\{(1, 2, 2), (2, 1, -2), (1, -2, 2)\}$ for $\mathbf{R}^3$

10. $\{(0, \sin\theta, \cos\theta, 0), (0, \cos\theta, -\sin\theta, 0)\}$, where θ is any real number for the subspace of $\mathbf{R}^4$ of all vectors whose first and last components are zero

11. $\{(\sqrt{2}/2, 0, \sqrt{2}/2)\}$ for the solution space of $A\mathbf{x} = \mathbf{0}$, where

$$A = \begin{bmatrix} 1 & 0 & -1 \\ 0 & 1 & 0 \\ 1 & 1 & -1 \end{bmatrix}$$

12. $\left\{(\sqrt{2}/2, 0, \sqrt{2}/2)\right\}$ for the solution space of $A\mathbf{x} = \mathbf{0}$, where

$$A = \begin{bmatrix} 1 & 0 & -1 \\ 1 & 1 & -1 \\ 2 & 1 & -2 \end{bmatrix}$$

13. $\{1, x, x^2\}$ in $\mathbf{P}_2$; $(\mathbf{p}, \mathbf{q}) = \int_0^1 \mathbf{p}(x)\mathbf{q}(x)\,dx$

14. $\left\{\frac{x}{2}, x^2 - 1\right\}$ in $\mathbf{P}_2$; $(\mathbf{p}, \mathbf{q}) = \mathbf{p}(0)\mathbf{q}(0) + \mathbf{p}(1)\mathbf{q}(1) + \mathbf{p}(-1)\mathbf{q}(-1)$

15. $\left\{\begin{bmatrix} \frac{1}{3} & -\frac{2}{3} \\ \frac{2}{3} & 0 \end{bmatrix}, \begin{bmatrix} \frac{2}{3} & -\frac{1}{3} \\ -\frac{2}{3} & 0 \end{bmatrix}, \begin{bmatrix} \frac{2}{3} & \frac{2}{3} \\ \frac{1}{3} & 0 \end{bmatrix}\right\}$ in $\mathbf{M}^{2,2}$;

$(A, B) = a_{11}b_{11} + a_{12}b_{12} + a_{21}b_{21} + a_{22}b_{22}$

16. $\left\{\frac{2}{\pi}, \left(\frac{\pi}{2}\right)\cos x\right\}$ in $\mathbf{C}\left[0, \frac{\pi}{2}\right]$, $(\mathbf{f}, \mathbf{g}) = \int_0^{\frac{\pi}{2}} \mathbf{f}(x)\mathbf{g}(x)\,dx$

In Exercises 17 and 18 express the given vector x as a linear combination of the vectors in the orthonormal basis $\left\{\left(\frac{1}{3}, -\frac{2}{3}, \frac{2}{3}\right), \left(-\frac{2}{3}, \frac{1}{3}, \frac{2}{3}\right), \left(\frac{2}{3}, \frac{2}{3}, \frac{1}{3}\right)\right\}$.

17. $\mathbf{x} = (1, 2, 3)$

18. $\mathbf{x} = (-1, 0, 1)$

In Exercises 19 and 20 the given set is an ortho*gonal* basis for a certain subspace of $\mathbf{R}^4$. Find an ortho*normal* basis for this subspace.

19. $\{(4, 5, 0, -2), (-2, 2, 0, 1), (1, 0, 0, 2)\}$

20. $\{(1, -1, 1, -1), (1, 1, 1, 1), (1, 0, -1, 0)\}$

In Exercises 21–26 the given set is a basis for a certain subspace of $\mathbf{R}^m$. Use the Gram–Schmidt process to find an orthonormal basis for this subspace.

21. $\{(1, -1, 0), (0, 1, -1)\}$

22. $\{(1, 0, -1), (1, 0, 1)\}$

23. $\{(-1, 1, 0, 0), (1, -1, 1, 0), (0, 0, 1, 2)\}$

24. $\{(-1, -1, 2, 1), (2, 1, -1, -2), (1, 1, 0, 1)\}$

25. $\{(1, 0, 0, 0), (1, 1, 0, 0), (1, 1, 1, 0), (1, 1, 1, 1)\}$

26. $\{(1, 1, 1, 0), (1, -1, 1, -1), (1, 0, 1, 1), (0, 1, 1, 1)\}$

In Exercises 27–30 apply the Gram–Schmidt process to find an orthonormal basis for the subspace of C[*a*, *b*] (under the inner product (f, g) = $\int_a^b$ f(x)g(x) *dx*) spanned by the given set $\mathcal{T}$.

27. $\mathcal{S} = \{1, x, x^2, x^3\}, [a, b] = [-1, 1]$

28. $\mathcal{S} = \{1, x, x^2\}, [a, b] = [0, 1]$

29. $\mathcal{S} = \{1, e^x\}, [a, b] = [0, 1]$

30. $\mathcal{S} = \{1, e^{-x}\}, [a, b] = [0, 1]$

31. Show that the standard basis for $\mathbf{R}^m$, $\{\mathbf{e}_1, \mathbf{e}_2, \ldots, \mathbf{e}_m\}$ is an orthonormal set.

32. Prove that, for all vectors **v** in an inner product space **V**, we have $(\mathbf{0}, \mathbf{v}) = (\mathbf{v}, \mathbf{0}) = 0$ (part (b) of Theorem 1).

33. Using the fact that $\mathcal{T}'$ of Theorem 6 is an ortho*gonal* basis for **V**, show that the set $\mathcal{T}''$ of this theorem is an ortho*normal* basis for **V**.

Computational Exercises

34–39. Redo Exercises 21–26 using a computer algebra system.

6.5 Vector Spaces over Other Fields

The definition of a vector space given in Section 6.1 requires that the scalars belong to the set of real numbers. But the definition of a vector space makes sense when the scalars are chosen from other sets as well. What is required is that the scalars form what is called a *field*. To do this, they must satisfy the following rules:

DEFINITION

A set **F** together with operations called *addition* (+) and *multiplication* (*) is a **field** if the following properties are satisfied for every choice of elements *a*, *b*, and *c* in **F**:

a.	$a + b$ and $a * b$ are in **F**.	(**F** is closed under addition and multiplication.)
b.	$a + b = b + a$	(Addition is commutative.)
c.	$(a + b) + c = a + (b + c)$	(Addition is associative.)
d.	There is an element, denoted by 0, in **F** such that $a + 0 = a$ for all a in **F**.	(0 is the additive identity of **F**.)
e.	For each a in **F**, there is an element, denoted by $-a$, such that $a + (-a) = 0$.	($-a$ is the negative of a.)
f.	$a * b = b * a$	(Multiplication is commutative.)
g.	$(a * b) * c = a * (b * c)$	(Multiplication is associative.)
h.	There is an element, denoted by 1, in **F** such that $a * 1 = a$ for all a in **F**.	(1 is the multiplicative identity of **F**.)
i.	For each $a \neq 0$ in **F**, there is an element, denoted by a^{-1}, such that $a * a^{-1} = 1$.	(a^{-1} is the inverse of a.)
j.	$a * (b + c) = a * b + a * c$	(Multiplication is distributive over addition.)

For every field **F** and every positive integer m, there is an associated vector space $\mathbf{F}^m$ that consists of all m-tuples from **F**, where addition is component-wise, and scalar multiplication (by elements in **F**) is defined by multiplying each coordinate by the scalar. This definition coincides with our previous definition of Euclidean n-space when $\mathbf{F} = \mathbf{R}$. Many of the notions (such as *dimension*, *subspaces*, and *independence*) and results from vector spaces over **R** carry over almost word for word to other fields.

The real numbers satisfy all the rules of a field. Are there others? Indeed there are. The rational numbers **Q** form a field as well. They inherit rules (b), (c), (f), (g), and (j) automatically. They contain 0 and 1, so rules (d) and (h) hold as well. They are closed under addition and multiplication, so (a) is true. Finally, if $a = p/q$ is a rational number with p and q integers, then $-a = (-p)/q$ is an integer, and if $a \neq 0$, then $p \neq 0$ and so $a^{-1} = q/p$ is also a rational number. These show that (e) and (i) hold, so all the rules have been verified.

There are other fields that have finitely many elements, unlike the reals and the rationals. There is, in fact, a finite field with n elements just in case n is a power of a prime number. In particular, for any prime p, there is a unique field with exactly p elements, and this field is fairly easy to describe. Given a prime p, the field $\mathbf{Z}_p$ consists of the integers $\{0, 1, \ldots, p - 1\}$, with addition and multiplication *modulo* p. That is, we add and multiply as usual, but we always reduce the answer by taking the remainder upon division by p. So, for example, in the field $\mathbf{Z}_{11}$, $5 \times 6 = 8$ since $5 \times 6 = 30$ and the remainder when you divide 30 by 11 is 8 (as $30 = 2 \times 11 + 8$).

One field that has many applications, especially in computers, is the *binary field* $\mathbf{Z}_2$. It is ideally suited for electronics, as the field has only two elements, 0 and 1, and these correspond nicely with the "off/on" of electric circuits. Binary strings of some fixed length n form a vector space, and this has broad applications to data encryption and compression.

There is one important field that contains the real numbers—the complex numbers. A **complex number** is one of the form $z = a + bi$ where a and b are real numbers and $i = \sqrt{-1}$; a is called the **real part** of z whereas b is called the **imaginary part**. We denote the set of all complex numbers by **C**.

Addition of complex numbers is defined by adding real parts and imaginary parts; that is, $z = a + bi$ and $w = c + di$, then $z + w = (a + c) + (b + d)i$. Multiplication is a little more involved; you multiply complex numbers by multiplying all four possible combinations (as you do when you multiply the product of two linear polynomials):

$$zw = (a + bi)(c + di) = ac + adi + bci + bdi^2 = (ac - bd) + (ad + bc)i$$

(Here we used the fact that $i^2 = -1$.) With these definitions of addition and multiplication, the complex numbers form a field.

Two operations on complex numbers are very useful. The **absolute value** (or **modulus**) of a complex number $z = a + bi$ is denoted $|z|$ and is defined to be $|z| = \sqrt{a^2 + b^2}$. The complex conjugate (or just **conjugate**) of $z = a + bi$ is denoted by $\bar{z}$ and is given by $\bar{z} = a - bi$. It is not hard to check that for complex numbers z and w, we have that $\overline{z + w} = \bar{z} + \bar{w}$ and $\overline{zw} = \bar{z}\,\bar{w}$ (see Exercise 29). The absolute value of a complex number z is connected to the conjugate of z by the formula $|z|^2 = z\bar{z}$.

EXAMPLE 1

For $z = -1 + 2i$ and $w = 3 + 4i$, find $z + w$, zw, $|w|$, and $\bar{z}$.

Solution

$$z + w = (-1 + 3) + (2 + 4)i = 2 + 6i;$$

$$zw = (-1 + 2i)(3 + 4i) = -3 - 4i + 6i + 8(-1) = -11 + 2i;$$

$$|w| = \sqrt{3^2 + 4^2} = \sqrt{25} = 5;$$

$$\bar{z} = -1 + (-2)i = -1 - 2i.$$

■

Of course, the real numbers **R** form a subset of the set **C** of complex numbers since any x in **R** can be written in the form $x + 0i$. Moreover, when x is real, observe that $xz = (x + 0i)(a + bi) = xa + xbi + 0 + 0 = xa + xbi$. Also, the definition of absolute value for complex numbers is compatible with the definition for real numbers, since $|x + 0i| = \sqrt{x^2 + 0^2} = \sqrt{x^2} = |x|$. Moreover, a necessary and sufficient condition that a given complex number z be real is that $z = \bar{z}$ (since, if $z = a + bi$, $z = \bar{z}$ is equivalent to $a = a$ and $b = -b$, and the latter is true if and only if $b = 0$).

Complex numbers can be visualized in $\mathbf{R}^2$ by associating the vector (a, b) to the complex number $z = a + bi$. Complex addition corresponds to vector addition, and, from the previous paragraph, multiplication of a complex number by a real number x corresponds to multiplication of the vector by the scalar x. The absolute value of a complex number corresponds to the length of the vector. So the complex numbers are, from one viewpoint, much like the vector space $\mathbf{R}^2$, but they are indeed so much more.

This view of complex numbers as ordered pairs allows us to represent complex numbers in the plane. For example, the complex number $z = -1 + i$ is represented in the plane by the vector $(-1, 1)$, and the absolute value of z is the length of the vector $(-1, 1)$, which is $\sqrt{(-1)^2 + 1^2} = \sqrt{2}$ (Figure 5).

FIGURE 5

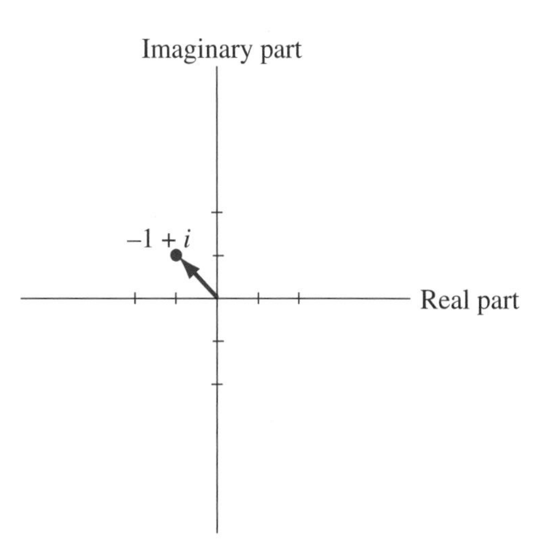

The set of all m-tuples of complex numbers, denoted $\mathbf{C}^m$, is called **complex *m*-space**. The definitions of addition and subtraction of vectors in $\mathbf{C}^m$ and multiplication of a vector in $\mathbf{C}^m$ by a scalar (here, a complex number) are identical to the $\mathbf{R}^m$ case, but the definitions of dot product and norm in $\mathbf{C}^m$ are not.

DEFINITION

Let $\mathbf{w} = (\mathbf{w}_1, \mathbf{w}_2, \ldots, \mathbf{w}_m)$ and $\mathbf{z} = (\mathbf{z}_1, \mathbf{z}_2, \ldots, \mathbf{z}_m)$ be vectors in $\mathbf{C}^m$. Then

a. $\mathbf{w} \cdot \mathbf{z} = w_1\bar{z}_1 + w_2\bar{z}_2 + \cdots + w_m\bar{z}_m$ (Dot product)

b. $\|\mathbf{w}\| = \sqrt{|w_1|^2 + |w_2|^2 + \cdots + |w_m|^2}$ (*Norm* or *length*)

NOTE These definitions reduce to the corresponding ones for $\mathbf{R}^m$ in the case when $\mathbf{w}$ and $\mathbf{z}$ are in $\mathbf{R}^m$.

Notice that the dot product, $\mathbf{w} \cdot \mathbf{z}$, involves complex conjugates. It is defined in this manner to preserve the property that $\mathbf{w} \cdot \mathbf{w} = \|\mathbf{w}\|^2$. If we had defined the dot product as $\mathbf{w} \cdot \mathbf{z} = w_1z_1 + w_2z_2 + \cdots + w_mz_m$, then this would not always be the case. With the definition just stated, however, we do have $\mathbf{w} \cdot \mathbf{w} = \|\mathbf{w}\|^2$.

EXAMPLE 2

Let $\mathbf{w} = (i, 1, 0)$ and $\mathbf{z} = (2i, 1 + i, 1 - i)$ be vectors in $\mathbf{C}^3$. Find $\mathbf{w} \cdot \mathbf{z}$, $\mathbf{z} \cdot \mathbf{w}$, and $\|\mathbf{w}\|$.

Solution

i. $\mathbf{w} \cdot \mathbf{z} = i(\overline{2i}) + 1(\overline{1 + i}) + 0(\overline{1 - i}) = i(-2i) + 1(1 - i)$
$= -2i^2 + 1 - i = 3 - i$

ii. $\mathbf{z} \cdot \mathbf{w} = 2i(\overline{i}) + (1 + i)\overline{1} + (1 - i)\overline{0} = 2i(-i) + (1 + i)1$
$= -2i^2 + 1 + i = 3 + i$

iii. $\|\mathbf{w}\| = \sqrt{|i|^2 + 1^2 + 0^2} = \sqrt{1^2 + 1^2} = \sqrt{2}$ ■

Notice that for the $\mathbf{w}$ and $\mathbf{z}$ of Example 2, we have $\mathbf{w} \cdot \mathbf{z} = \overline{\mathbf{z} \cdot \mathbf{w}}$. In fact, this equation holds for every $\mathbf{w}$ and $\mathbf{z}$ in $\mathbf{C}^m$. The next theorem presents some properties of the dot product in $\mathbf{C}^m$, including this one.

THEOREM 1

Let $\mathbf{w}$ and $\mathbf{z}$ be vectors in $\mathbf{C}^m$ and c be a scalar (a complex number). Then

a. $\mathbf{w} \cdot \mathbf{w} \geq 0$, $\mathbf{w} \cdot \mathbf{w} = 0$ if and only if $\mathbf{w} = \mathbf{0}$;

b. $\mathbf{w} \cdot \mathbf{z} = \overline{\mathbf{z} \cdot \mathbf{w}}$;

c. $(c\mathbf{w}) \cdot \mathbf{z} = c(\mathbf{w} \cdot \mathbf{z})$, $\mathbf{w} \cdot (c\mathbf{z}) = \overline{c}(\mathbf{w} \cdot \mathbf{z})$;

d. $\mathbf{w} \cdot \mathbf{w} = \|\mathbf{w}\|^2$.

Proof of (b) The proofs of (a), (c), and (d) are left for an exercise.
Let $\mathbf{w} = (w_1, w_2, \ldots, w_m)$ and $\mathbf{z} = (z_1, z_2, \ldots, z_m)$. Then

$$\begin{aligned}
\overline{\mathbf{z} \cdot \mathbf{w}} &= \overline{z_1\overline{w}_1 + z_2\overline{w}_2 + \cdots + z_m\overline{w}_m} \\
&= \overline{z_1\overline{w}_1} + \overline{z_2\overline{w}_2} + \cdots + \overline{z_m\overline{w}_m} \\
&= \overline{z}_1\overline{\overline{w}}_1 + \overline{z}_2\overline{\overline{w}}_2 + \cdots + \overline{z}_m\overline{\overline{w}}_m \\
&= \overline{z}_1 w_1 + \overline{z}_2 w_2 + \cdots + \overline{z}_m w_m \\
&= w_1\overline{z}_1 + w_2\overline{z}_2 + \cdots + w_m\overline{z}_m \\
&= \mathbf{w} \cdot \mathbf{z}.
\end{aligned}$$ ■

The results of previous sections on $\mathbf{R}^m$, such as those related to basis, dimension, spanning sets, linear independence, inner products, and orthogonality carry over to $\mathbf{C}^m$ in a natural way. We illustrate with an example.

EXAMPLE 3

Let S be the subspace of $\mathbf{C}^4$ generated by the basis

$$\{(1, 0, i, 0), (i, 0, 0, 1), (0, i, 0, -i)\}.$$

Find an orthonormal basis for S.

Solution We proceed via the Gram–Schmidt process. We set $\mathbf{u}_1 = (1, 0, i, 0)$, $\mathbf{u}_2 = (i, 0, 0, 1)$, and $\mathbf{u}_3 = (0, i, 0, -i)$. Then (recalling the definition of the dot product on complex space) we calculate the following:

$$\mathbf{v}_1 = \mathbf{u}_1 = (1, 0, i, 0)$$

$$\mathbf{v}_2 = \mathbf{u}_2 - \left(\frac{\mathbf{u}_2 \cdot \mathbf{v}_1}{\mathbf{v}_1 \cdot \mathbf{v}_1}\right)\mathbf{v}_1$$

$$= (i, 0, 0, 1) - \left(\frac{(i, 0, 0, 1) \cdot (1, 0, i, 0)}{(1, 0, i, 0) \cdot (1, 0, i, 0)}\right)(1, 0, i, 0)$$

$$= (i, 0, 0, 1) - \frac{i}{2}(1, 0, i, 0)$$

$$= \left(\frac{i}{2}, 0, \frac{1}{2}, 1\right)$$

$$\mathbf{v}_3 = \mathbf{u}_3 - \left(\frac{\mathbf{u}_3 \cdot \mathbf{v}_1}{\mathbf{v}_1 \cdot \mathbf{v}_1}\right)\mathbf{v}_1 - \left(\frac{\mathbf{u}_3 \cdot \mathbf{v}_2}{\mathbf{v}_2 \cdot \mathbf{v}_2}\right)\mathbf{v}_2$$

$$= (0, i, 0, -i) - \left(\frac{(0, i, 0, -i) \cdot (1, 0, i, 0)}{(1, 0, i, 0) \cdot (1, 0, i, 0)}\right)(1, 0, i, 0) - \left(\frac{(0, i, 0, -i) \cdot \left(\frac{i}{2}, 0, \frac{1}{2}, 1\right)}{\left(\frac{i}{2}, 0, \frac{1}{2}, 1\right) \cdot \left(\frac{i}{2}, 0, \frac{1}{2}, 1\right)}\right)\left(\frac{i}{2}, 0, \frac{1}{2}, 1\right)$$

$$= (0, i, 0, -i) - 0(1, 0, i, 0) - \left(-\frac{2i}{3}\right)\left(\frac{i}{2}, 0, \frac{1}{2}, 1\right)$$

$$= \left(-\frac{1}{3}, i, \frac{i}{3}, -\frac{i}{3}\right)$$

The set $\{\mathbf{v}_1, \mathbf{v}_2, \mathbf{v}_3\}$ is an orthogonal basis for S. Normalizing each vector, we arrive at an orthonormal basis $\{\mathbf{w}_1, \mathbf{w}_2, \mathbf{w}_3\}$, where

$$\mathbf{w}_1 = \left(\frac{\sqrt{2}}{2}, 0, \frac{\sqrt{2}i}{2}, 0\right)$$

$$\mathbf{w}_2 = \left(\frac{\sqrt{6}i}{6}, 0, \frac{\sqrt{6}}{6}, \frac{\sqrt{6}}{3}\right)$$

$$\mathbf{w}_3 = \left(\frac{-\sqrt{3}}{6}, \frac{\sqrt{3}i}{2}, \frac{\sqrt{3}i}{6}, -\frac{\sqrt{3}i}{6}\right).$$

■

6.5 Exercises

1. Given a set S and operations $+$ and $*$ on S, we can form a table of the operation $+$ by listing the elements of S along the top and down the left side, and filling in the table in the row labelled a (under column b) by the value of $a + b$. We can, of course, do a similar thing for the $*$ operation.
   (*a*) Form the tables for $\mathbf{Z}_2$, $\mathbf{Z}_3$, and $\mathbf{Z}_5$.
   (*b*) Explain how rules (a), (b), (d), (e), (f), (h), and (i) can be verified in general from the tables of operations.

2. Consider the subset S of $\mathbf{Z}_2^4$ that consists of all 4-tuples with an even number of 1's.
   (*a*) Prove that S is a subspace of $\mathbf{Z}_2^4$.
   (*b*) What is its dimension?

3. Find a field with four elements 0, 1, x, and $x + 1$. (*Hint*: Use $\mathbf{Z}_2$ addition and make $x^2 = x + 1$.)

For Exercises 4–8, let $z = 1 - i$ and $w = 2 + 3i$. Compute the following.

4. $z + w$ *5.* $w - z$ *6.* zw *7.* $\overline{z + w}$ *8.* $|z + 2w|$

For Exercises 9 and 10, let $\mathbf{u} = (1 + i, 1 - i, 2i)$ and $\mathbf{v} = (0, -1 + i, i)$. Compute the following.

9. $\mathbf{u} + \mathbf{v}$ *10.* $2\mathbf{u} - \mathbf{v}$

In Exercises 11–14, letting $\mathbf{w} = (i, 0, 1 - i, 1)$ and $\mathbf{z} = (-i, 2 + i, 0, -1)$, evaluate the expressions.

11. $\mathbf{w} \cdot \mathbf{z}$ *12.* $\mathbf{z} \cdot \mathbf{w}$ *13.* $\|\mathbf{w}\|$ *14.* $\mathbf{w} \cdot \mathbf{w}$

In Exercises 15–18 determine whether or not the given set is a basis for $\mathbf{C}^3$.

15. $\{(i, 0, -1), (1, 1, 1), (0, -i, i)\}$

16. $\{(i, -i, 1), (1, 1 + i, 1 - i), (0, 2 + i, 1)\}$

17. $\{(i, 1, 0), (0, 0, 1)\}$

18. $\{(0, i, 1), (i, 0, 0), (1 + i, -1, 0), (1 - i, 0, 1)\}$

For Exercises 19 and 20, show that the given set S is a subspace of $\mathbf{C}^4$ and determine a basis and dim S.

19. $S = \{(z_1, z_2, z_3, z_4) : 2z_1 - z_2 + iz_3 + z_4 = 0\}$

20. $S = \{(z_1, z_2, z_3, z_4) : z_1 = -z_2 \text{ and } z_3 = -iz_4\}$

21. Is the set $\{(1, 0, -i), (1, -1, 2i), (2 + i, -i, -2 - 2i)\}$ linearly independent in $\mathbf{C}^3$?

22. Is the set $\{(1, 0, -i), (1, -1, 2i), (2 + i, i, -2 - 2i)\}$ linearly independent in $\mathbf{C}^3$?

23. Find a basis for the span of the set $\{(1, 1 + i), (-1, 2i), (3 + i, 2), (-1 + i, 5 - 3i)\}$.

24. Find a basis for the span of the set $\{(i, 0, 0), (0, i, 0), (i, -i, 0), (0, i, -i)\}$.

25. Show that $\{(i, i, 0), (i, -i, 0), (0, 0, -i)\}$ is an orthogonal set and convert it to an orthonormal set.

26. Let S be the subspace of $\mathbf{C}^4$ generated by the basis

$$\{(1, 0, i, 0), (i, -1, 0, 1), (-1, 0, 0, i)\}.$$

Find an orthonormal basis for S.

27. Prove parts (a), (c), and (d) of Theorem 1.

28. Let $\mathbf{w}$ and $\mathbf{z}$ be in $\mathbf{C}^{mn}$. Define a scalar-valued product, $\mathbf{w} \circ \mathbf{z}$, by

$$\mathbf{w} \circ \mathbf{z} = w_1 z_1 + w_2 z_2 + \cdots + w_m z_m.$$

Show that $\mathbf{w} \circ \mathbf{w}$ is not necessarily equal to $\|\mathbf{w}\|^2$.

29. Let z_1 and z_2 be complex numbers. Show that

(a) $\overline{z_1 + z_2} = \bar{z}_1 + \bar{z}_2$

(b) $\overline{z_1 z_2} = \bar{z}_1 \bar{z}_2$

(c) $\bar{\bar{z}}_1 = z_1$

(d) $z_1 \bar{z}_1 = |z_1|^2$

6.6 APPLICATION—Approximation of Continuous Functions; Fourier Series

In Sections 2.5 and 3.6, we considered the problem of obtaining a polynomial approximation to a discrete set of data. Now we turn to the related idea of approximation of a vector by a vector in a subspace. The key to the development of this material is the topic of inner product spaces (Section 6.4).

We begin with a definition that extends Theorem 5 of Section 6.4.

DEFINITION

Let $\mathbf{V}$ be an inner product space and $\mathbf{U}$ a finite dimensional subspace of $\mathbf{V}$ with orthonormal basis $\{\mathbf{u}_1, \mathbf{u}_2, \ldots, \mathbf{u}_d\}$. Then the **orthogonal projection** of $\mathbf{v}$ onto $\mathbf{U}$ is given by

$$proj_{\mathbf{U}}\mathbf{v} = (\mathbf{v}, \mathbf{u}_1)\mathbf{u}_1 + (\mathbf{v}, \mathbf{u}_2)\mathbf{u}_2 + \cdots + (\mathbf{v}, \mathbf{u}_d)\mathbf{u}_d.$$

Note that the orthogonal projection of $\mathbf{v}$ onto $\mathbf{U}$ indeed lies in $\mathbf{U}$ because it is a linear combination of basis vectors of $\mathbf{U}$. We are now ready to prove a general result that will have a number of interesting applications.

THEOREM 1

Suppose $\mathbf{V}$ is an inner product space and $\mathbf{U}$ a finite dimensional subspace with orthonormal basis $\{\mathbf{u}_1, \mathbf{u}_2, \ldots, \mathbf{u}_d\}$. Let $\mathbf{v}$ be any vector in $\mathbf{V}$. Then there is a unique vector $\mathbf{p}$ in $\mathbf{U}$ (namely, the orthogonal projection of $\mathbf{v}$ onto $\mathbf{U}$) that minimizes $\|\mathbf{v} - \mathbf{u}\|^2$ over all choices of vectors $\mathbf{u}$ in $\mathbf{U}$.

Proof Set $\mathbf{p} = proj_{\mathbf{U}}\mathbf{v}$. As noted earlier, $\mathbf{p}$ lies in $\mathbf{U}$. Moreover, observe that $\mathbf{v} - \mathbf{p}$ is orthogonal to each $\mathbf{u}_i$ as $\{\mathbf{u}_1, \mathbf{u}_2, \ldots, \mathbf{u}_d\}$ is an orthonormal set:

$$\begin{aligned}(\mathbf{v} - \mathbf{p}, \mathbf{u}_i) &= (\mathbf{v} - ((\mathbf{v}, \mathbf{u}_1)\mathbf{u}_1 + (\mathbf{v}, \mathbf{u}_2)\mathbf{u}_2 + \cdots + (\mathbf{v}, \mathbf{u}_d)\mathbf{u}_d), \mathbf{u}_i) \\ &= (\mathbf{v} - (\mathbf{v}, \mathbf{u}_1)\mathbf{u}_1 - (\mathbf{v}, \mathbf{u}_2)\mathbf{u}_2 - \cdots - (\mathbf{v}, \mathbf{u}_d)\mathbf{u}_d, \mathbf{u}_i) \\ &= (\mathbf{v}, \mathbf{u}_i) - (\mathbf{v}, \mathbf{u}_1)(\mathbf{u}_1, \mathbf{u}_i) - (\mathbf{v}, \mathbf{u}_2)(\mathbf{u}_2, \mathbf{u}_i) - \cdots - (\mathbf{v}, \mathbf{u}_d)(\mathbf{u}_d, \mathbf{u}_i) \\ &= (\mathbf{v}, \mathbf{u}_i) - (\mathbf{v}, \mathbf{u}_i) \\ &= 0\end{aligned}$$

It follows that $\mathbf{v} - \mathbf{p}$ is orthogonal to every vector in $\mathbf{U}$. If $\mathbf{u}$ is any vector in $\mathbf{U}$, then $\mathbf{p} - \mathbf{u}$ is also in $\mathbf{U}$ since $\mathbf{p}$ lies in $\mathbf{U}$. Thus $\mathbf{v} - \mathbf{p}$ is orthogonal to $\mathbf{p} - \mathbf{u}$, and so

$$\begin{aligned}\|\mathbf{v} - \mathbf{u}\|^2 &= (\mathbf{v} - \mathbf{u}, \mathbf{v} - \mathbf{u},) \\ &= ((\mathbf{v} - \mathbf{p}) + (\mathbf{p} - \mathbf{u}), (\mathbf{v} - \mathbf{p}) + (\mathbf{p} - \mathbf{u})) \\ &= (\mathbf{v} - \mathbf{p}, \mathbf{v} - \mathbf{p}) + 2(\mathbf{v} - \mathbf{p}, \mathbf{p} - \mathbf{u}) + (\mathbf{p} - \mathbf{u}, \mathbf{p} - \mathbf{u}) \\ &= \|\mathbf{v} - \mathbf{p}\|^2 + \|\mathbf{p} - \mathbf{u}\|^2.\end{aligned}$$

It follows that $\|\mathbf{v} - \mathbf{u}\|^2$ is at least $\|\mathbf{v} - \mathbf{p}\|^2$, with equality if and only if $\|\mathbf{p} - \mathbf{u}\|^2 = 0$ (that is, if and only if $\mathbf{p} = \mathbf{u}$), so that $\mathbf{p} = proj_{\mathbf{U}}\mathbf{v}$ is the unique vector in $\mathbf{U}$ that minimizes $\|\mathbf{v} - \mathbf{u}\|^2$. ■

We now turn to some far reaching applications of this far-reaching result.

Least Square Approximations to Continuous Functions

There are many ways of approximating a continuous function on an interval by a polynomial. If $\mathbf{f}(x)$ is the function and $[a, b]$ is the interval, we can evaluate $\mathbf{f}$ at n distinct points in $[a, b]$ to obtain the data set $(x_i, \mathbf{f}(x_i))$, $i = 1, 2, \ldots, n$, and then construct the interpolating, or least square, polynomial for the data (Sections 2.5 and 3.6). This approach makes use of only a small part of the given information, however. To involve all our knowledge of $\mathbf{f}(x)$, we employ the following approach.

Recall that the least square polynomial, $\mathbf{p}(x)$, for a finite set of data, $\{(x_i, y_i) \mid i = 1, 2, \ldots, n\}$, is the one that minimizes

$$S = (y_1 - \mathbf{p}(x_1))^2 + (y_2 - \mathbf{p}(x_2))^2 + \cdots + (y_n - \mathbf{p}(x_n))^2.$$

By letting $\mathbf{y} = (y_1, \ldots, y_n)$ and $\mathbf{z} = (\mathbf{p}(x_1), \ldots, \mathbf{p}(x_n))$ and making use of the dot product for $\mathbf{R}^n$, we may write this equation as

$$S = (\mathbf{y} - \mathbf{z}) \cdot (\mathbf{y} - \mathbf{z}) = \|\mathbf{y} - \mathbf{z}\|^2.$$

This suggests one way to extend the idea of the least square approximation to a continuous function $\mathbf{f}(x)$: Find a polynomial $\mathbf{p}(x)$ that minimizes

$$(\mathbf{f} - \mathbf{p}, \mathbf{f} - \mathbf{p}) = \|\mathbf{f} - \mathbf{p}\|^2,$$

where this inner product is the one on $\mathbf{C}[a, b]$ (the vector space of functions continuous on the interval $[a, b]$) given by

$$(\mathbf{g}, \mathbf{h}) = \int_a^b \mathbf{g}(x)\mathbf{h}(x)\,dx \tag{1}$$

(see Section 6.4).

The following immediate corollary of Theorem 1 provides the basis for many of the approximation techniques of this section.

COROLLARY

Let $\mathbf{f}(x)$ be a member of $\mathbf{C}[a, b]$, and let $\mathbf{S}$ be a finite dimensional subspace of $\mathbf{C}[a, b]$. Then there is a unique function $\mathbf{p}(x)$ in $\mathbf{S}$ that minimizes

$$\|\mathbf{f} - \mathbf{p}\|^2 = \int_a^b (\mathbf{f}(x) - \mathbf{p}(x))^2\,dx. \qquad \blacksquare$$

NOTE The function $\mathbf{p}$ of the corollary is called the **least square approximation** to $\mathbf{f}$ from $\mathbf{S}$. From a practical point of view, the subspace $\mathbf{S}$ usually consists of polynomials of some sort.

Procedure for Determining the Least Square Approximation to a Continuous Function

Let $\mathbf{f}$ be a function in $\mathbf{C}[a, b]$ and $\mathbf{S}$ be a finite dimensional subspace of approximating functions. To construct the least square approximation for $\mathbf{f}$ from $\mathbf{S}$,

i. Find an orthonormal basis, $\{\mathbf{p}_1, \mathbf{p}_2, \ldots, \mathbf{p}_n\}$ for $\mathbf{S}$, and

ii. Let $\mathbf{p} = c_1\mathbf{p}_1 + c_2\mathbf{p}_2 + \cdots + c_n\mathbf{p}_n$, where

$$c_i = (\mathbf{f}, \mathbf{p}_i) = \int_a^b \mathbf{f}(x)\mathbf{p}_i(x)\,dx.$$

Then $\mathbf{p}$, the orthogonal projection of $\mathbf{f}$ onto $\mathbf{S}$, is the desired least square approximation.

Least Square Polynomial Approximation

Perhaps the simplest choice of the subspace $\mathbf{S}$ of the corollary is $\mathbf{P}_n[a, b]$, the vector space of all polynomials of degree less than or equal to n with domain restricted to $[a, b]$. In this case, the desired approximating function is called the **least square polynomial** for $\mathbf{f}$ on $[a, b]$. To apply the procedure of the corollary to construct the least square polynomial, we first need an orthonormal basis for $\mathbf{P}_n[a, b]$. Since $\mathcal{B} = \{1, x, x^2, \ldots, x^n\}$ is a basis for this subspace, we can find an orthonormal basis by applying the Gram–Schmidt process of Section 6.4 to $\mathcal{B}$.

EXAMPLE 1

Find the least square quadratic polynomial approximating $\mathbf{f}(x) = x^3$ on $[-1, 1]$.

Solution We apply the procedure of the corollary. Here $\mathbf{S} = \mathbf{P}_2[-1, 1]$, which has a basis $\mathcal{B} = \{1, x, x^2\}$. By applying the Gram–Schmidt process to $\mathcal{B}$, we obtain the orthonormal basis (see Example 11 of Section 6.4) $\{\mathbf{p}_0, \mathbf{p}_1, \mathbf{p}_2\}$, where

$$\mathbf{p}_0(x) = \sqrt{\frac{1}{2}},$$

$$\mathbf{p}_1(x) = \left(\sqrt{\frac{3}{2}}\right)x,$$

and $$\mathbf{p}_2(x) = \left(\frac{1}{2}\right)\left(\sqrt{\frac{5}{2}}\right)(3x^2 - 1).$$

Therefore, the quadratic least square polynomial, $\mathbf{p}(x)$, for $\mathbf{f}(x) = x^3$ on $[-1, 1]$ is given by

$$\mathbf{p}(x) = c_0\mathbf{p}_0(x) + c_1\mathbf{p}_1(x) + c_2\mathbf{p}_2(x),$$

where

$$c_0 = (\mathbf{f}, \mathbf{p}_0) = \int_{-1}^{1} x^3\left(\sqrt{\frac{1}{2}}\right)dx = 0,$$

$$c_1 = (\mathbf{f}, \mathbf{p}_1) = \int_{-1}^{1} x^3\left(\sqrt{\frac{3}{2}}\right)x\,dx = \sqrt{\frac{3}{2}}\int_{-1}^{1} x^4 dx = \left(\frac{2}{5}\right)\sqrt{\frac{3}{2}},$$

and
$$c_2 = (\mathbf{f}, \mathbf{p}_2) = \int_{-1}^{1} x^3\left(\frac{1}{2}\right)\left(\sqrt{\frac{5}{2}}\right)(3x^2 - 1)dx$$

$$= \left(\frac{1}{2}\right)\sqrt{\frac{5}{2}}\int_{-1}^{1}(3x^5 - x^3)dx = 0.$$

Thus,

$$\mathbf{p}(x) = \left[\left(\frac{2}{5}\right)\sqrt{\frac{3}{2}}\right]\left(\sqrt{\frac{3}{2}}\right)x = \left(\frac{3}{5}\right)x.$$

(The given $\mathbf{f}(x)$ and approximating $\mathbf{p}(x)$ are shown in Figure 6.)

FIGURE 6

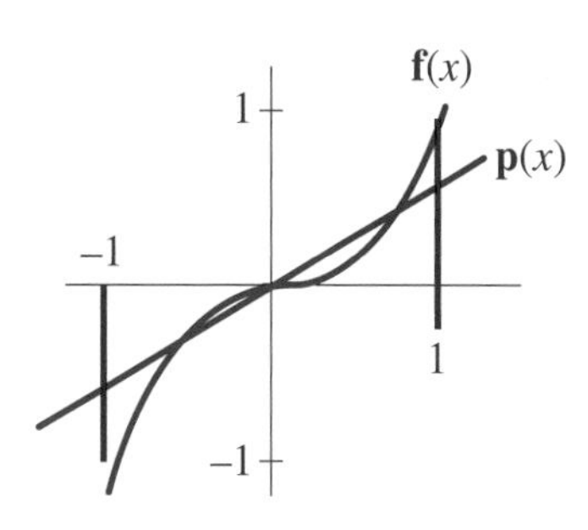

■

NOTE As we can see from Example 1, the least square approximation to **f** from $\mathbf{P}_2[a, b]$ does not need to have its degree exactly equal to 2.

The polynomials that form an orthonormal basis for $\mathbf{P}_n[-1, 1]$ are called *orthonormal Legendre* polynomials and are often listed in mathematical reference books. The next theorem demonstrates how we can construct a set of polynomials that are orthonormal on any interval $[a, b]$ in terms of the orthonormal Legendre polynomials.

THEOREM 2

Let $\{\mathbf{q}_0(t), \mathbf{q}_1(t), \ldots, \mathbf{q}_n(t)\}$ be an orthonormal basis for $\mathbf{P}_n[-1, 1]$. Then $\mathcal{B} = \{\mathbf{p}_0(x), \mathbf{p}_1(x), \ldots, \mathbf{p}_n(x)\}$ is an orthonormal basis for $\mathbf{P}_n[a, b]$, where

$$\mathbf{p}_i(x) = \left(\sqrt{\frac{2}{b - a}}\right)\mathbf{q}_i\left(\frac{2x - (a + b)}{b - a}\right).$$

Proof All we need to do is verify that $\mathcal{B}$ is an orthonormal *set* on $[a, b]$; that is,

$$\int_a^b (\mathbf{p}_i(x))^2 dx = 1$$

and
$$\int_a^b \mathbf{p}_i(x)\mathbf{p}_j(x)\,dx = 0, \qquad \text{if } i \neq j.$$

Then, by Theorem 3 of Section 6.4 and Theorem 7 of Section 6.2, $\mathscr{B}$ is an orthonormal *basis* for $\mathbf{P}_n[a, b]$. We leave the details of this verification as an exercise. ■

EXAMPLE 2

Find the least square linear polynomial, $\mathbf{p}(x)$, approximating

$$\mathbf{f}(x) = \sqrt{x}$$

on [0, 1].

Solution We again apply the procedure of the corollary. The approximation we seek is of the form

$$\mathbf{p} = c_0\mathbf{p}_0 + c_1\mathbf{p}_1,$$

where $\{\mathbf{p}_0, \mathbf{p}_1\}$ is an orthonormal basis for $\mathbf{P}_1[0, 1]$.

$$c_0 = \int_0^1 \mathbf{f}(x)\mathbf{p}_0(x)\,dx,$$

$$\text{and} \quad c_1 = \int_0^1 \mathbf{f}(x)\mathbf{p}_1(x)\,dx.$$

To find $\mathbf{p}_0$ and $\mathbf{p}_1$, we use Theorem 2. Since an orthonormal basis for $\mathbf{P}_1[-1, 1]$ is $\left\{\frac{1}{\sqrt{2}}, \left(\sqrt{\frac{3}{2}}\right)t\right\}$, and for this example $a = 0$ and $b = 1$, we obtain

$$\mathbf{p}_0(x) = \sqrt{2}\left(\frac{1}{\sqrt{2}}\right) = 1$$

$$\text{and} \quad \mathbf{p}_1(x) = \sqrt{2}\left(\sqrt{\frac{3}{2}}\right)(2x - 1) = \sqrt{3}(2x - 1).$$

Consequently,

$$c_0 = \int_0^1 \sqrt{x}\,dx = \frac{2}{3}$$

$$\text{and} \quad c_1 = \int_0^1 \sqrt{x}\sqrt{3}(2x - 1)\,dx$$

$$= \sqrt{3}\int_0^1 (2x^{3/2} - x^{1/2})\,dx = \frac{2\sqrt{3}}{15},$$

$$\text{yielding} \quad \mathbf{p}(x) = c_0\mathbf{p}_0(x) + c_1\mathbf{p}_1(x) = \left(\frac{4}{5}\right)x + \frac{4}{15}.$$

(The given $\mathbf{f}(x)$ and approximating $\mathbf{p}(x)$ are shown in Figure 7.)

FIGURE 7

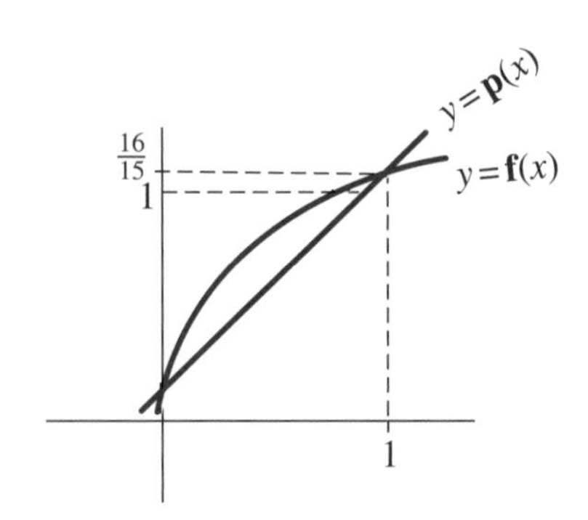

■

Fourier Series

Another type of relatively simple function used for approximation purposes is the *trigonometric polynomial.*

DEFINITION

A **trigonometric polynomial** is a function of the form

$$\mathbf{p}(x) = a_0 + a_1\cos x + a_2\cos 2x + \cdots + a_n\cos nx$$
$$+ b_1\sin x + b_2\sin 2x + \cdots + b_n\sin nx.$$

If a_n and b_n are not both zero, we say that $\mathbf{p}(x)$ has **order** n.

Joseph Fourier (1768–1830) was a mathematician and physicist who lived in France. His most important contribution to mathematics lay in his argument that a function of a single variable could be expanded into an infinite linear combination of sine and cosine functions. This result has had far-reaching consequences in many fields, including music.

The Bridgeman Art Library*

It can be shown that $\mathbf{T}_n[a, b]$, the set of all trigonometric polynomials on $[a, b]$ of order less than or equal to n, is a subspace of $\mathbf{C}[a, b]$. If we choose the subspace $\mathbf{S}$ of Theorem 1 to be $\mathbf{T}_n[-\pi, \pi]$, the resulting approximation is called the **Fourier series of order** n for the given function.

To obtain the Fourier series of order n for a continuous function $\mathbf{f}(x)$, we apply the procedure of the corollary with $\mathbf{S} = \mathbf{T}_n[-\pi, \pi]$. This subspace has the orthogonal basis

$$\mathcal{B} = \{1, \cos x, \cos 2x, \ldots, \cos nx, \sin x, \sin 2x, \ldots, \sin nx\}$$

(Exercise 17). Thus, to obtain an orthonormal basis for $\mathbf{T}_n[-\pi, \pi]$, we need to only normalize the vectors in $\mathcal{B}$. In other words, we seek constants $c_0, c_1, \ldots, c_n, d_1, \ldots, d_n$ so that

$$(c_0, c_0) = \int_{-\pi}^{\pi} c_0^2\, dx = 1,$$

$$(c_k\cos kx, c_k\cos kx) = \int_{-\pi}^{\pi} c_k^2\cos^2 kx\, dx = 1 \qquad (k = 1, 2, \ldots, n),$$

and $$(d_k\sin kx, d_k\sin kx) = \int_{-\pi}^{\pi} d_k^2\sin^2 kx\, dx = 1 \quad (k = 1, 2, \ldots, n).$$

Performing the indicated integrations and solving for the c_k and d_k yields

$$c_0 = \frac{1}{\sqrt{2\pi}},$$

$$c_k = d_k = \frac{1}{\sqrt{\pi}} \text{ for } k = 1, 2, \ldots, n.$$

Now that we have an orthonormal basis for $\mathbf{T}_n[-\pi, \pi]$, we can complete the procedure of the corollary and obtain the following theorem.

THEOREM 3

Let $\mathbf{f}(x)$ be a continuous function on $[-\pi, \pi]$. Then the Fourier series for $\mathbf{f}$ (the least square trigonometric polynomial approximating $\mathbf{f}$ on $[-\pi, \pi]$) of order n is

$$\mathbf{p}(x) = a_0 + a_1\cos x + a_2\cos 2x + \cdots + a_n\cos nx$$
$$+ b_1\sin x + b_2\sin 2x + \cdots + b_n\sin nx,$$

*Joseph Fourier (1768–1830) from 'Recueil de Portraits des Academiciens Celebres', 1823 (litho) (b/w photo), Boilly, Julien Leopold (1796–1874)/Bibliotheque de l'Institut de France, Paris, France/Giraudon/The Bridgeman Art Library

where

$$a_0 = \left(\frac{1}{2\pi}\right)\int_{-\pi}^{\pi} \mathbf{f}(x)\, dx,$$

$$a_k = \left(\frac{1}{\pi}\right)\int_{-\pi}^{\pi} \mathbf{f}(x) \cos kx\, dx \quad \text{for } k = 1, 2, \ldots, n,$$

and $$b_k = \left(\frac{1}{\pi}\right)\int_{-\pi}^{\pi} \mathbf{f}(x) \sin kx\, dx \quad \text{for } k = 1, 2, \ldots, n.$$ ■

EXAMPLE 3

Find the Fourier series of order 2 for $\mathbf{f}(x) = x$.

Solution Applying Theorem 3, the desired function has the form

$$\mathbf{p}(x) = a_0 + a_1 \cos x + a_2 \cos 2x + b_1 \sin x + b_2 \sin 2x,$$

where

$$a_0 = \left(\frac{1}{2\pi}\right)\int_{-\pi}^{\pi} x\, dx = 0,$$

$$a_1 = \left(\frac{1}{\pi}\right)\int_{-\pi}^{\pi} x \cos x\, dx = 0,$$

$$a_2 = \left(\frac{1}{\pi}\right)\int_{-\pi}^{\pi} x \cos 2x\, dx = 0,$$

$$b_1 = \left(\frac{1}{\pi}\right)\int_{-\pi}^{\pi} x \sin x\, dx = 2,$$

and $$b_2 = \left(\frac{1}{\pi}\right)\int_{-\pi}^{\pi} x \sin 2x\, dx = -1.$$

Therefore, $\mathbf{p}(x) = 2 \sin x - \sin 2x$. (The given $\mathbf{f}(x)$ and the approximating $\mathbf{p}(x)$ are shown in Figure 8.)

FIGURE 8

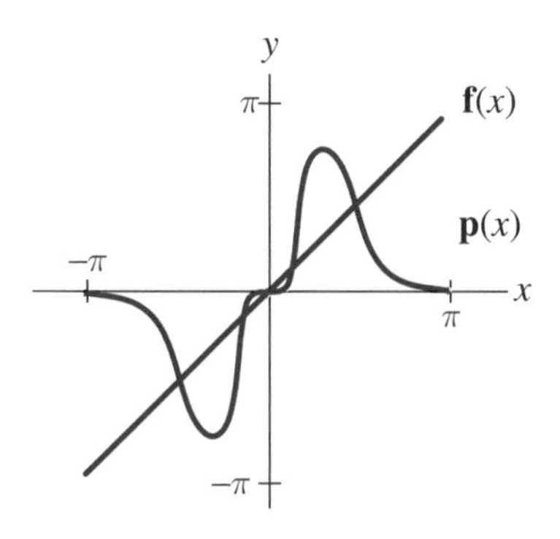

■

NOTE If a function **f** is continuous on $[-\pi, \pi]$, there is a Fourier series of order n for **f** for every natural number n. If this sequence of Fourier series is denoted by $\{\mathbf{p}_n\}$, then $\|\mathbf{f} - \mathbf{p}_n\| \to 0$ as $n \to \infty$. Therefore, the infinite series

$$a_0 + \sum_{k=1}^{\infty} (a_k \cos kx + b_k \sin kx)$$

converges to the function **f**. This series is called the **Fourier series** (without the qualifier *of order n*) for **f**.

Least Square Approximation to a Linear System

Consider the linear system $A\mathbf{x} = \mathbf{b}$. The system may have no solutions. We can, however, ask for a "best" solution, in the least square sense: we seek the minimum distance of $A\mathbf{x}$ from $\mathbf{b}$ over all choices of column vectors $\mathbf{x}$ in $\mathbf{R}^n$. This is equivalent to minimizing $\|A\mathbf{x} - \mathbf{b}\|^2$ over all $\mathbf{x}$ in $\mathbf{R}^n$.

THEOREM 4

Let A be an $m \times n$ matrix and $\mathbf{b}$ a column vector in $\mathbf{R}^m$. Then the vectors $\mathbf{x}$ in $\mathbf{R}^n$ that minimize $\|A\mathbf{x} - \mathbf{b}\|^2$ are the solutions to $(A^TA)\mathbf{x} = A^T\mathbf{b}$.

Proof The set $U = \{A\mathbf{x} : \mathbf{x} \text{ is in } \mathbf{R}^n\}$ is a subspace of $\mathbf{R}^n$, namely the column space of A (see Exercise 21). From Theorem 1, we see that the solution corresponds precisely to those vectors $\mathbf{x}$ for which $A\mathbf{x}$ is equal to the projection of $\mathbf{b}$ onto $\mathbf{U}$ (as the columns $\mathbf{a}^1, \mathbf{a}^2, \ldots, \mathbf{a}^n$ span $\mathbf{U}$, there are scalars $c_1, c_2, \ldots, c_n$ so that

$$proj_{\mathrm{u}}\mathbf{b} = c_1\mathbf{a}^1 + c_2\mathbf{a}^2 + \cdots + c_n\mathbf{a}^n = A\begin{bmatrix} c_1 \\ c_2 \\ \vdots \\ c_n \end{bmatrix}$$

and hence there is indeed at least one such $\mathbf{x}$). That is, the vectors $\mathbf{x}$ in $\mathbf{R}^n$ that minimize $\|A\mathbf{x} - \mathbf{b}\|^2$are those such that $\mathbf{b} - A\mathbf{x}$ is orthogonal to $\mathbf{U}$.

However, being orthogonal to $\mathbf{U}$ is equivalent to being orthogonal to each vector in a spanning set of $\mathbf{U}$. As the columns of A span the column space, $\mathbf{U}$, we conclude that the least square approximation $\mathbf{x}$ to the linear system $A\mathbf{x} = \mathbf{b}$ are those vectors $\mathbf{x}$ in $\mathbf{R}^n$ for which $(\mathbf{a}^i)^T(\mathbf{b} - A\mathbf{x}) = \mathbf{0}$ for each column $\mathbf{a}^i$ of A. But this is equivalent to $\mathbf{x}$ being a solution to $A^T(\mathbf{b} - A\mathbf{x}) = \mathbf{0}$, that is, $(A^TA)\mathbf{x} = A^T\mathbf{b}$. ■

EXAMPLE 4

Find the least square approximation to the linear system

$$\begin{aligned} -2x + y &= -1 \\ x - y &= -2 \\ x + y &= 0. \end{aligned}$$

Solution It is clear that the linear system has no solution; from the bottom two equations we derive that $x = -1$ and $y = 1$, and these values are not a solution to the first equation. To find a least square approximation, we write the system in the form $A\mathbf{x} = \mathbf{b}$, where

$$A = \begin{bmatrix} -2 & 1 \\ 1 & -1 \\ 1 & 1 \end{bmatrix} \text{ and } \mathbf{b} = \begin{bmatrix} -1 \\ -2 \\ 0 \end{bmatrix}.$$

From Theorem 4, the least square approximations $\mathbf{x}$ are the solutions to $(A^TA)\mathbf{x} = A^T\mathbf{b}$, that is,

$$\begin{bmatrix} 6 & -2 \\ -2 & 3 \end{bmatrix}\begin{bmatrix} x \\ y \end{bmatrix} = \begin{bmatrix} 0 \\ 1 \end{bmatrix}.$$

Solving this linear system, we find that $x = \frac{1}{7}$ and $y = \frac{3}{7}$, and this yields our least square best approximation to the linear system. ■

Theorem 4 can be used to provide a proof of Theorem 1 of Section 3.6. For a given set of points $(x_0, y_0), (x_1, y_1), \ldots, (x_m, y_m)$, it is not hard to show that seeking a polynomial

$$Q(x) = b_0 + b_1 x + b_2 x^2 + \cdots + b_n x^n$$

that minimizes

$$S = [y_0 - Q(x_0)]^2 + [y_1 - Q(x_1)]^2 + \cdots + [y_m - Q(x_m)]^2$$

is equivalent to seeking a least square approximation to the linear system $U\mathbf{b} = \mathbf{y}$, where $U = \begin{bmatrix} 1 & x_0 & x_0^2 & \cdots & x_0^n \\ 1 & x_1 & x_1^2 & \cdots & x_1^n \\ \vdots & \vdots & \vdots & & \vdots \\ 1 & x_m & x_m^2 & \cdots & x_m^n \end{bmatrix}$ and $y = \begin{bmatrix} y_0 \\ y_1 \\ \vdots \\ y_m \end{bmatrix}$. From Theorem 4, this corresponds to a solution of $(U^T U)\mathbf{b} = U^T\mathbf{y}$. The only remaining part to show is that the solution is unique, and this we leave to Exercise 22.

6.6 Exercises

In Exercises 1–4 find the *linear* least square polynomial approximating the given function on the indicated interval.

1. $f(x) = x^4$ on $[-1, 1]$

2. $f(x) = x^3 + 3x$ on $[-1, 1]$

3. $f(x) = 1/x$ on $[1, 2]$

4. $f(x) = \sqrt{x}$ on $[1, 4]$

5–8. Find the *quadratic* least square polynomial approximating the functions of Exercises 1–4 on the indicated intervals.

9–10. Find the *cubic* least square polynomial approximating the functions of Exercises 1 and 2 on the indicated intervals.

In Exercises 11–16 find the Fourier series of order *n* (for the indicated *n*) for the given function.

11. $f(x) = x + 1 \quad (n = 1)$

12. $f(x) = x^2 \quad (n = 1)$

13. $f(x) = x^3 + 1 \quad (n = 1)$

14. $f(x) = e^x \quad (n = 1)$

15. $f(x) = x - 1 \quad (n = 2)$

16. $f(x) = \sin 3x \quad (n = 2)$

17. Show that $\{1, \cos x, \cos 2x, \ldots, \cos nx, \sin x, \sin 2x, \ldots, \sin nx\}$ is an orthogonal basis for $\mathbf{T}_n[-\pi, \pi]$, the set of all trigonometric polynomials on $[-\pi, \pi]$ of order less than or equal to n.

For Exercises 18–20, compute a least square approximation to the linear system.

18.
$$\begin{aligned} x - y + z &= -1 \\ x - y + z &= -2 \\ -x + 2y - z &= 3 \\ 2x - y + 2z &= 0 \end{aligned}$$

19. $A\mathbf{x} = \mathbf{b}$, where $A = \begin{bmatrix} 2 & 1 \\ 1 & 0 \\ 0 & 2 \end{bmatrix}$ and $\mathbf{b} = \begin{bmatrix} 2 \\ -3 \\ 4 \end{bmatrix}$

20. $A\mathbf{x} = \mathbf{b}$, where $A = \begin{bmatrix} 2 & 1 & -1 \\ 0 & 4 & 2 \\ 1 & -1 & 1 \\ -1 & 0 & 1 \end{bmatrix}$ and $\mathbf{b} = \begin{bmatrix} 0 \\ 2 \\ 1 \\ 1 \end{bmatrix}$

21. Prove that the set $U = \{A\mathbf{x} : \mathbf{x} \text{ is in } \mathbf{R}^n\}$ is the column space of A.

22. Prove that the solution to $(U^TU)\mathbf{b} = U^T\mathbf{y}$ is unique. [*Hint*: First show that the matrix U has rank $n + 1$ by drawing a connection to Vandermonde matrices (page 172). Then argue that U^TU also has rank $n + 1$, and hence is invertible.]

6.7 APPLICATION—Music, Fourier Transforms, and *A Hard Day's Night*

Music is very closely tied to Fourier series. Pure musical tones are derived from some medium (usually air molecules) oscillating back and forth. The oscillation takes the form of $A\sin(kx + \phi)$ for some numbers A, k, and ϕ. A is called the amplitude and corresponds roughly to the loudness of the note. Our ears can rarely discern changes in ϕ, the *phase shift*, and it often can be safely ignored. The periodic properties of the trigonometric sine function show that the function $A\sin(kx + \phi)$ repeats $\frac{k}{2\pi}$ times every second; this is called the *frequency* of the note, and we hear it as the note's pitch. Higher values of the frequency are heard as higher pitches. Moreover, doubling the frequency moves a note up one octave.

Complex musical tones correspond to linear combinations of such functions. Our ears and minds have an innate ability to find the Fourier series approximations guaranteed by Theorem 3 of Section 6.6 to incoming sound waves—we can decompose the incoming sound waves into their component pure tones. Digitized musical tones are samples of the amplitude of the sound at equally spaced times; for CDs, the sampling rate is 44,100 Hz, that is, 44,100 times per second.

There is a process of fitting a linear combination of sine functions of the form $\sin(kx)$ to the data points, which is called a *Fourier transform* (it is the analogue of fitting a trigonometric polynomial to a function that passes through the data points). Computer programs can help us apply Fourier transforms to data points. This can allow us to reconstruct the original frequencies that made up the original sound. The following describes, in part, a mathematical deduction carried out on a mystery of pop music.

There has long been a controversy over how The Beatles played the opening guitar chord of *A Hard Day's Night*, certainly one of the most famous (and instantly recognizable) sounds in rock 'n' roll. We can use Fourier transforms to uncover what frequencies made up the sound, and at what amplitudes. Figure 9 shows what the sound wave for a portion of the opening chord looks like.

A Fourier transform was applied to the data, decomposing the sound into many, many frequencies—29,375 frequencies, to be exact. Anything that rattles, shakes, or rolls in the studio could have added frequencies to the sound, along with additional

FIGURE 9

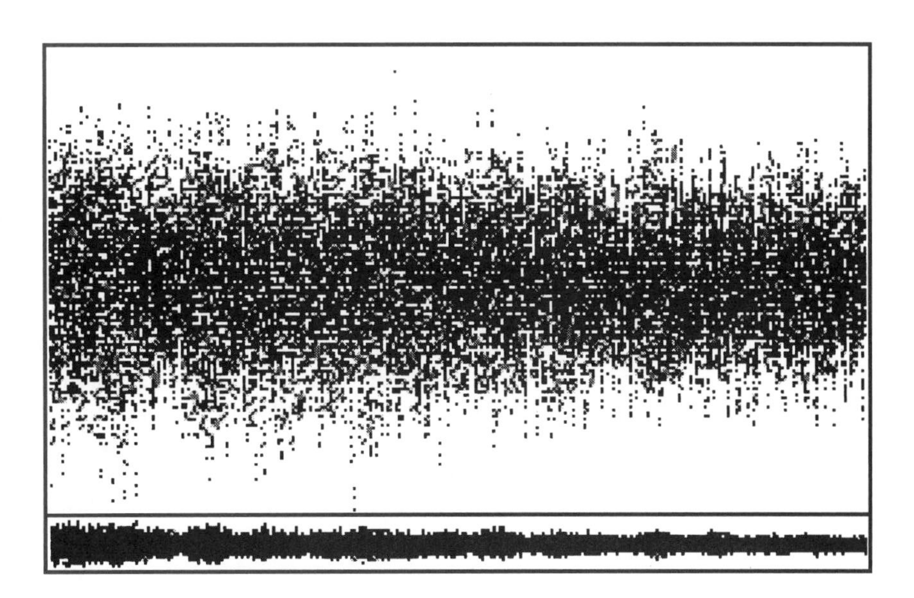

TABLE 1

Freq. (Hz)	Ampl.	Freq. (Hz)	Ampl.	Freq. (Hz)	Ampl.	Freq. (Hz)	Ampl.
110.340	0.0600967	299.494	0.0298296	1050.86	0.0687151	2368.93	0.0221358
145.619	0.0254850	392.570	0.0309716	1185.97	0.0372155	2371.19	0.0212846
148.621	0.0264278	438.358	0.0286329	1286.55	0.0231789	2371.19	0.0212846
149.372	0.0656018	524.678	0.0680974	1314.32	0.0381900	2372.69	0.0360420
150.123	0.1751490	587.730	0.0206130	1320.33	0.0223535	2637.65	0.0261839
174.142	0.0275547	588.480	0.0310337	1321.08	0.0494908	2638.40	0.0237794
174.893	0.0380282	589.231	0.0231753	1488.47	0.0241328	2754.00	0.0200010
175.643	0.0407103	785.141	0.0323532	1632.58	0.0205742	2763.76	0.0493617
195.159	0.0405164	786.642	0.0251928	1750.43	0.0234704	3083.52	0.0332062
218.428	0.0448308	787.393	0.0268553	2359.93	0.0366079	3147.32	0.0293723
261.964	0.0302402	960.784	0.0228509	2367.43	0.0267098	3148.07	0.0418507
262.714	0.0234502	981.801	0.0224200	2368.18	0.0755327	3158.58	0.0285631

tones, called *harmonics*, that appear in conjunction with other notes (and appear, in general, as integer multiples of the original frequencies).

What is clear from this plot is that, while many frequencies are present, there are some that appear with much larger amplitude than the rest. Among these are very likely the frequencies of the notes played by the musicians themselves. Some of the 48 frequencies of largest amplitude are listed in Table 1. The plot in Figure 10 shows the various frequencies, with their corresponding amplitudes.

FIGURE 10

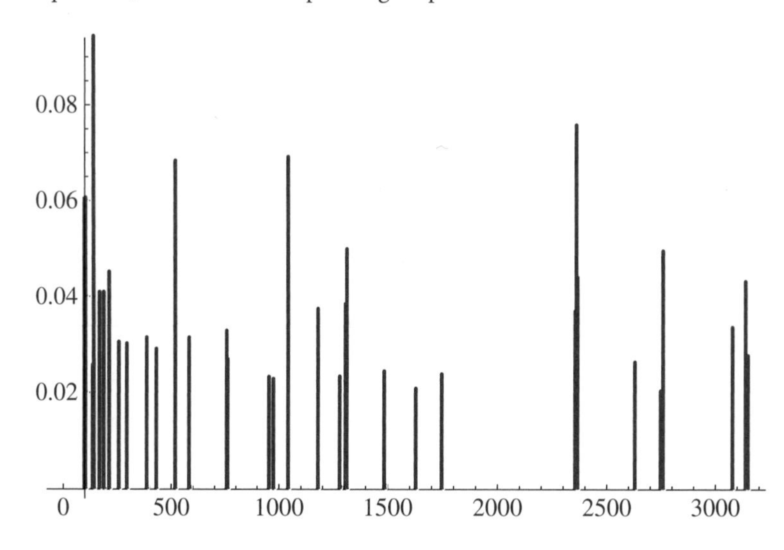

The note A below middle C has frequency 220 Hz and is written as A3 (the number refers to the octave of the tone, with middle C written as C4). If we round each frequency to the nearest note, we get the following list of notes:

A2, D3, D3, D3, D3, F3, F3, F3, G3, A3, C4, C4, D4, G4, A4, C5, D5, D5, D5, G5, G5, G5, B5, B5, C6, D6, E6, E6, E6, E6, F#6, G#6, A6, D7, D7, D7, D7, D7, D7, D7, E7, E7, F7, F7, G7, G7, G7, G7

We can safely ignore some of the higher notes as they are beyond the scope of the musical instruments played.

The data are quite illuminating. For instance, we see that there is one note, a D3 (with frequency 150.123 Hz) that has a much larger amplitude than any other note. This is almost certainly a bass note (played by Paul McCartney). What is also interesting is that there are multiple copies of some lower notes, with similar but slightly different frequencies. What does this mean? It means that these notes arose from different places, perhaps different guitar strings.

It is known that George Harrison, The Beatles' lead guitarist, played a brand new 12-string guitar on the song. The guitar strings occurred in octave pairs and were (from lowest to highest)

E2 E3 A2 A3 D3 D4 G3 G4 B3 B3 E4 E4.

Trying to figure out how he might have played the lower notes listed, even with Paul McCartney playing bass and John Lennon playing another guitar, poses problems. Among these notes are the three F3's—there is no way to account for them on George Harrison's and John Lennon's guitars, or by the stringing of George's 12-string. If he played an F3, then an F4 should be heard as well, but isn't. What is the solution?

The solution is that, indeed, there is no solution with the two guitars. But a piano is strung quite differently from a guitar. Each note of a piano typically consists of three strings, tuned to the same note, but often a little out of tune with each other. This offers the explanation of how the chord was played and why it remained a mystery for so long—the guitar chord wasn't just a guitar chord! The Beatles' producer, George Martin, played piano later on the track, so it is likely that the piano mixed into the chord is by his hands.

Further examination led to the following musical notation for the chord (Figure 11).

FIGURE 11

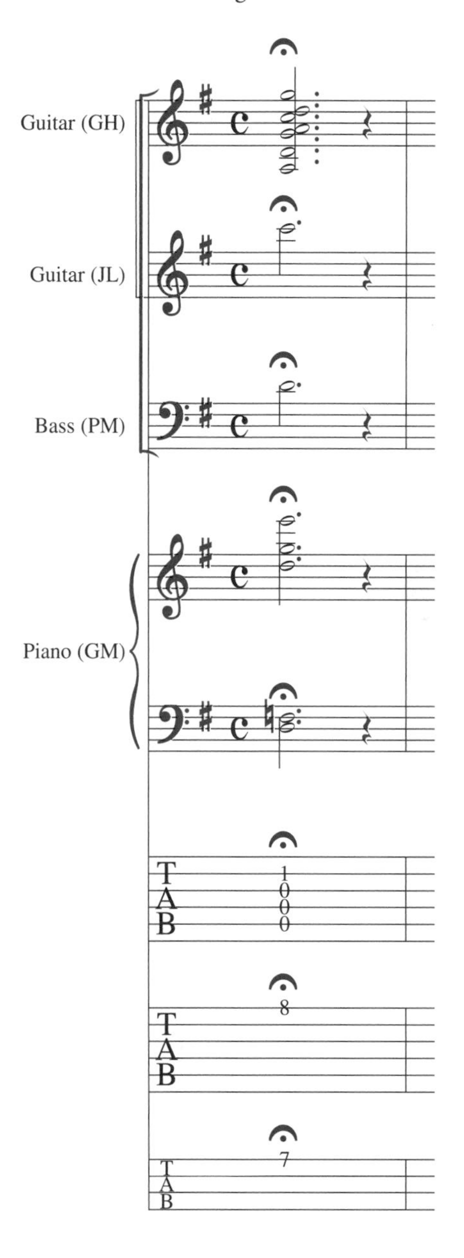

Mathematics can be a powerful tool in uncovering musical mysteries. For more information, read one of the two articles on Fourier transforms and *A Hard Day's Night* by Jason Brown in *Guitar Player* magazine (January 2005 and November 2006).

6.7 Exercises

1. The number of semitones a note is above A3 (with frequency 220 Hz) is given by the function $f(x) = 12 \log_2(x/220)$. Use the function to calculate the number of semitones each frequency in Table 1 is above and below A3. What does this tell you about the tuning of The Beatles' instruments?

2. Explain how the data in Table 1 explain the near perfect mixing of the piano and guitars on the opening chord.

6.8 APPLICATION—Fractals I: Self-Similarity, Projections, and Seeds

One of the key developers of the field of fractals was Benoit Mandelbrot (1924–2010), who was born in Poland. He fled the oncoming German invasion by moving to France with his family in 1936, and then worked at IBM's Thomas J. Watson Research Center in the United States for much of his career. He coined the term *fractal* in 1975. He investigated a variety of fractals, including the Mandelbrot set, named after him, that described the set of complex numbers *c* such that no matter how you repeatedly apply the function $f(z) = z^2 + c$ to the number 0, the results are always bounded.

Gamma-Rapho via Getty Images

Many of the objects in nature have what is called a "self-similar" structure, in that a small portion of the object looks very much like the object as a whole. For example, the craggy coastline of a country, upon magnification, again looks like a craggy coastline. Blowing up a picture of the side of a mountain again looks like the side of a mountain. Clouds, lightning, and trees also exhibit such self-similarity. Even some of the food we eat is self-similar. Pull apart a stalk of broccoli or cauliflower, and see how a part resembles the whole.

Contrast this with many of the geometric objects that you are accustomed to working with mathematically. Figures such as squares and circles are definitely not self-similar, as parts do not resemble the whole. For example, blowing up a part of a circle looks like a straight line.

As humans, what we often find most beautiful are objects that have self-similarity—trees, mountains, and so on. Figures that exhibit self-similarity are called *fractals.* For those working in computer animation, the generation of pictures of such objects is difficult using the basic shapes of geometry—lines, circles, ellipses, squares, and polygons. But the study of self-similar objects and how they can be generated has found many applications, including computer-generated landscapes. In the movie *Star Trek II—The Wrath of Khan*, the rebirth of a planet was created with great effect using fractals.

The fractal in Figure 12, which is known as *Sierpinski's Triangle*, was created using a mathematical program called *Maple.* Notice how the whole object is made up of three smaller Sierpinski's triangles, which are identical to the whole triangle, although smaller by half. Each of these three triangles is made up of three smaller ones, and so on.

One way to create fractals from line art is by a process of replacement (there are other methods as well—see Section 7.7). Suppose that we start off with four straight line segments of the same length, joined as follows (Figure 13).

The two slanted line segments make a 120° angle with the adjacent horizontal line segments, so that, with the missing piece, they would form an isosceles triangle.

Now the process is to replace each of the four line segments by a scaled copy of the entire figure. It is not clear exactly how the replacement should be done; to what side should the "notch" be pointed on each segment? What we will decide is to point the notch always to the left of the segment as we travel from left to right (although certainly there are other possibilities). Figure 14 shows what it would look like.

We iterate the process, replacing each of the line segments by a copy of the original figure. Figure 15 shows the next four iterations. The final iteration looks very much like a coastline.

FIGURE 12

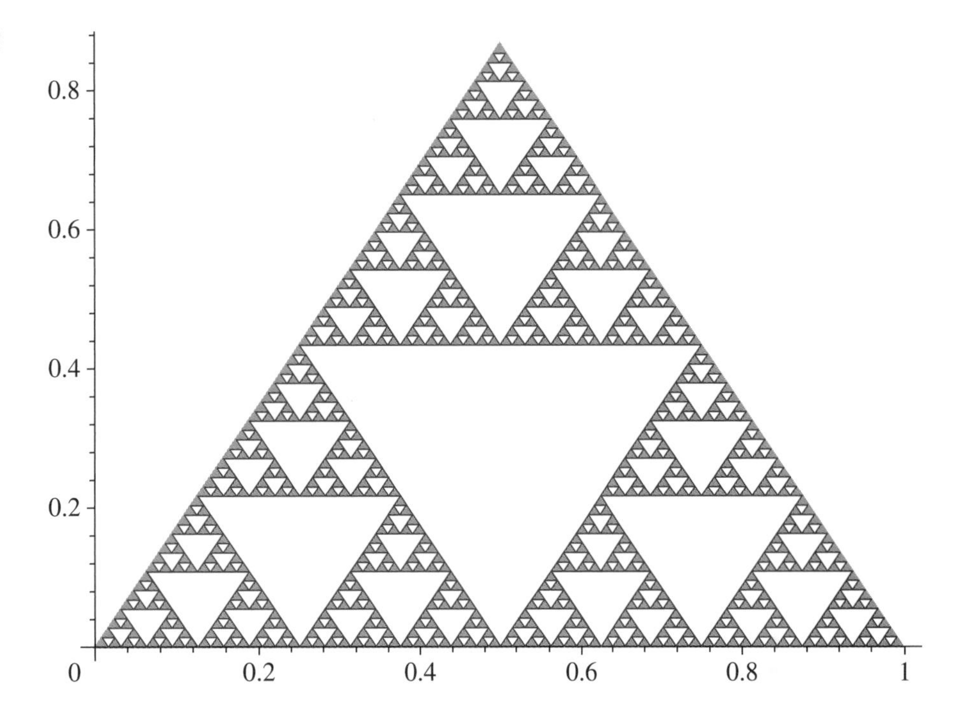

FIGURE 13

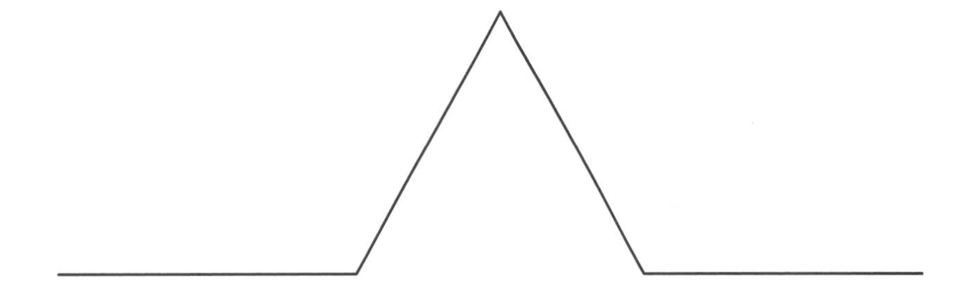

FIGURE 14

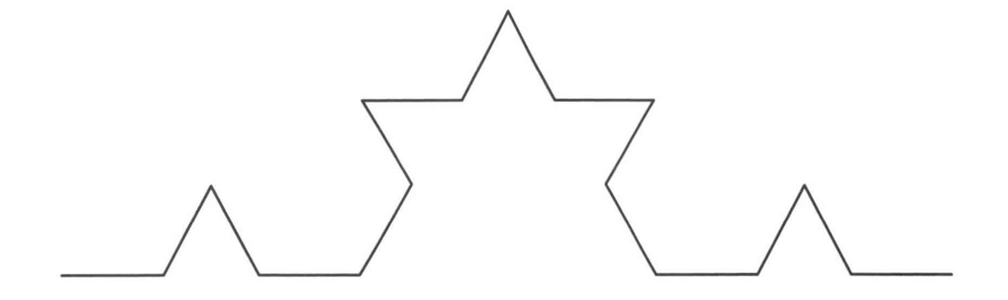

Further iterations won't produce visibly different pictures, as we are limited by the resolution of the printer (and our eyes). It should be clear that we can, in principle, continue the process indefinitely, and the actual fractal is the "limit" (in a sense we won't define here) of the figures. This curve, called the *Koch* curve, was introduced in 1904 by the mathematician Helge von Koch.

The general process involves taking a seed figure, consisting of a set of line segments, and then iterating the process of replacing each of the line segments by a scaled version of the seed. How can vectors and their orthogonal projections be used to carry out the process? To find out, we will make use of the following theorems (we leave the proof of the second for an exercise).

FIGURE 15

THEOREM 1

Let $\mathbf{v}_1$ be a nonzero vector in $\mathbf{R}^2$ and $\mathbf{v}_2$ be a nonzero vector perpendicular to $\mathbf{v}_1$. Then any vector $\mathbf{u}$ in $\mathbf{R}^2$ can be uniquely expressed as the sum of its orthogonal projections along $\mathbf{v}_1$ and $\mathbf{v}_2$:

$$\mathbf{u} = \text{proj}_{\mathbf{v}_1}\mathbf{u} + \text{proj}_{\mathbf{v}_2}\mathbf{u}$$

(The vector $\text{proj}_{\mathbf{v}_2}\mathbf{u}$ is called the orthogonal projection of $\mathbf{u}$ *orthogonal* to $\mathbf{v}_1$, in addition to being called the orthogonal projection of $\mathbf{u}$ *onto* $\mathbf{v}_2$.)

Proof From Theorem 3 of Section 6.4, $\{\mathbf{v}_1, \mathbf{v}_2\}$ is an orthogonal basis for the two-dimensional vector space $\mathbf{R}^2$. Thus, we can write vector $\mathbf{u}$ as a linear combination of $\mathbf{v}_1$ and $\mathbf{v}_2$, say

$$\mathbf{u} = c_1\mathbf{v}_1 + c_2\mathbf{v}_2$$

for some scalars c_1 and c_2.

If we take the inner product of both sides with $\mathbf{v}_1$, then we get

$$\begin{aligned}\mathbf{u} \bullet \mathbf{v}_1 &= (c_1\mathbf{v}_1 + c_2\mathbf{v}_2) \bullet \mathbf{v}_1 \\ &= c_1\mathbf{v}_1 \bullet \mathbf{v}_1 + c_2\mathbf{v}_2 \bullet \mathbf{v}_1 \\ &= c_1\mathbf{v}_1 \bullet \mathbf{v}_1 + c_2\,(0) \\ &= c_1\mathbf{v}_1 \bullet \mathbf{v}_1\end{aligned}$$

so that

$$c_1 = \frac{\mathbf{u} \cdot \mathbf{v}_1}{\mathbf{v}_1 \cdot \mathbf{v}_1}.$$

But then $c_1\mathbf{v}_1$ is equal to $\left(\frac{\mathbf{u} \cdot \mathbf{v}_1}{\mathbf{v}_1 \cdot \mathbf{v}_1}\right)\mathbf{v}_1$, which by Theorem 5 of Section 6.4 is nothing more than the orthogonal projection of $\mathbf{u}$ onto $\mathbf{v}_1$, $\text{proj}_{\mathbf{v}_1} \mathbf{u}$. Similarly,

$$c_2 = \frac{\mathbf{u} \cdot \mathbf{v}_2}{\mathbf{v}_2 \cdot \mathbf{v}_2}$$

and $c_2\mathbf{v}_2$ is the orthogonal projection of $\mathbf{u}$ onto $\mathbf{v}_2$, $\text{proj}_{\mathbf{v}_2} \mathbf{u}$. As $\mathbf{u} = c_1 \mathbf{v}_1 + c_2 \mathbf{v}_2$, we conclude that, indeed, $\mathbf{u} = \text{proj}_{\mathbf{v}_1}\mathbf{u} + \text{proj}_{\mathbf{v}_2}\mathbf{u}$. ■

THEOREM 2

Let $\mathbf{v}_1 = (a, b)$ be a nonzero vector in $\mathbf{R}^2$. Then $\mathbf{v}_2 = (-b, a)$ is a nonzero vector that is orthogonal to $\mathbf{v}_1$ and, moreover, it always points to the *left* of $\mathbf{v}_1 = (a, b)$.

Proof We leave the proof for Exercise 1. ■

EXAMPLE 1

Suppose that we have as our seed Figure 16, consisting of four line segments of length 1, with the inside two forming an equilateral triangle with the "missing" middle line segment.

FIGURE 16

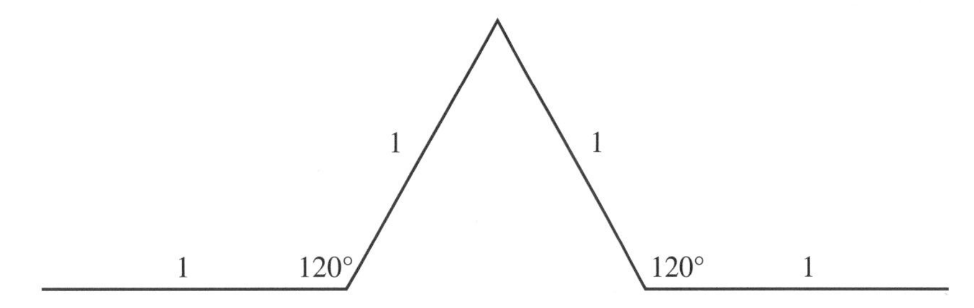

We calculate the points for the replacement of each of the four line segments of the seed by copies of the seed, with the "notch" always pointing to the left as we travel from left to right along the seed.

Solution We consider first the general problem of how to replace a line segment from point $P = (a, b)$ to $Q = (c, d)$ by a scaled copy of the seed. There are two possible orientations of the seed for each segment, depending on which endpoint of the segment we identify with the leftmost point of the seed. Thus, what we need to do is assign a direction to each of the segments that we are replacing. We also need to represent the seed as a sequence of vectors, lined up head to tail. Let's look at the general vector $\overrightarrow{PQ}$, along with the original directed seed (Figure 17).

FIGURE 17

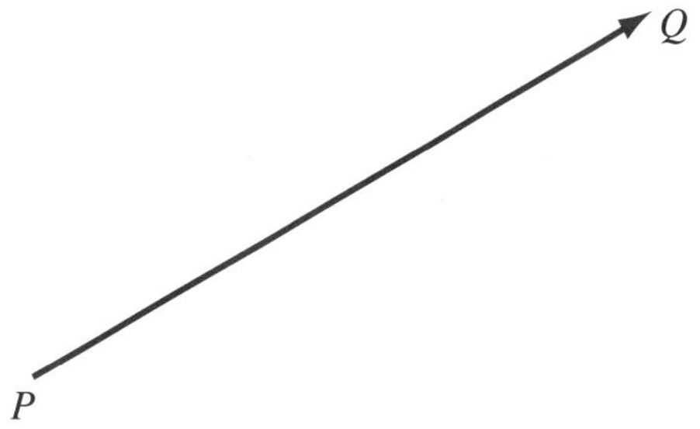

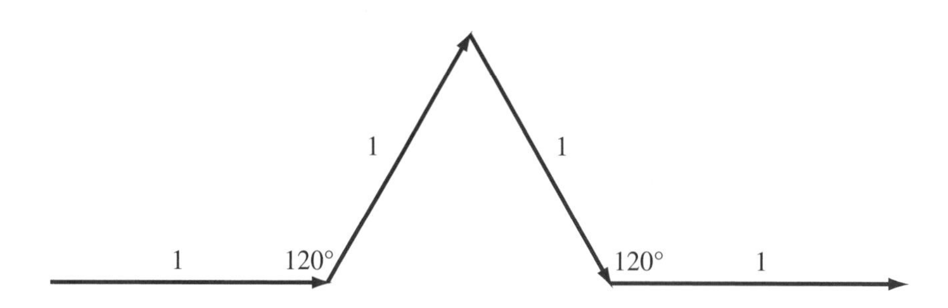

What we are after is shown in Figure 18.

FIGURE 18

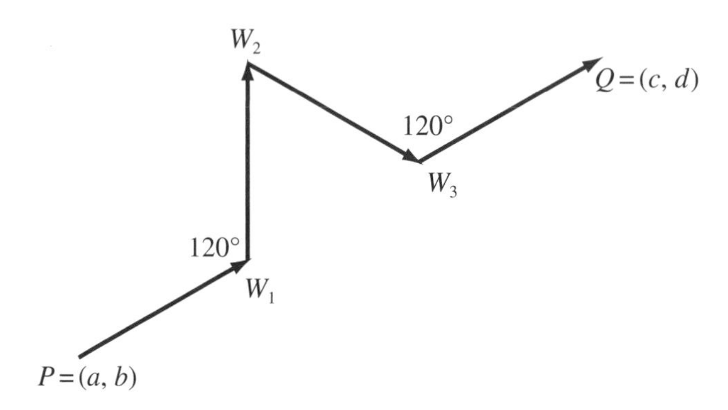

We need to find the locations of the points W_1, W_2, and W_3, which we will do by finding the vectors $\overrightarrow{PW_1}$, $\overrightarrow{W_1W_2}$, and $\overrightarrow{W_2W_3}$.

We'll start by representing the seed as a list of orthogonal projections along a vector that runs from the start to the end of the seed. If we align the seed so that it's leftmost point is at the origin, $O = (0, 0)$, then its right point will be at point $V = (3, 0)$ (Figure 19). We take the vector $\overrightarrow{OV}$ as our *reference vector* for the seed. We need to write each part of the seed in terms of the reference vector.

FIGURE 19

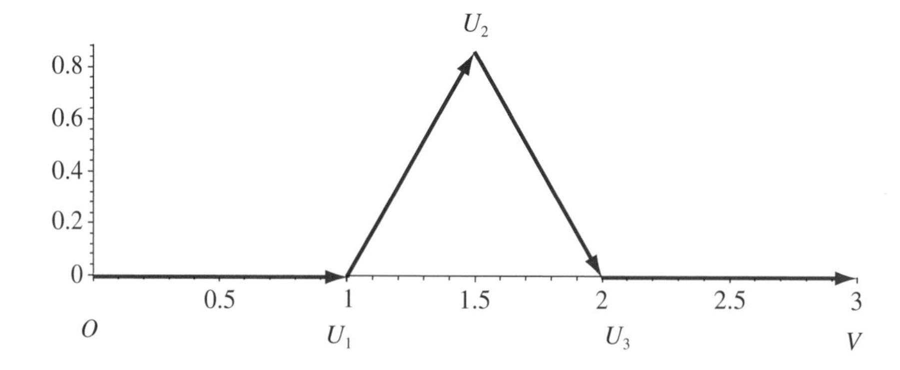

We can write the vector $\overrightarrow{OU_1}$ in terms of the orthogonal projection onto the reference vector $\overrightarrow{OV}$ and perpendicular (or orthogonal) to the reference vector. The calculation of the orthogonal projection of vector **u** onto vector **v** uses Theorem 5 of Section 6.4:

$$proj_{\mathbf{v}}\mathbf{u} = \left(\frac{\mathbf{u} \cdot \mathbf{v}}{\mathbf{v} \cdot \mathbf{v}}\right)\mathbf{v}$$

Thus, the orthogonal projection of $\mathbf{u}_1 = \overrightarrow{OU_1} = (1, 0)$ onto $\mathbf{v} = \overrightarrow{OV} = (3, 0)$ is

$$\begin{aligned} proj_{\mathbf{v}}\mathbf{u}_1 &= \left(\frac{(1, 0) \cdot (3, 0)}{(3, 0) \cdot (3, 0)}\right)\mathbf{v} \\ &= \frac{1}{3}\mathbf{v}. \end{aligned}$$

This corresponds to the obvious fact that $\overrightarrow{OU_1}$ is one-third of the way toward $\overrightarrow{OV}$.

What about the orthogonal projection of $\mathbf{u}_1$ orthogonal to $\mathbf{v}$? We will use a vector orthogonal to $\mathbf{v}$, but we do have a bit of choice here. Not only can we choose its magnitude, but we can also choose its direction—there are exactly two directions for a vector orthogonal to a given vector in $\mathbf{R}^2$ (namely one direction and its negative). Theorem 2 allows us to consistently choose a direction for the vector orthogonal to the reference vector $\mathbf{v}$, namely $\mathbf{v}' = (-0, 3) = (0, 3)$. The orthogonal projection of $\mathbf{u}_1$ onto $\mathbf{v}'$ is then

$$\begin{aligned} proj_{\mathbf{v}'}\mathbf{u}_1 &= \left(\frac{(1, 0) \cdot (0, 3)}{(0, 3) \cdot (0, 3)}\right)\mathbf{v}' \\ &= 0\mathbf{v}' \\ &= \mathbf{0}, \end{aligned}$$

which is what we expect (there is no part of $\mathbf{u}_1$ orthogonal to $\mathbf{v}$ as, in fact, $\mathbf{u}_1 = \frac{1}{3}\mathbf{v}$). Thus, we keep track of the fact that the pair of projections for $\mathbf{u}_1$ is $(proj_{\mathbf{v}}\mathbf{u}_1, proj_{\mathbf{v}'}\mathbf{u}_1) = \left(\frac{1}{3}\mathbf{v}, 0\mathbf{v}'\right)$.

We proceed onto the next vector, $\mathbf{u}_2 = \overrightarrow{U_1U_2}$, and calculate its orthogonal projections onto $\mathbf{v}$ and $\mathbf{v}'$. The points U_1 and U_2 have coordinates $(1, 0)$ and $\left(\frac{3}{2}, \frac{\sqrt{3}}{2}\right)$ respectively (see Exercise 2). Thus, the vector $\mathbf{u}_2$ is equal to $\left(\frac{3}{2} - 1, \frac{\sqrt{3}}{2} - 0\right) = \left(\frac{1}{2}, \frac{\sqrt{3}}{2}\right)$. We calculate its orthogonal projections onto $\mathbf{v}$ and $\mathbf{v}'$ as follows:

$$\begin{aligned} proj_{\mathbf{v}}\mathbf{u}_2 &= \left(\frac{(1/2, \sqrt{3}/2) \cdot (3, 0)}{(3, 0) \cdot (3, 0)}\right)\mathbf{v} \\ &= \frac{1}{6}\mathbf{v} \\ proj_{\mathbf{v}'}\mathbf{u}_2 &= \left(\frac{(1/2, \sqrt{3}/2) \cdot (0, 3)}{(0, 3) \cdot (0, 3)}\right)\mathbf{v}' \\ &= \frac{\sqrt{3}}{6}\mathbf{v}' \end{aligned}$$

Thus, the ordered pair of orthogonal projections associated with $\mathbf{u}_2$ is $(proj_{\mathbf{v}}\mathbf{u}_2, proj_{\mathbf{v}'}\mathbf{u}_2) = \left(\frac{1}{6}\mathbf{v}, \frac{\sqrt{3}}{6}\mathbf{v}'\right)$. Similarly, the ordered pair of orthogonal projections associated with $\mathbf{u}_3 = \overrightarrow{U_2U_3} = \left(\frac{1}{2}, -\frac{\sqrt{3}}{2}\right)$ is $(proj_{\mathbf{v}}\mathbf{u}_3, proj_{\mathbf{v}'}\mathbf{u}_3) = \left(\frac{1}{6}\mathbf{v}, -\frac{\sqrt{3}}{6}\mathbf{v}'\right)$. For $\mathbf{u}_4 = \overrightarrow{U_3V} = (1, 0)$, it is $(proj_{\mathbf{v}}\mathbf{u}_4, proj_{\mathbf{v}'}\mathbf{u}_4) = \left(\frac{1}{3}\mathbf{v}, 0\mathbf{v}'\right)$ (the same as for $\mathbf{u}_1$).

We are now ready for our replacement scheme. We replace $\mathbf{v}$ by $\mathbf{w} = (c - a, d - b)$ and $\mathbf{v}'$ by $\mathbf{w}' = (-(d - b), c - a)$ in all our formulas for the orthogonal projections.

Starting at P, we use the orthogonal projections $(proj_{\mathbf{v}}\mathbf{u}_1, proj_{\mathbf{v}'}\mathbf{u}_1) = \left(\frac{1}{3}\mathbf{v}, 0\mathbf{v}'\right)$ for the first segment $\mathbf{u}_1$ to calculate $\overrightarrow{PW_1}$ as

$$\overrightarrow{PW_1} = \frac{1}{3}\mathbf{w} + 0\mathbf{w}' = \frac{1}{3}\mathbf{w}.$$

Similarly,

$$\begin{aligned}
\overrightarrow{W_1W_2} &= \frac{1}{6}\mathbf{w} + \frac{\sqrt{3}}{6}\mathbf{w}', \\
\overrightarrow{W_2W_3} &= \frac{1}{6}\mathbf{w} - \frac{\sqrt{3}}{6}\mathbf{w}', \\
\text{and} \quad \overrightarrow{W_3Q} &= \frac{1}{3}\mathbf{w}.
\end{aligned}$$

This gives us the locations of each of the points W_1, W_2, and W_3 (see figure 18):

$$\begin{aligned}
\overrightarrow{OW_1} &= \overrightarrow{OP} + \overrightarrow{PW_1} \\
&= (a, b) + \frac{1}{3}\mathbf{w} \\
\overrightarrow{OW_2} &= \overrightarrow{OW_1} + \overrightarrow{W_1W_2} \\
&= \overrightarrow{OW_1} + \frac{1}{6}\mathbf{w} + \frac{\sqrt{3}}{6}\mathbf{w}' \\
\overrightarrow{OW_3} &= \overrightarrow{OW_2} + \overrightarrow{W_2W_3} \\
&= \overrightarrow{OW_2} + \frac{1}{6}\mathbf{w} - \frac{\sqrt{3}}{6}\mathbf{w}'
\end{aligned}$$

So, for example, when we replace the first line segment from $P = (0, 0)$ to $Q = (1, 0)$ by a copy of the seed, the points we insert are defined as follows (here $\mathbf{w}$ is the vector from P to Q, that is, $(1, 0)$):

$$\begin{aligned}
\overrightarrow{OW_1} &= \overrightarrow{OP} + \overrightarrow{PW_1} \\
&= (0, 0) + \frac{1}{3}\mathbf{w} \\
&= (0, 0) + \frac{1}{3}(1, 0) \\
&= \left(\frac{1}{3}, 0\right) \\
\overrightarrow{OW_2} &= \overrightarrow{OW_1} + \overrightarrow{W_1W_2} \\
&= \left(\frac{1}{3}, 0\right) + \frac{1}{6}(1, 0) + \frac{\sqrt{3}}{6}(0, 1) \\
&= \left(\frac{1}{2}, \frac{\sqrt{3}}{6}\right) \\
\overrightarrow{OW_3} &= \overrightarrow{OW_2} + \overrightarrow{W_2W_3} \\
&= \left(\frac{1}{2}, \frac{\sqrt{3}}{6}\right) + \frac{1}{6}(1, 0) - \frac{\sqrt{3}}{6}(0, 1) \\
&= \left(\frac{2}{3}, 0\right)
\end{aligned}$$

Thus, the line segment from (0, 0) to (1, 0) gets replaced by four line segments, running from $(0,0)$ to $\left(\frac{1}{3}, 0\right)$ to $\left(\frac{1}{2}, \frac{\sqrt{3}}{6}\right)$ to $\left(\frac{2}{3}, 0\right)$ to $(1,0)$. Verify that the other points are as follows: $\left(\frac{7}{6}, \frac{\sqrt{3}}{6}\right)$, $\left(1, \frac{\sqrt{3}}{3}\right)$, $\left(\frac{4}{3}, \frac{\sqrt{3}}{6}\right)$, $\left(\frac{3}{2}, \frac{\sqrt{3}}{2}\right)$, $\left(\frac{5}{3}, \frac{\sqrt{3}}{3}\right)$, $\left(2, \frac{\sqrt{3}}{3}\right)$, $\left(\frac{11}{6}, \frac{\sqrt{3}}{6}\right)$, (2, 0), $\left(\frac{7}{3}, 0\right)$, $\left(\frac{5}{2}, \frac{\sqrt{3}}{6}\right)$, $\left(\frac{8}{3}, 0\right)$, and (3, 0).

FIGURE 20

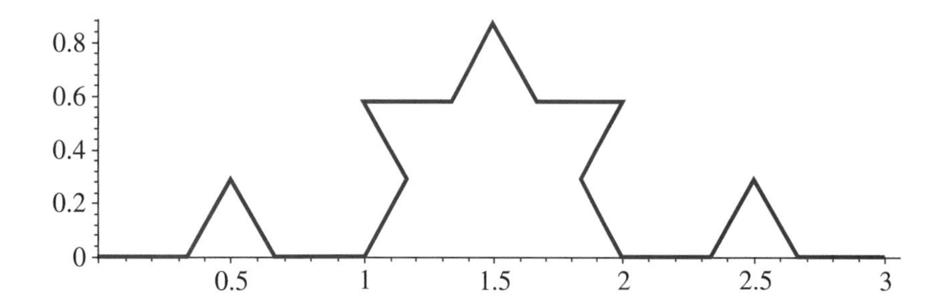

The process for replacement by a seed may seem long, but it is fairly straightforward to program in a language such as *Maple.* ■

The orientation of the line segments can make a big difference. For example, if we orient the third edge of the seed in the opposite direction, then the first replacement looks like Figure 21.

FIGURE 21

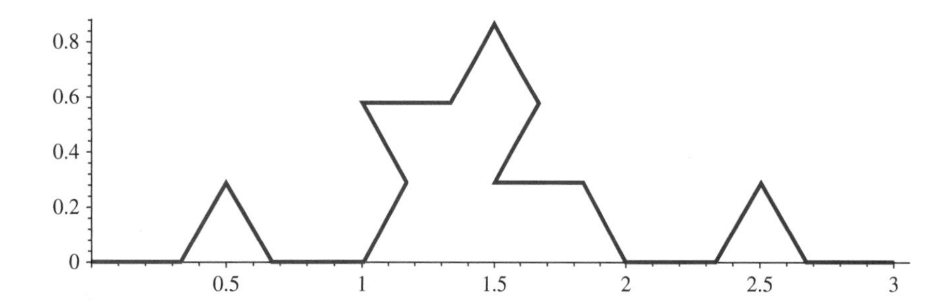

Three iterations later, the curve looks like Figure 22.

FIGURE 22

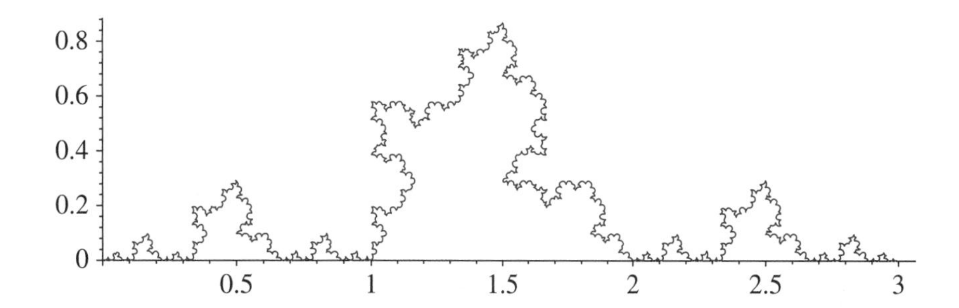

As another example of what can be done, consider Figure 23.

FIGURE 23

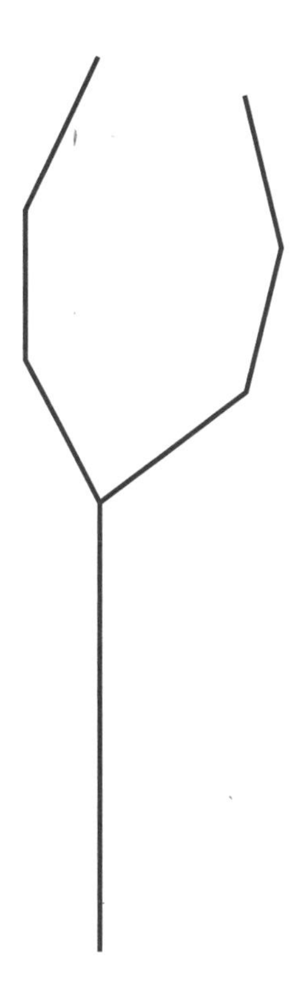

Figure 23 is a fairly simple seed. What we choose to do is replace all the branches, except for the bottom half of the vertical branch, by a copy of the seed. Figure 24 shows what we get under three successive iterations.

FIGURE 24

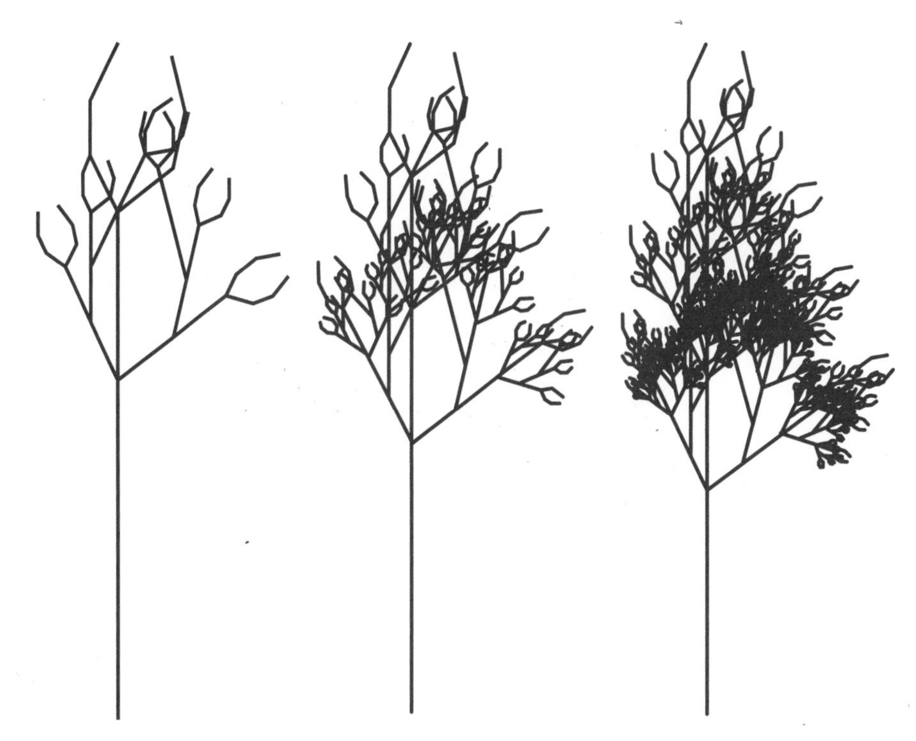

This is a good approximation to a leafless tree, all carried out with only some linear algebra and a computer; there was no scanning of a real tree involved.

There are several advantages to using fractals to generate realistic-looking images. First, high resolution images are large files. Fractals have a small description (namely the vectors for the seed and the line segments to replace, along with the procedure for generating them) so such images are greatly compressed, only requiring suitable computing power to regenerate "on the fly." Also, now that the image is represented in the computer, it can be manipulated and animated in a variety of mathematical ways. The sky's the limit!

6.8 Exercises

1. Prove Theorem 2.

2. In Example 2, show that U_1 and U_2 have coordinates $(1, 0)$ and $\left(\frac{3}{2}, \frac{\sqrt{3}}{2}\right)$, respectively.

3. Let $\mathbf{v} = (1, 2)$ and $\mathbf{v}' = (-2, 1)$. For the following vectors $\mathbf{u}$, determine the orthogonal projection of $\mathbf{u}$ onto $\mathbf{v}$ and onto $\mathbf{v}'$.

(a) $\mathbf{u} = (1, 2)$
(b) $\mathbf{u} = (-2, 1)$
(c) $\mathbf{u} = (2, -1)$
(d) $\mathbf{u} = (3, -7)$
(e) $\mathbf{u} = (1, 1)$

4. *(a)* Show that, in Example 1, the length of the curve under an iteration is 4/3 times the previous length.
(b) Explain why this implies that the Koch curve can be said to have infinite length even though it is contained in a bounded rectangle. (The Koch curve has many interesting properties, including being continuous everywhere but differentiable nowhere.)

5. Our bones, lungs, and brains all have fractal properties. Explain why. Can you give any biological advantages for their having fractal properties?

6. Explain how you would alter the procedure for replacement by a seed to allow only some segments to be replaced and for figures where the points are not arranged to form a polygon (such as in Figure 24).

7. Calculate the set of points for the seed of the tree, with the base of the tree at the origin $O = (0, 0)$ and the top of the tree at $V = (0, 3)$. Then calculate the projection of each line segment of the seed in terms of the reference vector $\mathbf{v} = \overrightarrow{OV}$.

CHAPTER SUMMARY

Vector Spaces and Subspaces

(a) DEFINITION: A *vector space* consists of a given set of objects and two given operations (called *addition* and *scalar multiplication*) defined on that set and satisfying certain special properties. (See Section 6.1 for the list of properties—the *vector space axioms.*)

(b) If **V** is a vector space and **W** is a subset of **V**, then **W** is a *subspace* of **V** if either

- **W** is itself a vector space under the operations of **V** (DEFINITION) or, equivalently,
- **W** is nonempty and closed under the operations of **V**.

(c) The following are vector spaces under the usual operations of addition and scalar multiplication:

- The set $\mathbf{M}^{m,n}$ of all $m \times n$ matrices
- The set **P** of all polynomials
- The set $\mathbf{P}_n$ of all polynomials of degree less than or equal to n

Basis and Dimension

(a) A set of vectors, $\{\mathbf{v}_1, \mathbf{v}_2, \ldots, \mathbf{v}_n\}$, in a vector space **V** is *linearly independent* if either

- The vector equation $c_1\mathbf{v}_1 + c_2\mathbf{v}_2 + \cdots + c_n\mathbf{v}_n = \mathbf{0}$ has only the trivial solution with all $c_i = 0$ (DEFINITION) or, equivalently,
- None of its vectors can be written as a linear combination of the others.

DEFINITION: A set of vectors is *linearly dependent* if and only if it is not linearly dependent.

(b) DEFINITION: A set of vectors $\mathcal{S}$ in a vector space **V** *spans* **V** if every vector in **V** can be expressed as a finite linear combination of those in $\mathcal{S}$.

(c) DEFINITION: A set of vectors $\mathcal{S}$ in a vector space **V** is a *basis* for **V** if $\mathcal{S}$ is linearly independent and spans **V**. (Every basis for **V** has the same number of vectors.)

(d) DEFINITION: The *dimension* of a vector space **V** is equal to the number of vectors in a basis for **V**.

(e) Let $\mathcal{S}$ be a subset of a vector space **V** of dimension n. If $\mathcal{S}$ contains s vectors, and

i. If $s > n$, then $\mathcal{S}$ is linearly dependent;

ii. If $s < n$, then $\mathcal{S}$ does not span **V**;

iii. If $s = n$, then $\mathcal{S}$ is linearly independent if and only if it spans **V**.

(f) The dimension of $\mathbf{M}^{m,n}$ is mn, and the dimension of $\mathbf{P}_n$ is $n + 1$.

Coordinate Vectors

(a) DEFINITION: Let $\mathcal{B} = \{\mathbf{b}_1, \mathbf{b}_2, \ldots, \mathbf{b}_n\}$ be a basis for a vector space **V**, and let **v** be a vector in **V**. Then there is a unique vector $\mathbf{x} = (x_1, x_2, \ldots, x_n)$ such that $\mathbf{v} = x_1\mathbf{b}_1 + x_2\mathbf{b}_2 + \cdots + x_n\mathbf{b}_n$. The vector **x** is called the *coordinate vector* of **v** with respect to the basis $\mathcal{B}$.

(b) Coordinate vectors (which are vectors in $\mathbf{R}^n$) for a given set of vectors in a vector space **V** may be used instead of these vectors to answer questions about linear independence, span, and basis.

Inner Product Spaces

(a) DEFINITION: Let **V** be a vector space (over the real numbers). An *inner product* on **V** is a function that associates a real number, denoted $(\mathbf{u}, \mathbf{v})$, with every pair of vectors **u** and **v** in **V** such that

i. $(\mathbf{u}, \mathbf{v}) = (\mathbf{v}, \mathbf{u})$,

ii. $(\mathbf{u}, \mathbf{v} + \mathbf{w}) = (\mathbf{u}, \mathbf{v}) + (\mathbf{u}, \mathbf{w})$,

iii. $c(\mathbf{u}, \mathbf{v}) = (c\mathbf{u}, \mathbf{v})$,

iv. $(\mathbf{u}, \mathbf{u}) \geq 0$; $(\mathbf{u}, \mathbf{u}) = 0$ if and only if $\mathbf{u} = \mathbf{0}$.

The vector space **V**, together with the inner product, is called an *inner product space.*

(b) DEFINITION: Let **C** be an inner product space. Then the *norm* of **u**, denoted $\|\mathbf{u}\|$, is given by $\|\mathbf{u}\| = \sqrt{(\mathbf{u}, \mathbf{u})}$.

(c) *Cauchy–Schwarz inequality*: $|(\mathbf{u}, \mathbf{v})| \leq \|\mathbf{u}\| \, \|\mathbf{v}\|$.

(d) DEFINITION: Vectors **u** and **v** in an inner product space are *orthogonal* if $(\mathbf{u}, \mathbf{v}) = 0$. A set of vectors $\mathcal{S}$ in an inner product space is *orthogonal* if every pair of vectors in $\mathcal{S}$ is orthogonal; an orthogonal set $\mathcal{S}$ is *orthonormal* if every vector in $\mathcal{S}$ has norm equal to 1.

(e) Let **V** be an inner product space with basis $\mathcal{B} = \{\mathbf{b}_1, \mathbf{b}_2, \ldots, \mathbf{b}_n\}$. Then, by the *Gram–Schmidt process*, we can construct an *orthonormal basis*—a basis that is an orthonormal set—from the $\mathbf{b}_i$.

Vector Spaces over Other Fields

(a) DEFINITION: A set **F**, together with operations called *addition* (+) and *multiplication* (∗), is a **field** if the following properties are satisfied for every choice of elements *a*, *b*, and *c* in **F**:

i. $a + b$ and $a * b$ are in **F**.

ii. $a + b = b + a$

iii. $(a + b) + c = a + (b + c)$

iv. There is an element in **F**, denoted by 0, such that $a + 0 = a$ for all a in **F**.

v. For each a in **F**, there is an element, denoted by $-a$, such that $a + (-a) = 0$.

vi. $a * b = b * a$

vii. $(a * b) * c = a * (b * c)$

viii. There is an element in **F**, denoted by 1, such that $a * 1 = a$ for all a in **F**.

ix. For each $a \neq 0$ in **F**, there is an element, denoted by a^{-1}, such that $a * a^{-1} = 1$.

x. $a * (b + c) = a * b + a * c$

(b) DEFINITION: For every field **F** and every positive integer *m*, there is an associated vector space $\mathbf{F}^m$ that consists of all *m*-tuples from **F**.

(c) DEFINITION: *Complex m-space*, $\mathbf{C}^m$, is the set of all *m*-tuples of complex numbers.

KEYWORDS

vector space
scalar
subspace
linearly dependent
linearly independent
span
basis
dimension
coordinate vector
inner product
inner product space
norm
Cauchy–Schwarz inequality
orthogonal
orthonormal
projection
Gram–Schmidt process
complex space
complex number
field
least square approximation
trigonometric polynomial
Fourier series
frequency
fractal
seed

SELF-TEST

In Exercises 1–4 determine whether or not the given set with the usual operations is a vector space. If it is not a vector space, give one property that fails to hold and verify that it fails with a *specific example.* [Section 6.1]

1. The set of all ordered pairs (a, b) with both a and b real numbers greater than or equal to 0

2. The set of all 2×2 matrices A with real entries and $a_{12} = 0$

3. The set of all polynomials with no x^2 term

4. The set of all polynomials with degree exactly 4, together with the zero polynomial

5. Verify explicitly that the set $\mathcal{T} = \{1 + x + x^2, x - x^2, 1 - x^2, 1 + x - x^2\}$ is linearly dependent in $\mathbf{P}_2$ by expressing $1 + x - x^2$ as a linear combination of the other three polynomials. [Section 6.2]

6. Verify that the set $\mathcal{T}$ of Exercise 5 is linearly dependent, using no calculations other than counting. [Section 6.2]

7. Use coordinate vectors to calculate the coefficients needed to express

$$\begin{bmatrix} 1 & 1 \\ 1 & 1 \end{bmatrix}$$

as a linear combination of the basis for $\mathbf{M}^{2,2}$ given by

$$\left\{\begin{bmatrix} 0 & 1 \\ 1 & 1 \end{bmatrix}, \begin{bmatrix} 1 & 0 \\ 1 & 1 \end{bmatrix}, \begin{bmatrix} 1 & 1 \\ 0 & 1 \end{bmatrix}, \begin{bmatrix} 1 & 1 \\ 1 & 0 \end{bmatrix}\right\}.$$ [Section 6.3]

8. Prove that in any inner product space $\mathbf{V}$, if $\mathbf{v}$ and $\mathbf{w}$ are orthogonal, then any scalar multiple of $\mathbf{v}$ is orthogonal to any scalar multiple of $\mathbf{w}$. [Section 6.4]

9. Verify that $f(x) = x^2 - 2x$ and $g(x) = x - 1$ are orthogonal in the inner product space of $\mathbf{P}_2$, with

$$(f, g) = f(0)g(0) + f(1)g(1) + f(2)g(2).$$ [Section 6.4]

10. The vectors $\mathbf{v}_1 = (1, 1, 1, 0)$ and $\mathbf{v}_2 = (1, 0, -1, 1)$ are orthogonal in $\mathbf{R}^4$, and $\mathbf{v}_3 = (1, 1, 1, 1)$ is linearly independent from them. Use the Gram–Schmidt process to replace $\mathbf{v}_3$ with a third vector $\mathbf{w}$ that is orthogonal to both $\mathbf{v}_1$ and $\mathbf{v}_2$ and that, together with them, generates the same subspace as the original three vectors. [Section 6.4]

11. Find an orthonormal basis for the subspace of $\mathbf{C}^3$ generated by

$$\{(i, 0, -i), (0, 2, 0), (-i, 2, i)\}.$$

[Section 6.5]

12. Are the vectors $(z_1, z_2 z_3 z_4)$ that satisfy the equations

$$\begin{aligned} -2z_1 + z_2 \qquad\qquad &= 0 \\ z_2 - 3z_3 + z_4 &= i \\ z_1 - \qquad\qquad z_4 &= 1 \end{aligned}$$

a subspace of $\mathbf{C}^4$? Explain your answer.

[Section 6.5]

REVIEW EXERCISES

In Exercises 1–6 determine whether or not the given set is a vector space under the indicated operations. If it is not a vector space, list each property that fails to hold.

1. The set of all polynomials of degree less than 3 whose coefficients have a sum of 1: the usual operations

2. The set of all polynomials of degree less than 3 whose coefficients have a sum of 0: the usual operations

3. Those functions in $\mathbf{C}[0, 3]$ that have value 0 at every point in the subinterval $[1, 2]$: the operations of Example 6 of Section 6.1

4. Those functions in $\mathbf{C}[0, 3]$ that have value *not* equal to 0 at every point in the subinterval $[1, 2]$: the operations of Example 6 of Section 6.1

5. The set of all 2×3 matrices of the form

$$\begin{bmatrix} a & b & a-b \\ a+2b & ab & b-a \end{bmatrix}$$

6. The set of all 2×3 matrices of the form

$$\begin{bmatrix} a & b & a-b \\ a+2b & a+b & b-a \end{bmatrix}$$

In Exercises 7–10 determine whether or not the given set is linearly independent. If the set is linearly dependent, express one of its vectors as a linear combination of the others. (*Note*: It is easier to use coordinate vectors with respect to the appropriate standard basis.)

7. $\{x + 2x^2 + 3x^3, -x + 2x^2 - x^3, x + x^3\}$ in $\mathbf{P}_3$

8. $\left\{\begin{bmatrix} 2 & 0 \\ 0 & 1 \end{bmatrix}, \begin{bmatrix} 1 & -2 \\ 0 & 0 \end{bmatrix}, \begin{bmatrix} 4 & -4 \\ 0 & 1 \end{bmatrix}, \begin{bmatrix} 1 & 1 \\ 0 & 1 \end{bmatrix}\right\}$ in $\mathbf{M}^{2,2}$

9. $\{x - \cos x, 3x + x^2 + \cos x + 2\sin x, -2x + 2\cos x\}$ in $\mathbf{C}^\infty$

10. $\left\{\begin{bmatrix} -1 & 1 \\ 2 & -1 \end{bmatrix}, \begin{bmatrix} 1 & 0 \\ -1 & 1 \end{bmatrix}, \begin{bmatrix} -1 & 2 \\ 3 & -1 \end{bmatrix}, \begin{bmatrix} 1 & 1 \\ 0 & 1 \end{bmatrix}\right\}$ in $\mathbf{M}^{2,2}$

In Exercises 11–16 determine whether or not the given set is orthogonal in the indicated inner product space.

11. $\{(2, -1, 1, 0), (0, 1, 1, 0), (1, 1, -1, -2)\}$ in $\mathbf{R}^4$: the usual dot product

12. The same set as in Exercise 11 but with the inner product of Example 2 of Section 6.4, with $m = 4$

13. $\{1, 1 - 2x\}$ in $\mathbf{P}_1$: $(\mathbf{p}, \mathbf{q}) = \int_0^1 \mathbf{p}(x)\mathbf{q}(x)\,dx$

14. $\{1, 1 - 2x, 3x^2 - 1\}$ in $\mathbf{P}_2$: $(\mathbf{p}, \mathbf{q}) = \int_0^1 \mathbf{p}(x)\mathbf{q}(x)\,dx$

15. $\{x - 2, x^2 - x\}$ in $\mathbf{P}_2$: $(\mathbf{p}, \mathbf{q}) = \mathbf{p}(0)\mathbf{q}(0) + \mathbf{p}(1)\mathbf{q}(1) + \mathbf{p}(2)\mathbf{q}(2)$

16. $\{x, \sin x\}$ in $\mathbf{C}^{\infty}\left[-\frac{\pi}{2}, \frac{\pi}{2}\right]$: $(\mathbf{f}, \mathbf{g}) = \int_{-\frac{\pi}{2}}^{\frac{\pi}{2}} \mathbf{f}(x)\mathbf{g}(x)\,dx$

In Exercises 17–20 determine whether or not each of the following is an inner product for the indicated vector space.

17. $(\mathbf{p}, \mathbf{q}) = \mathbf{p}(1)\mathbf{q}(1) + \mathbf{p}(2)\mathbf{q}(2)$ on $\mathbf{P}_2$

18. $(\mathbf{p}, \mathbf{q}) = \mathbf{p}(1)\mathbf{q}(1) + \mathbf{p}(2)\mathbf{q}(2) + \mathbf{p}(3)\mathbf{q}(3)$ on $\mathbf{P}_2$

19. $(A, B) = \sum_{i=1}^{3} a_{ii}b_{ii}$ on $\mathbf{M}^{3,3}$ (the sum of the products of the diagonal entries of A and B)

20. $(A, B) = \sum_{i=1}^{2}\sum_{j=1}^{2} ija_{ij}b_{ij}$ on $\mathbf{M}^{2,2}$

In Exercises 21 and 22 use the Gram–Schmidt process to find an orthonormal basis for the space generated by the set in the given inner product space.

21. $\{(1, 1, 1, 0), (1, 1, 0, 1), (1, 0, 1, 1)\}$ in $\mathbf{R}^4$

22. $\{e^x, e^{-x}\}$ in $\mathbf{C}[0, 1]$ with $(\mathbf{f}, \mathbf{g}) = \int_0^1 \mathbf{f}(x)\mathbf{g}(x)\,dx$

23. Prove that $(A, B) = \sum_{i=1}^{m}\sum_{j=1}^{n} ija_{ij}b_{ij}$ is an inner product on $\mathbf{M}^{m,n}$.

24. Show, by example, that vectors in a vector space may be orthogonal with respect to one inner product and not orthogonal with respect to another.

25. Consider the vector space $\mathbf{Z}_2^n$. Are the vectors whose coordinates have an odd sum a subspace of $\mathbf{Z}_2^4$? Explain your answer.

26. Is the set of vectors $\{(0,1,1,1), (1,0,0,1), (1,1,1,0)\}$ linearly independent over $\mathbf{R}$? Over $\mathbf{Z}_2$? Explain your answers.

27. Find a basis for the subspace of all vectors in $\mathbf{Z}_2^4$ that has an even number of 1's.

28. For vectors $\mathbf{u} = (u_1, \ldots, u_n)$ and $\mathbf{v} = (v_1, \ldots, v_n)$ in $\mathbf{Z}_2^n$, we define $\mathbf{u} \bullet \mathbf{v} = u_1v_1 + \cdots + u_nv_n$ (addition in $\mathbf{Z}_2$). Is this an inner product on $\mathbf{Z}_2^n$?

29. Consider the set $S = \{(i, 0, i), (i, 0, -i), (3, 0, 1), (0, 1, 0)\}$. Find a basis for the subspace of $\mathbf{C}^3$ spanned by S.

30. Prove that the set of vectors (z_1, z_2, z_3, z_4) that satisfies the equations

$$\begin{aligned} z_1 + \quad\quad\quad\quad z_4 &= 0 \\ z_2 - \quad\quad iz_4 &= 0 \\ z_3 - z_4 &= 0 \end{aligned}$$

form a subspace of $\mathbf{C}^4$. Find an orthonormal basis for the subspace and the dimension.

CHALLENGE YOURSELF

31. *(a)* Prove that the set of all 3×3 matrices A that commutes with

$$B = \begin{bmatrix} 1 & 1 & 1 \\ 0 & 1 & 1 \\ 0 & 1 & 1 \end{bmatrix}$$

(such that $AB = BA$) forms a subspace of $\mathbf{M}^{3,3}$.

(b) Find a basis for this subspace.

32. *(a)* Let n be any positive integer, and let r be any real number. Prove that the polynomials in $\mathbf{P}^n$ that have a root at r, form a subspace of $\mathbf{P}^n$.
(b) Find a basis for this subspace.

33. Suppose that we have vectors $\mathbf{v}_1, \mathbf{v}_2, \ldots, \mathbf{v}_k$ in $\mathbf{Q}^m$; that is, all their coordinates are *rational* numbers. Prove that if $\mathbf{v}_1, \mathbf{v}_2, \ldots, \mathbf{v}_k$ are linearly independent vectors in $\mathbf{Q}^m$, then they are linearly independent vectors in $\mathbf{R}^m$ as well.

34. Suppose that $\mathbf{W}$ is a subspace of $\mathbf{R}^m$ with an orthogonal basis $\mathbf{v}_1, \mathbf{v}_2, \ldots, \mathbf{v}_k$. The orthogonal projection of vector $\mathbf{u}$ onto $\mathbf{W}$ is the vector $\mathbf{u}'$ in $\mathbf{W}$ such that $\mathbf{u} - \mathbf{u}'$ is orthogonal to *every* vector in $\mathbf{W}$. Prove that the orthogonal projection of $\mathbf{u}$ onto $\mathbf{W}$ always exists and is unique.

35. Let $n \geq 2$. Suppose that we have subsets $A_1, A_2, \ldots, A_m$ and $B_1, B_2, \ldots, B_m$ of the set $\{1, 2, \ldots, n\}$ such that for all i and j, A_i and B_j have no elements in common if $i = j$ and have exactly one element in common otherwise. Prove that $m \leq n$. (*Hint*: For each subset A_i, associate the vector $\mathbf{a}_i$ in $\mathbf{R}^n$ whose kth component is 1 if k is in set A_i, and 0 otherwise. Define vectors $\mathbf{b}_j$ similarly. Then, using the dot product, prove that the vectors $\mathbf{a}_1, \mathbf{a}_2, \ldots, \mathbf{a}_m$ are linearly independent.)

SPOTLIGHT

Vera Bogaerts/Shutterstock

Among the computer-generated graphics we've seen on television or in the movies, an object spins and rockets through space, greenery grows at an amazing rate on a make-believe planet, and a spiralling picture of a television news program's logo swoops (almost) into your living room. Each of these special effects is made up of a sequence of discrete pictures, like the frames in movie film, mathematically generated by a computer.

How is this sequence of images generated? One way involves first creating a single, detailed picture made up of many individual points, or vectors, in $\mathbf{R}^3$—so many points that the mind visualizes the resulting image as a solid object. This object can then be given motion, rotated, resized, and even reshaped by performing *linear transformations* on its constituent vectors. To create images that make the object appear to move continuously before your eyes, a sequence of transformations is performed, each of which changes the object by a very small increment. Although each of the individual transformations is simple enough, an enormous number of them are required to make an effective graphic, necessitating the use of extremely fast computers containing very large amounts of memory.

In this chapter, we discuss linear transformations in general. In Section 7.1, we present examples of the kinds of rotating and scaling transformations that are fundamental in the use of computer graphics. Linear transformations also are an essential in the mathematical description of real-life objects. We have seen, in Chapter 6, how fractals can be built from a seed of vectors under replacement. In Section 7.7, we discuss another way to construct fractals from certain linear transformations called *contractions.* These are transformations that always shrink distances between points. The process, often carried out with a computer, starts with *any* filled polygon, and repeatedly applies all transformations to the set.

Another procedure uses randomness to successively apply a randomly chosen transformation to a point. The images are produced much quicker and are drawn (or *rendered*) on the screen much faster, although they lack the crispness of the images produced under the nonrandom procedure. Fairly realistic images of real-life objects can be achieved.

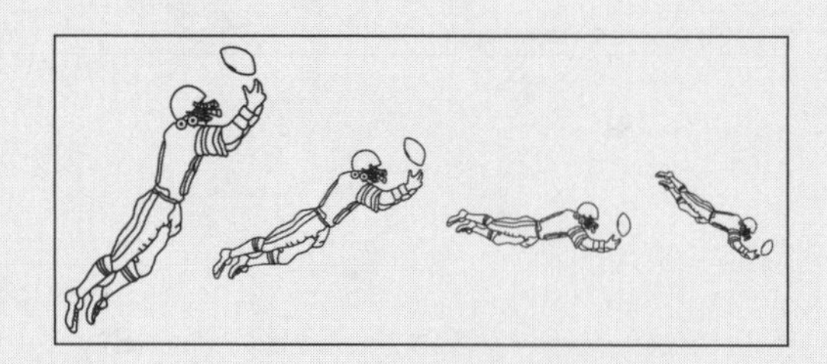

CHAPTER 7

Linear Transformations

OBJECTIVES

You will learn to:

- Determine whether or not a given function is a linear transformation.
- Find the matrix that represents a linear transformation from $\mathbf{R}^n$ to $\mathbf{R}^m$ and, more generally, from one vector space to another.
- Find the sum, difference, and product of two linear transformations.
- Find the kernel and image of a linear transformation.
- Determine whether or not a given linear transformation is one-to-one, is onto, or has an inverse.
- Find the change-of-basis matrix needed to express a given vector in $\mathbf{R}^n$ as the coordinate vector with respect to a given basis.

In this chapter, we discuss the concept of a *linear transformation*—a function from one vector space to another that possesses certain special properties. You will see that matrices play a key role in this subject and that many questions concerning a linear transformation can be answered by considering a corresponding matrix.

7.1 Definition of a Linear Transformation

Let us first recall some terminology about *functions*. A **function** or **map** from one set to another is an association of each element of the first set with a unique element of the second. If $\mathcal{A}$ and $\mathcal{B}$ are sets, the notation

$$f\colon \mathcal{A} \to \mathcal{B}$$

denotes a function from the **domain** $\mathcal{A}$ to the **codomain** $\mathcal{B}$. For a in $\mathcal{A}$, the unique element of $\mathcal{B}$ associated with a is called the **image** of a under f and is denoted by $f(a)$.

For our purposes, the sets $\mathcal{A}$ and $\mathcal{B}$ are vector spaces $\mathbf{V}$ and $\mathbf{W}$. We are not interested in all functions from $\mathbf{V}$ to $\mathbf{W}$, but only those that are *linear* in the following sense.

DEFINITION

Let **V** and **W** be vector spaces. A **linear transformation** from **V** to **W** is a function $T: \mathbf{V} \to \mathbf{W}$ that satisfies the following two conditions. For each **u** and **v** in **V** and scalar a,

i. $T(a\mathbf{u}) = aT(\mathbf{u})$ (Scalars factor out.)
ii. $T(\mathbf{u} + \mathbf{v}) = T(\mathbf{u}) + T(\mathbf{v})$ (T is additive.)

Linear transformations are precisely the functions that preserve what is essential about vector spaces, namely, vector addition and scalar multiplication.

EXAMPLE 1

Let $T: \mathbf{R}^2 \to \mathbf{R}^3$ be described by

$$T(x, y) = (x + y, x, 2x - y).$$

Show that T is a linear transformation from $\mathbf{R}^2$ to $\mathbf{R}^3$.

Solution To check conditions (i) and (ii) of the definition, let $\mathbf{u} = (x_1, y_1)$, $\mathbf{v} = (x_2, y_2)$, and a be a scalar.

i.
$$\begin{aligned} T(a\mathbf{u}) &= T(a(x_1, y_1)) \\ &= T((ax_1, ay_1)) \\ &= (ax_1 + ay_1, ax_1, 2(ax_1) - ay_1) \\ &= a(x_1 + y_1, x_1, 2x_1 - y_1) \\ &= aT((x_1, y_1)) = aT(\mathbf{u}) \end{aligned}$$

ii.
$$\begin{aligned} T(\mathbf{u} + \mathbf{v}) &= T((x_1, y_1) + (x_2, y_2)) \\ &= T((x_1 + x_2, y_1 + y_2)) \\ &= ((x_1 + x_2) + (y_1 + y_2), x_1 + x_2, 2(x_1 + x_2) - (y_1 + y_2)) \\ &= ((x_1 + y_1) + (x_2 + y_2), x_1 + x_2, (2x_1 - y_1) + (2x_2 - y_2)) \\ &= (x_1 + y_1, x_1, 2x_1 - y_1) + (x_2 + y_2, x_2, 2x_2 - y_2) \\ &= T(\mathbf{u}) + T(\mathbf{v}) \end{aligned}$$

■

EXAMPLE 2

Let $T: \mathbf{V} \to \mathbf{W}$ be described by $T(\mathbf{u}) = \mathbf{0}$, for all **u** in **V**. Show that T is a linear transformation.

Solution

i. If a is a scalar and **u** is in **V**, $T(a\mathbf{u}) = \mathbf{0} = a\mathbf{0} = aT(\mathbf{u})$.
ii. If **u** and **v** are in **V**, $T(\mathbf{u} + \mathbf{v}) = \mathbf{0} = \mathbf{0} + \mathbf{0} = T(\mathbf{u}) + T(\mathbf{v})$.

Thus, T is a linear transformation. ■

DEFINITION

The transformation of Example 2 is called the **zero transformation** from **V** to **W** and is denoted by *0*.

Another simple linear transformation, this time from a vector space to *itself*, is the one that leaves every element fixed; that is, it does not change anything. This is just the *identity function* on the space. That this function is a linear transformation is straightforward and is to be verified in Exercise 26.

DEFINITION

Let $\mathbf{V}$ be a vector space. The linear transformation $I: \mathbf{V} \rightarrow \mathbf{V}$ described by $I(\mathbf{v}) = \mathbf{v}$, for every $\mathbf{v}$ in $\mathbf{V}$, is called the **identity transformation**.

EXAMPLE 3

Define $T: \mathbf{M}^{2,3} \rightarrow \mathbf{P}_2$ by

$$T(A) = T\begin{bmatrix} a_{11} & a_{12} & a_{13} \\ a_{21} & a_{22} & a_{23} \end{bmatrix} = (a_{11} + a_{13})x^2 + (a_{21} - a_{22})x + a_{23}.$$

Show that T is a linear transformation.

Solution To verify condition (i) of the definition, let A be in $\mathbf{M}^{2,3}$ and c be a scalar. Applying T to cA, we have

$$\begin{aligned} T(cA) &= T\begin{bmatrix} ca_{11} & ca_{12} & ca_{13} \\ ca_{21} & ca_{22} & ca_{23} \end{bmatrix} \\ &= (ca_{11} + ca_{13})x^2 + (ca_{21} - ca_{22})x + ca_{23} \\ &= c[(a_{11} + a_{13})x^2 + (a_{21} - a_{22})x + a_{23}] = cT(A). \end{aligned}$$

To verify condition (ii), let A and B be in $\mathbf{M}^{2,3}$. Applying T to $A + B$, we have

$$\begin{aligned} T(A + B) &= T\begin{bmatrix} a_{11} + b_{11} & a_{12} + b_{12} & a_{13} + b_{13} \\ a_{21} + b_{21} & a_{22} + b_{22} & a_{23} + b_{23} \end{bmatrix} \\ &= [(a_{11} + b_{11}) + (a_{13} + b_{13})]x^2 \\ &\quad + [(a_{21} + b_{21}) - (a_{22} + b_{22})]x + [a_{23} + b_{23}] \\ &= (a_{11} + a_{13})x^2 + (a_{21} - a_{22})x + a_{23} \\ &\quad + (b_{11} + b_{13})x^2 + (b_{21} - b_{22})x + b_{23} \\ &= T(A) + T(B). \end{aligned}$$

Thus, T is a linear transformation. ■

The next example is very important in a more advanced study of linear algebra.

EXAMPLE 4

Let A be a square matrix of order n. Define $T: \mathbf{P} \rightarrow \mathbf{M}^{n,n}$ as follows. For each polynomial

$$\mathbf{p}(x) = a_m x^m + a_{m-1}x^{m-1} + \cdots + a_1 x + a_0,$$

define $T(\mathbf{p})$ to be the matrix

$$T(\mathbf{p}) = a_m A^m + a_{m-1}A^{m-1} + \cdots + a_1 A + a_0 I,$$

where I is the identity matrix of order n. Show that T is a linear transformation.

Solution To show condition (i), let $\mathbf{p}$ be the polynomial

$$\mathbf{p}(x) = a_m x^m + \cdots + a_1 x + a_0,$$

and c be a scalar. Then $c\mathbf{p}$ is the polynomial

$$(c\mathbf{p})(x) = (ca_m)x^m + \cdots + (ca_1)x + ca_0.$$

Applying T to $c\mathbf{p}$, we have

$$\begin{aligned} T(c\mathbf{p}) &= (ca_m)A^m + \cdots + (ca_1)A + (ca_0)I \\ &= c(a_m A^m + \cdots + a_1 A + A_0 I) = cT(\mathbf{p}). \end{aligned}$$

To show condition (ii), let **p** and **q** be the polynomials $\mathbf{p}(x) = a_m x^m + \cdots + a_1 x + a_0$ and $\mathbf{q}(x) = b_m x^m + \cdots + b_1 x + b_0$. (By using zero coefficients, if necessary, **p** and **q** can be made to appear as if they have the same degree.) Then $\mathbf{p} + \mathbf{q}$ is the polynomial

$$(\mathbf{p} + \mathbf{q})(x) = (a_m + b_m)x^m + \cdots + (a_1 + b_1)x + (a_0 + b_0).$$

Applying T to $\mathbf{p} + \mathbf{q}$, we have

$$\begin{aligned} T(\mathbf{p} + \mathbf{q}) &= (a_m + b_m)A^m + \cdots + (a_1 + b_1)A + (a_0 + b_0)I \\ &= (a_m A^m + \cdots + a_1 A + a_0 I) + (b_m A^m + \cdots + b_1 A + b_0 I) \\ &= T(\mathbf{p}) + T(\mathbf{q}). \end{aligned}$$

Thus, T is a linear transformation. ■

To show that a given function is *not* a linear transformation, we need to only show that *one* of the two conditions of the definition is violated in a *single* instance. This is illustrated in the next example.

EXAMPLE 5

Let $T: \mathbf{R}^2 \to \mathbf{R}^2$ be defined by $T(\mathbf{u}) = \mathbf{u} + (1, 2)$, for all **u** in $\mathbf{R}^2$. Show that T is *not* a linear transformation.

Solution To show that condition (i) is violated by this function, we let $a = 0$ and $\mathbf{u} = (0, 0)$. Then

$$T(a\mathbf{u}) = T((0, 0)) = (0, 0) + (1, 2) = (1, 2),$$

but

$$aT(\mathbf{u}) = 0\,T((0, 0)) = 0((0, 0) + (1, 2)) = 0(1, 2) = (0, 0).$$

Thus, it is not always true that $T(a\mathbf{u}) = aT(\mathbf{u})$, and condition (i) is violated: T is not linear. (It can be shown that condition (ii) does not hold either.) ■

Notice that the function described in Example 5 is a *translation* of the plane $\mathbf{R}^2$ by the vector (1, 2). For each point **u**, the line segment $\overrightarrow{\mathbf{u}T(\mathbf{u})}$ is equivalent to the vector (1, 2), as shown in Figure 1. In particular, the origin **0** gets mapped by T to the point (1, 2), $T(\mathbf{0}) = (1, 2)$. This fact, together with part (a) of the following theorem, provides another way to show that this function is not linear.

FIGURE 1

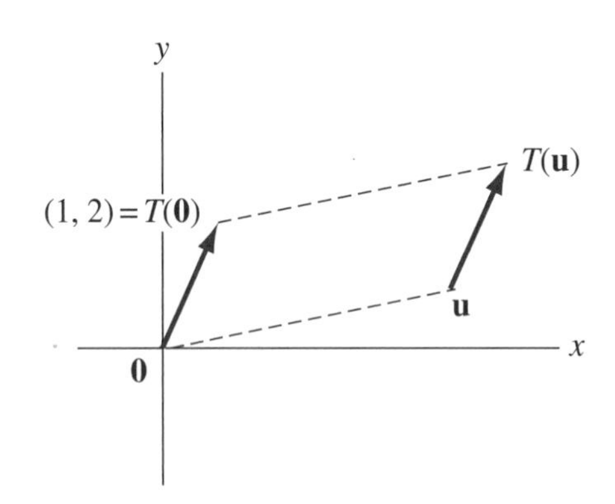

THEOREM 1

Let $T: \mathbf{V} \to \mathbf{W}$ be a linear transformation. Then

a. $T(\mathbf{0}) = \mathbf{0}$

b. $T(-\mathbf{u}) = -T(\mathbf{u})$ for each **u** in **V**

c. $T(\mathbf{u} - \mathbf{v}) = T(\mathbf{u}) - T(\mathbf{v})$ for each **u** and **v** in **V**

NOTE The vector $\mathbf{0}$ of $T(\mathbf{0})$ is an element of $\mathbf{V}$, and the vector $\mathbf{0}$ on the right side of the equation given in part (a) of Theorem 1 is an element of $\mathbf{W}$. We use the same notation for each, but you should always be aware of where the vector is located.

Proof

a. Since $\mathbf{0} = \mathbf{0} + \mathbf{0}$ in $\mathbf{V}$,

$$\begin{aligned} T(\mathbf{0}) &= T(\mathbf{0} + \mathbf{0}) && \text{(In } \mathbf{W}) \\ &= T(\mathbf{0}) + T(\mathbf{0}) && (T \text{ is additive}) \end{aligned}$$

Since $T(\mathbf{0})$ is in $\mathbf{W}$, using $\mathbf{0}$ in $\mathbf{W}$, we have

$$T(\mathbf{0}) + \mathbf{0} = T(\mathbf{0}) + T(\mathbf{0}).$$

Adding $-T(\mathbf{0})$ to both sides, we have

$$\mathbf{0} = T(\mathbf{0}).$$

b. By part (a) of Theorem 1, and the fact that $\mathbf{0} = \mathbf{u} + (-\mathbf{u})$, we have

$$\mathbf{0} = T(\mathbf{0}) = T((\mathbf{u} + (-\mathbf{u})) = T(\mathbf{u}) + T(-\mathbf{u}).$$

That is,

$$\mathbf{0} = T(\mathbf{u}) + T(-\mathbf{u}).$$

Adding $-T(\mathbf{u})$ to both sides, we have

$$-T(\mathbf{u}) = T(-\mathbf{u}).$$

c. The proof of part (c) is Exercise 20 at the end of this section. ■

WARNING In algebra, we define a linear *function* from $\mathbf{R}$ to $\mathbf{R}$ to be one of the form $f(x) = mx + b$, where m and b are arbitrary constants. As you can see from part (a) of Theorem 1, however, a linear function is not a linear transformation unless $b = 0$. More generally, the function $f\colon \mathbf{R}^m \to \mathbf{R}^m$ given by $f(\mathbf{x}) = m\mathbf{x} + \mathbf{b}$ (where m is a scalar and $\mathbf{x}$ and $\mathbf{b}$ are in $\mathbf{R}^m$) is a linear transformation if and only if $\mathbf{b} = \mathbf{0}$.

EXAMPLE 6

Let A be the matrix

$$A = \begin{bmatrix} 2 & 0 & 1 \\ 1 & 1 & -1 \end{bmatrix},$$

and let $\mathbf{x}$ be an element of $\mathbf{R}^3$. Let $T(\mathbf{x}) = A\mathbf{x}$, where $A\mathbf{x}$ denotes matrix multiplication. Show that T is a linear transformation.

Solution We let $\mathbf{x} = (x, y, z)$ and compute $T(\mathbf{x})$:

$$\begin{aligned} T(\mathbf{x}) = A\mathbf{x} &= \begin{bmatrix} 2 & 0 & 1 \\ 1 & 1 & -1 \end{bmatrix} \begin{bmatrix} x \\ y \\ z \end{bmatrix} \\ &= \begin{bmatrix} 2x + z \\ x + y - z \end{bmatrix} = (2x + z, x + y - z) \end{aligned}$$

Thus, we have the rule $T(x, y, z) = (2x + z, x + y - z)$, and we can follow the procedure of Example 1 to verify that T is linear. (You should do this for the practice.) ■

Example 6 is a special case of the following theorem.

THEOREM 2

Let A be an $m \times n$ matrix. Then $T: \mathbf{R}^n \rightarrow \mathbf{R}^m$, defined by $T(\mathbf{x}) = A\mathbf{x}$, is a linear transformation.

Proof First note that $\mathbf{x}$ is to be written as a column vector so that the matrix multiplication is defined: A is $m \times n$, and $\mathbf{x}$ is $n \times 1$. Thus, the product is defined, and the result is an $m \times 1$ matrix. The latter may be viewed as an element of $\mathbf{R}^m$.

To check property (i) of the definition of linear transformation, let a be a scalar and $\mathbf{x}$ be an element of $\mathbf{R}^n$. Then

$$\begin{aligned} T(a\mathbf{x}) &= A(a\mathbf{x}) \\ &= aA\mathbf{x} \qquad \text{(Part (n) of Theorem 1 of Section 3.1)} \\ &= aT(\mathbf{x}). \end{aligned}$$

For property (ii), let $\mathbf{x}$ and $\mathbf{y}$ be elements of $\mathbf{R}^n$. Then

$$\begin{aligned} T(\mathbf{x} + \mathbf{y}) &= A(\mathbf{x} + \mathbf{y}) \\ &= A\mathbf{x} + A\mathbf{y} \qquad \text{(Part (g) of Theorem 1 of Section 3.1)} \\ &= T(\mathbf{x}) + T(\mathbf{y}). \end{aligned}$$

Thus, T is a linear transformation. 

Examples of linear transformations given by Theorem 2 are of the utmost importance. In fact, it will be shown in Section 7.4 that, in some sense, all linear transformations are of this type. For the time being, such functions provide an infinite supply of linear transformations for which we need not check the definition; that has already been done in Theorem 2. We continue with some interesting special cases.

EXAMPLE 7

Let θ be any real number, and view θ as the radian measure of an angle. Let T_θ be given by $T_\theta(\mathbf{x}) = A_\theta\mathbf{x}$, where

$$A_\theta = \begin{bmatrix} \cos\theta & -\sin\theta \\ \sin\theta & \cos\theta \end{bmatrix}.$$

Then T_θ is the **rotation** by θ. In other words, T_θ is the linear transformation that rotates each vector of $\mathbf{R}^2$ counterclockwise through an angle of θ radians.

Solution To see that this is so, we need only show that, for all $\mathbf{x}$ in $\mathbf{R}^2$, $\mathbf{x}$ and $T_\theta(\mathbf{x})$ have the same length, and if $\mathbf{x} \neq \mathbf{0}$ the angle between these vectors is θ (Figure 2).

FIGURE 2

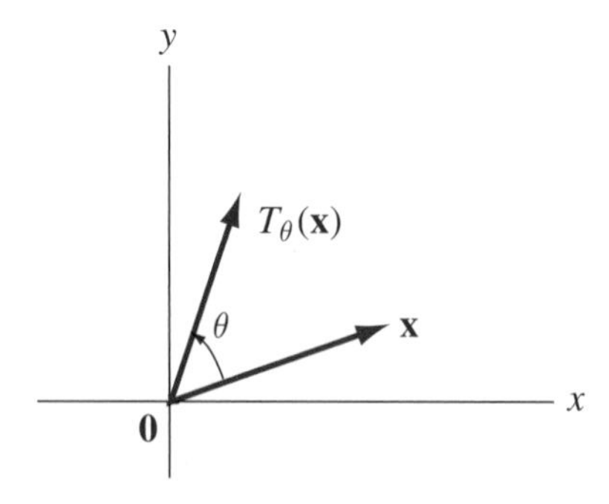

To show that $\mathbf{x}$ and $T_\theta(\mathbf{x})$ have the same length, let $\mathbf{x} = (x, y)$. Then $T_\theta(\mathbf{x}) = (x\cos\theta - y\sin\theta, x\sin\theta + y\cos\theta)$, so that

$$\|T_\theta(\mathbf{x})\| = \|(x\cos\theta - y\sin\theta, x\sin\theta + y\cos\theta)\|$$

$$= \sqrt{(x\cos\theta - y\sin\theta)^2 + (x\sin\theta + y\cos\theta)^2}$$
$$= \sqrt{x^2\cos^2\theta - 2xy\cos\theta\sin\theta + y^2\sin^2\theta + x^2\sin^2\theta + 2xy\sin\theta\cos\theta + y^2\cos^2\theta}$$
$$= \sqrt{x^2(\cos^2\theta + \sin^2\theta) + y^2(\sin^2\theta + \cos^2\theta)}$$
$$= \sqrt{x^2 + y^2} \qquad (\text{Since } \sin^2\theta + \cos^2\theta = 1)$$
$$= \|\mathbf{x}\|.$$

By Theorem 2 of Section 1.2, for $\mathbf{x} \neq \mathbf{0}$ the cosine of the angle ψ between $\mathbf{x}$ and $T_\theta(\mathbf{x})$ is given by

$$\cos\psi = \frac{\mathbf{x} \cdot T_\theta(\mathbf{x})}{\|\mathbf{x}\|\,\|T_\theta(\mathbf{x})\|}$$
$$= \frac{x^2\cos\theta - xy\sin\theta + xy\sin\theta + y^2\cos\theta}{\|\mathbf{x}\|\,\|\mathbf{x}\|}$$
$$= \frac{(x^2 + y^2)}{\|\mathbf{x}\|^2}\cos\theta$$
$$= \frac{\|\mathbf{x}\|^2}{\|\mathbf{x}\|^2}\cos\theta = \cos\theta.$$

Thus, $\cos\psi = \cos\theta$. If $0 \leq \theta \leq \pi$, this implies that $\psi = \theta$. If, on the other hand, $\pi < \theta < 2\pi$, then $\psi = 2\pi - \theta$. In either case, T_θ rotates $\mathbf{x}$ through an angle θ (see Figure 3).

FIGURE 3

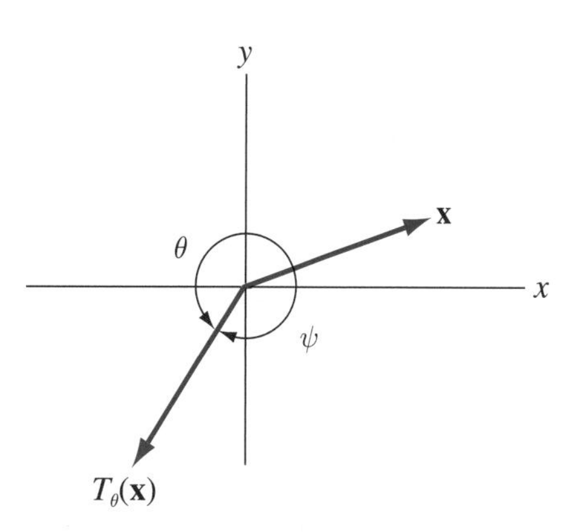

As a specific instance of Example 7, let $\theta = \pi/6$ radians. Then $\cos \pi/6 = \sqrt{3}/2$ and $\sin \pi/6 = 1/2$, so that

$$A = \begin{bmatrix} \sqrt{3}/2 & -1/2 \\ 1/2 & \sqrt{3}/2 \end{bmatrix}.$$

This matrix determines the linear transformation that rotates each vector in $\mathbf{R}^2$ by $\pi/6$ and leaves its length unchanged.

EXAMPLE 8

Let c be a positive scalar and $A = cI$, where I is the $n \times n$ identity matrix. Then T_c given by $T_c(\mathbf{x}) = A\mathbf{x}$ is the **dilation of $\mathbf{R}^n$** by c if $c > 1$ and is the **contraction** by c if $0 < c < 1$. In other words, T_c is the linear transformation that "stretches" or "shrinks" each vector of $\mathbf{R}^n$ by a factor of c but leaves its direction unchanged. Of course, if $c = 1$, T_c is the identity transformation. Figures 4 and 5 indicate the result when $c = 3$ and $c = \frac{1}{2}$, respectively.

Solution To verify the assertions, note that $T_c(\mathbf{x}) = c\mathbf{x}$. Thus, T_c is just multiplication by the scalar c, and, by Section 1.1, both the length and the direction are as stated.

FIGURE 4

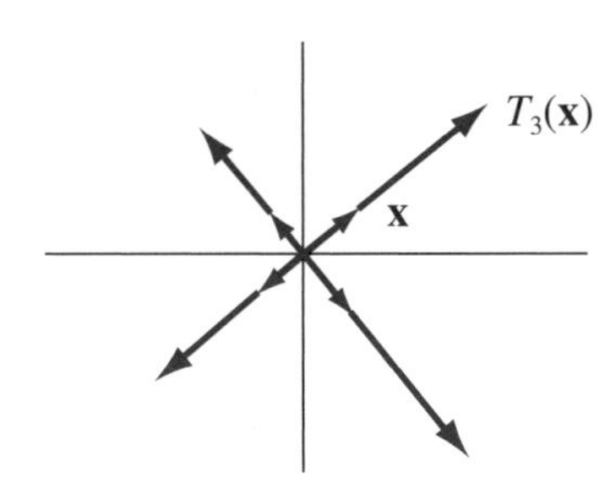

FIGURE 5

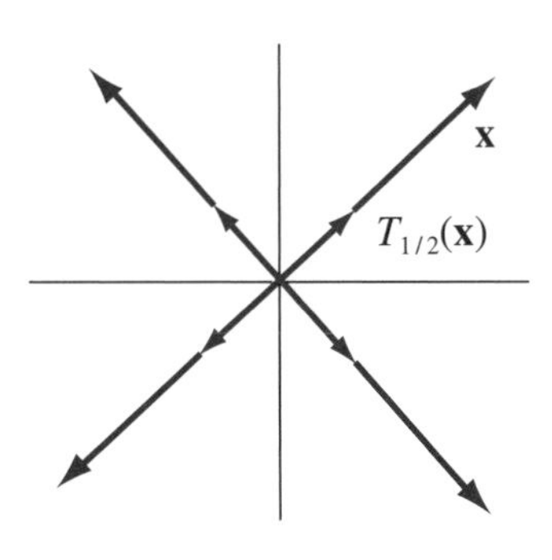

■

EXAMPLE 9

Let θ be any real number, and view θ as the radian measure of an angle. Let T^{θ} be given by $T^{\theta}(\mathbf{x}) = A\mathbf{x}$, where

$$A = \begin{bmatrix} \cos 2\theta & \sin 2\theta \\ \sin 2\theta & -\cos 2\theta \end{bmatrix}.$$

Then T^{θ} is the **reflection** in the line l through the origin that forms angle θ with the positive x-axis. For each $\mathbf{x}$ in $\mathbf{R}^2$, $T^{\theta}(\mathbf{x})$ is situated so that the perpendicular bisector of the segment from the endpoint of $\mathbf{x}$ to that of $T^{\theta}(\mathbf{x})$ is the line l (Figure 6). We leave the verification of this statement for Exercise 27 at the end of the section.

FIGURE 6

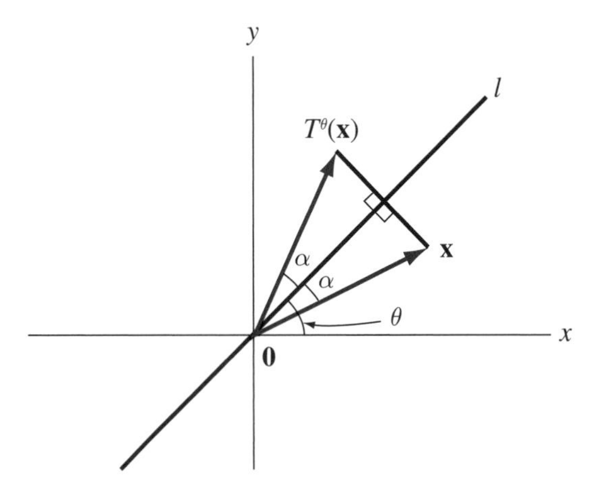

■

NOTE The linear transformations of Examples 7, 8, and 9 are all **linear operators**—they are all linear transformations that map the space back into itself. They are, in fact, more special than that. They are **similarities**; they preserve the shape of geometric figures. Examples 7 and 9 are also **isometries**, which preserve not only shape but also size (distance between any two points).

In the preceding examples, we checked the effect or "action" that a linear transformation T: $\mathbf{V} \to \mathbf{W}$ has on an arbitrary vector $\mathbf{x}$ in $\mathbf{V}$. Determining the effect of a linear transformation can also be done by checking its action on a specific, finite set of vectors. The next definition and theorem make this precise.

DEFINITION

Let S: $\mathbf{V} \rightarrow \mathbf{W}$ and T: $\mathbf{V} \rightarrow \mathbf{W}$ be linear transformations. Then S **equals** T if $S = T$ as functions; that is, if $S(\mathbf{v}) = T(\mathbf{v})$ for all $\mathbf{v}$ in $\mathbf{V}$.

THEOREM 3

If S: $\mathbf{V} \rightarrow \mathbf{W}$ and T: $\mathbf{V} \rightarrow \mathbf{W}$ are linear transformations, and $\mathscr{B} = \{\mathbf{b}_1, \mathbf{b}_2, \ldots, \mathbf{b}_n\}$ is a basis for $\mathbf{V}$ and $S(\mathbf{b}_i) = T(\mathbf{b}_i)$ for each i, then $S = T$.

Proof Since $\mathscr{B}$ is a basis for $\mathbf{V}$, for each $\mathbf{v}$ in $\mathbf{V}$ there exist scalars $c_1, \ldots, c_n$ with $\mathbf{v} = c_1\mathbf{b}_1 + c_2\mathbf{b}_2 + \cdots + c_n\mathbf{b}_n$. Then

$$\begin{aligned} S(\mathbf{v}) &= S(c_1\mathbf{b}_1 + c_2\mathbf{b}_2 + \cdots + c_n\mathbf{b}_n) \\ &= c_1S(\mathbf{b}_1) + c_2S(\mathbf{b}_2) + \cdots + c_nS(\mathbf{b}_n) && \text{(Exercise 29)} \\ &= c_1T(\mathbf{b}_1) + c_2T(\mathbf{b}_2) + \cdots + c_nT(\mathbf{b}_n) && \text{(Hypothesis)} \\ &= T(c_1\mathbf{b}_1 + c_2\mathbf{b}_2 + \cdots + c_n\mathbf{b}_n) && \text{(Exercise 29)} \\ &= T(\mathbf{v}). \end{aligned}$$

Thus, $S = T$. ▣

NOTE In words, Theorem 3 is stated as follows. A linear transformation from one vector space to another is determined by its *action* on a basis.

EXAMPLE 10

Describe the action of the linear transformation T: $\mathbf{R}^2 \rightarrow \mathbf{R}^2$ given by $T(x, y) = (2x - y, x + y)$.

Solution Theorem 3 applies only to *linear transformations*, so we must verify that T is a linear transformation. We leave this for Exercise 1 at the end of Section 7.1. We proceed, instead, to check the action of T on a basis for $\mathbf{R}^2$.

For the basis, we choose the most convenient one—the standard basis $\{\mathbf{e}_1, \mathbf{e}_2\}$. We have

$$T(\mathbf{e}_1) = T(1, 0) = (2, 1) \quad \text{and} \quad T(\mathbf{e}_2) = T(0, 1) = (-1, 1).$$

Next we sketch $\{\mathbf{e}_1, \mathbf{e}_2\}$ and $[T(\mathbf{e}_1), T(\mathbf{e}_2)]$ on separate copies of the space $\mathbf{R}^2$ (Figures 7 and 8).

FIGURE 7

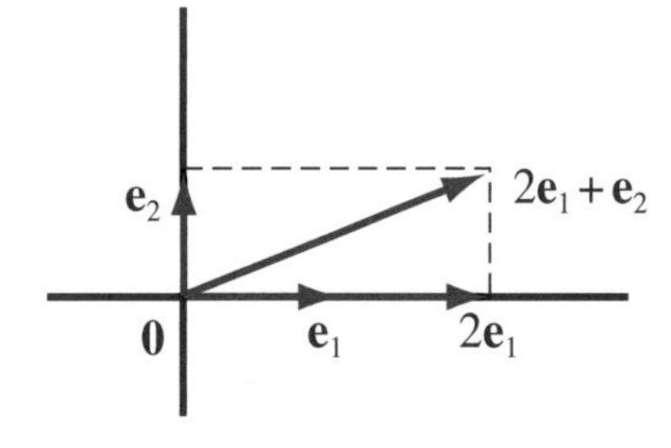

Here T is seen to rotate and stretch the vectors, but unevenly; $\mathbf{e}_1$ is not rotated as much as $\mathbf{e}_2$, and $\mathbf{e}_2$ is not stretched as much as $\mathbf{e}_1$. The importance of Theorem 3 is that it shows us that the effect of T on all other vectors is determined by its effect on these two. To be precise, for any $\mathbf{x} = (x, y)$ in $\mathbf{R}^2$, since $\mathbf{x} = x\mathbf{e}_1 + y\mathbf{e}_2$, we have $T(\mathbf{x}) = xT(\mathbf{e}_1) + yT(\mathbf{e}_2)$. For example, consider the point (2, 1). Since $(2, 1) = 2\mathbf{e}_1 + \mathbf{e}_2$, the point (2, 1) completes the parallelogram determined by $2\mathbf{e}_1$ and $\mathbf{e}_2$. Then $T(2, 1)$ must be the point that completes the parallelogram determined by $2T(\mathbf{e}_1)$ and $T(\mathbf{e}_2)$ (Figure 8). ▣

FIGURE 8

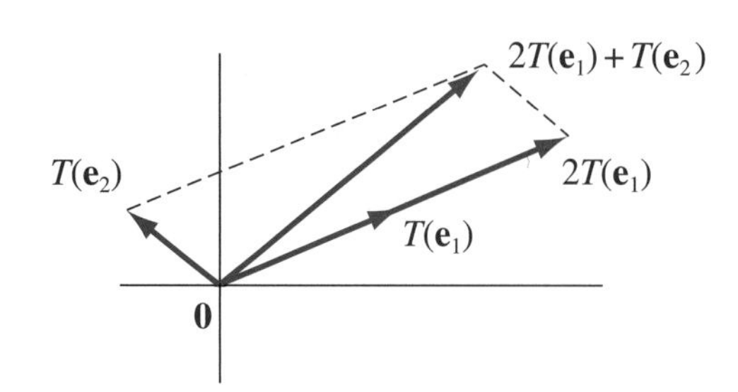

7.1 Exercises

In Exercises 1–8 show that each function is a linear transformation.

1. $S(x, y) = (2x - y, x + y)$ $(S\colon \mathbf{R}^2 \rightarrow \mathbf{R}^2)$
2. $T(x, y) = (2x, x + y, x - 2y)$ $(T\colon \mathbf{R}^2 \rightarrow \mathbf{R}^3)$
3. $U(x, y, z) = x + y + z$ $(U\colon \mathbf{R}^3 \rightarrow \mathbf{R})$
4. $I(x_1, \ldots, x_n) = (x_1, \ldots, x_n)$ $(I\colon \mathbf{R}^n \rightarrow \mathbf{R}^n)$
5. $T\colon \mathbf{R}^3 \rightarrow \mathbf{M}^{2,2}$ given by
$$T(x, y, z) = \begin{bmatrix} y & z \\ -x & 0 \end{bmatrix}$$
6. $T\colon \mathbf{P}_3 \rightarrow \mathbf{P}_3$ given by
$$T(a_0 + a_1x + a_2x^2 + a_3x^3) = (a_0 + a_2) - (a_1 + 2a_3)x^2$$
7. $T\colon \mathbf{P}_2 \rightarrow \mathbf{M}^{2,2}$ given by
$$T(a_0 + a_1x + a_2x^2) = \begin{bmatrix} a_0 & -a_2 \\ -a_2 & a_0 - a_1 \end{bmatrix}$$
8. $T\colon \mathbf{P}_2 \rightarrow \mathbf{P}_2$ given by
$$T(a_0 + a_1x + a_2x^2) = a_1 - (a_0 + a_2)x + (a_1 + a_2)x^2$$

In Exercises 9–12 determine whether or not the given function is a linear transformation, and justify your answer.

9. $F(x, y, z) = (0, 2x + y)$ $(F\colon \mathbf{R}^3 \rightarrow \mathbf{R}^2)$
10. $G(x, y, z) = (xy, y, x - z)$ $(G\colon \mathbf{R}^3 \rightarrow \mathbf{R}^3)$
11. $H(x, y) = \sqrt{x^2 + y^2}$ $(H\colon \mathbf{R}^2 \rightarrow \mathbf{R})$
12. $K(x, y) = (x, \sin y, 2x + y)$ $(K\colon \mathbf{R}^2 \rightarrow \mathbf{R}^3)$

In Exercises 13–16 let $T\colon \mathbf{R}^n \rightarrow \mathbf{R}^m$ be the linear transformation given by $T(\mathbf{x}) = A\mathbf{x}$, multiplication by the matrix A. Find n and m and $T(\mathbf{x}_0)$ for each given vector $\mathbf{x}_0$.

13. $A = \begin{bmatrix} 2 & 3 & 1 \\ 1 & -1 & 0 \end{bmatrix}$, $\mathbf{x}_0 = (1, 4, 2)$

14. $A = \begin{bmatrix} -1 & 1 & 2 \\ 2 & 4 & -1 \\ 1 & 0 & 1 \end{bmatrix}$, $\mathbf{x}_0 = (1, 4, 2)$

15. $A = \begin{bmatrix} 1 & 2 \\ 0 & 1 \\ 1 & 0 \end{bmatrix}$, $\mathbf{x}_0 = (2, -3)$

16. $A = \begin{bmatrix} 0 & 1 & 0 \\ 1 & 0 & 0 \\ 0 & 0 & 1 \end{bmatrix}$, $\mathbf{x}_0 = (x, y, z)$

17. The function Π_i: $\mathbf{R}^n \rightarrow \mathbf{R}$ defined by $\Pi_i(x_1, \ldots, x_n) = x_i$ is called the ith *projection map*. Prove that each projection map is a linear transformation.

18. If A is an $m \times n$ matrix, prove that the function T: $\mathbf{R}^m \rightarrow \mathbf{R}^n$ defined by $T(\mathbf{x}) = \mathbf{x}A$ is a linear transformation, where $\mathbf{x}$ is viewed as a matrix with one row. (*Hint*: Follow Theorem 2.)

19. For a linear transformation T: $\mathbf{R}^n \rightarrow \mathbf{R}^m$, the **negative** of T, $-T$: $\mathbf{R}^n \rightarrow \mathbf{R}^m$, is defined by $(-T)(\mathbf{x}) = -(T(\mathbf{x}))$. Prove that $-T$ is a linear transformation.

20. Prove part (c) of Theorem 1.

21. Find the matrix for the linear transformation $T_{3\pi/4}$, the rotation by $3\pi/4$.

22. Find the matrix for the linear transformation $T_{-\pi/2}$, the rotation by $-\pi/2$.

23. Find the matrix for the linear transformation $T^{\pi/4}$, the reflection in the line $x = y$.

24. Find the matrix for the linear transformation T^0, the reflection in the x-axis.

25. Let D: $\mathbf{P}_3 \rightarrow \mathbf{P}_2$ be defined as follows. For each

$$\mathbf{p}(x) = a_3x^3 + a_2x^2 + a_1x + a_0,$$

define

$$D(\mathbf{p}) = 3a_3x^2 + 2a_2x + a_1.$$

Prove that D is a linear transformation. (*Note from calculus*: This linear transformation gives the *derivative* of any polynomial in $\mathbf{P}_3$.)

26. Prove that the identity function on a vector space $\mathbf{V}$ is a linear transformation.

27. Complete the verification of Example 9.

28. Let F: $\mathbf{V} \rightarrow \mathbf{W}$ be a function. Prove that F is a linear transformation if and only if $F(a\mathbf{u} + \mathbf{v}) = aF(\mathbf{u}) + F(\mathbf{v})$ for each $\mathbf{u}$ and $\mathbf{v}$ in $\mathbf{V}$ and a in $\mathbf{R}$.

29. Let T: $\mathbf{V} \rightarrow \mathbf{W}$ be a linear transformation. Prove that for any linear combination $c_1\mathbf{v}_1 + \cdots + c_k\mathbf{v}_k$ in $\mathbf{V}$, $T(c_1\mathbf{v}_1 + \cdots + c_k\mathbf{v}_k) = c_1T(\mathbf{v}_1) + \cdots + c_kT(\mathbf{v}_k)$.

30. Let T: $\mathbf{R}^n \rightarrow \mathbf{R}^m$ be a linear transformation. If $\mathcal{S}$ is a subset of $\mathbf{R}^n$, then $T[\mathcal{S}]$ is the set of all images $T(\mathbf{s})$ for $\mathbf{s}$ in $\mathcal{S}$. If l is a line in $\mathbf{R}^n$, show that $T[l]$ is either a line or a point in $\mathbf{R}^m$. (*Hint*: Use Exercise 28 and the point-parallel or two-point form for a line.)

31. Extend Exercise 30 to include the case where the given set l is a plane.

Exercises 32–33 deal with the ideas of rotating and "resizing" (contracting or dilating) screen images as described in the Spotlight at the beginning of the chapter. The screen image is a flat, or two-dimensional entity, so the matrices that transform

a fixed image that spirals into the screen, say a football player making a spectacular catch, are substantially simpler than those of rotating a three-dimensional object.

SPOTLIGHT **32.** Give matrices for a sequence of rotation–contraction matrices of the type in Examples 7 and 8 that eventually rotate the plane two full revolutions while contracting a 10-unit line segment to 1/100th of a unit in steps of 1° with each transformation. *Note*: This can be done with a *single* matrix, with the multiplication repeated 720 times. The rotation part is almost obvious. The correct **contraction factor** is the 720th root of 1000.

In the three-dimensional case, the matrices for such transformations are the easy part. Far more involved is the process of "projecting" the three-dimensional representation onto the two-dimensional screen. Even a simple "wire figure" of the edges of a box turns out to be an interesting problem. Which edges get seen? If they all get seen, how do you decide which edges are from the front and which are from the back? How do you approach the problem for general solid figures? We'll ignore those problems entirely! Example 8 dealt with contraction–dilation of $\mathbf{R}^n$ for any n, so 3 is no special problem. Rotations always have an *axis of rotation*, a line of vectors that are unmoved by the linear transformation, and all others rotate uniformly about this axis. Any rotation of the x,y-plane can be made into a rotation of x,y,z-space by leaving the z-component unchanged. In this case, the axis of rotation is just the z-axis, and the matrix that represents such a rotation by angle θ is given by

$$A_\theta = \begin{bmatrix} \cos\theta & -\sin\theta & 0 \\ \sin\theta & \cos\theta & 0 \\ 0 & 0 & 1 \end{bmatrix}$$

Moreover, any vector in the x,y-plane of x,y,z-space (that is, a vector orthogonal to the z-axis) is rotated by angle θ. The same situation occurs for any rotation, but usually the axis of rotation is not so easily identified. If A is the matrix that represents the rotation, then there must be a nonzero solution to $A\mathbf{x} = \mathbf{x}$. Such a vector $\mathbf{x}$ establishes the axis of rotation, and any vector $\mathbf{v}$ orthogonal to it is rotated by the fixed angle of rotation. By calculating the angle between $\mathbf{v}$ and $A\mathbf{v}$, the angle of rotation can be determined.

SPOTLIGHT **33.** Use a computer software package or graphing calculator to find the axis of rotation and the angle of rotation of the following rotation matrix P.

$$P = \begin{bmatrix} 0.991895 & -0.039117 & -0.120889 \\ 0.024443 & 0.992404 & -0.120570 \\ 0.124687 & 0.116638 & -0.985317 \end{bmatrix}$$

7.2 Algebra of Linear Transformations

In this section, we define operations on linear transformations that have the same properties as the analogous ones for matrices. Then, restricting consideration to Euclidean spaces, we show that every linear transformation can be represented by a matrix, and that there is a very simple correspondence between an operation on the transformation and on its matrix.

Operations on Linear Transformations

Sums, differences, negatives, and scalar multiples of linear transformations between vector spaces $\mathbf{V}$ and $\mathbf{W}$ can be defined in a natural way.

DEFINITION

Let S and T be linear transformations from $\mathbf{V}$ to $\mathbf{W}$ and c be a scalar.

i. The **sum** $S + T$ is the function given by
$$(S + T)(\mathbf{x}) = S(\mathbf{x}) + T(\mathbf{x})$$
for all $\mathbf{x}$ in $\mathbf{V}$.

ii. The **difference** $S - T$ is the function given by
$$(S - T)(\mathbf{x}) = S(\mathbf{x}) - T(\mathbf{x})$$
for all $\mathbf{x}$ in $\mathbf{V}$.

iii. The **negative** $-S$ is the function given by
$$(-S)(\mathbf{x}) = -S(\mathbf{x})$$
for all $\mathbf{x}$ in $\mathbf{V}$.

iv. The **scalar multiple** cS is the function given by
$$(cS)(\mathbf{x}) = (c)(S(\mathbf{x}))$$
for all $\mathbf{x}$ in $\mathbf{V}$.

EXAMPLE 1

Let S and T be linear transformations from $\mathbf{R}^3$ to $\mathbf{R}^2$ defined by $S(x, y, z) = (x + y - z, 3y + z)$ and $T(x, y, z) = (2x + z, y + z)$. Describe the transformation $3S - 2T$.

Solution For any $\mathbf{x} = (x, y, z)$, we follow this definition:

$$\begin{aligned}(3S - 2T)(x, y, z) &= (3S)(x, y, z) - (2T)(x, y, z)\\ &= 3(S(x, y, z)) - 2(T(x, y, z))\\ &= 3(x + y - z, 3y + z) - 2(2x + z, y + z)\\ &= (3x + 3y - 3z, 9y + 3z) - (4x + 2z, 2y + 2z)\\ &= (-x + 3y - 5z, 7y + z)\end{aligned}$$
■

It is easy to verify directly that the resulting function $3S - 2T$ of Example 1 is a linear transformation, but the following theorem makes this unnecessary.

THEOREM 1

Let S and T be linear transformations from $\mathbf{V}$ to $\mathbf{W}$ and c be a scalar. Then $S + T$, $S - T$, $-S$, and cS are all linear transformations from $\mathbf{V}$ to $\mathbf{W}$.

Proof We prove only the case of $S + T$ and leave the remaining proofs for an exercise. We must check the two defining properties of linear transformation. If $\mathbf{x}$ and $\mathbf{y}$ are vectors in $\mathbf{V}$, then

$$\begin{aligned}(S + T)(\mathbf{x} + \mathbf{y}) &= S(\mathbf{x} + \mathbf{y}) + T(\mathbf{x} + \mathbf{y}) && \text{(Definition of } S + T)\\ &= S(\mathbf{x}) + S(\mathbf{y}) + T(\mathbf{x}) + T(\mathbf{y}) && \text{(Linearity of } S \text{ and } T)\\ &= S(\mathbf{x}) + T(\mathbf{x}) + S(\mathbf{y}) + T(\mathbf{y}) && \text{(Vector algebra)}\\ &= (S + T)(\mathbf{x}) + (S + T)(\mathbf{y}) && \text{(Definition of } S + T)\end{aligned}$$

If c is any scalar,

$$\begin{aligned}(S + T)(c\mathbf{x}) &= S(c\mathbf{x}) + T(c\mathbf{x}) && \text{(Definition of } S + T)\\ &= cS(\mathbf{x}) + cT(\mathbf{x}) && \text{(Linearity of } S \text{ and } T)\\ &= c(S(\mathbf{x}) + T(\mathbf{x})) && \text{(Vector algebra)}\\ &= c((S + T)(\mathbf{x})) && \text{(Definition of } S + T)\end{aligned}$$

Thus, the sum of two linear transformations is a linear transformation. ■

In general, if $f: \mathscr{A} \to \mathscr{B}$ and $g: \mathscr{B} \to \mathscr{C}$ are functions, the **composition** of f and g is the function $g \circ f: \mathscr{A} \to \mathscr{C}$ defined by $(g \circ f)(a) = g(f(a))$ for all a in $\mathscr{A}$. That is, first apply f to obtain $f(a)$, and then apply g to the result. As usual, we are concerned only with functions that are linear transformations between vector spaces.

EXAMPLE 2

Let $S: \mathbf{R}^2 \to \mathbf{R}^3$ be defined by $S(x, y) = (x - y, x + y, 2y)$ and $T: \mathbf{R}^3 \to \mathbf{R}^3$ be defined by $T(x, y, z) = (x, 2x - y, x + y + z)$. Determine $T \circ S$.

Solution Before computing the general result, let us find $(T \circ S)(1, 2)$. First, $S(1, 2) = (-1, 3, 4)$. Now apply T to this result:

$$(T \circ S)(1, 2) = T(S(1, 2)) = T(-1, 3, 4) = (-1, -5, 6).$$

The general result is computed in exactly the same way, but, unless we change variable names, it is easy to become confused. Let us rewrite T as $T(s, t, u) = (s, 2s - t, s + t + u)$. Now, to evaluate T at $S(x, y) = (x - y, x + y, 2y)$, we have $s = x - y$, $t = x + y$, and $u = 2y$:

$$\begin{aligned}(T \circ S)(x, y) &= T(S(x, y)) = T(x - y, x + y, 2y) \\ &= (x - y, 2(x - y) - (x + y), (x - y) + (x + y) + 2y) \\ &= (x - y, x - 3y, 2x + 2y).\end{aligned}$$

The resulting description of $T \circ S$ is thus

$$(T \circ S)(x, y) = (x - y, x - 3y, 2x + 2y).$$ ■

In Example 2, you can check to see that S, T, and $T \circ S$ are all linear transformations. No mention was made of the composition in reverse order, $S \circ T$, since the image of any vector $T(\mathbf{x})$ is in $\mathbf{R}^3$, and S is only defined for vectors in $\mathbf{R}^2$. These observations are stated more generally in the following theorem.

THEOREM 2

Let $S: \mathbf{U} \to \mathbf{V}$ and $T: \mathbf{V} \to \mathbf{W}$ be linear transformations. Then the composition $T \circ S: \mathbf{U} \to \mathbf{W}$ is a linear transformation.

The theory of linear transformations was developed by Arthur Cayley (1821–1895). In the course of working on the composition of linear transformations, he introduced the definition of matrix multiplication. Cayley wrote over 900 papers in various areas of mathematics, 250 of them while holding down a job as a lawyer!

Hulton Archive/Getty Images

Proof For any $\mathbf{u}$ in $\mathbf{V}$ and scalar a,

$$\begin{aligned}(T \circ S)(a\mathbf{u}) &= T(S(a\mathbf{u})) && \text{(Definition of composition)} \\ &= T(aS(\mathbf{u})) && \text{(}S\text{ is linear)} \\ &= aT(S(\mathbf{u})) && \text{(}T\text{ is linear)} \\ &= a(T \circ S)(\mathbf{u}) && \text{(Definition of composition)}\end{aligned}$$

$(T \circ S)(\mathbf{u} + \mathbf{v}) = (T \circ S)(\mathbf{u}) + (T \circ S)(\mathbf{v})$ for all $\mathbf{u}$ and $\mathbf{v}$ is done in a similar manner; this is left for Exercise 22. ■

It is common to speak of the *product* of two linear transformations instead of their composition. In this vein, for $S: \mathbf{U} \to \mathbf{V}$ and $T: \mathbf{V} \to \mathbf{W}$, the composition $T \circ S$ is often abbreviated to TS. In general, a linear transformation cannot be composed with itself. In other words, for $T: \mathbf{V} \to \mathbf{W}$, $TT = T \circ T$ is not defined unless $\mathbf{V} = \mathbf{W}$. If $\mathbf{V} = \mathbf{W}$ (that is, if T is a linear operator), the composition is defined and can be repeated as often as we wish. The convenient notation $T^2, T^3, \ldots$ symbolizes $T \circ T$, $T \circ (T \circ T)$, and so on.

DEFINITION

Let $S: \mathbf{U} \to \mathbf{V}$, $T: \mathbf{V} \to \mathbf{W}$ be linear transformations and $L: \mathbf{V} \to \mathbf{V}$ be a linear operator.

i. The **product** TS is the linear transformation $T \circ S$.

ii. The **powers** of L are the linear transformations $L^0 = I$, $L^1 = L$, $L^2 = LL, \ldots, L^{k+1} = L(L^k)$ for all $k = 2, 3, \ldots$

We now present those properties of linear transformations that correspond to the analogous ones for matrices (Theorem 1 of Section 3.1).

THEOREM 3

Let S, T, and U be linear transformations and a and b be scalars. Assume that the linear transformations are such that each of the following operations is defined. Then

a. $S + T = T + S$ (Commutative law for addition)

b. $S + (T + U) = (S + T) + U$ (Associative law for addition)

c. $S + 0 = S$

d. $S + (-S) = 0$

e. $S(TU) = (ST)U$ (Associative law for multiplication)

f. $SI = S$, $IS = S$

g. $\left.\begin{aligned} S(T + U) &= ST + SU \\ (T + U)S &= TS + US \end{aligned}\right\}$ (Distributive laws)

h. $a(T + U) = aT + aU$

i. $(a + b)U = aU + bU$

j. $(ab)U = a(bU)$

k. $1S = S$

l. $S0 = 0$, $0T = 0$

m. $a0 = 0$

n. $a(ST) = (aS)T$

Proof We prove parts (a), (e), (g), and (j) and leave the other proofs for exercises. To show that two linear transformations are equal, we show that they are equal as functions—that they agree at any vector for which they are defined. In these arguments, when we say "for each **x**," we really mean "for each **x** for which the functions are defined."

a. For each **x**,

$$\begin{aligned} (S + T)(\mathbf{x}) &= S(\mathbf{x}) + T(\mathbf{x}) && \text{(Definition of sum)} \\ &= T(\mathbf{x}) + S(\mathbf{x}) && \text{(Definition of vector space)} \\ &= (T + S)(\mathbf{x}) && \text{(Definition of sum)} \end{aligned}$$

Therefore $S + T = T + S$.

e. For each **x**,

$$\begin{aligned} S(TU)(\mathbf{x}) &= S(TU(\mathbf{x})) && \text{(Definition of product)} \\ &= S(T(U(\mathbf{x}))) && \text{(Definition of product)} \\ &= ST(U(\mathbf{x})) = (ST)U(\mathbf{x}) && \text{(Definition of product)} \end{aligned}$$

Therefore $S(TU) = (ST)U$.

g. For each **x**,

$$\begin{aligned} (S(T + U))(\mathbf{x}) &= S((T + U)(\mathbf{x})) && \text{(Definition of product)} \\ &= S(T(\mathbf{x}) + U(\mathbf{x})) && \text{(Definition of sum)} \\ &= S(T(\mathbf{x})) + S(U(\mathbf{x})) && \text{(}S\text{ is a linear transformation)} \end{aligned}$$

$$= ST(\mathbf{x}) + SU(\mathbf{x}) \qquad \text{(Definition of product)}$$
$$= (ST + SU)(\mathbf{x}) \qquad \text{(Definition of sum)}$$

Therefore, $S(T + U) = ST + SU$. (The other case of part (g) is similarly handled.)

j. For each $\mathbf{x}$,

$$(ab)U(\mathbf{x}) = (ab)(U(\mathbf{x})) \qquad \text{(Definition of multiplication by scalar)}$$
$$= a(bU(\mathbf{x})) \qquad \text{(Definition of vector space)}$$
$$= (a(bU))(\mathbf{x}) \qquad \text{(Definition of multiplication by scalar)}$$

Therefore, $(ab)U = a(bU)$. ■

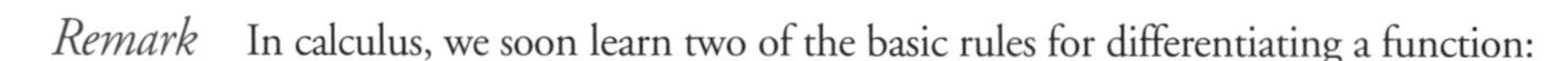

Remark In calculus, we soon learn two of the basic rules for differentiating a function:

i. The derivative of a sum is the sum of the derivatives.

ii. The derivative of a constant times a function is the constant times the derivative of the function.

If we denote the derivative of a function f by $D(f)$, then these rules can be more formally stated:

i. $D(f + g) = D(f) + D(g)$

ii. $D(cf) = cD(f)$, where c is a constant.

But these are just the properties we need to claim that a derivative is a linear operator. (For simplicity's sake, we can take its domain and codomain spaces as $\mathbf{C}^{\infty}(-\infty, \infty)$; see the note following Example 6 of Section 6.1.)

Since the powers of a linear operator are just that transformation applied over and over again, we have

$$D^2(f) = D(D(f)),$$

which is the derivative of the derivative—the *second derivative.* We also have

$$D^3(f) = D(D^2(f)),$$

which is the derivative of the second derivative—the *third derivative.* And, in general, $D^n(f)$ is the nth derivative of the function f.

Matrix of a Linear Transformation from $\mathbf{R}^n$ to $\mathbf{R}^m$

In the first section of this chapter, matrices were used to provide examples of linear transformations, whereas other linear transformations were presented by different means. In this section, we show that *every* linear transformation from $\mathbf{R}^n$ to $\mathbf{R}^m$ may be viewed as multiplication by an associated matrix. (In Section 7.4, we see that matrices can be used to represent linear transformations between general finite dimensional vector spaces.) We begin with a theorem that gives the procedure for finding the matrix just described.

THEOREM 4

Let $T: \mathbf{R}^n \to \mathbf{R}^m$ be a linear transformation and A be an $m \times n$ matrix with the ith column given by $\mathbf{a}^i = T(\mathbf{e}_i)$. Then A is the unique matrix for which $A\mathbf{x} = T(\mathbf{x})$ for all $\mathbf{x}$ in $\mathbf{R}^n$.

Proof Since $\mathbf{a}^i = (a_{1i}, \ldots, a_{mi}) = T(\mathbf{e}_i)$ for each $i = 1, \ldots, n$, we have the following computation:

$$A\mathbf{e}_i = \begin{bmatrix} a_{11} & \cdots & a_{1i} & \cdots & a_{1n} \\ a_{21} & \cdots & a_{2i} & \cdots & a_{2n} \\ \vdots & & \vdots & & \vdots \\ a_{m1} & \cdots & a_{mi} & \cdots & a_{mn} \end{bmatrix} \begin{bmatrix} 0 \\ \vdots \\ 0 \\ 1 \\ 0 \\ \vdots \\ 0 \end{bmatrix}$$

$$= \begin{bmatrix} a_{1i} \\ a_{2i} \\ \vdots \\ a_{mi} \end{bmatrix} = (a_{1i}, \ldots, a_{mi}) = T(\mathbf{e}_i)$$

Thus, multiplication on the left by A agrees with the linear transformation T on a basis for $\mathbf{R}^n$ (the standard one) and is itself a linear transformation (Theorem 2 of Section 7.1). By Theorem 3 of Section 7.1, $A(\mathbf{x}) = T(\mathbf{x})$ for all $\mathbf{x}$ in $\mathbf{R}^n$.

To show uniqueness, suppose that B is a matrix such that $B\mathbf{x} = T(\mathbf{x})$ for all $\mathbf{x}$ in $\mathbf{R}^n$. Then, in particular, $B\mathbf{e}_i = T(\mathbf{e}_i)$ for $i = 1, \ldots, n$. But $B\mathbf{e}_i = \mathbf{b}^i$ and $T(\mathbf{e}_i) = \mathbf{a}^i$, which means the ith column of B is the ith column of A for each i. Thus, $B = A$. ■

The matrix A of Theorem 4 is called the matrix **associated with** T, or the matrix that **represents** T.

EXAMPLE 3

Let T: $\mathbf{R}^2 \to \mathbf{R}^3$ be the linear transformation of Example 1 of Section 7.1. For each $\mathbf{x} = (x, y)$ in $\mathbf{R}^2$, $T(x, y) = (x + y, x, 2x - y)$. Find the matrix that represents T.

Solution We have already shown that T is linear, so we proceed with the computation of the associated matrix A. By Theorem 4, A is the 3×2 matrix with

$$\mathbf{a}^1 = T(\mathbf{e}_1) = T(1, 0) = (1, 1, 2)$$
$$\text{and} \quad \mathbf{a}^2 = T(\mathbf{e}_2) = T(0, 1) = (1, 0, -1).$$

Thus, the matrix A is given by

$$A = \begin{bmatrix} 1 & 1 \\ 1 & 0 \\ 2 & -1 \end{bmatrix}.$$

By Theorem 4, we do not need to verify that this matrix A represents T. To emphasize the point, however, we do it anyway. For any $\mathbf{x} = (x, y)$ in $\mathbf{R}^2$,

$$A\mathbf{x} = \begin{bmatrix} 1 & 1 \\ 1 & 0 \\ 2 & -1 \end{bmatrix} \begin{bmatrix} x \\ y \end{bmatrix} = \begin{bmatrix} x + y \\ x \\ 2x - y \end{bmatrix}.$$

The resulting vector is exactly $T(x, y)$. ■

WARNING Before applying Theorem 4, it is essential to check that the function T is indeed a linear transformation. Following the given procedure always produces a matrix, even for nonlinear functions. Unless the function is linear, however, the resulting matrix does not represent the function (see Exercise 28).

EXAMPLE 4

Find the matrix associated with the identity transformation I: $\mathbf{R}^n \to \mathbf{R}^n$.

Solution By definition, $I(\mathbf{x}) = \mathbf{x}$ for each $\mathbf{x}$ in $\mathbf{R}^n$. Then, in particular, $I(\mathbf{e}_i) = \mathbf{e}_i = \mathbf{a}^i$ for each i, and thus the matrix A associated with I is the matrix with ith column equal to $\mathbf{e}_i$ for each i. Thus, $A = I$, the $n \times n$ identity matrix. ■

NOTE The identity *transformation* I: $\mathbf{R}^n \to \mathbf{R}^n$ is represented by the $n \times n$ identity *matrix* I. Because of their close relationship, using the same symbol for both presents no difficulties. Similarly, the zero transformation 0: $\mathbf{R}^n \to \mathbf{R}^m$ is represented by the $m \times n$ zero matrix, 0. Again, the slight ambiguity is no problem.

Since the sum, difference, negative, and scalar product functions are linear transformations, there is a matrix that represents each of them. These associated matrices are compatible with the corresponding operation in the following sense.

THEOREM 5

Let S and T be linear transformations from $\mathbf{R}^n$ to $\mathbf{R}^m$ and their associated matrices be A and B, respectively. Then the associated matrix of the linear transformation

i. $S + T$ is $A + B$,

ii. $S - T$ is $A - B$,

iii. $-S$ is $-A$,

iv. cS is cA.

Proof Again we prove only the case of $S + T$ and leave the remaining proofs for an exercise. For any $\mathbf{x}$ in $\mathbf{R}^n$,

$$\begin{aligned}(A + B)\mathbf{x} &= A\mathbf{x} + B\mathbf{x} && \text{(part (g) of Theorem 1 of Section 3.1)}\\ &= S(\mathbf{x}) + T(\mathbf{x}) && (A \text{ and } B \text{ represent } S \text{ and } T)\\ &= (S + T)(\mathbf{x}) && (\text{Definition of } S + T)\end{aligned}$$

Therefore, the matrix $A + B$ represents the linear transformation $S + T$. ■

Let S and T be the linear transformations of Example 1. Let A, B, and C be the matrices that represent S, T, and $3S - 2T$, respectively. Verify directly that $C = 3A - 2B$.

Solution Following Theorem 4, we have from Example 1 that

$$A = \begin{bmatrix} 1 & 1 & -1 \\ 0 & 3 & 1 \end{bmatrix}, \quad B = \begin{bmatrix} 2 & 0 & 1 \\ 0 & 1 & 1 \end{bmatrix}, \quad \text{and} \quad C = \begin{bmatrix} -1 & 3 & -5 \\ 0 & 7 & 1 \end{bmatrix}.$$

We now compute $3A - 2B$ directly:

$$3A - 2B = 3\begin{bmatrix} 1 & 1 & -1 \\ 0 & 3 & 1 \end{bmatrix} - 2\begin{bmatrix} 2 & 0 & 1 \\ 0 & 1 & 1 \end{bmatrix} = \begin{bmatrix} -1 & 3 & -5 \\ 0 & 7 & 1 \end{bmatrix}$$

The resulting matrix is C, as required. ■

Since the composition of two linear transformations is a linear transformation, and since linear transformations can be represented by matrices (Theorem 4), it is natural to inquire about the nature of the matrix that represents the composition. The next theorem describes this very important situation.

THEOREM 6

Let $S: \mathbf{R}^n \to \mathbf{R}^m$ and $T: \mathbf{R}^m \to \mathbf{R}^l$ be linear transformations, let A be the matrix that represents S, and B be the matrix that represents T. Then the matrix product BA represents the composition $T \circ S$.

Proof Since B is $l \times m$, and A is $m \times n$, the matrix product BA is defined and is an $l \times n$ matrix as required ($T \circ S$ is a linear transformation from $\mathbf{R}^n$ to $\mathbf{R}^l$). For any $\mathbf{x}$ in $\mathbf{R}^n$,

$$\begin{aligned}(BA)\mathbf{x} &= (B)(A\mathbf{x}) && \text{(Matrix multiplication is associative)}\\ &= (B)(S(\mathbf{x})) && (A \text{ represents } S)\\ &= T(S(\mathbf{x})) && (B \text{ represents } T)\\ &= (T \circ S)(\mathbf{x}) && \text{(Definition of composition)}\end{aligned}$$

Thus, multiplication on the left by BA has exactly the same effect as mapping by the linear transformation $T \circ S$, and BA represents $T \circ S$. ■

EXAMPLE 6

Let S and T be as in Example 2. Find the matrices A, B, and C that represent S, T, and $T \circ S$, respectively, and verify that $C = BA$.

Solution From Example 2,

$$\begin{aligned} S(x, y) &= (x - y, x + y, 2y),\\ T(x, y, z) &= (x, 2x - y, x + y + z),\\ \text{and} \quad (T \circ S)(x, y) &= (x - y, x - 3y, 2x + 2y). \end{aligned}$$

Following the procedure described in Theorem 4 for each of these linear transformations,

$$A = \begin{bmatrix} 1 & -1 \\ 1 & 1 \\ 0 & 2 \end{bmatrix}, \quad B = \begin{bmatrix} 1 & 0 & 0 \\ 2 & -1 & 0 \\ 1 & 1 & 1 \end{bmatrix}, \quad \text{and} \quad C = \begin{bmatrix} 1 & -1 \\ 1 & -3 \\ 2 & 2 \end{bmatrix}.$$

The essence of Theorem 6 is that it is unnecessary to evaluate $T \circ S$ first and then compute its corresponding matrix; exactly the same result may be obtained by computing A and B and then $C = BA$. Performing this multiplication, we do indeed have

$$BA = \begin{bmatrix} 1 & 0 & 0 \\ 2 & -1 & 0 \\ 1 & 1 & 1 \end{bmatrix} \begin{bmatrix} 1 & -1 \\ 1 & 1 \\ 0 & 2 \end{bmatrix} = \begin{bmatrix} 1 & -1 \\ 1 & -3 \\ 2 & 2 \end{bmatrix} = C.$$ ■

EXAMPLE 7

Use Theorem 6 to show that the composition of a rotation of $\mathbf{R}^2$ by an angle θ followed by a rotation by an angle ψ is a rotation by the angle $\theta + \psi$. (Obviously, this is true geometrically.)

Solution From Example 7 of Section 7.1, we know that the rotation by θ is represented by the matrix

$$A_\theta = \begin{bmatrix} \cos\theta & -\sin\theta \\ \sin\theta & \cos\theta \end{bmatrix},$$

and the rotation by ψ is represented by the matrix

$$A_\psi = \begin{bmatrix} \cos\psi & -\sin\psi \\ \sin\psi & \cos\psi \end{bmatrix}.$$

By Theorem 6, the matrix that represents the composition is the product $A_\psi A_\theta$. Computing the product, we have

$$A_\psi A_\theta = \begin{bmatrix} \cos\psi\cos\theta - \sin\psi\sin\theta & -\cos\psi\sin\theta - \sin\psi\cos\theta \\ \sin\psi\cos\theta + \cos\psi\sin\theta & -\sin\psi\sin\theta + \cos\psi\cos\theta \end{bmatrix}$$
$$= \begin{bmatrix} \cos(\theta+\psi) & -\sin(\theta+\psi) \\ \sin(\theta+\psi) & \cos(\theta+\psi) \end{bmatrix} = A_{\theta+\psi}.$$

It is left for an exercise to show that the composition of two *reflections* is also a *rotation.* ■

EXAMPLE 8

Let $T: \mathbf{R}^3 \to \mathbf{R}^3$ be defined by $T(x, y, z) = (x + 2y - z, 3x - z, y + z)$. Describe the transformation $2T^2 - T + 3I$.

Solution We use the algebra of the corresponding matrices to simplify the computation. Let A represent T, and note that the identity matrix I of order 3 represents I. Then $2A^2 - A + 3I$ represents $2T^2 - T + 3I$. By Theorem 4, A is given by

$$A = \begin{bmatrix} 1 & 2 & -1 \\ 3 & 0 & -1 \\ 0 & 1 & 1 \end{bmatrix},$$

and hence

$$2A^2 - A + 3I = 2\begin{bmatrix} 1 & 2 & -1 \\ 3 & 0 & -1 \\ 0 & 1 & 1 \end{bmatrix}^2 - \begin{bmatrix} 1 & 2 & -1 \\ 3 & 0 & -1 \\ 0 & 1 & 1 \end{bmatrix} + 3\begin{bmatrix} 1 & 0 & 0 \\ 0 & 1 & 0 \\ 0 & 0 & 1 \end{bmatrix}$$
$$= \begin{bmatrix} 14 & 2 & -8 \\ 6 & 10 & -8 \\ 6 & 2 & 0 \end{bmatrix} - \begin{bmatrix} 1 & 2 & -1 \\ 3 & 0 & -1 \\ 0 & 1 & 1 \end{bmatrix} + \begin{bmatrix} 3 & 0 & 0 \\ 0 & 3 & 0 \\ 0 & 0 & 3 \end{bmatrix}$$
$$= \begin{bmatrix} 16 & 0 & -7 \\ 3 & 13 & -7 \\ 6 & 1 & 2 \end{bmatrix}.$$

Finally,

$$(2T^2 - T + 3I)(x, y, z) = \begin{bmatrix} 16 & 0 & -7 \\ 3 & 13 & -7 \\ 6 & 1 & 2 \end{bmatrix}\begin{bmatrix} x \\ y \\ z \end{bmatrix} = \begin{bmatrix} 16x \qquad - 7z \\ 3x + 13y - 7z \\ 6x + \quad y + 2z \end{bmatrix},$$

so that $(2T^2 - T + 3I)(x, y, z) = (16x - 7z, 3x + 13y - 7z, 6x + y + 2z)$. ■

7.2 Exercises

The following functions are all linear transformations. Use them in Exercises 1–20.

$$K(x, y, z) = (x, x + y, x + y + z)$$
$$L(x, y, z) = (2x - y, x + 2y)$$
$$S(x, y, z) = (z, y, x)$$
$$T(x, y) = (2x + y, x + y, x - y, x - 2y)$$

$$G: \mathbf{R}^3 \longrightarrow \mathbf{M}^{2,2} \text{ defined by } G(x, y, z) = \begin{bmatrix} y & z \\ -x + y & -x \end{bmatrix}$$

$$H: \mathbf{M}^{2,2} \longrightarrow \mathbf{P}_2 \text{ defined by } H\left(\begin{bmatrix} a & b \\ c & d \end{bmatrix}\right) = (a + b) + (a - 2d)x + dx^2$$

In Exercises 1–12 find the indicated linear transformation if it is defined. If it is not defined, explain why it is not.

1. $LK(=L \circ K)$

2. TL

3. S^2

4. $K + S$

5. T^2

6. GS

7. HG

8. $L + S$

9. $3K - 2S$

10. $2S^3 - S^2 + 3S - 4I$

11. $3H$

12. HGK

13–16. Find the matrix that represents each of the linear transformations K, L, S, and T.

17–20. Find the matrix that represents each of the linear transformations in Exercises 1–4 in two ways, first using the result of the exercise and then using the result of Exercises 13–16.

21. Show that the composition of one reflection T^θ followed by another T^ψ is the rotation by the angle $2(\psi - \theta)$.

22. Finish the proof of Theorem 2 by showing that if S and T are linear transformations such that $T \circ S$ is defined, then $(T \circ S)(\mathbf{u} + \mathbf{v}) = (T \circ S)\mathbf{u} + (T \circ S)\mathbf{v}$.

23. Complete the proof of Theorem 1.

24. Complete the proof of Theorem 3.

25. Complete the proof of Theorem 5.

26. Show that the composition of a contraction/dilation by a contraction/dilation is a contraction/dilation.

27. If $S: \mathbf{R}^n \rightarrow \mathbf{R}^n$ is a linear transformation, and $T_c: \mathbf{R}^n \rightarrow \mathbf{R}^n$ is a contraction/dilation, prove that $ST_c = T_cS$. In other words, show that a contraction/dilation *commutes* with any linear operator.

28. Let $f: \mathbf{R}^2 \rightarrow \mathbf{R}^2$ be the *nonlinear* function defined by $f(x, y) = (xy, x^2)$. Let A be the 2×2 matrix constructed as in Theorem 4. Show that A does *not* represent f. In other words, show that for some $\mathbf{x}$ in $\mathbf{R}^2$, $f(\mathbf{x}) \neq A\mathbf{x}$.

7.3 Kernel and Image

In this section, we investigate further properties of linear transformations, as well as two very important subspaces associated with them.

Kernel of a Linear Transformation

If T is a linear transformation, Theorem 1 of Section 7.1 showed us that $T(\mathbf{0}) = \mathbf{0}$. There may or may not be other vectors $\mathbf{x}$ such that $T(\mathbf{x}) = \mathbf{0}$.

DEFINITION

Let $T\colon \mathbf{V} \to \mathbf{W}$ be a linear transformation. The set of all vectors $\mathbf{x}$ in $\mathbf{V}$ that satisfy $T(\mathbf{x}) = \mathbf{0}$ is called the **kernel** (or **null space**) of T and is denoted by $\operatorname{Ker}(T)$.

EXAMPLE 1

Compute $\operatorname{Ker}(T)$ for $T\colon \mathbf{R}^2 \to \mathbf{R}^3$ given by $T(x, y) = (x + y, x, 2x - y)$.

Solution That the transformation T is linear has been shown in Example 1 of Section 7.1. To find $\operatorname{Ker}(T)$, we must solve the equation $T(x, y) = (0, 0, 0)$ for all possible ordered pairs (x, y). That is, we solve

$$(x + y, x, 2x - y) = (0, 0, 0).$$

Equating components, we have

$$\begin{aligned} x + y &= 0 \\ x \quad\; &= 0 \\ 2x - y &= 0. \end{aligned}$$

This **homogeneous** system of linear equations can easily be solved by inspection. The second equation implies that $x = 0$, which, in turn, implies that $y = 0$, from the first equation. Thus, $\operatorname{Ker}(T) = \{\mathbf{0}\}$, the set that consists solely of the zero vector. ■

EXAMPLE 2

Let O be the zero transformation from $\mathbf{V}$ to $\mathbf{W}$; that is, $O(\mathbf{x}) = \mathbf{0}$ for each $\mathbf{x}$ in $\mathbf{V}$. Compute $\operatorname{Ker}(O)$.

Solution That O is a linear transformation was verified in Example 2 of Section 7.1. Since every $\mathbf{x}$ in $\mathbf{V}$ gets mapped into $\mathbf{0}$, $\operatorname{Ker}(O) = \mathbf{V}$. ■

EXAMPLE 3

Describe $\operatorname{Ker}(T)$, where $T\colon \mathbf{R}^3 \to \mathbf{R}^2$ is the linear transformation given by $T(\mathbf{x}) = A\mathbf{x}$ and

$$A = \begin{bmatrix} 2 & 0 & 1 \\ 1 & 1 & -1 \end{bmatrix}.$$

Solution Since T is given by matrix multiplication, it is linear (by Theorem 2 of Section 7.1). To compute $\operatorname{Ker}(T)$, we let $\mathbf{x} = (x, y, z)$, determine $T(\mathbf{x})$, and set it equal to zero:

$$T(\mathbf{x}) = \begin{bmatrix} 2 & 0 & 1 \\ 1 & 1 & -1 \end{bmatrix} \begin{bmatrix} x \\ y \\ z \end{bmatrix} = \begin{bmatrix} 2x \quad + z \\ x + y - z \end{bmatrix} = \begin{bmatrix} 0 \\ 0 \end{bmatrix}$$

Equating components, we have

$$\begin{aligned} 2x \quad\;\; + z &= 0 \\ x + y - z &= 0. \end{aligned}$$

We solve the resulting homogeneous system by the methods of Section 2.3, forming the augmented matrix and transforming it to row-reduced echelon form:

$$\left[\begin{array}{ccc|c} 2 & 0 & 1 & 0 \\ 1 & 1 & -1 & 0 \end{array}\right] \to \left[\begin{array}{ccc|c} 1 & 0 & \frac{1}{2} & 0 \\ 0 & 1 & -\frac{3}{2} & 0 \end{array}\right]$$

Thus,

$$\begin{aligned} x &= -\tfrac{1}{2}t, \\ y &= \tfrac{3}{2}t, \\ \text{and} \quad z &= t, \end{aligned}$$

and we have a one-parameter family of solutions. Expressing this result in vector form,

$$\mathbf{x} = t\left(-\tfrac{1}{2}, \tfrac{3}{2}, 1\right),$$

we see that Ker (T) is the line through the origin determined by the vector $\left(-\tfrac{1}{2}, \tfrac{3}{2}, 1\right)$ (or more conveniently, by $(-1, 3, 2)$). ■

EXAMPLE 4

Compute Ker (T) for T: $\mathbf{P}_2 \rightarrow \mathbf{M}^{2,2}$ given by

$$T(a_0 + a_1 x + a_2 x^2) = \begin{bmatrix} a_1 & a_1 - a_0 \\ a_1 + a_0 & 2a_1 - a_2 \end{bmatrix}.$$

(That T is a linear transformation is left to the reader.)

Solution By definition of Ker (T), we must solve the equation

$$T(a_0 + a_1 x + a_2 x^2) = \begin{bmatrix} 0 & 0 \\ 0 & 0 \end{bmatrix}.$$

From the rule for T, this is equivalent to

$$\begin{bmatrix} a_1 & a_1 - a_0 \\ a_1 + a_0 & 2a_1 - a_2 \end{bmatrix} = \begin{bmatrix} 0 & 0 \\ 0 & 0 \end{bmatrix}$$

or, equating coefficients,

$$\begin{aligned} a_1 \qquad &= 0 \\ -a_0 + a_1 \qquad &= 0 \\ a_0 + a_1 \qquad &= 0 \\ 2a_1 - a_2 &= 0. \end{aligned}$$

Solving this system, we see that the only polynomials $\mathbf{p}(x) = a_0 + a_1 x + a_2 x^2$ in $\mathbf{P}_2$ for which $T(\mathbf{p}) = \mathbf{0}$ are those for which $a_0 = a_1 = a_2 = 0$. That is, $\mathbf{p} = \mathbf{0}$, so Ker $(T) = \{\mathbf{0}\}$. ■

In each of these four examples, the kernel has been a subspace of **V**. The following theorem shows that the kernel of a linear transformation is *always* a subspace. The dimension of the kernel is called the **nullity** of the linear transformation.

THEOREM 1

Let T: $\mathbf{V} \rightarrow \mathbf{W}$ be a linear transformation. Then the kernel of T is a subspace of **V**.

Proof That Ker (T) is a nonempty *subset* of **V** is clear from the definition and from the fact that $T(\mathbf{0}) = \mathbf{0}$.

For **u** and **v** in Ker (T), we compute $T(\mathbf{u} + \mathbf{v})$:

$$T(\mathbf{u} + \mathbf{v}) = T(\mathbf{u}) + T(\mathbf{v}) = \mathbf{0} + \mathbf{0} = \mathbf{0}$$

Thus, $\mathbf{u} + \mathbf{v}$ is in Ker (T) as well.

For **u** in Ker (T) and c any scalar,

$$T(c\mathbf{u}) = cT(\mathbf{u}) = c\mathbf{0} = \mathbf{0}.$$

Thus, $c\mathbf{u}$ is also in Ker (T).

These two computations confirm that the kernel of a linear transformation is a subspace of **V**. ■

If the linear transformation is given by a matrix as in Example 3, computation of the kernel is particularly easy. Note that, in this example, we obtained the kernel by augmenting the matrix A by a column of zeros and row-reducing. That this is always the case is simply a statement of the fact that $T(\mathbf{x}) = \mathbf{0}$ if and only if $A\mathbf{x} = \mathbf{0}$. The latter equation may be solved by row-reducing the augmented matrix $[A \mid \mathbf{0}]$, and a basis for the kernel may be produced by following the procedure of Theorem 4 of Section 5.3. This observation is stated as the next theorem. Recall that the *rank* of a matrix is the number of nonzero rows in its row-reduced echelon form.

THEOREM 2

Let $T: \mathbf{R}^n \rightarrow \mathbf{R}^m$ be a linear transformation and A be the matrix that represents T. Then the kernel of T is the solution space of the vector equation $A\mathbf{x} = \mathbf{0}$. Thus, dim (Ker (T)), the nullity of T, is $n - r$, where r is the rank of the matrix A. ■

Procedure for Computing a Basis for the Kernel of a Transformation Given by a Matrix A

Solve $A\mathbf{x} = \mathbf{0}$ by transforming the matrix A to row-reduced echelon form, and then follow the procedure of Theorem 4 of Section 5.3. The result is a basis for Ker (T).

EXAMPLE 5

Let $T: \mathbf{R}^5 \rightarrow \mathbf{R}^4$ be given by $T(\mathbf{x}) = A\mathbf{x}$, where A is the matrix

$$A = \begin{bmatrix} 1 & 2 & 0 & 1 & 0 \\ 2 & 4 & 1 & 0 & 0 \\ 0 & 0 & 1 & -2 & 1 \\ 1 & 2 & 1 & -1 & 1 \end{bmatrix}.$$

Find the nullity of T and a basis for Ker (T).

Solution The row-reduced echelon form of A is the matrix

$$B = \begin{bmatrix} 1 & 2 & 0 & 1 & 0 \\ 0 & 0 & 1 & -2 & 0 \\ 0 & 0 & 0 & 0 & 1 \\ 0 & 0 & 0 & 0 & 0 \end{bmatrix}.$$

Since B has three nonzero rows, the rank of A is $r = 3$. Since $n = 5$ in this example, the nullity is $n - r = 2$. Following the procedure of Theorem 4 of Section 5.3, we determine that a basis for Ker (T) is $\{(-2, 1, 0, 0, 0), (-1, 0, 2, 1, 0)\}$. ■

A function $f: \mathcal{A} \rightarrow \mathcal{B}$ is **one-to-one** if, whenever a_1 and a_2 satisfy $f(a_1) = f(a_2)$, then $a_1 = a_2$. In other words, no two elements of $\mathcal{A}$ are mapped to the same element of $\mathcal{B}$. It is particularly easy to check to see if a *linear transformation* is one-to-one.

THEOREM 3

Let $T: \mathbf{V} \rightarrow \mathbf{W}$ be a linear transformation. Then T is one-to-one if and only if Ker (T) = $\{\mathbf{0}\}$.

Proof If T is one-to-one, then $T(\mathbf{x}) = \mathbf{0}$ can have only one solution, namely $\mathbf{x} = \mathbf{0}$. That is, Ker (T) = $\{\mathbf{0}\}$. Conversely, suppose that Ker (T) = $\{\mathbf{0}\}$ and $T(\mathbf{u}) = T(\mathbf{v})$. We need to show that $\mathbf{u} = \mathbf{v}$. Since $T(\mathbf{u}) = T(\mathbf{v})$,

$$T(\mathbf{u}) - T(\mathbf{v}) = \mathbf{0},$$

and therefore $T(\mathbf{u} - \mathbf{v}) = \mathbf{0}$ (part (c) of Theorem 1 of Section 7.1). Thus, $\mathbf{u} - \mathbf{v}$ is in Ker (T). But then $\mathbf{u} - \mathbf{v} = \mathbf{0}$ since $\mathbf{0}$ is the *only* element of Ker (T). Hence $\mathbf{u} = \mathbf{v}$, and T is one-to-one. ■

From the computations given in Example 1, we see that the linear transformation $T(x, y) = (x + y, x, 2x - y)$ is one-to-one. On the other hand, the linear transformation $T(\mathbf{x}) = A\mathbf{x}$ given in Example 5 is not.

Image of a Linear Transformation

Let $T\colon \mathbf{V} \to \mathbf{W}$ be a linear transformation. The set of all vectors $T(\mathbf{x})$ is called the **image** (or **range**) of T and is denoted by $\text{Im}(T)$.

EXAMPLE 6

Describe $\text{Im}(0)$, where 0 is the zero transformation from $\mathbf{V}$ to $\mathbf{W}$.

Solution Since $0(\mathbf{x}) = 0$ for all $\mathbf{x}$ in $\mathbf{V}$, $\text{Im}(0) = \{\mathbf{0}\}$. ■

EXAMPLE 7

Describe $\text{Im}(T)$, where $T\colon \mathbf{R} \to \mathbf{R}^2$ is given by $T(x) = (x, 2x)$ for each x in $\mathbf{R}$.

Solution Letting $T(x) = (x, y)$, we have $y = 2x$ for each x in the real numbers $\mathbf{R}$. Thus, $\text{Im}(T)$ is the set of points on the line in $\mathbf{R}^2$ that passes through the origin and has slope 2. ■

EXAMPLE 8

Describe $\text{Im}(T)$ where $T\colon \mathbf{P}_2 \to \mathbf{M}^{2,2}$ is given by

$$T(a_0 + a_1x + a_2x^2) = \begin{bmatrix} a_1 & a_1 - a_0 \\ a_1 + a_0 & 2a_1 - a_2 \end{bmatrix},$$

which is the transformation of Example 4.

Solution We express $T(a_0 + a_1x + a_2x^2)$ in the form

$$\begin{bmatrix} a_1 & a_1 - a_0 \\ a_1 + a_0 & 2a_1 - a_2 \end{bmatrix} = a_0\begin{bmatrix} 0 & -1 \\ 1 & 0 \end{bmatrix} + a_1\begin{bmatrix} 1 & 1 \\ 1 & 2 \end{bmatrix} + a_2\begin{bmatrix} 0 & 0 \\ 0 & -1 \end{bmatrix}.$$

Since a_0, a_1, and a_2 are arbitrary, $\text{Im}(T)$ is the set spanned by $\mathcal{S}$, where

$$\mathcal{S} = \left\{\begin{bmatrix} 0 & -1 \\ 1 & 0 \end{bmatrix}, \begin{bmatrix} 1 & 2 \\ 1 & 2 \end{bmatrix}, \begin{bmatrix} 0 & 0 \\ 0 & -1 \end{bmatrix}\right\}.$$

Since $\mathcal{S}$ is linearly independent (the computations in Example 4 establish this), $\text{Im}(T)$ is a three-dimensional subspace of the four-dimensional vector space $\mathbf{M}^{2,2}$. ■

Just as the kernel of a linear transformation $T\colon \mathbf{V} \to \mathbf{W}$ is a subspace of $\mathbf{V}$ (Theorem 1), the image of T is a subspace of $\mathbf{W}$. The dimension of the image is called the **rank** of T.

THEOREM 4

Let $T\colon \mathbf{V} \to \mathbf{W}$ be a linear transformation. Then the image of T is a subspace of $\mathbf{W}$.

Proof That $\text{Im}(T)$ is a nonempty subset of $\mathbf{W}$ is clear from the definition of image.

For $\mathbf{u}$ and $\mathbf{v}$ in $\text{Im}(T)$, we must show that $\mathbf{u} + \mathbf{v}$ is also in $\text{Im}(T)$. Since $\mathbf{u}$ is in $\text{Im}(T)$, there is some $\mathbf{x}$ in $\mathbf{V}$ such that $T(\mathbf{x}) = \mathbf{u}$. Similarly, there is some $\mathbf{y}$ in $\mathbf{V}$ such that $T(\mathbf{y}) = \mathbf{v}$. Let $\mathbf{z} = \mathbf{x} + \mathbf{y}$. Then

$$T(\mathbf{z}) = T(\mathbf{x} + \mathbf{y}) = T(\mathbf{x}) + T(\mathbf{y}) = \mathbf{u} + \mathbf{v},$$

and $\mathbf{u} + \mathbf{v}$ is the image of $\mathbf{z}$ under T. That is, $\mathbf{u} + \mathbf{v}$ is in $\text{Im}(T)$.

For any scalar c, $T(c\mathbf{x}) = cT(\mathbf{x}) = c\mathbf{u}$. Thus, $c\mathbf{u}$ is the image of $c\mathbf{x}$ under T. That is, $c\mathbf{u}$ is in $\text{Im}(T)$.

These two computations confirm that the image of a linear transformation is a subspace of $\mathbf{W}$. ■

EXAMPLE 9

Describe Im (T), where T: $\mathbf{R}^3 \to \mathbf{R}^3$ is given by $T(\mathbf{x}) = A\mathbf{x}$ and

$$A = \begin{bmatrix} 2 & 1 & 1 \\ 1 & -1 & 0 \\ 1 & 2 & 1 \end{bmatrix}.$$

Solution Letting $\mathbf{x} = (x, y, z)$, we compute $T(\mathbf{x})$:

$$T(\mathbf{x}) = A\mathbf{x} = \begin{bmatrix} 2 & 1 & 1 \\ 1 & -1 & 0 \\ 1 & 2 & 1 \end{bmatrix}\begin{bmatrix} x \\ y \\ z \end{bmatrix} = \begin{bmatrix} 2x + y + z \\ x - y \\ x + 2y + z \end{bmatrix}$$

$$= x\begin{bmatrix} 2 \\ 1 \\ 1 \end{bmatrix} + y\begin{bmatrix} 1 \\ -1 \\ 2 \end{bmatrix} + z\begin{bmatrix} 1 \\ 0 \\ 1 \end{bmatrix}$$

Thus, we see that for any $\mathbf{x}$ in $\mathbf{R}^3$, $T(\mathbf{x})$ is a linear combination of the vectors (2, 1, 1), (1, −1, 2), and (1, 0, 1). Furthermore, any given linear combination of these,

$$\mathbf{y} = a(2, 1, 1) + b(1, -1, 2) + c(1, 0, 1),$$

can be seen to be an image of some element $\mathbf{x}$ in $\mathbf{R}^3$ under T by letting $\mathbf{x} = (a, b, c)$:

$$T(a, b, c) = a(2, 1, 1) + b(1, -1, 2) + c(1, 0, 1).$$

Therefore, the set of vectors {(2, 1, 1), (1, −1, 2), (1, 0, 1)} generates Im (T). Knowing a spanning set for Im (T), we can produce a basis for it, following Theorem 1 of Section 5.3. We form the matrix with the spanning set as columns and row-reduce:

$$\begin{bmatrix} 2 & 1 & 1 \\ 1 & -1 & 0 \\ 1 & 2 & 1 \end{bmatrix} \to \begin{bmatrix} 1 & 0 & \frac{1}{3} \\ 0 & 1 & \frac{1}{3} \\ 0 & 0 & 0 \end{bmatrix}.$$

Thus, a basis for Im (T) is given by the first two columns, {(2, 1, 1), (1, −1, 2)}. From the geometric point of view, Im (T) is a plane in $\mathbf{R}^3$. ■

For the remainder of this section, we make use of the matrix of a linear transformation, so the discussion is restricted to Euclidean m-space. Then, in Section 7.4, we extend all the results obtained here to general vector spaces.

You probably noticed that the matrix of columns we needed to row-reduce in Example 9 was the matrix A that defined the linear transformation T at the beginning of the example. The following theorem shows that this is no coincidence. Recall that the *column space* of a matrix A is the set of all linear combinations of its columns (see Section 5.4).

THEOREM 5

Let T: $\mathbf{R}^n \to \mathbf{R}^m$ be a linear transformation and A be the matrix that represents T. Then Im (T) is the column space of A, and dim (Im (T)), the rank of the transformation T, is equal to the rank of the matrix A.

Procedure for Computing a Basis for the Image of a Transformation Given by a Matrix A

Transform the matrix A to its row-reduced echelon matrix form, B. A basis for Im (T) consists of those columns of A that are transformed into the distinct elementary columns of B.

Proof Let $\mathbf{x} = (x_1, \ldots, x_n)$. Then we have

$$T(\mathbf{x}) = A\mathbf{x} = \begin{bmatrix} a_{11} & a_{12} & \cdots & a_{1n} \\ a_{21} & a_{22} & \cdots & a_{2n} \\ \vdots & \vdots & & \vdots \\ a_{m1} & a_{m2} & \cdots & a_{mn} \end{bmatrix} \begin{bmatrix} x_1 \\ x_2 \\ \vdots \\ x_n \end{bmatrix}$$

$$= \begin{bmatrix} a_{11}x_1 + a_{12}x_2 + \cdots + a_{1n}x_n \\ a_{21}x_1 + a_{22}x_2 + \cdots + a_{2n}x_n \\ \vdots \qquad \vdots \qquad \vdots \\ a_{m1}x_1 + a_{m2}x_2 + \cdots + a_{mn}x_n \end{bmatrix}$$

$$= x_1 \begin{bmatrix} a_{11} \\ a_{21} \\ \vdots \\ a_{m1} \end{bmatrix} + x_2 \begin{bmatrix} a_{12} \\ a_{22} \\ \vdots \\ a_{m2} \end{bmatrix} + \cdots + x_n \begin{bmatrix} a_{1n} \\ a_{2n} \\ \vdots \\ a_{mn} \end{bmatrix}$$

$$= x_1\mathbf{a}^1 + x_2\mathbf{a}^2 + \cdots + x_n\mathbf{a}^n,$$

where $\mathbf{a}^i$ denotes the *i*th column of A. In other words, for each $\mathbf{x}$ in $\mathbf{R}^n$, $T(\mathbf{x})$ is a linear combination of the columns of A. Conversely, if $\mathbf{y}$ in $\mathbf{R}^m$ is a linear combination of the columns of A,

$$\mathbf{y} = c_1\mathbf{a}^1 + c_2\mathbf{a}^2 + \cdots + c_n\mathbf{a}^n,$$

then $\mathbf{y} = T(\mathbf{c})$, where $\mathbf{c} = (c_1, \ldots, c_n)$. That is, $\mathbf{y}$ is in Im (T) if and only if $\mathbf{y}$ is a linear combination of the columns of A. ■

The preceding procedure is essentially that of Theorem 1 of Section 5.3. It is appropriate here as a consequence of Theorem 5.

EXAMPLE 10

For the matrix A, let $T: \mathbf{R}^5 \to \mathbf{R}^4$ be defined by $T(\mathbf{x}) = A\mathbf{x}$ for each $\mathbf{x}$ in $\mathbf{R}^5$. Find the rank of T and a basis for Im (T).

$$A = \begin{bmatrix} 1 & 2 & 0 & 1 & 0 \\ 2 & 4 & 1 & 0 & 0 \\ 0 & 0 & 1 & -2 & 1 \\ 1 & 2 & 1 & -1 & 1 \end{bmatrix}$$

Solution We row-reduce A to obtain

$$B = \begin{bmatrix} 1 & 2 & 0 & 1 & 0 \\ 0 & 0 & 1 & -2 & 0 \\ 0 & 0 & 0 & 0 & 1 \\ 0 & 0 & 0 & 0 & 0 \end{bmatrix}.$$

Since B has three columns that contain leading ones (the first, third, and fifth), the rank of T is 3. A basis for Im (T) is given by the columns of A that correspond to these, $\{\mathbf{a}^1, \mathbf{a}^3, \mathbf{a}^5\}$. That is, $\{(1, 2, 0, 1), (0, 1, 1, 1), (0, 0, 1, 1)\}$ is a basis for Im (T). ■

A function $f: \mathcal{A} \to \mathcal{B}$ is **onto** if Im (f), the set of all images $f(a)$, is the set $\mathcal{B}$. An immediate consequence of part (c) of Theorem 7 of Section 6.2 is that a *linear transformation* $T: \mathbf{V} \to \mathbf{W}$ is onto if and only if dim (Im (T)) = dim $\mathbf{W}$. In case $T: \mathbf{R}^n \to \mathbf{R}^m$ is represented by a matrix, the following theorem gives an easy test to see whether or not the linear transformation is onto.

THEOREM 6

Let $T: \mathbf{R}^n \rightarrow \mathbf{R}^m$ be a linear transformation and A be the matrix that represents T. Then T is onto if and only if the rank of A is equal to m.

Proof Since $\text{Im}(T)$ is a subspace of $\mathbf{R}^m$, we know that $\text{Im}(T) = \mathbf{R}^m$ if and only if $\dim(\text{Im}(T)) = m$. By Theorem 5, $\dim(\text{Im}(T))$ is equal to the rank of A, so we have the desired result. ■

As examples of Theorem 6, note that the linear transformation of Example 3 is onto, since $\dim(\text{Im}(T)) = 2$, which is the rank of the matrix, but that of Example 5 is not, since $\dim(\text{Im}(T)) = 3$, which is less than 4.

Combining Theorem 2 with Theorem 5 yields an interesting result. Assume that A is the $m \times n$ matrix that represents the linear transformation T, and let r be the rank of A. As a consequence of Theorem 2, the dimension of the kernel (the nullity of A) is $n - r$. As a consequence of Theorem 5, the dimension of the image (the rank of A) is r. Combining these facts, we obtain the following theorem.

THEOREM 7

Let $T: \mathbf{R}^n \rightarrow \mathbf{R}^m$ be a linear transformation. Then the sum of the dimension of the kernel of T and the dimension of the image of T is equal to the dimension of the domain of T. That is,

$$\dim(\text{Ker}(T)) + \dim(\text{Im}(T)) = n,$$

$$\text{or} \qquad \text{nullity} + \text{rank} = n.$$

In general, a *function* may be one-to-one without being onto, or it may be onto without being one-to-one. As the next theorem tells us, however, this cannot be the case for a *linear operator* on $\mathbf{R}^n$, a linear transformation from $\mathbf{R}^n$ into itself.

THEOREM 8

Let $T: \mathbf{R}^n \rightarrow \mathbf{R}^n$ be a linear transformation. Then T is one-to-one if and only if it is onto.

Proof From Theorem 7, we have

$$\dim(\text{Ker}(T)) + \dim(\text{Im}(T)) = n.$$

By Theorem 3, T is one-to-one if and only if $\text{Ker}(T) = \{\mathbf{0}\}$; that is, if and only if $\dim(\text{Ker}(T)) = 0$. But then the preceding equation implies that $\dim(\text{Im}(T)) = n$; that is, T is onto $\mathbf{R}^n$ by Theorem 6. ■

Let A be the matrix that represents a linear transformation T from $\mathbf{R}^n$ into itself. Then A is an order n square matrix. Since T is one-to-one if and only if it is onto, by Theorems 3 and 6 it is one-to-one if and only if the columns of A are a basis for $\mathbf{R}^n$. By Theorem 6 of Section 5.3 and Theorem 5 of Section 4.2, this occurs if and only if A is *invertible*. In this situation, A^{-1} also gives rise to a linear transformation T^{-1} on $\mathbf{R}^n$ defined by $T^{-1}(\mathbf{x}) = A^{-1}\mathbf{x}$, for each $\mathbf{x}$. The linear transformation T^{-1} is called the **inverse** of T, and we say that T is **invertible**. It is easy to confirm that $T^{-1}T = TT^{-1} = I$, since for each $\mathbf{x}$ in $\mathbf{R}^n$,

$$T^{-1}T(\mathbf{x}) = T^{-1}(T(\mathbf{x})) = T^{-1}(A\mathbf{x}) = A^{-1}A\mathbf{x} = I\mathbf{x} = I(\mathbf{x})$$

and, similarly, $TT^{-1}(\mathbf{x}) = I(\mathbf{x})$.

The following theorem is an immediate consequence of the preceding remarks.

THEOREM 9

Let $T\colon \mathbf{R}^n \to \mathbf{R}^n$ be a linear transformation. The following are equivalent:

a. T is one-to-one,

b. T is onto,

c. T is invertible,

d. $\det A \neq 0$, where A is the matrix that represents T. 

EXAMPLE 11

Let $T\colon \mathbf{R}^3 \to \mathbf{R}^3$ be defined by $T(x, y, z) = (2x - y, x + y + z, 2y - z)$ for each $\mathbf{x} = (x, y, z)$ in $\mathbf{R}^3$. Show that T is invertible, and compute T^{-1}.

Solution We first compute the matrix A that represents T,

$$A = \begin{bmatrix} 2 & -1 & 0 \\ 1 & 1 & 1 \\ 0 & 2 & -1 \end{bmatrix}.$$

Since $\det A = -7 \neq 0$, A is invertible. Its inverse is

$$A^{-1} = \frac{1}{7}\begin{bmatrix} 3 & 1 & 1 \\ -1 & 2 & 2 \\ -2 & 4 & -3 \end{bmatrix}.$$

Thus, T^{-1} is given by

$$T^{-1}(\mathbf{x}) = \mathrm{A}^{-1}\mathbf{x} = \frac{1}{7}\begin{bmatrix} 3 & 1 & 1 \\ -1 & 2 & 2 \\ -2 & 4 & -3 \end{bmatrix}\begin{bmatrix} x \\ y \\ z \end{bmatrix} = \begin{bmatrix} (\ 3x + \ y + \ z)/7 \\ (\ -x + 2y + 2z)/7 \\ (-2x + 4y - 3z)/7 \end{bmatrix}$$

or

$$T^{-1}(x, y, z) = \left(\frac{3x + y + z}{7}, \frac{-x + 2y + 2z}{7}, \frac{-2x + 4y - 3z}{7}\right). \quad \blacksquare$$

7.3 Exercises

In Exercises 1–6 determine whether or not $\mathbf{v}_1$ or $\mathbf{v}_2$ is in the kernel of the given linear transformation.

1. $T\colon \mathbf{R}^4 \to \mathbf{R}^3$ given by $T(\mathbf{x}) = A\mathbf{x}$, where

$$\mathrm{A} = \begin{bmatrix} 1 & 2 & -1 & 1 \\ 1 & 0 & 1 & 1 \\ 2 & -4 & 6 & 2 \end{bmatrix} \text{ and } \mathbf{v}_1 = (-2, 0, 0, 2), \quad \mathbf{v}_2 = (-2, 2, 2, 0).$$

2. $T\colon \mathbf{R}^3 \to \mathbf{R}^4$ given by $T(\mathbf{x}) = A\mathbf{x}$, where

$$A = \begin{bmatrix} 1 & 1 & 0 \\ 2 & 0 & 3 \\ -1 & 1 & 1 \\ 1 & 1 & 2 \end{bmatrix} \text{ and } \mathbf{v}_1 = (-3, 3, 2), \quad \mathbf{v}_2 = (0, 0, 0).$$

3. $T\colon \mathbf{R}^3 \to \mathbf{R}^2$ given by $T(x, y, z) = (3x + y - z, x + 2y + z)$ and $\mathbf{v}_1 = (3, -4, 5)$, $\mathbf{v}_2 = (1, 4, 7)$.

4. $T: \mathbf{R}^4 \to \mathbf{R}$ given by $T(x_1, x_2, x_3, x_4) = x_1 + 2x_2 + 3x_3 + 4x_4$ and $\mathbf{v}_1 = (1, 2, 1, -2)$, $\mathbf{v}_2 = (2, 3, 0, -2)$.

5. $T: \mathbf{P}_2 \to \mathbf{M}^{2,2}$ of Example 4 and $\mathbf{v}_1 = 1 + x - 2x^2$, $\mathbf{v}_2 = \mathbf{0}$.

6. $T: \mathbf{M}^{2,3} \to \mathbf{P}_2$ of Example 3 of Section 7.1 and

$$\mathbf{v}_1 = \begin{bmatrix} -1 & 3 & 1 \\ 2 & 2 & 0 \end{bmatrix}, \quad \mathbf{v}_2 = \begin{bmatrix} 2 & 1 & -2 \\ 1 & 1 & 0 \end{bmatrix}.$$

In Exercises 7–12 determine whether or not $\mathbf{w}_1$ or $\mathbf{w}_2$ is in the image of the corresponding linear transformation of Exercises 1–6.

7. $\mathbf{w}_1 = (1, 3, 1)$, $\mathbf{w}_2 = (-1, -1, -2)$

8. $\mathbf{w}_1 = (0, 0, -8, -4)$, $\mathbf{w}_2 = (-1, 13, 0, 7)$

9. $\mathbf{w}_1 = (1, 1)$, $\mathbf{w}_2 = (-3, 1)$

10. $\mathbf{w}_1 = -13$, $\mathbf{w}_2 = 0$

11. $\mathbf{w}_1 = \begin{bmatrix} 2 & 1 \\ 3 & 4 \end{bmatrix}$, $\mathbf{w}_2 = \begin{bmatrix} 3 & 2 \\ 3 & 4 \end{bmatrix}$

12. $\mathbf{w}_1 = x^2 + 2x + 1$, $\mathbf{w}_2 = x - 2$

13–18. For each of the linear transformations T in Exercises 1–6, find the nullity of T and give a basis for $\text{Ker}(T)$. Which of the transformations T are one-to-one?

19–24. For each of the linear transformations T of Exercises 1–6, find the rank of T and give a basis for $\text{Im}(T)$. Which of the transformations T are onto?

25. If $T: \mathbf{R}^n \to \mathbf{R}^m$ is a linear transformation, prove that $\dim(\text{Im}(T)) \leq n$.

26. If $T: \mathbf{R}^n \to \mathbf{R}^m$ is a linear transformation and $n < m$, prove that T is not onto.

27. If $T: \mathbf{R}^n \to \mathbf{R}^m$ is a linear transformation and $n > m$, prove that T is not one-to-one.

28. If $T: \mathbf{R}^n \to \mathbf{R}^m$ is a linear transformation that is one-to-one and $\mathcal{B}$ is a basis for $\mathbf{R}^n$, prove that $T[\mathcal{B}]$, the set of all images $T(\mathbf{b})$ for $\mathbf{b}$ in $\mathcal{B}$, is a basis for $\text{Im}(T)$.

In Exercises 29–35 determine whether or not the given linear transformation is invertible. If it is invertible, compute its inverse.

29. $T: \mathbf{R}^3 \to \mathbf{R}^3$ given by $T(x, y, z) = (x + z, x - y + z, y + 2z)$

30. $T: \mathbf{R}^2 \to \mathbf{R}^2$ given by $T(x, y) = (3x + 2y, -6x - 4y)$

31. $T: \mathbf{R}^3 \to \mathbf{R}^3$ given by $T(\mathbf{x}) = A\mathbf{x}$, where

$$A = \begin{bmatrix} 3 & 1 & 2 \\ 5 & -3 & 4 \\ 1 & -2 & 1 \end{bmatrix}.$$

32. $T: \mathbf{R}^4 \to \mathbf{R}^4$ given by $T(\mathbf{x}) = A\mathbf{x}$, where

$$A = \begin{bmatrix} -1 & 2 & 1 & 0 \\ -1 & 1 & 0 & -1 \\ 2 & -1 & 0 & 4 \\ 1 & -2 & 0 & 0 \end{bmatrix}.$$

33. $T_\theta: \mathbf{R}^2 \to \mathbf{R}^2$, the rotation of the plane given by Example 7 of Section 7.1.

34. $T^{\theta}\colon \mathbf{R}^2 \to \mathbf{R}^2$, the reflection of the plane given by Example 9 of Section 7.1.

35. $T_c\colon \mathbf{R}^n \to \mathbf{R}^n$, the contraction/dilation given by Example 8 of Section 7.1.

Computational Exercises

In Exercises 36–38 use a computer algebra system to find the values of the variables for which *T* is one-to-one or onto, or both.

36. $T\colon \mathbf{R}^3 \to \mathbf{R}^2$, given by $T(\mathbf{x}) = A\mathbf{x}$, where

$$A = \begin{bmatrix} 1 & a & 2 \\ 2a & 1 & a+b \end{bmatrix}$$

37. $T\colon \mathbf{R}^4 \to \mathbf{R}^4$, given by $T(\mathbf{x}) = A\mathbf{x}$, where

$$A = \begin{bmatrix} 1 & -1 & a & 1 \\ a & a+b & 2 & 1 \\ -1 & 3a & 0 & 2b \\ a & 2b & b & c \end{bmatrix}$$

38. $T\colon \mathbf{R}^3 \to \mathbf{R}^4$, given by $T(\mathbf{x}) = A\mathbf{x}$, where

$$A = \begin{bmatrix} 1 & -1 & a \\ a & a+b & 2 \\ -1 & 3a & 0 \\ a & 2b & b \end{bmatrix}$$

7.4 Matrix of a General Linear Transformation

In this section, we show how the matrix representation of linear transformations from $\mathbf{R}^n$ to $\mathbf{R}^m$ (Section 7.2) can be extended to incorporate linear transformations between any two finite dimensional vector spaces.

Let $\mathbf{V}$ be a vector space with basis $\mathcal{B} = \{\mathbf{v}_1, \mathbf{v}_2, \ldots, \mathbf{v}_n\}$. Recall, from Section 6.3, that the *coordinate vector* of a vector $\mathbf{v}$ in $\mathbf{V}$ is the unique $\mathbf{x}$ in $\mathbf{R}^n$ of coefficients needed to express $\mathbf{v}$ as a linear combination of elements of $\mathcal{B}$:

$$\mathbf{v} = x_1\mathbf{v}_1 + x_2\mathbf{v}_2 + \cdots + x_n\mathbf{v}_n.$$

The coordinate vector can be used to replace the original vector in many calculations. In the case of a linear transformation $T\colon \mathbf{V} \to \mathbf{W}$, there are two spaces involved, $\mathbf{V}$ and $\mathbf{W}$. If $\mathbf{V} = \mathbf{W}$, it is possible to start with the same basis for both spaces, but, in general, the two spaces might be completely different, so that a basis for one looks nothing like that for the other. It is still possible, however, to use a matrix to represent a linear transformation, but a fixed basis for *each* space must be chosen before the matrix can be constructed, and the elements must be kept in the same order.

Procedure for Constructing a Matrix That Represents a Linear Transformation $T\colon \mathbf{V} \to \mathbf{W}$.

i. Choose $\mathcal{B} = \{\mathbf{b}_1, \ldots, \mathbf{b}_n\}$ and $\mathcal{B}' = \{\mathbf{b}'_1, \ldots, \mathbf{b}'_m\}$ as bases of $\mathbf{V}$ and $\mathbf{W}$, respectively.

ii. Express each $T(\mathbf{b}_i)$ as a linear combination of $\mathcal{B}'$.

iii. Let $\mathbf{a}^i$ be the coefficients of the $\mathbf{b}'_j$; that is, $\mathbf{a}^i$ is the coordinate vector of $T(\mathbf{b}_i)$ with respect to $\mathcal{B}'$.

iv. A is the matrix with ith column $\mathbf{a}^i$.

EXAMPLE 1

Let L: $\mathbf{M}^{2,3} \rightarrow \mathbf{P}_2$ be as in Example 3 of Section 7.1.

$$T\begin{bmatrix} a_{11} & a_{12} & a_{13} \\ a_{21} & a_{22} & a_{23} \end{bmatrix} = (a_{11} + a_{13})x^2 + (a_{21} - a_{22})x + a_{23}$$

Find the matrix A in $\mathbf{M}^{3,6}$ that represents T with respect to the bases $\mathcal{B} = \{E^{ij} \mid i = 1, 2; j = 1, 2, 3\}$ and $\mathcal{B}' = \{1, x, x^2\}$.

Solution Following Theorem 1, we compute $T(E^{ij})$ for each i, j and represent the result by means of the coordinate vector with respect to $\mathcal{B}'$:

$$\begin{aligned}
T(E^{11}) &= (1 + 0)x^2 + (0 - 0)x + 0 = 0(1) + 0(x) + 1(x^2) \leftrightarrow (0, 0, 1),\\
T(E^{12}) &= (0 + 0)x^2 + (0 - 0)x + 0 = 0(1) + 0(x) + 0(x^2) \leftrightarrow (0, 0, 0),\\
T(E^{13}) &= (0 + 1)x^2 + (0 - 0)x + 0 = 0(1) + 0(x) + 1(x^2) \leftrightarrow (0, 0, 1),\\
T(E^{21}) &= (0 + 0)x^2 + (1 - 0)x + 0 = 0(1) + 1(x) + 0(x^2) \leftrightarrow (0, 1, 0),\\
T(E^{22}) &= (0 + 0)x^2 + (0 - 1)x + 0 = 0(1) - 1(x) + 0(x^2) \leftrightarrow (0, -1, 0),\\
\text{and} \quad T(E^{23}) &= (0 + 0)x^2 + (0 - 0)x + 1 = 1(1) + 0(x) + 0(x^2) \leftrightarrow (1, 0, 0).
\end{aligned}$$

The resulting matrix is then

$$A = \begin{bmatrix} 0 & 0 & 0 & 0 & 0 & 1 \\ 0 & 0 & 0 & 1 & -1 & 0 \\ 1 & 0 & 1 & 0 & 0 & 0 \end{bmatrix}.$$

■

EXAMPLE 2

Let the derivative transformation D: $\mathbf{P}_3 \rightarrow \mathbf{P}_2$ be defined as follows. For each

$$\mathbf{p}(x) = a_3x^3 + a_2x^2 + a_1x + a_0,$$

define

$$D(\mathbf{p}) = 3a_3x^2 + 2a_2x + a_1.$$

Find the matrix A that represents D with respect to the bases $\mathcal{B} = \{1, x, x^2, x^3\}$ and $\mathcal{B}' = \{1, x, x^2\}$.

Solution We have shown, in the Remark after Theorem 3 of Section 7.2, that D is a linear transformation. Now

$$\begin{aligned}
D(1) &= 0 &&= 0(1) + 0(x) + 0(x^2) \leftrightarrow (0, 0, 0),\\
D(x) &= 1 &&= 1(1) + 0(x) + 0(x^2) \leftrightarrow (1, 0, 0),\\
D(x^2) &= 2x &&= 0(1) + 2(x) + 0(x^2) \leftrightarrow (0, 2, 0),\\
\text{and} \quad D(x^3) &= 3x^2 &&= 0(1) + 0(x) + 3(x^2) \leftrightarrow (0, 0, 3).
\end{aligned}$$

The resulting matrix is then

$$A = \begin{bmatrix} 0 & 1 & 0 & 0 \\ 0 & 0 & 2 & 0 \\ 0 & 0 & 0 & 3 \end{bmatrix}.$$

■

NOTE (*From calculus*) Using the result of Example 2, we can give the derivative of any polynomial of degree less than or equal to 3 by multiplication by A. That is,

$$\frac{d}{dx}(a_3x^3 + a_2x^2 + a_1x + a_0) \leftrightarrow \begin{bmatrix} 0 & 1 & 0 & 0 \\ 0 & 0 & 2 & 0 \\ 0 & 0 & 0 & 3 \end{bmatrix}\begin{bmatrix} a_0 \\ a_1 \\ a_2 \\ a_3 \end{bmatrix}.$$

For example, if $\mathbf{p}(x) = 1 - 3x^2 + x^3$, then $\frac{d}{dx}(\mathbf{p}(x))$ can be obtained by computing

$$\begin{bmatrix} 0 & 1 & 0 & 0 \\ 0 & 0 & 2 & 0 \\ 0 & 0 & 0 & 3 \end{bmatrix} \begin{bmatrix} 1 \\ 0 \\ -3 \\ 1 \end{bmatrix} = \begin{bmatrix} 0 \\ -6 \\ 3 \end{bmatrix}.$$

Now $(0, -6, 3)$ is the coordinate vector of $\frac{d}{dx}(\mathbf{p}(x)) = -6x + 3x^2$.

The preceding note shows why we say that the matrix constructed *represents* the linear transformation. Instead of applying the original linear transformation to the original vector space, we multiply the constructed matrix times each coordinate vector to obtain a result that can be interpreted in the second vector space. In general, let $T: \mathbf{V} \to \mathbf{W}$ be a linear transformation from an n-dimensional space $\mathbf{V}$ to an m-dimensional space $\mathbf{W}$, and A be the matrix that represents T with respect to a basis of each $\mathbf{V}$ and $\mathbf{W}$. Since A is an $m \times n$ matrix, multiplication by A is a linear transformation, $T_A: \mathbf{R}^n \to \mathbf{R}^m$, from $\mathbf{R}^n$ to $\mathbf{R}^m$. For each vector $\mathbf{v}$ in $\mathbf{V}$, the coordinate vector $\mathbf{x}$ with respect to the basis of $\mathbf{V}$ is in $\mathbf{R}^n$, so $T_A(\mathbf{x}) = A\mathbf{x}$ makes sense and is some element of $\mathbf{R}^m$. Each vector in $\mathbf{R}^m$ can be interpreted as the coordinate vector of some element of $\mathbf{W}$ with respect to the basis of $\mathbf{W}$. The next theorem asserts that the vector in $\mathbf{W}$ of which $A\mathbf{x}$ is the coordinate vector is exactly $T(\mathbf{v})$.

THEOREM 1

Let $T: \mathbf{V} \to \mathbf{W}$ be a linear transformation, A be the matrix that represents T with respect to the bases $\mathcal{B}$ of $\mathbf{V}$ and $\mathcal{B}'$ of $\mathbf{W}$, $\mathbf{v}$ be a vector in $\mathbf{V}$ with coordinate vector $\mathbf{x}$, and $\mathbf{w}$ be a vector in $\mathbf{W}$ with coordinate vector $\mathbf{y}$. Then $T(\mathbf{v}) = \mathbf{w}$ if and only if $T_A(\mathbf{x}) = \mathbf{y}$; that is, if and only if $A\mathbf{x} = \mathbf{y}$ (see Figure 9).

Proof Since $T: \mathbf{V} \to \mathbf{W}$ and $T_A: \mathbf{R}^n \to \mathbf{R}^m$ are linear transformations, it suffices to show that each acts appropriately on a basis. The basis we choose for $\mathbf{V}$ is $\mathcal{B} = \{\mathbf{b}_1, \ldots, \mathbf{b}_n\}$, and the corresponding basis for $\mathbf{R}^n$ is the standard basis for $\mathbf{R}^n$, $\{\mathbf{e}_1, \ldots, \mathbf{e}_n\}$, since $\mathbf{e}_i$ is the coordinate vector of $\mathbf{b}_i$ with respect to the basis $\mathcal{B}$. For each i, $T(\mathbf{b}_i)$ has as its coordinate vector $\mathbf{a}^i$, the ith column of A by the construction procedure for A. But $T_A(\mathbf{e}_i) = A\mathbf{e}_i = \mathbf{a}^i$, so T_A acting on the coordinate vector of $\mathbf{b}_i$ is the coordinate vector of $T(\mathbf{b}_i)$. Thus, the two linear transformations agree in their actions on their respective bases, and the proof is complete. ■

FIGURE 9

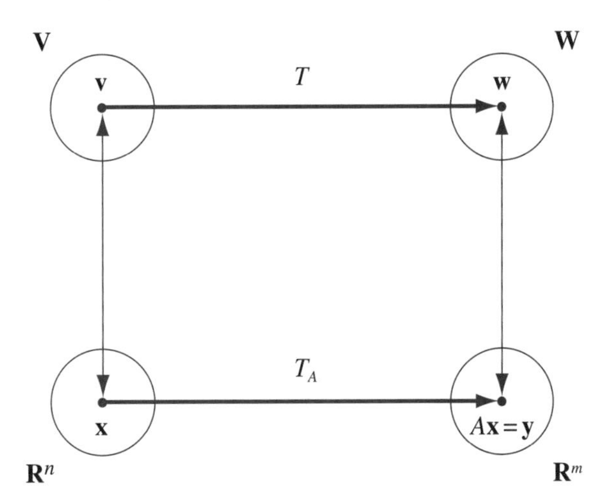

NOTE Theorems 5 and 6 of Section 7.2, which demonstrate that the sum, scalar multiple, and composition of linear transformations correspond naturally to the analogous operations on their matrices, can easily be extended to this more general case. The proofs for the general case are almost identical to those given in Section 7.2.

The matrix that represents a linear transformation can be used to characterize the *kernel* and *image* of the linear transformation and thereby determine whether or not the transformation is one-to-one or onto. The following theorems explain this correspondence. They are almost immediate corollaries to Theorem 1, and their proofs are left for Exercises 18 and 19.

THEOREM 2

Let $T: \mathbf{V} \rightarrow \mathbf{W}$ be a linear transformation and A be a matrix that represents T. Then a vector $\mathbf{v}$ is in $\text{Ker}(T)$ if and only if its coordinate vector $\mathbf{x}$ is in $\text{Ker}(T_A)$; that is, if and only if $A\mathbf{x} = \mathbf{0}$. ■

THEOREM 3

Let $T: \mathbf{V} \rightarrow \mathbf{W}$ be a linear transformation and A be a matrix that represents T. Then a vector $\mathbf{w}$ is in $\text{Im}(T)$ if and only if its coordinate vector $\mathbf{y}$ is in $\text{Im}(T_A)$. ■

EXAMPLE 3

Let $T: \mathbf{M}^{2,3} \rightarrow \mathbf{P}_2$ be as in Example 1.

$$T\begin{bmatrix} a_{11} & a_{12} & a_{13} \\ a_{21} & a_{22} & a_{23} \end{bmatrix} = (a_{11} + a_{13})x^2 + (a_{21} - a_{22})x + a_{23}.$$

Find the kernel and image of T using the matrix that represents T with respect to the standard bases.

Solution From Example 1, we have the appropriate matrix:

$$A = \begin{bmatrix} 0 & 0 & 0 & 0 & 0 & 1 \\ 0 & 0 & 0 & 1 & -1 & 0 \\ 1 & 0 & 1 & 0 & 0 & 0 \end{bmatrix}.$$

By inspection, we see that the rank of A is 3. Since $T_A: \mathbf{R}^6 \rightarrow \mathbf{R}^3$, by Theorem 6 of Section 7.3 we know that T_A is onto. But then, by Theorem 3, T must be onto as well. Thus, the image of T is $\mathbf{P}_2$.

To find $\text{Ker}(T)$, we also use one of the procedures of Section 7.3—this time the procedure to compute a basis for $\text{Ker}(T_A)$. Let B be the row-reduced echelon form for A. Then

$$B = \begin{bmatrix} 1 & 0 & 1 & 0 & 0 & 0 \\ 0 & 0 & 0 & 1 & -1 & 0 \\ 0 & 0 & 0 & 0 & 0 & 1 \end{bmatrix}.$$

Interpreting B as the matrix of coefficients of a homogeneous linear system in the variables x_{11}, x_{12}, x_{13}, x_{21}, x_{22}, and x_{23} (corresponding to the order of the E^{ij}), we have

$$\begin{aligned} x_{11} + x_{13} \qquad\qquad &= 0 \\ x_{21} - x_{22} \quad &= 0 \\ x_{23} &= 0. \end{aligned}$$

Parameterizing x_{12}, x_{13}, and x_{22} with a, b, and c, we have

$$\begin{aligned} x_{11} &= 0a - 1b + 0c = -b, \\ x_{21} &= 0a + 0b + 1c = c, \\ \text{and} \quad x_{23} &= 0a + 0b + 0c = 0. \end{aligned}$$

Thus, we have Ker(T) as those matrices of the form

$$\begin{bmatrix} -b & a & b \\ c & c & 0 \end{bmatrix}.$$

■

With Theorems 1 through 3, it is a straightforward task to extend several theorems of Section 7.3 to the general case. These are summarized in the following theorem.

THEOREM 4

Let T: $\mathbf{V} \rightarrow \mathbf{W}$ be a linear transformation and A be a matrix that represents T with respect to bases for $\mathbf{V}$ and $\mathbf{W}$. Then

a. T is a one-to-one if and only if $A\mathbf{x} = \mathbf{0}$ has only the trivial solution;

b. T is onto if and only if rank $(A) = \dim(\mathbf{W})$;

c. $\dim(\text{Ker}(T)) + \dim(\text{Im}(T)) = \dim(\mathbf{V})$;

d. If $\dim(\mathbf{V}) = \dim(\mathbf{W})$, then the following are equivalent:

- **i.** T is one-to-one,
- **ii.** T is onto,
- **iii.** $\det A \neq 0$.

■

EXAMPLE 4

Let T: $\mathbf{M}^{2,3} \rightarrow \mathbf{P}_2$ be as in Examples 1 and 3:

$$T\begin{bmatrix} a_{11} & a_{12} & a_{13} \\ a_{21} & a_{22} & a_{23} \end{bmatrix} = (a_{11} + a_{13})x^2 + (a_{21} - a_{22})x + a_{23}.$$

Is T one-to-one or onto, or both?

Solution That T is not one-to-one is immediately apparent from the result of Example 3. For example, with $a = 1$, $b = 2$, and $c = 3$, the final matrix of Example 3 yields the following nonzero matrix in Ker(T):

$$\begin{bmatrix} -2 & 1 & 2 \\ 3 & 3 & 0 \end{bmatrix}$$

Part (a) of Theorem 4, however, gives an even easier solution to the question of whether or not T is one-to-one. The matrix A of T, calculated in Example 1, has more columns than rows. Hence, $A\mathbf{x} = \mathbf{0}$ has nontrivial solutions by Corollary 2 of Section 3.3, and therefore T is not one-to-one.

Yet a third, and even easier, solution follows from Theorem 7 of Section 7.3, restated here as part (c) of Theorem 4. The dimension of $\mathbf{M}^{2,3}$ is 6 and the dimension of $\mathbf{P}_2$ is 3, so we know that $\dim(\text{Ker}(T)) \geq 3$ without even looking at T itself!

Since this same matrix A has rank 3 (its row-reduced echelon form is matrix B in Example 3), and since $\dim(\mathbf{P}_2) = 3$, T is onto $\mathbf{P}_2$, by part (b) of Theorem 4. ■

EXAMPLE 5

Let D: $\mathbf{P}_3 \rightarrow \mathbf{P}_3$ be given by the same rule as in Example 2. For each $\mathbf{p}(x) = a_3x^3 + a_2x^2 + a_1x + a_0$, define

$$D(\mathbf{p}) = 3a_3x^2 + 2a_2x + a_1.$$

Is D one-to-one or onto, or both?

Solution Since D is from $\mathbf{V} = \mathbf{P}_3$ to $\mathbf{W} = \mathbf{P}_3$, we have dim $(\mathbf{V})$ = dim $(\mathbf{W})$, as required for part (d) of Theorem 4. We do need a matrix to represent the transformation. To construct one, we use the same basis $\mathcal{B} = \{1, x, x^2, x^3\}$ for both $\mathbf{V}$ and $\mathbf{W}$. This gives almost the same matrix as was computed in Example 2; this time, it has an extra zero row:

$$A = \begin{bmatrix} 0 & 1 & 0 & 0 \\ 0 & 0 & 2 & 0 \\ 0 & 0 & 0 & 3 \\ 0 & 0 & 0 & 0 \end{bmatrix}$$

Because of the zero row, det $A = 0$. By part (d) of Theorem 4, D is neither one-to-one nor onto. ■

7.4 Exercises

In Exercises 1–4 find the matrix *A* that represents the linear transformation *T* with respect to the bases $\mathcal{B}$ and $\mathcal{B}'$. (See Exercises 5–8 of Section 7.1.)

1. T: $\mathbf{R}^3 \rightarrow \mathbf{M}^{2,2}$ given by

$$T(x, y, z) = \begin{bmatrix} y & z \\ -x & 0 \end{bmatrix},$$

where $\mathcal{B} = \{\mathbf{e}_1, \mathbf{e}_2, \mathbf{e}_3\}$ and $\mathcal{B}' = \{E^{ij} \mid i = 1, 2; j = 1, 2\}$.

2. T: $\mathbf{P}_3 \rightarrow \mathbf{P}_3$ given by

$$T(a_0 + a_1x + a_2x^2 + a_3x^3) = (a_0 + a_2) - (a_1 + 2a_3)x^2,$$

where $\mathcal{B} = \mathcal{B}' = \{1, x, x^2, x^3\}$.

3. T: $\mathbf{P}_2 \rightarrow \mathbf{M}^{2,2}$ given by

$$T(a_0 + a_1x + a_2x^2) = \begin{bmatrix} a_0 & -a_2 \\ -a_2 & a_0 - a_1 \end{bmatrix},$$

where $\mathcal{B} = \{1, x, x^2\}$ and $\mathcal{B}' = \{E^{ij} \mid i = 1, 2; j = 1, 2\}$.

4. T: $\mathbf{P}_2 \rightarrow \mathbf{P}_2$ given by

$$T(a_0 + a_1x + a_2x^2) = a_1 - (a_0 + a_2)x + (a_1 + a_2)x^2,$$

where $\mathcal{B} = \{1, x, x^2\}$ and $\mathcal{B}' = \{x + 1, x - 1, x^2 + 1\}$.

5–8. Use Theorem 2 to compute Ker (T) for each linear transformation in Exercises 1–4. Which transformations are one-to-one?

9–12. Use Theorem 3 to compute Im (T) for each linear transformation in Exercises 1–4. Which transformations are onto?

13. Find the 4×5 matrix associated with the derivative transformation

$$D: \mathbf{P}_4 \rightarrow \mathbf{P}_3$$

with respect to $\mathcal{B} = \{1, x, x^2, x^3, x^4\}$ and $\mathcal{B}' = \{1, x, x^2, x^3\}$ (see Example 2).

14. Use the matrix of Exercise 13 to find the derivative of

$$\mathbf{p}(x) = 1 + x^2 - 3x^3 + 2x^4.$$

15. Let S: $\mathbf{P}_2 \rightarrow \mathbf{P}_3$ be defined as follows. For each $\mathbf{P}(x) = a_2x^2 + a_1x + a_0$, define $S(\mathbf{p}) = \frac{a_2}{3}x^3 + \frac{a_1}{2}x^2 + a_0x$. Find the matrix A that represents S with respect to the bases $\mathcal{B} = \{1, x, x^2\}$ and $\mathcal{B}' = \{1, x, x^2, x^3\}$. (The linear transformation S gives the *integral* of $\mathbf{p}(x)$, with the constant term equal to zero.)

16. Use the matrix of Exercise 15 to find the integral of $\mathbf{p}(x) = 1 - x + 2x^2$.

17. Let T: $\mathbf{V} \rightarrow \mathbf{W}$ be a linear transformation that is both one-to-one and onto. Prove that T is *invertible*. That is, prove that there exists a transformation S: $\mathbf{W} \rightarrow \mathbf{V}$ such that $S \circ T$ is the identity on $\mathbf{V}$ and $T \circ S$ is the identity on $\mathbf{W}$. (*Hint*: See Theorem 9 of Section 7.3.)

18. Prove Theorem 2.

19. Prove Theorem 3.

7.5 Change of Basis

The matrix associated with a linear transformation T: $\mathbf{R}^n \rightarrow \mathbf{R}^m$ of Section 7.2 was computed using the standard basis vectors $\mathbf{e}_i$ (see Theorem 1 of Section 7.2). There are times when it is more convenient to compute the matrix of a linear transformation using a different basis (see Section 8.5). The procedure of Section 7.4 can be applied to construct this new matrix, but a more efficient procedure is described in this section.

The components of any vector $\mathbf{x}$ in $\mathbf{R}^n$ are its coordinates with respect to the standard basis $\{\mathbf{e}_1, \ldots, \mathbf{e}_n\}$, since $\mathbf{x} = x_1\mathbf{e}_1 + \cdots + x_n\mathbf{e}_n$. When we find the coordinate vector for $\mathbf{x}$ with respect to some other basis, we are, in effect, making a **change of basis**. For example, let $n = 2$. Then any vector $\mathbf{x} = (x, y)$ can be written as $\mathbf{x} = x\mathbf{e}_1 + y\mathbf{e}_2$; the ordered pair (x, y) expresses $\mathbf{x}$ in terms of the standard basis. Now, if $\mathcal{B} = \{\mathbf{b}_1, \mathbf{b}_2\}$ is some other basis, there exist unique scalars s and t such that $\mathbf{x} = s\mathbf{b}_1 + t\mathbf{b}_2$ (Figure 10). The coordinate vector $(s, t)_{\mathcal{B}}$ of $\mathbf{x}$ with respect to $\mathcal{B}$ represents the same point as $\mathbf{x}$ (albeit with different coordinates), as long as it is understood that $(s, t)_{\mathcal{B}}$ means the point obtained by adding the vectors $s\mathbf{b}_1$ and $t\mathbf{b}_2$. When this convention is made, the usual coordinate axes are replaced by the lines determined by $\mathbf{b}_1$ and $\mathbf{b}_2$, and each point is expressed in terms of these. For instance, $(1, 0)_{\mathcal{B}}$ is the endpoint of $\mathbf{b}_1$ and $(1, 1)_{\mathcal{B}}$ is the endpoint of $\mathbf{b}_1 + \mathbf{b}_2$ in the new coordinates. The following theorem shows that the conversion of coordinates from the standard basis to some other basis can be effected by multiplication by a matrix, instead of by solving for the coordinate vector.

FIGURE 10

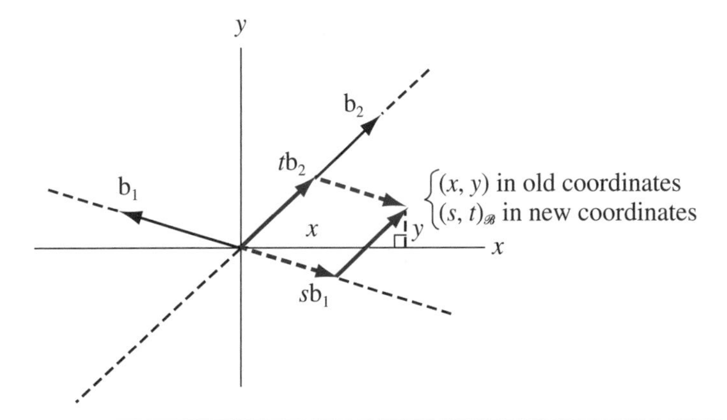

THEOREM 1

Let $\mathcal{B} = \{\mathbf{b}_1, \ldots, \mathbf{b}_n\}$ be a basis for $\mathbf{R}^n$ and P be the matrix with ith column $\mathbf{p}^i = \mathbf{b}_i$. Then P is invertible, and P^{-1} is the change-of-basis matrix. In other words, for each $\mathbf{x}$ in $\mathbf{R}^n$, $P^{-1}\mathbf{x}$ is the coordinate vector of $\mathbf{x}$ with respect to $\mathcal{B}$, $\mathbf{y}_{\mathcal{B}} = P^{-1}\mathbf{x}$.

Proof The matrix P is invertible, since it is square of order n with n independent columns. Since a linear transformation is determined by its action on a basis (Theorem 3 of Section 7.1), we need only confirm that multiplication by P^{-1} "acts right" on a basis. Note that each basis element $\mathbf{b}_i$ has coordinate vector $\mathbf{e}_i$ with respect to $\mathscr{B}$, since

$$\mathbf{b}_i = 0\mathbf{b}_1 + \cdots + 0\mathbf{b}_{i-1} + 1\mathbf{b}_i + 0\mathbf{b}_{i+1} + \cdots + 0\mathbf{b}_n.$$

Thus, the transformation must carry $\mathbf{b}_i$ into $\mathbf{e}_i$ for each i. But this is exactly the effect of multiplication on the left by P^{-1}, since $P^{-1}P = I$, or by columns, $P^{-1}\mathbf{p}^i = P^{-1}\mathbf{b}_i = \mathbf{e}_i$. ■

EXAMPLE 1

For $\mathscr{B} = \{(2, 1, 3), (1, 0, 1), (-1, 1, 1)\}$, a basis for $\mathbf{R}^3$, compute the change-of-basis matrix and use it to find the coordinate vector of $\mathbf{x} = (2, -1, 1)$.

Solution Following Theorem 1, we let

$$P = \begin{bmatrix} 2 & 1 & -1 \\ 1 & 0 & 1 \\ 3 & 1 & 1 \end{bmatrix}.$$

Then the change-of-basis matrix is

$$P^{-1} = \begin{bmatrix} 1 & 2 & -1 \\ -2 & -5 & 3 \\ -1 & -1 & 1 \end{bmatrix}.$$

Finally, the required coordinate vector is

$$P^{-1}(2, -1, 1) = \begin{bmatrix} 1 & 2 & -1 \\ -2 & -5 & 3 \\ -1 & -1 & 1 \end{bmatrix}\begin{bmatrix} 2 \\ -1 \\ 1 \end{bmatrix} = \begin{bmatrix} -1 \\ 4 \\ 0 \end{bmatrix}_{\mathscr{B}} = (-1, 4, 0)_{\mathscr{B}}.$$

■

If P^{-1} is a change-of-basis matrix for $\mathbf{R}^n$, then any vector $\mathbf{x}$ is represented by $P^{-1}\mathbf{x}$ with respect to the new basis. In other words, $\mathbf{x}$ and $P^{-1}\mathbf{x}$ each represent the same point but express it in terms of different bases. If T: $\mathbf{R}^n \rightarrow \mathbf{R}^n$ is a linear transformation, the point represented by $T(\mathbf{x})$ can also be represented in terms of the new basis as $P^{-1}T(\mathbf{x})$.

EXAMPLE 2

Let T: $\mathbf{R}^3 \rightarrow \mathbf{R}^3$ be defined by $T(x, y, z) = (x + y, z, 2x - y)$. Find the coordinate vector of $T(2, -1, 1)$ with respect to the basis

$$\mathscr{B} = \{(2, 1, 3), (1, 0, 1), (-1, 1, 1)\}.$$

Solution From Example 1, we have the change-of-basis matrix P^{-1}, so the result is

$$P^{-1}T(2, -1, 1) = \begin{bmatrix} 1 & 2 & -1 \\ -2 & -5 & 3 \\ -1 & -1 & 1 \end{bmatrix}\begin{bmatrix} 1 \\ 1 \\ 5 \end{bmatrix} = \begin{bmatrix} -2 \\ 8 \\ 3 \end{bmatrix}_{\mathscr{B}}.$$

That is, $P^{-1}T(2, -1, 1) = (-2, 8, 3)_{\mathscr{B}}$. ■

In Example 1, we computed the coordinates of the vector $(2, -1, 1)$ with respect to a basis $\mathscr{B}$. In Example 2, we computed the coordinates of the vector $T(2, -1, 1)$ with respect to the same basis. To make the computation, however, we essentially went back to the expression for the point in terms of the standard basis. Theorem 2 shows that it is possible to go directly from the coordinate vector of $\mathbf{x}$ to the coordinate vector of $T(\mathbf{x})$ by means of multiplication by the appropriate matrix.

DEFINITION

The matrix $A_{\mathcal{B}}$ **represents** T with **respect to the basis** $\mathcal{B}$ if

$$P^{-1}(T(\mathbf{x})) = A_{\mathcal{B}}\,(P^{-1}\mathbf{x})$$

for every $\mathbf{x}$ in $\mathbf{R}^n$, where P^{-1} is the change-of-basis matrix for $\mathcal{B}$. In other words, multiplication of the coordinate vector of $\mathbf{x}$ with respect to $\mathcal{B}$ by $A_{\mathcal{B}}$ gives the coordinate vector of $T(\mathbf{x})$ with respect to $\mathcal{B}$.

NOTE Let A be the matrix associated with a linear transformation T: $\mathbf{R}^n \to \mathbf{R}^n$. Then A is the matrix that represents T with respect to the *standard* basis, since $P^{-1} = P = I$ is the change-of-basis matrix for the standard basis.

THEOREM 2

Let T: $\mathbf{R}^n \to \mathbf{R}^n$ be a linear transformation with associated matrix A, $\mathcal{B}$ be a basis for $\mathbf{R}^n$, and P^{-1} be the change-of-basis matrix for $\mathcal{B}$. Then $P^{-1}AP$ is the matrix that represents T with respect to the basis $\mathcal{B}$, $A_{\mathcal{B}} = P^{-1}AP$.

Proof For each $\mathbf{x}$ in $\mathbf{R}^n$, $P^{-1}\mathbf{x}$ is the coordinate vector of the same point with respect to the basis $\mathcal{B}$. Checking the action of $P^{-1}AP$ on any such vector, we have

$$(P^{-1}AP)(P^{-1}\mathbf{x}) = P^{-1}AI\mathbf{x} = P^{-1}A\mathbf{x} = P^{-1}T(\mathbf{x}),$$

since A is the matrix associated with T. But this last vector, $P^{-1}\mathrm{T}(\mathbf{x})$, is the coordinate vector of $T(\mathbf{x})$ with respect to the basis $\mathcal{B}$. ■

EXAMPLE 3

Let T: $\mathbf{R}^3 \to \mathbf{R}^3$ be given by $T(x, y, z) = (x + y, z, 2x - y)$, and let $\mathcal{B} = \{(2, 1, 3), (1, 0, 1), (-1, 1, 1)\}$. Calculate $A_{\mathcal{B}}$ for T, and use it to find the coordinate vector for the vector $\mathbf{y}$ with coordinate vector $(-1, 4, 0)_{\mathcal{B}}$.

Solution To use Theorem 2, we need the matrix A associated with T,

$$A = \begin{bmatrix} 1 & 1 & 0 \\ 0 & 0 & 1 \\ 2 & -1 & 0 \end{bmatrix}.$$

From Example 1, we have P and P^{-1}. So, applying Theorem 2, we have

$$\begin{aligned} A_{\mathcal{B}} = P^{-1}AP &= \begin{bmatrix} 1 & 2 & -1 \\ -2 & -5 & 3 \\ -1 & -1 & 1 \end{bmatrix}\begin{bmatrix} 1 & 1 & 0 \\ 0 & 0 & 1 \\ 2 & -1 & 0 \end{bmatrix}\begin{bmatrix} 2 & 1 & -1 \\ 1 & 0 & 1 \\ 3 & 1 & 1 \end{bmatrix} \\ &= \begin{bmatrix} 6 & 1 & 5 \\ -12 & -1 & -14 \\ -3 & 0 & -4 \end{bmatrix}. \end{aligned}$$

Finally, we use $A_{\mathcal{B}}$ to compute the coordinate vector for the vector $\mathbf{y}$ with coordinate vector $(-1, 4, 0)_{\mathcal{B}}$:

$$\begin{bmatrix} 6 & 1 & 5 \\ -12 & -1 & -14 \\ -3 & 0 & -4 \end{bmatrix}\begin{bmatrix} -1 \\ 4 \\ 0 \end{bmatrix}_{\mathcal{B}} = \begin{bmatrix} -2 \\ 8 \\ 3 \end{bmatrix}_{\mathcal{B}}.$$

As it must, the result agrees with Example 2. ■

The relationship between the matrices $P^{-1}AP$ and A is important enough to warrant the following definition.

DEFINITION

Let A and B be square matrices of order n. We say that B is **similar** to A if there exists an invertible matrix P such that

$$B = P^{-1}AP.$$

Several remarks are in order. For one, P must also be of order n if the multiplication is to be defined. For another, $B = P^{-1}AP$ implies that $(P^{-1})^{-1}BP^{-1} = A$, so that if B is similar to A, then A is also similar to B (using P^{-1}). For a third, the matrix P is not unique. To see this, let $A = B = I$, the identity matrix of order n. Then *any* invertible matrix of order n serves for P, since $I = P^{-1}IP$, no matter what P is.

By Theorem 2, if a matrix A is associated with a linear transformation T: $\mathbf{R}^n \rightarrow \mathbf{R}^n$, if $\mathcal{B}$ is a basis for $\mathbf{R}^n$, and if B is the matrix that represents T with respect to $\mathcal{B}$, then B is similar to A. It is natural to ask if the converse is true. That is, let B be similar to A, and T be the linear transformation $T(\mathbf{x}) = A\mathbf{x}$. Is there a basis $\mathcal{B}$ such that B represents T with respect to $\mathcal{B}$? The answer is yes.

THEOREM 3

Let A and B be square matrices of order n. Suppose that B is similar to A, and let T: $\mathbf{R}^n \rightarrow \mathbf{R}^n$ be the linear transformation given by $T(\mathbf{x}) = \text{A}\mathbf{x}$, for each $\mathbf{x}$ in $\mathbf{R}^n$. Then if P is an invertible matrix such that $B = P^{-1}AP$, the set $\mathcal{B}$ of columns $\mathbf{b}_i = \mathbf{p}^i$, $i = 1, \ldots, n$, is a basis for $\mathbf{R}^n$ such that B represents T with respect to $\mathcal{B}$, $A_{\mathcal{B}} = B$.

Proof First note that $\mathcal{B}$ is a basis, since the columns of P are linearly independent (Theorem 5 of Section 3.3 and Theorem 4 of Section 5.1), and there are n of them.

Since P is the matrix with columns $\mathbf{b}_i = \mathbf{p}^i$, by Theorem 1, $P^{-1}\mathbf{x}$ is the coordinate vector of $\mathbf{x}$ with respect to $\mathcal{B}$.

Finally, by Theorem 2, $B = P^{-1}AP$ represents T with respect to $\mathcal{B}$. ∎

EXAMPLE 4

Let T: $\mathbf{R}^3 \rightarrow \mathbf{R}^3$ be given by multiplication by the following matrix A. Using the given matrices P and B, show that B is similar to A, and find a basis for $\mathbf{R}^3$ with respect to which B represents the linear transformation T.

$$A = \begin{bmatrix} 2 & 1 & 3 \\ 1 & -2 & 1 \\ 0 & 2 & 1 \end{bmatrix}, \quad P = \begin{bmatrix} 1 & 0 & 1 \\ 2 & 1 & 2 \\ 0 & 2 & 1 \end{bmatrix}, \quad B = \begin{bmatrix} -22 & -25 & -30 \\ -11 & -14 & -16 \\ 26 & 32 & 37 \end{bmatrix}$$

Solution First compute P^{-1} to be

$$P^{-1} = \begin{bmatrix} -3 & 2 & -1 \\ -2 & 1 & 0 \\ 4 & -2 & 1 \end{bmatrix}.$$

Then direct multiplication confirms that $B = P^{-1}AP$, so B is similar to A. Finally, Theorem 3 assures us that the columns of P provide a basis for $\mathbf{R}^3$ with the required property,

$$\mathcal{B} = \{(1, 2, 0), (0, 1, 2), (1, 2, 1)\}.$$

Note that we only used P^{-1} to confirm that B is similar to A, $B = P^{-1}AP$. Multiplying both sides on the left by P, we see that $B = P^{-1}AP$ if and only if $PB = AP$. Thus, an easier solution is to observe that P is invertible ($\det(P) = 1 \neq 0$, for example) and simply compute PB and AP. ∎

7.5 Exercises

In Exercises 1–6 verify that the given set $\mathcal{B}$ is a basis for $\mathbf{R}^n$. Compute the change-of-basis matrix for each of the bases, and use it to find the coordinate vector of v with respect to $\mathcal{B}$.

1. $\mathcal{B} = \{(1, 2), (1, -2)\}$, $\mathbf{v} = (-1, 3)$, $(n = 2)$

2. $\mathcal{B} = \{(2, 0), (1, 1)\}$, $\mathbf{v} = (3, -1)$, $(n = 2)$

3. $\mathcal{B} = \{(1, 1, 1), (1, 0, 1), (-1, 1, 0)\}$, $\mathbf{v} = (3, -1, 1)$, $(n = 3)$

4. $\mathcal{B} = \{(1, 2, 1), (-1, 1, 2), (1, 2, 3)\}$, $\mathbf{v} = (2, 1, 0)$, $(n = 3)$

5. $\mathcal{B} = \{(1, 2, 1, -1), (0, 2, 1, 0), (1, 1, 0, 1), (1, 3, 0, 0)\}$, $\mathbf{v} = (-1, 1, -1, 1)$, $(n = 4)$

6. $\mathcal{B} = \{(-1, 1, -1, 1), (2, 1, 0, 1), (1, 0, 0, 2), (0, 0, 3, 2)\}$, $\mathbf{v} = (2, 1, 2, 3)$, $(n = 4)$

In Exercises 7 and 8 let $T\colon \mathbf{R}^2 \to \mathbf{R}^2$ be given by $T(\mathbf{x}) = A\mathbf{x}$, where A is the matrix

$$A = \begin{bmatrix} 1 & 1 \\ 0 & -1 \end{bmatrix}.$$

7. Find the matrix $A_{\mathcal{B}}$ that represents T with respect to the basis $\mathcal{B}$ of Exercise 1.

8. Find the matrix $A_{\mathcal{B}}$ that represents T with respect to the basis $\mathcal{B}$ of Exercise 2.

In Exercises 9 and 10 let $T\colon \mathbf{R}^3 \to \mathbf{R}^3$ be given by $T(\mathbf{x}) = A\mathbf{x}$, where A is the matrix

$$A = \begin{bmatrix} 1 & 1 & 3 \\ 2 & -1 & 1 \\ 1 & 2 & 0 \end{bmatrix}.$$

9. Find the matrix $A_{\mathcal{B}}$ that represents T with respect to the basis $\mathcal{B}$ of Exercise 3.

10. Find the matrix $A_{\mathcal{B}}$ that represents T with respect to the basis $\mathcal{B}$ of Exercise 4.

In Exercises 11 and 12 let $T\colon \mathbf{R}^3 \to \mathbf{R}^3$ be given by $T(\mathbf{x}) = T(x, y, z) = (2x - y, x + y + z, y - x)$.

11. Find the matrix that represents T with respect to the basis $\mathcal{B}$ of Exercise 3.

12. Find the matrix that represents T with respect to the basis $\mathcal{B}$ of Exercise 4.

In Exercises 13 and 14 let $T\colon \mathbf{R}^4 \to \mathbf{R}^4$ be given by $T(\mathbf{x}) = T(w, x, y, z) = (w + y - z, 2w - y, w + 2x + y + z, y - z)$.

13. Find the matrix that represents T with respect to the basis $\mathcal{B}$ of Exercise 5.

14. Find the matrix that represents T with respect to the basis $\mathcal{B}$ of Exercise 6.

15. If A is similar to B, prove that $\det A = \det B$.

16. Let A and B be square matrices of the same order with $\det A = \det B$. Show that it is not necessarily true that A is similar to B.

17. If A is similar to B, prove that A^2 is similar to B^2.

18. If A is similar to B, prove that $\operatorname{rank} A = \operatorname{rank} B$. (*Hint*: Use Theorem 3.)

7.6 APPLICATION—Linear Differential Equations

In this section, we use linear transformations to discuss the topic of linear differential equations. We begin by giving some definitions basic to the study of differential equations. A **differential equation** is an equation involving an "unknown" function and at least one of its derivatives. The **order** of a differential equation is that of its highest-order derivative. For example,

$$\mathbf{y}'' + 3\mathbf{y}' - 5\mathbf{y} = x^3$$

is a differential equation of second order. This particular equation symbolizes the problem of determining a function $\mathbf{y} = \mathbf{f}(x)$ such that $\mathbf{f}''(x) + 3\mathbf{f}'(x) - 5\mathbf{f}(x) = x^3$. Such a function is called a **solution** of the differential equation.

We consider only *ordinary linear* differential equations. An **ordinary differential equation** is one in which the unknown is a function of a single variable. (If it is a function of several variables, we have a *partial* differential equation.) An (ordinary) **linear differential equation** of order n is one of the form

$$a_n\mathbf{y}^{(n)} + a_{n-1}\mathbf{y}^{(n-1)} + \cdots + a_1\mathbf{y}' + a_0\mathbf{y} = \mathbf{b}, \qquad (1)$$

where, in general, the a_i and $\mathbf{b}$ are functions of the independent variable. We consider only the case for which the a_i are constants—the *constant coefficient* case—and $a_n \neq 0$.

EXAMPLE 1

The differential equation

$$\mathbf{y}''' - 4\mathbf{y}' = \mathbf{0}$$

is a linear differential equation of order 3. Comparing it with the general form of Equation (1), we have $n = 3$, $a_3 = 1$, $a_2 = 0$, $a_1 = -4$, $a_0 = 0$, and $\mathbf{b} = \mathbf{0}$. One solution of this differential equation is $\mathbf{y} = e^{2x}$. This is easily verified by differentiating to get $\mathbf{y}' = 2e^{2x}$ and $\mathbf{y}''' = 8e^{2x}$ so that

$$\mathbf{y}''' - 4\mathbf{y}' = 8e^{2x} - 4(2e^{2x}) = \mathbf{0},$$

as desired. (We will soon see how to determine *all* solutions of this differential equation.) ■

Leonard Euler (1707–1783) was the first to notice that the solutions to homogeneous linear differential equations could be written as linear combinations of linearly independent basic solutions. Euler was one of the most prolific mathematicians of all time, carrying out fundamental research in many areas of mathematics and physics, including calculus, differential equations, number theory, and graph theory. What is even more remarkable is that even though he became totally blind in 1766, his mathematical productivity didn't go down!

We can further divide the study of linear differential equations into two cases: *homogeneous* and *nonhomogeneous* equations. A **homogeneous** (ordinary) linear differential equation has the form of Equation (1) with $\mathbf{b} = \mathbf{0}$; it is **nonhomogeneous** if the function $\mathbf{b}$ (which we assume to be infinitely differentiable) takes on nonzero values. For example, the equation $\mathbf{y}'' - \mathbf{y} = \mathbf{0}$ is a linear homogeneous one, where $\mathbf{y}'' - \mathbf{y}' = \sin x$ is nonhomogeneous.

Homogeneous Linear Differential Equations

We first consider the homogeneous version of Equation (1),

$$a_n\mathbf{y}^{(n)} + a_{n-1}\mathbf{y}^{(n-1)} + \cdots + a_1\mathbf{y}' + a_0\mathbf{y} = \mathbf{0}, \qquad (2)$$

with the a_i constant and $a_n \neq 0$. To help analyze the solutions of this differential equation, we restate the problem in terms of linear operators (linear transformations from a vector space into itself).

Let $\mathbf{C}^\infty$ $(=\mathbf{C}^\infty(-\infty, \infty))$ denote the vector space of all functions $\mathbf{y}(x)$ that have continuous derivatives of all orders for all values of x (see the Note following Example 6 of Section 6.1). Also, let D: $\mathbf{C}^\infty \to \mathbf{C}^\infty$ be the linear operator that maps each function into its derivative (see the Remark after Theorem 3 of Section 7.2). Notice that since $D(\mathbf{y}) = \mathbf{y}'$, we have $D^2\mathbf{y} = D(D(\mathbf{y})) = D\mathbf{y}' = \mathbf{y}''$ and, in general, $D^k(\mathbf{y})$ gives the kth derivative of $\mathbf{y}$. Finally, we define the linear operator L: $\mathbf{C}^\infty \to \mathbf{C}^\infty$ to be

$$L = a_nD^n + a_{n-1}D^{n-1} + \cdots + a_1D + a_0I,$$

so that Equation (2) may be written as simply $L(\mathbf{y}) = \mathbf{0}$ (and Equation (1) as $L(\mathbf{y}) = \mathbf{b}$). The verification that L is indeed a linear operator on $\mathbf{C}^\infty$ is left for an exercise.

Using this operator notation, we see that the problem of solving the homogeneous Equation (2) may be restated as that of determining the *kernel* of the operator L—those functions $\mathbf{y}$ for which $L(\mathbf{y}) = \mathbf{0}$ (Section 7.3). Before discussing how to find the kernel of L—the solutions of Equation (2)—we present several theorems that give *properties* of these solutions.

THEOREM 1

Let $\mathbf{y}_1$ and $\mathbf{y}_2$ be solutions of $L(\mathbf{y}) = \mathbf{0}$, and let c be a constant. Then

i. $\mathbf{y}_1 + \mathbf{y}_2$ is a solution of $L(\mathbf{y}) = \mathbf{0}$;

ii. $c\mathbf{y}_1$ and $c\mathbf{y}_2$ are solutions of $L(\mathbf{y}) = \mathbf{0}$.

Proof Since $\mathbf{y}_1$ and $\mathbf{y}_2$ are solutions of $L(\mathbf{y}) = \mathbf{0}$, they are both in the kernel of L. But, by Theorem 1 of Section 7.3, the kernel of L is a subspace of $\mathbf{C}^\infty$, and hence $\mathbf{y}_1 + \mathbf{y}_2$ and $c\mathbf{y}_i$ are also in the kernel. Thus, they are also solutions of $L(\mathbf{y}) = \mathbf{0}$. ■

We prove the following two theorems at the end of this section.

THEOREM 2

Let $\mathbf{y}_1, \mathbf{y}_2, \ldots, \mathbf{y}_n$ be solutions of $L(\mathbf{y}) = \mathbf{0}$. Then $\mathbf{y}_1, \mathbf{y}_2, \ldots, \mathbf{y}_n$ form a linearly dependent set if and only if

$$\det\begin{bmatrix} \mathbf{y}_1 & \mathbf{y}_2 & \cdots & \mathbf{y}_n \\ \mathbf{y}_1' & \mathbf{y}_2' & \cdots & \mathbf{y}_n' \\ \vdots & \vdots & & \vdots \\ \mathbf{y}_1^{(n-1)} & \mathbf{y}_2^{(n-1)} & \cdots & \mathbf{y}_n^{(n-1)} \end{bmatrix} = \mathbf{0};$$

that is, the determinant is identically zero (zero for all values of the variable). ■

NOTE The determinant of Theorem 2 is called the **Wronskian** of $\mathbf{y}_1, \mathbf{y}_2, \ldots, \mathbf{y}_n$ and is denoted by $W(\mathbf{y}_1, \mathbf{y}_2, \ldots, \mathbf{y}_n)$. We can also show that, if $\mathbf{y}_1, \mathbf{y}_2, \ldots, \mathbf{y}_n$ are solutions of $L(\mathbf{y}) = \mathbf{0}$, then $W(\mathbf{y}_1, \mathbf{y}_2, \ldots, \mathbf{y}_n)$ is either identically zero or else it is never zero.

EXAMPLE 2

Show that the solutions $\mathbf{y}_1 = e^{2x}$, $\mathbf{y}_2 = e^{-2x}$, and $\mathbf{y}_3 = 3$ of the differential equation $\mathbf{y}''' - 4\mathbf{y}' = \mathbf{0}$ are linearly independent.

Solution We compute the Wronskian

$$W(\mathbf{y}_1, \mathbf{y}_2, \mathbf{y}_3) = \det\begin{bmatrix} e^{2x} & e^{-2x} & 3 \\ 2e^{2x} & -2e^{-2x} & 0 \\ 4e^{2x} & 4e^{-2x} & 0 \end{bmatrix} = 48.$$

Thus, by Theorem 2, $\{\mathbf{y}_1, \mathbf{y}_2, \mathbf{y}_3\}$ is a linearly independent set. ■

THEOREM 3

The dimension of the kernel of the linear operator L is n, the order of the differential equation $L(\mathbf{y}) = \mathbf{0}$ (or, in other words, there exist exactly n linearly independent solutions of this differential equation). ■

EXAMPLE 3

Determine the general form for the solution of $\mathbf{u}''' - 4\mathbf{u}' = \mathbf{0}$.

Solution This is the differential equation of Example 1, although here the unknown function is called "**u**." Thus, we know that e^{2x}, e^{-2x}, and 3 are three linearly independent solutions of the given equation. By Theorem 3, this set of functions provides a basis for the kernel of $L = D^3 - 4D$, the subspace of all solutions of the given differential equation. Hence, these three functions span this subspace, and every solution can be written as a linear combination of them. Thus,

$$\mathbf{u}(x) = c_1e^{2x} + c_2e^{-2x} + c_3(3).$$ ■

NOTE The general form of the solution of $L(\mathbf{y}) = \mathbf{0}$ (or $L(\mathbf{y}) = \mathbf{b}$) is called the **general solution** of this equation. Thus, the general solution of $\mathbf{u}''' - 4\mathbf{u}' = \mathbf{0}$ is (from Example 3) $\mathbf{u}(x) = c_1e^{2x} + c_2e^{-2x} + c_3(3)$.

Solving Linear Homogeneous Differential Equations

The linear operator

$$L = a_nD^n + a_{n-1}D^{n-1} + \cdots + a_1D + a_0I$$

can be viewed as a polynomial of degree n in the "variable" D. This polynomial,

$$P(\lambda) = a_n\lambda^n + a_{n-1}\lambda^{n-1} + \cdots + a_1\lambda + a_0, \quad (3)$$

is called the **characteristic polynomial** for L (or for the associated differential equation (1)). The algebraic equation $P(\lambda) = 0$ is called the **characteristic equation** of (1). As we will see, the roots of the characteristic equation play a key role in the process of solving linear differential equations.

EXAMPLE 4

Find the roots of the characteristic equation for the differential equation $\mathbf{y}''' - 4\mathbf{y}' = \mathbf{0}$.

Solution The characteristic equation is easily obtained by replacing the derivatives of **y** with the corresponding powers of λ. Doing this, we get $\lambda^3 - 4\lambda = 0$. The left side of this equation factors as $\lambda^3 - 4\lambda = \lambda(\lambda - 2)(\lambda + 2)$, and so the roots of the characteristic equations are 0, 2, and -2. ■

To simplify matters, we first consider the special case of the second-order equation

$$\mathbf{y}'' + p\mathbf{y}' + q\mathbf{y} = \mathbf{0}. \quad (4)$$

Letting $L: \mathbf{C}^\infty \to \mathbf{C}^\infty$ be the linear operator

$$L = D^2 + pD + qI,$$

Equation (4) takes the form $L(\mathbf{y}) = \mathbf{0}$. Here the characteristic polynomial is a quadratic one,

$$P(\lambda) = \lambda^2 + p\lambda + q. \quad (5)$$

The roots of a quadratic Equation can always be found by the quadratic formula, which, using the notation in Equation (5), is

$$\lambda = \frac{-p \pm \sqrt{p^2 - 4q}}{2}.$$

If we denote the two roots of the characteristic polynomial (5) by α and β, we can restrict ourselves to three cases:

i. α and β are real numbers with $\alpha \neq \beta$;

ii. α and β are real numbers with $\alpha = \beta$;

iii. α and β are complex (but nonreal) numbers and are complex conjugates of each other—if $\alpha = \alpha_1 + \alpha_2 i$, then $\beta = \alpha_1 - \alpha_2 i$.

By Theorem 3, we know that the differential equation (4) has exactly two linearly independent solutions. The next theorem tells us how to find these solutions once we know the roots of the characteristic polynomial (5).

THEOREM 4

Let the two roots of the characteristic equation for $\mathbf{y}'' + p\mathbf{y}' + q\mathbf{y} = \mathbf{0}$ (p, q are constant) be denoted by α and β.

i. If α and β are real numbers with $\alpha \neq \beta$, then

$$e^{\alpha x} \quad \text{and} \quad e^{\beta x}$$

are linearly independent solutions.

ii. If α and β are real numbers with $\alpha = \beta$, then

$$e^{\alpha x} \quad \text{and} \quad xe^{\alpha x}$$

are linearly independent solutions.

iii. If α and β are complex conjugates with $\alpha = \alpha_1 + \alpha_2 i$, then

$$e^{\alpha_1 x}\sin \alpha_2 x \quad \text{and} \quad e^{\alpha_1 x}\cos \alpha_2 x$$

are linearly independent solutions.

Proof We prove only part (ii) and leave the other proofs for exercises. First, we must show that $e^{\alpha x}$ and $xe^{\alpha x}$ are solutions of the differential equation

$$\mathbf{y}'' + p\mathbf{y}' + q\mathbf{y} = \mathbf{0}.$$

Since the characteristic polynomial is $P(\lambda) = \lambda^2 + p\lambda + q = (\lambda - \alpha)(\lambda - \alpha) = \lambda^2 - 2\alpha\lambda + \alpha^2$, we may rewrite the differential equation in terms of α and β as

$$(D^2 - 2\alpha D + \alpha^2 I)(\mathbf{y}) = \mathbf{0} \qquad \text{or} \qquad (D - \alpha I)(D - \alpha I)(\mathbf{y}) = \mathbf{0}.$$

Since

$$\begin{aligned}(D - \alpha I)(D - \alpha I)(e^{\alpha x}) &= (D - \alpha I)(De^{\alpha x} - \alpha e^{\alpha x})\\ &= (D - \alpha I)(\alpha e^{\alpha x} - \alpha e^{\alpha x})\\ &= (D - \alpha I)(\mathbf{0}) = \mathbf{0},\end{aligned}$$

$e^{\alpha x}$ is a solution. Similarly,

$$\begin{aligned}(D - \alpha I)(D - \alpha I)(xe^{\alpha x}) &= (D - \alpha I)(D(xe^{\alpha x}) - \alpha xe^{\alpha x})\\ &= (D - \alpha I)(\alpha xe^{\alpha x} + e^{\alpha x} - \alpha xe^{\alpha x})\\ &= (D - \alpha I)(e^{\alpha x}) = \mathbf{0},\end{aligned}$$

and $xe^{\alpha x}$ is a solution as well.

To show that $e^{\alpha x}$ and $xe^{\alpha x}$ are linearly independent, we need only show that their Wronskian is nonzero (by Theorem 2). Now

$$\begin{aligned}W(e^{\alpha x}, xe^{\alpha x}) &= \det\begin{bmatrix} e^{\alpha x} & xe^{\alpha x} \\ \alpha e^{\alpha x} & \alpha xe^{\alpha x} + e^{\alpha x}\end{bmatrix}\\ &= e^{\alpha x}(\alpha xe^{\alpha x} + e^{\alpha x}) - \alpha xe^{\alpha x}e^{\alpha x}\\ &= e^{\alpha x}e^{\alpha x} = e^{2\alpha x},\end{aligned}$$

which is nonzero regardless of the values of α and x. ∎

EXAMPLE 5

Find the general solution of $\mathbf{y}'' - 4\mathbf{y}' + 4\mathbf{y} = \mathbf{0}$.

Solution The characteristic equation is $\lambda^2 - 4\lambda + 4 = 0$, which has the double root $\lambda = 2$ (case (ii) of Theorem 4). Hence, two linear independent solutions are e^{2x} and xe^{2x}, and, by Theorem 3, the general solution of the given differential equation is $\mathbf{y} = c_1e^{2x} + c_2xe^{2x}$. ■

EXAMPLE 6

Find the general solution of $\left(\dfrac{d^2\mathbf{s}}{dt^2}\right) + \mathbf{s} = \mathbf{0}$.

Solution Here the characteristic equation is $\lambda^2 + 1 = 0$, which has roots $\lambda_1 = i$ $(= 0 + 1i)$ and $\lambda_2 = -i\,(= 0 - 1i)$. Thus, by part (iii) of Theorem 4, two linearly independent solutions of the given differential equation are $e^{0t}\sin t = \sin t$ and $e^{0t}\cos t = \cos t$. By Theorem 3, its general solution is $\mathbf{s} = c_1 \sin t + c_2 \cos t$. ■

Higher-Order Homogeneous Equations

For the more general case of Equation (2),

$$a_n\mathbf{y}^{(n)} + a_{n-1}\mathbf{y}^{(n-1)} + \cdots + a_1\mathbf{y}' + a_0\mathbf{y} = \mathbf{0},$$

Theorem 3 tells us that there are n linearly independent solutions. To find them, we turn again to the characteristic equation. The problem here is that the latter now has degree n, and it is difficult (sometimes impossible) to find the roots of a general nth-degree polynomial equation. However, *if* we can determine all the roots of this equation, then the following procedure, which generalizes Theorem 4, provides the needed n linearly independent solutions.

Procedure for Solving a Linear Homogeneous Differential Equation
Use this procedure to solve the differential equation

$$a_n\mathbf{y}^{(n)} + a_{n-1}\mathbf{y}^{(n-1)} + \cdots + a_1\mathbf{y}' + a_0\mathbf{y} = \mathbf{0},$$

where the a_i are constant and $a_n \neq 0$:

i. Find the roots $\alpha_1, \alpha_2, \ldots, \alpha_n$ of the characteristic equation

$$a_n\lambda^n + a_{n-1}\lambda^{n-1} + \cdots + a_1\lambda + a_0 = 0.$$

ii. Form the following collection of solutions:

- **a.** If α is a real root, $e^{\alpha x}$ is a solution.
- **b.** If $\alpha = \alpha_1 + \alpha_2 i$ is a nonreal root, then $\overline{\alpha} = \alpha_1 - \alpha_2 i$ is also a root and $e^{\alpha_1 x}\sin \alpha_2 x$ and $e^{\alpha_1 x}\cos \alpha_2 x$ are solutions.
- **c.** If α is any root of multiplicity k, and $\overline{\mathbf{y}}$ is a solution constructed from α as described above, then so are $x\overline{\mathbf{y}}, x^2\overline{\mathbf{y}}, \ldots, x^{k-1}\overline{\mathbf{y}}$.

iii. The collection of distinct solutions $\mathbf{y}_1, \mathbf{y}_2, \ldots, \mathbf{y}_n$ described in (ii) forms a linearly independent set of n solutions, and the general solution is given by

$$\mathbf{y} = c_1\mathbf{y}_1 + c_2\mathbf{y}_2 + \cdots + c_n\mathbf{y}_n.$$

EXAMPLE 7

Find the general solution of $\mathbf{y}^{(\mathrm{v})} - 2\mathbf{y}^{(\mathrm{iv})} + 2\mathbf{y}''' = \mathbf{0}$.

Solution Following the preceding procedure, we first must solve the characteristic equation

$$\lambda^5 - 2\lambda^4 + 2\lambda^3 = 0$$

$$\text{or}\quad \lambda^3(\lambda^2 - 2\lambda + 2) = 0.$$

The first factor gives the triple root (a root of multiplicity 3) $\lambda = 0$, whereas the roots corresponding to the second factor can be determined by the quadratic formula to be $\lambda = 1 + i$ and $\lambda = 1 - i$.

Now the solution of the differential equation corresponding to the real root $\lambda = 0$ is $\mathbf{y} = e^{0x} = 1$, and since 0 is a triple root, $\mathbf{y} = x$ and $\mathbf{y} = x^2$ are also solutions. Corresponding to the complex root $1 + i$ (and $1 - i$) are the solutions $\mathbf{y} = e^x \sin x$ and $\mathbf{y} = e^x \cos x$. We have thus found the five linearly independent solutions of the given fifth-order differential equation; its general solution is

$$\mathbf{y} = c_1 + c_2x + c_3x^2 + c_4e^x \cos x + c_5e^x \sin x. \qquad \blacksquare$$

The Nonhomogeneous Equation

Although we do not go into the details of finding *particular* solutions of the nonhomogeneous linear differential equation (1), we give a theorem showing how its general solution can be found.

THEOREM 5

Let $\mathbf{y}_p$ be a given solution of the differential equation (1), $L(\mathbf{y}) = \mathbf{b}$. Then the general solution of $L(\mathbf{y}) = \mathbf{b}$ is

$$\mathbf{y} = \mathbf{y}_p + \mathbf{y}_c,$$

where $\mathbf{y}_c$ is the general solution of the *associated homogeneous equation* $L(\mathbf{y}) = \mathbf{0}$.

NOTE The functions $\mathbf{y}_p$ and $\mathbf{y}_c$ of Theorem 5 are often called a **particular solution** and the **complementary function** for $L(\mathbf{y}) = \mathbf{b}$, respectively. The function $\mathbf{y}_p + \mathbf{y}_c$ is said to be the **general solution** of this equation.

Proof First we show that $\mathbf{y} = \mathbf{y}_p + \mathbf{y}_c$ is a solution of $L(\mathbf{y}) = \mathbf{b}$. We have

$$L(\mathbf{y}) = L(\mathbf{y}_p + \mathbf{y}_c) = L(\mathbf{y}_p) + L(\mathbf{y}_c),$$

since L is linear. Now $L(\mathbf{y}_p) = \mathbf{b}$ and $L(\mathbf{y}_c) = \mathbf{0}$, so $L(\mathbf{y}) = \mathbf{b} + \mathbf{0} = \mathbf{b}$, as desired.

Now we show that, if $\mathbf{v}$ is an arbitrary solution of $L(\mathbf{y}) = \mathbf{b}$, then it can be written as $\mathbf{y}_p + \mathbf{y}_c$ for any particular solution $\mathbf{y}_p$ of this equation. Let $\mathbf{y}_k = \mathbf{v} - \mathbf{y}_p$. Then $L(\mathbf{y}_k) = L(\mathbf{v} - \mathbf{y}_p) = L(\mathbf{v}) - L(\mathbf{y}_p) = \mathbf{b} - \mathbf{b} = \mathbf{0}$. Therefore $\mathbf{y}_k$ is in the kernel of L, so it is a specific member of the general solution $\mathbf{y}_c$. Transposing, we have $\mathbf{v} = \mathbf{y}_p + \mathbf{y}_k$, and $\mathbf{v}$ is expressed in the required form. $\blacksquare$

As we said, we will not delve into the various techniques for finding a particular solution of $L(\mathbf{y}) = \mathbf{b}$. If we do know one particular solution, however, we can then use Theorem 5 to determine its general solution.

EXAMPLE 8

Given that $\mathbf{y}_p = e^x$ is a particular solution of $\mathbf{y}'' + \mathbf{y} = 2e^x$, find the general solution of this differential equation.

Solution Applying Theorem 5, we see that we must find the general solution $\mathbf{y}_c$ of the associated homogeneous equation $\mathbf{y}'' + \mathbf{y} = \mathbf{0}$. Since the characteristic equation of the latter is $\lambda^2 + 1 = 0$, which has roots $\lambda = i$ and $\lambda = -i$, we have $\mathbf{y}_c = c_1 \cos x + c_2 \sin x$. Therefore, the general solution of $\mathbf{y}'' + \mathbf{y} = 2e^x$ is $\mathbf{y} = \mathbf{y}_p + \mathbf{y}_c = e^x + c_1 \cos x + c_2 \sin x$. $\blacksquare$

Initial Value Problems

The general solution of an nth-order linear differential equation contains n arbitrary constants (which we have been calling $c_1, c_2, \ldots, c_n$). In applications, additional information about the solution is usually known, and this information can be used to determine the arbitrary constants. The case in which we know the value of the unknown function as well as its first $n - 1$ derivatives at a single point leads to an *initial value problem.*

DEFINITION

Given a differential equation $L(\mathbf{y}) = \mathbf{b}$, a point x_0 and numbers $y_0, y_1, \ldots, y_{n-1}$, the problem of determining a function $\mathbf{y}(x)$ that satisfies

$$L(\mathbf{y}) = \mathbf{b}$$

$$\text{and} \quad \mathbf{y}(x_0) = y_0, \mathbf{y}'(x_0) = y_1, \ldots, \mathbf{y}^{(n-1)}(x_0) = y_{n-1} \qquad \textbf{(6)}$$

is called an **initial value problem**. The equations in (6) are called the **initial conditions** for the problem.

EXAMPLE 9

Solve the initial value problem

$$\mathbf{y}''' - \mathbf{y}' = \mathbf{0},$$
$$\mathbf{y}(0) = 1, \quad \mathbf{y}'(0) = 3, \quad \text{and} \quad \mathbf{y}''(0) = -1.$$

Solution We first solve the differential equation, ignoring the initial conditions for now. Since its characteristic equation is $\lambda^3 - \lambda = 0$ with roots $\lambda = 0$, $\lambda = 1$, and $\lambda = -1$, its general solution is given by

$$\mathbf{y} = c_1 + c_2 e^x + c_3 e^{-x}.$$

Differentiating this solution, we have

$$\begin{aligned} \mathbf{y}' &= c_2 e^x - c_3 e^{-x} \\ \text{and} \quad \mathbf{y}'' &= c_2 e^x + c_3 e^{-x}. \end{aligned}$$

Thus, the initial conditions imply the following:

$$\begin{aligned} \mathbf{y}(0) &= 1 \rightarrow c_1 + c_2 + c_3 = 1, \\ \mathbf{y}'(0) &= 3 \rightarrow \quad c_2 - c_3 = 3, \\ \text{and} \quad \mathbf{y}''(0) &= -1 \rightarrow \quad c_2 + c_3 = -1 \end{aligned}$$

The solution of this system of linear (algebraic) equations is $c_1 = 2$, $c_2 = 1$, $c_3 = -2$. Hence, substituting these values into the general solution, we arrive at the solution of the initial value problem,

$$\mathbf{y} = 2 + e^x - 2e^{-x}. \qquad \blacksquare$$

EXAMPLE 10

A weight attached to a spring suspended from the ceiling moves up and down in such a way that its displacement **s** (in centimetres) from its equilibrium position (position while "at rest") is given by $(d^2\mathbf{s}/dt^2) + 4\mathbf{s} = \mathbf{0}$, where t represents time in seconds (Figure 11). If the initial position of the spring is $\mathbf{s} = \mathbf{0}$ (its equilibrium position) and its initial velocity, $d\mathbf{s}/dt$, is 2 cm/s in the upward direction, find the position $\mathbf{s}(t)$ of the spring at any time t.

FIGURE 11

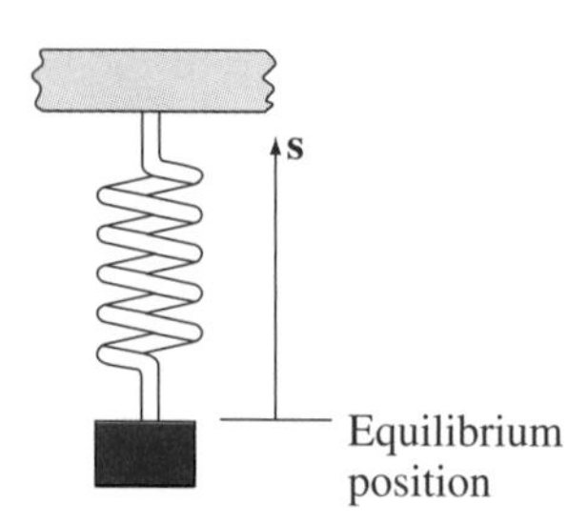

Solution Phrasing this as an initial value problem, we have the differential equation

$$\frac{d^2\mathbf{s}}{dt^2} + 4\mathbf{s} = \mathbf{0},$$

with initial conditions

$$\mathbf{s}(0) = 0 \quad \text{and} \quad \mathbf{s}'(0) = 2.$$

Solving the differential equation, we obtain the general solution

$$\mathbf{s}(t) = c_1 \cos 2t + c_2 \sin 2t,$$

so

$$\mathbf{s}'(t) = -2c_1 \sin 2t + 2c_2 \cos 2t.$$

Thus, the initial conditions yield

$$\begin{aligned} \mathbf{s}(0) = 0 &\rightarrow \quad c_1(1) + \quad c_2(0) = 0 \\ \text{and} \quad \mathbf{s}'(0) = 2 &\rightarrow -2c_1(0) + 2c_2(1) = 2. \end{aligned}$$

So $c_1 = 0$ and $c_2 = 1$, resulting in the solution of the initial value problem,

$$\mathbf{s}(t) = \sin 2t.$$

■

We now state (without proof) a basic theorem in the study of initial value problems. This theorem is used in the following proofs for Theorems 2 and 3.

THEOREM 6

The initial value problem

$$L(\mathbf{y}) = \mathbf{b} \qquad \text{(Equation (1))}$$

$$\mathbf{y}(x_0) = y_0, \ \ \mathbf{y}'(x_0) = y_1, \ldots, \mathbf{y}^{(n-1)}(x_0) = y_{n-1}$$

has a solution, and this solution is unique.

■

Proof of Theorem 2 We first assume that $\mathbf{y}_1, \mathbf{y}_2, \ldots, \mathbf{y}_n$ are linearly dependent and show that $W(\mathbf{y}_1, \mathbf{y}_2, \ldots, \mathbf{y}_n) = \mathbf{0}$. By definition, there exist n scalars, not all zero, such that

$$c_1\mathbf{y}_1 + c_2\mathbf{y}_2 + \cdots + c_n\mathbf{y}_n = \mathbf{0}.$$

This equation, together with the equations obtained by differentiating both sides of it $n - 1$ times, results in

$$\begin{array}{llll} c_1\mathbf{y}_1(x) & + c_2\mathbf{y}_2(x) & + \cdots + c_n\mathbf{y}_n(x) & = 0 \\ c_1\mathbf{y}_1'(x) & + c_2\mathbf{y}_2'(x) & + \cdots + c_n\mathbf{y}_n'(x) & = 0 \\ \vdots & \vdots & \vdots & \\ c_1\mathbf{y}_1^{(n-1)}(x) & + c_2\mathbf{y}_2^{(n-1)}(x) & + \cdots + c_n\mathbf{y}_n^{(n-1)}(x) & = 0, \end{array} \tag{7}$$

for each x. Thus, the vector $(c_1, c_2, \ldots, c_n)$ is a solution of a system of homogeneous linear equations. Since not all c_i are zero, this linear system has a nonzero solution, so its determinant, $W(\mathbf{y}_1(x), \mathbf{y}_2(x), \ldots, \mathbf{y}_n(x))$ is zero (by Theorem 6 of Section 4.2). Since this is true for each x, $W(\mathbf{y}_1,\mathbf{y}_2, \ldots, \mathbf{y}_n) = \mathbf{0}$, as desired.

We now assume that the Wronskian is zero for some fixed real number a and demonstrate the linear dependence of the $\mathbf{y}_i$. Since the Wronskian is zero, there exist scalars $c_1, c_2, \ldots, c_n$, not all of which are zero, such that the linear system (7) is satisfied at $x = a$. Now let $\mathbf{y}$ be the function

$$\mathbf{y} = c_1\mathbf{y}_1 + c_2\mathbf{y}_2 + \cdots + c_n\mathbf{y}_n.$$

By Theorem 1, $\mathbf{y}$ is a solution of $L(\mathbf{y}) = \mathbf{0}$, since all of the $\mathbf{y}_i$ are solutions. Moreover, $\mathbf{y}(x)$ satisfies the initial conditions

$$\mathbf{y}(a) = \mathbf{y}'(a) = \cdots = \mathbf{y}^{(n-1)}(a) = 0.$$

But this implies (Exercise 32) that $\mathbf{y} = \mathbf{0}$. Hence, we have found scalars, $c_1, c_2, \ldots, c_n$, not all zero, such that

$$c_1\mathbf{y}_1 + c_2\mathbf{y}_2 + \cdots + c_n\mathbf{y}_n = \mathbf{0},$$

and therefore the $\mathbf{y}_i$ are linearly dependent. ■

Proof of Theorem 3 We explicitly give a basis for the kernel of L, Ker (L), consisting of n functions, $\mathbf{y}_1, \mathbf{y}_2, \ldots, \mathbf{y}_n$. Let a be any fixed real number and $\mathbf{y}_i$ be the solutions of the n initial value problems

$$\begin{aligned} L(\mathbf{y}) &= \mathbf{0}, \\ \mathbf{y}_i(a) = \mathbf{y}_i'(a) &= \cdots = \mathbf{y}_i^{(i-2)}(a) = \mathbf{y}_i^{(i)}(a) = \cdots = \mathbf{y}_i^{(n-1)}(a) = 0, \\ \mathbf{y}_i^{(i-1)}(a) &= 1, \end{aligned}$$

for $i = 1, 2, \ldots, n$. The fact that these $\mathbf{y}_i$ exist (and are unique) is a consequence of Theorem 6.

We now show that the set $\mathcal{S} = \{\mathbf{y}_1, \mathbf{y}_2, \ldots, \mathbf{y}_n\}$ is linearly independent and that it spans Ker (L). For these functions $\mathbf{y}_i$, the Wronskian evaluated at $x = a$ is equal to det $(I) = 1 \neq 0$. Hence, by Theorem 2, $\mathcal{S}$ is a linearly independent set.

To show that $\mathcal{S}$ spans Ker (L), let $\mathbf{v}$ be any function in this space, and define the function $\mathbf{u}$ by

$$\mathbf{u} = \mathbf{v} - \mathbf{v}(a)\mathbf{y}_1 - \mathbf{v}'(a)\mathbf{y}_2 - \cdots - \mathbf{v}^{(n-1)}(a)\mathbf{y}_n.$$

Since $\mathbf{v}$ and each $\mathbf{y}_i$ are in Ker (L), $\mathbf{u}$ is in Ker (L). Differentiating both sides, we obtain

$$\mathbf{u}^{(k)} = \mathbf{v}^{(k)} - \mathbf{v}(a)\mathbf{y}_1^{(k)} - \mathbf{v}'(a)\mathbf{y}_2^{(k)} - \cdots - \mathbf{v}^{(n-1)}(a)\mathbf{y}_n^{(k)},$$

for $k = 1, 2, \ldots, n$. But direct substitution shows that

$$\mathbf{u}(a) = \mathbf{u}'(a) = \cdots = \mathbf{u}^{(n-1)}(a) = 0,$$

and since $L(\mathbf{u}) = \mathbf{0}$, we have (by Exercise 32) $\mathbf{u} = \mathbf{0}$. Thus, $\mathbf{v}$ can be written as a linear combination of the $\mathbf{y}_i$, and $\mathcal{S}$ is a spanning set. ■

7.6 Exercises

In Exercises 1–16 find the general solution of the given differential equation. (Assume that *x* is the independent variable unless otherwise specified.)

1. $\mathbf{y}'' - 3\mathbf{y}' + 2\mathbf{y} = \mathbf{0}$

2. $\mathbf{y}'' + 2\mathbf{y}' - 3\mathbf{y} = \mathbf{0}$

3. $\mathbf{y}'' + 2\mathbf{y}' - \mathbf{y} = \mathbf{0}$

4. $\mathbf{y}'' - 4\mathbf{y}' + \mathbf{y} = \mathbf{0}$

5. $\mathbf{u}'' + 4\mathbf{u}' + 4\mathbf{u} = \mathbf{0}$

6. $4\mathbf{u}'' - 4\mathbf{u}' + \mathbf{u} = \mathbf{0}$

7. $4\dfrac{d^2\mathbf{s}}{dt^2} + \mathbf{s} = \mathbf{0}$

8. $\dfrac{d^2\mathbf{s}}{dt^2} + 9\mathbf{s} = \mathbf{0}$

9. $\mathbf{y}'' + 2\mathbf{y}' + 5\mathbf{y} = \mathbf{0}$

10. $\mathbf{y}'' - 2\mathbf{y}' + 4\mathbf{y} = \mathbf{0}$

11. $\mathbf{u}''' + 2\mathbf{u}'' + \mathbf{u}' = \mathbf{0}$

12. $\mathbf{u}''' - 4\mathbf{u}'' + 4\mathbf{u}' = \mathbf{0}$

13. $\dfrac{d^4\mathbf{s}}{dt^4} - \mathbf{s} = \mathbf{0}$

14. $\dfrac{d^4\mathbf{s}}{dt^4} + \dfrac{d^2\mathbf{s}}{dt^2} = \mathbf{0}$

15. $\dfrac{d^4\mathbf{s}}{dt^4} - \dfrac{2d^3\mathbf{s}}{dt^3} + \dfrac{4d^2\mathbf{s}}{dt^2} = \mathbf{0}$

16. $\dfrac{d^4\mathbf{s}}{dt^4} + \dfrac{2d^2\mathbf{s}}{dt^2} + \mathbf{s} = \mathbf{0}$

In Exercises 17–20 a particular solution of the differential equation is given. Find its general solution.

17. $\mathbf{y}'' - 2\mathbf{y}' - 3\mathbf{y} = 10 \sin x, \quad \mathbf{y}_p = -2 \sin x + \cos x$

18. $\mathbf{y}'' - \mathbf{y} = x^2 + 1, \quad \mathbf{y}_p = -x^2 - 3$

19. $\mathbf{u}'' + \mathbf{u} = 2e^x + 2e^{-x}, \quad \mathbf{u}_p = e^x + e^{-x}$

20. $\mathbf{u}'' + \mathbf{u}' = 2 \cos x, \quad \mathbf{u}_p = \sin x - \cos x$

In Exercises 21–26 solve the given initial value problems.

21. $\mathbf{y}'' + 2\mathbf{y}' - 3\mathbf{y} = \mathbf{0}, \quad \mathbf{y}(0) = 1, \quad \mathbf{y}'(0) = 2$

22. $\mathbf{y}'' - \mathbf{y}' = \mathbf{0}, \quad \mathbf{y}(1) = 0, \quad \mathbf{y}'(1) = 1$

23. $\dfrac{d^2\mathbf{s}}{dt^2} + \mathbf{s} = \mathbf{0}, \quad \mathbf{s}\left(\dfrac{\pi}{2}\right) = 1, \quad \dfrac{d\mathbf{s}}{dt}\left(\dfrac{\pi}{2}\right) = 0$

24. $\dfrac{d^2\mathbf{s}}{dt^2} = \mathbf{0}, \quad \mathbf{s}(2) = 0, \quad \dfrac{d\mathbf{s}}{dt}(2) = 1$

25. $\mathbf{u}^{(\text{iv})} - \mathbf{u}'' = \mathbf{0}, \quad \mathbf{u}(0) = 0, \quad \mathbf{u}'(0) = 1, \quad \mathbf{u}''(0) = 2, \quad \mathbf{u}'''(0) = 4$

26. $\mathbf{u}''' + \mathbf{u}' = \mathbf{0}, \quad \mathbf{u}(0) = 2, \quad \mathbf{u}'(0) = 1, \quad \mathbf{u}''(0) = 0$

In Exercises 27 and 28 a weight is placed on a spring suspended from the ceiling, and it is then disturbed as described below. If its motion is represented by the differential equation

$$\frac{d^2\mathbf{s}}{dt^2} + 9\mathbf{s} = \mathbf{0},$$

where s is the displacement from its equilibrium position in centimetres and *t* is time in seconds, in each case find the formula s(*t*) for the motion of the spring.

27. The weight is pulled down 2 cm below its equilibrium position and released with an upward velocity of 6 cm/s.

28. The weight is pulled 2 cm below its equilibrium position and let go (initial velocity = 0).

In Exercises 29 and 30 a pendulum swings in such a way that its motion is described by the differential equation

$$\frac{d^2\boldsymbol{\theta}}{dt^2} + 16\boldsymbol{\theta} = \mathbf{0},$$

where *t* is time in seconds and θ is the angle in radians pictured in Figure 12 (θ is positive if the pendulum bob is to the right of the vertical). In each case, find the formula, θ(*t*), that describes the motion of the pendulum.

29. The pendulum bob, hanging vertically, is hit so that its initial angular velocity is 1 rad/s.

FIGURE 12

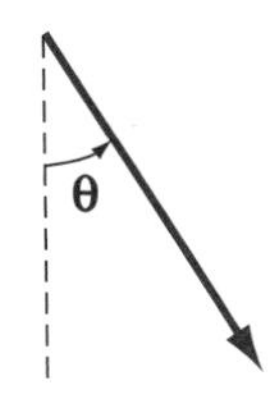

30. The pendulum bob is raised to an angle of $\pi/6$ rad (to the right of vertical) and released (zero initial velocity).

31. Prove parts (i) and (iii) of Theorem 4.

32. Use Theorem 6 to prove that if **f** is a solution of the initial value problem

$$a_n\mathbf{y}^{(n)} + a_{n-1}\mathbf{y}^{(n-1)} + \cdots + a_1\mathbf{y}' + a_0\mathbf{y} = \mathbf{0}$$
$$\mathbf{y}(x_0) = \mathbf{y}'(x_0) = \cdots = \mathbf{y}^{(n-1)}(x_0) = 0,$$

then **f** must be the zero function.

In Exercises 33 and 34 use Theorem 2 to show that the following sets of solutions of the indicated differential equation are linearly independent.

33. $\{1, x, e^x, e^{-x}\}$, $\mathbf{y}^{(iv)} - \mathbf{y}'' = \mathbf{0}$

34. $\{e^x, \cos x, \sin x\}$, $\mathbf{y}''' - \mathbf{y}'' + \mathbf{y}' - \mathbf{y} = \mathbf{0}$

7.7 APPLICATION—Fractals II: Contractive Affine Maps and IFSs

In Section 6.8, we introduced fractals, objects that exhibit self-similarity. We used linear algebra (in particular, projections) to create two-dimensional fractals from seed drawings. In this section, we are going to use certain maps, related to linear transformations, to create fractals in a different way.

We have seen, in this chapter, that each linear transformation $T: \mathbf{R}^n \rightarrow \mathbf{R}^n$ has an associated $n \times n$ matrix A such that, for all vectors $\mathbf{x} \in \mathbf{R}^n$, $T(\mathbf{x}) = A\mathbf{x}$. An *affine map* (also called an *affine function* or *affine transformation*) S on $\mathbf{R}^n$ has an associated $n \times n$ matrix A and a vector $\mathbf{b} \in \mathbf{R}^n$ such that, for all $\mathbf{x} \in \mathbf{R}^n$,

$$S(\mathbf{x}) = A\mathbf{x} + \mathbf{b}.$$

That is, an affine transformation is a linear transformation followed by a translation.

EXAMPLE 1

Consider the function

$$S(\mathbf{x}) = \begin{bmatrix} 1/2 & 0 \\ 0 & 1/2 \end{bmatrix}\mathbf{x} + \begin{bmatrix} 1 \\ 0 \end{bmatrix}.$$

This function is an affine transformation, with

$$A = \begin{bmatrix} 1/2 & 0 \\ 0 & 1/2 \end{bmatrix} \text{ and } \mathbf{b} = \begin{bmatrix} 1 \\ 0 \end{bmatrix}.$$

It takes the vector

$$\mathbf{y} = \begin{bmatrix} -1 \\ 2 \end{bmatrix}$$

to

$$S(y) = \begin{bmatrix} 1/2 & 0 \\ 0 & 1/2 \end{bmatrix}\begin{bmatrix} -1 \\ 2 \end{bmatrix} + \begin{bmatrix} 1 \\ 0 \end{bmatrix} = \begin{bmatrix} 1/2 \\ 1 \end{bmatrix}.$$

■

There are certain affine transformations that play a key role in fractals. An affine transformation S on $\mathbf{R}^n$ is a *contraction (mapping) with contraction factor* $s \in (0, 1)$ if, for all vectors $\mathbf{x}, \mathbf{y} \in \mathbf{R}^n$,

$$\|S(\mathbf{y}) - S(\mathbf{x})\| \leq s\|\mathbf{y} - \mathbf{x}\|;$$

that is, S shrinks the distance between two vectors by a factor of s (the fact that s is less than 1 ensures that S actually shrinks the distance). This notion of a contraction is a generalization of the one used in Example 8 of Section 7.1. A collection of contractions $\{S_1, \ldots, S_k\}$ on $\mathbf{R}^n$ is called an *iterated function system,* or an *IFS* for short.

EXAMPLE 2

The affine function

$$S(\mathbf{x}) = \begin{bmatrix} 1/2 & 0 \\ 0 & 1/2 \end{bmatrix}\mathbf{x} + \begin{bmatrix} 1/2 \\ 0 \end{bmatrix} = \left(\frac{1}{2}(x_1 + 1), \frac{1}{2}x_2\right)$$

from Example 1 is, in fact, a contraction with contraction factor $\frac{1}{2}$ since, for any vectors $\mathbf{x}, \mathbf{y} \in \mathbf{R}^2$,

$$\begin{aligned} S(\mathbf{y}) - S(\mathbf{x}) &= \left(\frac{1}{2}(y_1 + 1), \frac{1}{2}y_2\right) - \left(\frac{1}{2}(x_1 + 1), \frac{1}{2}x_2\right) \\ &= \left(\frac{1}{2}(y_1 - x_1), \frac{1}{2}(y_2 - x_2)\right), \end{aligned}$$

so

$$\begin{aligned} \|S(\mathbf{y}) - S(\mathbf{x})\| &= \sqrt{\left(\frac{1}{2}(y_1 - x_1)\right)^2 + \left(\frac{1}{2}(y_2 - x_2)\right)^2} \\ &= \sqrt{\frac{1}{4}(y_1 - x_1)^2 + \frac{1}{4}(y_2 - x_2)^2} \\ &= \sqrt{\frac{1}{4}\left((y_1 - x_1)^2 + (y_2 - x_2)^2\right)} \\ &= \frac{1}{2}\sqrt{(y_1 - x_1)^2 + (y_2 - x_2)^2} \\ &= \frac{1}{2}\|\mathbf{y} - \mathbf{x}\|. \end{aligned}$$

Similarly, the affine functions

$$R(\mathbf{x}) = \begin{bmatrix} 1/2 & 0 \\ 0 & 1/2 \end{bmatrix}\mathbf{x} = \left(\frac{1}{2}x_1, \frac{1}{2}x_2\right)$$

and $$T(\mathbf{x}) = \begin{bmatrix} 1/2 & 0 \\ 0 & 1/2 \end{bmatrix}\mathbf{x} + \begin{bmatrix} 1/4 \\ \sqrt{3}/4 \end{bmatrix} = \left[\frac{1}{2}\left(x_1 + \frac{1}{2}\right), \frac{1}{2}\left(x_2 + \frac{\sqrt{3}}{2}\right)\right]$$

are also contractions with contraction factor 1/2. It follows that $\{R, S, T\}$ is an IFS. ■

From an IFS $\{S_1, \ldots, S_k\}$ on $\mathbf{R}^n$, a fractal can be constructed using a *deterministic* algorithm, that is, a process that has precise rules, which produce the same outcome every time. (We will look at a different process, which uses randomness, later in this section.)

Start with *any* compact set C, that is, any subset of $\mathbf{R}^n$ that is *closed* (contains its boundary points) and *bounded* (sits within some rectangle). Usually a closed polygon, such as a triangle or a square, is used. Then the first iteration is the union of all the S_i's on C; this becomes the new set, replacing C. You then iterate the process, taking the union of all the S_i's on this new set, and so on. The fractal associated with the IFS is the object achieved *in the limit,* but any sufficiently large number of iterations will approximate the fractal.

EXAMPLE 3

For the IFS of Example 2, let's start with c being the unit square, with the bottom left corner at the origin and the upper right corner at (1, 1). For example, R moves the four corners of the square C, (0, 0), (1, 0), (1, 1), and (0, 1), as follows:

$$R((0,0)) = (0,0),\ R((1,0)) = (1/2, 0),\ R((1,1)) = (1/2, 1/2), \text{ and } R((0, 1)) = (0, 1/2).$$

So R takes C, the unit square, to the square with vertices (0, 0), (1/2, 0), (1/2, 1/2), and (0, 1/2). Continue in the same way for S and T. Figure 13 shows the union of $R(C) \cup S(C) \cup T(C)$ graphically. Note how the three squares form the general shape of a triangle.

FIGURE 13

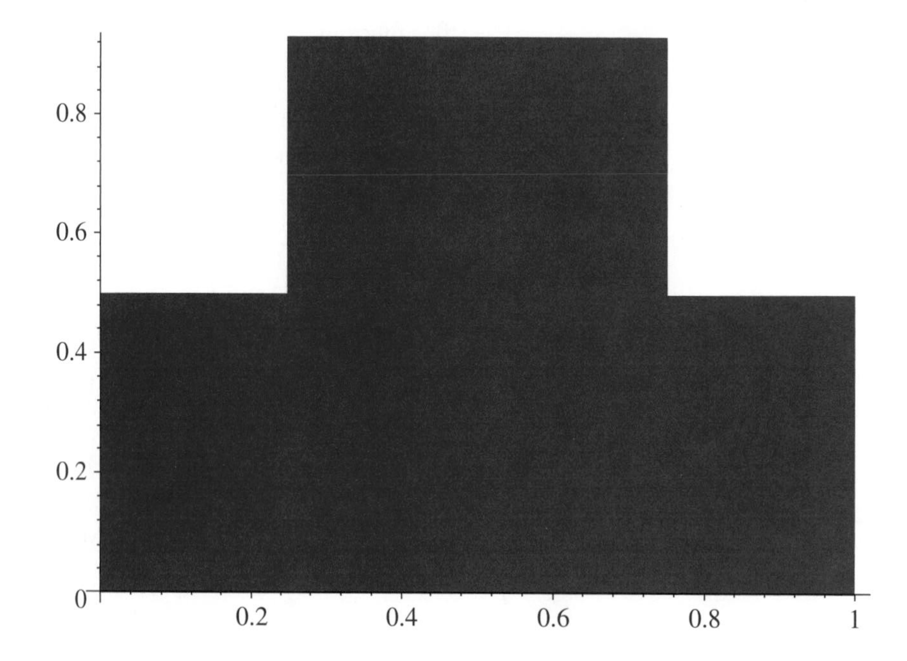

Figure 14 shows the next five iterations:

FIGURE 14

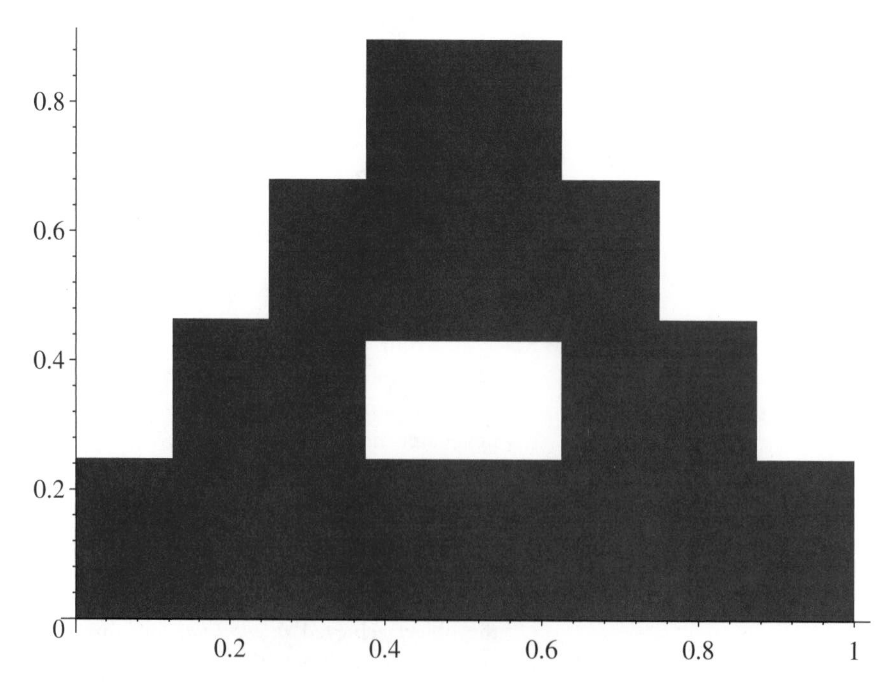

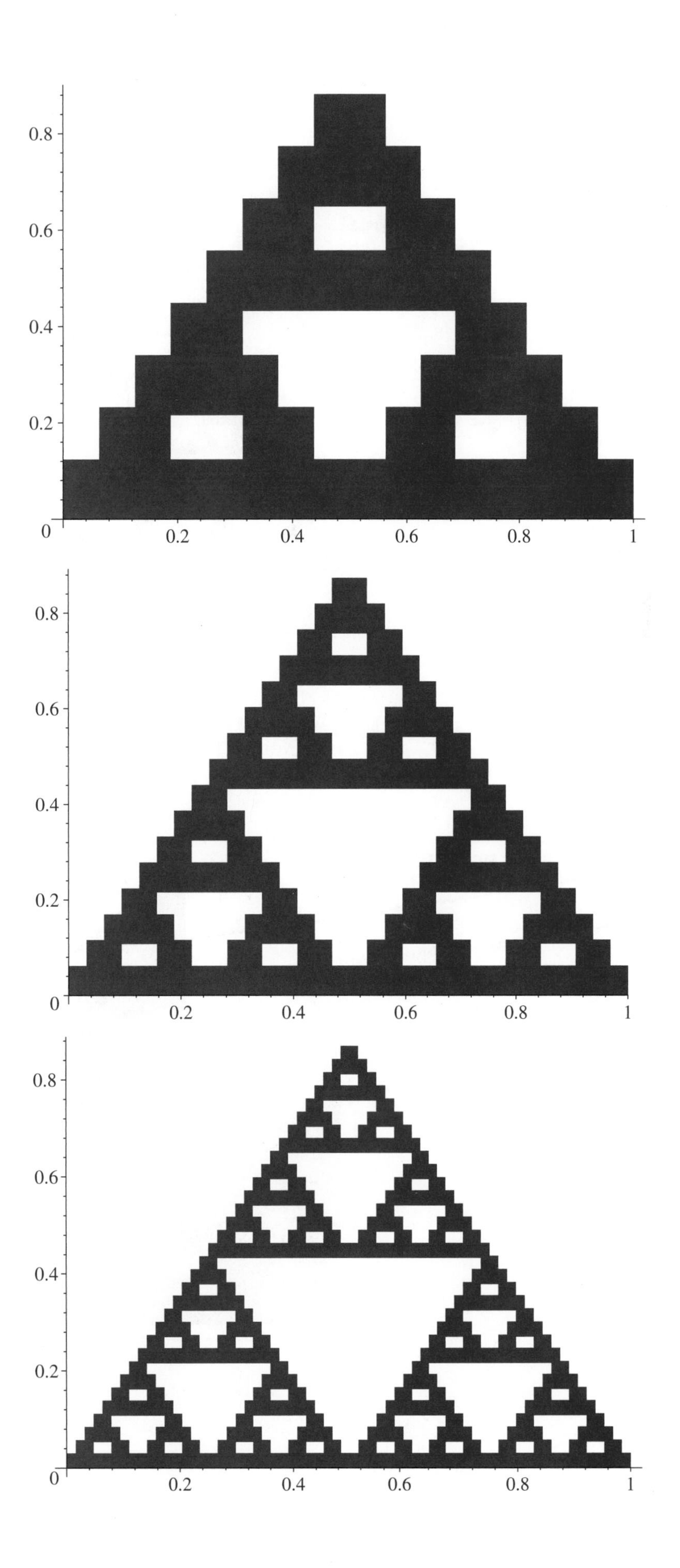
0.8
0.6
0.4
0.2
0
0.2
0.4
0.6
0.8
1
0.8
0.6
0.4
0.2
0
0.2
0.4
0.6
0.8
1
0.8
0.6
0.4
0.2
0
0.2
0.4
0.6
0.8
1

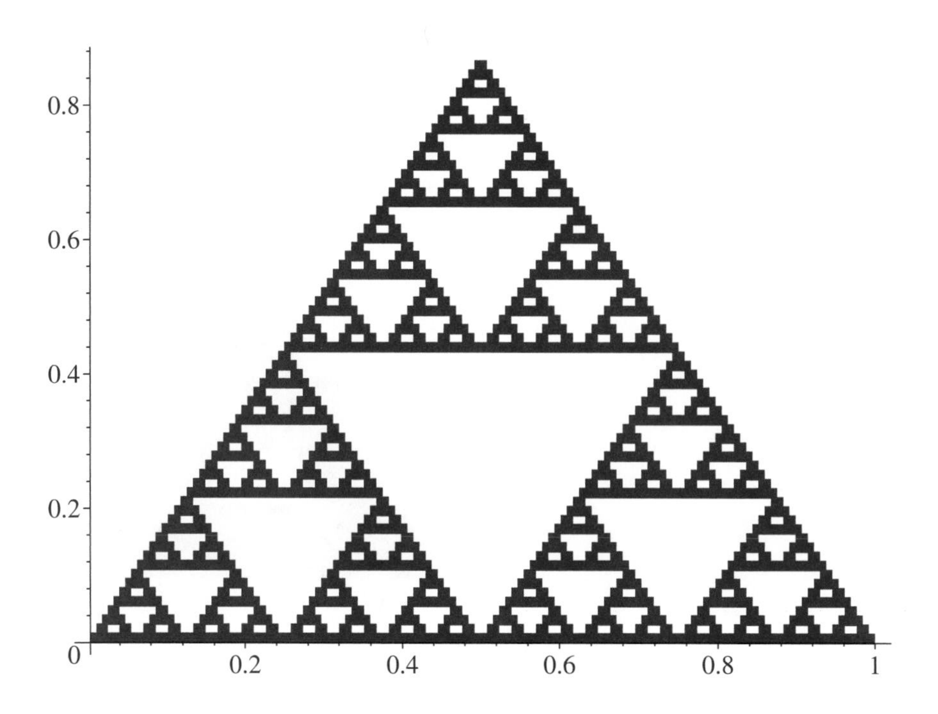

Notice that the final diagram in Figure 14 is unmistakably close to Sierpinski's triangle. In fact, the limiting object is indeed Sierpinski's triangle.

What if we start, instead, with a triangle for C, say the one with vertices (0, 0), (1, 0), and (0, 1)? Figure 15 shows what we get for the first two iterations with the same IFS.

FIGURE 15

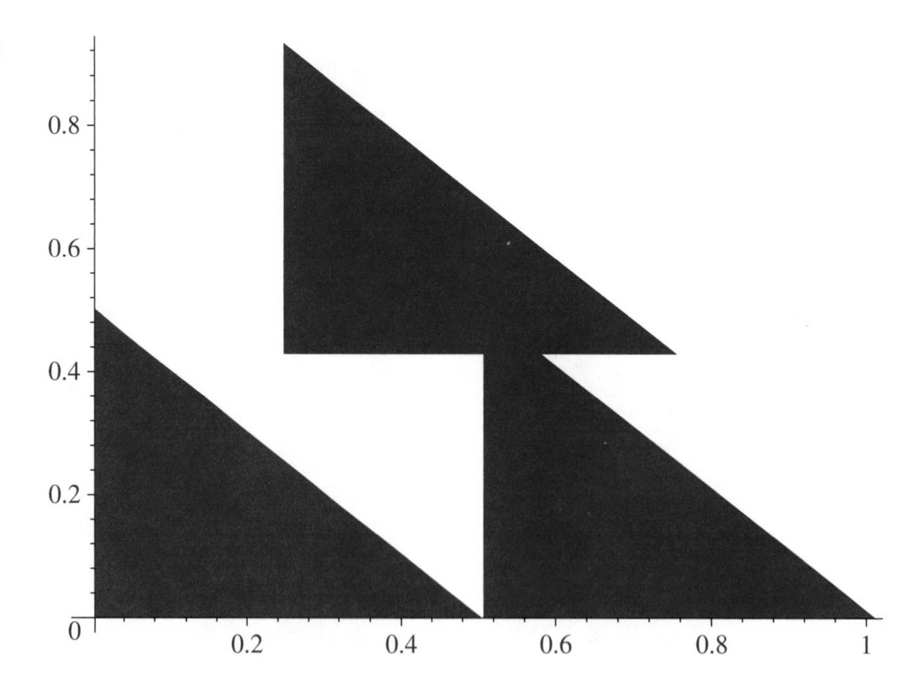

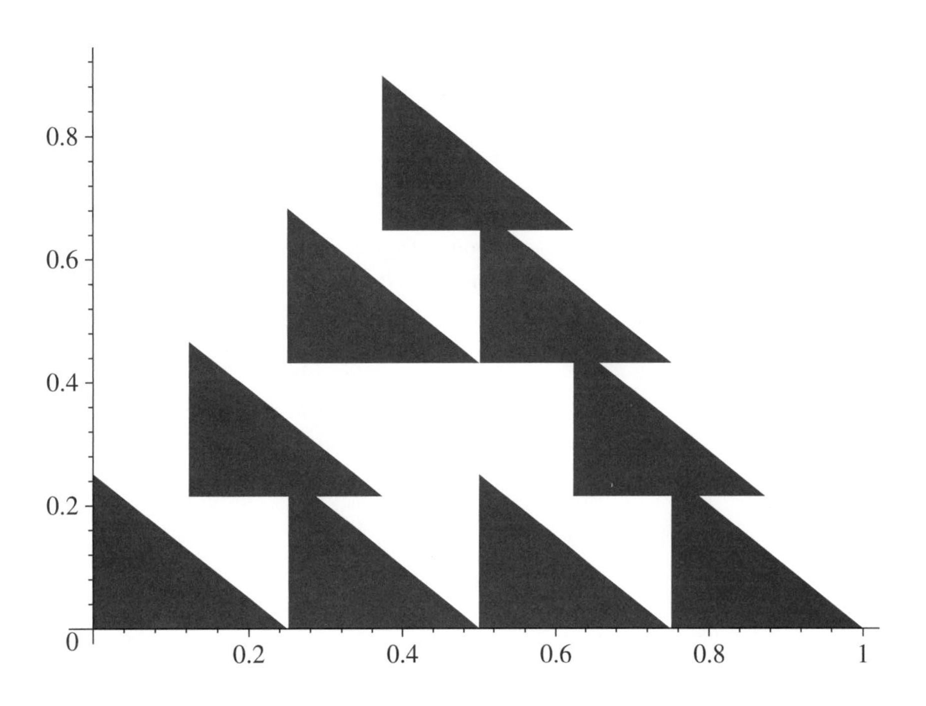

Figure 16 shows what we get after four more iterations:

FIGURE 16

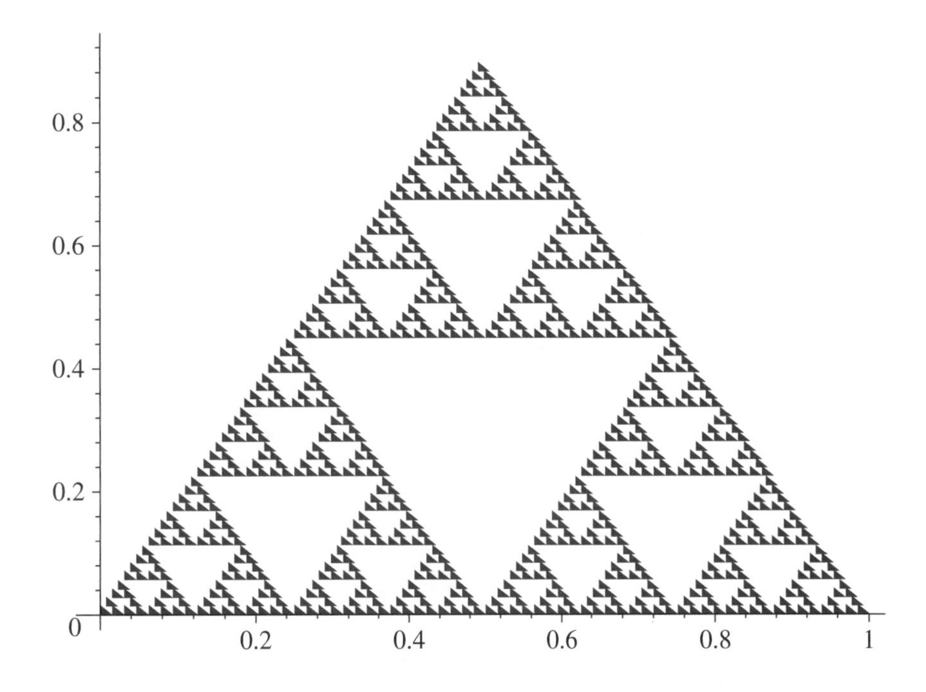

Again, we are approaching the same Sierpinski's triangle! It looks a little different on the microscopic scale, but still it seems to be (and indeed is) approaching the same limiting figure. ∎

The amazing result (which we won't prove here) is that, for *any* IFS, the deterministic algorithm applied to *any* compact set approaches the same limiting object, which will be a fractal. The deterministic algorithm produces crisp, attractive images. The drawback is that we may have to iterate many times, and the number of polygons produced grows exponentially. It may take a long time to render the image from the polygons, and it may require a fair amount of computer storage to store the polygons as well.

There is another way to construct fractals that, instead of requiring the use of all contractions, uses just one of them at a time. This *nondeterministic* (or random) algorithm starts with a single point and randomly picks one of the contractions, applies it, and repeats the process. The fractal produced is the collection of points from all the iterations, except that usually a certain number of the initial points are left out. This algorithm generally runs much faster than the deterministic one and uses less computer memory, as it does not need to carry along all the descriptions of the various polygons produced by the deterministic algorithm. Figure 17 was created by starting at the point (0.8, 0.1) and randomly applying one of the three contractions, R, S, and T from Example 2 (each equally likely), to the point 12,000 times, ignoring the first 200 points.

FIGURE 17

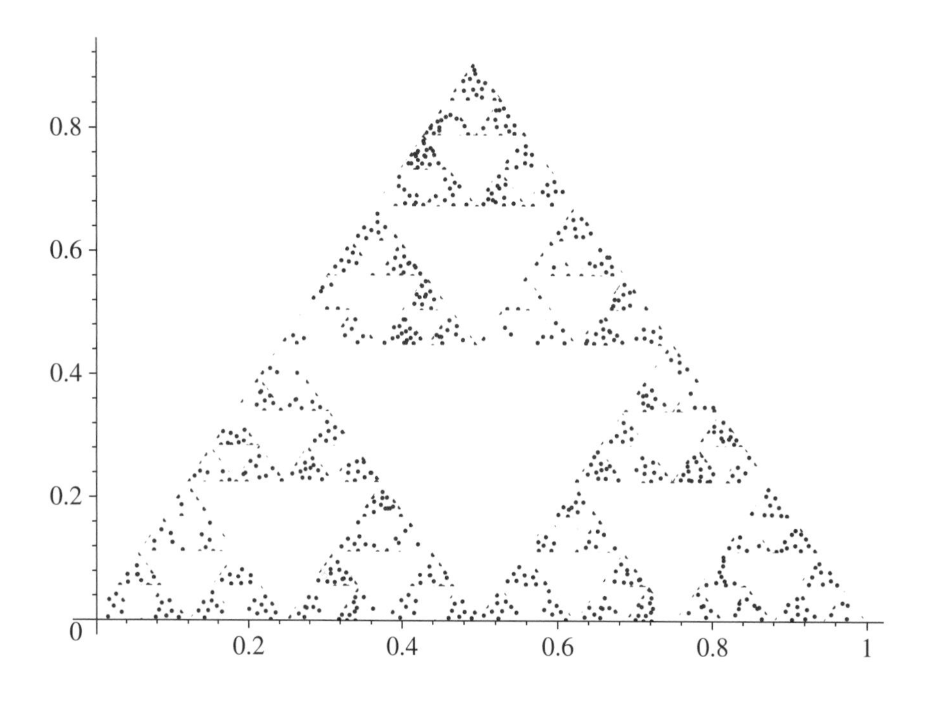

Michael Barnsley is one of the pioneers in the study of fractals. He discovered the Collage Theorem that allows you to produce a desired fractal image via iterated function systems. He also started up Iterated Systems Inc., a company that successfully applied fractals to problems such as image compression.

Courtesy of Dr. Michael Barnsley

There is some choice in how often you can choose each of the functions of an IFS under the nondeterministic algorithm, as varying the probabilities can dramatically affect how well the image "fills in." See Exercise 6 for more details.

There are two possible drawbacks to the nondeterministic algorithm. First, the images produced are never quite the same—they are close, to be sure, but not identical under different runs of the program, even when starting at the same point. Second, the images are not as sharp as those produced by the deterministic algorithm, with certain portions of the image taking longer to fill in than others. The two examples shown in Figure 18 were both generated by the nondeterministic algorithm from the same IFS of four contractions with the same starting point, the same number of initial points ignored, and the same number of total points. All that differed were the probabilities used for choosing each contraction; the first image had all the contractions equally likely to be chosen (each with probability $\frac{1}{4}$), while the second had one contraction chosen only 1% of the time, one chosen 85% of the time, and the other two chosen 7% of the time each. The probabilities used make quite a difference, don't they?

Is it possible to reverse the process and find an iterated function system that has a given image as its fractal? Indeed it is! The reverse process is called the *Collage Theorem*, but it is well beyond the scope of what is discussed here.

FIGURE 18

7.7 Exercises

1. Show that whether or not the affine function $S(\mathbf{x}) = A\mathbf{x} + \mathbf{b}$ is a contraction does not depend on $\mathbf{b}$.

2. Are dilations, rotations, or reflections contraction mappings?

3. For each of the following affine functions $S(\mathbf{x}) = A\mathbf{x} + \mathbf{b}$ on $\mathbf{R}^2$, is S a contraction mapping? If so, find contraction factor.

   (a) $A = \begin{bmatrix} 1/3 & 0 \\ 0 & 1/4 \end{bmatrix}, \ \mathbf{b} = \begin{bmatrix} -1 \\ 1 \end{bmatrix}$

   (b) $A = \begin{bmatrix} 0 & -1/2 \\ 1/2 & 0 \end{bmatrix}, \mathbf{b} = \begin{bmatrix} 10 \\ 100 \end{bmatrix}$

   (c) $A = \begin{bmatrix} 4/3 & 0 \\ 0 & 1/4 \end{bmatrix}, \ \mathbf{b} = \begin{bmatrix} 0 \\ 0 \end{bmatrix}$

4. Consider the following functions on $\mathbf{R}^2$. Write each function in matrix form, and use a computer program, such as *Maple,* to generate a picture under at least six iterations of the deterministic algorithm, applied to the triangle C with vertices (0, 0), (1, 0), and (0, 1). What does the image look like?

   (a) $f_1(x, y) = (0, 0.16y)$
   $f_2(x, y) = (0.85x + 0.04y, \ -0.04x + 0.85y + 1.6)$
   $f_3(x, y) = (0.2x - 0.26y, 0.23x + 0.22y + 1.6)$
   $f_4(x, y) = (-0.15x + 0.28y, 0.26x + 0.24y + 0.44)$

   (b) $f_1(x, y) = (0.333x, 0.333y)$
   $f_2(x, y) = (0.333x, 0.333y + 0.333)$
   $f_3(x, y) = (0.333x, 0.333y + 0.667)$
   $f_4(x, y) = (0.333x + 0.333, 0.333y)$
   $f_5(x, y) = (0.333x + 0.333, 0.333y + 0.667)$
   $f_6(x, y) = (0.333x + 0.667, 0.333y)$
   $f_7(x, y) = (0.333x + 0.667, 0.333y + 0.333)$
   $f_8(x, y) = (0.333x + 0.667, 0.333y + 0.667)$

5. Repeat Exercise 4, parts (a) and (b), with the nondeterministic algorithm, with each function of the IFS chosen equally likely.

6. Let $\{S_1, \ldots, S_k\}$ on $\mathbf{R}^n$ be an IFS, with $S_i(\mathbf{x}) = A_i\mathbf{x} + \mathbf{b}_i$ for $i = 1, \ldots, k$. It can be shown that the optimal way to choose the probabilities for the nondeterministic algorithm is to choose them in proportion to the absolute value of the determinants of the A_is. That is, if

$$d = |\det(A_1)| + |\det(A_2)| + \cdots + |\det(A_k)|,$$

   choose contraction A_i with probability

$$\frac{|\det(A_i)|}{d}.$$

   (a) Show that the sum of all such probabilities equals 1.

   (b) If $\det(A_i) = 0$ for some i, that function will never be chosen, and all functions in the IFS should be chosen. How do you think the probabilities should be altered in this case?

   (c) What should be the probabilities for each of the functions R, S, and T from Example 2 for Sierpinski's triangle?

   (d) What should be the probabilities for each of the functions in the IFS in Exercise 4, part (a)? Generate a corresponding picture of the fractal.

CHAPTER SUMMARY

Definition of a Linear Transformation

A *linear transformation* from one vector space $\mathbf{V}$ to another $\mathbf{W}$ is a function $T: \mathbf{V} \to \mathbf{W}$ that satisfies the following two conditions for every $\mathbf{u}$ and $\mathbf{v}$ in $\mathbf{V}$ and every scalar c:

i. $T(c\mathbf{u}) = cT(\mathbf{u})$

ii. $T(\mathbf{u} + \mathbf{v}) = \mathrm{T}(\mathbf{u}) + \mathrm{T}(\mathbf{v})$

Operations on Linear Transformations

(a) Let $S: \mathbf{V} \to \mathbf{W}$ and $T: \mathbf{V} \to \mathbf{W}$ be linear transformations, and c be a scalar. Then the following are linear transformations.

Sum: $(S + T)\mathbf{x} = S(\mathbf{x}) + T(\mathbf{x})$ for all $\mathbf{x}$ in $\mathbf{V}$

Difference: $(S - T)\mathbf{x} = S(\mathbf{x}) - T(\mathbf{x})$ for all $\mathbf{x}$ in $\mathbf{V}$

Scalar multiple: $(cS)(\mathbf{x}) = cS(\mathbf{x})$ for all $\mathbf{x}$ in $\mathbf{V}$

(b) Let $S: \mathbf{U} \to \mathbf{V}$ and $T: \mathbf{V} \to \mathbf{W}$ be linear transformations. The *product* TS, which represents the composition of T and S, is the linear transformation

$$TS(\mathbf{x}) = T(S(\mathbf{x})) \text{ for all } \mathbf{x} \text{ in } \mathbf{U}.$$

(c) Let $S: \mathbf{V} \to \mathbf{V}$ be a linear transformation. The *powers* of S are the linear transformations given by

$$S^0 = I \text{ (the identity matrix)}, S^1 = S, S^{k+1} = SS^k \text{ for } k = 1, 2, \ldots .$$

Matrix of a Linear Transformation

(a) DEFINITION: Let $T: \mathbf{R}^n \to \mathbf{R}^m$ be a linear transformation. Then the $m \times n$ matrix A with ith column given by $\mathbf{a}^i = T(\mathbf{e}_i)$ is called the matrix *associated* with T (or the matrix that *represents* T) and has the property that for all $\mathbf{x}$ in $\mathbf{R}^n$, $T(\mathbf{x}) = A\mathbf{x}$.

(b) If S and T are linear transformations, and if matrix A represents S and matrix B represents T, then (assuming that the given operation is defined)

$A + B$ represents $S + T$,
$A - B$ represents $S - T$,
cA represents cS (for any scalar c),
BA represents TS,
A^{-1} represents S^{-1}.

(c) Let $T: \mathbf{V} \to \mathbf{W}$ be a linear transformation, and $\mathcal{B}$ and $\mathcal{B}'$ be bases for $\mathbf{V}$ and $\mathbf{W}$, respectively. Suppose that $\mathbf{v}$ is an arbitrary vector in $\mathbf{V}$ with coordinate vector $\mathbf{x}$ (in $\mathbf{R}^n$) with respect to $\mathcal{B}$, and that $\mathbf{y}$ (in $\mathbf{R}^m$) is the coordinate vector of $T(\mathbf{v})$ with respect to $\mathcal{B}'$. Then the $m \times n$ matrix A that *represents* T with respect to these bases has the property that $\mathbf{y} = A\mathbf{x}$. Letting $\mathcal{B} = \{\mathbf{v}_1, \mathbf{v}_2, \ldots, \mathbf{v}_n\}$, the ith column of A is given by the coordinate vector of $T(\mathbf{v}_i)$.

Kernel and Image of a Linear Transformation

Let $T: \mathbf{V} \to \mathbf{W}$ be a linear transformation.

(a) DEFINITION: The *kernel* (*null space*) of T, Ker (T), is the set of all $\mathbf{x}$ in $\mathbf{V}$ for which $T(\mathbf{x}) = \mathbf{0}$. The kernel of T is a subspace of $\mathbf{V}$. T is one-to-one if and only if Ker $(T) = \{\mathbf{0}\}$.

(b) DEFINITION: The *image* of T, $\text{Im}(T)$, is the set of vectors $T(\mathbf{x})$ for all $\mathbf{x}$ in $\mathbf{V}$. The image of T is a subspace of $\mathbf{W}$. T is onto if and only if $\text{Im}(T) = \mathbf{W}$.

(c) DEFINITION: The *nullity* of T is the dimension of $\text{Ker}(T)$; the *rank* of T is the dimension of $\text{Im}(T)$. If $\dim(\mathbf{V}) = n$, then nullity + rank = n.

(d) If $\dim(\mathbf{V}) = \dim(\mathbf{W})$, then T is invertible if and only if T is one-to-one or T is onto. That is, T is invertible if and only if *either* $\text{Ker}(T) = \{\mathbf{0}\}$ or $\text{Im}(T) = \mathbf{W}$.

Change of Basis

(a) Let $\mathcal{B} = \{\mathbf{v}_1, \mathbf{v}_2, \ldots, \mathbf{v}_n\}$ be a basis for $\mathbf{R}^n$, and P be the $n \times n$ matrix with ith column given by $\mathbf{p}^i = \mathbf{v}_i$. Then P is invertible and $P^{-1}\mathbf{x}$ is the coordinate vector of $\mathbf{x}$ with respect to $\mathcal{B}$. P^{-1} is called the *change-of-basis matrix for* $\mathcal{B}$.

(b) DEFINITION: If A and B are square matrices, then B is *similar* to A if there is an invertible matrix P such that $B = P^{-1}AP$.

(c) Let $T: \mathbf{R}^n \to \mathbf{R}^n$ be a linear transformation with associated matrix A and $\mathcal{B}$ be a basis for $\mathbf{R}^n$. If P^{-1} is the change-of-basis matrix for $\mathcal{B}$, then $B = P^{-1}AP$ is the matrix that represents T with respect to the basis $\mathcal{B}$.

KEYWORDS

linear transformation
power of linear transformations
matrix representing a linear transformation
image
rank
linear differential equation
homogeneous
contraction
iterated function system (IFS)
product of linear transformations
matrix associated with a linear transformation
kernel
nullity
change-of-basis matrix
similar matrix
characteristic polynomial
affine map
contraction factor

SELF-TEST

In Problems 1 and 2 determine whether or not the given function is a linear transformation. Justify your answer.

1. $L: \mathbf{R}^2 \to \mathbf{R}^3$ given by $L(x, y) = (x - y, x + y, 1)$.

2. $K: \mathbf{P}_2 \to \mathbf{M}^{2,2}$ given by $K(a + bx + cx^2) = \begin{bmatrix} 0 & a - b \\ a + b & c \end{bmatrix}$. [Section 7.1]

In Problems 3–6 let $T: \mathbf{R}^3 \to \mathbf{R}^3$ be defined by $T(x, y, z) = (z, x + y, 0)$.

3. Find T^2 $(T \circ T)$. [Section 7.2]

4. Find the matrix that represents T. [Section 7.2]

5. *(a)* Find a basis for the kernel of T.
(b) Is T one-to-one? [Section 7.3]

6. *(a)* Find a basis for the image of T.
(b) Is T onto? [Section 7.3]

7. Suppose that a linear transformation L: $\mathbf{R}^2 \to \mathbf{R}^2$ is given by $L(\mathbf{x}) = A\mathbf{x}$, where

$$A = \begin{bmatrix} 2 & 5 \\ 1 & 3 \end{bmatrix}.$$

(a) Find $L(x, y)$ for arbitrary (x, y) in $\mathbf{R}^2$. [Section 7.1]

(b) Find $L^{-1}(x, y)$ for arbitrary (x, y) in $\mathbf{R}^2$. [Section 7.3]

In Problems 8 and 9 let S: $\mathbf{P}_2 \to \mathbf{R}^4$ be the linear transformation given by

$$S(a + bx + cx^2) = (c, a - b, 0, 4a).$$ [Section 7.4]

8. Find the matrix that represents S with respect to the standard bases for $\mathbf{P}_2$ and $\mathbf{R}^4$.

9. *(a)* Find the dimension of the kernel of S. Is S one-to-one?
(b) Find the dimension of the image of S. Is S onto?

10. Compute the change-of-basis matrix for the basis $\{(1, 0, 0), (1, 1, 0), (1, 1, 1)\}$ of $\mathbf{R}^3$. [Section 7.5]

REVIEW EXERCISES

In Exercises 1 and 2 show that the given function is a linear transformation.

1. S: $\mathbf{R}^3 \to \mathbf{R}^2$ given by $S(x, y, z) = (x - 2y, x + y + z)$.

2. T: $\mathbf{P}_3 \to \mathbf{M}^{2,2}$ given by

$$T(a_0 + a_1x + a_2x^2 + a_3x^3) = \begin{bmatrix} a_0 + a_2 & a_1 - a_3 \\ 2a_2 & 2a_0 + a_1 - a_3 \end{bmatrix}.$$

In Exercises 3 and 4 show that the given function is not a linear transformation.

3. S: $\mathbf{R}^2 \to \mathbf{R}^2$ given by $S(x, y) = (x + 2y, xy)$.

4. T: $\mathbf{R}^2 \to \mathbf{R}^3$ given by $T(x, y) = \left(x, x + y, x + |y|\right)$.

Assume that the following are linear transformations and use them in Exercises 5–16, 21 and 22.

D: $\mathbf{R}^2 \to \mathbf{R}^3$ given by $D(x, y) = (x + y, x - 2y, x)$
E: $\mathbf{R}^3 \to \mathbf{R}^3$ given by $E(x, y, z) = (x, x + y + z, 2x + y + z)$
F: $\mathbf{R}^2 \to \mathbf{R}^3$ given by $F(x, y) = (x, 2x + y, 2x - y)$

In Exercises 5–8, find the indicated linear transformation if it is defined. If it is not defined, explain why it is not defined.

5. FD

6. $D - 2F$

7. E^2

8. F^2

9. Find the matrix that represents D.

10. Find the matrix that represents E.

11. Is D one-to-one? Explain.

12. Is D onto? Explain.

13. Is E one-to-one? Explain.

14. Is E onto? Explain.

15. Find Ker (D).

16. Find Ker (E).

In Exercises 17–20, let T be the linear transformation of Exercise 2.

17. Find the matrix that represents T with respect to the standard bases for $\mathbf{P}_3$ and $\mathbf{M}^{2,2}$.

18. Is T one-to-one? Explain.

19. Is T onto? Explain.

20. Find Ker (T).

21. Directly find the matrix $A_{\mathcal{O}}$ that represents the linear transformation E with respect to the basis $\mathcal{B} = \{(1, 1, 0), (1, 0, 1), (0, 1, 1)\}$.

22. Again calculate $A_{\mathcal{O}}$ of Exercise 21, but this time use the result of Exercise 10 and the appropriate change-of-basis matrix.

CHALLENGE YOURSELF

23. Let $n \geq 2$, and consider the map D_n: $\mathbf{P}_n \rightarrow \mathbf{P}_{n-1}$ given by

$$D_n(a_0 + a_1 x + \cdots + a_n x^n) = a_1 + 2a_2 x + 3a_3 x^2 + \cdots + na_n x^{n-1}.$$

(a) Prove that D is a linear transformation.

(b) What is Ker (D)?

24. Let $n \geq 2$, and consider the map N: $\mathbf{P}_n \rightarrow \mathbf{P}_{n+1}$ given by

$$N_n(a_0 + a_1 x + \cdots + a_n x^n) = a_0 x + \frac{a_1}{2} x^2 + \frac{a_2}{3} x^3 + \cdots + \frac{a_n}{n+1} x^{n+1}.$$

(a) Prove that N is a linear transformation.

(b) What is Im (N)?

(c) For the transformation D of Problem 23, what is the linear transformation $D_{n+1} \circ N_n$? What is the transformation $N_{n-1} \circ D_n$?

25. Prove that the set of all transformations from $\mathbf{R}^m$ to $\mathbf{R}^m$ form a vector space. What is the dimension of this vector space?

26. Suppose that s is a positive real number less than 1, and that T: $\mathbf{R}^m \rightarrow \mathbf{R}^m$ is a linear transformation such that for every vector $\mathbf{v}$ in $\mathbf{R}^m$, $\|T(\mathbf{v})\| \leq s\|\mathbf{v}\|$. Prove that $\lim_{k \rightarrow \infty} T^k(\mathbf{v}) = \mathbf{0}$.

27. Suppose that T: $\mathbf{V} \rightarrow \mathbf{V}$ is a linear operator, and that $\mathbf{v}$ is a vector in $\mathbf{V}$ such that $T^{k-1}(\mathbf{v}) \neq \mathbf{0}$ but $T^k(\mathbf{v}) = \mathbf{0}$. Prove that the vectors $\mathbf{v}$, $T(\mathbf{v})$, $T^2(\mathbf{v})$, . . . , $T^{k-1}(\mathbf{v})$, are linearly independent.

SPOTLIGHT

Wesley Aston/Shutterstock

When studying physical, economic, and social systems, scientists are sometimes interested in the conditions that will lead to a *stable*, or *equilibrium*, state—one that does not change as time passes. Questions concerning stability arise when studying mechanical systems subject to vibration, electric circuits, subatomic particles, the production and distribution of goods, population growth, and migration patterns, to name just a few.

Let's look at a simple example. Suppose that we are studying a certain animal population and collect the following data on its female component: 40% of the females in this group survive for 1 year or more, and 30% of these live to be at least 2; however, none of the females live more than 3 years. Moreover, the average number of daughters born to a female in her first, second, and third years of life is 0.5, 2.5, and 1.5, respectively. (The reason for just studying females is that, normally, they alone determine how a population grows; the number of males is usually not relevant.)

To model this situation, for a given year, let $\mathbf{x} = (x_1, x_2, x_3)$ where x_1 is the number of females that are less than 1 year old, x_2 is the number between 1 and 2 years old, and x_3 is the number more than 2 years old. Then the number in each age group that will be present the *next* year is given by the components of the vector $L\mathbf{x}$, where

$$L = \begin{bmatrix} 0.5 & 2.5 & 1.5 \\ 0.4 & 0 & 0 \\ 0 & 0.3 & 0 \end{bmatrix}.$$

(We use "L" to represent the matrix because it is known as a *Leslie matrix*.) For example, if there are 100 females in each age group this year ($\mathbf{x} = (100, 100, 100)$), then next year there will be 450 less than one year old, 40 between 1 and 2 years old, and 30 over 2 years old (that is, $L\mathbf{x} = (450, 40, 30)$).

In a situation like this, we might take *stability* to mean that the number of females in each age group remains the same from year to year (that is, $L\mathbf{x} = \mathbf{x}$), a condition that is not likely to happen (unless $\mathbf{x}$ is equal to $\mathbf{0}$, perhaps!). Instead, we could ask the general question, "For what population distribution does the number of females in each age group change in the same proportion?" or, in other words, "For which (nonzero) $\mathbf{x}$, is there a constant λ such that $L\mathbf{x} = \lambda\mathbf{x}$?" In general terms, this is known as an *eigenvalue problem*: λ is an *eigenvalue* of L and $\mathbf{x}$ is the corresponding *eigenvector*.

This chapter deals with eigenvalue problems, providing techniques for solving them, as well as a few of their applications.

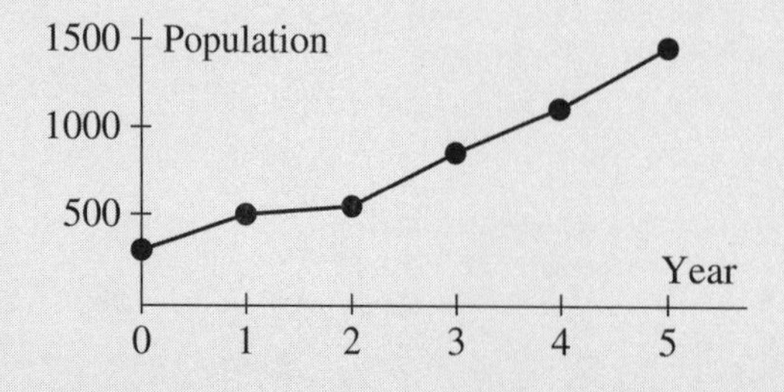

CHAPTER

8

Eigenvalues and Eigenvectors

OBJECTIVES

You will learn to:

- Find the eigenvalues and corresponding eigenvectors of simple linear operators on $\mathbf{R}^m$ and square matrices with real entries.
- Determine a basis for the eigenspace of a matrix.
- Diagonalize a square matrix with real entries (or demonstrate that it cannot be diagonalized).
- Orthogonally diagonalize a symmetric matrix.
- Factor a square matrix A as the product of an orthogonal matrix Q and an upper triangular matrix R: $A = QR$.
- Use the QR factorization technique to approximate the eigenvalues of a given matrix.
- Find the complex eigenvalues and corresponding eigenvectors (in complex m-space) of a square matrix.
- Diagonalize a square matrix with complex entries and unitarily diagonalize a hermitian matrix.

In this chapter, we discuss *eigenvalues* and *eigenvectors,* a concept that has many far-reaching applications. This topic was introduced in Chapter 4 for matrices, but now we extend our discussion to the context of linear transformations on general vector spaces. In Section 8.4, we introduce *complex m-space* and discuss eigenvalues that are complex numbers and eigenvectors with complex entries.

The use of the term *eigen* (German for "characteristic") is due to German mathematician David Hilbert (1862–1943). Hilbert was a very influential mathematician who, in 1902, posed 23 fundamental problems in mathematics, some of which still defy resolution.

8.1 Definitions and Examples

From a geometric point of view, a linear transformation T, of $\mathbf{R}^n$ into itself, maps one vector into another by altering the magnitude or direction (or both) of the former. The *eigenvectors* of T are those nonzero vectors $\mathbf{x}$ for which $T(\mathbf{x})$ and $\mathbf{x}$ are collinear. In other words, eigenvectors are characterized by the property that $T(\mathbf{x})$ is a scalar multiple of $\mathbf{x}$. The scalar involved is called the *eigenvalue* of T associated with the eigenvector $\mathbf{x}$. In more precise terms, we have the following definition. (It extends the definitions of *eigenvalue* and *eigenvector of a square matrix* from Section 4.5.)

DEFINITION

Let $\mathbf{V}$ be a vector space and $T: \mathbf{V} \to \mathbf{V}$ be a linear transformation. A scalar λ is called a real **eigenvalue** of T if there is a *nonzero* vector $\mathbf{x}$ in $\mathbf{V}$ such that $T(\mathbf{x}) = \lambda\mathbf{x}$. The vector $\mathbf{x}$ is said to be an **eigenvector** corresponding to the eigenvalue λ.

Eigenvalues are sometimes referred to as *characteristic values, proper values,* or *latent values.* Similarly, other names for eigenvectors are *characteristic vectors, proper vectors,* and *latent vectors.* In the first three sections of this chapter, we consider only eigenvalues that are real numbers; however, in Section 8.4, we will allow eigenvalues and the components of eigenvectors to be complex numbers.

EXAMPLE 1

Let $T: \mathbf{R}^3 \to \mathbf{R}^3$ be the linear transformation given by

$$T(x, y, z) = (0, y + z, 3y - z).$$

Show that $\lambda_1 = 0$ and $\lambda_2 = 2$ are eigenvalues of T with corresponding eigenvectors $\mathbf{x}_1 = (1, 0, 0)$ and $\mathbf{x}_2 = (0, 1, 1)$, respectively.

Solution We must verify that $T(\mathbf{x}_1) = \lambda_1\mathbf{x}_1$ and $T(\mathbf{x}_2) = \lambda_2\mathbf{x}_2$. Now

$$T(\mathbf{x}_1) = T(1, 0, 0) = (0, 0, 0) = 0(1, 0, 0) = \lambda_1\mathbf{x}_1$$

and

$$T(\mathbf{x}_2) = T(0, 1, 1) = (0, 2, 2) = 2(0, 1, 1) = \lambda_2\mathbf{x}_2,$$

as desired. ■

NOTE By definition, an eigen*vector* must be nonzero. A linear transformation may have a zero eigen*value*, however, as can be seen from Example 1.

It is easy to verify that, for the linear transformation of Example 1, any nonzero vector of the form $(s, 0, 0)$ is an eigenvector corresponding to $\lambda_1 = 0$, whereas any nonzero vector of the form $(0, t, t)$ is an eigenvector for $\lambda_2 = 2$. In general, if T is any linear transformation with eigenvalue λ_0 and corresponding eigenvector $\mathbf{x}_0$, then all vectors of the form $c\mathbf{x}_0$, $c \neq 0$, are eigenvectors of T corresponding to λ_0. This statement follows from the fact that

$$T(c\mathbf{x}_0) = cT(\mathbf{x}_0) = c(\lambda\mathbf{x}_0) = \lambda(c\mathbf{x}_0).$$

EXAMPLE 2

Let $T: \mathbf{R}^2 \to \mathbf{R}^2$ be the linear transformation that maps every vector into its reflection in the x-axis (see Example 9 of Section 7.1). Find the eigenvalues and eigenvectors of T.

Solution By the nature of the given transformation, the only nonzero vectors collinear with their images under T are those that lie along the coordinate axes. Thus, all eigenvectors of T are of the form $c_1(1, 0)$ or $c_2(0, 1)$, where c_1 and c_2 are nonzero scalars. Moreover, for any vector $\mathbf{x}$ lying on the x-axis, $T(\mathbf{x}) = \mathbf{x}$, whereas if $\mathbf{x}$ lies on the y-axis, we have $T(\mathbf{x}) = -\mathbf{x}$. So the eigenvectors of the form $c_1(1, 0)$ satisfy $T(\mathbf{x}) = \mathbf{x}$, whereas those of the form $c_2(0, 1)$ satisfy $T(\mathbf{x}) = -\mathbf{x}$. Comparing these equations to the defining equation for an eigenvalue, $T(\mathbf{x}) = \lambda\mathbf{x}$, we see the eigenvalue corresponding to $c_1(1, 0)$ is $\lambda_1 = 1$ and the eigenvalue corresponding to $c_2(0, 1)$ is $\lambda_2 = -1$ (Figure 1).

FIGURE 1

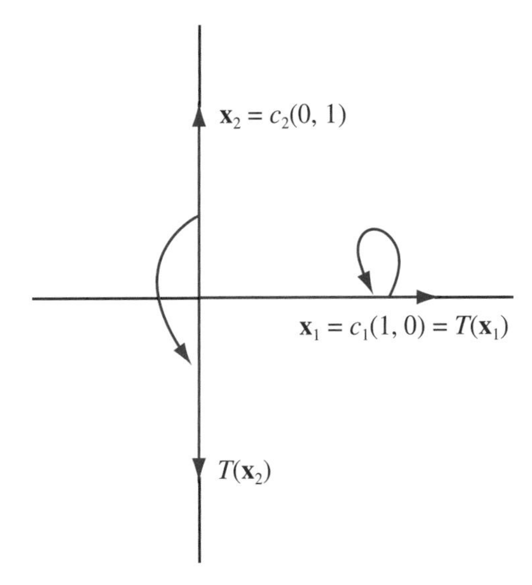

■

NOTE Let $T: \mathbf{V} \to \mathbf{V}$ be a linear transformation. For a scalar λ and a vector $\mathbf{x}$, we may rewrite the equation $T(\mathbf{x}) = \lambda\mathbf{x}$ as $T(\mathbf{x}) = \lambda I(\mathbf{x})$ or $(T - \lambda I)(\mathbf{x}) = \mathbf{0}$, where I is the identity transformation. This last equation tells us that a nonzero vector $\mathbf{x}$ is an eigenvector of T if and only if $\mathbf{x}$ lies on the kernel of the linear transformation $T - \lambda I$. From Theorem 1 of Section 7.3, we know that the kernel of this linear transformation is a subspace of $\mathbf{V}$. Thus, the set of all eigenvectors of T that correspond to λ, together with the zero vector, forms a subspace of $\mathbf{V}$, Ker $(T - \lambda I)$. This subspace is called the **eigenspace** of λ. For example, the eigenspaces of $\lambda_1 = 1$ and $\lambda_2 = -1$ in Example 2 are $\{c_1(1, 0)\}$ and $\{c_2(0, 1)\}$, respectively.

In Section 7.4, we demonstrated that any linear transformation $T: \mathbf{V} \to \mathbf{W}$ may be represented as multiplication by an $n \times m$ matrix A. (In dealing with eigenvalues and eigenvectors, $\mathbf{V} = \mathbf{W}$, so $n = m$, and the same basis can be used for each space.) This matrix A may then be used to calculate eigenvalues and eigenvectors of the transformation T. For this reason, we can restrict our attention to the matrix case. The following theorem makes this idea precise. The proof is an immediate consequence of Theorem 1 of Section 7.4 and the definition of eigenvalue and eigenvector.

THEOREM 1

Let $T: = \mathbf{V} \to \mathbf{V}$ be a linear transformation, $\mathcal{B}$ be a basis of $\mathbf{V}$, and A be the matrix that represents T with respect to $\mathcal{B}$. Let λ be a real number and $\mathbf{v}$ be a vector in $\mathbf{V}$ with coordinate vector $\mathbf{x}$ with respect to $\mathcal{B}$. Then λ is an eigenvalue of T with corresponding eigenvector $\mathbf{v}$ if and only if λ is an eigenvalue of A with corresponding eigenvector $\mathbf{x}$. ■

DEFINITION

The **eigenspace** of an eigenvalue λ_0 of a matrix A is the set of *all* solutions of $(A - \lambda_0 I)\mathbf{x} = \mathbf{0}$.

From Theorem 4 of Section 5.3, the dimension of the eigenspace for λ_0 is $n - r$, where r is the rank of $A - \lambda_0 I$. The procedure associated with this theorem provides a basis for the eigenspace.

EXAMPLE 3

Find the eigenvalues and corresponding eigenvectors, as well as a basis for each eigenspace, of the matrix

$$A = \begin{bmatrix} 4 & 1 & -3 \\ 0 & 0 & 2 \\ 0 & 0 & -3 \end{bmatrix}.$$

Solution Following the procedure for finding the eigenvalues of A (page 175), we compute

$$\det(A - \lambda I) = \det\begin{bmatrix} 4-\lambda & 1 & -3 \\ 0 & -\lambda & 2 \\ 0 & 0 & -3-\lambda \end{bmatrix} = (4-\lambda)(-\lambda)(-3-\lambda).$$

We now set this expression equal to zero, obtaining the equation

$$(4-\lambda)(-\lambda)(-3-\lambda) = 0.$$

The eigenvalues of A are the roots of the above equation:

$$\lambda_1 = 4, \qquad \lambda_2 = 0, \qquad \text{and} \qquad \lambda_3 = -3$$

To find the eigenvectors corresponding to $\lambda_1 = 4$, we form the matrix

$$A - 4I = \begin{bmatrix} 0 & 1 & -3 \\ 0 & -4 & 2 \\ 0 & 0 & -7 \end{bmatrix}$$

and solve the homogeneous linear system $(A - 4I)\mathbf{x} = \mathbf{0}$. Doing so, we see that all eigenvectors corresponding to $\lambda_1 = 4$ take the form $(t, 0, 0)$, where $t \neq 0$, and that a basis for the eigenspace of $\lambda_1 = 4$ is $\{(1, 0, 0)\}$.

To find the eigenspaces of $\lambda_2 = 0$ and $\lambda_3 = -3$, we work in a similar manner with the matrices $A - 0I\ (= A)$ and $A + 3I$, respectively. This yields eigenvectors of the form $(r, -4r, 0)$, $r \neq 0$, for λ_2 and $(11s, -14s, 21s)$, $s \neq 0$, for λ_3. The respective eigenspaces have bases given by $\{(1, -4, 0)\}$ and $\{(11, -14, 21)\}$. ■

WARNING The scalar 0 may or may not be an *eigenvalue* of a matrix A. By definition, however, the zero vector $\mathbf{0}$ is never an *eigenvector* of A. On the other hand, $\mathbf{0}$ must be in every *eigenspace* of A because the zero vector is an element of every subspace of $\mathbf{R}^m$.

Since an nth-degree polynomial equation has at most n roots, we see that an $n \times n$ matrix has *at most n* eigenvalues. As the next example illustrates, a matrix may have *fewer* than n eigenvalues.

EXAMPLE 4

Find the eigenvalues and bases for the corresponding eigenspaces of

$$A = \begin{bmatrix} -1 & 1 & 1 & 0 \\ 0 & 0 & 0 & -1 \\ 0 & 1 & 0 & 0 \\ 0 & 1 & 1 & -1 \end{bmatrix}$$

Solution Here the characteristic polynomial is of fourth degree and is computed (by expanding down the first column of $A - \lambda I$) to be

$$\det(A - \lambda I) = (-1-\lambda)\det\begin{bmatrix} -\lambda & 0 & -1 \\ 1 & -\lambda & 0 \\ 1 & 1 & -1-\lambda \end{bmatrix}$$
$$= (-1-\lambda)[-\lambda^3 - \lambda^2 - \lambda - 1].$$

Thus, one of the eigenvalues of A is $\lambda = -1$ (from the first factor), and the others are the real roots of the cubic equation $\lambda^3 + \lambda^2 + \lambda + 1 = 0$ (from the second factor). Since

$$\lambda^3 + \lambda^2 + \lambda + 1 = \lambda^2(\lambda + 1) + (\lambda + 1) = (\lambda^2 + 1)(\lambda + 1).$$

we see that -1 also satisfies this equation. We can find the remaining eigenvalues of A by solving the quadratic equation $\lambda^2 + 1 = 0$. Since the latter has no real roots (try, for example, the quadratic formula), A has only one eigenvalue, $\lambda_1 = -1$, which appears twice as a root of the characteristic polynomial.

To find the eigenspace of this eigenvalue, we form the matrix

$$A - (-1)I = A + I = \begin{bmatrix} 0 & 1 & 1 & 0 \\ 0 & 1 & 0 & -1 \\ 0 & 1 & 1 & 0 \\ 0 & 1 & 1 & 0 \end{bmatrix}$$

and solve the system $(A + I)\mathbf{x} = \mathbf{0}$. Doing so, we see that the eigenspace of $\lambda_1 = -1$ is the set of all vectors of the form

$$(s, t, -t, t) = s(1, 0, 0, 0) + t(0, 1, -1, 1).$$

A basis for this eigenspace is $\{(1, 0, 0, 0), (0, 1, -1, 1)\}$. (Note that this is quite different from Example 3, where the dimension of each eigenspace is 1; here, the dimension of the eigenspace corresponding to eigenvalue -1 is 2, and we notice that eigenvalue -1 appears more than once as a root of the characteristic polynomial.) ■

8.1 Exercises

In Exercises 1–8 find the eigenvalues and eigenvectors of the given linear transformations.

1. The identity transformation, I: $\mathbf{R}^n \to \mathbf{R}^n$ defined by $I(\mathbf{x}) = \mathbf{x}$

2. The zero transformation, 0: $\mathbf{R}^n \to \mathbf{R}^n$ defined by $0(\mathbf{x}) = \mathbf{0}$

3. The transformation $T_{\pi/2}$: $\mathbf{R}^2 \to \mathbf{R}^2$, which rotates every vector $\mathbf{x}$ in $\mathbf{R}^2$ by $\pi/2$

4. The transformation T_{π}: $\mathbf{R}^2 \to \mathbf{R}^2$, which rotates every vector $\mathbf{x}$ in $\mathbf{R}^2$ by π

5. The transformation $T^{\pi/2}$: $\mathbf{R}^2 \to \mathbf{R}^2$, which reflects every vector $\mathbf{x}$ in $\mathbf{R}^2$ in the y-axis

6. The transformation $T^{\pi/4}$: $\mathbf{R}^2 \to \mathbf{R}^2$, which reflects every vector $\mathbf{x}$ in $\mathbf{R}^2$ in the line $y = x$

7. The transformation T: $\mathbf{P}_3 \to \mathbf{P}_3$ given by

$$T(a_0 + a_1x + a_2x^2 + a_3x^3) = (a_0 + a_2) + (a_1 + 2a_3)x^3$$

8. The transformation T: $\mathbf{M}^{2,2} \to \mathbf{M}^{2,2}$ given by

$$T\left(\begin{bmatrix} a_{11} & a_{12} \\ a_{21} & a_{22} \end{bmatrix}\right) = \begin{bmatrix} a_{11} & -a_{21} \\ -a_{21} & a_{11} - a_{12} \end{bmatrix}$$

9–20. Find all eigenvectors and bases for the eigenspaces of the matrices in Exercises 1–12 of Section 4.4.

21. Let T_θ: $\mathbf{R}^2 \to \mathbf{R}^2$ be the rotation by θ (see Example 7 of Section 7.1). Find the eigenvalues and eigenvectors of T_θ by (a) using the definition and (b) working with the matrix associated with T_θ.

22. Let T^{θ}: $\mathbf{R}^2 \rightarrow \mathbf{R}^2$ be the reflection in the line through the origin that forms an angle θ with positive x-axis (see Example 9 of Section 7.1). Follow the instructions in Exercise 21 for T^{θ}.

23. Prove that the eigenvalues of a matrix are "invariant under similarity transformation." That is, if A is an $n \times n$ matrix and P is an invertible matrix of order n, then the eigenvalues of A and $P^{-1}AP$ are identical. (*Hint*: $\det[P^{-1}AP - \lambda I] = \det[P^{-1}(AP - \lambda PI)] = \det[P^{-1}(A - \lambda I)P]$.)

24–27. If $P(x) = a_n x^n + \cdots + a_1 x + a_0$ is a polynomial with coefficients a_i, $i = 0, 1, \ldots, n$, with coefficients a_i being real numbers, and A is an order n square matrix, then $P(A)$ is defined to be the matrix

$$P(A) = a_n A^n + \cdots + a_1 A + a_0 I.$$

The *Cayley–Hamilton theorem* states that if $P(x)$ is the characteristic polynomial of a square matrix A, then $P(A) = 0$. Verify the Cayley–Hamilton theorem for each matrix in Exercises 3–6 of Section 4.4.

Exercises 28 and 29 deal with the *Leslie population model* introduced in the Spotlight at the beginning of this chapter. If we divide a given female population into *n* age groups, the corresponding *Leslie matrix* is a square matrix of order *n* of the form

$$L = \begin{bmatrix} d_1 & d_2 & d_3 & \cdots & d_{n-1} & d_n \\ p_1 & 0 & 0 & \cdots & 0 & 0 \\ 0 & p_2 & 0 & \cdots & 0 & 0 \\ 0 & 0 & p_3 & \cdots & 0 & 0 \\ \vdots & \vdots & & \ddots & & \vdots \\ 0 & 0 & 0 & & p_{n-1} & 0 \end{bmatrix}.$$

Here, d_k is the average number of daughters per female in group *k* and p_k is the probability that a female in group *k* survives to become a member of group $k + 1$.

SPOTLIGHT *28.* Suppose that the female population of a certain country is divided into two age groups:

(*i*) Group A, those females who are under age 30;
(*ii*) Group B, those females who are of age 30 or older.

Further suppose that

- The average number of daughters per female in group A is 0.5,
- The average number of daughters per female in group B is 0.25,
- The probability that a female in group A will live longer than 30 years is 0.95.

(*a*) Find the Leslie matrix for this population.
(*b*) Find the positive eigenvalue, λ, for the Leslie matrix.
(*c*) What is the significance of λ, viewed in the light of the Leslie population model?

SPOTLIGHT *29.* Let A be a 3×3 Leslie matrix of the form

$$A = \begin{bmatrix} a & b & c \\ p & 0 & 0 \\ 0 & q & 0 \end{bmatrix},$$

where a, b, c, p, and q are nonnegative. Assuming that $a > 0$, show that A has a single positive eigenvalue. (*Hint*: Descartes's "rule of signs" may come in handy here.)

Computational Exercises

In Exercises 30 and 31 find numerical approximations for the eigenvalues using a linear algebra software package.

30. $\begin{bmatrix} 1 & 2 & 3 & 4 \\ -1 & 0 & 1 & 3 \\ 2 & 1 & 0 & 1 \\ 1 & 1 & 3 & -1 \end{bmatrix}$

31. $\begin{bmatrix} 1 & \frac{1}{2} & \frac{1}{3} & \frac{1}{4} \\ \frac{1}{2} & \frac{1}{3} & \frac{1}{4} & \frac{1}{5} \\ \frac{1}{3} & \frac{1}{4} & \frac{1}{5} & \frac{1}{6} \\ \frac{1}{4} & \frac{1}{5} & \frac{1}{6} & \frac{1}{7} \end{bmatrix}$

32, 33. Redo Exercises 30 and 31 using a computer algebra system.

34. Find a reasonable conjecture for the eigenvalues of a matrix of order n that has zeros down the main diagonal and ones everywhere else. (*Hint*: Start with $n = 2$, then $n = 3$, etc., using a computer algebra system.)

35. Prove your result for Exercise 34.

SPOTLIGHT **36.** Use a graphing calculator or computer software package to find the positive eigenvalue and corresponding eigenvectors of the matrix

$$L = \begin{bmatrix} 0.5 & 2.5 & 1.5 \\ 0.4 & 0 & 0 \\ 0 & 0.3 & 0 \end{bmatrix},$$

which arose in the Spotlight at the beginning of the chapter. Viewed in the context of the Spotlight, what do this eigenvalue and its eigenvectors represent?

8.2 Diagonalization

As we know from Section 7.4, the matrix of a linear transformation T from $\mathbf{V}$ into itself depends on the choice of basis for $\mathbf{V}$. Moreover (from Section 7.5), A and B are both matrices of a transformation T (corresponding to possibly different bases for $\mathbf{V}$) if and only if A and B are *similar*; that is, if and only if there is a nonsingular matrix P such that $B = P^{-1}AP$. Consequently, the problem investigated in this section may be phrased in two equivalent ways:

i. If T is a linear transformation of $\mathbf{V}$ into itself, can we determine a basis for $\mathbf{V}$ for which the matrix associated with T is diagonal?

ii. If A is a square matrix, can we determine a nonsingular matrix P such that $P^{-1}AP$ is a diagonal matrix?

In either formulation, we refer to this problem as **diagonalization** of the transformation T or the matrix A. As in Section 8.1, we concentrate our efforts on the *matrix* rather than the transformation.

DEFINITION

A square matrix A is **diagonalizable** (or can be **diagonalized**) if it is similar to a diagonal matrix.

The following procedure is justified by Theorems 1 through 3, given later in this section.

Procedure for Diagonalizing a Matrix

Let A be a square matrix of order n.

i. Calculate the eigenvalues of A.

ii. Calculate a basis for the eigenspace of each eigenvalue.

iii. Let $\mathcal{B} = \{\mathbf{x}_1, \mathbf{x}_2, \ldots, \mathbf{x}_m\}$ be the union of all the bases of (ii).

Then,

iv. If $m < n$, A is not diagonalizable;

v. If $m = n$, $\mathcal{B}$ is a basis for $\mathbf{R}^n$ and A is diagonalizable. A diagonalization of A is given by the $n \times n$ matrix D with ith diagonal element $d_{ii} = \lambda_i$, where λ_i is the eigenvalue to which $\mathbf{x}_i$ corresponds. (If P is the matrix with ith column $\mathbf{p}^i = \mathbf{x}_i$, then P is invertible and $D = P^{-1}AP$.)

EXAMPLE 1

Let

$$A = \begin{bmatrix} 1 & 0 & 0 \\ 0 & 0 & 1 \\ 0 & 1 & 0 \end{bmatrix}.$$

Show that A is diagonalizable, find a matrix P so that $P^{-1}AP$ is a diagonal matrix, and determine $P^{-1}AP$.

Solution To diagonalize A, we first calculate the eigenvalues of A. The characteristic polynomial for A is

$$\det(A - \lambda I) = (1 - \lambda)(\lambda^2 - 1) = -(\lambda - 1)^2(\lambda + 1).$$

Thus, the eigenvalues of A are $\lambda_1 = 1$ and $\lambda_2 = -1$. After some computation, we see that the bases for their respective eigenspaces are $\{(1, 0, 0), (0, 1, 1)\}$, and $\{(0, 1, -1)\}$. The eigenvectors $(1, 0, 0)$, $(0, 1, 1)$, and $(0, 1, -1)$ form a basis for $\mathbf{R}^3$, and consequently A is diagonalizable. Moreover, the matrix P that we are seeking may be taken to be the one whose columns are these three eigenvectors, so that

$$P = \begin{bmatrix} 1 & 0 & 0 \\ 0 & 1 & 1 \\ 0 & 1 & -1 \end{bmatrix}.$$

Finally, to find the diagonal matrix $P^{-1}AP$, we could perform the indicated inversion and multiplications, but it is much easier to apply step (v) of the procedure. Since the eigenvalues corresponding to the eigenvectors $(1, 0, 0)$, $(0, 1, 1)$, and $(0, 1, -1)$ are 1, 1, and -1, respectively, we have

$$P^{-1}AP = \begin{bmatrix} 1 & 0 & 0 \\ 0 & 1 & 0 \\ 0 & 0 & -1 \end{bmatrix}.$$

■

NOTE Reordering the eigenvectors produces a different diagonal matrix, so if A is diagonalizable, it is usually similar to several diagonal matrices. Of course, P changes accordingly.

EXAMPLE 2

Determine whether or not the matrix

$$A = \begin{bmatrix} 1 & 0 & 0 \\ 0 & 0 & -1 \\ 0 & 1 & 0 \end{bmatrix}$$

is diagonalizable.

Solution The characteristic polynomial for A is $\det(A - \lambda I) = (1 - \lambda)(\lambda^2 + 1)$, so A has only one real number eigenvalue $\lambda = 1$. Moreover, a basis for the eigenspace of $\lambda = 1$ is given by $\{(1, 0, 0)\}$ and is consequently one-dimensional. Thus, A has no set of three independent eigenvectors and, by the procedure, it is not diagonalizable. ■

In Example 2, two of the roots of the characteristic equation were imaginary numbers. The next example illustrates that even if *all* its characteristic roots are real, a matrix still may not be diagonalizable.

EXAMPLE 3

Determine whether or not the matrix

$$A = \begin{bmatrix} 1 & 2 & -1 \\ 0 & -1 & 3 \\ 0 & 0 & 1 \end{bmatrix}$$

is diagonalizable.

Solution Since A is upper triangular, its eigenvalues are given by the diagonal entries. Therefore, A has two distinct eigenvalues, $\lambda_1 = 1$ and $\lambda_2 = -1$. Computing bases for the corresponding eigenspaces, we see that $\{(1, 0, 0)\}$ is a basis for the eigenspace of λ_1 and $\{(-1, 1, 0)\}$ is a basis for the eigenspace of λ_2. Thus, all eigenvectors are nonzero scalar multiples of these two vectors, and A has no set of three independent eigenvectors. Consequently, by the procedure, A is not diagonalizable. ■

We now state and prove the theorems that support the preceding procedure. The first theorem gives the reason why $\mathcal{B}$ is a basis when it has "enough" vectors in step (v) of the procedure for diagonalizing a matrix.

THEOREM 1

For a linear transformation T: $\mathbf{V} \to \mathbf{V}$, nonzero eigenvectors that correspond to *distinct* eigenvalues are linearly independent. Moreover, if $\lambda_1 \neq \lambda_2$ are eigenvalues, if $\mathcal{T}_1$ is a linearly independent set of vectors that corresponds to λ_1, and if $\mathcal{T}_2$ is a linearly independent set of vectors that corresponds to λ_2, then $\mathcal{T} = \mathcal{T}_1 \cup \mathcal{T}_2$ is linearly independent. More generally, if λ_i, $i = 1, \ldots, n$ are distinct eigenvalues of T and $\mathcal{T}_i$ is a set of linearly independent eigenvectors corresponding to λ_i for each $i = 1, \ldots, n$, then $\mathcal{T} = \mathcal{T}_1 \cup \mathcal{T}_2 \cup \cdots \cup \mathcal{T}_n$ is linearly independent.

Proof Let $\mathcal{T}_1 = \{\mathbf{x}_1, \ldots, \mathbf{x}_k\}$ and $\mathcal{T}_2 = \{\mathbf{x}_{k+1}, \ldots, \mathbf{x}_m\}$, and suppose that $\mathbf{0}$ is a linear combination of the vectors in $\mathcal{T} = \mathcal{T}_1 \cup \mathcal{T}_2$. Then

$$a_1\mathbf{x}_1 + \cdots + a_k\mathbf{x}_k + a_{k+1}\mathbf{x}_{k+1} + \cdots + a_m\mathbf{x}_m = \mathbf{0}. \tag{1}$$

Applying T to both sides and using the fact that each $\mathbf{x}_i$ is an eigenvector of one of λ_1 or λ_2, we have

$$T(a_1\mathbf{x}_1 + \cdots + a_k\mathbf{x}_k + a_{k+1}\mathbf{x}_{k+1} + \cdots + a_m\mathbf{x}_m) = T(\mathbf{0}) = \mathbf{0},$$

so

$$a_1\lambda_1\mathbf{x}_1 + \cdots + a_k\lambda_1\mathbf{x}_k + a_{k+1}\lambda_2\mathbf{x}_{k+1} + \cdots + a_m\lambda_2\mathbf{x}_m = \mathbf{0}. \tag{2}$$

Multiplying Equation (1) by λ_1, we have

$$a_1\lambda_1\mathbf{x}_1 + \cdots + a_k\lambda_1\mathbf{x}_k + a_{k+1}\lambda_1\mathbf{x}_{k+1} + \cdots + a_m\lambda_1\mathbf{x}_m = \mathbf{0}. \tag{3}$$

Subtracting Equation (3) from Equation (2), we have

$$a_{k+1}(\lambda_2 - \lambda_1)\mathbf{x}_{k+1} + \cdots + a_m(\lambda_2 - \lambda_1)\mathbf{x}_m = \mathbf{0}. \tag{4}$$

But the left side of Equation (4) is a linear combination of vectors in $\mathcal{T}_2$, a linearly independent set. Therefore, all the coefficients are equal to 0; that is,

$$a_{k+1}(\lambda_2 - \lambda_1) = \cdots = a_m(\lambda_2 - \lambda_1) = 0.$$

Since the eigenvalues are distinct, $\lambda_2 - \lambda_1 \neq 0$, so we have $a_{k+1} = \cdots = a_m = 0$.

Returning to Equation (1), we now have

$$a_1\mathbf{x}_1 + \cdots + a_k\mathbf{x}_k = \mathbf{0}. \quad (5)$$

But the left side of Equation (5) is a linear combination of vectors in $\mathcal{T}_1$, also linearly independent. Therefore, the rest of the coefficients must also equal 0, $a_1 = \cdots = a_k = 0$.

In conclusion, any linear combination of vectors in $\mathcal{T}$ can equal $\mathbf{0}$ only if the coefficients are all 0. That is, $\mathcal{T} = \mathcal{T}_1 \cup \mathcal{T}_2$ is linearly independent.

The extension to the case of n distinct eigenvalues is a straightforward proof by induction using exactly the same idea. ■

THEOREM 2

A square matrix of order n is diagonalizable if and only if A has n independent eigenvectors; that is, if and only if there exists a basis for $\mathbf{R}^n$ of eigenvectors of A.

Proof Suppose that A is diagonalizable, D is a diagonal matrix, and P is the matrix of similarity, such that

$$D = P^{-1}AP.$$

Then $AP = PD$, and letting $\lambda_1, \ldots, \lambda_n$ be the diagonal entries of $D(d_{ii} = \lambda_i)$, we have for each column $\mathbf{p}^i$ of P, $A\mathbf{p}^i = \lambda_i\mathbf{p}^i$. But this implies that each column of P is an eigenvector of A corresponding to the eigenvalue λ_i. By Theorem 3 of Section 5.1, these columns are independent (since P is invertible), and thus A has n independent eigenvectors.

Conversely, suppose that A possesses n independent eigenvectors $\mathbf{x}_1, \ldots, \mathbf{x}_n$ that correspond respectively to (not necessarily distinct) eigenvalues $\lambda_1, \ldots, \lambda_n$. Following the procedure, we let P be the matrix of columns $\mathbf{p}^i = \mathbf{x}_i$ and D be the diagonal matrix whose ith diagonal entry, d_{ii} is λ_i. By Theorem 3 of Section 5.1, P is invertible, since its columns are linearly independent. Since λ_i is an eigenvalue of A with corresponding eigenvector $\mathbf{x}_i$, we have

$$A\mathbf{p}^i = A\mathbf{x}_i = \lambda_i\mathbf{x}_i = \lambda_i\mathbf{p}^i.$$

Thus, the ith column of AP is $\lambda_i\mathbf{p}^i$, so that the ith column of $P^{-1}AP$ is

$$P^{-1}(\lambda_i\mathbf{p}^i) = \lambda_i(P^{-1}\mathbf{p}^i) = \lambda_i\mathbf{e}_i.$$

But $\lambda_i\mathbf{e}_i$ is just the ith column of the diagonal matrix D. So, taking all the columns collectively, we conclude that

$$D = P^{-1}AP.$$

■

In one important case, we can find a diagonalization D of a matrix (although not the matrix of similarity P) without calculating a basis of eigenvectors for each eigenspace, as the procedure generally requires. In this case, only steps (i) and (v) of the procedure are necessary. The following theorem describes this fortunate situation.

THEOREM 3

Let A be an $n \times n$ matrix. If A has n distinct eigenvalues, then A is diagonalizable.

Proof Choosing a nonzero eigenvector for each eigenvalue, we have n linearly independent eigenvectors, by Theorem 1. By Theorem 2, A is diagonalizable. ■

EXAMPLE 4

Determine whether or not the matrix

$$A = \begin{bmatrix} 1 & 1 \\ 1 & 1 \end{bmatrix}$$

is diagonalizable.

Solution The characteristic polynomial $\det(A - \lambda I)$ is given by $\lambda^2 - 2\lambda$, which has zeros $\lambda_1 = 0$ and $\lambda_2 = 2$. Thus, the 2×2 matrix A has two distinct eigenvalues, and we know, from Theorem 3, that A is diagonalizable. ■

Notice that Theorem 3 gives only a *sufficient* condition that a matrix be diagonalizable. Consequently, if an $n \times n$ matrix A has fewer than n eigenvalues, then A *may* or *may not* be diagonalizable. For example, the matrices of Examples 1 and 3 have fewer than three eigenvalues, yet the first of these is diagonalizable, whereas the second is not.

Diagonalization of Symmetric Matrices

We now consider the class of *symmetric* matrices. These matrices can easily be identified by inspection. As we will see, all symmetric matrices are diagonalizable. Recall the following definition from Section 3.1:

DEFINITION

A square matrix A is **symmetric** if it is equal to its transpose A^T.

It is very easy to determine whether or not a given matrix is symmetric. Simply check for symmetry with respect to the main diagonal. For example, the matrices

$$\begin{bmatrix} 0 & 1 \\ 1 & 0 \end{bmatrix} \quad \text{and} \quad \begin{bmatrix} 1 & -1 & 3 \\ -1 & 2 & 0 \\ 3 & 0 & -5 \end{bmatrix}$$

are symmetric, but

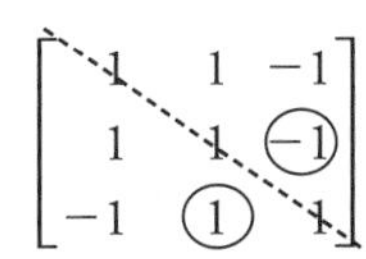

is not.

We now proceed to investigate the diagonalization of symmetric matrices. First, recall from Section 2.1 that two vectors $\mathbf{x}$ and $\mathbf{y}$ in $\mathbf{R}^n$ are *orthogonal* if and only if their dot product, $\mathbf{x} \cdot \mathbf{y}$, is equal to zero.

THEOREM 4

Let λ_1 and λ_2 be distinct eigenvalues of a symmetric matrix A with corresponding eigenvectors $\mathbf{x}_1$ and $\mathbf{x}_2$. Then $\mathbf{x}_1$ and $\mathbf{x}_2$ are orthogonal.

Proof We must show that $\mathbf{x}_1 \cdot \mathbf{x}_2 = 0$. By definition, $A\mathbf{x}_1 = \lambda_1\mathbf{x}_1$, so

$$\lambda_1(\mathbf{x}_1 \cdot \mathbf{x}_2) = (\lambda_1\mathbf{x}_1) \cdot \mathbf{x}_2 = (A\mathbf{x}_1) \cdot \mathbf{x}_2,$$

which, by Exercise 28, is equal to $\mathbf{x}_1 \cdot (A^T\mathbf{x}_2)$. Since A is symmetric, the last expression is equal to $\mathbf{x}_1 \cdot (A\mathbf{x}_2)$, which is the same as $\mathbf{x}_1 \cdot (\lambda_2\mathbf{x}_2)$ or $\lambda_2(\mathbf{x}_1 \cdot \mathbf{x}_2)$. Thus, $\lambda_1(\mathbf{x}_1 \cdot \mathbf{x}_2) = \lambda_2(\mathbf{x}_1 \cdot \mathbf{x}_2)$. Now, if $\mathbf{x}_1 \cdot \mathbf{x}_2$ were not 0, then dividing by this quantity, we would get $\lambda_1 = \lambda_2$, a contradiction. Therefore, $\mathbf{x}_1 \cdot \mathbf{x}_2$ must be 0, as desired. ■

EXAMPLE 5

Verify Theorem 4 for the matrix

$$A = \begin{bmatrix} 3 & 2 \\ 2 & 0 \end{bmatrix}.$$

Solution Since $\det(A - \lambda I) = \lambda^2 - 3\lambda - 4$, the eigenvalues of A are $\lambda_1 = 4$ and $\lambda_2 = -1$. Theorem 4 tells us that since A is symmetric and λ_1 and λ_2 are distinct, any pair of eigenvectors corresponding to these eigenvalues should be orthogonal. The eigenspaces of λ_1 and λ_2 are easily computed to be $\{(2s, s)\}$ and $\{(-t, 2t)\}$, respectively. To confirm that Theorem 4 holds in this case, we form the dot product

$$(2s, s) \cdot (-t, 2t) = -2st + 2st,$$

which is indeed equal to 0. ■

Since the matrix A of Example 5 has distinct eigenvalues, Theorem 3 implies that A is diagonalizable. In other words, there is a 2×2 invertible matrix P such that $P^{-1}AP = D$, a diagonal matrix. We can then apply the diagonalization procedure of this section to construct P by taking its columns to be linearly independent eigenvectors of A. By choosing these eigenvectors to have length 1 (that is, *normalizing* them), we obtain a matrix with particularly interesting properties. The matrix of normalized eigenvectors is

$$P = \begin{bmatrix} \frac{2}{\sqrt{5}} & \frac{-1}{\sqrt{5}} \\ \frac{1}{\sqrt{5}} & \frac{2}{\sqrt{5}} \end{bmatrix}.$$

We not only have $\mathbf{p}^1 \cdot \mathbf{p}^2 = 0$
(reflecting the orthogonality of the eigenvectors) and $\mathbf{p}^1 \cdot \mathbf{p}^1 = \mathbf{p}^2 \cdot \mathbf{p}^2 = 1$ (reflecting the fact that we chose eigenvectors of length 1), but, computing P^{-1}, we see that

$$P^{-1} = \begin{bmatrix} \frac{2}{\sqrt{5}} & \frac{1}{\sqrt{5}} \\ \frac{-1}{\sqrt{5}} & \frac{2}{\sqrt{5}} \end{bmatrix},$$

which is just P^T. As we shall see, *every* symmetric matrix A (whether or not it has distinct eigenvalues) is diagonalizable by a matrix P having these properties.

DEFINITION

A nonsingular matrix P is **orthogonal** if $P^{-1} = P^T$.

EXAMPLE 6

Let

$$P = \begin{bmatrix} \frac{1}{3} & \frac{2\sqrt{2}}{3} & 0 \\ \frac{2}{3} & \frac{-\sqrt{2}}{6} & \frac{\sqrt{2}}{2} \\ -\frac{2}{3} & \frac{\sqrt{2}}{6} & \frac{\sqrt{2}}{2} \end{bmatrix}.$$

Show that P is orthogonal.

Solution We could compute P^{-1} and then verify that $P^{-1} = P^T$, but there is an easier way. We simply compute the product P^TP. Since $P^TP = I$, by Theorem 2 of Section 3.2, $P^{-1} = P^T$, and thus P is orthogonal by definition. ■

NOTE It is easy to see that the columns of the matrix P of Example 6 have the following properties:

i. $\mathbf{p}^i \cdot \mathbf{p}^j = 0$, if $i \neq j$ $\quad (i, j = 1, 2, 3)$,

ii. $\mathbf{p}^i \cdot \mathbf{p}^i = 1$ $\quad (i = 1, 2, 3)$.

Recall that a set of vectors that possesses these two properties is called an *orthonormal set*. The proof of the next theorem is Exercise 29.

THEOREM 5

An $n \times n$ matrix P is orthogonal if and only if its columns

$$\{\mathbf{p}^1, \mathbf{p}^2, \ldots, \mathbf{p}^n\}$$

form an orthonormal set. ■

DEFINITION

A square matrix A is **orthogonally diagonalizable** if there exists an orthogonal matrix P such that $P^{-1}AP\,(=P^TAP)$ is a diagonal matrix.

The next theorem states that *every* symmetric matrix is orthogonally diagonalizable. The proof is omitted.

THEOREM 6

Let A be a symmetric matrix. Then there exists an orthogonal matrix P such that $D = P^{-1}AP$ is a diagonal matrix. ■

NOTE If A is a symmetric matrix, then it is diagonalizable (by Theorem 6) and, for each i, $A\mathbf{p}^i = PP^{-1}A\mathbf{p}^i = P\mathbf{d}^i = d_{ii}\mathbf{p}^i$. Thus, we have

i. The columns of P are linearly independent eigenvectors of A, $\mathbf{x}_i = \mathbf{p}^i$, $i = 1, \ldots, n$;

ii. The ith diagonal entry $\lambda_i = d_{ii}$ of D is the eigenvalue of A corresponding to the eigenvector $\mathbf{x}_i$.

Moreover, by Theorem 5, the columns of P form an orthonormal set. Consequently, $\{\mathbf{x}_1, \mathbf{x}_2, \ldots, \mathbf{x}_n\}$ is an orthonormal set of eigenvectors.

Procedure for Orthogonally Diagonalizing a Symmetric Matrix

Let A be a symmetric matrix of order n.

i. Calculate the eigenvalues of A.

ii. Calculate a basis for the eigenspace of each eigenvalue.

iii. With the aid of the Gram–Schmidt process (Section 6.4), replace each basis by an orthonormal basis. The union of all these bases is an orthonormal basis for $\mathbf{R}^n$.

iv. A diagonalization of A is the diagonal matrix D with $d_{ii} = \lambda_i$, the ith eigenvalue of A. The orthogonal matrix P has ith column $\mathbf{p}^i = \mathbf{x}_i$, the ith vector in the basis.

EXAMPLE 7 Let

$$A = \begin{bmatrix} -1 & 0 & 0 \\ 0 & 0 & 2 \\ 0 & 2 & 3 \end{bmatrix}.$$

Is the matrix A orthogonally diagonalizable? If so, find an orthogonal matrix P and diagonal matrix D such that $D = P^{-1}AP$.

Solution Since A is symmetric, by Theorem 6, A is orthogonally diagonalizable.

To find the matrices P and D, we apply the procedure just described. We first find the eigenvalues of A by solving the characteristic equation $\det(A - \lambda I) = 0$, where

$$\det(A - \lambda I) = \det\begin{bmatrix} -1-\lambda & 0 & 0 \\ 0 & -\lambda & 2 \\ 0 & 2 & 3-\lambda \end{bmatrix} = -(\lambda + 1)^2(\lambda - 4).$$

Thus, the eigenvalues of A are $\lambda_1 = -1$ and $\lambda_2 = 4$.

We now determine bases for the eigenspaces $\mathbf{S}_1$ and $\mathbf{S}_2$ of λ_1 and λ_2 by considering the linear homogeneous systems $(A + I)\mathbf{x} = \mathbf{0}$ and $(A - 4I)\mathbf{x} = \mathbf{0}$, respectively. We see that a basis for $\mathbf{S}_1$ is $\mathcal{T}_1 = \{(1, 0, 0), (0, -2, 1)\}$, whereas a basis for $\mathbf{S}_2$ is $\mathcal{T}_2 = \{(0, 1, 2)\}$.

Applying the Gram–Schmidt process to $\mathcal{T}_1$, we obtain the orthonormal basis $\mathcal{B}_1 = \{(1, 0, 0), (0, -2/\sqrt{5}, 1/\sqrt{5})\}$ for the eigenspace of λ_1. (Since $\mathcal{T}_1$ was orthogonal, all we actually needed to do was to normalize each vector in this set.) Then, normalizing the vector $(0, 1, 2)$ of $\mathcal{T}_2$, we obtain an orthonormal basis $\mathcal{B}_2 = \{(0, 1/\sqrt{5}, 2/\sqrt{5})\}$ for $\mathbf{S}_2$. Finally, taking the union of $\mathcal{B}_1$ and $\mathcal{B}_2$, we obtain an orthonormal set consisting of the three eigenvectors $\mathbf{x}_1 = (1, 0, 0)$, $\mathbf{x}_2 = (0, -2/\sqrt{5}, 1/\sqrt{5})$, and $\mathbf{x}_3 = (0, 1/\sqrt{5}, 2/\sqrt{5})$.

Thus, the matrix P is given by $\mathbf{p}^j = \mathbf{x}_j$, or

$$P = \begin{bmatrix} 1 & 0 & 0 \\ 0 & \frac{-2}{\sqrt{5}} & \frac{1}{\sqrt{5}} \\ 0 & \frac{1}{\sqrt{5}} & \frac{2}{\sqrt{5}} \end{bmatrix}.$$

To construct D, we simply form the 3×3 diagonal matrix whose diagonal entries are the eigenvalues corresponding to the eigenvectors $\mathbf{x}_1$, $\mathbf{x}_2$, and $\mathbf{x}_3$.

Therefore,

$$D = \begin{bmatrix} -1 & 0 & 0 \\ 0 & -1 & 0 \\ 0 & 0 & 4 \end{bmatrix}.$$

■

We now know that every symmetric matrix is orthogonally diagonalizable. A natural question to ask at this point is whether any other types of matrices are orthogonally diagonalizable. The next theorem, which is the converse of Theorem 6, provides a negative answer to this question. Its proof is Exercise 30.

THEOREM 7

If a square matrix A is orthogonally diagonalizable, it is symmetric. ■

8.2 Exercises

In Exercises 1–12 determine whether or not the given matrix A is diagonalizable. If it is, find a matrix P such that $P^{-1}AP$ is diagonal, and find this diagonal matrix.

1. $\begin{bmatrix} 1 & 4 \\ 2 & -1 \end{bmatrix}$

2. $\begin{bmatrix} 0 & 1 \\ 1 & 0 \end{bmatrix}$

3. $\begin{bmatrix} 1 & 2 \\ -2 & -1 \end{bmatrix}$

4. $\begin{bmatrix} 0 & 1 \\ -1 & 2 \end{bmatrix}$

5. $\begin{bmatrix} 0 & -1 & 0 \\ -1 & 0 & -1 \\ 0 & -1 & 0 \end{bmatrix}$

6. $\begin{bmatrix} 2 & -1 & 0 \\ -1 & 2 & -1 \\ 0 & -1 & 2 \end{bmatrix}$

7. $\begin{bmatrix} 1 & 1 & 1 \\ 1 & 1 & 1 \\ 1 & 1 & 1 \end{bmatrix}$

8. $\begin{bmatrix} 1 & 1 & -1 \\ -1 & -1 & 1 \\ 2 & 2 & -2 \end{bmatrix}$

9. $\begin{bmatrix} 1 & 0 & 3 \\ 0 & -2 & 1 \\ 0 & 0 & 1 \end{bmatrix}$

10. $\begin{bmatrix} 1 & 0 & -1 \\ 1 & 1 & 2 \\ 2 & 1 & 1 \end{bmatrix}$

11. $\begin{bmatrix} 1 & 1 & 1 & 1 \\ 1 & 1 & 1 & 1 \\ 1 & 1 & 1 & 1 \\ 1 & 1 & 1 & 1 \end{bmatrix}$

12. $\begin{bmatrix} 1 & 0 & 0 & 0 \\ 0 & 2 & 0 & 0 \\ 0 & 1 & -1 & 0 \\ -2 & 0 & 0 & 2 \end{bmatrix}$

In Exercises 13–16 determine whether or not the given matrix is orthogonal.

13. $\begin{bmatrix} \frac{1}{3} & \frac{2}{3} & \frac{2}{3} \\ \frac{2}{3} & \frac{1}{3} & -\frac{2}{3} \\ \frac{2}{3} & -\frac{2}{3} & \frac{1}{3} \end{bmatrix}$

14. $\begin{bmatrix} 0 & 1 & 0 \\ 1 & 0 & 0 \\ 0 & 0 & 1 \end{bmatrix}$

15. $\begin{bmatrix} \frac{1}{\sqrt{5}} & \frac{-2}{\sqrt{5}} & 1 \\ \frac{-2}{\sqrt{5}} & \frac{1}{\sqrt{5}} & -1 \\ 0 & 0 & 1 \end{bmatrix}$

16. $\begin{bmatrix} 0 & 1 & 0 \\ 1 & 0 & 1 \\ 0 & 1 & 0 \end{bmatrix}$

In Exercises 17–24 find an orthogonal matrix P such that $P^{-1}AP$ is diagonal, where A is the given symmetric matrix.

17. $\begin{bmatrix} 0 & 2 \\ 2 & 0 \end{bmatrix}$

18. $\begin{bmatrix} 1 & 1 \\ 1 & 1 \end{bmatrix}$

19. $\begin{bmatrix} 0 & -1 & 0 \\ -1 & 0 & -1 \\ 0 & -1 & 0 \end{bmatrix}$

20. $\begin{bmatrix} 2 & -1 & 0 \\ -1 & 2 & -1 \\ 0 & -1 & 2 \end{bmatrix}$

21. $\begin{bmatrix} 1 & 1 & 1 \\ 1 & 1 & 1 \\ 1 & 1 & 1 \end{bmatrix}$

22. $\begin{bmatrix} 0 & 1 & 0 \\ 1 & 0 & 0 \\ 0 & 0 & 1 \end{bmatrix}$

23. $\begin{bmatrix} 0 & 1 & 0 & 0 \\ 1 & 0 & 0 & 0 \\ 0 & 0 & 0 & 1 \\ 0 & 0 & 1 & 0 \end{bmatrix}$

24. $\begin{bmatrix} 1 & 1 & 1 & 1 \\ 1 & 1 & 1 & 1 \\ 1 & 1 & 1 & 1 \\ 1 & 1 & 1 & 1 \end{bmatrix}$

25. Show that the rotation matrix

$$A = \begin{bmatrix} \cos\theta & -\sin\theta \\ \sin\theta & \cos\theta \end{bmatrix}$$

is orthogonal.

26. Let A be an orthogonal matrix. Show that $\det(A) = \pm 1$. (*Hint*: $\det(A^TA) = \det(I) = 1$.)

27. For the matrix

$$A = \begin{bmatrix} a & b \\ c & d \end{bmatrix},$$

obtain conditions on a, b, c, and d such that (a) A is diagonalizable and (b) A is not diagonalizable. (*Hint*: See Exercise 21 of Section 4.4.)

28. Let A be an $n \times n$ matrix and $\mathbf{x}_1$ and $\mathbf{x}_2$ be in $\mathbf{R}^n$. Prove that $(A\mathbf{x}_1) \cdot \mathbf{x}_2 = \mathbf{x}_1 \cdot (A^T\mathbf{x}_2)$. (*Hint*: Viewing $\mathbf{u}$ and $\mathbf{v}$ as column vectors, the dot product is given by $\mathbf{u} \cdot \mathbf{v} = \mathbf{v}^T\mathbf{u}$.)

29. Prove that an $n \times n$ matrix P is orthogonal if and only if $\{\mathbf{p}^1, \mathbf{p}^2, \ldots, \mathbf{p}^n\}$ is an orthonormal set (Theorem 5).

30. Prove that if a square matrix is orthogonally diagonalizable, then it is symmetric (Theorem 7). (*Hint*: Consider $(P^{-1}AP)^T = D^T$.)

31. Let P and Q be orthogonal matrices of the same order. Show that PQ and QP are also orthogonal. (*Hint*: Show that $(PQ)(PQ)^T = I$.)

32. Let θ be a real number and s and t be integers with $1 \leq s < t \leq n$. Define an $n \times n$ matrix P by

$$p_{ss} = p_{tt} = \cos\theta, \qquad p_{ts} = -p_{st} = \sin\theta,$$
$$\text{and} \quad p_{ii} = 1, \quad p_{ij} = 0 \quad (i \neq j), \qquad \text{otherwise.}$$

Show that P is orthogonal. (The matrix P is called a *rotation matrix for the s,t-plane.*)

33. Let P be a third-order rotation in the 1,3-plane (see Exercise 32); that is

$$P = \begin{bmatrix} \cos\theta & 0 & -\sin\theta \\ 0 & 1 & 0 \\ \sin\theta & 0 & \cos\theta \end{bmatrix}.$$

Let A be an arbitrary 3×3 symmetric matrix. Show that if we choose θ so that $\tan 2\theta = 2a_{13}/(a_{11} - a_{33})$, then the matrix $B = P^TAP\ (=P^{-1}AP)$ satisfies $b_{13} = b_{31} = 0$. (*Note*: In general, if P is an nth-order rotation in the s,t-plane and A is symmetric, then, by choosing θ so that $\tan 2\theta = 2a_{st}/(a_{ss} - a_{tt})$, both the st and the ts entries of P^TAP are equal to 0.)

COMPUTATIONAL NOTE

Jacobi's method approximates the eigenvalues of a symmetric matrix A by forming the sequence of products

$$(P_k^T P_{k-1}^T \cdots P_1^T)A(P_1P_2 \cdots P_k), \quad k = 1, 2, 3, \ldots.$$

In this expression, each P_i is a rotation matrix (see Exercise 32) with θ chosen to "annihilate" (send to zero) the largest off-diagonal entry of

$$(P_{i-1}^T \cdots P_1^T)A(P_1 \cdots P_{i-1})$$

(see Exercise 33). Unfortunately, as one off-diagonal entry is annihilated, a previously annihilated one may become nonzero. Nevertheless, the size of all off-diagonal entries does approach zero as the process is continued, and the diagonal entries, in turn, get closer and closer to the eigenvalues of A. Jacobi's method is perhaps the most straightforward way of approximating the eigenvalues of a symmetric matrix, but there are more sophisticated techniques that are much more efficient. (See the last part of Section 8.3 for another method of this type.)

Computational Exercises

In Exercises 34 and 35 use a computer algebra system to find the orthogonal diagonalization of the symmetric matrix.

34. $\begin{bmatrix} 1 & 2 & 3 & 4 \\ 2 & 1 & 3 & 4 \\ 3 & 3 & 1 & 4 \\ 4 & 4 & 4 & 1 \end{bmatrix}$

35. $\begin{bmatrix} 0 & 1 & 0 & 2 \\ 1 & 0 & 1 & 3 \\ 0 & 1 & 0 & 4 \\ 2 & 3 & 4 & 0 \end{bmatrix}$

An eigenvalue of a matrix is called a dominant eigenvalue of the matrix if its absolute value is larger than the absolute values of the remaining eigenvalues. If a matrix A is diagonalizable with a dominant eigenvalue, then, for an arbitrary vector $\mathbf{v}$, the vector $A^n\mathbf{v}$ is usually a good approximation of a dominant eigenvector of A when the exponent is large. The dominant eigenvalue can be approximated by

$$\frac{(A^{n+1}\mathbf{v}) \bullet (A^n\mathbf{v})}{(A^n\mathbf{v}) \bullet (A^n\mathbf{v})} \approx \frac{(A\mathbf{x}) \bullet \mathbf{x}}{\mathbf{x} \bullet \mathbf{x}} = \frac{(\lambda\mathbf{x}) \bullet \mathbf{x}}{\mathbf{x} \bullet \mathbf{x}} = \frac{\lambda(\mathbf{x} \bullet \mathbf{x})}{\mathbf{x} \bullet \mathbf{x}} = \lambda,$$

where $\mathbf{x}$ is the dominant eigenvector.

36. Let

$$A = \begin{bmatrix} 3 & 1 \\ 2 & 5 \end{bmatrix}.$$

Evaluate $A^{10}\mathbf{v}$, $A^{20}\mathbf{v}$, $A^{30}\mathbf{v}$ for

$$\mathbf{v} = \begin{bmatrix} 1 \\ 1 \end{bmatrix} \quad \text{and} \quad \mathbf{v} = \begin{bmatrix} 2 \\ -1 \end{bmatrix}$$

and normalize the resulting vectors. Is there a dominant eigenvector? If so, find the dominant eigenvalues.

37. Let

$$A = \begin{bmatrix} 2 & -8 & 0 & 3 \\ 1 & -2 & 0 & 0 \\ 0 & 0 & -1 & 2 \\ 0 & 0 & -1 & 1 \end{bmatrix}.$$

Evaluate $A^{10}\mathbf{v}$, $A^{20}\mathbf{v}$, and $A^{30}\mathbf{v}$ for

$$\mathbf{v} = \begin{bmatrix} 1 \\ 1 \\ 1 \\ 1 \end{bmatrix} \quad \text{and} \quad \mathbf{v} = \begin{bmatrix} 1 \\ 0 \\ 1 \\ 0 \end{bmatrix},$$

and normalize the resulting vectors. Is there a dominant eigenvector? If so, find the dominant eigenvalues.

38. Let

$$A = \begin{bmatrix} 1 & 2 & 4 \\ -5 & 1 & 2 \\ 6 & -3 & 0 \end{bmatrix}.$$

Evaluate $A^{10}\mathbf{v}$, $A^{20}\mathbf{v}$, and $A^{30}\mathbf{v}$ for

$$\mathbf{v} = \begin{bmatrix} 1 \\ 0 \\ 1 \end{bmatrix} \quad \text{and} \quad \mathbf{v} = \begin{bmatrix} 2 \\ 1 \\ 1 \end{bmatrix},$$

and normalize the resulting vectors. Is there a dominant eigenvector? If so, find the dominant eigenvalues.

39. Use a computer algebra system to compute the eigenvalues and eigenvectors for the matrices in Exercises 36–38. Do your results agree with your findings?

8.3 The *QR* Method for Approximating Eigenvalues

In Section 3.4, we discussed the LU decomposition of a matrix—its factorization as the product of lower- and upper-triangular matrices—and applied this technique to the solution of systems of linear equations. Here, we consider another general way to decompose a matrix—as the product of orthogonal and upper-triangular matrices. As we will see, this factorization leads to a useful method for approximating the eigenvalues of a square matrix.

The *QR* Factorization of a Matrix

Recall that a square matrix M is *upper triangular* (Section 4.1) if all entries below its main diagonal are zero and is *orthogonal* (Section 8.2) if it is nonsingular with $M^{-1} = M^T$. Both types of matrices have special properties that make them important in linear algebra and its applications. In particular,

- If Q is an orthogonal matrix, then its columns form an orthonormal set;
- If R is an upper-triangular matrix, then the determinant of R is the product of its main diagonal entries, which, in turn, implies that the eigenvalues of R *are* its main diagonal entries.

The following procedure describes how to factor a square nonsingular matrix A as the product of orthogonal and upper-triangular matrices. This process is justified by Theorem 1, given later in this section.

NOTE Theorem 3 of Section 5.1 tells us that a square matrix A is nonsingular if and only if its columns form a linearly independent set, which would then be a basis for the *column space* of A (Section 5.4). Therefore, if A is nonsingular, we can use the Gram–Schmidt process (Section 6.4) to find an orthonormal basis for this column space.

A variant of QR factorization can also be performed on *rectangular* matrices with linearly independent columns, a technique that can be applied to solving systems of linear equations with more equations than unknowns. In this text, however, we discuss only QR factorization of square matrices and its application to approximating eigenvalues.

Procedure for Finding the QR Factorization of a Matrix

Let A be a nonsingular $n \times n$ matrix.

i. Apply the Gram–Schmidt process of Section 6.4 to find an orthonormal basis, $\{\mathbf{q}_1, \mathbf{q}_2, \ldots, \mathbf{q}_n\}$, for the column space of A.

ii. Define Q to be the $n \times n$ matrix whose ith column is $\mathbf{q}_i$.

iii. Define R to be the upper-triangular matrix (with zero entries below the main diagonal) for which $r_{ij} = \mathbf{a}^j \cdot \mathbf{q}^i$ (the dot product of the jth column of A with the ith column of Q) when $j \geq i$.

EXAMPLE 1

Find the QR factorization of the matrix

$$A = \begin{bmatrix} 1 & -1 & 0 \\ -1 & 1 & -1 \\ 0 & -1 & 1 \end{bmatrix}.$$

Solution To obtain the QR factorization, we follow the procedure just described. First, we apply the Gram–Schmidt process (Section 6.4) to the columns of A, $\{(1, -1, 0), (-1, 1, -1), (0, -1, 1)\}$, which gives the following *orthogonal* basis for the column space of A: $\{(-1, 1, 0), (0, 0, 1), (1, 1, 0)\}$. Dividing each of these vectors by its length yields the *orthonormal* sets $\mathbf{q}_1 = (-1/\sqrt{2}, 1/\sqrt{2}, 0)$, $\mathbf{q}_2 = (0, 0, 1)$, and $\mathbf{q}_3 = (1/\sqrt{2}, 1/\sqrt{2}, 0)$. These vectors become the columns of the matrix Q:

$$Q = \begin{bmatrix} \frac{-1}{\sqrt{2}} & 0 & \frac{1}{\sqrt{2}} \\ \frac{1}{\sqrt{2}} & 0 & \frac{1}{\sqrt{2}} \\ 0 & 1 & 0 \end{bmatrix}$$

The upper-triangular matrix R has the following nonzero entries:

$$\begin{aligned} r_{11} &= \mathbf{a}^1 \cdot \mathbf{q}^1 = (1, -1, 0) \cdot (-1/\sqrt{2}, 1/\sqrt{2}, 0) = -\sqrt{2} \\ r_{12} &= \mathbf{a}^2 \cdot \mathbf{q}^1 = (-1, 1, -1) \cdot (-1/\sqrt{2}, 1/\sqrt{2}, 0) = \sqrt{2} \\ r_{13} &= \mathbf{a}^3 \cdot \mathbf{q}^1 = (0, -1, 1) \cdot (-1/\sqrt{2}, 1/\sqrt{2}, 0) = -1/\sqrt{2} \\ r_{22} &= \mathbf{a}^2 \cdot \mathbf{q}^2 = (-1, 1, -1) \cdot (0, 0, 1) = -1 \\ r_{23} &= \mathbf{a}^3 \cdot \mathbf{q}^2 = (0, -1, 1) \cdot (0, 0, 1) = 1 \\ \text{and } r_{33} &= \mathbf{a}^3 \cdot \mathbf{q}^3 = (0, -1, 1) \cdot (1/\sqrt{2}, 1/\sqrt{2}, 0) = -1/\sqrt{2} \end{aligned}$$

$$\text{Thus,} \quad R = \begin{bmatrix} -\sqrt{2} & \sqrt{2} & \frac{-1}{\sqrt{2}} \\ 0 & -1 & 1 \\ 0 & 0 & \frac{-1}{\sqrt{2}} \end{bmatrix}.$$

■

We now state and prove a theorem that justifies the factorization procedure used in the previous example.

THEOREM 1

If A is a nonsingular $n \times n$ matrix, then there exist $n \times n$ matrices Q and R, where Q is orthogonal and R is upper triangular, such that $A = QR$.

Proof Construct the matrices Q and R as in the procedure above. Since the columns of Q form a basis for the column space of A, we can write each column of A, $\mathbf{a}^j$, as a linear combination of the $\mathbf{q}^k$.

$$\mathbf{a}^j = c_1\mathbf{q}^1 + c_2\mathbf{q}^2 + \cdots + c_n\mathbf{q}^n \tag{1}$$

Now, taking the dot product of each $\mathbf{q}^i$ with the given $\mathbf{a}^j$, we have

$$\mathbf{a}^j \cdot \mathbf{q}^i = c_1\mathbf{q}^1 \cdot \mathbf{q}^i + c_2\mathbf{q}^2 \cdot \mathbf{q}^i + \cdots + c_n\mathbf{q}^n \cdot \mathbf{q}^i = c_i\mathbf{q}^i \cdot \mathbf{q}^i = c_i,$$

because the $\mathbf{q}^k$ form an orthonormal set. Thus, for each i, $c_i = \mathbf{a}^j \cdot \mathbf{q}^i$. Substituting these expressions into Equation (1), we obtain

$$\mathbf{a}^j = (\mathbf{a}^j \cdot \mathbf{q}^1)\mathbf{q}^1 + (\mathbf{a}^j \cdot \mathbf{q}^2)\mathbf{q}^2 + \cdots + (\mathbf{a}^j \cdot \mathbf{q}^n)\mathbf{q}^n. \tag{2}$$

Setting the ith component of $\mathbf{a}^j$ equal to the ith component of the expression on the right (and remembering that $\mathbf{a}^j$ and the $\mathbf{q}^k$ are just columns of their respective matrices), Equation (2) becomes

$$a_{ij} = (\mathbf{a}^j \cdot \mathbf{q}^1)q_{i1} + (\mathbf{a}^j \cdot \mathbf{q}^2)q_{i2} + \cdots + (\mathbf{a}^j \cdot \mathbf{q}^n)q_{in}. \tag{3}$$

Now, recall from the Gram–Schmidt process that $\mathbf{a}^j$ is orthogonal to $\mathbf{q}^k$ whenever $j < k$. Thus, the dot products in Equation (3) give the entries in the jth column of R; in other words,

$$a_{ij} = q_{i1}r_{1j} + q_{i2}r_{2j} + \cdots + q_{in}r_{nj}.$$

This states exactly what we are trying to prove: A is the product of Q and R. ■

NOTE The proof of Theorem 1 requires that the columns of the given matrix A be linearly independent to guarantee that we can carry out the Gram–Schmidt process. This condition is not a necessary one, however; a matrix might have a QR factorization even if it is singular. Exercise 14, at the end of this section, provides such an example.

The *QR* Algorithm for Finding Eigenvalues

The QR factorization process provides an *iterative* procedure for approximating the eigenvalues of an $n \times n$ matrix A—it can be used to construct a sequence of matrices whose diagonal entries get closer and closer to the eigenvalues of A. This sequence of matrices, $A_1, A_2, A_3, \ldots$, is generated in the following way:

- Define $A_1 = A$ and factor A_1 as the product of an orthogonal matrix Q_1 and an upper-triangular matrix R_1: $A_1 = Q_1R_1$
- Define $A_2 = R_1Q_1$ (note the reversal of order), and then determine Q_2 and R_2 by factoring: $A_2 = Q_2R_2$
- Define $A_3 = R_2Q_2$ and factor it, obtaining Q_3 and R_3: $A_3 = Q_3R_3$

and so on.

The sequence of matrices generated in this way has two properties that ensure the success of the procedure for approximating the eigenvalues of A.

i. Each A_k is similar to A, and consequently has the same eigenvalues as A. To see that A_2 is similar to A_1 $(=A)$, multiply both sides of $A_1 = Q_1R_1$ by Q_1^{-1}, obtaining $R_1 = Q_1^{-1}A_1$. Thus, $A_2 = R_1Q_1 = Q_1^{-1}A_1Q_1$, as desired. Similarly, A_3 is similar to A_2, which makes it similar to A_1. Continuing in this way, we see that each A_k is similar to $A_1 = A$.

ii. If A has n (real) eigenvalues, as k increases, the entries below the main diagonal of the A_k get closer and closer to 0. In other words, as k increases, the A_k more and more resemble upper-triangular matrices, with eigenvalues that get closer and closer to their main diagonal entries.

Putting these two facts together, we see that as k increases, the diagonal entries of the A_k get closer and closer to (*approach*) the eigenvalues of A. This gives us a means—by generating more A_k—of obtaining approximations that are as close as we like to the actual eigenvalues of A. We summarize this discussion in the following procedure.

The QR Algorithm for Approximating the Eigenvalues of a Matrix

Given an $n \times n$ matrix A with n (real) eigenvalues,

i. Construct the sequence of matrices $A_1, A_2, A_3, \ldots$ as follows:

- Set $A_1 = A$.
- For $k = 1, 2, 3, \ldots$, find the QR factorization of A_k, $A_k = Q_kR_k$, and then multiply to get $A_{k+1} = R_kQ_k$.

ii. The diagonal entries of the A_k approach the eigenvalues of A.

EXAMPLE 2

Perform one iteration of the QR algorithm (that is, find A_2) on the matrix

$$A = \begin{bmatrix} 1 & -1 & 0 \\ -1 & 1 & -1 \\ 0 & -1 & 1 \end{bmatrix}.$$

Solution Notice that this is the matrix of Example 1, in which we found the QR factorization of A:

$$Q_1 = \begin{bmatrix} \frac{-1}{\sqrt{2}} & 0 & \frac{1}{\sqrt{2}} \\ \frac{1}{\sqrt{2}} & 0 & \frac{1}{\sqrt{2}} \\ 0 & 1 & 0 \end{bmatrix} \quad \text{and} \quad R_1 = \begin{bmatrix} -\sqrt{2} & \sqrt{2} & \frac{-1}{\sqrt{2}} \\ 0 & -1 & 1 \\ 0 & 0 & \frac{-1}{\sqrt{2}} \end{bmatrix}$$

Setting $A_2 = R_1 Q_1$, we have

$$A_2 = \begin{bmatrix} 2 & \frac{-1}{\sqrt{2}} & 0 \\ \frac{-1}{\sqrt{2}} & 1 & \frac{-1}{\sqrt{2}} \\ 0 & \frac{-1}{\sqrt{2}} & 0 \end{bmatrix}.$$

■

Be aware that one iteration of the QR algorithm (as in Example 2) is not enough to draw any conclusions about the eigenvalues of a matrix. For example, the matrix A_2 of Example 2 has diagonal entries 2, 1, and 0, but its eigenvalues are $\sqrt{2} + 1$, 1, and $1 - \sqrt{2}$. However, performing just two additional iterations of the QR algorithm results in diagonal entries for A_4 that approximate the eigenvalues of A to the nearest tenth.

COMPUTATIONAL NOTE

The QR algorithm is the basis for a practical, efficient way to find the eigenvalues of a square matrix. First, a similarity transformation is performed on the given matrix to transform it into a matrix H in which all entries below the subdiagonal, lying just under the main diagonal, are zero ($h_{ij} = 0$ whenever $i > j + 1$). Such a matrix is said to be in *Hessenberg form.* (The matrix A of Example 2 is in Hessenberg form.) Then the QR algorithm (or a variation of it) is performed on the matrix H, generating approximations to its eigenvalues until the desired accuracy is achieved.

8.3 Exercises

In Exercises 1–6 find the *QR* factorization of each matrix.

1. $\begin{bmatrix} 0 & 3 \\ 1 & 2 \end{bmatrix}$

2. $\begin{bmatrix} 3 & 2 \\ 4 & 1 \end{bmatrix}$

3. $\begin{bmatrix} 2 & -1 & 0 \\ -1 & 2 & -1 \\ 0 & -1 & 2 \end{bmatrix}$

4. $\begin{bmatrix} 1 & 0 & 0 \\ -1 & 2 & 1 \\ 0 & 0 & 1 \end{bmatrix}$

5. $\begin{bmatrix} 1 & 1 & 1 \\ 0 & 1 & 1 \\ 0 & 0 & 1 \end{bmatrix}$

6. $\begin{bmatrix} 1 & \frac{1}{2} & \frac{1}{3} \\ \frac{1}{2} & \frac{1}{3} & \frac{1}{4} \\ \frac{1}{3} & \frac{1}{4} & \frac{1}{5} \end{bmatrix}$

7–12. Perform one iteration of the QR algorithm on the matrices of Exercises 1–6.

13. Determine which matrices of Exercises 1–6 are in Hessenberg form.

14. Find a QR factorization of the singular matrix

$$\begin{bmatrix} 1 & 1 \\ 0 & 0 \end{bmatrix}.$$

Hint: You may be able to determine Q and R by inspection.

Computational Exercises

15–20. If you have access to a software package (such as *Maple* or *Matlab*) that performs a QR factorization, use it to perform several iterations of the

QR algorithm on the matrices of Exercises 1–6 to approximate the eigenvalues of each matrix. Then check your answers by using the software package to compute the eigenvalues of the given matrix.

8.4 Complex Eigenvalues and Eigenvectors

Except for Section 6.5, all scalars, including components of vectors and entries of matrices, have been real numbers. In this section, we allow scalars to be complex numbers. This generalization of the concept of scalar alters very little of the material of the first seven chapters and makes the subject of eigenvalues and eigenvectors more complete.

Matrices with Entries in $\mathbf{C}$

In this section, we allow the entries of matrices to be complex numbers. All definitions and theorems concerning matrices and determinants given in Chapters 3 and 4 remain completely unchanged, with one minor exception. The product of an $m \times n$ matrix A and an $n \times p$ matrix B, whether the entries are complex or real, is defined to be the $m \times p$ matrix C in which

$$c_{ij} = a_{i1}b_{1j} + a_{i2}b_{2j} + \cdots + a_{in}b_{nj}.$$

Because the complex dot product involves complex conjugates, however, this expression can no longer be given as simply $c_{ij} = \mathbf{a}_i \cdot \mathbf{b}^j$, as it was in Section 3.1.

EXAMPLE 1

Let

$$A = \begin{bmatrix} i & 1-i & 2 \\ 0 & -2i & 1+i \\ 1 & 0 & 3-i \end{bmatrix} \quad \text{and} \quad B = \begin{bmatrix} 1-i & i \\ -i & 0 \\ 0 & 2+i \end{bmatrix}.$$

Find $AB - 2B$.

Solution

$$AB - 2B = \begin{bmatrix} i & 1-i & 2 \\ 0 & -2i & 1+i \\ 1 & 0 & 3-i \end{bmatrix}\begin{bmatrix} 1-i & i \\ -i & 0 \\ 0 & 2+i \end{bmatrix} - 2\begin{bmatrix} 1-i & i \\ -i & 0 \\ 0 & 2+i \end{bmatrix}$$

$$= \begin{bmatrix} 0 & 3+2i \\ -2 & 1+3i \\ 1-i & 7+2i \end{bmatrix} - \begin{bmatrix} 2-2i & 2i \\ -2i & 0 \\ 0 & 4+2i \end{bmatrix}$$

$$= \begin{bmatrix} -2+2i & 3 \\ -2+2i & 1+3i \\ 1-i & 3 \end{bmatrix}$$ ■

Before proceeding to consider complex eigenvalues and eigenvectors, we should point out that all definitions and theorems in Chapters 5 through 7 (linear independence, basis, linear transformations, and so on) are exactly the same for the $\mathbf{C}^n$ case as they were in $\mathbf{R}^n$ (except that the dot product involves conjugates, as per Section 6.5).

Complex Eigenvalues and Eigenvectors in $\mathbf{C}^n$

Although much of the material dealing with complex eigenvalues and eigenvectors is completely analogous to the real case (Sections 4.4 and 8.1), we will also see some interesting differences.

DEFINITION

Let A be an $n \times n$ matrix with complex entries. A complex number λ is an **eigenvalue** of A if there is a nonzero vector $\mathbf{z}$ in $\mathbf{C}^n$ such that $A\mathbf{z} = \lambda\mathbf{z}$. The vector $\mathbf{z}$ is said to be an **eigenvector** corresponding to λ. The **eigenspace** of λ is the set of all solutions of the homogeneous linear system $(A - \lambda I)\mathbf{z} = \mathbf{0}$.

The computation of complex eigenvalues and their corresponding eigenvectors proceeds in a manner similar to that of the real case (see Theorem 1 of Section 4.4 and its accompanying procedure). However, since the characteristic equation, $\det(A - \lambda I) = 0$, is a polynomial equation in the *complex* numbers, it must have at least one complex root, by the *Fundamental Theorem of Algebra*. Consequently, we now have the following theorem.

THEOREM 1

Every $n \times n$ matrix has at least one (complex) eigenvalue. 

EXAMPLE 2

Find the eigenvalues and associated eigenspaces of the matrix

$$A = \begin{bmatrix} i & 0 & 0 \\ 0 & 0 & 1 \\ 0 & -1 & 0 \end{bmatrix}.$$

Solution To find the eigenvalues of A, we compute the characteristic polynomial

$$\det(A - \lambda I) = \det\begin{bmatrix} i - \lambda & 0 & 0 \\ 0 & -\lambda & 1 \\ 0 & -1 & -\lambda \end{bmatrix} = (i - \lambda)(\lambda^2 + 1).$$

Thus, the eigenvalues of A, the roots of the characteristic equation, are $\lambda_1 = i$ and $\lambda_2 = -i$ (since $\lambda^2 + 1 = 0$ has the solution $\lambda = \pm i$).

To compute the eigenspaces of λ_1 and λ_2, we solve the homogeneous linear systems $(A - iI)\mathbf{z} = \mathbf{0}$ and $(A + iI)\mathbf{z} = \mathbf{0}$, respectively. Doing so, we see that the eigenspace of $\lambda_1 = i$ is the set of all vectors of the form $(r, -is, s)$, whereas the eigenspace of $\lambda_2 = -i$ consists of all vectors of the form $(0, it, t)$, where r, s, and t are scalars (complex numbers). (*Note*: Bases for the eigenspaces of λ_1 and λ_2 are, respectively, $\{(1, 0, 0), (0, -i, 1)\}$, and $\{(0, i, 1)\}$.) ■

Diagonalization of Complex Matrices

The basic theory concerning diagonalization of matrices carries over to the case of matrices with complex entries. Specifically, Theorems 1, 2, and 3 of Section 8.2 continue to hold, and the proofs are essentially the same. We restate these theorems here in a more concise form.

THEOREM 2

Let A be an $n \times n$ matrix with complex entries.

a. Nonzero eigenvectors that correspond to distinct eigenvalues of A are linearly independent.

b. A is diagonalizable if and only if A has n linearly independent eigenvectors in $\mathbf{C}^n$.

c. If A has n distinct eigenvalues, then A is diagonalizable. 

The next example illustrates the diagonalization procedure of Section 8.2 in the case of complex matrices.

EXAMPLE 3

Determine whether or not the matrix

$$A = \begin{bmatrix} i & 0 & 0 \\ 0 & 0 & 1 \\ 0 & -1 & 0 \end{bmatrix}$$

is diagonalizable. If it is, find an invertible matrix P such that $P^{-1}AP$ is diagonal. Also find this diagonal matrix.

Solution Notice that the given matrix is the same as the one of Example 2. Since it has only two distinct eigenvalues, $\lambda_1 = i$ and $\lambda_2 = -i$, we cannot apply Theorem 2(c). However, the union of the bases for the eigenspaces of λ_1 and λ_2 (computed in Example 2) is $\{(1, 0, 0), (0, -i, 1), (0, i, 1)\}$, a set of three linearly independent eigenvectors. Therefore, by Theorem 2(b), A is diagonalizable.

Moreover, following the procedure of Section 8.2, P is the matrix whose columns are these eigenvectors:

$$P = \begin{bmatrix} 1 & 0 & 0 \\ 0 & -i & i \\ 0 & 1 & 1 \end{bmatrix}$$

Finally, the diagonal matrix $P^{-1}AP$ is the one with diagonal entries that are the eigenvalues of A. Thus,

$$P^{-1}AP = \begin{bmatrix} i & 0 & 0 \\ 0 & i & 0 \\ 0 & 0 & -i \end{bmatrix}.$$

■

We next consider the complex analog of the orthogonal diagonalization of real symmetric matrices.

DEFINITION

The **conjugate transpose** of a matrix A, denoted A^H, is the transpose of the matrix obtained from A by replacing each entry with its complex conjugate. We say that a square matrix A is **hermitian** if $A = A^H$.

For example, the matrix

$$A = \begin{bmatrix} 1 & 1+i & 0 \\ 1-i & -3 & -2i \\ 0 & 2i & 0 \end{bmatrix}$$

is hermitian, since

$$A^H = \left(\begin{bmatrix} 1 & 1-i & 0 \\ 1+i & -3 & 2i \\ 0 & -2i & 0 \end{bmatrix}\right)^T = A.$$

NOTE The *diagonal* entries of a hermitian matrix must be real, since they are equal to their complex conjugates. Moreover, if *all* entries of a square matrix A are real, then $A^H = A^T$. Thus, *hermitian* is the complex generalization of *symmetric*.

The reason we consider *conjugate transpose* instead of just *transpose* in the complex case is to preserve the validity of theorems such as Theorems 4, 6, and 7 of Section 8.2 when we replace *symmetric* with *hermitian*. For example, if a square matrix A has imaginary entries, it is *not* true that $A = A^T$ implies that A is diagonalizable (see Exercise 31). The underlying reason for the need for *conjugate* transpose in these situations is that complex dot products involve conjugates. Notice how this factor enters into the proofs of the following two theorems. (Theorem 4 is a complex analogue of Theorem 4 of Section 8.2.)

THEOREM 3

Let A be a hermitian matrix. Then all eigenvalues of A are real numbers.

Proof Let λ be an eigenvalue of A with corresponding eigenvector $\mathbf{z}$. Then $A\mathbf{z} = \lambda\mathbf{z}$ and

$$\lambda(\mathbf{z} \cdot \mathbf{z}) = (\lambda\mathbf{z}) \cdot \mathbf{z} = (A\mathbf{z}) \cdot \mathbf{z} = \mathbf{z} \cdot (A^H\mathbf{z}),$$

by Exercise 32. Moreover, A is hermitian, so

$$\mathbf{z} \cdot (A^H\mathbf{z}) = \mathbf{z} \cdot (A\mathbf{z}) = \mathbf{z} \cdot (\lambda\mathbf{z}) = \bar{\lambda}(\mathbf{z} \cdot \mathbf{z}),$$

by Theorem 1(c) of Section 6.5. Hence $\lambda(\mathbf{z} \cdot \mathbf{z}) = \bar{\lambda}(\mathbf{z} \cdot \mathbf{z})$. But $\mathbf{z}$ is an eigenvector, and therefore $\mathbf{z} \neq \mathbf{0}$, which, in turn, implies that $\mathbf{z} \cdot \mathbf{z} \neq 0$ (Theorem 1(a) of Section 6.5). Consequently $\lambda = \bar{\lambda}$, and we conclude that λ is real. ■

THEOREM 4

Let λ_1 and λ_2 be distinct eigenvalues of a hermitian matrix A with corresponding eigenvectors $\mathbf{z}_1$ and $\mathbf{z}_2$, respectively. Then $\mathbf{z}_1$ and $\mathbf{z}_2$ are *orthogonal.*

Proof The proof is virtually identical to that of Theorem 4 of Section 8.2. We have

$$\lambda_1(\mathbf{z}_1 \cdot \mathbf{z}_2) = (\lambda_1\mathbf{z}_1) \cdot \mathbf{z}_2 = (A\mathbf{z}_1) \cdot \mathbf{z}_2 = \mathbf{z}_1 \cdot (A^H\mathbf{z}_2),$$

by Exercise 32. But A is hermitian, so

$$\mathbf{z}_1 \cdot (A^H\mathbf{z}_2) = \mathbf{z}_1 \cdot (A\mathbf{z}_2) = \mathbf{z}_1 \cdot (\lambda_2\mathbf{z}_2) = \bar{\lambda}_2(\mathbf{z}_1 \cdot \mathbf{z}_2),$$

by Theorem 1(c) of Section 6.5. Thus, $\lambda_1(\mathbf{z}_1 \cdot \mathbf{z}_2) = \bar{\lambda}_2(\mathbf{z}_1 \cdot \mathbf{z}_2) = \lambda_2(\mathbf{z}_1 \cdot \mathbf{z}_2)$, by Theorem 3.

Now, if $\mathbf{z}_1 \cdot \mathbf{z}_2$ were not equal to zero, we would have $\lambda_1 = \lambda_2$, a contradiction. Thus, $\mathbf{z}_1 \cdot \mathbf{z}_2 = 0$, as desired. ■

To state the main theorem concerning the diagonalization of hermitian matrices, we must first introduce the complex analog of an orthogonal matrix.

DEFINITION

A nonsingular matrix P with complex entries is **unitary** if $P^{-1} = P^H$.

For example, the matrix

$$P = \begin{bmatrix} \frac{i}{\sqrt{2}} & \frac{i}{\sqrt{2}} \\ \frac{i}{\sqrt{2}} & \frac{-i}{\sqrt{2}} \end{bmatrix}$$

is unitary. It is an easy matter to check that $P^HP = I$, which implies that $P^{-1} = P^H$.

The next theorem tells us that the columns of unitary matrices behave in exactly the same manner with respect to the complex dot product, as do those of orthogonal matrices relative to the real dot product. Its proof is Exercise 33.

THEOREM 5

An $n \times n$ matrix P with complex entries is unitary if and only if its columns $\{\mathbf{p}^1, \mathbf{p}^2, \ldots, \mathbf{p}^n\}$ form an orthonormal set; that is, if and only if $\mathbf{p}^i \cdot \mathbf{p}^j = 0$ if $i \neq j$ while $\mathbf{p}^i \cdot \mathbf{p}^i = 1$. ■

Our last theorem for this section gives a generalization of Theorems 6 and 7 of Section 8.2 to the complex case. We omit the proof.

THEOREM 6

Let A be a square matrix with complex entries. Then there exists a unitary matrix P such that $P^{-1}AP$ $(= P^H AP)$ is diagonal with real entries if and only if A is hermitian. ■

EXAMPLE 4

Let

$$A = \begin{bmatrix} 0 & -i & 0 \\ i & 0 & 0 \\ 0 & 0 & 2 \end{bmatrix}.$$

Find a unitary matrix P such that $P^{-1}AP$ is diagonal.

Solution To find the matrix P, we follow the procedure given in Section 8.2 for orthogonally diagonalizing a real symmetric matrix. Since the characteristic polynomial of A is det $(A - \lambda I) = (2 - \lambda)(\lambda^2 - 1)$, the eigenvalues of A are $\lambda_1 = -1$, $\lambda_2 = 1$, and $\lambda_3 = 2$. Bases for the eigenspaces corresponding to these eigenvalues are then computed to be

$$\begin{aligned} &\{(i, 1, 0)\} && \text{for } \lambda_1 = -1, \\ &\{(-i, 1, 0)\} && \text{for } \lambda_2 = 1, \\ \text{and } &\{(0, 0, 1)\} && \text{for } \lambda_3 = 2. \end{aligned}$$

Finally, to obtain the columns of P, we normalize these three eigenvectors to get $(i/\sqrt{2}, 1/\sqrt{2}, 0)$, $(-i/\sqrt{2}, 1/\sqrt{2}, 0)$, and $(0, 0, 1)$. Hence,

$$P = \begin{bmatrix} \frac{i}{\sqrt{2}} & \frac{-i}{\sqrt{2}} & 0 \\ \frac{1}{\sqrt{2}} & \frac{1}{\sqrt{2}} & 0 \\ 0 & 0 & 1 \end{bmatrix}.$$

■

NOTE In general, to obtain a unitary matrix that diagonalizes a hermitian one, it may be necessary to employ the Gram–Schmidt process in the same manner as in Section 8.2.

8.4 Exercises

In Exercises 1–9 let

$$A = \begin{bmatrix} 1 & 0 & -i \\ 0 & 1+i & 1 \\ i & -1 & 2+i \end{bmatrix} \quad \text{and} \quad B = \begin{bmatrix} i & -i & 0 \\ i & i & 1 \\ 0 & 1 & i \end{bmatrix}.$$

1. Solve $A\mathbf{z} = \mathbf{b}$, where $\mathbf{b} = (i, 0, -1)$.

2. Solve $B\mathbf{z} = \mathbf{c}$, where $\mathbf{c} = (1 + i, 1 - i, 1)$.

3. Find AB.

4. Find BA.

5. Find A^{-1}.

6. Find B^{-1}.

7. Find $\det(A)$.

8. Find $\det(B)$.

9. Is either A or B hermitian?

10. Show that the dimension of $\mathbf{C}^n$ is equal to n.

In Exercises 11 and 12 find bases for the given subspace of $\mathbf{C}^n$.

11. The kernel of the transformation $T: \mathbf{C}^2 \rightarrow \mathbf{C}^2$ given by $T(z_1, z_2) = (z_1 - z_2, 0)$

12. The kernel of the transformation $S: \mathbf{C}^3 \rightarrow \mathbf{C}^2$ given by $S(\mathbf{z}) = A\mathbf{z}$, where

$$A = \begin{bmatrix} i & 0 & 1 \\ 1 & -i & 0 \end{bmatrix}.$$

In Exercises 13–20 find the eigenvalues and associated eigenspaces for each matrix. Also find a basis for each eigenspace.

13. $\begin{bmatrix} 0 & 3 + 4i \\ 3 - 4i & 0 \end{bmatrix}$

14. $\begin{bmatrix} 0 & -2 \\ 2 & 0 \end{bmatrix}$

15. $\begin{bmatrix} 2 & 0 & -4 \\ 0 & 1 & 0 \\ 2 & 0 & -2 \end{bmatrix}$

16. $\begin{bmatrix} 1 & i & 0 \\ i & 1 & -3i \\ 0 & i & 1 \end{bmatrix}$

17. $\begin{bmatrix} i & 0 & 1 \\ 0 & i & 0 \\ 0 & 0 & i \end{bmatrix}$

18. $\begin{bmatrix} 1 + i & 0 & 0 \\ -1 & 1 - i & 0 \\ 0 & -1 & 0 \end{bmatrix}$

19. $\begin{bmatrix} 1 & 0 & 0 \\ 0 & 0 & -i \\ 0 & i & 0 \end{bmatrix}$

20. $\begin{bmatrix} 0 & 0 & -1 \\ 0 & 1 & 0 \\ 1 & 0 & 0 \end{bmatrix}$

21–28. Determine whether or not each matrix in Exercises 13–20 is diagonalizable. If it is diagonalizable, find an invertible matrix P that performs the diagonalization, as well as the resultant diagonal matrix. If the given matrix is hermitian, find a unitary matrix P that performs the diagonalization, as well as the resultant diagonal matrix.

29. Which, if any, of the matrices in Exercises 13–20 are unitary?

30. Let A be a unitary matrix. Show that $\det(A) = \pm 1$. (See Exercise 26 of Section 8.2.)

31. Let

$$A = \begin{bmatrix} 1 & 0 & 0 \\ 0 & 1 & i \\ 0 & i & 3 \end{bmatrix}.$$

Show that A is not diagonalizable. (Notice that $A = A^T$.)

32. Let A be an $n \times n$ matrix with complex entries and $\mathbf{w}$ and $\mathbf{z}$ be in $\mathbf{C}^n$. Show that $(A\mathbf{w}) \bullet \mathbf{z} = \mathbf{w} \bullet (A^H\mathbf{z})$. (*Hint*: See Exercise 28 of Section 8.3.)

33. Prove that an $n \times n$ matrix P with complex entries is unitary if and only if its columns form an orthonormal set, under the complex dot product (Theorem 5).

8.5 APPLICATION—Powers of Matrices and Recurrence Relations

The Power of Matrices

There are many applications where one needs to raise a square matrix to higher and higher powers (one example is transition matrices from Markov chains—see Section 4.5). While it seems difficult to raise matrices to large powers, there is one class of matrices for which it is quite simple—diagonal matrices. Note that the product of two diagonal matrices A and B of the same size is again a diagonal matrix where the diagonal entries are products of the corresponding diagonal entries of A and B:

$$\begin{bmatrix} \alpha_1 & 0 & 0 & \dots & 0 \\ 0 & \alpha_2 & 0 & \dots & 0 \\ \vdots & \vdots & \vdots & \vdots & \vdots \\ 0 & 0 & 0 & \dots & \alpha_n \end{bmatrix} \begin{bmatrix} \beta_1 & 0 & 0 & \dots & 0 \\ 0 & \beta_2 & 0 & \dots & 0 \\ \vdots & \vdots & \vdots & \vdots & \vdots \\ 0 & 0 & 0 & \dots & \beta_n \end{bmatrix} = \begin{bmatrix} \alpha_1\beta_1 & 0 & 0 & \dots & 0 \\ 0 & \alpha_2\beta_2 & 0 & \dots & 0 \\ \vdots & \vdots & \vdots & \vdots & \vdots \\ 0 & 0 & 0 & \dots & \alpha_n\beta_n \end{bmatrix}.$$

It follows that

$$\begin{bmatrix} \alpha_1 & 0 & 0 & \cdots & 0 \\ 0 & \alpha_2 & 0 & \cdots & 0 \\ \vdots & \vdots & \vdots & \vdots & \vdots \\ 0 & 0 & 0 & \cdots & \alpha_n \end{bmatrix}^2 = \begin{bmatrix} \alpha_1\alpha_1 & 0 & 0 & \cdots & 0 \\ 0 & \alpha_2\alpha_2 & 0 & \cdots & 0 \\ \vdots & \vdots & \vdots & \vdots & \vdots \\ 0 & 0 & 0 & \cdots & \alpha_n\alpha_n \end{bmatrix} = \begin{bmatrix} \alpha_1^2 & 0 & 0 & \dots & 0 \\ 0 & \alpha_2^2 & 0 & \dots & 0 \\ \vdots & \vdots & \vdots & \vdots & \vdots \\ 0 & 0 & 0 & \dots & \alpha_n^2 \end{bmatrix}$$

and more generally, for any natural number k,

$$\begin{bmatrix} \alpha_1 & 0 & 0 & \dots & 0 \\ 0 & \alpha_2 & 0 & \dots & 0 \\ \vdots & \vdots & \vdots & \vdots & \vdots \\ 0 & 0 & 0 & \dots & \alpha_n \end{bmatrix}^k = \begin{bmatrix} \alpha_1^k & 0 & 0 & \dots & 0 \\ 0 & \alpha_2^k & 0 & \dots & 0 \\ \vdots & \vdots & \vdots & \vdots & \vdots \\ 0 & 0 & 0 & \dots & \alpha_n^k \end{bmatrix}.$$

EXAMPLE 1

If $A = \begin{bmatrix} 3 & 0 & 0 \\ 0 & -1 & 0 \\ 0 & 0 & 2 \end{bmatrix}$, then what is A^3? For positive integer k, what is A^k?

Solution As A is diagonal,

$$A^3 = \begin{bmatrix} 3^3 & 0 & 0 \\ 0 & (-1)^3 & 0 \\ 0 & 0 & 2^3 \end{bmatrix} = \begin{bmatrix} 27 & 0 & 0 \\ 0 & -1 & 0 \\ 0 & 0 & 8 \end{bmatrix}$$

while

$$A^k = \begin{bmatrix} 3^k & 0 & 0 \\ 0 & (-1)^k & 0 \\ 0 & 0 & 2^k \end{bmatrix}.$$

■

While raising nondiagonal square matrices seems harder, if a matrix is diagonalizable, we can use the diagonalization to make short work of powers, and indeed find explicit formulas for powers of a matrix. Suppose we have diagonalized a square matrix A as $D = P^{-1}AP$, where D is a diagonal matrix. Then observe that for any positive integer k,

$$\begin{aligned} D^k &= (P^{-1}AP)^k \\ &= (P^{-1}AP)(P^{-1}AP)\cdots(P^{-1}AP) \\ &= P^{-1}A(PP^{-1})A(PP^{-1})A\cdots A(PP^{-1})AP \\ &= P^{-1}AIAI\cdots IAP \\ &= P^{-1}A^kP. \end{aligned}$$

Thus we can solve for A^k as $A^k = PD^kP^{-1}$.

EXAMPLE 2

If $A = \begin{bmatrix} \frac{1}{2} & -\frac{3}{2} & -\frac{7}{2} \\ -\frac{3}{2} & \frac{1}{2} & -\frac{3}{2} \\ 0 & 0 & 4 \end{bmatrix}$, then what is A^5? For positive integer k, what is A^k?

Solution Raising A to some low powers, we calculate that

$$A^2 = \begin{bmatrix} \frac{1}{2} & -\frac{3}{2} & -\frac{7}{2} \\ -\frac{3}{2} & \frac{1}{2} & -\frac{3}{2} \\ 0 & 0 & 4 \end{bmatrix}\begin{bmatrix} \frac{1}{2} & -\frac{3}{2} & -\frac{7}{2} \\ -\frac{3}{2} & \frac{1}{2} & -\frac{3}{2} \\ 0 & 0 & 4 \end{bmatrix} = \begin{bmatrix} \frac{5}{2} & -\frac{3}{2} & -\frac{27}{2} \\ -\frac{3}{2} & \frac{5}{2} & -\frac{3}{2} \\ 0 & 0 & 16 \end{bmatrix}$$

while

$$A^3 = AA^2 = \begin{bmatrix} \frac{1}{2} & -\frac{3}{2} & -\frac{7}{2} \\ -\frac{3}{2} & \frac{1}{2} & -\frac{3}{2} \\ 0 & 0 & 4 \end{bmatrix}\begin{bmatrix} \frac{5}{2} & -\frac{3}{2} & -\frac{27}{2} \\ -\frac{3}{2} & \frac{5}{2} & -\frac{3}{2} \\ 0 & 0 & 16 \end{bmatrix} = \begin{bmatrix} \frac{7}{2} & -\frac{9}{2} & -\frac{121}{2} \\ -\frac{9}{2} & \frac{7}{2} & -\frac{9}{2} \\ 0 & 0 & 64 \end{bmatrix}.$$

You might observe some patterns among the entries, but certainly not for all of them. Let's diagonalize the matrix, if we can. By solving $\det(A - \lambda I) = 0$, we find (see Exercise 13) that the eigenvalues are 2, 4, and -1, with corresponding eigenvectors $(1, -1, 0)$, $(-1, 0, 1)$, and $(1, 1, 0)$. This means that $D = P^{-1}AP$ where

$$D = \begin{bmatrix} 2 & 0 & 0 \\ 0 & 4 & 0 \\ 0 & 0 & -1 \end{bmatrix},$$

$$P = \begin{bmatrix} 1 & -1 & 1 \\ -1 & 0 & 1 \\ 0 & 1 & 0 \end{bmatrix},$$

and hence

$$P^{-1} = \begin{bmatrix} \frac{1}{2} & -\frac{1}{2} & \frac{1}{2} \\ 0 & 0 & 1 \\ \frac{1}{2} & \frac{1}{2} & \frac{1}{2} \end{bmatrix}.$$

It follows that for all positive integers k, $A^k = PD^kP^{-1}$, and we calculate

$$A^5 = PD^5P^{-1}$$
$$= \begin{bmatrix} 1 & -1 & 1 \\ -1 & 0 & 1 \\ 0 & 1 & 0 \end{bmatrix} \begin{bmatrix} 2^5 & 0 & 0 \\ 0 & 4^5 & 0 \\ 0 & 0 & (-1)^5 \end{bmatrix} \begin{bmatrix} \frac{1}{2} & -\frac{1}{2} & \frac{1}{2} \\ 0 & 0 & 1 \\ \frac{1}{2} & \frac{1}{2} & \frac{1}{2} \end{bmatrix}$$
$$= \begin{bmatrix} \frac{31}{2} & -\frac{33}{2} & -\frac{2017}{2} \\ -\frac{33}{2} & \frac{31}{2} & -\frac{33}{2} \\ 0 & 0 & 1024 \end{bmatrix}$$

More generally, in the same way, we can even find a formula for A^k:

$$A^k = PD^kP^{-1}$$
$$= \begin{bmatrix} 1 & -1 & 1 \\ -1 & 0 & 1 \\ 0 & 1 & 0 \end{bmatrix} \begin{bmatrix} 2^k & 0 & 0 \\ 0 & 4^k & 0 \\ 0 & 0 & (-1)^k \end{bmatrix} \begin{bmatrix} \frac{1}{2} & -\frac{1}{2} & \frac{1}{2} \\ 0 & 0 & 1 \\ \frac{1}{2} & \frac{1}{2} & \frac{1}{2} \end{bmatrix}$$
$$= \begin{bmatrix} \frac{2^k + (-1)^k}{2} & \frac{-2^k + (-1)^k}{2} & \frac{2^k - 2(4^k) + (-1)^k}{2} \\ \frac{-2^k + (-1)^k}{2} & \frac{2^k + (-1)^k}{2} & \frac{-2^k + (-1)^k}{2} \\ 0 & 0 & 4^k \end{bmatrix}.$$

You can check that for $k = 5$ this gives the same result as before. ■

The formula from Example 2 shows much more. We see that all of the entries in A^k are linear combinations of the powers of the eigenvalues, and this is true in general if we can diagonalize the matrix.

Moreover, if all we are interested in is what happens eventually to the matrix, we can ignore eigenvalues where absolute values are small, as powers of these (even multiplied by constants) will matter little in comparison to the powers of eigenvalues where absolute values are larger (the largest of which is called the **dominant eigenvalue** of the matrix). For example, for large k, the matrix A^k of Example 2 is approximately

$$\begin{bmatrix} 0 & 0 & -4^k \\ 0 & 0 & 0 \\ 0 & 0 & 4^k \end{bmatrix},$$

as the dominant eigenvalue is 4, and if we remove all terms involving other eigenvalues, we arrive at the matrix above. So, for example, for the matrix of Example 2,

$$A^5 = \begin{bmatrix} 15.5 & -16.5 & -1008.5 \\ -16.5 & 15.5 & -16.5 \\ 0 & 0 & 1024 \end{bmatrix},$$

which is not so different from

$$\begin{bmatrix} 0 & 0 & 4^5 \\ 0 & 0 & 0 \\ 0 & 0 & 4^5 \end{bmatrix} = \begin{bmatrix} 0 & 0 & 1024 \\ 0 & 0 & 0 \\ 0 & 0 & 1024 \end{bmatrix}$$

(the other entries are small in comparison to 1024).

The fact that when a matrix is diagonalizable, powers of the matrix have entries that are linear combinations of the powers of the eigenvalues can tell you a lot. For example, if the matrix has all its eigenvalues in absolute value less than 1, the powers of the matrix converge to the zero matrix (that is, all of the entries converge to 0), as the powers of numbers that are in absolute value less than 1 tend to 0.

EXAMPLE 3

If $A = \begin{bmatrix} \frac{1}{12} & \frac{7}{12} \\ \frac{7}{12} & \frac{1}{12} \end{bmatrix}$, then we find that the eigenvalues of A are $-\frac{1}{2}$ and $\frac{2}{3}$. As both of these are less than 1 in absolute value, the entries of A^k, being linear combinations of these, tend to 0, so A itself tends to the zero matrix. As an example,

$$A^{10} \approx \begin{bmatrix} 0.009 & 0.008 \\ 0.008 & 0.009 \end{bmatrix}.$$

■

It can be shown (see Exercise 14) that if a matrix A has 1 as its dominant eigenvalue and the absolute values of all other eigenvalues are less than 1, then powers of the matrix converge to some fixed matrix. This fact plays a key role in the power method for finding the stationary distribution (see Sections 4.5 and 4.6).

Diagonalization and Recurrence Relations

We end this section with an application of powers of matrices and diagonalization to certain types of recurrence relations, which are discrete versions of differential equations. Suppose you want to calculate a sequence of numbers, $f_0, f_1, f_2, \ldots, f_n$ where each f_i is not defined explicitly, but rather is given in terms of some previous ones. A **homogeneous linear recurrence relation** (or **difference equation**) is a relation of the form

$$f_n = c_{n-1}f_{n-1} + c_{n-2}f_{n-2} + \ldots c_{n-k}f_{n-k}$$

where $k \geq 1$ is a fixed positive integer and the coefficients $c_{n-1}, c_{n-2}, \ldots, c_{n-k}$ are fixed numbers; n is allowed to be any positive integer of at least k. You are also given **boundary conditions** consisting of values for the first k values $f_0, f_1, f_2, \ldots, f_{k-1}$. We'll illustrate the process of finding an explicit formula for the terms in the recurrence relation with an example of one of the most famous sequences of numbers in mathematics.

EXAMPLE 4

The **Fibonacci sequence**

$$0, 1, 1, 2, 3, 5, 8, 13, \ldots$$

is formed from 0, 1 by insisting that each subsequent number be the sum of the two previous ones. That is, it is the recurrence relation

$$f_0 = 0, f_1 = 1 \text{ and } f_n = f_{n-1} + f_{n-2} \text{ for } n \geq 2.$$

Find a formula for the general term f_n.

Solution We can model this recurrence with matrices by setting $\mathbf{g}_n = \begin{bmatrix} f_{n+1} \\ f_n \end{bmatrix}$ for $n \geq 1$. Then setting $A = \begin{bmatrix} 1 & 1 \\ 1 & 0 \end{bmatrix}$, find that the recurrence can be written as

$$\mathbf{g}_0 = \begin{bmatrix} 1 \\ 0 \end{bmatrix} \text{ and } \mathbf{g}_n = A\mathbf{g}_{n-1}$$

as

$$A\mathbf{g}_{n\text{-}1} = \begin{bmatrix} 1 & 1 \\ 1 & 0 \end{bmatrix}\begin{bmatrix} f_n \\ f_{n-1} \end{bmatrix} = \begin{bmatrix} f_n + f_{n-1} \\ f_n \end{bmatrix} = \begin{bmatrix} f_{n+1} \\ f_n \end{bmatrix} = \mathbf{g}_n.$$

It follows that

$$\begin{aligned} \mathbf{g}_1 &= A\mathbf{g}_0, \\ \mathbf{g}_2 &= A\mathbf{g}_1 = AA\mathbf{g}_0 = A^2\mathbf{g}_0, \\ &\vdots \\ \mathbf{g}_n &= A^n\mathbf{g}_0. \end{aligned}$$

Thus we are back to the same problem of raising a square matrix to powers. We diagonalize as before:

$$\det\left(\begin{bmatrix} 1-\lambda & 1 \\ 1 & -\lambda \end{bmatrix}\right) = 0$$

which implies (after a small bit of algebra) that $\lambda^2 - \lambda - 1 = 0$. The quadratic formula tells us that the eigenvalues are $\lambda_1 = \frac{1+\sqrt{5}}{2}$ and $\lambda_2 = \frac{1-\sqrt{5}}{2}$. The eigenvectors $\mathbf{x} = \begin{bmatrix} x \\ y \end{bmatrix}$ corresponding to eigenvalue λ_1 satisfy

$$\begin{bmatrix} 1-\lambda_1 & 1 \\ 1 & -\lambda_1 \end{bmatrix}\begin{bmatrix} x \\ y \end{bmatrix} = \begin{bmatrix} 0 \\ 0 \end{bmatrix}$$

and hence (from the bottom coordinates) we find that $x - \lambda_1 y = 0$. It follows that an eigenvector for λ_1 is $\begin{bmatrix} \lambda_1 \\ 1 \end{bmatrix} = \begin{bmatrix} \frac{1+\sqrt{5}}{2} \\ 1 \end{bmatrix}$. Similarly, an eigenvector for λ_2 is $\begin{bmatrix} \lambda_2 \\ 1 \end{bmatrix} = \begin{bmatrix} \frac{1-\sqrt{5}}{2} \\ 1 \end{bmatrix}$.

Forming the matrix of eigenvectors

$$P = \begin{bmatrix} \frac{1+\sqrt{5}}{2} & \frac{1-\sqrt{5}}{2} \\ 1 & 1 \end{bmatrix}$$

and its inverse,

$$P^{-1} = \frac{1}{\sqrt{5}}\begin{bmatrix} 1 & -\frac{1-\sqrt{5}}{2} \\ -1 & \frac{1+\sqrt{5}}{2} \end{bmatrix}$$

we find that $A = PDP^{-1}$ where $D = \begin{bmatrix} \frac{1+\sqrt{5}}{2} & 0 \\ 0 & \frac{1-\sqrt{5}}{2} \end{bmatrix}$. As before,

$$\begin{aligned}
A^k &= PD^kP^{-1}, \text{ and thus} \\
\mathbf{g}_n &= A^n\mathbf{g}_0 \\
&= PD^nP^{-1}\mathbf{g}_0 \\
&= \begin{bmatrix} \frac{1+\sqrt{5}}{2} & \frac{1-\sqrt{5}}{2} \\ 1 & 1 \end{bmatrix}\begin{bmatrix} \left(\frac{1+\sqrt{5}}{2}\right)^n & 0 \\ 0 & \left(\frac{1-\sqrt{5}}{2}\right)^n \end{bmatrix}\frac{1}{\sqrt{5}}\begin{bmatrix} 1 & -\frac{1-\sqrt{5}}{2} \\ -1 & \frac{1+\sqrt{5}}{2} \end{bmatrix}\begin{bmatrix} 1 \\ 0 \end{bmatrix} \\
&= \frac{1}{\sqrt{5}}\begin{bmatrix} \frac{1+\sqrt{5}}{2} & \frac{1-\sqrt{5}}{2} \\ 1 & 1 \end{bmatrix}\begin{bmatrix} \left(\frac{1+\sqrt{5}}{2}\right)^n & 0 \\ 0 & \left(\frac{1-\sqrt{5}}{2}\right)^n \end{bmatrix}\begin{bmatrix} 1 \\ -1 \end{bmatrix} \\
&= \frac{1}{\sqrt{5}}\begin{bmatrix} \frac{1+\sqrt{5}}{2} & \frac{1-\sqrt{5}}{2} \\ 1 & 1 \end{bmatrix}\begin{bmatrix} \left(\frac{1+\sqrt{5}}{2}\right)^n \\ \left(\frac{1-\sqrt{5}}{2}\right)^n \end{bmatrix} \\
&= \frac{1}{\sqrt{5}}\begin{bmatrix} \left(\frac{1+\sqrt{5}}{2}\right)^{n+1} & -\left(\frac{1-\sqrt{5}}{2}\right)^{n+1} \\ \left(\frac{1+\sqrt{5}}{2}\right)^n & -\left(\frac{1-\sqrt{5}}{2}\right)^n \end{bmatrix}
\end{aligned}$$

Thus as the bottom coordinate of $\mathbf{g}_n$ is f_n, we find that

$$f_n = \frac{1}{\sqrt{5}}\left(\left(\frac{1+\sqrt{5}}{2}\right)^n - \left(\frac{1-\sqrt{5}}{2}\right)^n\right) \text{ for all } n \geq 0.$$

Isn't this surprising—not only that we did, indeed, find an explicit formula for the nth Fibonnaci number, but it also turns out to involve the irrational number $\sqrt{5}$, when we know that the terms must all be integers! Also, the eigenvalue $\lambda_1 = \frac{1+\sqrt{5}}{2} \approx 1.618$ is clearly the dominant one, as the other is $\lambda_2 = \frac{1-\sqrt{5}}{2} \approx -0.618$. As the second is less than 1 in absolute value, higher powers of it tend to 0, so the nth Fibonnaci number f_n is approximately $\frac{1}{\sqrt{5}}\left(\frac{1+\sqrt{5}}{2}\right)^n$. For large n this is quite close to the real answer; for example, $f_{20} = 6765$, while $\frac{1}{\sqrt{5}}\left(\frac{1+\sqrt{5}}{2}\right)^{20} = 6764.999964\ldots$ (and the latter is much easier to calculate). ■

General recurrence relations $f_n = c_{n-1}f_{n-1} + c_{n-2}f_{n-2} + \cdots c_{n-k}f_{n-k}$ can be solved in much the same way, by calculating the eigenvalues of the matrix

$$A = \begin{bmatrix} c_{n-1} & c_{n-2} & c_{n-3} & \cdots & c_{n-k} \\ 1 & 0 & 0 & 0 & 0 \\ 1 & 0 & 0 & 0 & 0 \\ \vdots & \vdots & \vdots & \vdots & \vdots \\ 1 & 0 & 0 & \cdots & 0 \end{bmatrix}$$

This turns out to be the roots of the polynomial

$$p(\lambda) = \lambda^k - c_{n-1}\lambda^{k-1} - c_{n-2}\lambda^{k-2} - \cdots - c_{n-k+1}\lambda - c_{n-k}.$$

If all of the eigenvalues $\lambda_1, \lambda_2, \ldots, \lambda_k$ are distinct (even if some are complex), then the general solution is a linear combination of powers of the eigenvalues $\lambda_1^n, \lambda_2^n, \ldots, \lambda_k^n$, that is, of the form

$$a_1\lambda_1^n + a_2\lambda_2^n + \cdots + a_k\lambda_k^n.$$

We can find the coefficients by either finding the matrix P of eigenvectors, as in Example 4, or by substituting the values $n = 0, 1, \ldots, k-1$ into this form and using the boundary conditions. If some roots repeat, then for each root λ that repeats, say exactly l times, we find that each of λ^n, $n\lambda^n$, $n^2\lambda^n$, $\ldots$, $n^{l-1}\lambda^n$ is a solution to the recurrence, and the general solution is a linear combination of all of these (over all eigenvalues). We leave the details and proofs for another time!

8.5 Exercises

In Exercises 1–8 for the given matrix *A* calculate A^6 and A^n.

1. $\begin{bmatrix} 1 & 0 \\ 0 & -2 \end{bmatrix}$

2. $\begin{bmatrix} 1 & -1 \\ -1 & 1 \end{bmatrix}$

3. $\begin{bmatrix} 1 & 0 & -2 \\ 2 & -1 & -2 \\ 1 & -1 & -3 \end{bmatrix}$

4. $\begin{bmatrix} \frac{5}{4} & \frac{3}{4} & -1 \\ -\frac{1}{4} & \frac{1}{4} & 1 \\ \frac{3}{4} & \frac{3}{4} & -\frac{1}{2} \end{bmatrix}$

5. $\begin{bmatrix} \frac{1}{2} & 0 & 0 & 0 \\ 0 & -\frac{1}{2} & 0 & 0 \\ \frac{3}{4} & 0 & -\frac{1}{4} & 0 \\ \frac{1}{2} & 0 & -\frac{1}{2} & \frac{1}{4} \end{bmatrix}$

6. $\begin{bmatrix} 1 & 1 \\ -1 & 1 \end{bmatrix}$

7. Using the dominant eigenvalue, for each matrix A in Exercises 1–6, approximate A^{10}.

8. Which of the A in Exercises 1–6 have their powers tending to a matrix with finite entries?

For Exercises 9–12, solve the recurrence relation.

9. $f_0 = 1, f_1 = 1$, and $f_n = f_{n-1} + f_{n-2}$ for $n \geq 2$.

10. $f_0 = 0, f_1 = 1$, and $f_n = f_{n-1} + 2f_{n-2}$ for $n \geq 2$.

11. $f_0 = -1, f_1 = 1$, and $f_n = 2f_{n-1} - 2f_{n-2}$ for $n \geq 2$.

12. $f_0 = 0, f_1 = 4$, and $f_n = 2f_{n-1} - 2f_{n-2}$ for $n \geq 2$.

13. Fill in the details for Example 2.

14. Prove that if a matrix A has 1 as its dominant eigenvalue and the absolute values of all other eigenvalues are less than 1, then powers of the matrix converge to some fixed matrix.

8.6 APPLICATION—Quadric Surfaces

A common topic in analytic geometry is the reduction of a quadratic equation in two variables—the equation of a *conic section*—to standard form. In Section 8.6, we discuss an extension of this topic. With the aid of the techniques of Section 8.2, we reduce a quadratic equation in three variables—that of a *quadric surface*—to certain standard forms.

DEFINITION

A second-degree polynomial (or **quadratic**) **equation** in x, y, and z is one of the form

$$ax^2 + by^2 + cz^2 + dxy + exz + fyz + gx + hy + iz = k, \qquad (1)$$

where $a, b, \ldots, i$ and k are constants and $a, b, \ldots, f$ are not all zero. The graph of such an equation is called a **quadric** or **quadric surface**.

For example, the equation

$$x^2 + y^2 + z^2 - 2x + 4y = 4$$

is the special case of Equation (1), with $a = b = c = 1$, $d = e = f = 0$, $g = -2$, $h = 4$, $i = 0$, and $k = 4$. (Its graph is a *sphere* in 3-space, with the centre at the point $(1, -2, 0)$ and the radius equal to 3.)

There are nine basic types of *nondegenerate* quadric surfaces. These are listed in *standard form* in the next subsection. We will develop a *reduction procedure* to transform Equation (1) to one of these standard forms. If it is impossible to reduce the equation to one of these forms, the corresponding surface is called a *degenerate quadric*. Graphs of the latter include planes, single points, and no points at all. It is not always clear from the original equation whether or not a quadric is nondegenerate, but, as we will see, once the reduction is performed, the given surface can be easily identified.

Standard Forms

i. Ellipsoid (Figure 2):

$$\frac{x^2}{p^2} + \frac{y^2}{q^2} + \frac{z^2}{r^2} = 1 \quad (p, q, r > 0)$$

(A *sphere* is the special case of an ellipsoid for which $p = q = r$.)

FIGURE 2

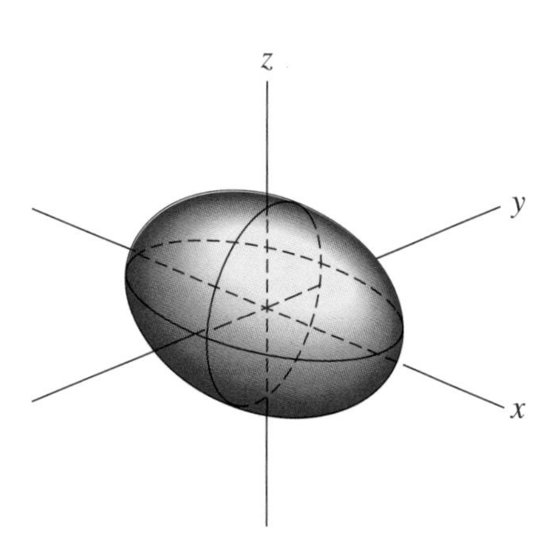

ii. Elliptic paraboloid (Figure 3):

$$z = \frac{x^2}{p^2} + \frac{y^2}{q^2} \quad (p, q > 0)$$

FIGURE 3

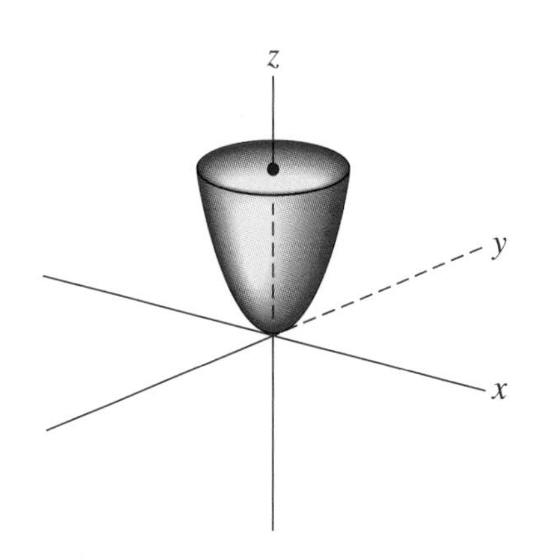

iii. **Hyperbolic paraboloid** (Figure 4):

$$z = \frac{x^2}{p^2} - \frac{y^2}{q^2} \quad (p, q > 0)$$

FIGURE 4

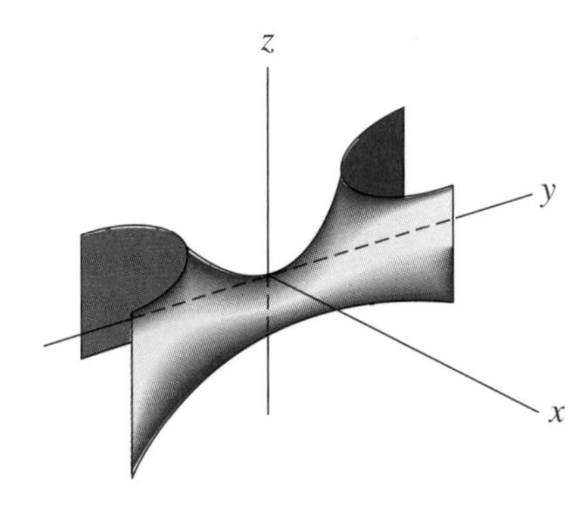

iv. **Hyperboloid of one sheet** (Figure 5):

$$\frac{x^2}{p^2} + \frac{y^2}{q^2} - \frac{z^2}{r^2} = 1 \quad (p, q, r > 0)$$

FIGURE 5

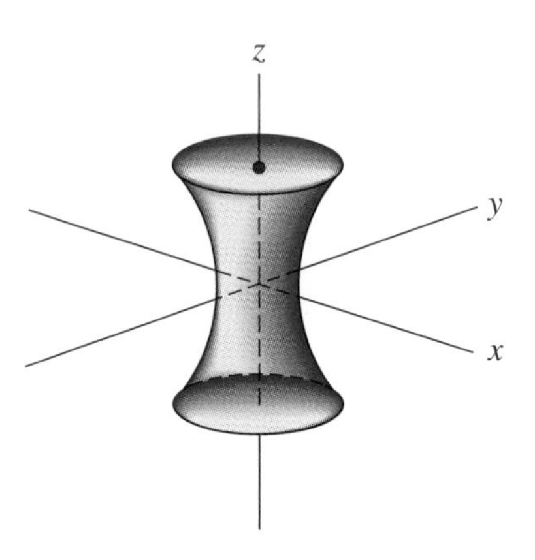

v. **Hyperboloid of two sheets** (Figure 6):

$$\frac{x^2}{p^2} - \frac{y^2}{q^2} - \frac{z^2}{r^2} = 1 \quad (p, q, r > 0)$$

FIGURE 6

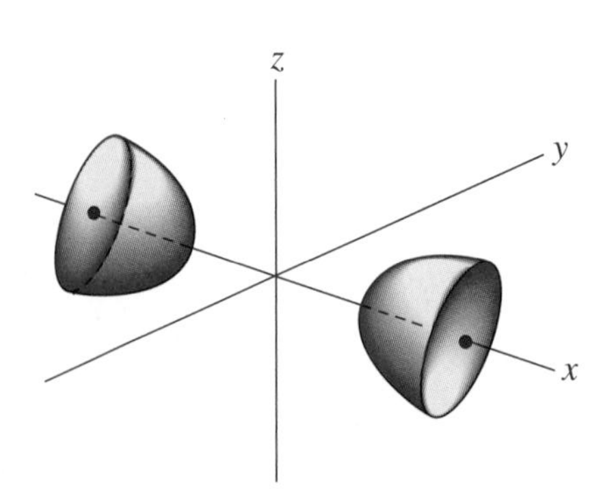

vi. Cone (Figure 7):

$$\frac{x^2}{p^2} + \frac{y^2}{q^2} - \frac{z^2}{r^2} = 0 \quad (p, q, r > 0)$$

FIGURE 7

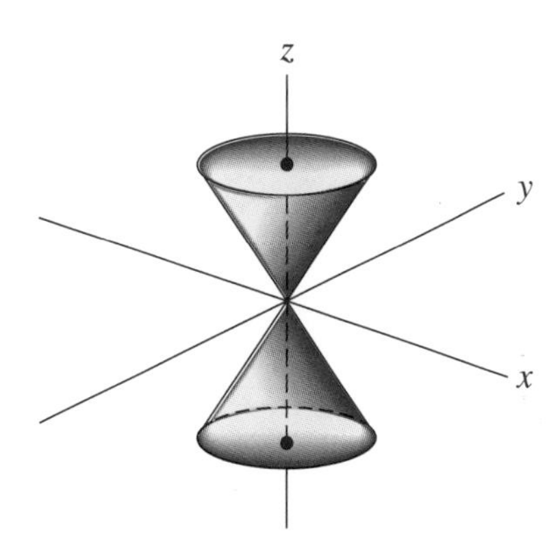

vii. Parabolic cylinder (Figure 8):

$$x^2 = py + qz \quad (p, q \text{ not both zero})$$

FIGURE 8

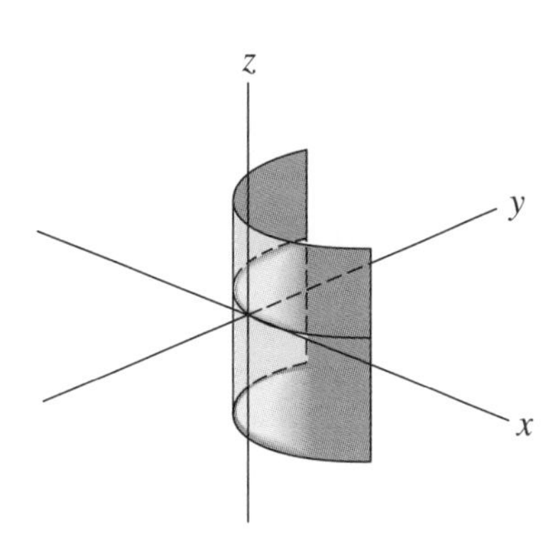

viii. Elliptic cylinder (Figure 9):

$$\frac{x^2}{p^2} + \frac{y^2}{q^2} = 1 \quad (p, q > 0)$$

FIGURE 9

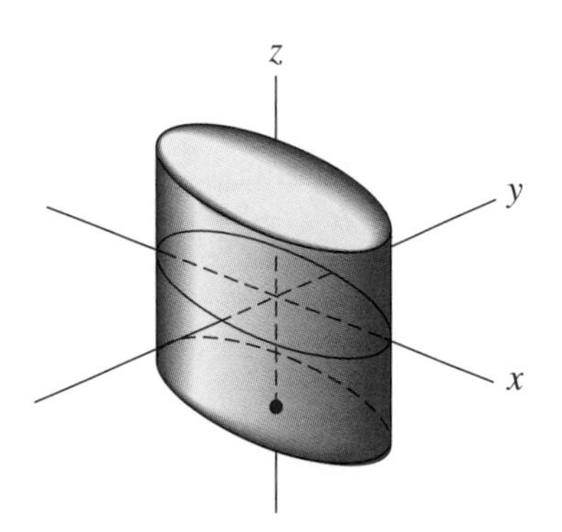

ix. Hyperbolic cylinder (Figure 10):

$$\frac{x^2}{p^2} - \frac{y^2}{q^2} = 1 \quad (p, q > 0)$$

FIGURE 10

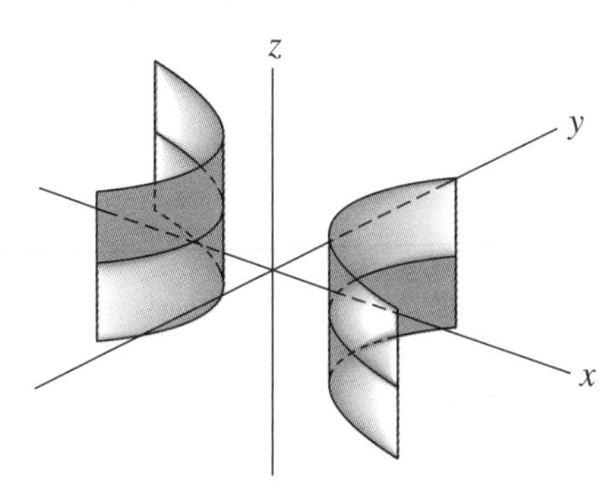

EXAMPLE 1

Identify the quadrics whose equations are

a. $\dfrac{x^2}{4} + \dfrac{y^2}{9} + z^2 = 1,$

b. $9x^2 - y^2 - 16z^2 = 144,$

c. $x^2 - y^2 - z = 0.$

Solution Equation (a) fits the standard form (i), with $p = 2$, $q = 3$, and $r = 1$; it represents an ellipsoid. Dividing Equation (b) by 144, we obtain one of the standard forms (v), with $p = 4$, $q = 12$, and $r = 3$; it represents a hyperboloid of two sheets. Finally, by transposing the z term of Equation (c), we see that it takes the standard form (iii), with $p = 1$ and $q = 1$; it represents a hyperbolic paraboloid. ■

The Reduction Procedure

We now derive a procedure for transforming (or *reducing*) those special cases of Equation (1) that represent nondegenerate quadric surfaces to one of the given standard forms. We begin by writing Equation (1) in the matrix form

$$\mathbf{x}^T A\mathbf{x} + B\mathbf{x} = k, \tag{2}$$

where A, B, and $\mathbf{x}$ are the matrices.

$$A = \begin{bmatrix} a & \frac{d}{2} & \frac{e}{2} \\ \frac{d}{2} & b & \frac{f}{2} \\ \frac{e}{2} & \frac{f}{2} & c \end{bmatrix}, \quad B = [g \quad h \quad i], \quad \text{and} \quad \mathbf{x} = \begin{bmatrix} x \\ y \\ z \end{bmatrix}. \tag{3}$$

To see that Equations (1) and (2) are equivalent, simply perform the operations indicated in Equation (2), and Equation (1) results. Notice that A is a *symmetric* matrix.

EXAMPLE 2

Find matrices A and B such that the equation $x^2 - 4y^2 - 2z^2 + 8xy - 6yz + 7x - 3y = 1$ takes the matrix form of Equation (2).

Solution Comparing the given equation to Equation (1), we see that $a = 1$, $b = -4$, $c = -2$, $d = 8$, $e = 0$, $f = -6$, $g = 7$, $h = -3$, $i = 0$, and $k = 1$. Therefore,

$$A = \begin{bmatrix} 1 & 4 & 0 \\ 4 & -4 & -3 \\ 0 & -3 & -2 \end{bmatrix} \quad \text{and} \quad B = [7 \quad -3 \quad 0].$$

■

The first step in the reduction procedure is the construction of a transformation that, when applied to Equation (1), results in the elimination of the *cross-product* (the xy, xz, and yz) terms. In relation to Equation (2), this is equivalent to diagonalizing the symmetric matrix A, since the off-diagonal entries of A are the ones that produce the coefficients of the cross-product terms.

Letting P be an orthogonal 3×3 matrix such that $D = P^TAP$ is diagonal (see Section 8.3), we make the change of variable

$$\mathbf{x} = P\mathbf{x}'$$

and substitute into Equation (2), obtaining

$$(P\mathbf{x}')^TA(P\mathbf{x}') + B(P\mathbf{x}') = k.$$

Now $(P\mathbf{x}')^T = (\mathbf{x}')^TP^T$, so this equation becomes

$$(\mathbf{x}')^T(P^TAP)\mathbf{x}' + BP\mathbf{x}' = k.$$

But $P^TAP = D$ (a diagonal matrix), so the preceding equation takes the form

$$(\mathbf{x}')^TD\mathbf{x}' + BP\mathbf{x}' = k. \quad \textbf{(4)}$$

Recalling, from Section 8.2, that the diagonal entries of D are just the eigenvalues of A, λ_1, λ_2, and λ_3, we need not actually perform the matrix multiplications P^TAP to obtain D; we simply set

$$D = \begin{bmatrix} \lambda_1 & 0 & 0 \\ 0 & \lambda_2 & 0 \\ 0 & 0 & \lambda_3 \end{bmatrix}.$$

Obtaining BP requires more effort, since the procedure for determining P involves finding an orthonormal basis of eigenvectors of A. Letting P be a matrix with these eigenvectors as columns, we form the product BP and obtain a matrix we denote by

$$BP = [a' \quad b' \quad c'].$$

Finally, letting $\mathbf{x}' = (x', y', z')$, we see that Equation (4) can be rewritten as the scalar equation

$$\lambda_1(x')^2 + \lambda_2(y')^2 + \lambda_3(z')^2 + a'x' + b'y' + c'z' = k. \quad \textbf{(5)}$$

Notice that Equation (5) contains no cross-product terms.

NOTE If Equation (1) contains no linear (x, y, or z) terms, then $B = 0$, and hence $BP = 0$. In this case, we do not have to bother with P at all to identify the quadric surface, since Equation (5) becomes simply

$$\lambda_1(x')^2 + \lambda_2(y')^2 + \lambda_3(z')^2 = k$$

and depends only on the eigenvalues of A. This is the case in the following example.

EXAMPLE 3

Reduce the equation $5x^2 + 11y^2 + 2z^2 - 16xy + 20xz + 4yz = 18$ to standard form and identify the quadric.

Solution We first write the given equation in the form

$$\mathbf{x}^TA\mathbf{x} = 18,$$

where $\mathbf{x} = (x, y, z)$ and

$$A = \begin{bmatrix} 5 & -8 & 10 \\ -8 & 11 & 2 \\ 10 & 2 & 2 \end{bmatrix}.$$

To diagonalize A, we compute its eigenvalues, $\lambda_1 = 9, \lambda_2 = 18, \lambda_3 = -9$, and set

$$D = \begin{bmatrix} 9 & 0 & 0 \\ 0 & 18 & 0 \\ 0 & 0 & -9 \end{bmatrix}.$$

Then the transformed equation, $(\mathbf{x}')^T D\mathbf{x}' = 18$, is simply

$$9(x')^2 + 18(y')^2 - 9(z')^2 = 18.$$

Finally, dividing both sides by 18, we obtain the standard form

$$\frac{(x')^2}{2} + \frac{(y')^2}{1} - \frac{(z')^2}{2} = 1.$$

Thus, the quadric is a hyperboloid of one sheet (see Figure 11).

FIGURE 11

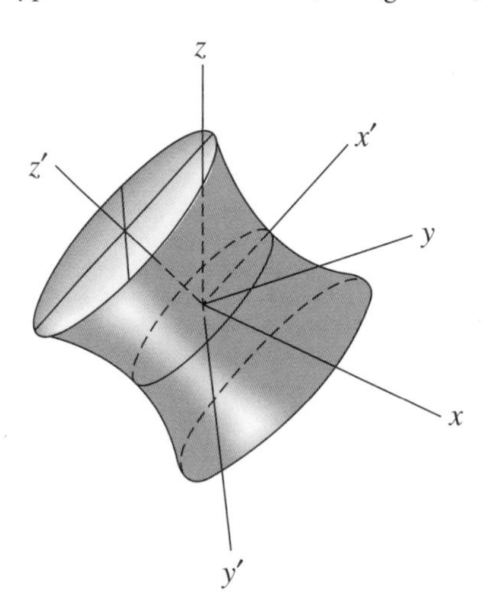

WARNING Unless the eigenvalues of A (the diagonal entries of D) are arranged in the proper order, it may not be possible to obtain a final equation that is *exactly* the same as one of the listed standard forms. It may have some of the variables interchanged. This corresponds to a permutation of the coordinate axes relative to those of the given standard form. For instance, if, in Example 3, we take

$$D = \begin{bmatrix} 9 & 0 & 0 \\ 0 & -9 & 0 \\ 0 & 0 & 18 \end{bmatrix},$$

the resulting equation is

$$\frac{(x')^2}{2} - \frac{(y')^2}{2} + (z')^2 = 1,$$

a hyperboloid of one sheet whose axis of symmetry is the y-axis.

We now know how to transform Equation (1) so that the cross-product terms are eliminated. In general, we also have to eliminate certain linear terms. This is accomplished with the aid of the process of *completing the square.* Recall that to complete the square in the expression $pu^2 + qu$, we proceed as follows:

$$\begin{aligned} pu^2 + qu &= p\left(u^2 + \left(\frac{q}{p}\right)u\right) \\ &= p\left(u^2 + \left(\frac{q}{p}\right)u + \frac{q^2}{4p^2}\right) - \frac{q^2}{4p} \\ &= p\left(u + \frac{q}{2p}\right)^2 - \frac{q^2}{4p} \end{aligned}$$

The manner in which completing the square is used in the reduction process is illustrated in the next example.

EXAMPLE 4

Reduce the equation $x^2 + 2y^2 - 2x + 8y - 4z = 3$ to standard form and identify the quadric.

Solution Notice that no cross-product terms appear in this equation. We complete the square in the quadratic variables, x and y, and transpose the z term to the right side:

$$(x^2 - 2x + 1) + 2(y^2 + 4y + 4) = 4z + 3 + 1 + 8$$
$$(x - 1)^2 + 2(y + 2)^2 = 4z + 12$$

We now divide both sides of this equation by the coefficient of z, obtaining

$$\frac{(x - 1)^2}{4} + \frac{(y + 2)^2}{2} = z + 3.$$

Finally, we make the change of variables $x' = x - 1$, $y' = y + 2$, and $z' = z + 3$. This yields

$$\frac{(x')^2}{4} + \frac{(y')^2}{2} = z',$$

which is the standard form of an equation (in variables x', y', and z') of an elliptic paraboloid (Figure 12).

FIGURE 12

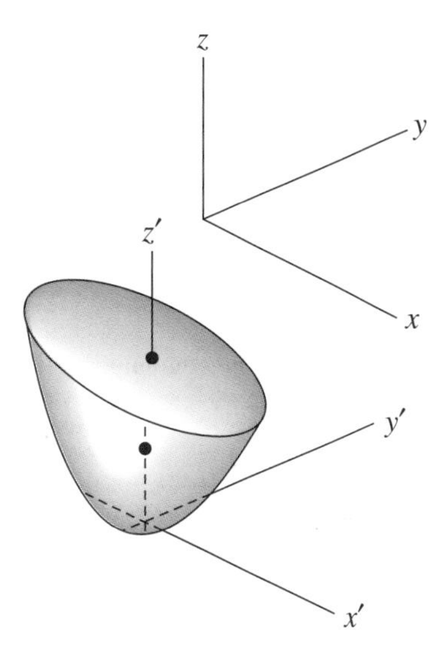

■

NOTE Geometrically, the change of variables of the type used in Example 4, $x' = x - x_0$, $y' = y - y_0$, $z' = z - z_0$, *translates* the xyz-coordinate system into the $x'y'z'$-system. The origin of the $x'y'z'$-coordinate system is the point (x_0, y_0, z_0) in the xyz-system. For example, the change of variables $x' = x - 1$, $y' = y + 2$, $z' = z + 3$ of Example 2 results in an $x'y'z'$ origin at $(1, -2, -3)$ of the xyz-system (see Figure 12).

Before giving another example, we present this technique in the form of a procedure.

Procedure to Reduce a Quadratic Equation in Three Variables to Standard Form

Given the equation

$$ax^2 + by^2 + cz^2 + dxy + exz + fyz + gx + hy + iz = k,$$

i. Determine the matrices A and B described by Equation (3);

ii. Find an orthogonal matrix P and a diagonal matrix D such that $P^TAP = D$;

iii. Make the change of variable $\mathbf{x} = P\mathbf{x}'$ by performing the indicated matrix operations in

$$(\mathbf{x}')^T D\mathbf{x}' + BP\mathbf{x}' = k$$

to eliminate the cross-product terms, and convert it to a scalar equation;

iv. Complete the square, translate the coordinate system by making a change of variables, and rearrange the terms in the scalar equation so that it is in standard form.

EXAMPLE 5

Reduce the equation $2x^2 + 2y^2 + 2z^2 + 2xz + 4\sqrt{2}x + 12y = -10$ to standard form and identify the quadric.

Solution The given equation has the matrix form $\mathbf{x}^T A\mathbf{x} + B\mathbf{x} = -10$, where

$$A = \begin{bmatrix} 2 & 0 & 1 \\ 0 & 2 & 0 \\ 1 & 0 & 2 \end{bmatrix} \quad \text{and} \quad B = [4\sqrt{2} \quad 12 \quad 0].$$

Since the eigenvalues and corresponding normalized eigenvectors of A are

$$\lambda_1 = 1, \quad \lambda_2 = 2, \quad \lambda_3 = 3,$$

$$\text{and} \quad \mathbf{x}_1 = \left(\frac{-1}{\sqrt{2}}, 0, \frac{1}{\sqrt{2}}\right), \quad \mathbf{x}_2 = (0, 1, 0), \quad \mathbf{x}_3 = \left(\frac{1}{\sqrt{2}}, 0, \frac{1}{\sqrt{2}}\right),$$

we may take

$$P = \begin{bmatrix} \frac{-1}{\sqrt{2}} & 0 & \frac{1}{\sqrt{2}} \\ 0 & 1 & 0 \\ \frac{1}{\sqrt{2}} & 0 & \frac{1}{\sqrt{2}} \end{bmatrix} \quad \text{and} \quad D = \begin{bmatrix} 1 & 0 & 0 \\ 0 & 2 & 0 \\ 0 & 0 & 3 \end{bmatrix}.$$

Thus, applying the transformation $\mathbf{x} = P\mathbf{x}'$, we obtain the equation

$$(\mathbf{x}')^T D\mathbf{x}' + BP\mathbf{x}' = -10,$$

$$\text{or} \quad (x')^2 + 2(y')^2 + 3(z')^2 - 4x' + 12y' + 4z' = -10.$$

We now complete the square in x', y', and z' and obtain

$$(x'-2)^2 + 2(y' + 3)^2 + 3\left(z' + \frac{2}{3}\right)^2 = \frac{40}{3}.$$

Finally, letting $x'' = x' - 2$, $y'' = y' + 3$, and $z'' = z' + \frac{2}{3}$, and dividing both sides of the last equation by 40/3 yields the standard form

$$\frac{(x'')^2}{40/3} + \frac{(y'')^2}{20/3} + \frac{(z'')^2}{40/9} = 1.$$

The graph is an ellipsoid. ▣

NOTE The techniques of this section can also be applied to the simpler case of a quadratic equation in two variables

$$ax^2 + bxy + cy^2 + dx + ey = f,$$

which is the equation of a *conic section.* The reduction of a conic section to standard form is explored in Exercises 11 through 18.

8.6 Exercises

In Exercises 1–4 the given equation contains no linear terms. Reduce it to standard form and identify the quadric surface it represents.

1. $x^2 + y^2 + z^2 - 2xy - 2xz - 2yz = 2$

2. $3x^2 + 3y^2 + 3z^2 - 3xy - 4yz = 66$

3. $x^2 + y^2 + z^2 + 6xy + 8yz = 0$

4. $x^2 + 2\sqrt{2}xy = 4$

In Exercises 5 and 6 the given equation contains no cross-product terms. Reduce it to standard form and identify the quadric surface it represents.

5. $2x^2 + y^2 + z^2 - 4x + 8y + 6z = 9$

6. $2x^2 - 2y^2 - z^2 - 8x + 4y - 4z = 2$

In Exercises 7–10 reduce the given equation to standard form and identify the quadric surface it represents.

7. $2x^2 + y^2 + z^2 + 2\sqrt{6}xz + 2y + 2\sqrt{5}z = 2$

8. $x^2 + y^2 + z^2 + 2xy + 2xz + 2yz + \sqrt{6}y = 0$

9. $x^2 + y^2 - z^2 + 2xy - 2z = 3$

10. $3x^2 + 3y^2 + z^2 - 2xy + 12z = 7$

In Exercises 11–14 a method for reducing quadratic equations in two variables—those of *conic sections*—to standard form is investigated.

11. Show that equation $ax^2 + bxy + cy^2 + dx + ey = f$ can be written in the matrix form $\mathbf{x}^T A\mathbf{x} + B\mathbf{x} = f$, where

$$A = \begin{bmatrix} a & \frac{b}{2} \\ \frac{b}{2} & c \end{bmatrix}, \quad B = [d \quad e], \quad \text{and} \quad \mathbf{x} = \begin{bmatrix} x \\ y \end{bmatrix}.$$

12. Use the result of Exercise 11 to write the equation

$$3x^2 + 4xy - y^2 + 5x - 2y = 1$$

in matrix form.

13. Given the matrix equation $\mathbf{x}^T A\mathbf{x} + B\mathbf{x} = f$, let P be an orthogonal matrix such that $P^T AP = D$, a diagonal matrix. Show that the change of variable $\mathbf{x} = P\mathbf{x}'$ transforms the given equation into one of the form

$$(\mathbf{x}')^T D\mathbf{x}' + BP\mathbf{x}' = f.$$

14. Show that the matrix equation $(\mathbf{x}')^T D\mathbf{x}' + BP\mathbf{x}' = f$ of Exercise 13 is equivalent to the scalar equation $a'(x')^2 + c'(y')^2 + d'x' + e'y' = f$, where a' and c' are the diagonal entries of D and d' and e' are the entries of BP. (Notice that there is no *cross-product*—that is, $x'y'$—term in the scalar equation.)

In Exercises 15 and 16 use the results of Exercises 11, 13, and 14 to transform the given equation into one in which no cross-product terms appear.

15. $x^2 + 4xy + y^2 + 5\sqrt{2}x + \sqrt{2}y = 0$

16. $x^2 - 4xy + 4y^2 + 10\sqrt{5}x = 15$

17, 18. With the aid of the *completing-the-square* process reduce the transformed equations obtained in Exercises 15 and 16 to one of the following *standard forms* for a *nondegenerate conic section*:

$$\textit{ellipse}, \frac{x^2}{p^2} + \frac{y^2}{q^2} = 1 \quad (p, q > 0)$$

$$\textit{parabola}, y^2 = px \quad (p \neq 0)$$

$$\text{or} \quad \textit{hyperbola}, \frac{x^2}{p^2} - \frac{y^2}{q^2} = 1 \quad (p, q > 0)$$

8.7 APPLICATION—Fractals III: When Is an Affine Map Contractive?

Recall from Section 7.7 that an affine map S on $\mathbf{R}^n$ has an $n \times n$ matrix A and a vector $\mathbf{b} \in \mathbf{R}^n$ so that for all $\mathbf{x} \in \mathbf{R}^n$, $S(\mathbf{x}) = A\mathbf{x} + \mathbf{b}$. Moreover, the map is a *contraction* with contraction factor $s \in (0, 1)$ if, for all vectors, $\mathbf{x}, \mathbf{y} \in \mathbf{R}^n$,

$$\|S(\mathbf{y}) - S(\mathbf{x})\| \leq s\|\mathbf{y} - \mathbf{x}\|,$$

so that S shrinks the distance between two vectors by a factor of s. In Section 7.7, we discussed how contractions could be used to build fractals (by either the deterministic or nondeterministic procedure). But it is essential to check that the affine maps are indeed contractive. How can this be done? For some maps, it can be fairly straightforward to show that the map is a contraction. The affine transformation

$$S(x_1, x_2) = \left(\frac{1}{2}(x_1 + 1), \frac{1}{2}x_2\right)$$

was shown to be a contraction directly in Example 2 of Section 7.7. For many affine maps, it can be difficult to manipulate the inequalities to show whether they are indeed contractions. For example, is the affine map

$$g(x, y) = (0.85x + 0.04y, -0.04x + 0.85y + 1.6)$$

a contraction? If so, what is the contraction factor? We'll learn, in this section, how eigenvalues can direct us to the answer.

Let $S(\mathbf{x}) = A\mathbf{x} + \mathbf{b}$ be an affine map on $\mathbf{R}^n$. We begin by recalling that if we write all vectors as column vectors, then the dot product can be written in matrix notation as

$$\mathbf{x} \bullet \mathbf{y} = \mathbf{x}^T\mathbf{y}.$$

It follows that the distance between vectors $\mathbf{x}$ and $\mathbf{y}$ can be written as

$$\begin{aligned}\text{dist}(\mathbf{x}, \mathbf{y}) &= \|\mathbf{y} - \mathbf{x}\| \\ &= \sqrt{(\mathbf{y} - \mathbf{x}) \bullet (\mathbf{y} - \mathbf{x})} \\ &= \sqrt{(\mathbf{y} - \mathbf{x})^T(\mathbf{y} - \mathbf{x})}.\end{aligned}$$

Similarly,

$$\begin{aligned}\text{dist}(S(\mathbf{x}), S(\mathbf{y})) &= \text{dist}(A\mathbf{x} + \mathbf{b}, A\mathbf{y} + \mathbf{b}) \\ &= \|(A\mathbf{y} + \mathbf{b}) - (A\mathbf{x} + \mathbf{b})\| \\ &= \|A(\mathbf{y} - \mathbf{x})\| \\ &= \sqrt{(A(\mathbf{y} - \mathbf{x}))^T(A(\mathbf{y} - \mathbf{x}))} \\ &= \sqrt{(\mathbf{y} - \mathbf{x})^T A^T A(\mathbf{y} - \mathbf{x})}.\end{aligned}$$

Let's set $\mathbf{z} = \mathbf{y} - \mathbf{x}$. Then $\text{dist}(S(\mathbf{x}), S(\mathbf{y})) = \sqrt{\mathbf{z}^T A^T A\mathbf{z}}$. We shall find an upper bound for $\mathbf{z}^T A^T A\mathbf{z}$ in terms of $\mathbf{z}^T\mathbf{z} = (\mathbf{y} - \mathbf{x})^T(\mathbf{y} - \mathbf{x})$.

The matrix A^TA has all real entries, and, moreover, it is symmetric, as

$$(A^TA)^T = A^T(A^T)^T = A^TA.$$

From Section 8.2, the symmetric matrix A^TA is orthogonally diagonalizable. Let the eigenvalues of A^TA be $\lambda_1 \geq \lambda_2 \geq \ldots \geq \lambda_n$. We can choose an orthonomal basis of eigenvectors $\mathbf{x}_1, \mathbf{x}_2, \ldots, \mathbf{x}_n$, so that

$$A^TA\mathbf{x}_i = \lambda_i\mathbf{x}_i$$

for each i. Note that by multiplying on the left by $\mathbf{x}_i^T$, we get

$$\begin{aligned} \mathbf{x}_i^T A^TA\mathbf{x}_i &= \mathbf{x}_i^T\lambda_i\mathbf{x}_i \\ (A\mathbf{x}_i) \bullet (A\mathbf{x}_i) &= \lambda_i\mathbf{x}_i^T\mathbf{x}_i \\ \|A\mathbf{x}_i\|^2 &= \lambda_i\|\mathbf{x}_i\|^2 \end{aligned}$$

so

$$\lambda_i = \frac{\|A\mathbf{x}_i\|^2}{\|\mathbf{x}_i\|^2} \geq 0.$$

This means that all the eigenvalues are nonnegative numbers, so that $\lambda_1 \geq \lambda_2 \geq \ldots \geq \lambda_n \geq 0$.

We are finally ready to reconsider $\mathbf{z}^TA^TA\mathbf{z}$. As $\{\mathbf{x}_1, \mathbf{x}_2, \ldots, \mathbf{x}_n\}$ is an orthonormal basis of $\mathbf{R}^n$, we can find real numbers $c_1, c_2, \ldots, c_n$ such that

$$\mathbf{z} = c_1\mathbf{x}_1 + c_2\mathbf{x}_2 + \cdots + c_n\mathbf{x}_n.$$

Then $\mathbf{z}^T\mathbf{z} = c_1^2 + c_2^2 + \cdots + c_n^2$. Using the fact that $A^TA\mathbf{x}_i = \lambda_i\mathbf{x}_i$ and $\mathbf{x}_i \bullet \mathbf{x}_j = 0$ if $i \neq j$ and $\mathbf{x}_i \bullet \mathbf{x}_j = 1$ if $i = j$, we derive the following:

$$\begin{aligned} \mathbf{z}^TA^TA\mathbf{z} &= (c_1\mathbf{x}_1 + c_2\mathbf{x}_2 + \cdots + c_n\mathbf{x}_n)^TA^TA(c_1\mathbf{x}_1 + c_2\mathbf{x}_2 + \cdots + c_n\mathbf{x}_n) \\ &= (c_1\mathbf{x}_1^T + c_2\mathbf{x}_2^T + \cdots + c_n\mathbf{x}_n^T)A^TA(c_1\mathbf{x}_1 + c_2\mathbf{x}_2 + \cdots + c_n\mathbf{x}_n) \\ &= (c_1\mathbf{x}_1^T + c_2\mathbf{x}_2^T + \ldots + c_n\mathbf{x}_n^T)(c_1A^TA\mathbf{x}_1 + c_2A^TA\mathbf{x}_2 + \cdots + c_nA^TA\mathbf{x}_n) \\ &= (c_1\mathbf{x}_1^T + c_2\mathbf{x}_2^T + \cdots + c_n\mathbf{x}_n^T)(c_1\lambda_1\mathbf{x}_1 + c_2\lambda_2\mathbf{x}_2 + \cdots + c_n\lambda_n\mathbf{x}_n) \\ &= c_1^2\lambda_1 + c_2^2\lambda_2 + \cdots + c_n^2\lambda_n \\ &\leq c_1^2\lambda_1 + c_2^2\lambda_1 + \cdots + c_n^2\lambda_1 \\ &= \lambda_1(c_1^2 + c_2^2 + \cdots + c_n^2) \\ &= \lambda_1\mathbf{z}^T\mathbf{z} \end{aligned}$$

It follows that

$$\text{dist}(S(\mathbf{x}), S(\mathbf{y})) \leq \sqrt{\lambda_1}\,\text{dist}(\mathbf{x}, \mathbf{y}) \quad (*)$$

The value $\sqrt{\lambda_1}$ of the square root of the largest eigenvalue of the matrix A^TA is given a special name, the *spectral norm* of matrix A. Equation (*) implies the following theorem.

THEOREM 1

Let $S(\mathbf{x}) = A\mathbf{x} + \mathbf{b}$ be an affine map on $\mathbf{R}^n$. Then S is a contraction if the spectral norm λ of A is less than 1. ■

It can be shown that if the spectral norm is greater than or equal to 1, then the affine map is not contractive (see Exercise 2). Thus, Theorem 1 is all we need to determine if an affine map is contractive, and hence able to be used to generate a fractal.

EXAMPLE 1

Consider the affine function

$$S(\mathbf{x}) = \begin{bmatrix} \frac{1}{2} & 0 \\ 0 & \frac{1}{2} \end{bmatrix}\mathbf{x} + \begin{bmatrix} \frac{1}{2} \\ 0 \end{bmatrix} = \left[\tfrac{1}{2}(x_1 + 1), \tfrac{1}{2}x_2\right].$$

To calculate the spectral norm of the matrix $A = \begin{bmatrix} \frac{1}{2} & 0 \\ 0 & \frac{1}{2} \end{bmatrix}$, we form

$$A^TA = \begin{bmatrix} \frac{1}{2} & 0 \\ 0 & \frac{1}{2} \end{bmatrix}\begin{bmatrix} \frac{1}{2} & 0 \\ 0 & \frac{1}{2} \end{bmatrix} = \begin{bmatrix} \frac{1}{4} & 0 \\ 0 & \frac{1}{4} \end{bmatrix}.$$

This matrix is diagonal and clearly has just one eigenvalue, $\frac{1}{4}$. Thus, the spectral norm of A is the square root of the largest of these, which is obviously $\sqrt{\frac{1}{4}} = \frac{1}{2}$. As this is less than 1, Theorem 1 tells us that the affine map is indeed contractive, with contraction factor $\sqrt{\frac{1}{4}} = \frac{1}{2}$. ■

EXAMPLE 2

Consider the affine function

$$g(x, y) = (0.85x + 0.04y, -0.04x + 0.85y + 1.6).$$

We can rewrite g in matrix form:

$$g(\mathbf{x}) = \begin{bmatrix} 0.85 & 0.04 \\ -0.04 & 0.85 \end{bmatrix}\mathbf{x} + \begin{bmatrix} 0 \\ 1.6 \end{bmatrix}$$

To calculate the spectral norm of the matrix $A = \begin{bmatrix} 0.85 & -0.04 \\ -0.04 & 0.85 \end{bmatrix}$, we form

$$A^TA = \begin{bmatrix} 0.85 & -0.04 \\ 0.04 & 0.85 \end{bmatrix}\begin{bmatrix} 0.85 & 0.04 \\ -0.04 & 0.85 \end{bmatrix} = \begin{bmatrix} 0.7241 & 0 \\ 0 & 0.7241 \end{bmatrix}.$$

This matrix is diagonal, and it has just one eigenvalue, 0.7241. Thus, the spectral norm of A is $\sqrt{0.7241} \approx 0.85$. As this is less than 1, Theorem 1 tells us that the affine map is indeed contractive, with contraction factor approximately 0.85. ■

EXAMPLE 3

Consider the affine function

$$h(x, y) = (0.8x + 0.4y + 0.1, 0.6x - 0.9y - 0.1).$$

We can rewrite h in matrix form as

$$h(\mathbf{x}) = \begin{bmatrix} 0.8 & 0.4 \\ 0.6 & -0.9 \end{bmatrix}\mathbf{x} + \begin{bmatrix} 0.1 \\ -0.1 \end{bmatrix}.$$

To calculate the spectral norm of the matrix $A = \begin{bmatrix} 0.8 & 0.4 \\ 0.6 & -0.9 \end{bmatrix}$, we form

$$A^TA = \begin{bmatrix} 0.8 & 0.6 \\ 0.4 & -0.9 \end{bmatrix}\begin{bmatrix} 0.8 & 0.4 \\ 0.6 & -0.9 \end{bmatrix} = \begin{bmatrix} 1 & -0.22 \\ -0.22 & 0.97 \end{bmatrix}.$$

To calculate the eigenvalues of $A^TA = \begin{bmatrix} 1 & -0.22 \\ -0.22 & 0.97 \end{bmatrix}$, we form the characteristic polynomial of this matrix:

$$\begin{aligned} \det\begin{bmatrix} 1 - \lambda & -0.22 \\ -0.22 & 0.97 - \lambda \end{bmatrix} &= (1 - \lambda)(0.97 - \lambda) - (-0.22)(-0.22) \\ &= \lambda^2 - 1.97\lambda + 0.9216 \end{aligned}$$

We can find the roots of this matrix using the quadratic formula, and we get λ to be approximately 0.76 and 1.21. Thus, the spectral norm is the square root of the largest of these, approximately 1.1. As this is larger than 1, we conclude that the affine map h is *not* a contraction. ■

8.7 Exercises

1. In the argument that showed $\mathbf{z}^T A^T A\mathbf{z} \leq \lambda_1 \mathbf{z}^T\mathbf{z}$, justify each of the lines.
2. If the spectral norm of matrix A is greater than 1, show that affine map S: $\mathbf{R}^n \rightarrow \mathbf{R}^n$ defined by $S(\mathbf{x}) = A\mathbf{x} + \mathbf{b}$ is not contractive by finding vectors $\mathbf{x}$ and $\mathbf{y}$ such that dist($S(\mathbf{x})$, $S(\mathbf{y})$) $\geq$ dist($\mathbf{x}$, $\mathbf{y}$).
3. Confirm that all the affine maps in Exercise 4 of Section 7.7 are indeed contractions and find contraction factors.

CHAPTER SUMMARY

Eigenvalues and Eigenvectors

(a) DEFINITION: The scalar λ is an *eigenvalue,* and the nonzero vector $\mathbf{x}$ is an *eigenvector* of

- The linear transformation T: $\mathbf{V} \rightarrow \mathbf{V}$ if $T(\mathbf{x}) = \lambda\mathbf{x}$,
- The square matrix A if $A\mathbf{x} = \lambda\mathbf{x}$.

(b) The eigenvalues of the matrix A are the roots of the *characteristic equation* $\det(A - \lambda I) = 0$.

(c) The eigenvectors corresponding to eigenvalue λ_0 are the nonzero solutions of the homogeneous linear system $(A - \lambda_0 I)\mathbf{x} = \mathbf{0}$. *All* solutions of $(A - \lambda_0 I)\mathbf{x} = \mathbf{0}$ form the *eigenspace* of λ_0.

Diagonalization

(a) DEFINITION: A square matrix A is *diagonalizable* if it is similar to a diagonal matrix D; that is, if there exists an invertible matrix P such that $D = P^{-1}AP$ is a diagonal matrix.

(b) To diagonalize the $n \times n$ matrix A, take the union of the bases of all eigenspaces of A: $\mathcal{B} = \{\mathbf{x}_1, \mathbf{x}_2, \ldots, \mathbf{x}_m\}$. Then

- If $m < n$, then A is not diagonalizable
- If $m = n$, then $P^{-1}AP = D$, where $\mathbf{p}^i = \mathbf{x}_i$ and the diagonal entries of D are the eigenvalues of A.

(c) A square matrix A is diagonalizable if and only if A has n linearly independent eigenvectors.

Diagonalization of Symmetric Matrices

(a) DEFINITION: A square matrix A is *symmetric* if $A = A^T$. An invertible matrix P is *orthogonal* if $P^{-1} = P^T$.

(b) If A is a symmetric matrix, then there exists an orthogonal matrix P such that $P^{-1}AP = D$, a diagonal matrix; that is, every symmetric matrix is *orthogonally diagonalizable.*

QR Factorization

(a) Every nonsingular square matrix can be factored as the product of an orthogonal matrix Q and an upper-triangular matrix: R: $A = QR$.

(b) The matrix Q is obtained by performing the Gram–Schmidt process on the columns of the given matrix A; the nonzero entries of R are obtained by $r_{ij} = \mathbf{a}^j \cdot \mathbf{q}^i$.

(c) The QR algorithm approximates the eigenvalues of a matrix A by constructing, in the following way, a sequence of matrices whose diagonal entries approach the eigenvalues:

i. Set $A_1 = A$.

ii. For $k = 1, 2, 3, \ldots$ find the QR factorization of A_k, $A_k = Q_kR_k$, and multiply to get $A_{k+1} = R_kQ_k$.

Complex Eigenvalues and Eigenvectors

(a) Every square matrix with complex entries has at least one complex eigenvalue; that is, given a square matrix A, there exists a complex scalar λ and a nonzero vector $\mathbf{z}$ in $\mathbf{C}^n$ such that $A\mathbf{z} = \lambda\mathbf{z}$.

(b) DEFINITION: A square matrix A is *hermitian* if $A = A^H$, where A^H denotes the transpose of the matrix obtained from A by replacing each entry by its complex conjugate. An invertible matrix P is *unitary* if $P^{-1} = P^H$.

(c) If A is an Hermitian matrix with complex entries, then there exists a unitary matrix P such that $P^{-1}AP = D$, a diagonal matrix with real entries.

KEYWORDS

eigenvalue
eigenvector
characteristic equation
eigenspace
diagonalizable
orthogonal matrix
orthogonally diagonalizable
QR factorization
complex *m*-space
hermitian
unitary
quadric surface
ellipsoid
elliptic paraboloid
hyperboloid of one sheet
hyperboloid of two sheets
cone
parabolic cylinder
elliptic cylinder
hyperbolic cylinder
spectral norm

SELF-TEST

1. Find the eigenvalues and corresponding eigenvectors of the linear transformation $T: \mathbf{R}^2 \rightarrow \mathbf{R}^2$ that projects every vector in $\mathbf{R}^2$ onto the x-axis; that is, for which $T(x, y) = (x, 0)$. [Section 8.1]

For each matrix in Exercises 2–4 find

(a) Its characteristic equation
(b) All real eigenvalues
(c) Bases for the corresponding eigenspaces [Section 8.1]

2. $\begin{bmatrix} 1 & 1 \\ 0 & 1 \end{bmatrix}$ *3.* $\begin{bmatrix} 2 & 2 & 2 \\ 2 & 2 & 2 \\ 2 & 2 & 2 \end{bmatrix}$ *4.* $\begin{bmatrix} 2 & -1 & 0 \\ -1 & 2 & -1 \\ 0 & -1 & 2 \end{bmatrix}$

In Exercises 5 and 6 determine whether or not the matrix A is diagonalizable. If it is, find matrices P and D such that $D = P^{-1}AP$ is a diagonal matrix.

5. $A = \begin{bmatrix} 2 & 2 & 2 \\ 2 & 2 & 2 \\ 2 & 2 & 2 \end{bmatrix}$ *6.* $A = \begin{bmatrix} 1 & 0 & 0 \\ 0 & 0 & 1 \\ 0 & -1 & 2 \end{bmatrix}$ [Section 8.2]

7. Let $B = \begin{bmatrix} 0 & 0 & 1 \\ 0 & 1 & 0 \\ 1 & 0 & 1 \end{bmatrix}$.

(a) Is B a symmetric matrix?
(b) Is B an orthogonal matrix? [Section 8.2]

8. Find the QR factorization of matrix B of Exercise 7. [Section 8.3]

9. Perform one iteration of the QR algorithm on matrix B of Exercise 7. [Section 8.3]

10. Let $A = \begin{bmatrix} i & -i \\ i & i \end{bmatrix}$, $\mathbf{w} = (i, 1, 1 + i)$, and $\mathbf{z} = (-i, i, -i)$.

(a) Find $\mathbf{w} \cdot \mathbf{z}$.
(b) Find $\|\mathbf{w}\|$.
(c) Find A^2.
(d) Find A^{-1}.
(e) Is A hermitian?
(f) Is A unitary? [Section 8.4]

11. Let $B = \begin{bmatrix} i & 0 & 0 \\ 0 & 0 & i \\ 0 & i & 0 \end{bmatrix}$.

Find all complex eigenvalues of B and bases for the corresponding eigenspaces. [Section 8.4]

12. Is matrix B of Exercise 11 diagonalizable? If so, find matrices P and D such that $D = P^{-1}BP$ is a diagonal matrix. [Section 8.4]

REVIEW EXERCISES

Use the following matrices for Exercises 1–6 and 9–13.

$A = \begin{bmatrix} 2 & 1 \\ 1 & 2 \end{bmatrix}$ $B = \begin{bmatrix} 0 & 4 \\ 1 & 0 \end{bmatrix}$

$C = \begin{bmatrix} 1 & 0 & 0 \\ -1 & 2 & 1 \\ 0 & 0 & 1 \end{bmatrix}$ $D = \begin{bmatrix} -6 & 2 & 7 \\ 0 & 2 & 0 \\ -9 & 2 & 10 \end{bmatrix}$

$E = \begin{bmatrix} 1 & 1 & -1 & 2 \\ 0 & 3 & 0 & 1 \\ 0 & 0 & 5 & 1 \\ 0 & 0 & 0 & 7 \end{bmatrix}$ $F = \begin{bmatrix} 2 & 1 & 1 \\ 1 & 1 & 0 \\ 1 & 0 & 1 \end{bmatrix}$

1. Find a basis of eigenvectors for each eigenvalue in matrix A.

2. Find a basis of eigenvectors for each eigenvalue in matrix C.

3. Identify which of matrices A through F can be diagonalized. Explain each of your answers.

4. Find a matrix P such that $P^{-1}CP$ is a diagonal matrix.

5. Find a matrix P such that $P^{-1}FP$ is a diagonal matrix.
6. Identify which of matrices A through F are orthogonally diagonalizable. Explain each of your answers.
7. Is the following matrix orthogonal? Verify.

$$\begin{bmatrix} \frac{1}{\sqrt{5}} & 0 & \frac{-2}{\sqrt{5}} \\ 0 & 1 & 0 \\ \frac{2}{\sqrt{5}} & 0 & \frac{1}{\sqrt{5}} \end{bmatrix}$$

8. Is the following matrix orthogonal? Verify.

$$\begin{bmatrix} -1 & 0 & 2 \\ 0 & 1 & 0 \\ 2 & 0 & 1 \end{bmatrix}$$

9. For each matrix H identified in Exercise 6, find an orthogonal matrix P such that $P^{-1}HP$ is a diagonal matrix.

10–13. For the corresponding matrices A through D,
   (a) Find the QR factorization;
   (b) Perform one iteration of the QR algorithm.

In Exercises 14 and 15 determine whether or not the given set is a basis for $\mathbf{C}^3$.

14. $\{(1 + i, 1, 2 - i), (1, 1 + i, 1 - i), (1, 3 + 2i, 2 - i)\}$
15. $\{(1, i, -i), (1, 1 + i, 1 - i), (1, 1 - i, 1 + i)\}$
16. Find the kernel of the linear transformation $T: \mathbf{C}^3 \rightarrow \mathbf{C}^3$ given by $T(\mathbf{z}) = G\mathbf{z}$, where

$$G = \begin{bmatrix} i & -i & 1 \\ 1 & 1 + i & 1 - i \\ 0 & 2 + i & 1 \end{bmatrix}.$$

17. Is the linear transformation of Exercise 16 one-to-one? Explain.
18. Is the linear transformation of Exercise 16 onto? Explain.

In Exercises 19 and 20 find the eigenvalues and associated eigenspaces, as well as a basis for each eigenspace.

19. $\begin{bmatrix} 1 & 0 & i \\ 0 & 1 & 0 \\ -i & 0 & 2 \end{bmatrix}$

20. $\begin{bmatrix} 1 & 2 & 1 \\ 0 & i & -i \\ 0 & 0 & 3 \end{bmatrix}$

21. Both of the matrices in Exercises 19 and 20 are diagonalizable. Explain why.
22. One of the matrices in Exercises 19 and 20 is orthogonally diagonalizable. Which matrix, and why?
23. Find a unitary matrix P such that $P^{-1}AP$ is diagonal, where A is the matrix of Exercise 22.

CHALLENGE YOURSELF

24. Suppose that A is an $n \times n$ matrix with distinct eigenvalues $\lambda_1, \lambda_2, \ldots, \lambda_n$. If k is a positive integer, what are the eigenvalues of A^k?
25. Let A be an $n \times n$ diagonalizable matrix. Show that if all the eigenvalues of A lie strictly between -1 and 1, then for any vector $\mathbf{v}$ in $\mathbf{R}^n$,

$$\lim_{k \to \infty} A^k\mathbf{v} = \mathbf{0}.$$

26. Let A be an $n \times n$ diagonalizable matrix with an eigenvalue of 1 and all other eignevalues λ satisfying $|\lambda| < 1$. Show that for *any* vector $\mathbf{u}$ there is a unique vector $\mathbf{v}$ such that,

$$\lim_{k\to\infty} A^k\mathbf{u} = \mathbf{v}.$$

(This fact is the basis of Google's ranking of web pages—see Section 4.6.)

27. Suppose that A is an $n \times n$ matrix such that A^k is an $n \times n$ zero matrix, for some positive integer k. Prove that the only possible eigenvalue for A is 0.

SPOTLIGHT

Danish Khan/Shutterstock

Many situations involve applications in which relationships vary in approximately linear ways. For example, the fabric requirements for making garments could safely be assumed to be linear. Double the garment production and, assuming a fairly constant mix of garments, the fabric requirements would have to double as well. Although linear, the manufacturer's need to have enough of each component material is better stated as an inequality, greater than or equal to, than as an equation. Such kinds of constraints should be built into a mathematical model that purports to analyze the situation. Similarly, a given factory may be able to produce various garments with different production schedules simultaneously, but these are also better viewed as inequalities, less than or equal to this time, or the factory would be assumed to be operating at maximum capacity at all times. At first glance, that might seem optimal but, if economic conditions are such that the factory is losing money, producing at maximum capacity might be the worst thing the manufacturer could do. Similarly, the manufacturer's revenue may be assumed to vary linearly with production: double the sales and double the revenue. Although it is easy to create situations in which such an assumption is wrong or even ridiculous, it is also fair to assume that, in some situations, it is eminently reasonable. Determining an optimal strategy in situations in which the various constraints are close to being linear is what linear programming is about.

We start with a simple example that can be solved by graphing in $\mathbf{R}^2$. Similar solution techniques apply in $\mathbf{R}^3$. Although it is difficult to visualize the three-dimensional regions that result, they resemble the figure shown here. For $\mathbf{R}^m$ with $m > 3$, we lose the geometric picture but not the ideas. Amazingly, the techniques developed here are applied to systems of linear inequalities that may run into the thousands of variables!

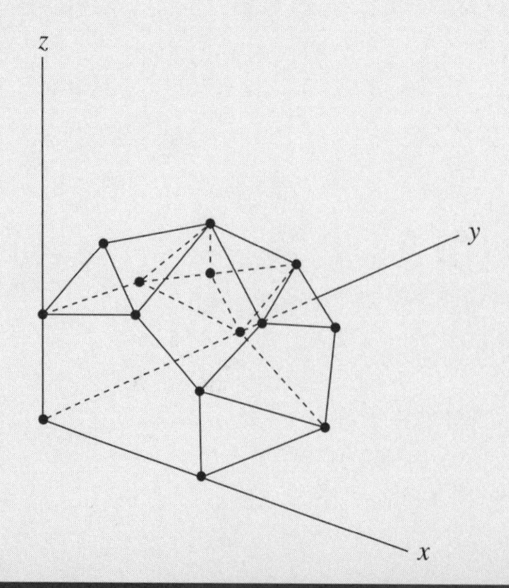

CHAPTER 9

Linear Programming

OBJECTIVES

You will learn to:

- Represent linear programming problems geometrically and algebraically.
- Translate systems of linear inequalities into a system of linear equations.
- Apply the simplex algorithm.
- Extend the methods to different settings.

The subject of **linear programming** deals with the problem of optimizing (maximizing or minimizing) a linear function, subject to the constraints imposed by a system of linear inequalities.

After introducing the topic in the first section, we demonstrate the basic solution technique in Section 9.2 and then extend this technique to a more general situation in Section 9.3. The proofs of the major theorems are delayed until Section 9.4. Adequate preparation for this chapter is given by Chapter 3.

9.1 Introduction and Terminology

In this section, you will see how linear programming problems might arise and how very simple ones can be solved geometrically. We also introduce some of the basic terminology and a standard form for linear programming problems.

A Simple Example

Suppose that a manufacturer produces two types of liquids: X and Y. Because of past sales experience, the market researcher estimates that at least twice as much Y as X is needed. The manufacturing capacity of the plant allows for a total of nine units to be manufactured. If each unit of liquid X results in a profit of \$2 and the profit for each unit of Y is \$1, how much of each should be produced to maximize the profit?

Let x be the amount of X to be produced and y be the amount of Y. Then the condition that there be at least twice as much Y as X becomes

$$2x \leq y.$$

The manufacturing capacity of nine units implies that

$$x + y \leq 9.$$

Let $z = z(x, y)$ be the profit associated with the production schedule of producing x units of X and y units of Y. Then the *profit function* or *objective function* is given by

$$z = 2x + y.$$

There is also a further condition implied but not stated in the problem. The variables must be nonnegative, since we cannot produce negative amounts of a product. Thus, we have

$$x \geq 0, \qquad y \geq 0.$$

Collecting all this information together, we may state the given problem mathematically as follows: Maximize $z = 2x + y$ subject to the constraints

$$\begin{aligned} 2x - y &\leq 0 \\ x + y &\leq 9 \\ x \geq 0, \qquad y &\geq 0. \end{aligned}$$

In the next section, we develop a technique (the *simplex algorithm*) for solving such problems in general, but we can solve this particular one geometrically. First, we sketch the set of points in $\mathbf{R}^2$ that satisfy the set of constraints (shaded part of Figure 1). This region is called the *feasible region.* If a point is in the region, it satisfies all the constraints and is called *feasible*; if it is not in the region, it violates at least one of the constraints and is called *infeasible.*

Different points in the feasible region give different values of the objective function. For example, the origin (0, 0) is feasible and gives a profit of $2(0) + (0) = 0$. Similarly, (1, 5) is feasible and gives a profit of $2(1) + (5) = 7$, which is better than 0. We seek the point or points of the feasible region that yield the maximal profit.

Corresponding to a *fixed* value of z, the set of solutions to the equation $z = 2x + y$ is the line (called the *objective line*) with slope -2 and y-intercept z. In other words, all points along the line $y = -2x + z$ correspond to the same z value. Remember that we are interested only in those points that lie in the feasible region, and we want z to be as large as possible (Figure 2).

FIGURE 1

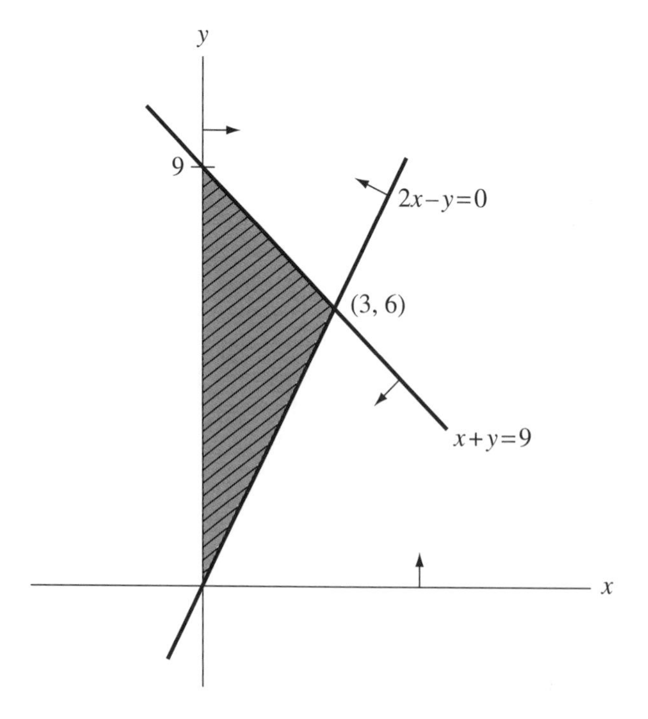

FIGURE 2

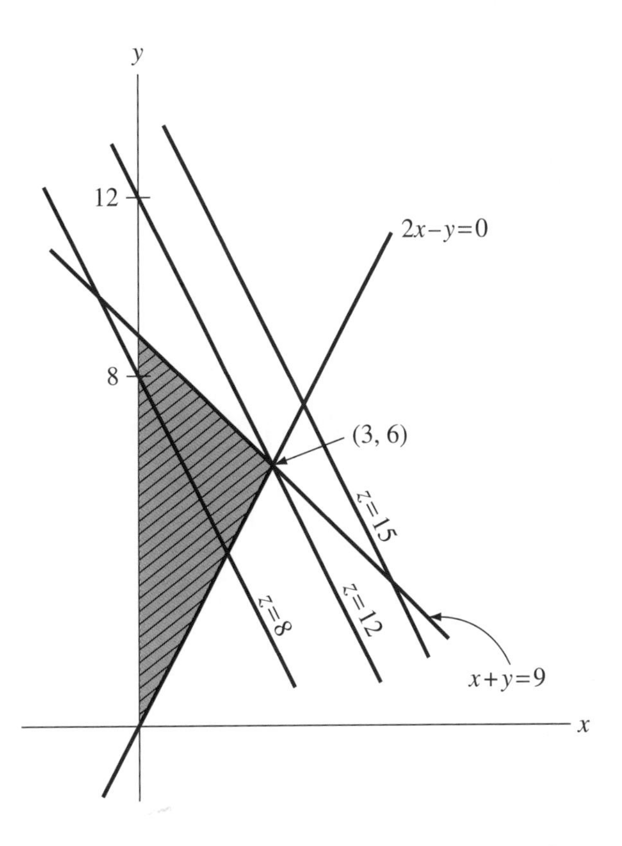

From Figure 2, it is clear that the optimal value occurs when $z = 12$, since moving the objective line up increases its y-intercept (the value of z), but moving beyond the point (3, 6) gives a line that does not intersect the feasible region at all.

NOTE In the preceding discussion, we obtained a unique solution; the maximum value $z = 12$ occurs only at the point (3, 6). If, however, the objective function is changed to $z = 2x + 2y$, all the points in the feasible region that lie on the edge determined by $x + y = 9$ are optimal, with value $z = 18$ (Figure 3). Thus, the point at which the optimal value occurs need not be unique.

FIGURE 3

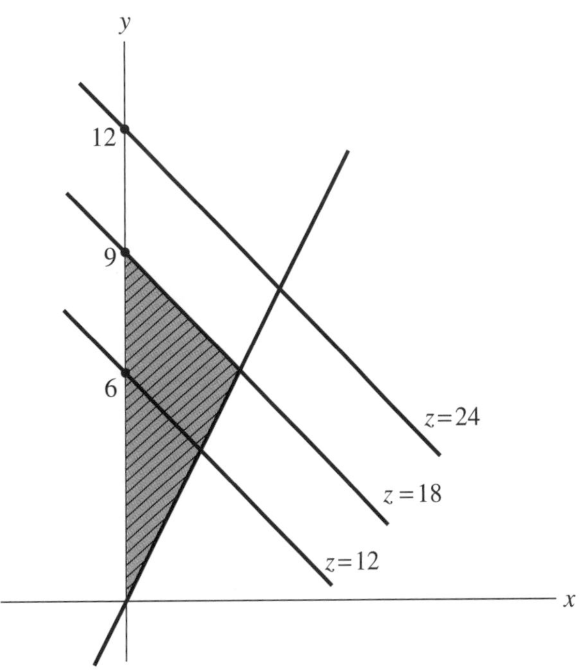

Moreover, if the feasible region is *unbounded*, as would be the case in the preceding discussion if the constraint $x + y \leq 9$ were removed, an optimal value may not exist. That is, it is possible that no matter which point is chosen, a "better" one always exists.

The General Problem

The previous example is a simple case of the general *linear programming problem.* We describe this problem mathematically after introducing the following notation for comparing certain vectors.

DEFINITION

Let $\mathbf{u}$ and $\mathbf{v}$ be vectors in $\mathbf{R}^m$. We say **u is less than v** and write $\mathbf{u} < \mathbf{v}$ if $u_i < v_i$ for each $i = 1, \ldots, m$. Likewise, $\mathbf{u} \leq \mathbf{v}$ means $u_i \leq v_i$ for each i. Similar definitions hold for $\mathbf{u} > \mathbf{v}$ and $\mathbf{u} \geq \mathbf{v}$. The vector $\mathbf{v}$ is **positive** if $\mathbf{v} > \mathbf{0}$ and **nonnegative** if $\mathbf{v} \geq \mathbf{0}$.

EXAMPLE 1

Let $\mathbf{u} = (1, 2, -3)$, $\mathbf{v} = (2, 4, 1)$, and $\mathbf{w} = (2, 0, 1)$. Then $\mathbf{u} < \mathbf{v}$ and $\mathbf{w} \leq \mathbf{v}$, but no special relationship exists between $\mathbf{u}$ and $\mathbf{w}$. As well, $\mathbf{v} > \mathbf{0}$ and $\mathbf{w} \geq \mathbf{0}$. ■

Given a system of inequalities,

$$\begin{aligned} a_{11}x_1 + a_{12}x_2 + \cdots + a_{1n}x_n &\leq b_1 \\ a_{21}x_1 + a_{22}x_2 + \cdots + a_{2n}x_n &\leq b_2 \\ \vdots \qquad\quad \vdots \qquad\qquad\quad \vdots \qquad &\quad \vdots \\ a_{m1}x_1 + a_{m2}x_2 + \cdots + a_{mn}x_n &\leq b_m, \end{aligned}$$

we can use the vector inequality notation to write the system as

$$A\mathbf{x} \leq \mathbf{b},$$

where, as was the case with linear equations,

$$A = \begin{bmatrix} a_{11} & \cdots & a_{1n} \\ \vdots & & \vdots \\ a_{m1} & \cdots & a_{mn} \end{bmatrix}, \quad \mathbf{b} = \begin{bmatrix} b_1 \\ \vdots \\ b_m \end{bmatrix}, \quad \text{and} \quad \mathbf{x} = \begin{bmatrix} x_1 \\ \vdots \\ x_n \end{bmatrix}.$$

EXAMPLE 2

Graph the set of solutions of the system of inequalities

$$A\mathbf{x} \leq \mathbf{b}, \qquad \mathbf{x} \geq \mathbf{0},$$

where

$$A = \begin{bmatrix} 2 & -1 \\ 1 & 1 \end{bmatrix} \quad \text{and} \quad \mathbf{b} = \begin{bmatrix} 0 \\ 9 \end{bmatrix}.$$

Solution The matrix inequality,

$$A\mathbf{x} = \begin{bmatrix} 2 & -1 \\ 1 & 1 \end{bmatrix} \begin{bmatrix} x \\ y \end{bmatrix} \leq \begin{bmatrix} 0 \\ 9 \end{bmatrix}, \quad \mathbf{x} \geq \mathbf{0},$$

is equivalent to the system of inequalities

$$\begin{aligned} 2x - y &\leq 0 \\ x + y &\leq 9 \\ x \geq 0, \qquad y &\geq 0. \end{aligned}$$

This is the same set of inequalities as was derived at the beginning of this section. Thus, the set of solutions is just that of Figure 1. ■

Consider the system of linear inequalities

$$A\mathbf{x} \leq \mathbf{b}, \qquad \mathbf{b} \geq \mathbf{0}, \qquad \mathbf{x} \geq \mathbf{0}.$$

The set of all solutions to this system is called the **feasible region**. Notice that $\mathbf{0}$ is always in the feasible region of this system, since $A\mathbf{0} = \mathbf{0} \leq \mathbf{b}$, and usually there are infinitely many feasible points. As in the example presented earlier, we are seeking solutions that *optimize* (*maximize* or *minimize*) the value z of a linear function

$$z = \mathbf{c} \bullet \mathbf{x} = c_1x_1 + c_2x_2 + \cdots + c_nx_n,$$

where the c_i are scalars. The function $z = \mathbf{c} \bullet \mathbf{x}$ is called the **objective function**.

DEFINITION

A linear programming problem is in **standard form** if it is stated as follows: Optimize the function $z = \mathbf{c} \bullet \mathbf{x}$, subject to the conditions

$$A\mathbf{x} \leq \mathbf{b}, \qquad \mathbf{b} \geq \mathbf{0}, \qquad \text{and} \qquad \mathbf{x} \geq \mathbf{0}.$$

To apply our knowledge of linear *equations* to the study of linear *inequalities*, we convert the system of linear inequalities to a system of linear equations by adding a nonnegative variable, called a **slack variable**, to each constraint. For example, the inequality

$$2x + 3y \leq 5$$

is equivalent to the equation

$$2x + 3y + s = 5,$$

with the restriction that s be nonnegative. Notice that $s = 5 - (2x + 3y)$ is the "slack," or difference, between $2x + 3y$ and 5. There is a tradeoff in this process—in order to work with equations instead of inequalities, the number of variables increases by one for each inequality.

The mathematical description of this conversion process is as follows. Let

$$A\mathbf{x} \leq \mathbf{b}, \qquad \mathbf{b} \geq \mathbf{0}, \qquad \mathbf{x} \geq \mathbf{0}$$

be a system of m linear inequalities in n unknowns. The corresponding system of equations is

$$A'\mathbf{x}' = \mathbf{b}, \qquad \mathbf{b} \geq \mathbf{0}, \qquad \mathbf{x}' \geq \mathbf{0},$$

where A' is an $m \times (n + m)$ matrix whose first n columns are those of A and last m columns are those of the identity matrix of order m. The first n components of the vector $\mathbf{x}'$ correspond to those of $\mathbf{x}$, and the last m components are the slack variables of the system.

EXAMPLE 3

Convert the following system of linear inequalities to the corresponding system of linear equations.

$$\begin{aligned} 2x_1 - x_2 + x_3 &\leq 5 \\ x_2 - 2x_3 &\leq 2 \\ x_1 - x_2 + 2x_3 &\leq 4 \\ x_i \geq 0, \quad i = 1, 2, 3 & \end{aligned}$$

Solution The resulting system in equation form is

$$\begin{aligned} 2x_1 - x_2 + x_3 + x_4 \qquad\qquad &= 5 \\ x_2 - 2x_3 \quad + x_5 \qquad &= 2 \\ x_1 - x_2 + 2x_3 \qquad\quad + x_6 &= 4 \\ x_i \geq 0, \quad i = 1, \ldots, 6. \end{aligned}$$

The matrices A and A' are given by

$$A = \begin{bmatrix} 2 & -1 & 1 \\ 0 & 1 & -2 \\ 1 & -1 & 2 \end{bmatrix} \quad \text{and} \quad A' = \begin{bmatrix} 2 & -1 & 1 & 1 & 0 & 0 \\ 0 & 1 & -2 & 0 & 1 & 0 \\ 1 & -1 & 2 & 0 & 0 & 1 \end{bmatrix}.$$ ■

NOTE We are not interested in the *values* of the slack variables. Thus, in a linear programming problem with objective function $z = \mathbf{c} \bullet \mathbf{x}$, in this conversion process we take as the new objective function $z' = \mathbf{c}' \bullet \mathbf{x}'$, where $c_j' = c_j$ for $j = 1, \ldots, n$ and $c_j' = 0$ for $j = n + 1, \ldots, n + m$.

A system of linear equations that corresponds to the linear inequalities of a linear programming problem in standard form is always *row-reduced* in the following sense.

DEFINITION

A matrix of rank r is **row-reduced** if it contains r nonzero rows, the rest are zero, and there exists a subset of r columns that are distinct elementary columns.

NOTE A matrix in row-reduced echelon form is, of course, row-reduced in this sense. The columns that contain the leading ones provide the set of elementary columns.

EXAMPLE 4

Which of the following are row-reduced?

$$B = \begin{bmatrix} 1 & 3 & 1 & 0 \\ -1 & 2 & 0 & 1 \\ 0 & 0 & 0 & 0 \end{bmatrix}, \quad C = \begin{bmatrix} 2 & 0 & 3 & 0 \\ 1 & 1 & 2 & 0 \\ 3 & 0 & 4 & 1 \end{bmatrix}, \quad D = \begin{bmatrix} 2 & 0 & 0 & 2 & 1 \\ 1 & 1 & 0 & 1 & 0 \\ -1 & 0 & 1 & 4 & 0 \end{bmatrix}.$$

Solution The matrix B is row-reduced, since $r = 2$ and $\mathbf{b}^3 = \mathbf{e}_1$ and $\mathbf{b}^4 = \mathbf{e}_2$. D is also row-reduced, since $r = 3$ and $\mathbf{d}^5 = \mathbf{e}_1$, $\mathbf{d}^2 = \mathbf{e}_2$, and $\mathbf{d}^3 = \mathbf{e}_3$. C, however, is not row-reduced, since $r = 3$, but there is no column equal to $\mathbf{e}_1$. ■

For an $m \times n$ matrix of rank m, being row-reduced is tantamount to possessing a subset of m columns that may be rearranged to form an identity matrix. It is sometimes convenient to speak of an *embedded identity matrix* when referring to the set of elementary columns. For example, in the matrix A' of a linear system of equations that corresponds to a system of linear inequalities, the last m columns form an embedded identity matrix.

As in the case of a system in row-reduced echelon form, it is easy to identify a particular solution, called a **basic solution**. To obtain the corresponding basic solution, let the variables that correspond to the elementary columns assume the corresponding $\mathbf{b}$ column value (these are called the **basic variables** of the solution), and let all the others be zero (these are the **nonbasic variables**). Basic solutions correspond to the vertices of the feasible region in $\mathbf{R}^n$.

EXAMPLE 5

Find a basic solution to the following system of equations and identify the basic and nonbasic variables.

$$\begin{aligned} 2x_2 + x_3 \quad\quad - x_5 &= 2 \\ x_1 - x_2 \quad\quad + 3x_5 &= 3 \\ x_2 \quad + x_4 \quad\quad &= 4 \end{aligned}$$

Solution Forming the corresponding augmented matrix, we see that it is already in row-reduced form:

$$\left[\begin{array}{ccccc|c} 0 & 2 & 1 & 0 & -1 & 2 \\ 1 & -1 & 0 & 0 & 3 & 3 \\ 0 & 1 & 0 & 1 & 0 & 4 \end{array}\right]$$

A particular solution is then obtained by letting $x_3 = 2$ (because $\mathbf{a}^3 = \mathbf{e}_1$), $x_1 = 3$ (because $\mathbf{a}^1 = \mathbf{e}_2$), $x_4 = 4$ (because $\mathbf{a}^4 = \mathbf{e}_3$), and $x_2 = x_5 = 0$. The basic variables are x_1, x_3, and x_4, and the nonbasic variables are x_2 and x_5. ■

9.1 Exercises

In Exercises 1–4 sketch the feasible region in $\mathbf{R}^2$ described by the constraints.

1. $\begin{aligned} x - y &\le 1 \\ x + 2y &\le 8 \\ x, y &\ge 0 \end{aligned}$

2. $\begin{aligned} x - y &\le 2 \\ 3x - y &\le 6 \\ x, y &\ge 0 \end{aligned}$

3. $\begin{aligned} x + y &\le 2 \\ -x + 3y &\le 9 \\ 2x - y &\le 6 \\ x, y &\ge 0 \end{aligned}$

4. $\begin{aligned} x - y &\le 1 \\ 2x - 2y &\le 5 \\ 2x - y &\ge 6 \\ x \ge 1, \quad y &\le 2 \end{aligned}$

5. Maximize the objective function $z = 2x + y$ on the feasible region of Exercise 1, or show that no maximum exists.

6. Minimize the objective function $z = 3x - y$ on the feasible region of Exercise 2, or show that no minimum exists.

7. Maximize the function $z = x - 3y$ on the feasible region of Exercise 3, or show that no maximum exists.

8–10. Convert each system of constraints in Exercises 1–3 into equation form.

In Exercises 11–14 decide whether or not the system of linear equations is row-reduced. If it is row-reduced, identify the columns that yield an embedded identity matrix, and find the corresponding basic solution.

11. $\begin{aligned} 2x_1 \quad\quad + x_3 - 3x_4 \quad\quad &= 2 \\ x_1 + x_2 \quad\quad - x_4 \quad\quad &= 1 \\ 3x_1 \quad\quad\quad + 2x_4 + x_5 &= 4 \end{aligned}$

12. $\begin{aligned} x_1 \quad\quad\quad - x_4 &= 1 \\ x_2 \quad + x_4 &= 3 \\ x_3 + 2x_4 &= 5 \end{aligned}$

13. $\begin{aligned} x_1 \quad\quad + x_3 \quad\quad\quad &= 2 \\ -x_1 + x_2 \quad\quad + x_4 \quad &= 6 \\ 2x_1 \quad\quad\quad\quad + x_5 &= 1 \end{aligned}$

(*Note*: There are two possible answers.)

14.
$$\begin{aligned} 2x_2 - x_3 + 2x_4 + x_5 + 2x_6 &= 2 \\ x_1 - x_2 \qquad\qquad\qquad - 3x_6 &= 2 \end{aligned}$$

In Exercises 15 and 16 convert the given information to a linear programming problem in standard form and then to equation form. You do *not* need to solve the problems.

15. A man wishes to invest a portion of \$50,000 in savings, stocks, and property. He feels that no more than one-fifth of the total amount should be in savings and no more than two-fifths in savings and stocks together. If he estimates an 8% return from savings, 12% from real estate, and 15% from stocks, how should he invest his money?

16. A toy manufacturer has available 200 kg of wood, 400 kg of mouldable plastic, and 50 kg of metal to produce toys *A*, *B*, *C*, and *D*, each of which requires some or all of the materials for component parts. Table 1 gives the requirements for each toy, the price of each toy, and the price of the materials used in production. How many of each toy should the manufacturer produce to maximize profit?

TABLE 1

	Wood	Plastic	Metal	\$/Toy
A	0.2	0.3	0.1	5
B	0	0.5	0.2	4
C	0	0.9	0.1	3
D	1	0	0.1	6
\$/kg	1	0.5	2	

9.2 The Simplex Algorithm

The simplex method was developed by George Dantzig (1914–2005), an American mathematician. While he worked in a number of areas of mathematics, including statistics, he is most well known for his work in linear programming. His simplex algorithm, developed while working for the U.S. Air Force in the 1940s, remains one of the most important procedures in mathematics.

With permission from News Service Library, Stanford University

In this section, we introduce a general procedure for solving linear programming problems. Now that the translation from inequalities to equations has been described (Section 9.1), we assume that the linear programming problem is in equation form and row-reduced.

Let the linear programming problem be given as follows: Optimize the linear function

$$z = \mathbf{c} \cdot \mathbf{x},$$

subject to the system of constraints

$$A\mathbf{x} = \mathbf{b}, \qquad \mathbf{b} \geq \mathbf{0}, \qquad \mathbf{x} \geq \mathbf{0}, \tag{1}$$

where A has an embedded identity matrix.

Let $\mathbf{c}^*$ be the $1 \times m$ matrix that consists of the components of $\mathbf{c}$ associated with the basic variables in the current basic solution. Then the value associated with the current solution is $\mathbf{c}^*\mathbf{b}$ (the nonbasic variables are zero). The n-vector $\mathbf{c}^*A - \mathbf{c}$ is called the vector of **simplex indicators**. (Here $\mathbf{c}$ is viewed as a $1 \times n$ matrix so that the operations are defined.) We call the jth component of this vector, $\mathbf{c}^*\mathbf{a}^j - c_j$, the jth **simplex indicator**.

Associated with System (1), we prepare the array shown in Figure 4, called the **tableau**, associated with the current solution. For convenience, tableaux will sometimes be presented with the row of variables at the top and the basic variables and the vector $\mathbf{c}^*$ on the left (see, for example, Figure 5). The basic variables are always listed in the order that the corresponding elementary columns form the identity matrix.

FIGURE 4

A	$\mathbf{b}$
$\mathbf{c}^*A - \mathbf{c}$	$\mathbf{c}^*\mathbf{b}$

FIGURE 5

	$\mathbf{c}^*$	x_1	x_2	x_3	x_4	x_5	$\mathbf{b}$
x_3	4	0	2	1	0	−1	2
x_1	2	1	−1	0	0	3	3
x_4	0	0	1	0	1	0	4
		0	7	0	0	4	14

EXAMPLE 1

Let $A\mathbf{x} = \mathbf{b}$, $\mathbf{b} \geq \mathbf{0}$, $\mathbf{x} \geq \mathbf{0}$ be the system of Example 5 of Section 9.1,

$$\begin{aligned} 2x_2 + x_3 \qquad - x_5 &= 2 \\ x_1 - x_2 \qquad + 3x_5 &= 3 \\ x_2 \quad + x_4 \qquad &= 4, \end{aligned}$$

with objective function $z = 2x_1 - x_2 + 4x_3 - 2x_5$. Prepare the corresponding tableau.

Solution Since the basic solution is $x_3 = 2$, $x_1 = 3$, and $x_4 = 4$, we have $\mathbf{c}^* = (c_3, c_1, c_4) = (4, 2, 0)$. The tableau is given in Figure 5.

Notice, in Figure 5, that the second simplex indicator has been computed as $\mathbf{c}^*\mathbf{a}^2 - c_2 = (4, 2, 0) \bullet (2, -1, 1) - (-1) = 7$. The rightmost bottom row entry is $\mathbf{c}^*\mathbf{b} = 14$, the value associated with the basic solution $x_3 = 2$, $x_1 = 3$, and $x_4 = 4$. ■

Conditions for Terminating the Algorithm

The next theorem implies that the solution of Example 1 is maximal. In other words, there is no solution with value greater than 14, which occurs when $x_3 = 2$, $x_1 = 3$, $x_4 = 4$, and $x_2 = x_5 = 0$. The proof of this and subsequent theorems is delayed until Section 9.4.

THEOREM 1

Let a linear programming problem be given as follows: Optimize $z = \mathbf{c} \bullet \mathbf{x}$, subject to the constraints

$$A\mathbf{x} = \mathbf{b}, \qquad \mathbf{b} \geq \mathbf{0}, \qquad \mathbf{x} \geq \mathbf{0},$$

where A has an embedded identity matrix. If no simplex indicator of the associated tableau is negative (that is, $\mathbf{c}^*A - \mathbf{c} \geq \mathbf{0}$), then the corresponding basic solution is maximal. If no simplex indicator is positive, the corresponding basic solution is minimal. ■

Theorem 1 gives us conditions under which we have solved the problem and can terminate the procedure. The next theorem also gives us conditions under which we can cease our search, for a different reason—no optimal solution exists.

THEOREM 2

Let a linear programming problem be given as follows: Optimize $z = \mathbf{c} \bullet \mathbf{x}$, subject to the constraints

$$A\mathbf{x} = \mathbf{b}, \qquad \mathbf{b} \geq \mathbf{0}, \qquad \mathbf{x} \geq \mathbf{0},$$

where A has an embedded identity matrix. Let j be a column number such that the simplex indicator is negative ($\mathbf{c}^*\mathbf{a}^j - c_j < 0$). If every element of $\mathbf{a}^j$ is less than or equal to zero ($\mathbf{a}^j \leq \mathbf{0}$), then no maximal solution exists. If everything is the same except that the simplex indicator is positive, then no minimal solution exists. ■

EXAMPLE 2

Show that the following linear programming problem has no maximal solution. Subject to the constraints

$$\begin{aligned} x_1 \quad + 2x_3 - x_4 \quad + x_6 &= 2 \\ - x_3 \quad + x_5 + 3x_6 &= 1 \\ x_2 + x_3 - 2x_4 \quad + x_6 &= 4 \\ x_j \geq 0, \quad j = 1, \ldots, 6, \end{aligned}$$

maximize the function

$$z = x_1 + 3x_2 - x_3 + 3x_4 - 2x_5.$$

Solution We prepare the corresponding tableau, Figure 6, using $\mathbf{c} = (1, 3, -1, 3, -2, 0)$ and $\mathbf{c}^* = (1, -2, 3)$, since the basic variables are x_1, x_5, and x_2.

FIGURE 6

	$\mathbf{c}^*$	x_1	x_2	x_3	x_4	x_5	x_6	$\mathbf{b}$
x_1	1	1	0	2	−1	0	1	2
x_5	−2	0	0	−1	0	1	3	1
x_2	3	0	1	1	−2	0	1	4
		0	0	8	−10	0	−2	12

We have $\mathbf{c}^*\mathbf{a}^4 - c_4 = -10$ and $\mathbf{a}^4 = (-1, 0, -2) \leq \mathbf{0}$, so, by Theorem 2, no maximal solution exists. ■

Finding a Better Solution

We now have a criterion for optimality (Theorem 1) and a criterion under which no optimal solution exists (Theorem 2). Next we introduce a procedure that takes us from a basic solution that meets neither criterion to a new solution that might meet one of them. This process is called **pivoting**; it is just the row-reduction technique of Chapter 2. Once the pivot is chosen, divide that row by the pivot value (to obtain a 1 in the pivot position), and use the pivot row to obtain zeros in the rest of the pivot column.

EXAMPLE 3

In the following tableau (that of Example 2 without the bottom row), pivot on the 1, 3 position:

1	0	(2)	−1	0	1	2
0	0	−1	0	1	3	1
0	1	1	−2	0	1	4

Solution We first divide the top row by the circled entry in the 1, 3 position to obtain

$\frac{1}{2}$	0	1	$-\frac{1}{2}$	0	$\frac{1}{2}$	1
0	0	−1	0	1	3	1
0	1	1	−2	0	1	4

Now we add the top row to the second, and the negative of the top row to the third, to create zeros in the rest of the third column. This yields the new tableau:

$\frac{1}{2}$	0	1	$-\frac{1}{2}$	0	$\frac{1}{2}$	1
$\frac{1}{2}$	0	0	$-\frac{1}{2}$	1	$\frac{7}{2}$	2
$-\frac{1}{2}$	1	0	$-\frac{3}{2}$	0	$\frac{1}{2}$	3

The new tableau is row-reduced and corresponds to the basic solution $x_3 = 1, x_5 = 2, x_2 = 3$, and $x_1 = x_4 = x_6 = 0$. The new $\mathbf{c}^*$ vector is $\mathbf{c}^* = (-1, -2, 3)$. Completing the tableau to check for optimality, we have the tableau of Figure 7.

FIGURE 7

	$\mathbf{c}^*$	x_1	x_2	x_3	x_4	x_5	x_6	$\mathbf{b}$
x_3	-1	$\frac{1}{2}$	0	1	$-\frac{1}{2}$	0	$\frac{1}{2}$	1
x_5	-2	$\frac{1}{2}$	0	0	$-\frac{1}{2}$	1	$\frac{7}{2}$	2
x_2	3	$-\frac{1}{2}$	1	0	$-\frac{3}{2}$	0	$\frac{1}{2}$	3
		-4	0	0	-6	0	-6	4

By Theorem 1, this new basic solution is minimal. ■

Notice that, in Example 3, we could have computed the bottom row of the new tableau (Figure 7) by continuing the pivot procedure to the bottom row. In other words, the bottom row of the tableau of Figure 6 was

$$[0 \quad 0 \quad 8 \; -10 \quad 0 \; -2 \mid 12].$$

If we add -8 times the pivot row to it, we obtain

$$[-4 \quad 0 \quad 0 \; -6 \quad 0 \; -6 \mid 4],$$

which is the bottom row of the new tableau of Figure 7. This technique always works, and we state it as the next theorem.

THEOREM 3

The bottom row of a new tableau may be computed from the preceding tableau by using the pivot row and row operations to obtain a zero in the pivot column position of the bottom row. ■

In this pivoting process, we must check two conditions. The simplex indicator (bottom row entry) determines whether the value of the solution increases or decreases in the next tableau. That is, if the simplex indicator is *negative*, that column can be chosen to *increase* the value in the next tableau; if it is *positive*, that column causes the value to *decrease*. Care must also be taken, however, to ensure that the next solution remains *feasible*; in other words, that all entries in the new **b** column are nonnegative.

Once an appropriate column has been chosen, a row is chosen as follows. Compute the quotients of the **b** column entries divided by the corresponding *positive* entries in the desired column. Choose a row that gives the smallest such quotient, because it will lead to a new solution that is feasible. In Example 3, we chose to pivot on the 1, 3 position by the following procedure. To minimize, we first chose the third column, since the simplex indicator of this column, 8, is greater than zero. Then we computed the quotients $b_1/a_{13} = 2/2 = 1$ and $b_3/a_{33} = 4/1 = 4$ (b_2/a_{23} was ignored because $a_{23} \leq 0$), using the smaller of the two quotients (the first one) as the basis for choosing the first row. The justification for the process just described is the following theorem.

!

THEOREM 4

Let a linear programming problem be given as follows: Optimize $z = \mathbf{c} \cdot \mathbf{x}$, subject to the constraints

$$A\mathbf{x} = \mathbf{b}, \qquad \mathbf{b} \geq \mathbf{0}, \qquad \mathbf{x} \geq \mathbf{0},$$

where A has an embedded identity matrix. Let j be a column number such that the simplex indicator is negative (that is, $\mathbf{c}^*\mathbf{a}^j - c_j < 0$). If a_{ij} is positive for at least one row number $i = 1, \ldots, m$, let k be a row number such that $a_{kj} > 0$ and $b_k/a_{kj} \leq b_i/a_{ij}$ for all $i = 1, \ldots, m$ with $a_{ij} > 0$. Then pivoting on the k, j position leads to a new basic solution with value greater than or equal to the former. If everything is the same, except that the simplex indicator is positive, the new solution has value less than or equal to the former. ■

The Simplex Algorithm

Combining Theorems 1 through 4, we have the following procedure for *maximizing* a linear programming problem. To *minimize*, consideration of negative simplex indicators should be replaced by that of *positive* simplex indicators.

i. Convert the given linear programming problem to the following form: Optimize $z = \mathbf{c} \cdot \mathbf{x}$, subject to the constraints

$$A\mathbf{x} = \mathbf{b}, \quad \mathbf{x} \geq \mathbf{0}, \quad \mathbf{b} \geq \mathbf{0},$$

where A has an embedded identity matrix.

ii. Prepare the first tableau.

iii. If no simplex indicator is negative, then the solution is maximal and the algorithm terminates.

iv. If some simplex indicator is negative and no entry in that column is positive, then there is no maximal solution and the algorithm terminates.

v. If at least one simplex indicator is negative and each such column has a positive entry, choose any such column j and pivot on the k, j position, where $b_k/a_{kj} = \min\{b_i/a_{ij} \mid a_{ij} > 0\}$.

vi. Repeat step (v) until either step (iii) or step (iv) is satisfied.

Because of Theorem 3, we only need to construct $\mathbf{c}^*$ for the first tableau. For each succeeding tableau, the vector of simplex indicators $\mathbf{c}^*A - \mathbf{c}$ is obtained by pivoting. Furthermore, we do not need to worry about which variables are basic and which are nonbasic until the final tableau. For these reasons, tableaux subsequent to the first do not include the columns that indicate the basic variables or $\mathbf{c}^*$.

EXAMPLE 4

Use the simplex algorithm to solve the motivating example of Section 9.1. Maximize the function $z = 2x + y$, subject to the constraints

$$\begin{aligned} 2x - y &\leq 0 \\ x + y &\leq 9 \\ x, y &\geq 0. \end{aligned}$$

Solution We first convert it to equation form by using slack variables s and t, as follows. Maximize the function $z = 2x + y$, subject to the constraints

$$\begin{aligned} 2x - y + s \qquad &= 0, \\ x + y \qquad + t &= 9, \\ x, y, s, t &\geq 0. \end{aligned}$$

We now have the problem in the desired form, and we proceed to prepare the first tableau with $\mathbf{c} = (2, 1, 0, 0)$ and $\mathbf{c}^* = (0, 0)$, since the embedded identity matrix columns correspond to the slack variables s and t:

	$\mathbf{c}^*$	x	y	s	t	$\mathbf{b}$
s	0	2	−1	1	0	0
t	0	1	1	0	1	9
		−2	−1	0	0	0

Here we could choose either column 1 or column 2 as the pivot column, since the simplex indicators are negative for these columns (−2 and −1, respectively). We choose column 1. Both entries in this column are positive, so, to determine the pivot row, we need to form both quotients $0/2 = 0$ and $9/1 = 9$. Since $0 < 9$, the pivot row is row 1. Pivoting on the 1, 1 position, we have

1	$-\frac{1}{2}$	$\frac{1}{2}$	0	0
0	$\frac{3}{2}$	$-\frac{1}{2}$	1	9
0	−2	1	0	0

Now the only negative simplex indicator is in the second column, so this is the pivot column. There is only one positive entry in the second column ($\frac{3}{2} > 0$), so the new pivot row is the second row. Pivoting on the 2, 2 position yields

1	0	$\frac{1}{3}$	$\frac{1}{3}$	3
0	1	$-\frac{1}{3}$	$\frac{2}{3}$	6
0	0	$\frac{1}{3}$	$\frac{4}{3}$	12

Since all the simplex indicators are nonnegative, the resulting basic solution $x = 3$, $y = 6$, and $s = t = 0$ is maximal, and the value of the objective function at this point is 12. Of course, this is the result that was obtained geometrically in Section 9.1. ■

NOTE From now on, we proceed directly from one tableau to the next without explanation, only circling the pivot entry at each stage. For example, the simplex algorithm for Example 4 is shown in Figure 8.

FIGURE 8

	$\mathbf{c}^*$	x	y	s	t	$\mathbf{b}$
s	0	(2)	−1	1	0	0
t	0	1	1	0	1	9
		−2	−1	0	0	0
		1	$-\frac{1}{2}$	$\frac{1}{2}$	0	0
		0	($\frac{3}{2}$)	$-\frac{1}{2}$	1	9
		0	−2	1	0	0
		1	0	$\frac{1}{3}$	$\frac{1}{3}$	3
		0	1	$-\frac{1}{3}$	$\frac{2}{3}$	6
		0	0	$\frac{1}{3}$	$\frac{4}{3}$	12

EXAMPLE 5

Minimize $z = 3x_1 - x_2 - 2x_3 + x_4$, subject to the constraints

$$\begin{aligned} 2x_1 - 4x_2 - x_3 + x_4 &\leq 8 \\ x_1 + x_2 + 2x_3 - 3x_4 &\leq 7 \\ x_1 - x_2 - x_3 + x_4 &\leq 3 \\ x_j \geq 0, \quad j &= 1, \ldots, 4. \end{aligned}$$

Solution We first convert the problem to equation form. Minimize $z = 3x_1 - x_2 - 2x_3 + x_4 + 0x_5 + 0x_6 + 0x_7$, subject to

$$\begin{aligned} 2x_1 - 4x_2 - x_3 + x_4 + x_5 \qquad\qquad &= 8 \\ x_1 + x_2 + 2x_3 - 3x_4 \qquad + x_6 \qquad &= 7 \\ x_1 - x_2 - x_3 + x_4 \qquad\qquad + x_7 &= 3 \\ x_j \geq 0, \quad j = 1, \ldots, 7. \end{aligned}$$

We then prepare the first tableau, Figure 9, and use it to start the simplex algorithm. The pivot used is circled in the tableau. From the fourth column of the last tableau, we see that no minimal solution exists.

FIGURE 9

	$\mathbf{c}^*$	x_1	x_2	x_3	x_4	x_5	x_6	x_7	$\mathbf{b}$
x_5	0	2	−4	−1	1	1	0	0	8
x_6	0	1	(1)	2	−3	0	1	0	7
x_7	0	1	−1	−1	1	0	0	1	3
		−3	1	2	−1	0	0	0	0
		6	0	7	−11	1	4	0	36
		1	1	2	−3	0	1	0	7
		2	0	1	−2	0	1	1	10
		−4	0	0	2	0	−1	0	−7

■

EXAMPLE 6

A factory makes three products: *A*, *B*, and *C*. The company has available each week 200 h of lathe time, 100 h of milling time, 100 h of grinding time, and 80 h of packing time. The time requirements for the production of one unit of each product are given in Table 2.

TABLE 2

	Lathe	Mill	Grind	Pack
A	3	1	2	1
B	1	2	1	2
C	4	2	1	2

If each unit of *A* yields a profit of \$20, each unit of *B* yields a profit of \$16, and each unit of *C* yields a profit of \$24, find a week's production schedule that maximizes profit.

Solution Let x_1 be the number of units of *A* to be produced, x_2 the number of *B*, and x_3 the number of *C*. Knowing that the total lathe time cannot exceed 200 h, we have

$$3x_1 + x_2 + 4x_3 \leq 200.$$

Similarly, the milling, grinding, and packing limitations yield the constraints

$$\begin{aligned} x_1 + 2x_2 + 2x_3 &\le 100 \\ 2x_1 + x_2 + x_3 &\le 100 \\ x_1 + 2x_2 + 2x_3 &\le 80. \end{aligned}$$

The profit function is $z = 20x_1 + 16x_2 + 24x_3$, and obviously each variable must be nonnegative. Putting all these conditions together, we have the mathematical problem: Maximize $z = 20x_1 + 16x_2 + 24x_3$, subject to the constraints

$$\begin{aligned} 3x_1 + x_2 + 4x_3 &\le 200 \\ x_1 + 2x_2 + 2x_3 &\le 100 \\ 2x_1 + x_2 + x_3 &\le 100 \\ x_1 + 2x_2 + 2x_3 &\le 80 \\ x_j \ge 0, \quad j &= 1, 2, 3. \end{aligned}$$

We restate the problem in equation form: Maximize

$$z = 20x_1 + 16x_2 + 24x_3 + 0x_4 + 0x_5 + 0x_6 + 0x_7,$$

subject to the constraints

$$\begin{aligned} 3x_1 + x_2 + 4x_3 + x_4 \qquad\qquad\qquad &= 200 \\ x_1 + 2x_2 + 2x_3 \qquad + x_5 \qquad\qquad &= 100 \\ 2x_1 + x_2 + x_3 \qquad\qquad + x_6 \qquad &= 100 \\ x_1 + 2x_2 + 2x_3 \qquad\qquad\qquad + x_7 &= 80 \\ x_j \ge 0, \quad j = 1, \ldots, 7. \end{aligned}$$

We prepare the first tableau (Figure 10) and use it to start the simplex algorithm. The pivot used at each iteration is circled.

FIGURE 10

	$\mathbf{c}^*$	x_1	x_2	x_3	x_4	x_5	x_6	x_7	$\mathbf{b}$
x_4	0	3	1	4	1	0	0	0	200
x_5	0	1	2	2	0	1	0	0	100
x_6	0	2	1	1	0	0	1	0	100
x_7	0	1	2	(2)	0	0	0	1	80
		-20	-16	-24	0	0	0	0	0
		1	-3	0	1	0	0	-2	40
		0	0	0	0	1	0	-1	20
		($\frac{3}{2}$)	0	0	0	0	1	$-\frac{1}{2}$	60
		$\frac{1}{2}$	1	1	0	0	0	$\frac{1}{2}$	40
		-8	8	0	0	0	0	12	960
		0	-3	0	1	0	$-\frac{2}{3}$	$-\frac{5}{3}$	0
		0	0	0	0	1	0	-1	20
		1	0	0	0	0	$\frac{2}{3}$	$-\frac{1}{3}$	40
		0	1	1	0	0	$-\frac{1}{3}$	$\frac{2}{3}$	20
		0	8	0	0	0	$\frac{16}{3}$	$\frac{28}{3}$	1280

Since no simplex indicator of the last tableau is negative, the solution is maximal. The conclusion is that the company should produce 40 units of *A*, 20 units of *C*, and 0 units of *B*. The associated profit is \$1280 per week. ■

9.2 Exercises

In Exercises 1 and 2 complete the first tableau for the given linear programming problem.

1. Objective function: $z = 3x_1 - 2x_2 + 2x_4 - x_5$

	$\mathbf{c}^*$	x_1	x_2	x_3	x_4	x_5	$\mathbf{b}$
x_3	0	3	1	1	0	0	1
x_4	2	−1	2	0	1	0	5
x_5	−1	4	5	0	0	1	9

2. Objective function: $z = 2x_1 + x_2 - 3x_3 + x_4$

	$\mathbf{c}^*$	x_1	x_2	x_3	x_4	x_5	x_6	x_7	x_8	$\mathbf{b}$
		2	1	1	0	3	0	−2	0	10
		1	0	−1	0	1	0	1	1	9
		1	0	1	0	2	1	2	0	7
		3	0	2	1	0	0	−1	0	2

In Exercises 3–6 prepare the first tableau for each linear programming problem.

3. Maximize $z = x_1 - 2x_2 + x_3 + 2x_5$, subject to

$$\begin{aligned} 3x_1 + x_2 + 2x_3 &= 2 \\ x_1 - 3x_3 + x_5 &= 5 \\ 2x_1 - x_3 + x_4 &= 8 \\ x_j \geq 0, \quad j = 1, \ldots, 5. \end{aligned}$$

4. Minimize $z = 2x_1 - x_2 + x_3$, subject to

$$\begin{aligned} x_1 + x_2 + x_3 &\leq 10 \\ 2x_1 - x_2 - x_3 &\leq 5 \\ x_1 + 3x_2 - 4x_3 &\leq 1 \\ x_j \geq 0, \quad j = 1, 2, 3. \end{aligned}$$

5. Minimize $z = x_1 - 2x_2 + x_3 + 4x_4$, subject to

$$\begin{aligned} 2x_1 + x_2 + x_3 + 3x_4 &\leq 12 \\ x_1 - 3x_2 - x_3 + x_4 &\leq 8 \\ x_j \geq 0, \quad j = 1, \ldots, 4. \end{aligned}$$

6. Maximize $z = x_1 - x_2 + 2x_3$, subject to

$$\begin{aligned} x_1 - 3x_2 - x_3 &\leq 6 \\ x_1 + 3x_2 - 4x_3 &\leq 8 \\ x_1 - x_2 - 3x_3 &\leq 10 \\ x_j \geq 0, \quad j = 1, 2, 3. \end{aligned}$$

In Exercises 7–10 explain why each tableau is the final tableau for a linear programming problem. If the solution is optimal, give the optimal solution and corresponding value of the objective function.

7. Maximize:

x_1	x_2	x_3	x_4	x_5	**b**
−3	1	0	−2	0	2
−1	0	0	3	1	4
−2	0	1	1	0	1
2	0	0	3	0	5

8. Tableau in Exercise 7, but *minimized*

9. Minimize:

x_1	x_2	x_3	x_4	x_5	x_6	**b**
0	1	1	−3	1	0	8
1	2	0	−2	−3	4	11
0	−2	0	−1	−2	−5	−8

10. Tableau in Exercise 9, but *maximized*

11–14. Solve Exercises 3–6 completely.

15, 16. Solve Exercises 15 and 16 of Section 9.1.

9.3 Nonstandard LP Problems

In this section, we show how other types of problems can be put in the form needed for the simplex algorithm.

The Two-Phase Problem

So far, we have considered only problems in which the constraints are of the type

$$A\mathbf{x} \leq \mathbf{b}, \qquad \mathbf{b} \geq \mathbf{0}, \qquad \text{and} \qquad \mathbf{x} \geq \mathbf{0}.$$

What if we have constraints of the type $\mathbf{a}_i \bullet \mathbf{x} \geq b_i$ for $b_i \geq 0$? To handle this situation, we introduce a slack variable, as before, but this time we *subtract* it. That is, $\mathbf{a}_i \bullet \mathbf{x} \geq b_i$ is equivalent to the equation $\mathbf{a}_i \bullet \mathbf{x} - x_{n+1} = b_i$, where $x_{n+1} \geq 0$ (and $\mathbf{x} = (x_1, \ldots, x_n)$). Unfortunately, a problem arises here. We no longer have an embedded identity matrix to start the solution process. The same problem arises if we have equations as well as inequalities in the system.

EXAMPLE 1

Convert the following problem to equation form:

$$\begin{aligned} 3x_1 - 2x_2 + 5x_3 &\leq 14 \\ x_1 - x_2 - 2x_3 &\geq 1 \\ x_2 - x_3 &= 5 \\ x_j \geq 0, \quad j &= 1, 2, 3. \end{aligned}$$

Solution We add a slack variable to the first equation and subtract a slack variable from the second equation to obtain this system:

$$\begin{aligned} 3x_1 - 2x_2 + 5x_3 + x_4 \quad\quad &= 14 \\ x_1 - x_2 - 2x_3 \quad\quad - x_5 &= 1 \\ x_2 - x_3 \quad\quad &= 5 \\ x_j \geq 0, \quad j = 1, \ldots, 5. \end{aligned}$$

■

In Example 1, we have only one elementary column (the fourth), and the simplex algorithm cannot be initiated. There is, however, a device that allows us to use the same method. As needed, we add new variables, called **artificial variables**, to create an embedded identity matrix. In this case, we add one artificial variable to each of the second and third constraints, to obtain

$$\begin{aligned} 3x_1 - 2x_2 + 5x_3 + x_4 \quad\quad\quad &= 14 \\ x_1 - x_2 - 2x_3 \quad\quad - x_5 + x_6 \quad &= 1 \\ x_2 - x_3 \quad\quad\quad + x_7 &= 5 \\ x_j \geq 0, \quad j = 1, \ldots, 7. \end{aligned}$$

Now we have an embedded identity matrix: the entries of the fourth, sixth, and seventh columns.

Unfortunately, the new system is *not* equivalent to the original. In the previous equation, for example, if x_7 is strictly positive, then $x_2 - x_3 < 5$ instead of $x_2 - x_3 = 5$. If $x_7 = 0$, however, we have $x_2 - x_3 + 0 = 5$, which is equivalent to the original. The idea, then, is to use the basic solution that has been "rigged" ($x_4 = 14$, $x_6 = 1$, $x_7 = 5$) to start the simplex algorithm, and then try to move to a basic solution in which the artificial variables are *not* basic. To this end, we introduce an objective function that is zero if the artificial variables are all zero and positive otherwise. This objective function is taken to be the sum of the artificial variables. In the case of Example 1, this is $z = x_6 + x_7$. We then use the simplex method to *minimize* this function. If the minimum value is zero, we have a solution—in fact, a *basic* solution—to the original problem. If the minimum value is positive, the original system can have *no* solutions; that is, the constraints are inconsistent.

EXAMPLE 2

Find a basic solution to the following system, or show that none exists.

$$\begin{aligned} 3x_1 - 2x_2 + 5x_3 &\leq 14 \\ x_1 - x_2 - 2x_3 &\geq 1 \\ x_2 - x_3 &= 5 \\ x_j \geq 0, \quad j = 1, 2, 3. \end{aligned}$$

Solution Including the slack variables x_4 and x_5, and artificial variables x_6 and x_7, we minimize the function $z = x_6 + x_7$. The tableaux are given in Figure 11.

From the final tableau of Figure 11, we see that x_6 and x_7 are nonbasic variables, so the basic solution that corresponds to the final tableau is a solution to the original problem. That is, $x_4 = 6$, $x_1 = 6$, $x_2 = 5$, and $x_3 = x_5 = 0$ is a basic solution to the original problem. Ignoring variables x_6 and x_7, the final tableau corresponds to the following system of equations:

$$\begin{aligned} 12x_3 + x_4 + 3x_5 &= 6 \\ x_1 \quad - 3x_3 \quad\quad - x_5 &= 6 \\ x_2 - x_3 \quad\quad\quad &= 5. \end{aligned} \tag{1}$$

FIGURE 11

	c^*	x_1	x_2	x_3	x_4	x_5	x_6	x_7	**b**
x_4	0	3	−2	5	1	0	0	0	14
x_6	1	①	−1	−2	0	−1	1	0	1
x_7	1	0	1	−1	0	0	0	1	5
		1	0	−3	0	−1	0	0	6
		0	1	11	1	3	−3	0	11
		1	−1	−2	0	−1	1	0	1
		0	①	−1	0	0	0	1	5
		0	1	−1	0	0	−1	0	5
		0	0	12	1	3	−3	−1	6
		1	0	−3	0	−1	1	1	6
		0	1	−1	0	0	0	1	5
		0	0	0	0	0	−1	−1	0

■

The process just described is called the **first phase** of the linear programming problem—locating a basic feasible solution. The **second phase** is taking the result of the first phase to construct the first tableau with the *original* objective function. From then on, the problem is optimized in the usual manner.

EXAMPLE 3

Maximize the function $z = x_1 - 2x_2 + 2x_3$, subject to the constraints

$$\begin{aligned} 3x_1 - 2x_2 + 5x_3 &\le 14 \\ x_1 - x_2 - 2x_3 &\ge 1 \\ x_2 - x_3 &= 5 \\ x_j \ge 0, \quad j = 1, 2, 3. \end{aligned}$$

Solution Since the constraints are exactly those of Example 1, the first phase has already been completed, and we start from there. The first tableau of the second phase is constructed from System (1) of Example 2 and the objective function $z = x_1 - 2x_2 + 2x_3 + 0x_4 + 0x_5$. Notice that except for the bottom row, the columns of the first tableau are exactly those of the final tableau of the first stage, without the columns that correspond to the artificial variables. The tableaux of the second phase are shown in Figure 12.

FIGURE 12

	c^*	x_1	x_2	x_3	x_4	x_5	**b**
x_4	0	0	0	12	1	③	6
x_1	1	1	0	−3	0	−1	6
x_2	−2	0	1	−1	0	0	5
		0	0	−3	0	−1	−4
		0	0	4	$\frac{1}{3}$	1	2
		1	0	1	$\frac{1}{3}$	0	8
		0	1	−1	0	0	5
		0	0	1	$\frac{1}{3}$	0	−2

Since no simplex indicators in the final tableau are negative, we have the maximal solution $x_1 = 8$, $x_2 = 5$, $x_5 = 2$, and $x_3 = x_4 = 0$. ■

We summarize the preceding discussion with the following procedure.

The Two-Phase Procedure

i. Add artificial variables to create an embedded identity matrix.

ii. Use the simplex algorithm to minimize this linear programming problem when the objective function is the sum of the artificial variables.

iii. If the minimum is not zero, there is no solution to the system of constraints.

iv. If the minimum is zero, use the resulting basic solution, and delete the columns that correspond to artificial variables to create a first tableau with the original objective function.

v. Use the simplex algorithm to optimize the resulting linear programming problem.

NOTE In some kinds of problems, knowing whether or not a set of constraints has *any* solution is the only important question. In this case, only the first phase is necessary.

EXAMPLE 4

Show that there is no solution to the set of constraints

$$\begin{aligned} 4x_1 + 2x_2 - x_3 &\leq 5 \\ 2x_1 + x_2 + x_3 &\geq 8 \\ -2x_1 - x_2 + 2x_3 &\leq 2 \\ x_j \geq 0, \quad j &= 1, 2, 3. \end{aligned}$$

Solution Adding slack variables to the first and third constraints, and subtracting one from the second, we obtain the equivalent set of equations,

$$\begin{aligned} 4x_1 + 2x_2 - x_3 + x_4 \qquad\qquad &= 5 \\ 2x_1 + x_2 + x_3 \qquad - x_5 \qquad &= 8 \\ -2x_1 - x_2 + 2x_3 \qquad\qquad + x_6 &= 2 \\ x_j \geq 0, \quad j = 1, \ldots, 6. & \end{aligned}$$

Adding an artificial variable to the second constraint, we obtain the (non-equivalent) first-phase problem: Minimize $z = x_7$, subject to the constraints

$$\begin{aligned} 4x_1 + 2x_2 - x_3 + x_4 \qquad\qquad\qquad &= 5 \\ 2x_1 + x_2 + x_3 \qquad - x_5 \qquad + x_7 &= 8 \\ -2x_1 - x_2 + 2x_3 \qquad\qquad + x_6 \qquad &= 2 \\ x_j \geq 0, \quad j = 1, \ldots, 7. & \end{aligned}$$

The tableaux for the simplex algorithm are shown in Figure 13. Since this solution is minimal, and the artificial variable is still in the corresponding basic solution ($x_1 = 2$, $x_3 = 3$, $x_7 = 1$), there is no solution to the original problem.

FIGURE 13

	$\mathbf{c}^*$	x_1	x_2	x_3	x_4	x_5	x_6	x_7	$\mathbf{b}$
x_4	0	④	2	-1	1	0	0	0	5
x_7	1	2	1	1	0	-1	0	1	8
x_6	0	-2	-1	2	0	0	1	0	2
		2	1	1	0	-1	0	0	8
		1	$\frac{1}{2}$	$-\frac{1}{4}$	$\frac{1}{4}$	0	0	0	$\frac{5}{4}$
		0	0	$\frac{3}{2}$	$-\frac{1}{2}$	-1	0	1	$\frac{11}{2}$
		0	0	($\frac{3}{2}$)	$\frac{1}{2}$	0	1	0	$\frac{9}{2}$
		0	0	$\frac{3}{2}$	$-\frac{1}{2}$	-1	0	0	$\frac{11}{2}$
		1	$\frac{1}{2}$	0	$\frac{1}{3}$	0	$\frac{1}{6}$	0	2
		0	0	0	-1	-1	-1	1	1
		0	0	1	$\frac{1}{3}$	0	$\frac{2}{3}$	0	3
		0	0	0	-1	-1	-1	0	1

■

9.3 Exercises

In Exercises 1–4 convert each problem to the form $A\mathbf{x} = \mathbf{b}$, $\mathbf{b} \geq 0$, $\mathbf{x} \geq 0$, where A has an embedded identity matrix, by use of slack or artificial variables (or both).

1.
$$\begin{aligned} 2x - 3y - z &\leq 5 \\ y + 2z &\geq 8 \\ x - 2z &= 3 \\ x, y, z &\geq 0. \end{aligned}$$

2.
$$\begin{aligned} x_1 + 2x_2 + x_4 &\geq 6 \\ x_2 + x_3 + 4x_4 &\geq 12 \\ x_1 - 2x_3 + x_4 &\leq 4 \\ x_j \geq 0, \quad j &= 1, \ldots, 4. \end{aligned}$$

3.
$$\begin{aligned} 2x - y &= 7 \\ x + 2y &\leq 18 \\ 3x - 5y &\geq -3 \\ x, y &\geq 0. \end{aligned}$$

4.
$$\begin{aligned} x_1 + 2x_2 - x_3 &= 5 \\ 4x_2 - x_3 + x_4 &\geq 8 \\ x_2 - 3x_3 + x_4 &\leq -2 \\ x_j \geq 0, \quad j &= 1, \ldots, 4. \end{aligned}$$

5–8. Find a basic solution to each of Exercises 1–4, or show that none exists by using the simplex algorithm with the objective function given by the sum of the artificial variables.

In Exercises 9–12 use the results of Exercises 5–8 and the simplex algorithm to optimize the given objective function, subject to the constraints imposed in Exercises 1–4.

9. Maximize $w = x + 3y - z$.

10. Minimize $z = x_1 + 2x_3 + x_4$.

11. Minimize $z = 2x - 3y$.

12. Minimize $z = x_1 + x_2 + x_3 + x_4$.

In Exercise 13 and 14 use the two-phase method to solve each linear programming problem.

13. Maximize the function $z = 2x_1 + x_2 + 2x_4$, subject to

$$\begin{aligned} x_1 + 2x_2 - x_3 &\geq 5 \\ x_2 + x_3 + x_4 &\leq 12 \\ 2x_1 - x_2 - 3x_3 - 2x_4 &\leq -10 \\ x_j \geq 0, \quad j &= 1, \ldots, 4. \end{aligned}$$

14. Maximize the function $z = x_1 - x_2 + x_3 - x_4$, subject to

$$\begin{aligned} x_2 + 3x_3 + x_4 &\leq 20 \\ x_1 - 3x_2 + x_3 - 2x_4 &\geq 5 \\ x_2 - x_3 + 3x_4 &\geq 6 \\ x_j \geq 0, \quad j = 1, \ldots, 4. \end{aligned}$$

15. *The diet problem*: A company makes dog food from a mixture of four different ingredients (A, B, C, and D) that contain three basic nutrients (P, Q, and R) in various quantities. Figure 14 gives the amount of each nutrient in 1 kg of each ingredient. The column at the right gives the price per kilogram (in cents) of each ingredient, and the row at the bottom gives the minimum daily requirement of each nutrient per kilogram of final product. Find the dog food mix that supplies all the needs at the cheapest price.

FIGURE 14

	P	*Q*	*R*	
A	3	2	3	40
B	3	1	1	20
C	0	4	4	10
D	1	2	3	20
	2	2	3	

16. *The transportation problem*: A company has a particular raw material, stored at warehouses W_1, W_2, and W_3, that is required by factories F_1 and F_2. Figure 15 gives the number of kilometres (in hundreds) between each warehouse and each factory. The column at the right gives the quantity of material available at each warehouse, and the row at the bottom gives the amount needed by each factory. Find a shipping formula that supplies the needs of each factory and minimizes the total shipping distance.

FIGURE 15

	F_1	F_2	
W_1	4	8	6
W_2	2	1	8
W_3	0	3	5
	7	4	

9.4 Theory of Linear Programming

We now present the proofs of Theorems 1–4 of Section 9.2, describing the simplex algorithm. We consider only the problem of *maximizing*. The problem of minimizing requires only minor modifications, and therefore is omitted.

THEOREM 1

Let a linear programming problem be given as follows: Maximize $z = \mathbf{c} \cdot \mathbf{x}$, subject to the constraints

$$A\mathbf{x} = \mathbf{b}, \quad \mathbf{b} \geq \mathbf{0}, \quad \mathbf{x} \geq \mathbf{0},$$

where A has an embedded identity matrix. If no simplex indicator is negative (that is, $\mathbf{c}^*A - \mathbf{c} \geq \mathbf{0}$), then the corresponding basic solution is maximal.

Proof As usual, assume that A is an $m \times n$ matrix. By renumbering the variables if necessary, we may assume that the embedded identity matrix occurs in the first m columns and in their natural order. $A\mathbf{x} = \mathbf{b}$ has the form

$$\begin{array}{lllll} x_1 & & + a_{1,m+1}x_{m+1} + \cdots + a_{1,n}x_n & = & b_1 \\ & x_2 & + a_{2,m+1}x_{m+1} + \cdots + a_{2,n}x_n & = & b_2 \\ & \ddots & \vdots \qquad\qquad\qquad \vdots & & \vdots \\ & x_m & + a_{m,m+1}x_{m+1} + \cdots + a_{m,n}x_n & = & b_m. \end{array} \quad (1)$$

The corresponding basic solution is obtained by letting $x_i = b_i$ for $i = 1, \ldots, m$ and $x_{m+1} = x_{m+2} = \cdots = x_n = 0$. With the truncated cost vector $\mathbf{c}^* = (c_1, \ldots, c_m)$, the value of the current basic solution is

$$\begin{aligned} z &= c_1b_1 + c_2b_2 + \cdots + c_mb_m + c_{m+1}(0) + \cdots + c_n(0) \\ &= c_1b_1 + c_2b_2 + \cdots + c_mb_m = \mathbf{c}^*\mathbf{b}. \end{aligned}$$

From the theory of linear equations of Section 3.3, we have the *general* solution to the system of linear equations, System (1), as a parameterized family by letting

$x_{m+1} = s_1, x_{m+2} = s_2, \ldots, x_n = s_{n-m}$, and then

$$\begin{aligned} x_1 &= b_1 - a_{1,m+1}s_1 - \cdots - a_{1,n}s_{n-m} \\ x_2 &= b_2 - a_{2,m+1}s_1 - \cdots - a_{2,n}s_{n-m} \\ &\ \vdots \\ x_m &= b_m - a_{m,m+1}s_1 - \cdots - a_{m,n}s_{n-m}. \\ x_{m+1} &= s_1 \\ &\ \vdots \\ x_n &= s_{n-m}. \end{aligned} \quad (2)$$

The value associated with this general solution $\mathbf{x}$ is given by

$$\begin{aligned} z = \mathbf{c} \cdot \mathbf{x} &= c_1x_1 + c_2x_2 + \cdots + c_nx_n \\ &= \quad c_1b_1 - c_1a_{1,m+1}s_1 - \cdots - c_1a_{1,n}s_{n-m} \\ &\quad + c_2b_2 - c_2a_{2,m+1}s_1 - \cdots - c_2a_{2,n}s_{n-m} \\ &\quad\ \vdots \\ &\quad + c_mb_m - c_ma_{m,m+1}s_1 - \cdots - c_ma_{m,n}s_{n-m} \\ &\quad + c_{m+1}s_1 + \cdots + c_ns_{n-m}. \end{aligned}$$

Simplifying by columns, we have

$$z = \mathbf{c}^*\mathbf{b} - (\mathbf{c}^*\mathbf{a}^{m+1} - c_{m+1})s_1 - (\mathbf{c}^*\mathbf{a}^{m+2} - c_{m+2})s_2 - \cdots - (\mathbf{c}^*\mathbf{a}^n - c_n)s_{n-m}.$$

Since $\mathbf{c}^*\mathbf{a}^{m+j} - c_{m+j} \geq 0$ and $s_j \geq 0$ for each $j = 1, \ldots, n - m$, the value of z is maximal when $s_1 = s_2 = \cdots = s_{n-m} = 0$. Thus, the corresponding basic solution is maximal. ■

An important special case of the final expression for z is the following corollary.

COROLLARY

Let $A\mathbf{x} = \mathbf{b}$, $\mathbf{b} \geq \mathbf{0}$, $\mathbf{x} \geq \mathbf{0}$, and $z = \mathbf{c} \cdot \mathbf{x}$ be as in Theorem 1, j be the number of a nonbasic variable, and $\mathbf{x}$ be the solution obtained from the basic solution by letting $x_j = s$ and the other nonbasic variables be zero. Then the value of the solution $\mathbf{x}$ is given by

$$z = \mathbf{c}^*\mathbf{b} - (\mathbf{c}^*\mathbf{a}^j - c_j)s.$$ ■

Thus, we see that if $\mathbf{c}^*\mathbf{a}^j - c_j < 0$, a larger value of z results in the new solution (if $s > 0$). That is, if the simplex indicator is negative, the new solution is better for the purpose of maximizing. In fact, the equation $z = \mathbf{c}^*\mathbf{b} - (\mathbf{c}^*\mathbf{a}^j - c_j)s$ expresses the objective function value z as a *linear* function of s. In other words, $-(\mathbf{c}^*\mathbf{a}^j - c_j)$ measures the *rate of increase* (or *decrease* if $\mathbf{c}^*\mathbf{a}^j - c_j$ is positive) in the value of z as a function of s. If there are no restrictions on the size of s, the value of z is unbounded, and no maximum occurs. This is the situation of Theorem 2. If s is restricted, we make s as large as the restriction allows. This yields a new basic solution—the same situation as Theorem 4. We restate both theorems and prove them simultaneously.

Let a linear programming problem be given as follows: Optimize $z = \mathbf{c} \cdot \mathbf{x}$, subject to the constraints

$$A\mathbf{x} = \mathbf{b}, \qquad \mathbf{b} \geq \mathbf{0}, \qquad \mathbf{x} \geq \mathbf{0},$$

where A has an embedded identity matrix. Let j be a column number such that the simplex indicator is negative (that is, $\mathbf{c}^*\mathbf{a}^j - c_j < 0$).

THEOREM 2

If no element of $\mathbf{a}^j$ is positive ($\mathbf{a}^j \leq 0$), then no maximal solution exists.

THEOREM 3

If a_{ij} is positive for at least one row number $i = 1, \ldots, m$, let k be a row number such that $a_{kj} > 0$ and $b_k/a_{kj} \leq b_i/a_{ij}$ for all $i = 1, \ldots, m$ with $a_{ij} > 0$. Then pivoting on the k, j position leads to a new basic solution with value greater than or equal to the former.

Proofs From System (2), the solution $\mathbf{x}$ obtained from the basic solution $x_1 = b_1, \ldots, x_m = b_m, x_{m+1} = \cdots = x_n = 0$ (by letting the nonbasic variable $x_j = s$) is given by

$$\begin{aligned}
x_1 &= b_1 - a_{1j}s \\
x_2 &= b_2 - a_{2j}s \\
\vdots\ &\quad \vdots \qquad \vdots \\
x_m &= b_m - a_{mj}s \\
x_j &= \qquad\quad s \\
x_i &= 0 \quad \text{if } i \neq 1, \ldots, m \quad \text{and} \quad i \neq j.
\end{aligned} \qquad \mathbf{(3)}$$

From this, we see that $\mathbf{x}$ is a solution only if $s \geq 0$ (since $x_j = s$), and, for $i = 1, \ldots, m$, $b_i - a_{ij}s \geq 0$. If $a_{ij} \leq 0$, then $b_i - a_{ij}s \geq b_i \geq 0$, so this condition is satisfied for any value of $s \geq 0$. If $\mathbf{a}^j \leq 0$ (that is, if every $a_{ij} \leq 0$), we have no constraints on s (except $s \geq 0$), and, by the corollary to Theorem 1, Theorem 2 is proved. If, however, $a_{ij} > 0$ for some i, the condition $b_i - a_{ij}s \geq 0$ implies that $s \leq b_i/a_{ij}$. Thus, s must be chosen less than or equal to b_i/a_{ij} for each $i = 1, \ldots, m$ such that $a_{ij} > 0$. By choosing $s = \min\ \{b_i/a_{ij} \mid a_{ij} > 0\}$, we obtain a new solution. Theorem 3 is proved when it is shown that this new solution is the result of pivoting on the k, j position, where $s = b_k/a_{kj}$. To this end, we put System (1) in augmented matrix form (emphasizing the pivot column j and pivot row k):

$$\left[\begin{array}{ccccccc|c}
1 & & & \cdots & a_{1j} & \cdots & & b_1 \\
& \ddots & & & \vdots & & & \vdots \\
& & 1 & \cdots & \textcircled{a_{kj}} & \cdots & & b_k \\
& & & \ddots & \vdots & & & \vdots \\
& & & 1\ \cdots & a_{mj} & \cdots & & b_m
\end{array}\right]. \qquad (4)$$

Pivoting on the k, j position, we obtain the new matrix

$$\left[\begin{array}{cccccccc|c} 1 & -a_{1j}/a_{kj} & & \cdots & 0 & \cdots & b_1 - a_{1j}(b_k/a_{kj}) \\ \ddots & \vdots & & & \vdots & & \vdots \\ & 1/a_{kj} & & \cdots & 1 & \cdots & b_k/a_{kj} \\ & \vdots & \ddots & & \vdots & & \vdots \\ & -a_{mj}/a_{kj} & 1 & \cdots & 0 & \cdots & b_m - a_{mj}(b_k/a_{kj}) \end{array}\right]. \quad (5)$$

Since $s = b_k/a_{kj}$, the corresponding basic solution is that of System (3), as desired. ■

The final theorem needed for the simplex algorithm states that a new bottom row may be computed by continuing row operations to the bottom row.

THEOREM 4

The bottom row of a new tableau may be computed from the preceding tableau by using the pivot row and row operations to obtain a zero in the pivot column position of the bottom row.

Proof Again we assume that the current basic solution is in terms of the first m variables and in their natural order. Then matrix (4) of the preceding proof is the "heart" of the simplex tableau for the current solution, and matrix (5) is that of the new tableau. The bottom row of the first tableau is given by $\mathbf{c}^*A - \mathbf{c}$ for the first n entries and $\mathbf{c}^*\mathbf{b}$ for the last, where $\mathbf{c}^* = (c_1, \ldots, c_m)$. Then the entry in the jth position (the pivot column) is $\mathbf{c}^*\mathbf{a}^j - c_j$. To obtain a zero in that position by use of the pivot row $\mathbf{a}_k$, we must add $-[(\mathbf{c}^*\mathbf{a}^j - c_j)/a_{kj}]\mathbf{a}_k$ to the bottom row. The result is that the first n positions in the bottom row are given by

$$(\mathbf{c}^*A - \mathbf{c}) - [(\mathbf{c}^*\mathbf{a}^j - c_j)/a_{kj}]\mathbf{a}_k,$$

and the last position is given by

$$\mathbf{c}^*\mathbf{b} - [(\mathbf{c}^*\mathbf{a}^j - c_j)/a_{kj}]b_k.$$

To complete the proof, we must show that this is also the bottom row if we construct the bottom row from matrix (5), $\mathbf{c}^{*\prime}A' - \mathbf{c}$, where $\mathbf{c}^{*\prime}$ is the $1 \times m$ matrix of components of $\mathbf{c}$ that corresponds to basic variables in the new solution and A' is the new tableau. In the new solution, x_j has replaced x_k, so $\mathbf{c}^{*\prime} = (c_1, \ldots, c_{k-1}, c_j, c_{k+1}, \ldots, c_m)$. The matrix A' from matrix (5) is given by its rows, as

$$\mathbf{a}_i' = \mathbf{a}_i - \frac{a_{ij}}{a_{kj}}\mathbf{a}_k \quad \text{for } i \neq k$$

and

$$\mathbf{a}_k' = \frac{1}{a_{kj}}\mathbf{a}_k.$$

We can then compute $\mathbf{c}^{*\prime}A' - \mathbf{c}$ by rows as follows:

$$\begin{aligned} \mathbf{c}^{*\prime}A' - \mathbf{c} &= c_1\left(\mathbf{a}_1 - \frac{a_{1j}}{a_{kj}}\mathbf{a}_k\right) + \cdots + c_j\left(\frac{1}{a_{kj}}\mathbf{a}_k\right) \\ &\quad + \cdots + c_m\left(\mathbf{a}_m - \frac{a_{mj}}{a_{kj}}\mathbf{a}_k\right) - \mathbf{c} \\ &= (c_1\mathbf{a}_1 + \cdots + c_m\mathbf{a}_m - c_k\mathbf{a}_k) \\ &\quad - \left(\frac{c_1a_{1j}}{a_{kj}} + \cdots - \frac{c_j}{a_{kj}} + \cdots + \frac{c_ma_{mj}}{a_{kj}}\right)\mathbf{a}_k - \mathbf{c} \\ &= \mathbf{c}^*A - \left[\frac{(c_1a_{1j} + \cdots + c_ma_{mj} - c_j)}{a_{kj}}\right]\mathbf{a}_k - \mathbf{c} \end{aligned}$$

$$= \mathbf{c}^*A - \left[\frac{(\mathbf{c}^*\mathbf{a}^j - c_j)}{a_{kj}}\right]\mathbf{a}_k - \mathbf{c}$$

$$= (\mathbf{c}^*A - \mathbf{c}) - \left[\frac{(\mathbf{c}^*\mathbf{a}^j - c_j)}{a_{kj}}\right]\mathbf{a}_k.$$

Thus, the first n positions in the bottom row are correct, and we need only check the last position. This position is just the value of the new basic solution, and by the corollary to Theorem 1 with $s = b_k/a_{kj}$, it must be

$$\mathbf{c}^*\mathbf{b} - (\mathbf{c}^*\mathbf{a}^j - c_j)\left(\frac{b_k}{a_{kj}}\right).$$

Since this agrees with the row operation result, the proof is complete. ■

One natural question that remains is whether or not the process must terminate. In other words, if we are maximizing the objective function, are we guaranteed that eventually some tableau must satisfy the condition of Theorem 2 or Theorem 3? The answer is, "Not quite, but close enough." Although examples have been constructed that *cycle* (in other words, repeat the same sequence of basic solutions ad infinitum without locating the optimal value), none has arisen by accident. In practical terms, it is not necessary to take such situations into account. In one important situation (the most common), it is easy to prove that the process must terminate, as the final theorem shows.

!

THEOREM 5

If, in each tableau of the simplex algorithm, the basic solution $\mathbf{b}$ is strictly positive, then the process terminates in a finite number of steps.

Proof Assume that the problem is to maximize the objective function. Then, at any stage in the algorithm, we choose a column j for which $\mathbf{c}^*\mathbf{a}^j - c_j < 0$ and let $s = \min\{b_i/a_{ij} \mid a_{ij} > 0\}$. Since $\mathbf{b} > \mathbf{0}$, each $b_i > 0$, and thus $s > 0$. By the corollary to Theorem 1, the value of the next solution is $\mathbf{c}^*\mathbf{b} - (\mathbf{c}^*\mathbf{a}^j - c_j)s$, which is strictly greater than the value of the current solution $\mathbf{c}^*\mathbf{b}$. That is, the value associated with each basic solution *strictly increases* with each iteration of the method. Thus, no solution can be repeated, or else the value drops back to its former level—a contradiction. The proof is then complete if we know that there are only finitely many basic solutions. This follows from the fact that any basic solution arises as a transformation of the original system $A\mathbf{x} = \mathbf{b}$ into a new row-equivalent system with elementary columns that correspond to the basic solution variables. Since A has only finitely many columns, and since any such row-reduced form is unique, there are only finitely many such matrices. Thus, there are only finitely many basic solutions. ■

9.4 Exercises

1. Each of the theorems in this section was stated as a maximizing problem. Prove the analogous results for minimizing problems by proving the following statement: A solution $\mathbf{x}_0$ maximizes a linear function $z = \mathbf{c} \bullet \mathbf{x}$, subject to some set of constraints, if and only if it minimizes the linear function $z' = (-\mathbf{c}) \bullet \mathbf{x}$, subject to the same set of constraints.

2. At some stage in the operation of the simplex algorithm, there may be several simplex indicators that improve the value of the objective function if chosen for the pivot column. Show that the best choice is the simplex indicator

$$\mathbf{c}^*\mathbf{a}^j - c_j \qquad \text{such that} \qquad |(\mathbf{c}^*\mathbf{a}^j - c_j)q_j|$$

is greatest where

$$q_j = \min\left\{\frac{b_i}{a_{ij}} \middle| a_{ij} > 0\right\}.$$

3. Theorem 5 guarantees that if $\mathbf{b} > \mathbf{0}$ at each stage, the simplex algorithm terminates in a finite number of iterations. If $\mathbf{b}$ has some zero entries, however, the simplex algorithm may not have a finite number of iterations. Demonstrate this fact by showing that the following problem cycles if the simplex indicator of proper sign and maximal absolute value are always chosen. If a choice between pivot rows occurs, always choose the first one. (The example is due to Beale.)

   Minimize the linear function

$$z = -0.75x_1 + 150x_2 - 0.02x_3 + 6x_4,$$

   subject to the constraints

$$\begin{aligned} 0.25x_1 - 60x_2 - 0.04x_3 + 9x_4 &\leq 0 \\ 0.50x_1 - 90x_2 - 0.02x_3 + 3x_4 &\leq 0 \\ x_3 \qquad &\leq 1 \\ x_j \geq 0, \quad j = 1, \ldots, 4. \end{aligned}$$

4. Find the optimal solution to the problem in Exercise 3 by row-reducing the equation form and obtaining a matrix with an embedded identity matrix in columns 1, 3, and 4. Form the corresponding tableau, and proceed in the usual manner.

9.5 APPLICATION—Supply and Demand

Linear programming is very useful in a variety of settings. In this section, we'll model a certain type of problem with linear programming; you can then use the simplex method to solve the original problem. There are applied problems that have *supply and demand* constraints; that is, some commodities in limited supply are required for production, and there is a certain level of demand for each of the finished products produced. What you want to do is either minimize cost or maximize profit.

EXAMPLE 1

An animal feed company produces two types of mixed grain for feeding livestock: premium quality and high quality. The grains are all produced by mixing two grains: wheat and barley. The price per tonne of wheat and barley is given in Table 3. (The premium quality feed requires premium grade wheat or barley, and the high quality feed requires high quality wheat or barley.)

TABLE 3

	Premium Quality	High Qualilty
Wheat (tonnes)	190	180
Barley (tonnes)	130	110

The market prices of the two types of feed (per tonne) are as follows: premium quality, \$285.00; high quality, \$140.00. The total amount of wheat and barley that the company can acquire is limited by its storage facilities as follows: wheat, 220 t; barley, 480 t. The company must meet a certain demand for its feed (or lose long-time customers): premium quality, 300 t, high quality, 400 t. How much of each type of feed should be made, and what is the maximum profit?

Solution Let the amount (in tonnes) of premium quality wheat purchased be x_1 and high quality wheat purchased be x_2. Similary, let the amount (in tonnes) of premium quality barley purchased be x_3 and high quality barley purchased be x_4. Then the *supply constraints* (corresponding to the upper limits on the available supply of wheat and barley) are

$$\begin{aligned} x_1 + x_2 &\leq 220 \\ x_3 + x_4 &\leq 480. \end{aligned}$$

The demand constraints (corresponding to the lower limits on how little of each type of feed can be produced) are

$$\begin{aligned} x_1 + x_3 &\geq 300 \\ x_2 + x_4 &\geq 400. \end{aligned}$$

The objective function is

$$z = 285(x_1 + x_3) + 140(x_2 + x_4) - 190x_1 - 180x_2 - 130x_3 - 110x_4,$$

which is

$$z = 95x_1 - 40x_2 + 155x_3 + 30x_4.$$

Thus, the problem we want to solve can be written as follows: Maximize $z = 95x_1 - 40x_2 + 155x_3 + 30x_4$, subject to the constraints

$$\begin{aligned} x_1 + x_2 &\leq 220 \\ x_3 + x_4 &\leq 480 \\ x_1 + x_3 &\geq 300 \\ x_2 + x_4 &\geq 400 \\ x_j \geq 0, \quad j &= 1, \ldots, 4. \end{aligned}$$

We need to use the two-phase method of Section 9.3 to solve this LP problem. We introduce slack variables x_5, x_6, x_7, and x_8, as well as artificial variables x_9 and x_{10}, and solve.

Minimize $z = x_9 + x_{10}$, subject to the constraints

$$\begin{aligned} x_1 + x_2 + x_5 &= 220 \\ x_3 + x_4 + x_6 &= 480 \\ x_1 + x_3 - x_7 + x_9 &= 300 \\ x_2 + x_4 - x_8 + x_{10} &= 400 \\ x_j \geq 0, \quad j &= 1, \ldots, 10. \end{aligned}$$

We solve this system as in Section 9.3, remembering that we want to find a basic solution where the artificial variables are *not* basic. If the value of the objective function is then 0, then the original system is feasible and has a basic solution (by ignoring the artificial variables); if it is not 0, then the original system is infeasible.

From the final tableau of Figure 16, we see that the artificial variables x_9 and x_{10} are nonbasic variables, so the basic solution that corresponds to the final tableau is a solution to the original problem. That is,

$$x_2 = 220, x_3 = 300, x_4 = 180, \text{ and } x_1 = x_5 = x_6 = x_7 = x_8 = 0$$

is a basic solution to the original problem. Ignoring variables x_9 and x_{10}, the final tableau corresponds to the following system of equations:

$$\begin{aligned} x_1 + x_2 + x_5 &= 220 \\ x_5 + x_6 + x_7 + x_8 &= 0 \\ x_1 + x_3 - x_7 &= 300 \\ -x_1 + x_4 - x_5 - x_8 &= 180. \end{aligned}$$

We now proceed with the second phase. We use the system above and the objective function $z = 95x_1 - 40x_2 + 155x_3 + 30x_4$. The tableaux of the second phase is shown in Figure 17.

FIGURE 16

	$\mathbf{c}^*$	x_1	x_2	x_3	x_4	x_5	x_6	x_7	x_8	x_9	x_{10}	$\mathbf{b}$
x_5	0	①	1	0	0	1	0	0	0	0	0	220
x_6	0	0	0	1	1	0	1	0	0	0	0	480
x_9	1	1	0	1	0	0	0	−1	0	1	0	300
x_{10}	1	0	1	0	1	0	0	0	−1	0	1	400
		1	1	1	1	0	0	−1	−1	0	0	700
		1	1	0	0	1	0	0	0	0	0	220
		0	0	1	1	0	1	0	0	0	0	480
		0	−1	①	0	−1	0	−1	0	1	0	80
		0	1	0	1	0	0	0	−1	0	1	400
		0	0	1	1	−1	0	−1	−1	0	0	480
		1	①	0	0	1	0	0	0	0	0	220
		0	1	0	1	1	1	1	0	−1	0	400
		0	−1	1	0	−1	0	−1	0	1	0	80
		0	1	0	1	0	0	0	−1	0	1	400
		0	1	0	1	0	0	0	−1	−1	0	400
		1	1	0	0	1	0	0	0	0	0	220
		−1	0	0	1	0	1	1	0	−1	0	180
		1	0	1	0	0	0	−1	0	1	0	300
		−1	0	0	①	−1	0	0	−1	0	1	180
		−1	0	0	1	−1	0	0	−1	−1	0	180
		1	1	0	0	1	0	0	0	0	0	220
		0	0	0	0	1	1	1	1	−1	−1	0
		1	0	1	0	0	0	−1	0	1	0	300
		−1	0	0	1	−1	0	0	−1	0	1	180
		0	0	0	0	0	0	0	0	−1	−1	0

FIGURE 17

	$\mathbf{c}^*$	x_1	x_2	x_3	x_4	x_5	x_6	x_7	x_8	$\mathbf{b}$
x_2	−40	1	1	0	0	1	0	0	0	220
x_6	0	0	0	0	0	1	1	①	1	0
x_3	155	1	0	1	0	0	0	−1	0	300
x_4	30	−1	0	0	1	−1	0	0	−1	180
		−10	0	0	0	−70	0	−155	−30	43,100
		①	1	0	0	1	0	0	0	220
		0	0	0	0	1	1	1	1	0
		1	0	1	0	1	1	0	1	300
		−1	0	0	1	−1	0	0	−1	180
		−10	0	0	0	85	155	0	125	43,100
		1	1	0	0	1	0	0	0	220
		0	0	0	0	1	1	1	1	0
		0	−1	1	0	0	1	0	1	80
		0	1	0	1	0	0	0	−1	400
		0	10	0	0	95	155	0	125	45,300

Since no simplex indicators in the final tableau are negative, we have a maximal solution $x_1 = 220$, $x_3 = 80$, $x_4 = 400$, and $x_2 = x_5 = x_6 = x_7 = x_8 = 0$. This means that the maximum profit is \$45,300, and is achieved by ordering 220 t of premium quality wheat, no high quality wheat, 80 t of premium quality barley, and 400 t of high quality barley. This makes 300 t of premium quality feed and 400 t of high quality feed. ■

The above supply-demand model is *balanced* in that the total supply is equal to the total demand (both are 700 t). It can be shown that every supply-demand model where the total supply equals the total demand has a feasible solution, and hence an optimal value.

9.5 Exercises

1. In the second-last tableau of Figure 16, why was the pivot chosen in the fourth row rather than in the third row?
2. Redo the problem in Example 1 if the market prices of the two types of feed (per tonne) are as follows: premium quality, \$225.00; high quality, \$170.00, and the costs per tonne are as shown below:

TABLE 4

	Premium Quality	High Qualilty
Wheat (tonnes)	190	150
Barley (tonnes)	130	110

3. If the supply of wheat in Example 1 is 240 t instead of 220 t, how might you balance the problem?
4. If the supply of premium quality wheat in Example 1 is 450 t instead of 300 t, would the problem be feasible? Explain your answer.
5. If a supply-demand problem is unbalanced, how might you balance it?
6. Prove that every balanced supply-demand problem is feasible.

CHAPTER SUMMARY

Linear Programming Problems

(a) DEFINITION: A *linear programming problem* involves finding the optimal value of a linear, real-valued function on a geometric region determined by a system of linear inequalities in multi-dimensional space.

(b) DEFINITION: The region in (a) is called the *feasible region*, and the function in (a) is called the *objective function.*

(c) In simple cases, the optimal value can be found by graphical means. This process is especially workable in the case of two or three variables and is aided by the knowledge that the optimal value must occur at some vertex of the feasible region or else the feasible region must be unbounded.

Simplex Algorithm

(a) A general solution to the problem is given by the *simplex algorithm.* In this method of solution, the variables are all nonnegative.

(b) The process begins with the objective function value at the most obvious vertex of the feasible region, the origin of the coordinate system.

(c) The mathematical representation of the situation at that point is called a *tableau.*

i. The rates of increase or decrease in objective function value that result in moving toward each adjacent vertex are the entries in the bottom row of the corresponding tableau.
ii. One of those directions is chosen, and the algorithm proceeds to convert the tableau into a new tableau for that new vertex. If no direction improves the objective function value, that vertex has optimal value.
iii. Since the region has only finitely many vertices, the process must terminate in an optimal solution or else determine a direction in which the function improves in an unbounded manner.

(d) Many applications are amenable to the simplex algorithm even if their original description does not make that fact obvious.

i. For example, if the situation involves variables that must be negative, they can be replaced by their additive inverses to force them to be positive.
ii. If they are unrestricted positive or negative, they can be replaced by the difference of two variables that are individually restricted to be nonnegative.
iii. Sometimes the most important problem is to determine whether or not any solutions exist at all. That is, is the feasible region empty? Again, the appropriate adjustment allows the simplex algorithm to provide the answer to that question, as well as find an optimal answer in case the existence question has been answered affirmatively.

KEYWORDS

linear programming
objective function
tableau
simplex indicator
artificial variable
supply and demand
feasible region
simplex algorithm
slack variable
pivoting
two-phase problem
balanced

SELF-TEST

In Exercises 1 and 2 sketch the feasible region in $\mathbf{R}^2$.

1. $\begin{aligned} x + 5y &\le 30 \\ x + y &\le 6 \\ 5x + y &\le 20 \\ x, y &\ge 0 \end{aligned}$

2. $\begin{aligned} x + 5y &\le 30 \\ x + 2y &\le 6 \\ -x + y &\ge 4 \\ x, y &\ge 0 \end{aligned}$ [Section 9.1]

3. Maximize the objective function $z = 2x + y$ on the feasible region of Exercise 1 using graphical means. [Section 9.1]

4. Minimize the objective function $z = 2x - y$ on the feasible region of Exercise 1 using graphical means. [Section 9.1]

5. Convert the system of Exercise 1 into equation form. [Section 9.1]

In Exercises 6–10 use the tableau in Figure 18.

6. Give the value of each of the variables and the objective function value in the current solution. [Section 9.2]

FIGURE 18

x_1	x_2	x_3	x_4	x_5	x_6	**b**
3	0	1	2	0	1	3
−1	1	0	−1	0	2	2
1	0	0	1	1	2	1
−1	0	0	−2	0	−3	10

7. Which nonbasic variables, if any, can be "brought in" to increase the value of the objective function? [Section 9.2]

8. Which nonbasic variables, if any, can be "brought in" to decrease the value of the objective function? [Section 9.2]

9. The current solution is either maximal or minimal. Which is it, and how can you tell? [Section 9.2]

10. Which nonbasic variable can be brought in to yield the maximum (in absolute value) rate of increase of the objective function value? Find the new tableau resulting from that choice. [Section 9.2]

REVIEW EXERCISES

In Exercises 1 and 2 sketch the feasible region in $\mathbf{R}^2$.

1.
$$\begin{aligned} x + 10y &\le 40 \\ x + y &\le 5 \\ 6x + y &\le 20 \\ x, y &\le 0 \end{aligned}$$

2.
$$\begin{aligned} x + 10y &\le 40 \\ x + y &\le 5 \\ -x + y &\ge 1 \\ x, y &\ge 0 \end{aligned}$$

3. Maximize and minimize the objective function $z = 2x + y$ on the feasible region of Exercise 1 using graphical means.

4. Maximize and minimize the objective function $z = 2x + y$ on the feasible region of Exercise 2 using graphical means.

5. Maximize and minimize the objective function $z = x - y$ on the feasible region of Exercise 1.

6. Maximize and minimize the objective function $z = x - y$ on the feasible region of Exercise 2.

7. Convert the system of Exercise 1 to equation form.

8. Convert the system of Exercise 2 to equation form.

9. Using Exercise 7, together with the objective function $z = 2x + y$, create the first tableau of the simplex algorithm.

10. Using Exercise 7, together with the objective function $z = x - y$, create the first tableau of the simplex algorithm.

11. Use Exercise 9 to redo Exercise 3 using the simplex algorithm.

12. Use Exercise 8, with the addition of an artificial variable and the appropriate objective function, to create the first tableau for the first stage of redoing Exercise 6 using the simplex algorithm.

13. Use Exercise 12 to redo Exercise 6 using the simplex algorithm.

14. Redo Exercise 11 assuming that only x is restricted to be nonnegative.

15. Redo Exercise 13 assuming that only y is restricted to be nonnegative.

In Exercises 16 and 17 convert each problem to the form $A\mathbf{x} = \mathbf{b}$, $\mathbf{b} \ge 0$, $\mathbf{x} \ge 0$, where A has an embedded identity matrix.

16.
$$\begin{aligned} x + 2y + 3z &\le 10 \\ 2x - 5y - z &\ge 5 \\ x + y + z &= 12 \\ y, z &\ge 0 \end{aligned}$$

17.
$$\begin{aligned} x_1 + 2x_2 - x_3 + 3x_4 &\ge 5 \\ x_1 - 2x_2 + x_3 - x_4 &= 10 \\ 2x_1 + x_2 + x_3 - x_4 &\ge 8 \\ x_2, x_3 &\ge 0 \end{aligned}$$

18. Write your own "diet problem" (like Exercise 15 of Section 9.3) that involves mixing three different ingredients, each containing four different nutrients. Set up a simplex algorithm solution for your problem.

CHALLENGE YOURSELF

19. Give an example of a feasible region and an objective function z that is not constant such that the maximum of z over the region is the same as the minimum over the region.

20. Three manufacturers of flour supply three bakeries with flour. Manufacturer 1 can supply 8000 kg of flour per week. Manufacturer 2 can supply 10,000 kg of flour per week. Manufacturer 3 can supply 12,000 kg of flour per week. Bakery A needs at least 10,000 kg of flour per week. Bakery B needs at least 13,000 kg of flour per week. Bakery C needs at least 7000 kg of flour per week. The costs (in dollars) of shipping 1000 kg of flour from each manufacturer to the three bakeries are shown below.

TABLE 5

	Bakery A	Bakery B	Bakery C
Manufacturer 1	1.25	1.50	1.75
Manufacturer 2	1.50	1.50	1.25
Manufacturer 3	2.25	1.25	1.50

(*a*) Express the problem as a linear programming problem.

(*b*) What can you say if bakery C needs at least 23,000 kg of flour per week? Why?

(*c*) Solve the original problem.

21. Consider the following problem: Maximize $z = \mathbf{c} \cdot \mathbf{x}$, subject to $A\mathbf{x} \leq \mathbf{b}$, $\mathbf{b} \geq \mathbf{0}$, $\mathbf{x} \geq \mathbf{0}$, if all the coordinates of $\mathbf{x}$ are integers. (This is an example of an *integer programming problem.*)

(*a*) The *LP relaxation* of the problem omits the last condition on $\mathbf{x}$ (that its coordinates must be integers). Show that if the LP relaxation has a solution, then this solution is greater than the solution to the original integer programming problem (if it exists).

(*b*) How does part (a) change if the original integer programming problem is to minimize z instead of maximizing it?

22. Consider the following integer programming problem: Maximize $z = 7x + 5y$, subject to

$$\begin{aligned} x + y &\leq 8 \\ 13x + 4y &\leq 52 \\ x, y &\geq 0,\ x \text{ and } y \text{ are integers.} \end{aligned}$$

(*a*) Solve the LP relaxation of the problem.

(*b*) Solve the following two related problems:

Problem A: Maximize $z = 7x + 5y$, subject to

$$\begin{aligned} x + y &\leq 8 \\ 13x + 4y &\leq 52 \\ x &\leq 2 \\ x, y &\geq 0. \end{aligned}$$

Problem B: Maximize $z = 7x + 5y$, subject to

$$\begin{aligned} x + \quad y &\leq 8 \\ 13x + 4y &\leq 52 \\ x \quad\quad &\geq 3 \\ x, y &\geq 0. \end{aligned}$$

(*c*) Use parts (a) and (b) to solve the original integer programming problem.

23. In the following network diagram (Figure 19), each line can carry water flow in the direction indicated, with a maximum capacity on the line as shown. At every point, other than the source and the sink, the total flow into the point must equal the total flow out of the point. The value of a flow is the total amount of water flowing out of the source. The *network flow problem* is to maximize the value of the flow. Model the problem as a linear programming problem.

FIGURE 19

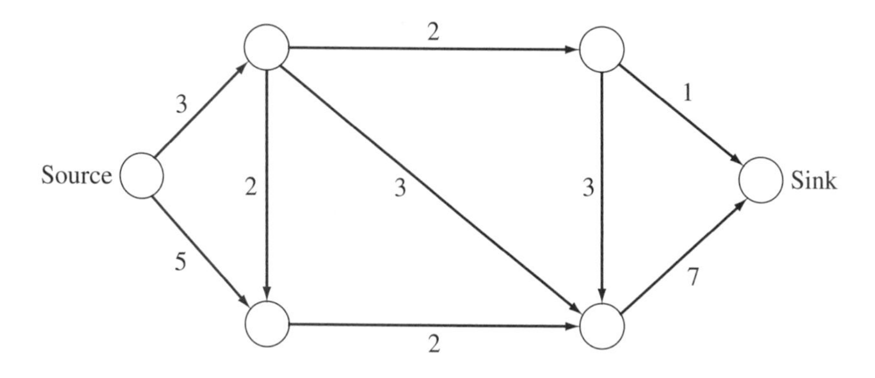

A Short Introduction to *Maple*

A.1 Getting Started

Maple is a programming language that is well-suited to mathematics. It can carry out both symbolic and numerical calculations, as well as provide plots and animations for visualization. This short section will introduce you to *Maple* and the commands you will need to use to investigate linear algebra with it.

Once you have installed *Maple* on your computer, you can start the program by choosing *Maple* from the Start Menu (Windows) or by double-clicking on its icon in the Applications window (Mac). The program should open up with a blank document that is in **Document Mode** or **Worksheet Mode**, depending on your settings for *Maple*. Select **Worksheet Mode** for now from the **New** submenu of the **File menu**. You should be presented with a window that looks like this:

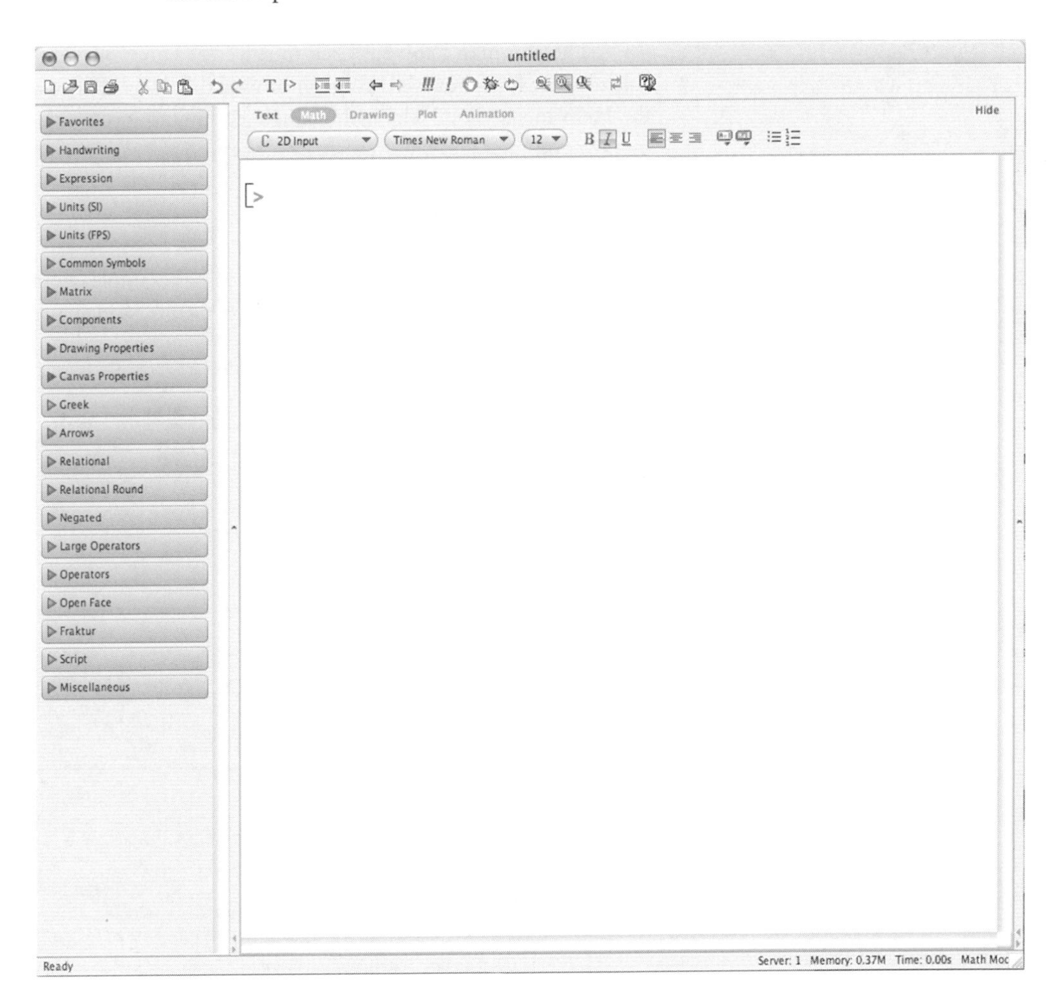

The red > sign is the prompt at which you type a *Maple* command (it is hidden in document mode). After each command, make sure that you type a semicolon followed by the Return/Enter key; this sends the command to the *Maple* engine that carries out calculations.

Let's type a few simple commands. First, let's type the command to calculate $2^5(41 \times 32 - 21)$. In *Maple*, you indicate powers by the caret (^) sign and multiplication by the asterisk (*) sign. You enter $2^5(41 \times 32 - 21)$ as **2^5*(41*32-21)**. *Maple* versions 10 and higher format your input in a mathematical, two-dimension style, so you need to use the down arrow key after typing the 5 to move the cursor from the superscript level back down to the main level. Remember to end your line with a semicolon, followed by the Return/Enter key.

> $2^5 \cdot (41 \cdot 32 - 21);$

$$41312$$

>

Let's try another example. Suppose that you want to find all the roots of the quadratic equation $x^2 + 15x - 4 = 0$. You can enter the equation in *Maple* and assign it a name for easy reference. Let's call the equation **eqn**. Assignments of names occur at the start of a line, followed by the assignment keystroke : = (a colon followed by an equal sign).

> $eqn := x^2 + 15 \cdot x - 4 = 0;$

$$eqn := x^2 + 15x - 4 = 0$$

The command **solve** is used to solve the equation. You need to specify the variable to solve for (x) as well. You can browse the *Maple* Help facilities by selecting from the **Help** menu. For Help with a single command, type a question mark (?) followed by the command and the Return/Enter key.

> *?solve*

The bottom of the ensuing Help screen has examples from which you can glean the structure of the command. Here is a portion of the *Maple* Help page for the **solve** command.

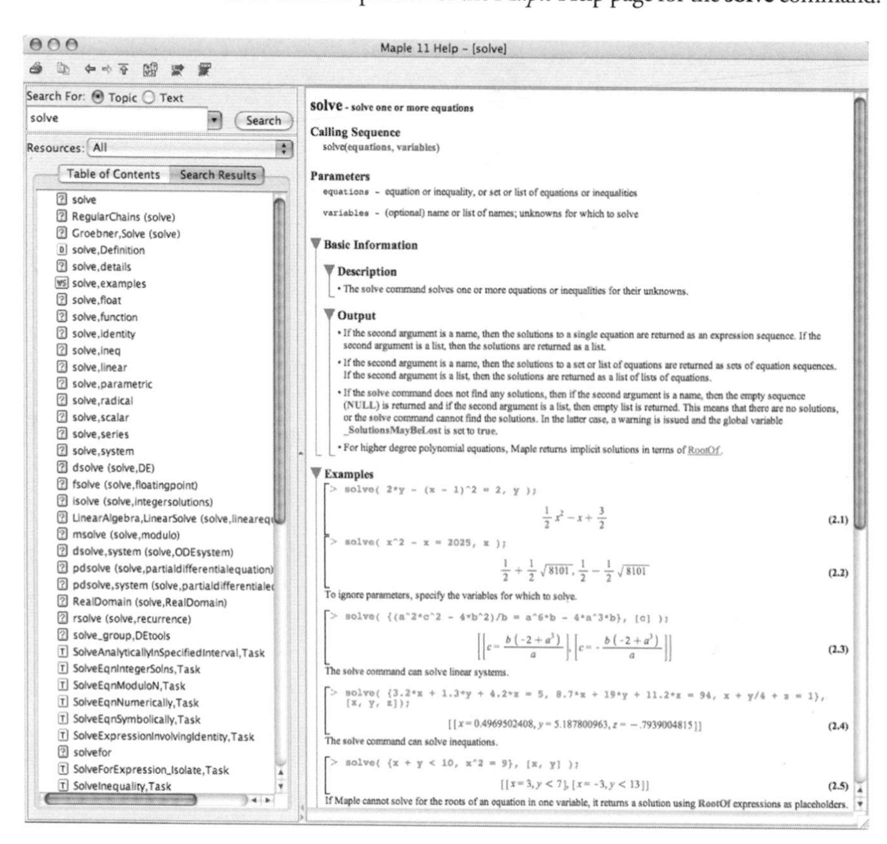

Issuing the **solve** command on **eqn** and ***x*** yields the following, solving the equation $x^2 + 15x - 4 = 0$ exactly for x:

> *solve* (*eqn*, *x*);

$$-\frac{15}{2} + \frac{1}{2}\sqrt{241}, -\frac{15}{2} - \frac{1}{2}\sqrt{241}$$

If approximations are desired for the roots, you can use either the **fsolve** command or the all-purpose **evalf** command, which evaluates an exact value to an approximation. The percent symbol is also a handy one to know in *Maple*, as it refers to the previous calculation (rather than the previous line) carried out (and %% represents the second last calculation carried out, and so on).

> *solve* (*eqn*, *x*);

$$-\frac{15}{2} + \frac{1}{2}\sqrt{241}, -\frac{15}{2} - \frac{1}{2}\sqrt{241}$$

> *fsolve* (*eqn*, *x*);

$$-15.26208735, 0.2620873481$$

> *evalf*(%%);

$$0.262087350, -15.26208735$$

You can also ask *Maple* to provide a variety of plots for visualization. In the example below, the **plot** command plots the left-hand side of **eqn** *f*or $x = -16$ to 3. (For a full description of the **plot** command, see the *Maple* Help page for **plot**.)

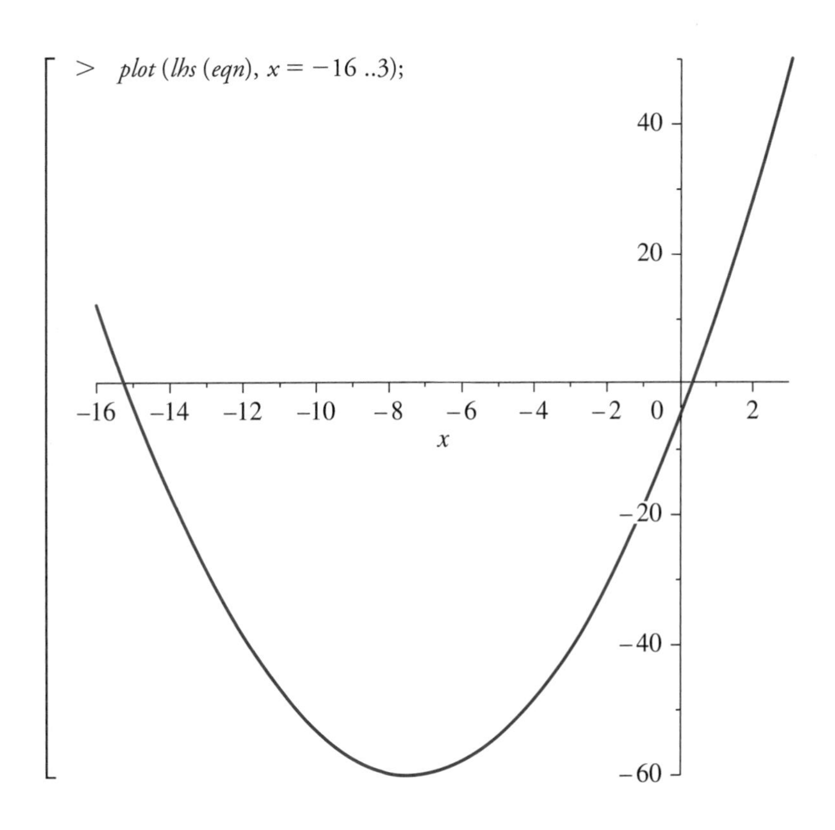

In addition to having *Maple* carry out single commands, you can instruct *Maple* to delay several calculations. You do this by typing a Shift-Return or Shift-Enter keystroke (that is, holding down the Shift key as you type Return or Enter) at the end of a line, hitting the Return/Enter key by itself only on the last line of a section. For example, the following command adds the first 1000 positive integers by adding each successive number to the name **s** which starts off at 0. The **for** loop runs a counter from the start value to the end value, carrying out the command between the **do** and **od** keywords for each iteration. Because we don't want to see the intermediate results of the calculations, we also use the colon (:) instead of the semicolon (;) as the former carries out the same calculation, but does not display the result. Make sure to hit the Return/Enter key only at the end of the block to have the commands sent off to *Maple* for processing.

```
> s := 0:
  for i from 1 to 1000 do
    s := s + i:
  od:
```

There is a slight problem here; the use of the colon hid all the results, so we do not know the value of *s*! Not to worry—you can see the value of *s* by typing the name followed by a semicolon and the Return/Enter key:

```
> s := 0:
  for i from 1 to 1000 do
    s := s + i:
  od:
> s;
                              500500
```

A.2 *Maple* and Linear Algebra

Maple has a built-in library for linear algebra. (A library is a set of routines that can be called upon to make available specific commands.) This library is called **linalg**, and you include it with the following command:

```
> with(linalg);
[BlockDiagonal, GramSchmidt, JordanBlock, LUdecomp, QRdecomp, Wronskian, addcol,
  addrow, adj, adjoint, angle, augment, backsub, band, basis, bezout, blockmatrix,
  charmat, charpoly, cholesky, col, coldim, colspace, colspan, companion, concat, cond,
  copyinto, crossprod, curl, definite, delcols, delrows, det, diag, diverge, dotprod,
  eigenvals, eigenvalues, eigenvectors, eigenvects, entermatrix, equal, exponential,
  extend, ffgausselim, fibonacci, forwardsub, frobenius, gausselim, gaussjord, geneqns,
  genmatrix, grad, hadamard, hermite, hessian, hilbert, htranspose, ihermite, indexfunc,
  innerprod, intbasis, inverse, ismith, issimilar, iszero, jacobian, jordan, kernel,
  laplacian, leastsqrs, linsolve, matadd, matrix, minor, minpoly, mulcol, mulrow,
  multiply, norm, normalize, nullspace, orthog, permanent, pivot, potential, randmatrix,
  randvector, rank, ratform, row, rowdim, rowspace, rowspan, rref, scalarmul,
  singularvals, smith, stackmatrix, submatrix, subvector, sumbasis, swapcol, swaprow,
  sylvester, toeplitz, trace, transpose, vandermonde, vecpotent, vectdim, vector,
  wronskian]
```

You'll notice a number of words that are specific to linear algebra echoed back by *Maple*. These are linear algebra commands that are now available for you to use in a *Maple* session. (You may wonder why these aren't available upon starting *Maple*. To include all possible commands upon start-up would tax your computer's RAM and processing capabilities greatly—and perhaps even crash your system—so *Maple* wisely chooses to have most commands bundled in packages that can be called upon and read into memory only when needed.)

You can now start working with matrices. There are many ways to input a matrix. One useful way is to use the **matrix** command, but first we need to talk about lists. A list is a sequence of items, separated by commas and enclosed in square brackets. Here is a list of the first five positive integers, under the name **lst**:

> $lst := [1, 2, 3, 4, 5];$

$$lst := [1, 2, 3, 4, 5]$$

A matrix is a list of lists. Here is a *Maple* command to initiate two matrices:

> $A := matrix([[1, -1], [-2, 5]]);$
> $B := matrix([[0, 21], [-11, 0]]);$

$$A := \begin{bmatrix} 1 & -1 \\ -2 & 5 \end{bmatrix}$$

$$B := \begin{bmatrix} 0 & 21 \\ -11 & 0 \end{bmatrix}$$

You can select a certain item in the matrix by using the name of the matrix followed by the row and column of the entry, enclosed in square brackets.

> $A[1, 2];$

$$-1$$

You can't use the plain arithmetic commands on matrices. But you can use the command **evalm** along with the usual operations (sum, difference, scalar multiplication, powers) to generate the appropriate matrix.

> evalm (A + B);

$$\begin{bmatrix} 1 & 20 \\ -13 & 5 \end{bmatrix}$$

> evalm (A − B);

$$\begin{bmatrix} 1 & -22 \\ 9 & 5 \end{bmatrix}$$

> evalm (3·B);

$$\begin{bmatrix} 0 & 63 \\ -33 & 0 \end{bmatrix}$$

> $evalm(A^2);$

$$\begin{bmatrix} 3 & -6 \\ -12 & 27 \end{bmatrix}$$

Matrix multiplication is just a little more complicated. The operation * is only used (in evalm) for scalar multiplication. For matrix multiplication, use **&*** instead.

> *evalm* (*A*&· *B*);

$$\begin{bmatrix} 11 & 21 \\ -55 & -42 \end{bmatrix}$$

Maple's linalg package has built-in commands for many matrix operations, such as determinants, inverses, and eigenvalues. Some of the operations will work even if the matrix has variable entries.

> *C* := *matrix* ([[1, *x*], [*x*, 2 · *x*]]);

$$C := \begin{bmatrix} 1 & x \\ x & 2x \end{bmatrix}$$

> *det* (*C*);

$$2x - x^2$$

> *inverse* (*C*);

$$\begin{bmatrix} -\frac{2}{-2+x} & \frac{1}{-2+x} \\ \frac{1}{-2+x} & -\frac{1}{x(-2+x)} \end{bmatrix}$$

> *E* := *matrix* ([[0.4, 0.4, 0.2], [0.4, 0.3, 0.3], [0.2, 0.3, 0.5]]);
F := *evalm* (*transpose* (*E*) & · *E*);

$$E := \begin{bmatrix} 0.4 & 0.4 & 0.2 \\ 0.4 & 0.3 & 0.3 \\ 0.2 & 0.3 & 0.5 \end{bmatrix}$$

$$F := \begin{bmatrix} 0.36 & 0.34 & 0.30 \\ 0.34 & 0.34 & 0.32 \\ 0.30 & 0.32 & 0.38 \end{bmatrix}$$

> *eigenvals* (*F*);

0.005358983849, 0.07464101615, 1.000000000

Maple has a built-in routine for Gauss–Jordan elimination as well:

> G := matrix ([[1, 2, 0, 0, 1], [1, 2, 3, 1, 0], [−1, −1, 1, 1, −2], [0, 1, 1, 1, −1], [0, −1, 2, 0, 0]]);

$$G := \begin{bmatrix} 1 & 2 & 0 & 0 & 1 \\ 1 & 2 & 3 & 1 & 0 \\ -1 & -1 & 1 & 1 & -2 \\ 0 & 1 & 1 & 1 & -1 \\ 0 & -1 & 2 & 0 & 0 \end{bmatrix}$$

> *gaussjord* (G);

$$\begin{bmatrix} 1 & 0 & 0 & -\frac{4}{3} & \frac{7}{3} \\ 0 & 1 & 0 & \frac{2}{3} & -\frac{2}{3} \\ 0 & 0 & 1 & \frac{1}{3} & -\frac{1}{3} \\ 0 & 0 & 0 & 0 & 0 \\ 0 & 0 & 0 & 0 & 0 \end{bmatrix}$$

While *Maple* can handle vectors as matrices of the appropriate dimensions, the linalg package has a built-in data structure for vectors. You input a vector between the $<$ and $>$ symbols, separating items by the | symbol. The items can be involved in matrix multiplication, being interpreted as a column or row vector as needed (but being displayed as a row vector in any event).

There are many commands for working with vectors. Here are but a few:

> $vect1 := \langle 1|0|-1\rangle$;

$$vect1 := [1 \;\; 0 \;\; -1]$$

> $H := matrix([[1, 2, 3], [-1, 0, 1], [0, 1, 3]])$;

$$H := \begin{bmatrix} 1 & 2 & 3 \\ -1 & 0 & 1 \\ 0 & 1 & 3 \end{bmatrix}$$

> $evalm\,(H\&\cdot vect1)$;

$$[-2 \;\; -2 \;\; -3]$$

> $evalm\,(vect1\&\cdot H)$;

$$[1 \;\; 1 \;\; 0]$$

You can also find bases of vector spaces spanned by a set of vectors:

> $vect4 := \langle 1|0|1\rangle$;
$vect5 := \langle 1|-1|2\rangle$;
$vect6 := \langle 0|1|-1\rangle$;
$Basis\,(\{vect4, vect5, vect6\})$;

$$vect4 := [\,1 \;\; 0 \;\; 1\,]$$
$$vect5 := [\,1 \;\; -1 \;\; 2\,]$$
$$vect6 := [\,0 \;\; 1 \;\; -1\,]$$
$$\{[\,0 \;\; 1 \;\; -1\,], [\,1 \;\; 0 \;\; 1\,]\}$$

You can explore the many useful commands in *Maple* by looking at the examples in the online Help and trying out your own. (It's often helpful to copy some of the examples from the Help pages to your worksheet and then alter them.)

A.3 *Maple* and Linear Programming

Maple has another package, **Optimization**, that has useful commands for linear programming. In particular, the **LPSolve** command can be used to solve a linear programming problem. For example, suppose that you want to solve the following problem:

Maximize $z = 95x_1 - 40x_2 + 155x_3 + 30x_4$, subject to the constraints

$$\begin{aligned} x_1 + x_2 &\leq 220 \\ x_3 + x_4 &\leq 480 \\ x_1 + x_3 &\geq 300 \\ x_2 + x_4 &\geq 400 \\ x_j \geq 0, \quad j = 1, \ldots, 4 \end{aligned}$$

You can use the following *Maple* commands:

```
> with(Optimization);
  [ImportMPS, Interactive, LPSolve, LSSolve, Maximize, Minimize, NLPSolve, QPSolve]

> f := 95·x1 − 40·x2 + 155·x3 + 30·x4;
                    f := 95 x1 − 40 x2 + 155 x3 + 30 x4

> LPSolve(f, {x1 + x2 ≤ 220, x3 + x4 ≤ 480, x1 + x3 ≥ 300, x2 + x4 ≥ 400, x1 ≥ 0,
      x2 ≥ 0, x3 ≥ 0, x4 ≥ 0}, maximize);
[45300.0000000981, [x1 = 220.000000001032588, x2 = 0., x3 = 80.0000000000000284, x4 = 400.]]
```

Note that the answers are subject to rounding errors. You can find out more about the **LPSolve** command and its options from the *Maple* Help facilities.

APPENDIX B

Answers to Selected Exercises

Chapter 1

Section 1.1

1. $(-4, -8, 12)$
3. $(1, -1)$
5. $(5, 1, 1)$
7. $(-3, -2, 2)$
9. $(-9, -3, 1)$
11. $\sqrt{14}$
13. $\sqrt{17}$
15. $\sqrt{14} + \sqrt{5}$
17. $\left(\frac{3}{5}, -\frac{4}{5}\right)$
19. $\left(-\frac{2}{3}, \frac{1}{3}, -\frac{2}{3}\right)$
21. $(4, 2)$
23. $(1, -4, -7)$
25. $2\mathbf{i} - 3\mathbf{k}$
27. $\sqrt{14}$
33. **i.** $(x_1 - x_2)^2 + (y_1 - y_2)^2 = r^2$, where $\mathbf{p} = (x_2, y_2)$.
ii. $(x_1 - x_2)^2 + (y_1 - y_2)^2 + (z_1 - z_2)^2 = r^2$, where $\mathbf{p} = (x_2, y_2, z_2)$.
35. $\frac{2}{3}\mathbf{p} + \frac{1}{3}\mathbf{q}$
37. $\mathbf{q} - \mathbf{p} = (1, 2, -1)$ and $\mathbf{s} - \boldsymbol{r} = -2(\mathbf{q} - \mathbf{p})$
39. 659.00 km/h, $\theta = 4.92°$

Section 1.2

1. 4
3. 4
5. $(11, -7, -10)$
7. $(-22, 14, 20)$
9. -38
11. Obtuse
13. $\frac{-8}{\sqrt{145}}$
15. $-\frac{1}{2}$
17. $(-4, 7, -1)$
19. $\frac{1}{2}\sqrt{14}$
23. $-\sqrt{5}$
25. $\mathbf{0}$
27. The dot product of a vector with a scalar is not defined.
29. The cross product is not associative.
33. Supplementary
35. u_1, u_2, u_3 are the cosines of the angles formed by $\mathbf{v}$ with $\mathbf{i}, \mathbf{j}, \mathbf{k}$, where $\mathbf{u} \cdot \mathbf{i} = u_1$, $\mathbf{u} \cdot \mathbf{j} = u_2$, and $\mathbf{u} \cdot \mathbf{k} = u_3$
37. $V = Ah = |(\mathbf{u} \times \mathbf{v}) \cdot \mathbf{w}|$
41. All three vectors $\mathbf{v}_1$, $\mathbf{v}_2$, and $(\mathbf{v}_1 + \mathbf{v}_2)$ lie on the same plane.
43. 100 Nm
45. 34.47 Nm

Computational Exercises

47. *(a)* $\theta \approx 2.2974$ radian
(b) so $\theta \approx 0.9303$ radian

Section 1.3

1. Point-parallel form: $\boldsymbol{x}(t) = (2 + t, 1 + 2, -3 + 2)$
Parametric equations: $(x, y, z) = (2 + t, 1 + 2t, -3 + 2t)$ or, $x = 2 + t$, $y = 1 + 2t$, $z = -3 + 2t$.

3. Point-parallel form: $\boldsymbol{x}(t) = (2 + t, -3, 1)$
Parametric equations: $(x, y, z) = (2 + t, -3, 1)$ or, $x = 2 + t$, $y = -3$, $z = 1$.

5. The two-point form: $\boldsymbol{x}(t) = (2 - t, 4t, -2 + 4t)$
Parametric equations: $(x, y, z) = (2 - t, 4t, -2 + 4t)$ or, $x = 2 - t$, $y = 4t$, $z = -2 + 4t$.

7. Point-parallel form: $\boldsymbol{x}(t) = (2 + 5t, 4 - 5t, 5 - 10t)$
Parametric equations: $(x, y, z) = (2 + 5t, 4 - 5t, 5 - 10t)$ or, $x = 2 + 5t$, $y = 4 - 5t$, $z = 5 - 10t$.

9. Point-parallel form: $\boldsymbol{x}(t) = (1 + t, -1 - 3t)$
Parametric equations: $(x, y) = (1 + t, -1 - 3t)$ or, $x = 1 + t$, $y = -1 - 3t$.

11. $2x + y + 3z = 0$

13. $3x + 6y + z = 12$

15. $(2, 3, 0) \cdot \left(\mathbf{x} - \left(\frac{1}{2}, 0, 0\right)\right) = 0$

17. $(2, 1) \cdot (\mathbf{x} - (-1, 2)) = 0$

19. $(-2, 1) \cdot (\mathbf{x} - (-2, 5)) = 0$

21. Point-parallel form: $(-7, -5, 0) \cdot (\mathbf{x} - (-1, 2, 3)) = 0$

Standard form: $-7x - 5y = -3$

Parametric form of the plane is $\Pi = (-1, 2, 3) + s(1, -1, 4) + t(2, -2, 1)$, where s and t are real numbers.

23. $(-3, -1) \cdot (\mathbf{x} - (2, 3)) = 0$

25. $\frac{\sqrt{6}}{2}$

27. $\frac{9}{\sqrt{5}}$

29. $(x, y, z) = \left(\frac{13}{5}, 1, -\frac{7}{5}\right)$

31. $(2, -1, -2) \cdot (1, 0, 1) = 0$ so the given line and plane are parallel.

33. $\mathbf{n}_1 = (1, -1, -2)$ and $\mathbf{n}_2 = (3, -2, 1)$, they are not multiples of each other so the two planes are not parallel.

35. 1.16 radian

37. $\theta = 90°$

Section 1.4

1. Suppose $a = c$. Then if $d < b$, the mouse is directly below the center of the dial, so the dial is to be rotated by 180°. If $d \geq b$, the mouse is directly above the center of the dial, so the dial should not be rotated.

3. Using the notation of Example 1, set $\mathbf{g} = \mathbf{m} + (10, 20)$.

Self-Test

1. *(a)* -2 *(b)* $\mathbf{y} \times \mathbf{z} = (-5, 1, 13)$ *(c)* $3\sqrt{14}$ *(d)* $3\sqrt{74}$ *(e)* $\left(\frac{-3}{14}, \frac{-2}{14}, \frac{-1}{14}\right)$ *(f)* $\left(\frac{3}{14}, \frac{-3}{14}, \frac{1}{14}\right)$

3. $(1/2, 3, 7)$

5. $\mathbf{x}(t) = (2, 1, -3) + t(2, -3, 1)$

7. Not parallel

Review Exercises

1. $(4, 2, -9)$

3. $\sqrt{34}$

5. $2\sqrt{5}$

7. $(6, -3, 2)$

9. -1

11. $(1/\sqrt{10}, 0, -3/\sqrt{10})$

13. $z - 3 = 0$.

15. $(3, 1, -2) \cdot (\mathbf{x} - (0, 1, 0)) = 0$

17. $\mathbf{x}(t) = (1 + t)(1, 0, 2)$

19. $x + y + z = d$, where d is an arbitrary constant.

Review Exercises—Challenge Problems

21. $P(2, 1, -1)$

23. $(-b, a)$ points to the left of (a, b)

25. $\mathbf{u}_1 = (-1, 1, 1)$ and $\mathbf{u}_2 = (2, 0, -1)$. $\mathbf{x}(s, t) = (0, 2, 1) + s\mathbf{u}_1 + t\mathbf{u}_2$

Chapter 2

Section 2.1

1. $(1, 2, 5, 1)$

3. $(12, -6, 14, 40)$

5. 2

7. $6\sqrt{11}$

9. $\sqrt{31}$

11. $2\sqrt{47}$

13. $\mathbf{u}_1$ and $\mathbf{u}_3$ are orthogonal, $\mathbf{u}_2$ and $\mathbf{u}_3$ are orthogonal, $\mathbf{u}_2$ and $\mathbf{u}_4$ are orthogonal, $\mathbf{u}_3$ and $\mathbf{u}_4$ are orthogonal.

15. $\frac{1}{\|\mathbf{u}\|}\mathbf{u}$ is a unit vector.

17. $\left(\frac{2}{\sqrt{31}}, \frac{1}{\sqrt{31}}, \frac{-1}{\sqrt{31}}, 0, \frac{3}{\sqrt{31}}, \frac{4}{\sqrt{31}}\right)$

19. $-2x_1 - x_2 + x_4 = t$, $t \in \mathbf{R}$

21. $x = -1, y = 0, z = 3 + t, w = 2 + 3t$

23. $x_1 - x_2 + x_3 - x_4 + x_5 = 5$

25. $(2, -3, 0, 1, -1, 0) \cdot (\mathbf{x} - (1, 0, 0, 0, 0, 0)) = 0$

27. Not linear

29. Not linear

31. *(a)* $\mathbf{u} + \mathbf{v} = \mathbf{v} + \mathbf{u}$ *(b)* $(\mathbf{u} + \mathbf{v}) + \mathbf{w} = \mathbf{u} + (\mathbf{v} + \mathbf{w})$
(c) $\mathbf{u} + \mathbf{0} = \mathbf{u}$ *(d)* $\mathbf{u} + (-\mathbf{u}) = \mathbf{0}$
(e) $(cd)\mathbf{u} = c(d\mathbf{u})$ *(f)* $(c + d)\mathbf{u} = c\mathbf{u} + d\mathbf{u}$
(g) $c(\mathbf{u} + \mathbf{v}) = c\mathbf{u} + c\mathbf{v}$ *(h)* $1\mathbf{u} = \mathbf{u}$
(i) $(-1)\mathbf{u} = -\mathbf{u}$ *(j)* $0\mathbf{u} = \mathbf{0}$

35. $\mathbf{x} = (3, 1, 0, 2, 1) + t_1(-1, 0, 4, 0, -1) + t_2(-4, 1, 1, 1, 0) + t_3(-3, 1, 0, -1, -1)$, where t_1, t_2, and t_3 are any scalars.

39. $V(2, 0, 2, -2)$

Section 2.2

1. Linear 3. Not linear 5. Not linear

7. $\begin{aligned} y + z &= 6 \\ x + y - z &= 0 \\ x - y - z &= -3 \end{aligned}$

9. $\begin{aligned} \mathbf{x}_1 + \mathbf{x}_2 + \mathbf{x}_3 - \mathbf{x}_4 &= 0 \\ \mathbf{x}_1 \quad\quad\quad\quad &= 1 \\ - \mathbf{x}_2 \quad + \mathbf{x}_4 &= 3 \end{aligned}$

15. $(x, y) = (1 + 4t, t)$, where the parameter t is any scalar.

17. $(2 - 3t, -1 - s + 2t, s, t)$

21. $u_6 = 60$ and $u_7 = 70$ represent the temperatures at internal points 6 and 7 on the grid.

Computational Exercises

23. $x = \frac{147}{16},\ y = -\frac{231}{16}$

25. No solution

27. $c_1 = \frac{3 + k^2}{3 - k},\ c_2 = \frac{k^2 + 2k - 3}{3 - k},\ c_3 = -\frac{2(3 + k^2)}{3 - k}$, no solution when $k = 3$.

29. $c_1 = \frac{k^2 - 2k + 21}{k - 7},\ c_2 = -\frac{k^2 + 7}{k - 7},\ c_3 = \frac{2(k^2 - 4k + 35)}{k - 7}$, no solution when $k = 7$.

Section 2.3

1. $\left[\begin{array}{ccc|c} 2 & -3 & 1 & 0 \\ 1 & 0 & -2 & 1 \\ 0 & -4 & 1 & -1 \end{array}\right]$

3. $\left[\begin{array}{ccc|c} 1 & -1 & 0 & 0 \\ 0 & 1 & -1 & 0 \end{array}\right]$

5. No 7. Yes 9. No 11. $x_1 = 0, x_2 = 2, x_3 = -1$

13. No solution 15. $(4 - 2s - 3t, s, t)$ 17. $(1 + 2s, s)$ 19. $\left(-\frac{3}{4}, -\frac{5}{4}, \frac{13}{4}\right)$

21. $\left(2s, \frac{5s - 1}{3}, s\right)$ 23. No solution 25. $\left(\frac{1}{2} + s, 1 + 2s - t, s, t\right)$. 27. $(1 + 2s, s)$

29. $x = -\frac{3}{4}, y = -\frac{5}{4}, z = \frac{13}{4}$ 31. $\left(2s, \frac{5s - 1}{3}, s\right)$ 33. No solution

35. $\left(\frac{1}{2} + s, 2 + 1s - t, s, t\right)$ 37. $\left(-\frac{21}{5}, \frac{13}{5}, -\frac{12}{5}\right)$ 39. $(1 + t, 1 + 2t, t)$ 41. $c = -1$

43. No value of k such that the system will have infinitely many solutions.

45. *(a)* $\left[\begin{array}{cccc|c} 4 & -1 & 0 & 0 & 100 \\ 1 & -4 & 1 & 0 & -120 \\ 0 & 1 & -4 & 1 & -160 \\ 0 & 0 & 1 & -4 & -320 \end{array}\right]$

(b) $u_1 = 40$, $u_2 = 60$, $u_3 = 80$, and $u_4 = 100$

Computational Exercises

47. $(1 + s, s)$, where s is any scalar.

49. $(-0.75, -1.25, 3.25)$

51. $(-2s, 1.666666667 - 0.3333333333s, s)$ where s is any scalar.

53. No solution

55. $(0.5 + s, 1 + 2s - t, s, t)$, where s and t are any scalars.

57. $(35.99835642, -629.9538302, 3359.69089, -7559.13460, -2771.661717)$

59. *(a)* The solution is $(-1149.42064, -704.6287859, 1775.5)$.

(b) The solution is $(2.242677824, -1.138075314, -1, -4.686192469)$.

(c) The solution is $(-0.4403384846, 0.7565217391, 0.4373082689)$.

61. The solution in each case is found to be unique.

Section 2.4

1. $I_1 = \frac{11}{12}, I_2 = \frac{5}{12}, I_3 = \frac{1}{2}$

3. $I_1 = \frac{90}{13} \approx 0.80, I_2 = \frac{46}{113} \approx 0.41, I_3 = \frac{122}{113} \approx 1.08, I_4 = \frac{78}{113} \approx 0.69, I_5 = \frac{44}{13}$

5. $R_c = s, R_e = t, R_a = \frac{4}{5} - \frac{2}{5}s + \frac{2}{5}t, R_b = \frac{2}{3} + \frac{2}{3}s - \frac{2}{3}t, R_d = \frac{1}{2} - \frac{1}{2}t$

7. $x_1 = 36.75, x_2 = 49.29, x_3 = 17.95, x_4 = 50.71, x_5 = 18.80, x_6 = -1.42, x_7 = 17.38$

9. $k_c = s, k_e = t, k_a = \frac{4}{5} - \frac{2}{5}s + \frac{2}{5}t, k_b = \frac{2}{3} + \frac{2}{3}s - \frac{2}{3}t, k_d = \frac{1}{2} - \frac{1}{2}t$, where $k_a, k_b, k_c, k_d, k_e \geq 0$.

Computational Exercises

11. $I_1 = -\frac{416}{137}, I_2 = -\frac{412}{137}, I_3 = -\frac{348}{137}, I_4 = -\frac{126}{137}, I_5 = -\frac{22}{137}, I_6 = -\frac{86}{137},$

$I_7 = -\frac{90}{137}, I_8 = \frac{200}{137}, I_9 = \frac{290}{137}, I_{10} = \frac{64}{137}, I_{11} = -\frac{222}{137}, I_{12} = \frac{4}{137}$

Section 2.5

1. $P(x) = -3 + 2x + x^2$

3. $P(x) = -1 - x^2 + x^3$

5. $P(x) = 1 - 2x^2 + x^3$

7. $P(x) = -0.4259 + 0.48835x - 0.06245x^2$

$\log_{10}(2.5) \approx P(2.5) = 0.4046625$

9.
$$a_1 = \frac{-2x_0y_0^2 + x_0^2y_1 + x_0^2y_2 + 3x_0x_1y_1 - 2x_0x_1y_2 + x_0x_2y_0 - x_0x_2y_1 - x_0x_1y_0}{(x_1 - x_0)(x_2 - x_0)(x_2 - x_1)} + \frac{x_1^2y_1 - x_1^2y_2 - x_1x_2y_1 + x_1x_2y_0 - x_2^2y_0 + x_2^2y_1}{(x_1 - x_0)(x_2 - x_0)(x_2 - x_1)}$$

Computational Exercises

11. $P(x) = \frac{7}{2} - x - \frac{1}{2}x^2$

13. $P(x) = 1 - 2x + x^2$

15. $P(x) = -3 - 8x^2 + 2x^4$

17.

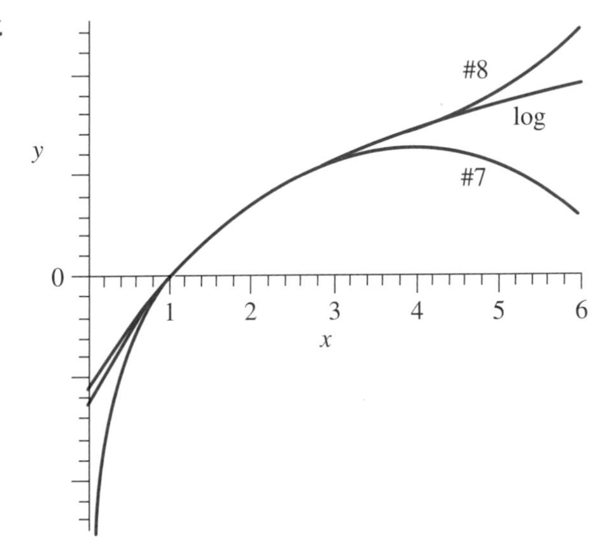

19. When $s = 0$ the conic is degenerate and when $k = 2$ or $k = 3$ the points lie on the conic.

21. $4x^2 + 6xy + 7y^2 - 22x - 69y + 74 = 0$

Self-Test

1. *(a)* -8

(b) $2\sqrt{7}$

3. $(1, -1, 2, -1) \bullet (\mathbf{x} - (0, 0, 0, 6)) = 0$

5. $\left[\begin{array}{ccc|c} 1 & 0 & -\frac{1}{5} & \frac{1}{5} \\ 0 & 1 & -\frac{4}{5} & -\frac{1}{5} \end{array}\right]$

7. $(1 + 3s - 7t, 2s - 4t, s, t)$

Review Exercises

1. $(4, -3, 3, -3, -10)$

3. $2\sqrt{14}$

5. 7

7. $x + 2y + 3z + 4u + 5v = 15$

9. $t(1, 0, 2, 0)$

11. Expressions, (b), (c), (f) are not defined.

13. $(-1 + 2s, s)$, where s is a scalar.

15. $(2s - t, s, t)$, where s and t are scalars.

17. $(1, 0, -1)$

19. No solution

Review Exercises—Challenge Problems

21. $x_1 + x_3 - x_4 = 0$

23. $(1, 2, 0, 0) + t(0, -2, 1, 0)$

27. $x = \frac{-1}{2} - \frac{i}{2}, y = \frac{1}{2} + \frac{i}{2}$

Chapter 3

Section 3.1

1. $\begin{bmatrix} -1 & 4 & -1 \\ 5 & 0 & 5 \\ -1 & 1 & -1 \end{bmatrix}$

3. $\begin{bmatrix} -2 & -12 & 8 \\ -5 & -5 & 0 \\ 3 & 2 & -7 \end{bmatrix}$

5. $\begin{bmatrix} -1-\lambda & 0 & 1 \\ 2 & -1-\lambda & 3 \\ 0 & 1 & -2-\lambda \end{bmatrix}$

7. $\begin{bmatrix} -1 & 2 & 0 \\ 0 & -1 & 1 \\ 1 & 3 & -2 \end{bmatrix}$

9. $\begin{bmatrix} -1 & -4 & 3 \\ -6 & 7 & -3 \\ 5 & 1 & 0 \end{bmatrix}$

11. $\begin{bmatrix} -1 & -6 & 5 \\ -4 & 7 & 1 \\ 3 & -3 & 0 \end{bmatrix}$

13. $\begin{bmatrix} 2 & 1 \\ -1 & 0 \\ \frac{1}{2} & -2 \end{bmatrix}$

15. $\mathbf{a}_1 = (2, 1)$, $\mathbf{a}_2 = (1, 0, -2)$, and $\mathbf{a}_3 = \left(\frac{1}{2}, -2\right)$

17. $AB = \begin{bmatrix} 7 & -1 & 1 \\ 0 & -4 & -2 \\ 6 & -6 & 0 \end{bmatrix}$, $BA = \begin{bmatrix} -1 & 2 & 4 \\ 5 & 6 & -1 \\ 6 & 4 & -2 \end{bmatrix}$

19. AB: Not defined, BA: Not defined.

21. $AB = \begin{bmatrix} 2 & 1 & 0 \\ -1 & 4 & 3 \end{bmatrix}$, BA: Not defined.

23. $\begin{bmatrix} -9 & 1 & 8 \\ -2 & -3 & -10 \end{bmatrix}$

25. $A^2 = \begin{bmatrix} 2 & -3 \\ -3 & 5 \end{bmatrix}$, $A^3 = \begin{bmatrix} 5 & -8 \\ -8 & 13 \end{bmatrix}$

27. $A^2 = \begin{bmatrix} 1 & 0 & 0 \\ 4 & 1 & 0 \\ 10 & 4 & 1 \end{bmatrix}$, $A^3 = \begin{bmatrix} 1 & 0 & 0 \\ 6 & 1 & 0 \\ 21 & 6 & 1 \end{bmatrix}$

35. $A^n = \begin{bmatrix} 1 & n \\ 0 & 1 \end{bmatrix}$

37. $\begin{bmatrix} -2 & -3 \\ 1 & 2 \end{bmatrix}$

Computational Exercises

41. Not equal

43. $A + B = \begin{bmatrix} 2.00 & 0.83 & 0.53 \\ 0.83 & 0.53 & 0.40 \\ 0.53 & 0.40 & 0.31 \end{bmatrix}$, $A - B = \begin{bmatrix} 0.00 & 0.17 & 0.13 \\ 0.17 & 0.13 & 0.10 \\ 0.13 & 0.10 & 0.09 \end{bmatrix}$

45. **(a)**

$$\begin{aligned} ax_{11} + bx_{21} + cx_{31} &= 1 \\ ax_{12} + bx_{22} + cx_{32} &= 0 \\ ax_{13} + bx_{23} + cx_{33} &= 0 \\ dx_{11} + ex_{21} + fx_{31} &= 0 \\ dx_{12} + ex_{22} + fx_{32} &= 1 \\ dx_{13} + ex_{23} + fx_{33} &= 0 \\ gx_{11} + hx_{21} + jx_{31} &= 0 \\ gx_{12} + hx_{22} + jx_{32} &= 0 \\ gx_{13} + hx_{23} + jx_{33} &= 1 \end{aligned}$$

(b)

$$x_{11} = \frac{ej - hf}{\Delta}, x_{12} = \frac{hc - bj}{\Delta}, x_{13} = \frac{-ec + bf}{\Delta}$$

$$x_{21} = \frac{gf - dj}{\Delta}, x_{22} = \frac{aj - gc}{\Delta}, x_{23} = \frac{-af + dc}{\Delta}, \text{ where } \Delta = gbf - gec - ahf - bdj + eaj + dhc.$$

$$x_{31} = \frac{dh - gc}{\Delta}, x_{31} = \frac{gb - ah}{\Delta}, x_{33} = \frac{-db + ae}{\Delta}$$

(c) No solution exactly when the denominator Δ is zero.

(d) For $k \neq \frac{12}{7}$, $AB = \begin{bmatrix} 1 & 0 & 0 \\ 0 & 1 & 0 \\ 0 & 0 & 1 \end{bmatrix}$ for some matrix B.

Section 3.2

1. $\begin{bmatrix} 1 & -1 & 3 \\ 1 & 0 & -1 \\ -2 & 1 & 0 \end{bmatrix}\begin{bmatrix} x \\ y \\ z \end{bmatrix} = \begin{bmatrix} 1 \\ 0 \\ -1 \end{bmatrix}$

3. $\begin{bmatrix} 1 & -1 & 1 \\ 1 & -1 & -1 \end{bmatrix}\begin{bmatrix} x \\ y \\ z \end{bmatrix} = \begin{bmatrix} 0 \\ 0 \end{bmatrix}$

5. A and B are inverses of each other.

7. A and B are inverses of each other.

9. $\begin{bmatrix} \frac{1}{2} & 0 \\ \frac{3}{2} & 1 \end{bmatrix}$

11. $\begin{bmatrix} 1 & -2 & 5 \\ 0 & 1 & -2 \\ 0 & 0 & 1 \end{bmatrix}$

13. $\begin{bmatrix} 2 & 1 & -1 \\ 1 & 1 & -1 \\ 1 & 1 & 0 \end{bmatrix}$

15. No inverse

17. If we let $c = -6$, then A has no inverse.

19. $\begin{bmatrix} -3 \\ -8 \end{bmatrix}$

21. $\begin{bmatrix} \frac{7}{2} \\ -\frac{3}{2} \\ -\frac{3}{2} \end{bmatrix}$

23. $\begin{bmatrix} 12 \\ 7 \end{bmatrix}$

25. $\begin{bmatrix} 4 \\ -4 \\ -3 \end{bmatrix}$

27. No inverse

29. $\frac{1}{ad - bc}\begin{bmatrix} d & -b \\ -c & a \end{bmatrix}$

31. $\begin{bmatrix} 1 & \frac{1}{4} \\ 0 & \frac{1}{4} \end{bmatrix}$

33. $\begin{bmatrix} \frac{1}{d_1} & 0 & \cdot & \cdot & 0 \\ 0 & \frac{1}{d_2} & \cdot & \cdot & 0 \\ \cdot & \cdot & \cdot & \cdot & \cdot \\ \cdot & \cdot & \cdot & \cdot & \cdot \\ 0 & \cdot & \cdot & \cdot & \frac{1}{d_n} \end{bmatrix}$

Computational Exercises

35. *(a)* $\begin{bmatrix} 1 \\ 3 \\ 5 \\ 2 \\ 4 \end{bmatrix}$ *(b)* $\begin{bmatrix} 125 \\ -2880 \\ 14490 \\ -24640 \\ 13230 \end{bmatrix}$

37. $\frac{1}{k^2 + k - 2}\begin{bmatrix} -1 & -1 & k+1 \\ -1 & k+1 & -1 \\ k+1 & -1 & -1 \end{bmatrix}$. No inverse when $k = 1$ or $k = -2$.

39. $A^{-1} = \frac{1}{n-1}B$, where B is the $n \times n$ matrix with $(n - 2)$'s on the main diagonal and -1's elsewhere.

Section 3.3

1. Homogeneous

3. 2

5. 3

7. *(a)* $\text{rank}(A) = 2$ provided $c \neq 0$, $d \neq 0$, and $ad - bc \neq 0$.
   (b) $\text{rank}(A) = 1$ and $\text{rank}(A|\mathbf{b}) = 2$ if and only if $ad = bc$ but $af \neq ce$.
   (c) $\text{rank}(A) = \text{rank}(A|\mathbf{b}) = 1$ if and only if $ad = bc$ and $af = ce$.

9. False

11. True

13. False

15. Infinitely many solutions

17. One solution or infinitely many solutions.

21. *(a)* No solution since $x = 0$, $y \neq 0$.
    (b) if $x \neq 0$ then the linear system has a unique solution.
    (c) Infinitely many solutions if $x = y = 0$.

Computational Exercises

23. 2

25. 3

Section 3.4

1. $L = \begin{bmatrix} 1 & 0 \\ -2 & 1 \end{bmatrix}$, $U = \begin{bmatrix} 2 & -1 \\ 0 & 3 \end{bmatrix}$

3. $PA = LU = \begin{bmatrix} 2 & 1 & 1 \\ 4 & 3 & 2 \\ 2 & 1 & -1 \end{bmatrix}$

5. $L = \begin{bmatrix} 1 & 0 & 0 & 0 \\ 3 & 1 & 0 & 0 \\ 2 & 3 & 1 & 0 \\ 2 & 3 & 2 & 1 \end{bmatrix}, U = \begin{bmatrix} 1 & 2 & 1 & 2 \\ 0 & -1 & -2 & -4 \\ 0 & 0 & 3 & 9 \\ 0 & 0 & 0 & -12 \end{bmatrix}, P = \begin{bmatrix} 1 & 0 & 0 & 0 \\ 0 & 1 & 0 & 0 \\ 0 & 0 & 1 & 0 \\ 0 & 0 & 0 & 1 \end{bmatrix}$

7. $L = \begin{bmatrix} 1 & 0 \\ 3 & 1 \end{bmatrix}, U = \begin{bmatrix} 2 & -1 \\ 0 & 5 \end{bmatrix}, P = \begin{bmatrix} 1 & 0 \\ 0 & 1 \end{bmatrix}$

9. $L = \begin{bmatrix} 1 & 0 & 0 \\ 2 & 1 & 0 \\ 1 & 3 & 1 \end{bmatrix}, U = \begin{bmatrix} 3 & 1 & 2 \\ 0 & -1 & -1 \\ 0 & 0 & 2 \end{bmatrix}, P = \begin{bmatrix} 1 & 0 & 0 \\ 0 & 0 & 1 \\ 0 & 1 & 0 \end{bmatrix}$

11. $L = \begin{bmatrix} 1 & 0 & 0 \\ 0 & 1 & 0 \\ 2 & 1 & 1 \end{bmatrix}, U = \begin{bmatrix} 1 & -1 & 2 & -1 & 0 \\ 0 & 1 & -1 & 1 & 1 \\ 0 & 0 & 0 & 3 & 1 \end{bmatrix}, P = \begin{bmatrix} 0 & 1 & 0 \\ 1 & 0 & 0 \\ 0 & 0 & 1 \end{bmatrix}$

13. $y_1 = \frac{3}{2}, y_2 = \frac{-1}{2}, y_3 = \frac{1}{2}$

15. $(-4, 0, -2, 3)$

17. $x = \frac{3}{4}, y = \frac{1}{4}, z = \frac{-1}{4}$

19. A admits no LU decomposition unless a permutation matrix P is involved.

21. $A = \begin{bmatrix} 1 & 1 & 3 & 4 \\ -1 & 0 & 1 & 3 \\ -2 & 1 & 2 & 4 \end{bmatrix}, L = \begin{bmatrix} 1 & 0 & 0 \\ -1 & 1 & 0 \\ -2 & 3 & 1 \end{bmatrix}, U = \begin{bmatrix} 1 & 1 & 3 & 4 \\ 0 & 1 & 4 & 7 \\ 0 & 0 & -4 & -9 \end{bmatrix}$

Computational Exercises

23. (a) $L = \begin{bmatrix} 1 & 0 & 0 \\ 0 & 1 & 0 \\ 2 & 1 & 1 \end{bmatrix}, U = \begin{bmatrix} 1 & -1 & 2 & -1 & 0 \\ 0 & 1 & -1 & 1 & 1 \\ 0 & 0 & 0 & 3 & 1 \end{bmatrix}, P = \begin{bmatrix} 0 & 1 & 0 \\ 1 & 0 & 0 \\ 0 & 0 & 1 \end{bmatrix}$

(b) $L = \begin{bmatrix} 1 & 0 & 0 \\ -1 & 1 & 0 \\ -2 & 3 & 1 \end{bmatrix}, U = \begin{bmatrix} 1 & 1 & 3 & 4 \\ 0 & 1 & 4 & 7 \\ 0 & 0 & -4 & -9 \end{bmatrix}, P = \begin{bmatrix} 1 & 0 & 0 \\ 0 & 1 & 0 \\ 0 & 0 & 1 \end{bmatrix}$

25. $L = \begin{bmatrix} 0 & 0 & 0 \\ 2 & 0 & 0 \\ 1 & 0 & 0 \end{bmatrix}, U = \begin{bmatrix} 3 & 1 & 2 \\ 0 & -1 & -1 \\ 0 & 0 & -4 \end{bmatrix}, P = \begin{bmatrix} 1 & 0 & 0 \\ 0 & 0 & 1 \\ 0 & 1 & 0 \end{bmatrix}$

Section 3.5

1. Interchange the 2nd and 3rd rows of the identity matrix A.

3. EA has the effect of adding 3 times row2 of A to row1 of A.

5. $\begin{bmatrix} \frac{1}{2} & 0 & 0 \\ 0 & 1 & 0 \\ 0 & 0 & 1 \end{bmatrix}$

7. $A^{-1}B^{-1} \neq (AB)^{-1}$

9. $E_1^{-1} = E_{23}(-c)$ is the inverse of $E_1 = E_{23}(c)$

11. P_{ij} is its own inverse.

13. 10

15. 34

17. $\frac{n}{6}(4n^2 - 3n - 1)$

19. $A = (E_n \cdots E_2E_1)^{-1} = E_1^{-1}E_2^{-1} \cdots E_n^{-1}$, a product of elementary matrices.

Section 3.6

1. $Q(x) = \frac{-4}{5} + \frac{13}{10}x$

3. $Q(x) = \frac{15}{2} - \frac{137}{10}x + 3x^2$

5.
$$\begin{aligned} 7b_0 \quad\quad + 28b_2 \quad\quad &= -28 \\ 28b_1 \quad\quad + 196b_3 &= 0 \\ 28b_0 \quad\quad + 196b_2 \quad\quad &= 20 \\ 196b_1 \quad\quad + 1588b_3 &= 0 \end{aligned}$$

7. $p(x) = 303 - 86x$

Computational Exercises

9. Exercise 5: $p(x) = \frac{11}{2}x^2 - \frac{72}{7}$

Exercise 6: $p(x) = \frac{1}{24}x^4 - \frac{61}{24}x^2 - \frac{103}{42}x + \frac{2}{21}$

11. (a) Degree 3 least square polynomial:

$p(x) = -0.399x^2 + 1.255x - 0.014$

Degree 5 least square polynomial:

$q(x) = 0.037x^4 - 0.232x^3 + 0.050x^2 + 0.987x$

(b) $p\left(\frac{\pi}{2}\right) = 1$, $q\left(\frac{\pi}{2}\right) = 0.999$, $\sin\left(\frac{\pi}{2}\right) = 1$

(c) The higher degree polynomial gives a slightly better approximation.

13. (a)

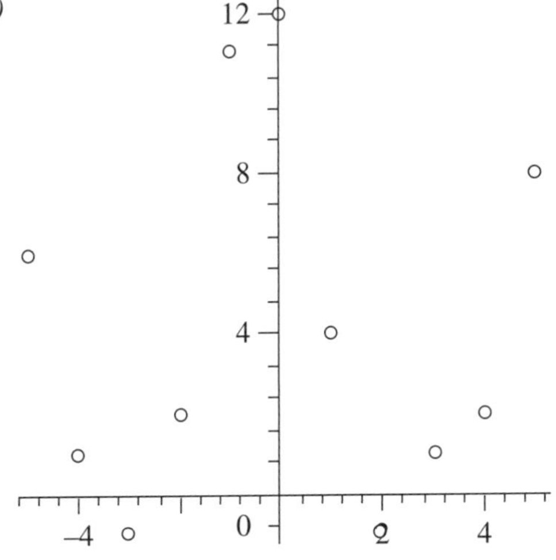

(b) $p(x) = -0.0046x^6 - 0.0067x^5 + 0.2294x^4 + 0.2475x^3 - 3.0334x^2 - 1.8317x + 10.5977$

(c)

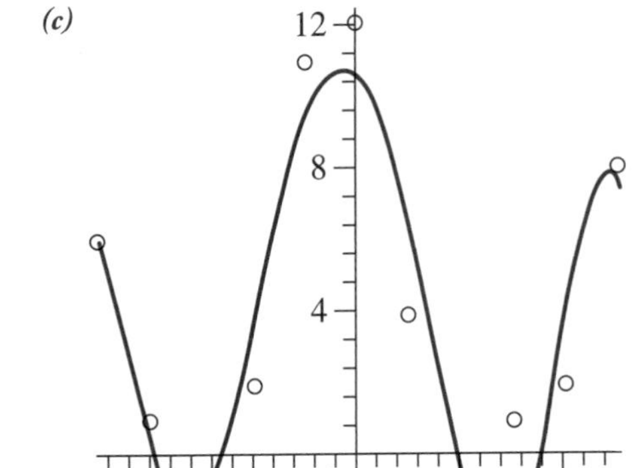

Section 3.7

1.

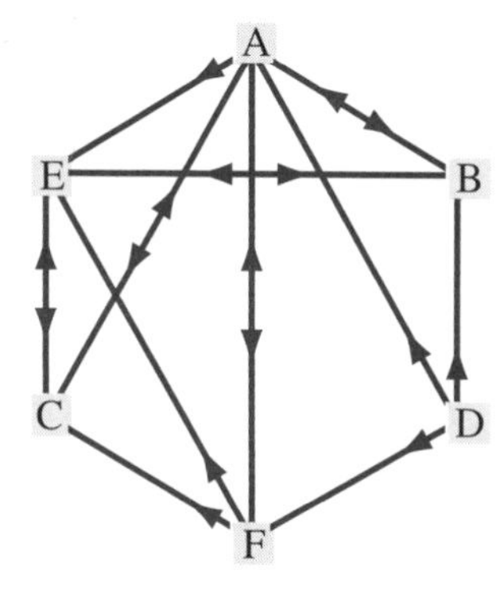

3. For a vertex, v, let indeg(v) and outdeg(v) be equal to the in-degrees and out-degrees of v respectively.

indeg(A) = 4 outdeg(A) = 4, indeg(B) = 3 outdeg(B) = 2, indeg(C) = 3 outdeg(C) = 2, indeg(D) = 0, outdeg(D) = 3, indeg(E) = 3, outdeg(E) = 2, indeg(F) = 2, outdeg(F) = 2

5. Not strongly connected

7.
$$\begin{bmatrix} 0 & 1 & 1 & 1 & 1 & 0 \\ 1 & 0 & 0 & 1 & 0 & 1 \\ 1 & 0 & 0 & 0 & 1 & 0 \\ 1 & 1 & 0 & 0 & 0 & 1 \\ 1 & 0 & 1 & 0 & 0 & 1 \\ 0 & 1 & 0 & 1 & 1 & 0 \end{bmatrix}$$

9. There are 93 walks of length at most 5 between u and v.

11. D

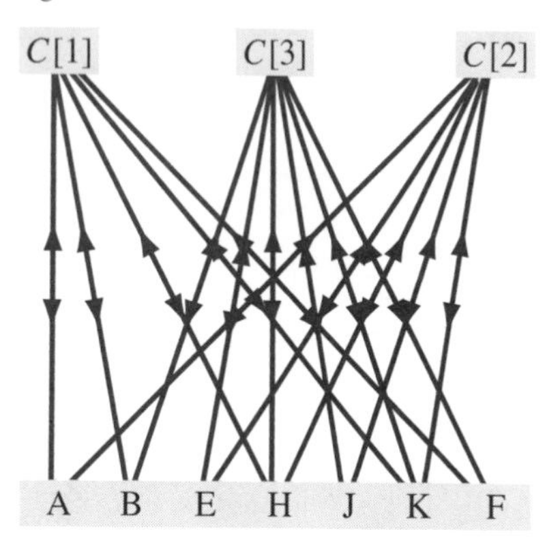

13.
$$\begin{bmatrix} 0 & 0 & 0 & 0 & 0 & 0 & 0 & 0 & 1 & 1 & 0 \\ 0 & 0 & 0 & 0 & 0 & 0 & 0 & 0 & 1 & 0 & 1 \\ 0 & 0 & 0 & 0 & 0 & 0 & 0 & 0 & 0 & 0 & 0 \\ 0 & 0 & 0 & 0 & 0 & 0 & 0 & 0 & 0 & 1 & 1 \\ 0 & 0 & 0 & 0 & 0 & 0 & 0 & 0 & 1 & 0 & 1 \\ 0 & 0 & 0 & 0 & 0 & 0 & 0 & 0 & 1 & 1 & 1 \\ 0 & 0 & 0 & 0 & 0 & 0 & 0 & 0 & 0 & 1 & 1 \\ 0 & 0 & 0 & 0 & 0 & 0 & 0 & 0 & 1 & 1 & 1 \\ 1 & 1 & 0 & 0 & 1 & 1 & 0 & 1 & 0 & 0 & 0 \\ 1 & 0 & 0 & 1 & 0 & 1 & 1 & 1 & 0 & 0 & 0 \\ 0 & 1 & 0 & 1 & 1 & 1 & 1 & 1 & 0 & 0 & 0 \end{bmatrix}$$
with the labels A, B, D, E, F, H, J, K, C_1, C_2, C_3

15. No cycles of odd length.

21. You can determine the diameter of a graph from its adjacency matrix, A by looking to see what for what value of k is A^k has all positive entries.

Section 3.8

1. $A = \frac{28s}{15}, \quad B = \frac{2s}{3}, \quad C = 15s$

3. $A = 2s,\ B = s,\ C = s$, for any $s > 0$

5. $A = 12.50,\ B = 18.83,\ C = 9.45$

7. A: 55.46; B: 45.21; C: 21.28

Self-Test

1. $\begin{bmatrix} 2 & 0 \\ -3 & 5 \end{bmatrix}$

3. $\begin{bmatrix} 2 & 3 \\ 0 & -1 \\ 1 & 0 \end{bmatrix}$

5. (1, 3)

7. ***(a)*** False ***(b)*** True

9. $(-8, 3, 1)$

Review Exercises

1. $\begin{bmatrix} 5 & -3 & -1 \\ -3 & -7 & 3 \\ -6 & 0 & 9 \end{bmatrix}$

3. $\begin{bmatrix} 1 & 1 & 0 \\ 7 & 3 & -2 \\ 6 & 0 & -3 \end{bmatrix}$

5. $\begin{bmatrix} 1 & 0 & -\frac{1}{3} \\ 0 & 1 & -1 \\ 0 & 0 & -\frac{1}{3} \end{bmatrix}$

7. $\begin{bmatrix} -1 & 1 & 1 \\ 1 & 3 & 1 \\ 8 & 10 & 1 \end{bmatrix}$, but the inverse does not exist.

9. 3

11. $\begin{bmatrix} -1 & 0 & -4 \\ 0 & -1 & -12 \\ 0 & 0 & -9 \end{bmatrix}$

13. A is not invertible.

15. A is not invertible.

17. $(1, 0, -1)$

19. Unique solution, **0**

21. Infinitely many solutions

25. $(2t - 1, t)$, where t is an arbitrary real number.

27. $(2s - t, s, t)$, where s and t are arbitrary real numbers.

29. $(1, 0, -1)$

31. **i.** $E_{12}(2) = \begin{bmatrix} 1 & 2 & 0 \\ 0 & 1 & 0 \\ 0 & 0 & 1 \end{bmatrix}$ **ii.** $P_{13} = \begin{bmatrix} 0 & 0 & 1 \\ 0 & 1 & 0 \\ 1 & 0 & 0 \end{bmatrix}$ **iii.** $E_{33}(2) = \begin{bmatrix} 1 & 0 & 0 \\ 0 & 1 & 0 \\ 0 & 0 & 2 \end{bmatrix}$

Review Exercises—Challenge Problems

35. The matrices that commute with A are of the form $\begin{bmatrix} -c + d & -c \\ c & d \end{bmatrix}$

Chapter 4

Section 4.1

1. The 3, 1 minor of A is -7 and the cofactor is -7. The 3, 2 minor of A is -1 and the cofactor is 1.

3. 3

5. -5

7. 15

9. 0

11. 120

13. a^3

15. $k^2(ad - bc)$

17. $t^2 - 3t + 3$

19. -8

21. 0

23. $aei - afh - bdi + bfg + cdh - ceg$

27. $x = 3$ and $x = -1$

29. $a_{11}a_{22}a_{33} + a_{12}a_{23}a_{31} + a_{13}a_{21}a_{32} - a_{13}a_{22}a_{31} - a_{11}a_{23}a_{32} - a_{12}a_{21}a_{33}$

35. 6

Section 4.2

1. 2

3. -24

5. $-\frac{1}{720}$

7. $x(-x - 1)(2x - 1)$

9. $\det(A) = -(\lambda - 1)(\lambda - 2)(\lambda + 1)(\lambda + 2)$

11. In Exercise 7: A is not invertible when x is $-1, 0,$ or $\frac{1}{2}$.

Exercise 8: A is not invertible when t is 3 or a root of $(t^2 - t - 8)$.
Exercise 9: A is not invertible when $\lambda \epsilon \{-2, -1, 1, 2\}$.

13. Has a unique solution.

15. Does not have a unique solution.

17. $\det(A) = 4 = \det(A^T)$, $\det(2A) = 32$

19. 0

21. If $k = 2m + 1$, where m is a positive integer then $\det(A') = (-1)^{2m+1} \det(A)$.

If $k = 2m$, where m is a positive integer then $\det(A') = (-1)^{2m} \det(A)$.

23. $\det(AB) = \det(BA) = \det(A)\det(B)$

27.
$$\begin{bmatrix} 1 & 0 & 0 & \cdots & 0 \\ a_{21} & 1 & 0 & \cdots & 0 \\ \vdots & \vdots & \vdots & \vdots & \vdots \\ a_{n1} & 0 & 0 & \cdots & 1 \end{bmatrix}\begin{bmatrix} 1 & a_{12} & a_{13} & \cdots & a_{1n} \\ 0 & a_{22} - a_{21}a_{12} & a_{23} - a_{21}a_{13} & \cdots & a_{2n} - a_{21}a_{1n} \\ \vdots & \vdots & \vdots & \vdots & \vdots \\ 0 & a_{n2} - a_{n1}a_{12} & a_{n3} - a_{n1}a_{13} & \cdots & a_{nm} - a_{n1}a_{1n} \end{bmatrix}$$

$$\begin{bmatrix} 1 & a_{12} & a_{13} & \cdots & a_{1n} \\ a_{21} & a_{22} & a_{23} & \cdots & a_{2n} \\ \vdots & \vdots & \vdots & \vdots & \vdots \\ a_{n1} & a_{n2} & a_{n3} & \cdots & a_{nm} \end{bmatrix}$$

31. $k_1x_1 - k_2x_2 + k_3x_3 - \ldots + (-1)^{1+n} k_nx_n = -(-1)^{n+2}k_{n+1}$

33. *(a)* $A^{-1} = C = B(AB)^{-1}$

(b) AB cannot be invertible

Computational Exercises

35. Approximately 0

37. $-(k + 2)(k - 1)^2$

39. *(a)* A^{-1} fails to exist exactly when $k = -2$ or $k = 1$.
(b) B^{-1} fails to exist exactly when x, y, and z are not all distinct.

41. $a = -997, b = -1122, c = 747, d = 17108, e = -1596, f = -59311$

Section 4.3

1. $x = 1, y - 1$

3. $\det A = 0$, Cramer's rule cannot be used to solve the given system.

5. $x = \frac{3}{5}, y = \frac{3}{2}, z = \frac{17}{10}$

7. $x = 2, y = 1, z = -1$

9. The adjoint of B is: $\begin{bmatrix} 4 & -2 \\ 3 & 1 \end{bmatrix}$, $B^{-1} = \begin{bmatrix} \frac{2}{5} & \frac{-1}{5} \\ \frac{3}{10} & \frac{1}{10} \end{bmatrix}$

11. The adjoint of A is: $\begin{bmatrix} 2 & 1 & 2 \\ 0 & 1 & 2 \\ 0 & 0 & 2 \end{bmatrix}$, $A^{-1} = \frac{1}{2}\begin{bmatrix} 2 & 1 & 2 \\ 0 & 1 & 2 \\ 0 & 0 & 2 \end{bmatrix}$.

13. The adjoint of A is: $\begin{bmatrix} -27 & 9 & 9 \\ 18 & -6 & -6 \\ 9 & -3 & -3 \end{bmatrix}$, the inverse doesn't exist.

15.
$$\frac{1}{\det A}\begin{bmatrix} a_{22}a_{33} - a_{23}a_{32} & -a_{12}a_{33} + a_{13}a_{32} & a_{12}a_{23} - a_{13}a_{22} \\ -a_{21}a_{33} + a_{23}a_{31} & a_{11}a_{33} - a_{13}a_{31} & -a_{11}a_{23} + a_{13}a_{21} \\ a_{21}a_{32} - a_{22}a_{31} & -a_{11}a_{32} + a_{12}a_{31} & a_{11}a_{22} - a_{12}a_{21} \end{bmatrix}$$

21. The unique solution is$(a_0', a_1', \ldots, a_n')$ and the unique polynomial of degree n passing through the given points is $P(x) = a_0' + a_1'x + a_2'x^2 + \cdots + a_n'x^n$.

Computational Exercises

23. $\text{Adj}(A) \approx \begin{bmatrix} 5.60 & -9.63 & 31.74 & 9.72 \\ 6.10 & 14.18 & -10.69 & -8.40 \\ -2.32 & 4.08 & 10.89 & 2.61 \\ -2.21 & 2.53 & -10.19 & 4.65 \end{bmatrix}$

25. $\det A = a_{33}a_{11}a_{22} - a_{33}a_{21}a_{12} - a_{31}a_{13}a_{22} - a_{32}a_{11}a_{23} + a_{32}a_{21}a_{13} + a_{31}a_{12}a_{23}$

$$\text{Adj}(A) = \begin{bmatrix} a_{22}a_{33} - a_{23}a_{32} & -a_{12}a_{33} + a_{13}a_{32} & a_{12}a_{23} - a_{13}a_{22} \\ -a_{21}a_{33} + a_{23}a_{31} & a_{11}a_{33} - a_{13}a_{31} & -a_{11}a_{23} + a_{13}a_{21} \\ a_{21}a_{32} - a_{22}a_{31} & -a_{11}a_{32} + a_{12}a_{31} & a_{11}a_{22} - a_{12}a_{21} \end{bmatrix}$$

Section 4.4

1. $\lambda = \pm 3$ *3.* $\lambda = 1$ *5.* $\lambda = 0, \lambda = \pm\sqrt{2}$ *7.* $\lambda = 0, \lambda = 3$

9. $\lambda = 1, \lambda = -2$ *11.* $\lambda = 0, \lambda = 4$ *13.* $\lambda = 1, \lambda = 3$, and $\lambda = 4$

15. $\lambda = 1, \lambda = 3, \lambda = 6$, and $\lambda = 10$ *19.* A and A^T have the same eigenvalues.

21. $\lambda = \dfrac{(a+d) \pm \sqrt{(a-d)^2 + 4bc}}{2}$

If $(a-d)^2 + 4bc > 0$, then there are two distinct eigenvalues λ_1 and λ_2.

If $(a-d)^2 + 4bc = 0$, then there is just one eigenvalue, $\lambda = \dfrac{a+d}{2}$.

If $(a-d)^2 + 4bc < 0$, then $\sqrt{(a-d)^2 + 4bc}$ does not exist as a real number and there are no eigenvalues.

Section 4.5

1. $t_{32} = 0.4$ *3.* $T = \begin{bmatrix} 0.2 & 0.1 & 0.2 \\ 0.5 & 0.8 & 0.4 \\ 0.3 & 0.1 & 0.4 \end{bmatrix}$ *5.* $x = \left(\frac{1}{3}, \frac{1}{3}, \frac{1}{3}\right)$ *7.* $\begin{bmatrix} 0.17 \\ 0.57 \\ 0.27 \end{bmatrix}$

9. $\begin{bmatrix} 0.15 \\ 0.62 \\ 0.23 \end{bmatrix}$ *11.* $T = \begin{bmatrix} 0 & \frac{1}{2} & \frac{1}{3} \\ \frac{2}{3} & 0 & \frac{1}{3} \\ \frac{1}{3} & \frac{1}{2} & \frac{1}{3} \end{bmatrix}$ *13.* $\left(\frac{5}{18}, \frac{1}{3}, \frac{7}{18}\right)$ *15.* $\frac{1}{81}$

17. $\left(\dfrac{1}{s+s+t}\right)(s, s, t) = \left(\dfrac{1}{2s+t}\right)(s, s, t)$; where $s, t \geq 0$, and $s + t > 0$.

Computational Exercises

21. $(-0.12, -0.94, -0.25)$ *23.* Most people would describe my personality as happy ☺.

Section 4.6

1. Pageranks: BGDEFACH. Make sense since H is the lowest rank. F has the most in links so pages that F link to should be ranked high, and B is the only website which F links to and B is the highest ranked.

3. $G_{0.5} = \begin{bmatrix} \frac{1}{6} & \frac{1}{6} & \frac{1}{3} \\ \frac{5}{12} & \frac{1}{6} & \frac{1}{3} \\ \frac{5}{12} & \frac{2}{3} & \frac{1}{3} \end{bmatrix}$, [0.2424 0.3030 0.4545] Pagerank: C, B, A

7.
[0.0851 0.2179 0.0674 0.1205 0.0940 0.1648 0.2179 0.0320]
[0.1241 0.2184 0.0933 0.1219 0.1114 0.1559 0.1431 0.0315]
[0.0925 0.2302 0.0936 0.1332 0.1199 0.1323 0.1662 0.0317]
[0.1035 0.2048 0.0981 0.1379 0.1178 0.1366 0.1680 0.0329]
[0.1048 0.2133 0.0914 0.1331 0.1134 0.1408 0.1695 0.0334]

Self-Test

1. 5

3. 36

5. $-\frac{3}{7}$

7. $t = \left(\frac{2 \pm \sqrt{6}}{2}\right)$

9. $\begin{bmatrix} 3 & 1 & 2 \\ 2 & 1 & 5 \\ 1 & 2 & 1 \end{bmatrix} \rightarrow \begin{bmatrix} 2 & -1 & 0 \\ 1 & 7 & -5 \\ -3 & -11 & 10 \end{bmatrix}$

11. *a)* The characteristic equation for A is $\lambda^2 - 2\lambda$.
b) The eigenvalues are 0 and 2.
c) The eigenvectors are $(-1, 1)$ and $(1, 1)$.

Review Exercises

1. -2

3. 6

5. $xy(x-1)(y-1)(y-x)$

7. 1

9. Exercises 1, 2, and 3 are invertible, but not 4.

11. $\det(A^T) = 2$ and $\det(A^T)^{-1} = \frac{1}{2}$

13. $A^{-1} = \left(-\frac{1}{2}\right)\text{Adj}(A)$, where $\text{Adj}(A) = \begin{bmatrix} -1 & -1 & 1 \\ -1 & 1 & -1 \\ 1 & -1 & -1 \end{bmatrix}$

15. $A^{-1} = \left(\frac{1}{6}\right)\text{Adj}(A)$, where $\text{Adj}(A) = \begin{bmatrix} 6 & 0 & 0 \\ -3 & 3 & 0 \\ 0 & -2 & 2 \end{bmatrix}$

17. $-\frac{1}{3}$

19. $x_1 = \frac{1}{2}, x_2 = \frac{1}{2}, x_3 = -\frac{1}{2}$

21. Cramer's rule does not apply.

23. $\lambda = -1, 3$

25. $\lambda = -2, 2, 4$

27. The eigenvector for $\lambda = -1$ is $(-s, s) = s(-1, 1)$.
The eigenvector for $\lambda = 3$ is $(s, s) = s(1, 1)$.

Review Exercises—Challenge Problems

29. A is invertible

31. $\det\begin{bmatrix} 1 & x & y \\ 1 & a & b \\ 1 & c & d \end{bmatrix} = 0$ goes through the points (a, b) and (c, d).

Chapter 5

Section 5.1

1. $\mathbf{w}_1 = 3\mathbf{u} + 2\mathbf{v}$

3. Only when $c = -2$ can $\mathbf{w}$ be written as a linear combination of $\mathbf{u}$ and $\mathbf{v}$.

5. $(0, 1, 2) = \frac{1}{2}(1, 2, 3) + \frac{1}{2}(-1, 0, 1)$

7. Linearly independent

9. Linearly independent

11. Linearly dependent

13. Linearly independent

15. Linearly dependent

17. Linearly independent

19. Linearly dependent

21. Linearly independent

23. Linearly dependent

25. Linearly dependent

27. Linearly dependent

29. Linearly independent

33. If $\mathbf{u} \neq \mathbf{0}$, $\mathbf{v} \neq \mathbf{0}$, and $c = d = 0$ in $c\mathbf{u} + d\mathbf{v} = 0$ then $\{\mathbf{u}, \mathbf{v}\}$ is linearly independent.

Computational Exercises

35. Linearly independent

37. Linearly independent

39. In summary, the given vectors are linearly independent if and only if a and b are not both 0.

41. $(1, 2, 3, 4)$ cannot be expressed as a linear combination of the given vectors.

Section 5.2

1. Let **S** be a set of all scalar multiples of the vector (1, 2, 3) in $\mathbf{R}^3$, then **S** is a subspace of $\mathbf{R}^3$.

3. Let **S** be a set of all linear combinations of the vectors (1, 0, 1, 0) and (0, 1, 0, 1) in $\mathbf{R}^4$, then **S** is a subspace of $\mathbf{R}^4$.

5. Let **S** be the set of all vectors of the form $(a, b, a - b, a + b)$ in $\mathbf{R}^4$, then **S** is a subspace of $\mathbf{R}^4$.

7. Let $\mathbf{S} = \{(1, 1, 1, 1)\} \subseteq \mathbf{R}^4$, then **S** does not form a subspace of $\mathbf{R}^4$.

9. Let T be the set of all m-vectors except the $m - 0$ vector in $\mathbf{R}^m$, then T is not a subspace of $\mathbf{R}^m$.

11. Let **S** be the set of all 3-vectors the sum of whose components is 1, then **S** does not form a subspace of $\mathbf{R}^3$.

13. Let $\mathbf{S} = \{(1, 2, 3), (-1, 0, 1), (0, 1, 2)\}$ then **S** does not span $\mathbf{R}^3$.

15. The set $\mathbf{S} = \{(1, 2, -1), (1, 0, 1)\}$ does not span $\mathbf{R}^3$.

17. Therefore, $\{\mathbf{e}_2, \mathbf{e}_3\}$ spans the subspace of $\mathbf{R}^4$ consisting of all vectors whose first and last components are zeros.

19. The set of all scalar multiples of (1, 2, 3) is, by definition, the span of {(1, 2, 3)}.

21. {(1, 0, 1, 0), (0, 1, 0, 1)} spans **S**.

23. The given subspace of $\mathbf{R}^5$ is spanned by the set $\{(1, 0, 1, 0, 1), (0, 1, 0, -1, 1)\}$.

25. The solution space of $A\mathbf{x} = \mathbf{0}$ consists of all vectors of the form: $\mathbf{x} = s(0, 1, 1)$.

Section 5.3

1. **S** is not a basis for $\mathbf{R}^3$.

3. $\{\mathbf{v}_1, \mathbf{v}_2\}$ is a basis for **S**.

5. $T = \{(1, 1, 0, 0), (0, 0, 1, 1)\}$ is not a basis for **S**.

7. $\{\mathbf{v}_1, \mathbf{v}_2\}$ is a basis for **S**.

9. $\{(1, -1, 1), (2, 0, 1)\}$ forms a basis for the subspace generated by $\{(1, -1, 1), (2, 0, 1), (1, 1, 0)\}$.

11. $\{(-1, 1, 2, -1), (1, 0, -1, 1)\}$ forms a basis for the subspace generated by $\{(-1, 1, 2, -1), (1, 0, -1, 1), (-1, 2, 3, -1), (1, 1, 0, 1)\}$.

13. $\{(1, 1, 0), (1, -1, 1), (1, 0, 0)\}$ forms a basis for $\mathbf{R}^3$.

15. $\{(1, 0, 0, 1), (0, 1, 1, 0), (1, 1, 1, 2), (0, 0, 1, 0)\}$ forms a basis for $\mathbf{R}^4$.

17. Linearly dependent

19. Linearly independent

21. $\{(1, 1, 1, 0), (-2, -1, 0, 1)\}$ is a basis for the solution space.

23. $\mathbf{v} = \{(1, 2, 0), (0, 0, 1)\}$ is a basis for the subspace of $\mathbf{R}^4$ of all vectors of the form $(x, 2x, z)$, so the dimension of the subspace of all vectors of the form $(x, 2x, z)$ is 2.

25. $\mathbf{v} = \{(1, 0, 1, 0), (0, -1, -2, 3)\}$ is a basis for the subspace of $\mathbf{R}^4$ of all vectors of the form $(x, -y, x - 2y, 3y)$ so the dimension of the subspace of all vectors of the form $(x, -y, x - 2y, 3y)$ is 2.

27. $\{(1, -1, 3, 1), (-2, 0, -2, 0)\}$ is a basis for the subspace **S** of $\mathbf{R}^4$ so spanned. Since it has a basis that contains two vectors, **S** has dimension 2

29. {**v**} is a basis for of **S** and since {**v**} has only one element, then the dimension of **S** is 1.

33. ***(a)*** Two degrees of freedom and the given subspace of $\mathbf{R}^4$ has dimension 2.
(b) Two degrees of freedom and the given subspace of $\mathbf{R}^4$ has dimension 2.

Computational Exercises

35. The given set of vectors is a basis for $\mathbf{R}^3$ whenever $x \neq 2$ and $x \neq 3$.

37. The set of vectors is a basis for $\mathbf{R}^4$ except when (x, y) lies on the hyperbola $-x^2 - 2xy + 5y^2 - 3x - 2y + 3 = 0$.

Section 5.4

1. 2

3. row rank = column rank = 1 for the given matrix A.

5. 3

7. (1, 0, 0) and (0, 1, 0) form a basis for the column space of A.

9. A basis of B is $\{(1, -1)\}$.

11. A basis for the column space of A is $\{(1, 0, 1, 2), (-1, 1, 1, -1), (3, 1, 0, 1)\}$.

13. $\{(1, 2, 0), (0, 0, 1)\}$ forms a basis for the row space of A.

15. The vector $(1, 0, -1)$ forms a basis for the row space of A.

17. $\{(1, -1, 3), (0, 1, 1), (1, 1, 0)\}$ forms a basis for the row space of A.

19. Linearly independent

23. The rows of A that correspond to a distinct set of elementary rows of B form a basis for the row space of A.

25. $\{(1, 2, 1), (2, 1, -1)\}$ forms a basis for the row space of A.

27. The vector $(-1, 2)$ forms a basis for the row space of A.

29. The set $\{(1, -1, 1, 0), (1, 1, 0, 0), (1, 0, 0, 1)\}$ forms a basis for the row space of B.

Section 5.5

1. Let $B_1 = \{(1, 1), (1, -1)\}$, then B_1 is a basis for $\mathbf{R}^2$

3. $(-7, -6.5, 2.7, 3) = a(1, 1, 1, 1) + b(1, 1, -1, -1) + c(1, -1, 0, 0) + d(0, 0, 1, -1)$

5. $x_1 = -4 - 2 - 1 = 17$
$x_2 = -4 - 2 + 1 = -5$
$x_3 = -4 + 2 - 4 = -6$
$x_4 = -4 + 2 + 4 = 2$

7. 10, 10, 10, 10, 10, 10, 4, 4, 5, 5, −1, −1, −1, −1, 5, 5, 5, 5, 3, 3

Self-Test

1. $(2, 1, 1, 1) = (2)(1, 0, 1, 0) + (1)(0, 1 -1, 1)$

3. $\mathbf{S}$ is a subsequence of $\mathbf{R}^4$.

5. $(a + b, a - b, c, 0) = a(1, 1, 0, 0) + b(1, -1, 0, 0) + c(0, 0, 1, 0)$

7. $(-s-2t, -s-t, s, t)$, when $s = 1, t = 0: (-1, -1, 1, 0)$
$s = 0, t = 1: (-2, -1, 0, 1)$

9. *(a)* $\{(1, 0, 1), (2, 1, 1)\}$, 2
(b) $\{(1, 2, 3, 4), (0, 1, 2, 3)\}$, 2

Review Exercises

1. Linearly independent

3. The given set is linearly dependent $(3, -1, 0, 1) = 2(1, 0, 1, 0) + (-1)(-1, 1, 2, -1)$

5. The given set of vectors is a subspace of $\mathbf{R}^4$.

7. The given set is not a subspace of $\mathbf{R}^3$.

9. $(-1, 1, 1)$ spans $\mathbf{S}$.

11. The vectors are linearly independent, and there are three of them so they form a basis for $\mathbf{R}^3$.

13. The given set is a basis for $\mathbf{S}$.

15. $\{(1, 1, 0, 0), (0, 1, 1, 0), (0, 1, 0, 1)\}$ is a basis for $\mathbf{S}$, and $\dim \mathbf{S} = 3$.

17. $\{(-2, 1, 0), (-3, 0, 1)\}$ is a basis of $\mathbf{S}$ and the dimension of $\mathbf{S}$ is 2.

19. $\{(1, 1, 1, 1, 1), (0, 1, 1, 1, 1), (0, 0, 1, 1, 1)\}$ is linearly independent, it is a basis for the row space, and the dimension of this space is 3.

21. $(1, -1, 2, 0)$ and $(-1, -1, 4, -2)$, form a basis for the subspace, $\mathbf{S}$, generated by the given vectors and the dimension of $\mathbf{S}$ is 2.

23. $\{(-1, 2)\}$ is a basis for the column space. $\{(-1, 1)\}$ is a basis for the row space. Rank of A is 1.

25. A basis for the column space of A is $\{(1, -2, 0, -1), (0, 01, 2)\}$. A basis for the row space of A is $\{(1, 0, 1), (-2, 0, -2)\}$. Rank of A is 2.

27. A basis for the column space of A is $\{(1, -1, 1, 1), (0, 1, 2, 0), (3, -2, 5, 3)\}$. A basis for the row space of A is $\{(1, -1, 1, 1), (0, 1, 2, 0), (0, 0, 1, 0)\}$.

Review Exercises—Challenge Problems

29. $\text{rank}(BA) \le \text{rank}(B)$ and $\text{rank}(BA) \le \text{rank}(A)$

31. $\{(0, -1, 1, 0, 0), (1, -1, 0, 1, 0), (-1, 1, 0, 0, 1)\}$, the dimension is 3.

33. *(a)* Either the rows or columns of A are linearly dependent.
(b) The rows and columns are both linearly independent.

Chapter 6

Section 6.1

1. *(a)* $A + B$ is in $\mathbf{V}$
(b) cA is in $\mathbf{V}$
(c) $A + B = B + A$
(d) $(A + B) + C = A + (B + C)$
(e) $A + \mathbf{0} = A$
(f) $A + (-A) = \mathbf{0}$
(g) $(cd)A = c(dA)$
(h) $(c + d)A = cA + dA$
(i) $c(A + B) = cA + cB$
(j) $1 \cdot A = A$

Therefore the given set is a vector space.

3. Not a vector space

5. Not a vector space

7. Not a vector space

9. Is a vector space

11. Is a vector space

13. $\mathbf{M}$ is a subspace of $\mathbf{M}^{2,2}$.

15. $\mathbf{M}$ is not a subspace of $\mathbf{M}^{2,2}$.

17. $\mathbf{N}$, the set of all diagonal matrices of order n, is a subspace of $\mathbf{M}^{n,n}$.

19. $\mathbf{S}$ is a subspace of $\mathbf{C}(-\infty, \infty)$.

21. $c_1v_1 + c_2v_2 + \cdots + c_nv_n$ is in $\mathbf{V}$.

Section 6.2

1. Linearly independent

3. Linearly dependent

5. Linearly independent

7. Linearly dependent

9. Linearly independent

11. Linearly dependent

13. $\mathbf{S}$ spans $\mathbf{P}_2$

15. $\{1, x - 1, x + 1\}$ spans $\mathbf{P}_1$

17. $\mathbf{M}$ spans $\mathbf{M}^2$

19. $\{1, \sin^2 x, \cos^2 x\}$ does not span $\mathbf{C}(-\infty, \infty)$.

21. $\mathbf{M}$ does not span the vector space of all 3×3 diagonal matrices.

23. $\mathbf{S}$ spans $\mathbf{V}$

25. $\mathbf{S}$ is a basis for $\mathbf{P}_2$.

27. Not a basis

29. $\mathbf{M}$ is a basis for $\mathbf{M}^{2,2}$.

31. Not a basis

33. Not basis

35. Not a basis

37. 12

39. 4

41. 1

43. A basis

45. Linearly dependent

Section 6.3

1. $(2, -1, 0, 3)$

3. $\left(-\frac{3}{2}, -\frac{7}{2}\right)$

5. $(4, 5, -\frac{9}{2})$

7. $\mathbf{S}_2$ is linearly independent in $\mathbf{P}_3$.

9. $\begin{bmatrix} 1 & 0 \\ 1 & 2 \end{bmatrix} = (1)\begin{bmatrix} 1 & 2 \\ -1 & 0 \end{bmatrix} + (2)\begin{bmatrix} 0 & -1 \\ 1 & 1 \end{bmatrix}$

11.
$f_1(x) = 2 + x - 3\sin x + \cos x \longleftrightarrow (2, 1, -3, 1)$
$f_2(x) = x + \sin x - 3\cos x \longleftrightarrow (0, 1, 1, -3)$
$f_3(x) = 1 - 2x + 3\sin x + \cos x \longleftrightarrow (1, -2, 3, 1)$
$f_4(x) = 2 - x - \sin x - \cos x \longleftrightarrow (2, -1, -1, -1)$
$f_5(x) = 2 + x - 3\cos x \longleftrightarrow (2, 1, 0, -3)$

$$f_5(x) = \frac{17}{28}f_1(x) + \frac{17}{14}f_2(x) + \frac{2}{7}f_3(x) + \frac{1}{4}f_4(x)$$

13. $x^2 - x + 1 = 1 - x + x^2 \longleftrightarrow (1, -1, 1)$

$x - 1 = -1 + x + 0x^2 \longleftrightarrow (-1, 1, 0)$

$x + 1 = 1 + x + 0x^2 \longleftrightarrow (1, 1, 0)$

There is no vector to be added to the given set to form an extended basis set.

15. $\begin{bmatrix} 1 & -1 \\ 1 & 0 \end{bmatrix} \longleftrightarrow (1, -1, 1, 0)$

$\begin{bmatrix} 1 & 1 \\ 0 & 0 \end{bmatrix} \longleftrightarrow (1, 1, 0, 0)$

$\begin{bmatrix} 0 & 0 \\ 1 & 1 \end{bmatrix} \longleftrightarrow (0, 0, 1, 1)$

$\begin{bmatrix} 1 & 0 \\ 0 & 0 \end{bmatrix}$, extends B to the basis $\left\{\begin{bmatrix} 1 & -1 \\ 1 & 0 \end{bmatrix}, \begin{bmatrix} 1 & 1 \\ 0 & 0 \end{bmatrix}, \begin{bmatrix} 0 & 0 \\ 1 & 1 \end{bmatrix}, \begin{bmatrix} 1 & 0 \\ 0 & 0 \end{bmatrix}\right\}$ of $\mathbf{M}^{2,2}$.

17. $B = \left\{\begin{bmatrix} 1 & 0 & -1 \\ 1 & -1 & 1 \end{bmatrix}, \begin{bmatrix} 0 & 1 & 1 \\ 2 & 1 & -1 \end{bmatrix}, \begin{bmatrix} 1 & -1 & 1 \\ 1 & -1 & 1 \end{bmatrix}, \begin{bmatrix} 0 & 1 & 0 \\ 0 & 1 & 0 \end{bmatrix}, \begin{bmatrix} 0 & 0 & 1 \\ 0 & 0 & 0 \end{bmatrix}, \begin{bmatrix} 0 & 0 & 0 \\ 1 & 0 & 0 \end{bmatrix}\right\}$ is a basis for $\mathbf{M}^{2,3}$.

19. Let $\mathbf{S} = \{x + 1, x - 1, x^2 + 1, x^2 - 1, x^2 + x, x^2 - x\} \subseteq \mathbf{P}_2$.

$\{x + 1, x - 1, 1 + x^2\}$ is an independent set that has the same span as $\mathbf{S}$.

21. $\left\{\begin{bmatrix} 2 & 1 \\ 0 & 1 \end{bmatrix}, \begin{bmatrix} -1 & 1 \\ 0 & 0 \end{bmatrix}, \begin{bmatrix} 1 & 2 \\ 0 & 2 \end{bmatrix}\right\}$ is linearly independent.

Computational Exercises

23. $f_4(x) = 0 \cdot f_1(x) + 3 \cdot f_2(x) + 2 \cdot f_3(x)$

25. $\mathbf{S}$ is linearly dependent if and only if: $a^2 + ab + 4a - 3b + 5 = 0$

Section 6.4

1. $(\mathbf{p}, \mathbf{q}) = a_0b_0 + a_1b_1$ forms an inner product space.

3. (A, B) is not an inner product on $\mathbf{M}^{2,2}$.

5. $(\mathbf{f}, \mathbf{g})$ is not an inner product.

7. $\{\mathbf{u}_1, \mathbf{u}_2, \mathbf{u}_3\}$ is orthonormal.

9. $\{(1, 2, 2), (1, -2, 2)\}$ is not orthonormal.

11. $\{\mathbf{v}\}$ is orthonormal

13. Let $\mathbf{Q} = \{1, x, x^2\}$ in $\mathbf{P}_2$, $\mathbf{Q}$ is not orthogonal, thus it is not orthonormal.

15. $\{A, B, C\}$ is orthogonal and orthonormal.

17. $\mathbf{x} = (1, 2, 3) = (1)\left(\frac{1}{3}, -\frac{2}{3}, \frac{2}{3}\right) + (2)\left(-\frac{2}{3}, \frac{1}{3}, \frac{2}{3}\right) + (3)\left(\frac{2}{3}, \frac{2}{3}, \frac{1}{3}\right)$

19. $\left\{\left(\frac{4}{3\sqrt{5}}, \frac{5}{3\sqrt{5}}, 0, -\frac{2}{3\sqrt{5}}\right), \left(-\frac{2}{3}, \frac{2}{3}, 0, \frac{1}{3}\right), \left(\frac{1}{\sqrt{5}}, 0, 0, \frac{2}{\sqrt{5}}\right)\right\}$

21. $\left\{\left(\frac{\sqrt{2}}{2}, -\frac{\sqrt{2}}{2}, 0\right), \left(\frac{\sqrt{6}}{6}, \frac{\sqrt{6}}{6}, -\frac{\sqrt{6}}{3}\right)\right\}$

23. $\left\{\left(-\frac{1}{\sqrt{2}}, \frac{1}{\sqrt{2}}, 0, 0\right), (0, 0, 1, 0), (0, 0, 0, 1)\right\}$ is an orthonormal basis.

25. $\{(1, 0, 0, 0), (0, 1, 0, 0), (0, 0, 1, 0), (0, 0, 0, 1)\}$

27. $\left\{\frac{1}{\sqrt{2}}, \frac{3}{\sqrt{6}}x\right\}, \frac{15}{2\sqrt{10}}\left(x^2 - \frac{1}{3}\right), \left[\frac{35}{2\sqrt{14}}\left(x^3 - \frac{3}{5}x\right)\right]$

29. $\tau' = \left\{1, \frac{\sqrt{2}(e^x - e + 1)}{\sqrt{4e - e^2 - 3}}\right\}$

31. $\{\mathbf{e}_1, \mathbf{e}_2, \dots, \mathbf{e}_m\}$ is an orthonormal set.

33. $\{\mathbf{w}_1, \mathbf{w}_2, \dots \mathbf{w}_n\}$

Computational Exercises

35. $\left\{\left(\frac{\sqrt{2}}{2}, 0, -\frac{\sqrt{2}}{2}\right), \left(\frac{\sqrt{2}}{2}, 0, \frac{\sqrt{2}}{2}\right)\right\}$

37. $\left\{\left(\frac{\sqrt{7}}{7}\right)(-1, -1, 2, 1), \left(\frac{\sqrt{3}}{3}\right)(1, 0, 1, -1), \left(\frac{\sqrt{35}}{35}\right)(3, 3, 1, 4)\right\}$

39. $\left\{\left(\frac{\sqrt{3}}{3}\right)(1, 1, 1, 0), \left(\frac{\sqrt{33}}{33}\right)(2, -4, 2, -3), \left(\frac{\sqrt{22}}{22}\right)(1, -2, 1, 4), \left(\frac{\sqrt{2}}{2}\right)(1, 0, 1, 0)\right\}$

Section 6.5

1. *(a)*

+	0	1
0	0	1
1	1	0

*	0	1
0	0	0
1	0	1

+	0	1	2
0	0	1	2
1	1	2	0
2	2	0	1

*	0	1	2
0	0	0	0
1	0	1	2
2	0	2	1

+	0	1	2	3	4
0	0	1	2	3	4
1	1	2	3	4	0
2	2	3	4	0	1
3	3	4	0	1	2
4	4	0	1	2	3

*	0	1	2	3	4
0	0	0	0	0	0
1	0	1	2	3	4
2	0	2	4	1	3
3	0	3	1	4	2
4	0	4	3	2	1

3.

+	0	1	x	$x+1$
0	0	1	x	$x+1$
1	1	0	$x+1$	x
x	x	$x+1$	0	1
$x+1$	$x+1$	x	1	0

*	0	1	x	$x+1$
0	0	0	0	0
1	0	1	x	$x+1$
x	0	x	$x+1$	1
$x+1$	0	$x+1$	1	x

To verify that **F** is a field, one must check that each property holds.

5. $2 + 4i$

7. $3 - 2i$

9. $(1 + i, 0, 3i)$

11. -2

13. 2

15. $B = \{(i, 0, -1), (1, 1, 1), (0, -i, i)\}$ is a basis for $\mathbb{C}^3$.

17. $B = \{(i, 1, 0), (0, 0, 1)\}$ is not a basis in $\mathbf{C}^3$.

19. Show **S** is closed under addition and scalar multiplication. A basis is $\{(1, 0, 0, -2), (0, 1, 0, -1), (0, 0, 1, i)\}$, so the dimension is 3.

21. Not linearly independent

23. $\{(1, 1 + i), (-1, 2i), (3 + i, 2), (-1 + i, 5 - 3i)\}$

25. $\left\{\left(\frac{i}{\sqrt{2}}, \frac{i}{\sqrt{2}}, 0\right), \left(\frac{i}{\sqrt{2}}, -\frac{i}{\sqrt{2}}, 0\right), (0, 0, -i)\right\}$

29. *(a)* $\overline{z_1 + z_2} = \overline{z_1} + \overline{z_2}$
(b) $\overline{z_1 z_2} = \overline{z_1} \cdot \overline{z_2}$
(c) $\overline{\overline{z_1}} = z_1$
(d) $z_1\overline{z_1} = |z_1|^2$

Section 6.6

1. $\left\{\frac{1}{\sqrt{2}}, \sqrt{\frac{3}{2}}x\right\}$ and $\mathbf{p}(x) = \frac{1}{5}$

3. $\left\{1, 2\sqrt{3}\left(x - \frac{3}{2}\right)\right\}$ and $\mathbf{p}(x) = (12 - 18\ln 2)x - (18 - 28\ln 2)$

5. $\mathbf{p}(x) = -\frac{3}{35} + \frac{6}{7}x^2$

7. $\mathbf{p}(x) = (390\ln 2 - 270)x^2 - (1188\ln 2 - 822)x + (873\ln 2 - 603)$

9. $\mathbf{p}(x) = -\frac{3}{35} + \frac{6}{7}x^2$

11. $\mathbf{p}(x) = 1 + 2\sin x$

13. $\mathbf{p}(x) = 1 + (2\pi^2 - 12)(\sin x) = 1 + 2(\pi^2 - 6)\sin x$

15. $\mathbf{p}(x) = 1 + 2\sin x - \sin 2x$

17. $\mathbf{S} = \{1, \cos x, \cos(2x), \ldots, \cos(nx), \sin x, \sin(2x), \ldots, \sin(nx)\}$ is an orthogonal basis for $\mathbf{T}_n[-\pi, \pi]$.

19. $x = -\frac{5}{7}, y = \frac{16}{7}$

Section 6.7

1. $f(110.34) = -11.95$ but some of the numbers are not close to an integer value which means by the last take of the song, some of the instruments should have been retuned!

Section 6.8

1. 0

3. (a) $proj_{\mathbf{v}}\mathbf{u} = \left(\frac{\mathbf{u}\cdot\mathbf{v}}{\mathbf{v}\cdot\mathbf{v}}\right)\mathbf{v} = \left(\frac{(1,2)\cdot(1,2)}{(1,2)\cdot(1,2)}\right)\mathbf{v} = \mathbf{v};$

$proj_{\mathbf{v}'}\mathbf{u} = \left(\frac{\mathbf{u}\cdot\mathbf{v}'}{\mathbf{v}'\cdot\mathbf{v}'}\right)\mathbf{v} = \left(\frac{(1,2)\cdot(-2,1)}{(-2,1)\cdot(-2,1)}\right)\mathbf{v}' = 0\mathbf{v}'$

(b) $proj_{\mathbf{v}}\mathbf{u} = \left(\frac{\mathbf{u}\cdot\mathbf{v}}{\mathbf{v}\cdot\mathbf{v}}\right)\mathbf{v} = \left(\frac{(-2,1)\cdot(1,2)}{(1,2)\cdot(1,2)}\right)\mathbf{v} = 0\mathbf{v};$

$proj_{\mathbf{v}'}\mathbf{u} = \left(\frac{\mathbf{u}\cdot\mathbf{v}'}{\mathbf{v}'\cdot\mathbf{v}'}\right)\mathbf{v} = \left(\frac{(-2,1)\cdot(-2,1)}{(-2,1)\cdot(-2,1)}\right)\mathbf{v}' = \mathbf{v}'$

(c) $proj_{\mathbf{v}}\mathbf{u} = \left(\frac{\mathbf{u}\cdot\mathbf{v}}{\mathbf{v}\cdot\mathbf{v}}\right)\mathbf{v} = \left(\frac{(2,-1)\cdot(1,2)}{(1,2)\cdot(1,2)}\right)\mathbf{v} = 0\mathbf{v};$

$proj_{\mathbf{v}'}\mathbf{u} = \left(\frac{\mathbf{u}\cdot\mathbf{v}'}{\mathbf{v}'\cdot\mathbf{v}'}\right)\mathbf{v} = \left(\frac{(2,-1)\cdot(-2,1)}{(-2,1)\cdot(-2,1)}\right)\mathbf{v}' = (-1)\mathbf{v}'$

(d) $proj_{\mathbf{v}}\mathbf{u} = \left(\frac{\mathbf{u}\cdot\mathbf{v}}{\mathbf{v}\cdot\mathbf{v}}\right)\mathbf{v} = \left(\frac{(3,-7)\cdot(1,2)}{(1,2)\cdot(1,2)}\right)\mathbf{v} = \left(-\frac{11}{5}\right)\mathbf{v};$

$proj_{\mathbf{v}'}\mathbf{u} = \left(\frac{\mathbf{u}\cdot\mathbf{v}'}{\mathbf{v}'\cdot\mathbf{v}'}\right)\mathbf{v} = \left(\frac{(3,-7)\cdot(-2,1)}{(-2,1)\cdot(-2,1)}\right)\mathbf{v}' = \left(-\frac{13}{5}\right)\mathbf{v}'$

(e) $proj_{\mathbf{v}}\mathbf{u} = \left(\frac{\mathbf{u}\cdot\mathbf{v}}{\mathbf{v}\cdot\mathbf{v}}\right)\mathbf{v} = \left(\frac{(1,1)\cdot(1,2)}{(1,2)\cdot(1,2)}\right)\mathbf{v} = \left(\frac{3}{5}\right)\mathbf{v};$

$proj_{\mathbf{v}'}\mathbf{u} = \left(\frac{\mathbf{u}\cdot\mathbf{v}'}{\mathbf{v}'\cdot\mathbf{v}'}\right)\mathbf{v} = \left(\frac{(1,1)\cdot(-2,1)}{(-2,1)\cdot(-2,1)}\right)\mathbf{v}' = \left(-\frac{1}{5}\right)\mathbf{v}'$

7.

Initial Point of Vector	Terminal Point of Vector	Projection onto $\overrightarrow{OV} = (0, 3)$	Projection onto $(-3, 0)$
$(0, 0)$	$(0, 1)$	$\frac{1}{3}$	0
$(0, 1)$	$(0, 2)$	$\frac{1}{3}$	0
$(0, 2)$	$(-\frac{1}{6}, \frac{7}{3})$	$\frac{1}{9}$	$\frac{1}{18}$
$(-\frac{1}{6}, \frac{7}{3})$	$(-\frac{1}{6}, \frac{8}{3})$	$\frac{1}{9}$	0
$(-\frac{1}{6}, \frac{8}{3})$	$(0, 3)$	$\frac{1}{9}$	$-\frac{1}{18}$
$(0, 2)$	$(\frac{1}{3}, \frac{9}{4})$	$\frac{1}{12}$	$-\frac{1}{9}$
$(\frac{1}{3}, \frac{9}{4})$	$(\frac{5}{12}, \frac{31}{12})$	$\frac{1}{9}$	$-\frac{1}{36}$
$(\frac{5}{12}, \frac{31}{12})$	$(\frac{1}{3}, \frac{35}{12})$	$\frac{1}{9}$	$\frac{1}{36}$

Self-Test

1. Not a vector space

3. Is a vector space

5. $1 + x - x^2 = (\tfrac{1}{3})(1 + x + x^2) + (\tfrac{2}{3})(x - x^2) + (\tfrac{2}{3})(1 - x^2)$

7. $A = \begin{bmatrix} 1 & 1 \\ 1 & 1 \end{bmatrix} \rightarrow (1, 1, 1, 1)$

$A_1 = \begin{bmatrix} 0 & 1 \\ 1 & 1 \end{bmatrix} \rightarrow (0, 1, 1, 1)$

$A_2 = \begin{bmatrix} 1 & 0 \\ 1 & 1 \end{bmatrix} \rightarrow (1, 0, 1, 1)$

$A_3 = \begin{bmatrix} 1 & 1 \\ 0 & 1 \end{bmatrix} \rightarrow (1, 1, 0, 1)$

$A_4 = \begin{bmatrix} 1 & 1 \\ 1 & 0 \end{bmatrix} \rightarrow (1, 1, 1, 0)$

$A = (\tfrac{1}{3})A_1 + (\tfrac{1}{3}3)A_2 + (1)A_3 + (\tfrac{1}{3})A_4$

9. $(\mathbf{f}, \mathbf{g}) = 0$

11. $\left\{\left(\frac{i}{\sqrt{2}} 0, \frac{-i}{\sqrt{2}}\right), (0, 1, 0)\right\}$

Review Exercises

1. Not a vector space

3. Is a vector space

5. Not a vector space

7. Linearly independent

9. Linearly dependent

11. Orthogonal

13. Orthogonal

15. Orthogonal

17. Not an inner product space

19. Not an inner product

21. $\left\{\left(\frac{1}{\sqrt{3}}\right)(1, 1, 1, 0), \left(\frac{1}{\sqrt{15}}\right)(1, 1, -2, 3), \left(\frac{1}{\sqrt{35}}\right)(1, -4, 3, 3)\right\}$

25. Not a subspace

27. $\{\mathbf{v}_1, \mathbf{v}_2, \mathbf{v}_3\}$ is a basis for **S** and the dimension of **S** is 3.

29. $\{(i, 0, i), (i, 0, -i), (0, 1, 0)\}$

Review Exercises—Challenge Problems

31. *(a)* **W** is a subspace of $\mathbf{M}^{3,3}$.

(b) A basis for **W** is $\left\{\begin{bmatrix} -1 & 1 & 1 \\ 0 & 0 & 0 \\ 0 & 0 & 0 \end{bmatrix}, \begin{bmatrix} 0 & 0 & 0 \\ 0 & 0 & 1 \\ 0 & 1 & 0 \end{bmatrix}, \begin{bmatrix} 1 & 0 & 0 \\ 0 & 1 & 0 \\ 0 & 0 & 1 \end{bmatrix}\right\}$.

33. Linearly independent

35. Linearly independent

Chapter 7

Section 7.1

1. S is a linear transformation.

3. U is a linear transformation.

5. T is a linear transformation.

7. T is a linear transformation.

9. F is a linear transformation.

11. H is not a linear transformation.

13. $(16, -3)$, $m = 2$

15. $(-4, -3, 2)$, $m = 3$

17. Π_i is a linear transformation.

19. T is a linear transformation.

21. When $\theta = \frac{3\pi}{4}$ we obtain, $A_{\frac{3\pi}{4}} = \begin{bmatrix} -\frac{\sqrt{2}}{2} & -\frac{\sqrt{2}}{2} \\ \frac{\sqrt{2}}{2} & -\frac{\sqrt{2}}{2} \end{bmatrix}$.

23. With $\theta = \frac{\pi}{4}$ we have $A^{\frac{\pi}{4}} = \begin{bmatrix} 0 & 1 \\ 1 & 0 \end{bmatrix}$.

31. If S is a plane in $\mathbf{R}^n$ and $T: \mathbf{R}^n \to \mathbf{R}^m$ is a linear transformation, then $T(S)$ is either a plane, a line, or a point in $\mathbf{R}^m$.

33. There is no axis of rotation corresponding to P.

Section 7.2

1. $(x - y, 3x + 2y)$

3. S^2 is the identity transformation on $\mathbf{R}^3$.

5. T^2 is not defined.

7. $(v + w) + (v + 2u)x - ux^2$

9. $(3x - 2z, 3x + y, x + 3y + 3z)$

11. $3(a + b) + 3(a - 2d)x + 3dx^2$

13. $A = \begin{bmatrix} 1 & 0 & 0 \\ 1 & 1 & 0 \\ 1 & 1 & 1 \end{bmatrix}$

15. $A = \begin{bmatrix} 0 & 0 & 1 \\ 0 & 1 & 0 \\ 1 & 0 & 0 \end{bmatrix}$

17. $A = \begin{bmatrix} 1 & -1 & 0 \\ 3 & 2 & 0 \end{bmatrix}$

19. $A = \begin{bmatrix} 1 & 0 & 0 \\ 0 & 1 & 0 \\ 0 & 0 & 1 \end{bmatrix}$

21. $A^{\psi}A^{\phi} = A_{2(\psi - \theta)}$ which is the rotation by the angle $2(\psi - \theta)$.

27. $ST_c = T_cS$

Section 7.3

1. $\mathbf{v}_1$ is in Ker(T) and $\mathbf{v}_2$ is also in Ker(T).

3. $\mathbf{v}_1$ is in the kernel of T, while $\mathbf{v}_2$ is not.

5. $\mathbf{v}_1$ is not in Ker(T) but $\mathbf{v}_2$ in Ker(T).

7. $\mathbf{w}_1$ is not in the image of T and $\mathbf{w}_2$ is in the image of T.

9. $\mathbf{w}_1$ is in Im (T) and $\mathbf{w}_2$ is in Im (T)

11. $\mathbf{w}_1$ is in the image of T and $\mathbf{w}_2$ is not in the image of T.

13. Nullity of A is 2
    Setting $s = 1$, $t = 0$, we obtain: $\mathbf{v}_1 = (-1, 1, 1, 0)$
    Settings $s = 0$, $t = 1$, we obtain: $\mathbf{v}_2 = (-1, 0, 0, 1)$
    Thus, the set $B = \{(-1, 1, 1, 0), (-1, 0, 0, 1)\}$ forms a basis for Ker(T), and since Ker(T) $\neq \{\mathbf{0}\}$, then T is not one-to-one.

15. The nullity of T is 1. A basis for Ker(T) is the set $B = \{(3, -4, 5)\}$. Since Ker(T) $\neq \{\mathbf{0}\}$, then T is not one-to-one.

17. The nullity of T is 0. Also since Ker(T) = 0, then T is one-to-one
    Since dim(Ker(T)) = 0, then a basis for Ker(T) is an empty set.

19. $\{(2, 0, -4), (1, 1, 2)\}$ is a basis for Im (T). Since the rank is 2 and $m = 3$ this transformation is not onto.

21. A basis for Im(T) is: $B = \{(3, 1), (1, 2)\}$.
    Since the image of T is of order 2 and the rank of A is 2, then T is onto.

23. rank(T) = 3 and T is not onto.

27. T is not one-to-one.

29. $T^{-1}(\mathbf{x}) = \left(\frac{3}{2}x - \frac{1}{2}y + \frac{1}{2}z, x - y, -\frac{1}{2}x + \frac{1}{2}y + \frac{1}{2}z\right)$

31. T is not invertible.

33. $T_\theta^{-1}(\mathbf{x}) = A_{-\theta}^{-1}(\mathbf{x}) = \begin{bmatrix} \cos(-\theta) & -\sin(-\theta) \\ \sin(-\theta) & \cos(-\theta) \end{bmatrix}\begin{bmatrix} x \\ y \end{bmatrix} = T_{-\theta}^{-1}(\mathbf{x})$

35. $T_c^{-1}: \mathbf{R}^2 \to \mathbf{R}^2$ exists and is given by, $T_c^{-1}(\mathbf{x}) = B\mathbf{x}$, where $B = A^{-1} = c^{-1}I$.

Computational Exercises

37. T is one-to-one and onto if and only if $(-4b^2a - 2b^3 - 6ac + 7b^2 - 3a^2b + 3a^3c - 2b^2a^2 + 3b + 2c + a^2c + bac + 4ab + 2a^3b - 3a^3 + 6a^2) \neq 0$.

Section 7.4

1. $\begin{bmatrix} 0 & 1 & 0 \\ 0 & 0 & 1 \\ -1 & 0 & 0 \\ 0 & 0 & 0 \end{bmatrix}$

3. $\begin{bmatrix} 1 & 0 & 0 \\ 0 & 0 & -1 \\ 0 & 0 & -1 \\ 1 & -1 & 0 \end{bmatrix}$

5. T is one-to-one.

7. Ker(T) consists of the single vector $\mathbf{v} = 0$ and so T is one-to-one.

9. $\begin{bmatrix} m_{11} \\ m_{12} \\ m_{21} \\ m_{22} \end{bmatrix}$, where m_{11}, m_{12}, m_{21} are any real numbers and $m_{22} = 0$.

11. $\text{Im}\,(T) = \left\{\begin{bmatrix} a & -c \\ -c & a-b \end{bmatrix}: a, b, c \text{ are scalars}\right\}$, T is not onto.

13. $\begin{bmatrix} 0 & 1 & 0 & 0 & 0 \\ 0 & 0 & 2 & 0 & 0 \\ 0 & 0 & 0 & 3 & 0 \\ 0 & 0 & 0 & 0 & 4 \end{bmatrix}$

15. $\begin{bmatrix} 0 & 0 & 0 \\ 1 & 0 & 0 \\ 0 & \frac{1}{2} & 0 \\ 0 & 0 & \frac{1}{3} \end{bmatrix}$

17. T is invertible.

Section 7.5

1. $\frac{1}{4}\begin{bmatrix} 2 & 1 \\ 2 & -1 \end{bmatrix}, \frac{1}{4}(1, -5)$

3. $\begin{bmatrix} 1 & 1 & -1 \\ -1 & -1 & 2 \\ -1 & 0 & 1 \end{bmatrix}, (1, 0, -2)$

5. $\frac{1}{5}\begin{bmatrix} 3 & -1 & 2 & -2 \\ -3 & 1 & 3 & 2 \\ 3 & -1 & 2 & 3 \\ -1 & 2 & -4 & -1 \end{bmatrix}, \frac{1}{5}(-8, 3, -3, 6)$

7. $\begin{bmatrix} 1 & 0 \\ 2 & -1 \end{bmatrix}$

9. $\begin{bmatrix} 4 & 6 & -4 \\ -1 & -5 & 5 \\ -2 & -3 & 1 \end{bmatrix}$

11. $\begin{bmatrix} 4 & 5 & -5 \\ -4 & -6 & 7 \\ -1 & -3 & 5 \end{bmatrix}$

13. $\frac{1}{5}\begin{bmatrix} 14 & 12 & 8 & 15 \\ 11 & 13 & 12 & 20 \\ 24 & 17 & 3 & 15 \\ -23 & -24 & -11 & -25 \end{bmatrix}$

15. $\det(A) = \det(B)$

Section 7.6

1. $\mathbf{y} = c_1e^x + c_2e^{2x}$

3. $\mathbf{y} = c_1e^{(-1+\sqrt{2})x} + c_2e^{(-1-\sqrt{2})x}$

5. $\mathbf{y} = c_1e^{-2x} + c_2xe^{-2x}$

7. $\mathbf{s} = c_1 \sin\left(\frac{1}{2}t\right) + c_2 \cos\left(\frac{1}{2}t\right)$

9. $\mathbf{y} = e^{-x}(c_1 \sin(2x) + c_2 \cos(2x))$

11. $\mathbf{u} = c_1 + c_2e^{-x} + c_3xe^{-x}$

13. $\mathbf{s} = c_1e^t + c_2e^{-t} + c_3 \sin t + c_4 \cos t$

15. $\mathbf{s} = c_1 + c_2t + c_3e^t \sin(\sqrt{3}t) + c_4e^t \cos(\sqrt{3}t)$

17. $\mathbf{y} = -2 \sin x + \cos x + c_1e^{3x} + c_2e^{-x}$

19. $\mathbf{u} = e^x + e^{-x} + c_1 \sin x + c_2 \cos x$

21. $\mathbf{y} = -\frac{1}{4}e^{-3x} + \frac{5}{4}e^x$

23. $\mathbf{s} = \sin t$

25. $\mathbf{y} = -2 - 3x + 3e^x - e^{-x}$

27. $\mathbf{s}(t) = 2 \sin 3t - 2 \cos 3t$

29. $\boldsymbol{\theta}(t) = \frac{1}{4} \sin 4t$

33. $\{1, x, e^x, e^{-x}\}$ is linearly independent.

Section 7.7

1. Depends only on $S(\mathbf{y}) - S(\mathbf{x})$, and does not depend on the vector $\mathbf{b}$.

3. *(a)* S is a contraction mapping with contraction factor $\frac{1}{3}$.
(b) S is a contraction mapping with contraction factor $\frac{1}{2}$.
(c) S is not a contraction.

5. 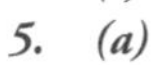 *(a)*

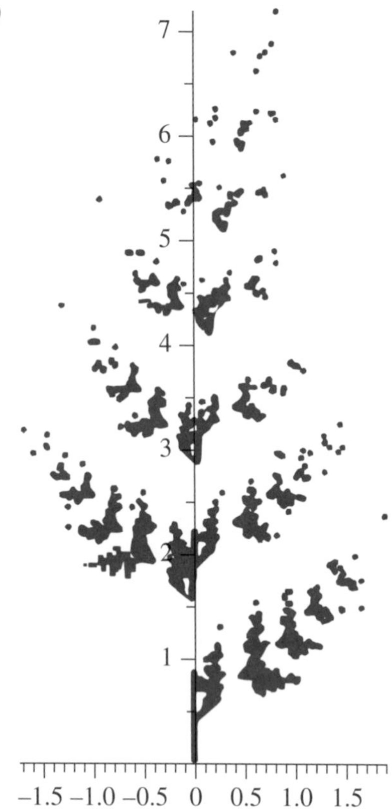

(b)

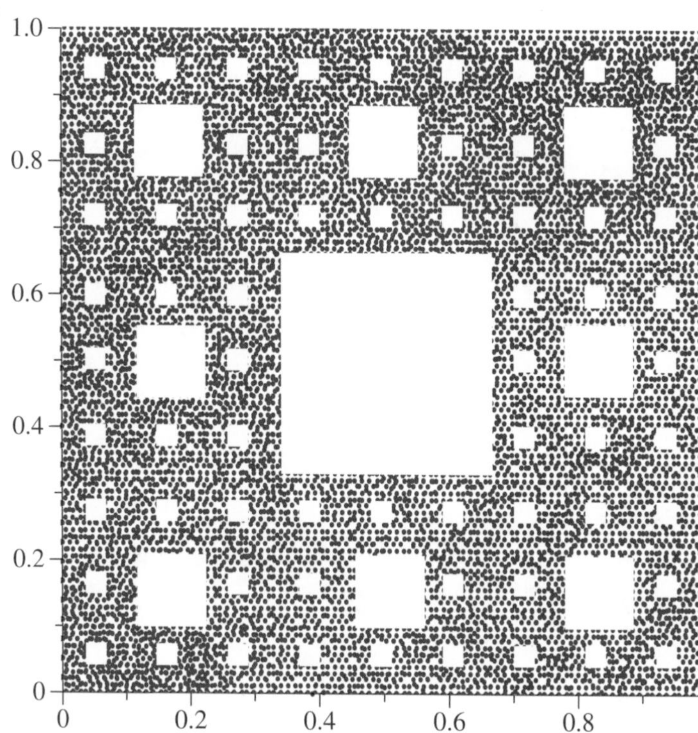

Self-Test

1. Not a linear transformation

3. $(0, x + y + z, 0)$

5. *(a)* The basis is the set containing the single vector $(-1, 1, 0)$.
(b) T is not one-to-one.

7. *(a)* $(2x + 5y, x + 3y)$
(b) $(3x - 5y, -x, + 2y)$

9. *(a)* S is one-to-one.
(b) S is not onto.

Review Exercises

1. Verify that $S(c\mathbf{x}) = cS(\mathbf{x})$ and that $S(\mathbf{x}_1 + \mathbf{x}_2) = S(\mathbf{x}_1) + S(\mathbf{x}_2)$.

3. S is not additive and scalars do not factor out.

5. FD is not defined.

7. $(x, 4x + 2y + 2z, 5x + 2y + 2z)$

9. $\begin{bmatrix} 1 & 1 \\ 1 & -2 \\ 1 & 0 \end{bmatrix}$

11. D is one-to-one.

13. E is not one-to-one.

15. D is one-to-one

17. $\begin{bmatrix} 1 & 0 & 1 & 0 \\ 0 & 1 & 0 & -1 \\ 0 & 0 & 2 & 0 \\ 2 & 1 & 0 & -1 \end{bmatrix}$

19. T is not onto.

21. $\begin{bmatrix} 0 & 0 & 0 \\ 1 & 1 & 0 \\ 2 & 2 & 2 \end{bmatrix}$

Review Exercises—Challenge Problems

23. *(a)* D_n is a linear transformation.
(b) The kernel of D_n is the set of all constant functions.

25. *(a)* The sum of two linear transformations on $\mathbf{R}^m$ and the product of a scalar times a linear transformation on $\mathbf{R}^m$ are linear transformations on $\mathbf{R}^m$.
(b) Let $B = \{T_{ij}: 1 \leq i, j \leq m\}$, where $T_{ij}: \mathbf{R}^m \rightarrow \mathbf{R}^m$ is the linear transformation defined by $T_{ij}(a_1\mathbf{v}_1 + a_2\mathbf{v}_2 + \cdots + a_m\mathbf{v}_m) = a_i\mathbf{v}_j$ for all i, j. B is a basis for the vector space of all linear transformations from $\mathbf{R}^m$ to $\mathbf{R}^m$.

27. $\{\mathbf{v}, T(\mathbf{v}), \ldots, T^{k-1}(\mathbf{v})\}$ is linearly independent.

Chapter 8

Section 8.1

1. $\lambda = 1$ is the eigenvalue for $I: \mathbf{R}^n \rightarrow \mathbf{R}^n$. The eigenvectors are all $\mathbf{x}$'s in $\mathbf{R}^n$ such that $\mathbf{x} \neq 0$.

3. No eigenvalues

5. Let $\mathbf{x} = s, s \neq 0$, then $(s, 0)$ is the eigenvector for eigenvalue $\lambda = -1$. Let $y = t, t \neq 0$, then the eigenvector for eigenvalue $\lambda = 1$ is $(0, t)$.

7. Eigenvalue 0; Eigenvectors: $\mathbf{v} = -s - 2tx + sx^2 + tx^3, st \neq 0$
Eigenvalue 1; Eigenvectors: $\mathbf{v} = s, s \neq 0$
Eigenvalue 2; Eigenvectors: $\mathbf{v} = sx^3, s \neq 0$

9. Eigenvector for eigenvalue $\lambda = 3$ is $(2c, c)$.
Eigenvector for eigenvalue $\lambda = -3$ is $(-t, t)$.

11. The eigenspace of eigenvalue $\lambda = 1$ is the set of all vectors of the form $(s, s) = s(1, 1)$. A basis for this eigenspace is $\{(1, 1)\}$.

13. The eigenvector for $\lambda = 0$ is $(-c, 0, c)$, c is nonzero.
A basis for the eigenspace for $\lambda = 0$ is $\{(-1, 0, 1)\}$.
The eigenvector for $\lambda = \sqrt{2}$ is $(c, -c\sqrt{2}, c)$.
A basis for the eigenspace for $\lambda = \sqrt{2}$ is $\{(1, -\sqrt{2}, 1)\}$.
The eigenvector for $\lambda = -\sqrt{2}$ is $(c, c\sqrt{2}, c)$.
$\{(1, \sqrt{2}, 1)\}$ is a basis for the eigenspace for $\lambda = -\sqrt{2}$.

15. The eigenspace of eigenvalue $\lambda = 0$ is the set of all vectors of the form$(s, t, -s - t) = s(1, 0, -1) + t(0, 1, -1)$. A basis for this eigenspace is $\{(1, 0, -1), (0, 1, -1)\}$. The eigenspace of eigenvector $\lambda = 3$ is the set of all vectors of the form $(s, s, s) = s(1, 1, 1)$. A basis for this eigenspace is $\{(1, 1, 1)\}$.

17. $(c, 0, 0), c \neq 0$ is the eigenvector for eigenvalue $\lambda = 1$. A basis for the eigenspace for $\lambda = 1$ is $\{(1, 0, 0)\}$.
$(0, c, 0), c \neq 0$ is the eigenvector for $\lambda = -2$. A basis for the eigenspace for $\lambda = -2$ is $\{(0, 1, 0)\}$.

19. The eigenspace of eigenvector $\lambda = 0$ is the set of all vectors of the form $(r, s, t, -r - s - t) = r(1, 0, 0, -1) + s(0, 1, 0, -1) + t(0, 0, 1, -1)$.
A basis for this eigenspace is $\{(1, 0, 0, -1), (0, 1, 0, -1), (0, 0, 1, -1)\}$.
The eigenspace of eigenvector $\lambda = 4$ is the set of all vectors of the form $(s, s, s, s) = s(1, 1, 1, 1)$. A basis for this eigenspace is $\{(1, 1, 1, 1)\}$.

21. Let $T_\theta: \mathbf{R}^2 \rightarrow \mathbf{R}^2$ be the rotation by θ radians.
(a) For values of θ other than integer multiples of π, $T_\theta(\mathbf{x})$ is not collinear with $\mathbf{x}$ for *any* non-zero vector $\mathbf{x}$, and thus there are no eigenvalues.
(b) The eigenspace is $\mathbf{R}^2$.

23. λ is an eigenvalue of B if and only if λ is an eigenvalue of A.

25. $\begin{bmatrix} 0 & 0 \\ 0 & 0 \end{bmatrix}$

27. $\begin{bmatrix} 0 & 0 & 0 \\ 0 & 0 & 0 \\ 0 & 0 & 0 \end{bmatrix}$

29. A has exactly one positive eigenvalue.

Computational Exercises

31. Eigenvalues: 1.5002, 0.1691, 0.0067, 0.0000

33. Eigenvalues: 1.5002, 0.1691, 0.0067, 0.0000

Section 8.2

1. Diagonalizable, $\begin{bmatrix} -3 & 0 \\ 0 & 3 \end{bmatrix}$

3. Not diagonalizable

5. Diagonalizable, $\begin{bmatrix} -\sqrt{2} & 0 & 0 \\ 0 & 0 & 0 \\ 0 & 0 & \sqrt{2} \end{bmatrix}$

7. Diagonalizable, $\begin{bmatrix} 0 & 0 & 0 \\ 0 & 0 & 0 \\ 0 & 0 & 3 \end{bmatrix}$

9. Not diagonalizable

11. Diagonalizable, $\begin{bmatrix} 0 & 0 & 0 & 0 \\ 0 & 0 & 0 & 0 \\ 0 & 0 & 0 & 0 \\ 0 & 0 & 0 & 4 \end{bmatrix}$

13. Orthogonal

15. Not orthogonal

17. $\left\{\left(\frac{1}{\sqrt{2}}, \frac{1}{\sqrt{2}}\right)\left(\frac{1}{\sqrt{2}}, -\frac{1}{\sqrt{2}}\right)\right\}$ is an orthonormal set and $P = \begin{vmatrix} \frac{1}{\sqrt{2}} & \frac{1}{\sqrt{2}} \\ \frac{1}{\sqrt{2}} & \frac{1}{\sqrt{2}} \end{vmatrix}$.

19. $\left\{\left(\frac{1}{\sqrt{2}}, 0, -\frac{1}{\sqrt{2}}\right), \left(\frac{1}{2}, -\frac{\sqrt{2}}{2}, \frac{1}{2}\right), \left(\frac{1}{2}, \frac{\sqrt{2}}{2}, \frac{1}{2}\right)\right\}$ forms an orthonormal basis for $\mathbf{R}^3$, and $P = \begin{bmatrix} \frac{\sqrt{2}}{2} & \frac{1}{2} & \frac{1}{2} \\ 0 & \frac{\sqrt{2}}{2} & \frac{\sqrt{2}}{2} \\ \frac{\sqrt{2}}{2} & \frac{1}{2} & \frac{1}{2} \end{bmatrix}$.

21. $\left\{\left(-\frac{1}{\sqrt{2}}, \frac{1}{\sqrt{2}}, 0\right)\left(-\frac{1}{\sqrt{6}}, -\frac{1}{\sqrt{6}}, \frac{2}{\sqrt{6}}\right)\left(\frac{1}{\sqrt{3}}, \frac{1}{\sqrt{3}}, \frac{1}{\sqrt{3}}\right)\right\}$ forms an orthonormal set for A, thus $P = \begin{vmatrix} -\frac{1}{\sqrt{2}} & -\frac{1}{\sqrt{6}} & \frac{1}{\sqrt{3}} \\ \frac{1}{\sqrt{2}} & -\frac{1}{\sqrt{6}} & \frac{1}{\sqrt{3}} \\ 0 & -\frac{1}{\sqrt{6}} & \frac{1}{\sqrt{3}} \end{vmatrix}$.

23. $\left\{\left(\frac{1}{\sqrt{2}}, \frac{1}{\sqrt{2}}, 0, 0\right), \left(0, 0, \frac{1}{\sqrt{2}}, \frac{1}{\sqrt{2}}\right), \left(-\frac{1}{\sqrt{2}}, \frac{1}{\sqrt{2}}, 0, 0\right), \left(0, 0, -\frac{1}{\sqrt{2}}, \frac{1}{\sqrt{2}}\right)\right\}$ forms an orthonormal basis for $\mathbf{R}^4$, and

$$P = \begin{bmatrix} \frac{1}{\sqrt{2}} & 0 & -\frac{1}{\sqrt{2}} & 0 \\ \frac{1}{\sqrt{2}} & 0 & \frac{1}{\sqrt{2}} & 0 \\ 0 & \frac{1}{\sqrt{2}} & 0 & -\frac{1}{\sqrt{2}} \\ 0 & \frac{1}{\sqrt{2}} & 0 & \frac{1}{\sqrt{2}} \end{bmatrix}.$$

25. Orthogonal

27. Not diagonalizable

33. $\begin{bmatrix} \cos\theta & 0 & \sin\theta \\ 0 & 1 & 0 \\ -\sin\theta & 0 & \cos\theta \end{bmatrix}\begin{bmatrix} a_{11} & a_{12} & a_{13} \\ a_{21} & a_{22} & a_{23} \\ a_{31} & a_{32} & a_{33} \end{bmatrix}\begin{bmatrix} \cos\theta & 0 & -\sin\theta \\ 0 & 1 & 0 \\ \sin\theta & 0 & \cos\theta \end{bmatrix}$, $\tan(2\theta) = \dfrac{2a_{13}}{a_{33} - a_{11}}$

Computational Exercises

35. $\begin{bmatrix} 6.07 & 0 & 0 & 0 \\ 0 & 0.10 & 0 & 0 \\ 0 & 0 & -1.33 & 0 \\ 0 & 0 & 0 & -4.84 \end{bmatrix}$

37. No dominant eigenvector

39. For Exercise 36:
eigenvalue $\lambda = 5.372$; eigenvectors $s(0.344, 0.939), s \neq 0$
eigenvalue $\lambda = 2.268$; eigenvectors $s(-1.283, 0.939), s \neq 0$
For Exercise 37:
eigenvalues: None
For Exercise 38:
eigenvalue $\lambda = 5.478$
eigenvectors: $s(0.565, -0.285, 0.775), s \neq 0$

Section 8.3

1. $Q = \begin{bmatrix} 0 & 1 \\ 1 & 0 \end{bmatrix}, R = \begin{bmatrix} 1 & 2 \\ 0 & 3 \end{bmatrix}$

3. $Q = \begin{bmatrix} \frac{2}{\sqrt{5}} & \frac{3}{\sqrt{70}} & \frac{1}{\sqrt{14}} \\ -\frac{1}{\sqrt{5}} & \frac{6}{\sqrt{70}} & \frac{2}{\sqrt{14}} \\ 0 & -\frac{5}{\sqrt{70}} & \frac{3}{\sqrt{14}} \end{bmatrix}, R = \begin{bmatrix} \frac{5}{\sqrt{5}} & -\frac{4}{\sqrt{5}} & \frac{1}{\sqrt{5}} \\ 0 & \frac{14}{\sqrt{70}} & -\frac{16}{\sqrt{70}} \\ 0 & 0 & \frac{4}{\sqrt{14}} \end{bmatrix}$

5. $Q = \begin{bmatrix} 1 & 0 & 0 \\ 0 & 1 & 0 \\ 0 & 0 & 1 \end{bmatrix}, R = \begin{bmatrix} 1 & 1 & 1 \\ 0 & 1 & 1 \\ 0 & 0 & 1 \end{bmatrix}$

7. $\begin{bmatrix} 2 & 1 \\ 3 & 0 \end{bmatrix}$

9. $\begin{bmatrix} 2.8000 & -0.7483 & 0.0001 \\ -0.7483 & 2.3428 & 0.6390 \\ 0 & 0.6388 & 0.8571 \end{bmatrix}$

11. $\begin{bmatrix} 1 & 1 & 1 \\ 0 & 1 & 1 \\ 0 & 0 & 1 \end{bmatrix}$

13. All the matrices of Exercises 1–4 are in Hessenberg form. In Exercise 5, Q and R are in Hessenberg form.

Computational Exercises

15. Eigenvalues: 3, −1

17. Eigenvalues: $2, 2 + \sqrt{2}, 2 - \sqrt{2}$

19. Eigenvalues: 1, 1, 1

Section 8.4

1. $z_1 = i, z_2 = 0, z_3 = 0$

3. $\begin{bmatrix} i & -2i & 1 \\ -1 + i & i & 1 + 2i \\ -1 - i & 3 & -2 + 2i \end{bmatrix}$

5. $\frac{1}{5}\begin{bmatrix} 8-i & 2+i & 1+3i \\ 2+i & 3-i & -1+2i \\ -1-3i & 1-2i & 3-i \end{bmatrix}$

7. $1 + 2i$

9. A is not Hermitian. B is not Hermitian.

11. A basis for Ker(T) is $\{(1, 1)\}$

13. The eigenspace for the eigenvalue $\lambda = 5$ is the set of all vectors $((3 + 4i)s, 5s) = s(3 + 4i, 5)$, where s is a complex scalar. A basis for the eigenspace is $\{(3 + 4i, 5)\}$.
The eigenspace for the eigenvalue $\lambda = -5$ is the set of all vectors $((3 + 4i)s, -5s) = s(3 + 4i, -5)$, where s is a complex scalar. A basis for the eigenspace is $\{(3 + 4i, -5)\}$.

15. The eigenspace for eigenvalue $\lambda = 1$ is $(0, s, 0) = (0, 1, 0)$.
A basis for that eigenspace is $\{(0, 1, 0)\}$.
The eigenspace for eigenvalue $\lambda = -2i$ is $(s(1 - i), 0, s)$.
A basis for the eigenspace is $\{(1 - i, 0, 1)\}$.
The eigenspace for eigenvalue $\lambda = 2i$ is $(-1(1 + i)t, 0, t)$.
A basis for the eigenspace is $\{(-1 - i, 0, 1)\}$.

17. The eigenspace for eigenvalue $\lambda = i$ is $(s, t, 0) = s(1, 0, 0) + t(0, 1, 0)$. A basis for the eigenspace is $\{(1, 0, 0), (0, 1, 0)\}$.

19. The eigenspace of the eigenvalue $\lambda = -1$ is $(0, ir, r)$. A basis for the eigenspace is $\{(0, i, 1)\}$.
The eigenspace of the eigenvalue $\lambda = 1$ is $(s, -it, t)$.A basis for the eigenspace is $\{(1, 0, 0), (0, -i, 1)\}$.

21. $P = \begin{bmatrix} 3+4i & 3+4i \\ 5 & -5 \end{bmatrix}$, $P^{-1}AP = \begin{bmatrix} 5 & 0 \\ 0 & -5 \end{bmatrix}$

23. Diagonizable, $\begin{bmatrix} 0 & 1-i & -1-i \\ 1 & 0 & 0 \\ 0 & 1 & 1 \end{bmatrix}$

25. Not diagonalizable

27. Diagonalizable, $P = \begin{bmatrix} 0 & 1 & 0 \\ \frac{i}{\sqrt{2}} & 0 & -\frac{i}{\sqrt{2}} \\ \frac{1}{\sqrt{2}} & 0 & \frac{1}{\sqrt{2}} \end{bmatrix}$, $D = \begin{bmatrix} -1 & 0 & 0 \\ 0 & 1 & 0 \\ 0 & 0 & 1 \end{bmatrix}$

29. Exercise 13: Not unitary
Exercise 14: Not unitary
Exercise 15: Not unitary
Exercise 16: Not unitary
Exercise 17: Not unitary
Exercise 18: Not unitary
Exercise 19: Unitary
Exercise 20: Unitary

Section 8.5

1. $\begin{bmatrix} 1 & 0 \\ 0 & 64 \end{bmatrix}$

3. $$\begin{bmatrix} \frac{-1}{4}(-3)^k + \frac{1}{2}(-1)^k + \frac{3}{4} & \frac{1}{4}(-3)^k - \frac{1}{2}(-1)^k + \frac{1}{4} & \frac{1}{2}(-3)^k - \frac{1}{2} \\ \frac{-1}{4}(-3)^k - \frac{1}{2}(-1)^k + \frac{3}{4} & \frac{1}{4}(-3)^k + \frac{1}{2}(-1)^k + \frac{1}{4} & \frac{1}{2}(-3)^k - \frac{1}{2} \\ \frac{-1}{2}(-3)^k + \frac{1}{2}(-1)^k & \frac{1}{2}(-3)^k - \frac{1}{2}(-1)^k & (-3)^k \end{bmatrix}$$

$$A^6 = \begin{bmatrix} -181 & 182 & 364 \\ -182 & 183 & 364 \\ -364 & 364 & 729 \end{bmatrix}$$

5. $$\begin{bmatrix} \left(\frac{1}{2}\right)^k & 0 & 0 & 0 \\ 0 & \left(\frac{-1}{2}\right)^k & 0 & 0 \\ \left(\frac{1}{2}\right)^k - \left(\frac{-1}{4}\right)^k & 0 & \left(\frac{-1}{4}\right)^k & 0 \\ \left(\frac{1}{4}\right)^k - \left(\frac{-1}{4}\right) & 0 & -\left(\frac{1}{2}\right)^k + \left(\frac{-1}{4}\right)^k & \left(\frac{1}{4}\right)^k \end{bmatrix}$$

$$A^6 = \begin{bmatrix} \frac{1}{64} & 0 & 0 & 0 \\ 0 & \frac{1}{64} & 0 & 0 \\ \frac{63}{4096} & 0 & \frac{1}{4096} & 0 \\ 0 & 0 & 0 & \frac{1}{4096} \end{bmatrix}$$

7. Exercise 1: $A^k = \begin{bmatrix} 0 & 0 \\ 0 & (-2)^k \end{bmatrix}$, so $A^{10} = \begin{bmatrix} 0 & 0 \\ 0 & 1024 \end{bmatrix}$

Exercise 2: $A^k = \begin{bmatrix} (-1)^{k+2}2^{k-1} & (-1)^{k+1}2^{k-1} \\ (-1)^{k+1}2^{k-1} & (-1)^{k}2^{k-1} \end{bmatrix}$, so $A^{10} = \begin{bmatrix} 512 & -512 \\ -512 & 512 \end{bmatrix}$

Exercise 3: $A^k = \begin{bmatrix} \frac{-1}{4}(-3)^k & \frac{1}{4}(-3)^k & \frac{1}{2}(-3)^k \\ \frac{-1}{4}(-3)^k & \frac{1}{4}(-3)^k & \frac{1}{2}(-3)^k \\ \frac{-1}{2}(-3)^k & \frac{1}{2}(-3)^k & (-3)^k \end{bmatrix}$, so $A^{10} = \begin{bmatrix} \frac{-59049}{4} & \frac{59049}{4} & \frac{59049}{2} \\ \frac{-59049}{4} & \frac{59049}{4} & \frac{59049}{2} \\ \frac{-59049}{4} & \frac{59049}{4} & 59049 \end{bmatrix}$

Exercise 4: $A^k = \begin{bmatrix} \frac{1}{2} & \frac{1}{2} & 0 \\ \frac{1}{2} & \frac{1}{2} & 0 \\ \frac{1}{2} & \frac{1}{2} & 0 \end{bmatrix}$, so $A^{10} = \begin{bmatrix} \frac{1}{2} & \frac{1}{2} & 0 \\ \frac{1}{2} & \frac{1}{2} & 0 \\ \frac{1}{2} & \frac{1}{2} & 0 \end{bmatrix}$

Exercise 5: $A^k = \begin{bmatrix} \left(\frac{1}{2}\right)^k & 0 & 0 & 0 \\ 0 & \left(\frac{-1}{2}\right)^k & 0 & 0 \\ \left(\frac{1}{2}\right)^k & 0 & 0 & 0 \\ 0 & 0 & 0 & 0 \end{bmatrix}$, so $A^{10} = \begin{bmatrix} \frac{1}{1024} & 0 & 0 & 0 \\ 0 & \frac{1}{1024} & 0 & 0 \\ \frac{1}{1024} & 0 & 0 & 0 \\ 0 & 0 & 0 & 0 \end{bmatrix}$

Exercise 6: $A^k = \begin{bmatrix} \frac{1}{2}(1+i)^k + \frac{1}{2}(1-i)^k & \frac{-1}{2}i(1+i)^k + \frac{1}{2}i(1-i)^k \\ \frac{1}{2}i(1+i)^k - \frac{1}{2}i(1-i)^k & \frac{1}{2}(1+i)^k + \frac{1}{2}(1-i)^k \end{bmatrix}$, so $A^{10} = \begin{bmatrix} 0 & 32 \\ -32 & 0 \end{bmatrix}$

9. $f_n = \left(\frac{\sqrt{5}}{10} + \frac{1}{2}\right)\left(\frac{1}{2} + \frac{\sqrt{5}}{2}\right)^n + \left(\frac{-\sqrt{5}}{10} + \frac{1}{2}\right)\left(\frac{1}{2} - \frac{\sqrt{5}}{2}\right)^n$

11. $f_n = \left(\frac{-1}{2} - i\right)(1+i)^n + \left(\frac{-1}{2} + i\right)(1-i)^n$

13. Eigenvalues of 2, 4, -1 $\lambda = 2$, eigenvector for 2 is $(-1, 1, 0)$ $\lambda = 4$ eigenvector for 4 is $(-1, 0, 1)$ $\lambda = -1$ eigenvalue for (-1) is $(1, 1, 0)$.

Section 8.6

1. $(x')^2 + (y')^2 - \frac{(z')^2}{2} = 1$, hyperboloid

3. $\frac{(x')^2}{1} + \frac{(y')^2}{\frac{1}{6}} - \frac{(z')^2}{\frac{1}{4}} = 0$, cone

5. $\frac{(x')^2}{18} + \frac{(y')^2}{36} + \frac{(z')^2}{36} = 1$, ellipsoid

7. $\frac{(x'')^2}{\frac{1}{2}} + \frac{(y'')^2}{\frac{1}{8}} - \frac{(z'')^2}{\frac{1}{2}} = 1$, hyperboloid

9. $2(x'')^2 - (z'')^2 = 2 \rightarrow (x'')^2 - \frac{(z'')^2}{2} = 1$, hyperbolic cylinder

13. $(\mathbf{x}')^T D\mathbf{x}' + BP\mathbf{x}' = f$

15. $3(x')^2 - (y')^2 + 6x' + 4y' = 0$

17. $-3(x'')^2 + (y'')^2 = 1$, hyperbola

Section 8.7

3. ***(i)*** $f_1(\mathbf{x})$ is a contraction with contraction factor 0.16.
(ii) $f_2(\mathbf{x})$ is a contraction with contraction factor 0.8509.
(iii) $f_3(\mathbf{x})$ is a contraction with contraction factor 0.3047.
(iv) $f_4(\mathbf{x})$ is a contraction with contraction factor 0.4690.
For Exercise 4(b) of Section 7.7 each affine function is a contraction with contraction factor 0.333.

Self-Test

1. $\lambda = 0$ is an eigenvalue with corresponding eigenvectors $(0, s)$ for $s \neq 0$ in **R**.

$\boldsymbol{\lambda} = 1$ is an eigenvalue and the corresponding eigenvectors are $(s, 0)$ for $s \neq 0$ in **R**.

3. *(a)* $\lambda^2(6 - \lambda)$
(b) $\boldsymbol{\lambda} = 0, 6$
(c) Eigenvalue $\lambda = 0$ has eigenvector $(-s - t, s, t)$, where s and t are in **R**. A basis is $\{(-1, 1, 0), (-1, 0, 1)\}$
Eigenvalue $\lambda = 6$ has eigenvector (s, s, s), where s is in **R**. A basis is $\{(1, 1, 1)\}$.

5. $P = \begin{bmatrix} -1 & -1 & 1 \\ 0 & 0 & 1 \\ 0 & 1 & 1 \end{bmatrix}, D = \begin{bmatrix} 0 & 0 & 0 \\ 0 & 0 & 0 \\ 0 & 0 & 6 \end{bmatrix}$

7. *(a)* Yes
(b) Not orthogonal

9. $\begin{bmatrix} 1 & 0 & 1 \\ 0 & 1 & 0 \\ 1 & 0 & 0 \end{bmatrix}$

11. Eigenvalue $\boldsymbol{\lambda} = i$ has eigenvector (s, t, t). A basis is $\{(1, 0, 0), (0, 1, 1)\}$.
Eigenvalue $\boldsymbol{\lambda} = -i$ has eigenvector. A basis is$\{(0, -1, 1)\}$.

Review Exercises

1. The eigenspace of $\lambda = 1$ is the set of all vectors of the form $(-s, s) = s(-1, 1)$. Thus, a basis for the eigenspace of $\lambda = 1$ is $\{(-1, 1)\}$.

The eigenspace of $\lambda = 3$ is the set of all vectors of the form $(s, s) = s(1, 1)$, which has a basis $\{(1, 1)\}$.

3. All of matrices A through F are diagonalizable.

5. $P = \begin{bmatrix} -1 & 0 & 2 \\ 1 & -1 & 1 \\ 1 & 1 & 1 \end{bmatrix}, P^{-1}FP = \begin{bmatrix} 0 & 0 & 0 \\ 0 & 1 & 0 \\ 0 & 0 & 3 \end{bmatrix}$

7. Orthogonal

9. Exercise 1, $P = \begin{bmatrix} -1/\sqrt{2} & 1/\sqrt{2} \\ 1/\sqrt{2} & 1/\sqrt{2} \end{bmatrix}$ is orthogonal and satisfies $P^TAP = \begin{bmatrix} 1 & 0 \\ 0 & 3 \end{bmatrix}$.

Exercise 5, $P = \begin{bmatrix} -1/\sqrt{3} & 0 & 2/\sqrt{6} \\ 1/\sqrt{3} & -1/\sqrt{2} & 1/\sqrt{6} \\ 1/\sqrt{3} & 1/\sqrt{2} & 1/\sqrt{6} \end{bmatrix}$ is orthogonal and satisfies

$P^{-1}FP = \begin{bmatrix} 0 & 0 & 0 \\ 0 & 1 & 0 \\ 0 & 0 & 3 \end{bmatrix}$.

11. $Q = \begin{bmatrix} 0 & 1 \\ 1 & 0 \end{bmatrix}, R = \begin{bmatrix} 1 & 0 \\ 0 & 4 \end{bmatrix}, B_2 = \begin{bmatrix} 0 & 1 \\ 4 & 0 \end{bmatrix}$

13. $Q = \begin{bmatrix} -\frac{2}{\sqrt{13}} & \frac{3}{\sqrt{182}} & \frac{3}{\sqrt{14}} \\ 0 & \frac{13}{\sqrt{182}} & -\frac{1}{\sqrt{14}} \\ -\frac{3}{\sqrt{13}} & -\frac{2}{\sqrt{182}} & -\frac{2}{\sqrt{14}} \end{bmatrix}, R = \begin{bmatrix} \frac{39}{\sqrt{13}} & -\frac{10}{\sqrt{13}} & -\frac{44}{\sqrt{13}} \\ 0 & \frac{28}{\sqrt{182}} & \frac{1}{\sqrt{182}} \\ 0 & 0 & \frac{1}{\sqrt{14}} \end{bmatrix} D_2 = \begin{bmatrix} \frac{54}{13} & \frac{75\sqrt{14}}{182} & \frac{215\sqrt{13}\sqrt{14}}{182} \\ -\frac{3\sqrt{14}}{182} & \frac{181}{91} & -\frac{15\sqrt{13}}{91} \\ -\frac{3\sqrt{13}\sqrt{14}}{182} & -\frac{\sqrt{13}}{91} & -\frac{1}{7} \end{bmatrix}$

15. Is a basis

17. Not one-to-one

19. The eigenspace of $\lambda = 1$is the set of all vectors $(0, s, 0) = s(0, 1, 0)$. A basis for the eigenspace is $\{(0, 1, 0)\}$.

The eigenspace of is $\lambda = \frac{3 + \sqrt{5}}{2}$the set of all vectors $((1 - \sqrt{5})i \cdot s, 0, -2s) = s((1 - \sqrt{5})i, 0, -2)$. A basis is $\{((1 - \sqrt{5})i, 0, -2)\}$.

The eigenspace of $\lambda = \frac{3 - \sqrt{5}}{2}$ is the set of all vectors $((1 + \sqrt{5})i \cdot s, 0, -2s) = s((1 + \sqrt{5})i, 0, -2)$. A basis is $\{((1 + \sqrt{5})i, 0, -2)\}$.

21. Exercise 19: Diagonalizable
Exercise 20: Diagonalizable

23. $P = \begin{bmatrix} 0 & \frac{1-\sqrt{5}}{\sqrt{10-2\sqrt{5}}}i & \frac{1+\sqrt{5}}{\sqrt{10+2\sqrt{5}}}i \\ 1 & 0 & 0 \\ 0 & -\frac{2}{\sqrt{10-2\sqrt{5}}} & -\frac{2}{\sqrt{10+2\sqrt{5}}} \end{bmatrix}$ is orthogonal, $P^{-1}AP = \begin{bmatrix} 1 & 0 & 0 \\ 0 & \frac{3+\sqrt{5}}{2} & 0 \\ 0 & 0 & \frac{3-\sqrt{5}}{2} \end{bmatrix}$.

Review Exercises—Challenge Problems

25. $\lim_{k\to\infty} A^k\mathbf{v} = \mathbf{0}$

Chapter 9

Section 9.1

1.

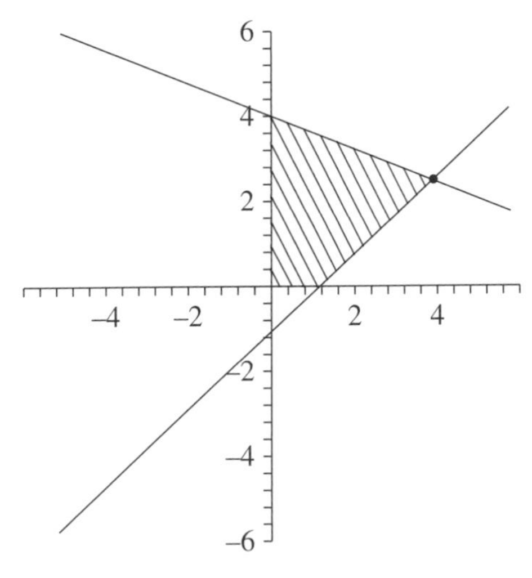

3.

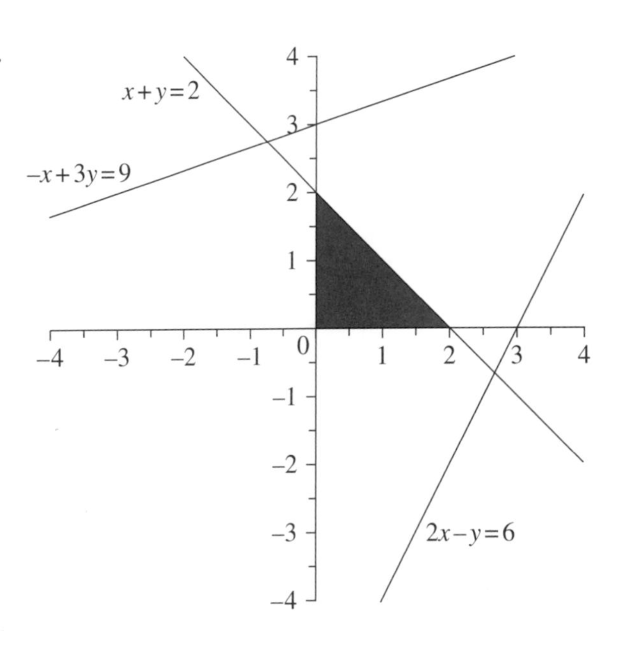

5. The point of intersection is $\left(\frac{10}{3}, \frac{7}{3}\right)$. The max value of z is 9.

7. The point of intersection is (2, 0). The max value of z is 2.

9. $x - y + s_1 = 2$
$3x - y + s_2 = 6$; where $x, y, s_1, s_2 \geq 0$.

11. $x_3 = 2, x_2 = 1, x_5 = 4$ and $x_1 = x_4 = 0$

13. ***i.*** $x_2 = 6, x_3 = 2, x_5 = 1, x_1 = x_4 = 0$
ii. $x_3 = 2, x_4 = 5, x_5 = 1, x_1 = x_2 = 0$

15. Maximize $z = 0.08a_1 + 0.15a_2 + 0.12a_3$ subject to the constraints

$$\begin{aligned} a_1 + a_2 + a_3 \qquad\quad &= 50{,}000 \\ a_1 \qquad\quad + s_1 \quad &= 10{,}000 \\ a_1 + a_2 \qquad\quad + s_2 &= 20{,}000 \\ a_1,\ a_2,\ a_3,\ s_1,\ s_2 &\geq 0 \end{aligned}$$

Section 9.2

1.

	c^*	x_1	x_2	x_3	x_4	x_5	**b**
x_3	0	3	1	1	0	0	1
x_4	2	−1	2	0	1	0	5
x_5	−1	4	5	0	0	1	9
		−9	1	0	0	0	1

3.

	c^*	x_1	x_2	x_3	x_4	x_5	**b**
x_2	−2	3	1	2	0	0	2
x_5	2	1	0	−3	0	1	5
x_4	0	2	0	−1	1	0	8
		−5	0	−11	0	0	6

5.

	c^*	x_1	x_2	x_3	x_4	x_5	x_6	**b**
x_5	0	2	1	1	3	1	0	12
x_6	0	1	−3	−1	1	0	1	8
		−1	2	−1	−4	0	0	0

7. 5, (0, 2, 1, 0, 4)

9. (11, 0, 8, 0, 0, 0)

11. $x_3 = 1, x_5 = 8, x_4 = 9$ and $x_1 = x_2 = 0$ which yields the maximum value $z = 17$.

13. $x_1 = x_3 = x_4 = 0, x_2 = 12$ with min value of $z = -24$.

15. The solution $z = \$6600$ is maximal and is achieved by investing $a_1 = \$0$ in savings, $a_2 = \$20{,}000$ in stock, and $a_3 = \$30{,}000$ in property.

Section 9.3

1.
$$\begin{aligned} 2x - 3y - z + s_1 \qquad\qquad &= 5 \\ y + 2z \quad - s_2 + a_1 \quad &= 8 \\ x \quad - 2z \qquad\qquad + a_2 &= 3 \end{aligned}$$

3.
$$\begin{aligned} 2x - y \qquad\qquad + a_1 &= 7 \\ x + 2y + s_1 \qquad &= 18 \\ -3x + 5y \quad + s_2 \qquad &= 3 \\ x, y, a_1, s_1\ s_2 \geq 0 \end{aligned}$$

5. $z = 3, y = 8, s_1 = 23, z = 0, s_2 = 0$

7. $x = \frac{7}{2}, s_1 = \frac{29}{2}, s_2 = \frac{27}{2}, y = 0, a_1 = 0$

9. No maximal solution

11. The solution $z = -\frac{5}{7}$ is minimal and occurs when $x = \frac{38}{7}, y = \frac{27}{7}$.

13. $x_1 = 7, x_4 = 12, x_2 = 0, x_3 = 0$, maximal value is 38.

15. $B = \frac{2}{3}, C = \frac{7}{12}, A = 0, D = 0$, the solution $z = \frac{115}{6} \approx \$19.17$ is minimal.

Section 9.4

3. The simplex algorithm cycles in this example.

Section 9.5

1. The artificial variable x_{10} does not leave the basis, and we must continue further.

3. Add an extra demand point for any excess wheat with a demand of 20 tonnes, with a maket price of \$0 per tonne.

5. If supply > demand, add a new demand point whose demand is equal to the excess supply over demand, with a cost of 0.
If demand > supply, you can try to add a new supply point with supply equal to the excess supply over demand, and attach a high cost to using any of this supply.

Self-Test

1.

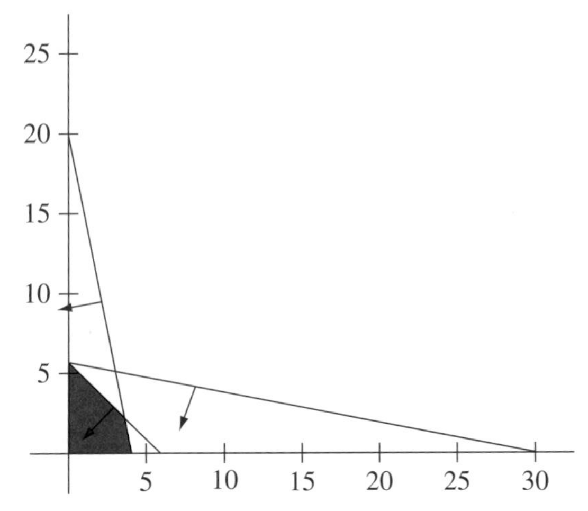

3. $x = \frac{7}{2}, y = \frac{5}{2}$, max value is $z = \frac{19}{2}$.

5.
$$\begin{aligned} x + 5y + s_1 \qquad &= 30 \\ x + y + \quad s_2 \quad &= 6 \\ 5x + y \qquad + s_3 &= 20 \end{aligned}$$

7. x_1, x_4, and x_6 will all increase the value of the objective function.

9. The current solution is minimal.

Review Exercises

1.

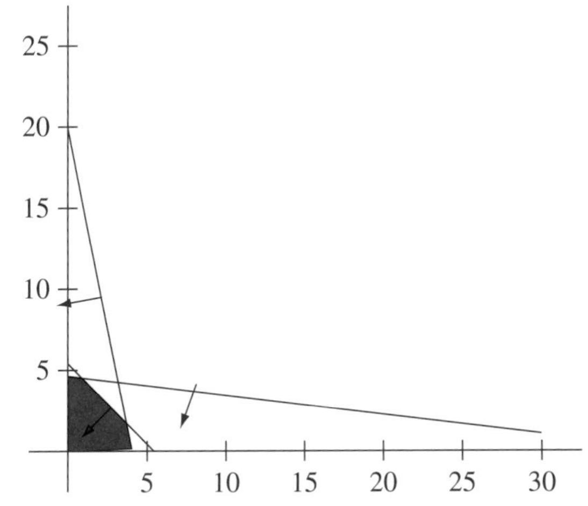

3. (0, 0) minimal value is 0, (3, 2) maximum value 8.

5. $(\frac{10}{3}, 0)$ maximal value $\frac{10}{3}$; (0, 4) minimum value -4.

7.
$$\begin{aligned} x + 10y + s_1 \qquad &= 40 \\ x + y + \quad s_2 \quad &= 5 \\ 6x + y \qquad + s_3 &= 20 \end{aligned}$$

9.

x	y	s_1	s_2	s_3	**b**
1	10	1	0	0	40
1	1	0	1	0	5
6	1	0	0	1	20
−2	1	0	0	0	0

11. (3, 2), maximum value of 8

13. (0, 1), maximal value -1; (0, 4) minimal value -4

15. (0, 1), maximal value -1; no minimal value exists

17.

y_1	y_2	x_2	x_3	x_4	s_1	s_2	a_1	a_2	a_3	**b**
0	0	-3	0	1	$-\frac{1}{2}$	1	$\frac{1}{2}$	$\frac{3}{2}$	-1	$\frac{19}{2}$
0	0	-2	1	-2	$\frac{1}{2}$	0	$-\frac{1}{2}$	$\frac{1}{2}$	0	$\frac{5}{2}$
1	-1	0	0	1	$-\frac{1}{2}$	0	$\frac{1}{2}$	$\frac{1}{2}$	0	$\frac{15}{2}$
0	0	0	0	0	0	0	-1	-1	-1	0

A feasible solution exists, and by discarding the columns of artificial variables and the bottom row, the starting tableau for a further optimization is available with the embedded identity matrix in the columns that correspond to y_1, x_3, and s_2.

Review Exercises—Challenge Problems

19. (0, 1), the maximum and minimum is 1.

21. *(a)* The maximum value of the objective function for the *LP* relaxation is at least as large as the maximum value of the associated integer programming problem (provided the values exist).
(b) The minimum value of the objective function for the *LP* relaxation is at most as large as the minimum value of the associated integer programming problem (provided the values exist).

23. Maximize $z = x_1 + x_2$ subject to

$$\begin{aligned} x_1 &= x_3 + x_4 + x_5 \\ x_2 + x_3 &= x_6 \\ x_4 &= x_7 + x_8 \\ x_5 + x_6 + x_8 &= x_9 \\ x_1 &\le 3 \\ x_2 &\le 5 \\ x_3 &\le 2 \\ x_4 &\le 2 \\ x_5 &\le 3 \\ x_6 &\le 2 \\ x_7 &\le 1 \\ x_8 &\le 3 \\ x_9 &\le 7 \\ x_i &\ge 0,\ i = 1, \ldots, 9 \end{aligned}$$

INDEX

V

W

Z